# Fundamentals of Atmospheric Modeling

This comprehensive text describes the atmospheric processes, numerical methods, and computational techniques required for a scientist to successfully study air pollution and meteorology.

Computer modeling has become a powerful tool in modern atmospheric sciences, combining the disciplines of meteorology, physics, mathematics, chemistry, computer sciences, and, to a lesser extent, geology, biology, microbiology, and oceanographic sciences. This text presents fundamental equations that describe physical, chemical, and dynamical processes in the atmosphere, and it provides numerical methods to solve these equations. Along with classic methods of simulating dynamical meteorology, the text contains several numerical techniques for simulating gas and aerosol processes not available in any other text.

The book has been developed from the author's graduate courses and research at Stanford University and contains homework and computer programming assignments. It is a valuable textbook for graduate and upper-level undergraduate courses in atmospheric sciences and meteorology. It will also be useful for courses in earth sciences, environmental sciences, and applied mathematics.

Mark Z. Jacobson is currently Professor of Civil and Environmental Engineering at Stanford University. He received a B.S. with distinction in Civil Engineering, a B.A. with distinction in Economics, an M.S. in Environmental Engineering from Stanford University in 1988, and an M.S. and PhD in Atmospheric Sciences from UCLA in 1991 and 1994, respectively. In addition to numerous awards and scholarships, he received an NCAA-ITCA scholar-athlete of the year award while at Stanford University, in 1985, 1986, and 1987.

*To Dionna and Daniel*

# Fundamentals of Atmospheric Modeling

MARK Z. JACOBSON

CAMBRIDGE
UNIVERSITY PRESS

PUBLISHED BY THE PRESS SYNDICATE OF THE UNIVERSITY OF CAMBRIDGE
The Pitt Building, Trumpington Street, Cambridge, United Kingdom

CAMBRIDGE UNIVERSITY PRESS
The Edinburgh Building, Cambridge CB2 2RU, UK http://www.cup.cam.ac.uk
40 West 20th Street, New York, NY 10011-4211, USA http://www.cup.org
10 Stamford Road, Oakleigh, Melbourne 3166, Australia



First published 1999

Printed in the United States of America

Typeset in Sabon 10.25/13 pt. and Melior in LaTeX 2ε [TB]

*A catalog record for this book is available from
the British Library*

*Library of Congress Cataloging-in-Publication Data*

Jacobson, Mark Z. (Mark Zachary)
Fundamentals of atmospheric modeling / Mark Z. Jacobson.
p. cm.
Includes bibliographical references and index.
ISBN 0-521-63143-2 (hardcover). – ISBN 0-521-63717-1 (pbk.)
1. Atmospheric physics – Mathematical models. 2. Air – Pollution – Mathematical models. I. Title.
QC861.2.J3 1999
551.5 – dc21 98-15180
CIP

ISBN 0 521 63143 2 hardback
ISBN 0 521 63717 1 paperback

# Contents

# Preface

Modern atmospheric science is a field that combines meteorology, physics, mathematics, chemistry, computer sciences, and to a lesser extent geology, biology, microbiology, and oceanographic sciences. Until the late 1940s scientific studies of the atmosphere were limited primarily to studies of the weather. At that time, heightened concern about air pollution caused a surge of atmospheric chemistry studies, and computer modeling of meteorology and air pollution commenced. Since the late 1940s, the number of meteorological and air-pollution studies has increased rapidly, and meteorological and air-pollution models have slowly merged.

## BRIEF HISTORY OF METEOROLOGICAL SCIENCES

The history of atmospheric sciences begins with weather forecasting. Forecasting originally grew out of three needs – for farmers to produce crops, sailors to survive at sea, and populations to avoid weather-related disasters such as floods. Every society has forecast wind, rain, and other weather events. Some forecasts are embodied in platitudes and lore. Virgil stated, "Rain and wind increase after a thunderclap." The Zuni Indians had a saying, "If the first thunder is from the east, winter is over." Human experiences with the weather have led to more recent forecast rhymes, such as, "Rainbow in morning, sailors take warning. Rainbow at night, a sailor's delight."

Primitive forecasts have also been made based on animal and insect behavior or the presence of a human ailment. Bird migration was thought to predict oncoming winds. This correlation has since proved unreliable. Rheumatism, arthritis, and gout have been associated with the onset of rain, but such ailments are usually unrelated to the weather. The presence of locusts has correctly been associated with rainfall in that locusts fly downwind until they reach an area of converging winds, where rain is likely to occur.

In the 1870s, forecasting based on observations and experience became a profession. Many felt that early professional forecasting was more of an art than a science, since it was not based on scientific theory. Although the number of data available to forecasters was large and increasing, the data were not always used. Data were gathered by observers who used instruments that measured winds, pressure, temperature, humidity, and rainfall. Many of these instruments had been developed over the previous two centuries, although ideas and crude technologies existed prior to that time.

The Greeks, around 430 B.C., may have been the first to measure winds. Yet, reliable instruments to measure wind force and direction were not developed until the seventeenth century. In 1667 Robert Hooke developed the **pressure-plate anemometer**, which measured the deflection and force of wind on a sheet of metal hanging vertically. This principle was used again in the **pressure-tube anemometer**, thought of earlier but not built until the 1740s. Windmills were used as early as 644 A.D. in Persia, but the first **cup anemometer**, which applies the principle of the windmill to measure wind speed, was not developed until the seventeenth century, in France. In the nineteenth century, additional work on the anemometer was carried out by T. R. Robinson and W. H. Dines.

The **mercury barometer**, used to measure air pressure, was invented in 1643 by Evangelista Torricelli (1608–1647), an associate of Galileo Galilei (1564–1642). Toricelli invented the barometer (*Encyclopedia Britannia* 1980)

> to make an instrument which might show the changes of the air, now heavier and coarser, now lighter and more subtle.

By 1663, the Royal Society of London had built its own barometer based on Torricelli's model. The **aneroid barometer**, which represented an advance over the mercury barometer, was not adequately developed until 1843. The aneroid barometer contains no fluid. Instead, it measures pressure by gauging the expansion and contraction of a tightly sealed metal cell that contains no air.

A third important invention for meteorologists was the thermometer. Prior to 1600, Galileo devised the **thermoscope**, which measured the expansion of air to estimate temperature changes. The instrument did not have a scale and was unreliable. Torricelli's mercury barometer, which contained fluid, led to the invention of the **liquid-in-glass thermometer** in Florence by the mid–seventeenth century. In the early eighteenth century, useful thermometer scales were developed by Gabriel Daniel Fahrenheit of Germany (1686–1736) and Anders Celsius of Sweden (1701–1744).

A fourth important invention was the **hygrometer**, which measures humidity. Leonardo da Vinci (1452–1519) was probably the first to implement a hygrometer. He based his idea on notes of Nicolas Cryfts, who suggested in 1450 that a hygroscope could be constructed with dried wool placed on a scale. The change in weight of the wool would give a rough idea of the change in humidity. Wood and seaweed were used later in place of wool. In the seventeenth century, gut, string, cord, and hair were also used to measure humidity, since the change in length of these materials with humidity could be measured crudely. The hair hygrometer is still used today, although another instrument, the **psychrometer**, is more accurate. A psychrometer consists of two liquid-in-glass thermometers mounted together, one with a dry bulb and the other with a bulb covered with a moistened cloth.

Following the inventions above, observations of pressure, temperature, humidity, wind force, wind direction, and rainfall became regular. By the nineteenth century, weather-station networks and meteorological tables were common. Observers gathered data and forecasters used the data to predict the weather, but neither observers nor forecasters applied significant theory in their work. Theoreticians studied physical laws of nature but did not take advantage of the abundance of data available.

One of the first weather theoreticians was Aristotle, who wrote *Meteorologica* about 340 B.C. In that text, Aristotle attempted to explain the cause of winds, clouds, rain, mist, dew, frost, snow, hail, thunder, lightning, thunderstorms, halos, rainbows, and mock suns. On the subject of winds, he wrote (Lee 1951)

> These, then are the most important different winds and their positions. There are two reasons for there being more winds from the northerly than from the southerly regions. First, our inhabited region lies towards the north; second, far more rain and snow is pushed up into this region because the other lies beneath the sun and its course. These melt and are absorbed by the earth and when subsequently heated by the sun and the earth's own heat cause a greater and more extensive exhalation.

On the subject of thunder, he wrote,

> Let us now explain lightning and thunder, and then whirlwinds, firewinds and thunderbolts: for the cause of all of them must be assumed to be the same. As we have said, there are two kinds of exhalation, moist and dry; and their combination (air) contains both potentially. It condenses into cloud, as we have explained before, and the condensation of clouds is thicker towards their farther limit. Heat when radiated disperses into the upper region. But any of the dry exhalation that gets trapped when the air is in process of cooling is forcibly ejected as the clouds condense and in its course strikes the surrounding clouds, and the noise caused by the impact is what we call thunder.

Aristotle's monograph established a method of qualitatively explaining meteorological problems. Since Aristotle was incorrect about nearly all his meteorological conclusions, *Meteorologica* was never regarded as a significant work. Aristotle made observations, as evidenced by diagrams and descriptions in *Meteorologica*, but he did not conduct experiments. Lacking experiments, his conclusions, while rational, were not scientifically based.

Aristotle's method of rationalizing observations with little or no experiment governed meteorological theory through the seventeenth century. In 1637, René Descartes (1596–1650) wrote *Les Météores*, a series of essays attached to *Discours de la Méthode*. In some parts of this work, Descartes improved upon Aristotle's treatise by discussing experiments. In other parts, Descartes merely expanded or reformulated many of Aristotle's explanations. On the subject of northerly winds, Descartes wrote (Olscamp 1965)

> We also observe that the north winds blow primarily during the day, that they come from above to below, and that they are very violent, cold and dry. You can see the explanation of this by considering that the earth EBFD [referring to a diagram] is covered with many clouds and mists near the poles E and F, where it is hardly heated by the sun at all; and that at B, where the sun is immediately overhead, it excites a

> quantity of vapors which are quite agitated by the action of its light and rise into the air very quickly, until they have risen so high that the resistance of their weight makes it easier for them to swerve, . . . .

Like Aristotle, Descartes was incorrect about many explanations. Despite some of the weaknesses of his work, Descartes is credited for being one of the first in meteorological sciences to form hypotheses and then to conduct experiments.

Between the seventeenth and mid–nineteenth centuries, knowledge of basic physics increased, but mathematics and physics were still not used rigorously to explain atmospheric behavior. In 1860, William Ferrel published a collection of papers that were the first to apply mathematical theory to fluid motions on a rotating earth. This work was the impetus behind the modern-day field of **dynamical meteorology**, which uses physics and mathematics to explain atmospheric motion.

Between 1860 and the early 1900s weather forecasting and theory advanced along separate paths. In 1903, Vilhelm Bjerknes of Norway (1862–1951) promulgated the idea that weather forecasting should be based on the laws of physics. This idea was not new, but Bjerknes advanced it further than others (Nebeker 1995). Bjerknes thought that weather could be described by seven primary variables – pressure, temperature, air density, air water content, and the three components of wind velocity. He also realized that many of the equations describing the change in these variables were physical laws already discovered. Such laws included the continuity equation for air, Newton's law of motion, the ideal-gas law, the hydrostatic equation, and the thermodynamic energy equation.

Bjerknes did not believe that prognostic meteorological equations could be solved analytically. He advocated the use of physical principles to operate on graphical observations to forecast the weather. This technique was called **graphical calculus**. Between 1913 and 1919, Lewis Fry Richardson (1881–1953), developed a different method of analyzing the equations describing the weather (Richardson 1922). The method involved simplifying the equations before solving them numerically by hand. Richardson was not satisfied, because data available to test his method were sparse, and predictions from his method were not accurate. Nevertheless, his was the first attempt to numerically predict the weather in detail (ibid.).

Until the 1940s, much of Richardson's work was ignored because of the lack of a means to carry out the large number of calculations required to implement his method. In 1946, John von Neumann (1903–1957), who was associated with work to build the first electronic computer, proposed a project to make weather forecasting its main application. The project was approved, and the first computer model of the atmosphere was planned. Among the workers on von Neumann's project was Jule Charney, who became director of the project in 1948. Charney made the first numerical forecast on the computer, ENIAC, with a one-dimensional model. Since that time, numerical models of weather prediction have become more elaborate, and computers have become faster.

## BRIEF HISTORY OF AIR-POLLUTION SCIENCES

Meteorological science is an old and established field; air-pollution science has a shorter history. Natural air pollution has occurred on earth since the planet's formation. Fires, volcanic eruptions, meteorite impacts, and high winds all cause natural air-pollution. Human-made air-pollution problems have existed on urban scales for centuries and have resulted from burning of wood, vegetation, coal, oil, natural gas, waste, and chemicals.

Before the twentieth century, air pollution was not treated as a science but as a regulatory problem (Boubel et al. 1994). In Great Britain, emissions from furnaces and steam engines led to the Public Health Act of 1848. Emissions of hydrogen chloride from soap making led to the Alkali Act of 1863. In both cases, pollution abatement was controlled by agencies. In the nineteenth century, pollution abatement in the United States was delegated to municipalities. Regulations did not reduce pollution much, but they led to pollution control technologies, such as the scrubber for removing effluent gases from smokestacks and, later, the electrostatic precipitator for reducing particulate emissions from them.

Inventions unrelated to air-pollution regulation reduced some pollution problems. In the early twentieth century, the advent of the electric motor centralized sources of combustion at electric utilities, reducing local air pollution caused by the steam engine.

At the same time, widespread use of automobiles and other combustion processes increased pollution, especially in urban regions. Most noticeable was a layer of pollution that formed almost daily in Los Angeles, California. This pollution became so serious that an Air Pollution Control District was formed in Los Angeles in 1947. In 1949, the first National Air Pollution Symposium was held in Los Angeles. In 1951, Arie Haagen-Smit produced ozone in a laboratory from oxides of nitrogen and reactive organic gases, in the presence of solar radiation, and he suggested that these gases were the main constituents of Los Angeles air pollution. Such pollution became known as **photochemical smog**. Photochemical smog, due primarily to automobile emissions, has since been observed in most cities of the world.

The term **smog** was first coined in 1905 by Harold Antoine Des Voeux, who described the combination of smoke and fog he observed in cities in Great Britain. The smoke was due to chimney and stack emissions of coal combustion products. In December 1952, such smog resulted in over 4000 deaths in London. This fatal episode was not the first in London. Pollution resulting from coal combustion in the presence of fog is commonly referred to as **London-type smog**.

## THE MERGING OF AIR-POLLUTION AND METEOROLOGICAL SCIENCES

In the 1950s, laboratory work was undertaken to better understand the formation of photochemical and London-type smog. Since the computer was already available, box models simulating atmospheric chemical reactions were immediately implemented. Between the 1950s and 1970s, air-pollution models, termed **air-quality models**, were expanded to three dimensions. Such models included treatment of transport, deposition, emissions, and gas chemistry. Most of these models used observed meteorological data as inputs. More recently, air quality models have used meteorological fields, either precalculated or calculated in real time, as inputs.

In the 1970s, atmospheric pollution problems, aside from urban air pollution, were increasingly recognized. Such problems included regional acid deposition, global ozone reduction, Antarctic ozone depletion, and global climate change. Initially, ozone depletion and climate change problems were treated separately by dynamical meteorologists and atmospheric chemists. More recently, computer models that incorporate atmospheric chemistry and dynamical meteorology have been used to study these problems.

The purposes of this book are to provide (1) a physical understanding of dynamical meteorology, gas chemistry, aerosol microphysics and chemistry, radiation, and cloud processes in the atmosphere, (2) a description of numerical methods and computational techniques used to simulate these processes, and (3) a catalog of steps required to construct, apply, and test a numerical model.

After the overview in the first chapter, atmospheric structure, composition, and thermodynamics are described in Chapter 2. In Chapters 3–5, basic equations describing dynamical meteorology are derived. In Chapter 6, numerical methods of solving partial differential equations are discussed. A finite-difference technique of solving dynamical meteorological equations is provided in Chapter 7. In Chapters 8 and 9, boundary-layer and cloud processes, respectively, are described. Chapter 10 introduces radiation. Chapters 11–13 focus on photochemistry and numerical methods of solving chemical equations. Chapters 14–19 describe aerosol physical and chemical processes. Chapter 20 describes sedimentation and dry deposition. Chapter 21 outlines computer model development, application, and testing.

The book is designed as a graduate, upper-level undergraduate, and research text. The text assumes students have a basic physical science, mathematical, and computational background. Both Système Internationale (SI) and centimeter-gram-second (CGS) units are used. Dynamical meteorologists often use SI units, and atmospheric chemists often use CGS units. Some chemical variables, such as gas concentrations, absorption cross sections, and rate coefficients, are most conveniently written in CGS units. Some meteorological variables, such as wind speed, geopotential, and energy, are most conveniently written in SI units. Thus, both unit systems are retained. Unit and variable conversions are given in Appendix A.

## ACKNOWLEDGMENTS

I would like to thank several colleagues who reviewed different sections of this text. In particular, I am indebted to (in alphabetical order) Akio Arakawa, Bob Chatfield, Frank Freedman, Ann Fridlind, James Holton, Daniel Jacob, Jinyou Liang, Jin-Sheng Lin, Gerard Ketefian, Doug Rotman, Roberto San Jose, Hanwant Singh, Amy Stuart, Azadeh Tabazadeh, and Don Wuebbles, who all provided significant comments, suggestions, and/or corrections relating to the text.

# 1

# Introduction

## 1.1. WEATHER, CLIMATE, AND AIR POLLUTION

A MODEL is a mathematical representation of a process. An **atmospheric computer model** is a computerized mathematical representation of dynamical, physical, chemical, and radiative processes in the atmosphere. In atmospheric models, time-dependent processes are mathematically represented by ordinary differential equations. Space- and time-dependent processes are represented by partial differential equations. Ordinary and partial differential equations are approximated with finite-difference or other methods, and solutions to the approximated equations are computerized.

Computer models also solve parameterized and empirical equations. A **parameterized equation** is an equation in which one parameter is expressed in terms of at least two other parameters. The equation of state, which relates pressure to temperature and air density, is a parameterized equation. An **empirical equation** is an equation in which one parameter is expressed as an empirical function (e.g., a polynomial fit) of at least one other parameter. Whereas parameterized equations are derived from insight, empirical equations do not always make physical sense. Instead, they reproduce observed results under a variety of conditions. In this text, computer modeling of the atmosphere is discussed. Such modeling requires solutions to ordinary differential equations, partial differential equations, parameterized equations, and empirical equations.

Since the advent of atmospheric computer modeling in 1948, models have been applied to study weather, climate, and air pollution on urban, regional, and global scales. **Weather** is the state of the atmosphere at a given time and place, and **climate** is the average of weather events over a long period. Some basic weather variables include wind speed, wind direction, pressure, temperature, relative humidity, and rainfall. Standard climate variables include mean annual temperatures and mean annual rainfall at a given location or averaged over a region.

**Air pollutants** are gases, liquids, or solids suspended in the atmosphere in high enough concentrations to affect human, animal, or vegetation health, or to erode structures. Standard air-pollution problems include urban smog, acid deposition, Antarctic ozone depletion, global ozone reduction, and global climate change. **Urban smog** is characterized by the local buildup of high concentrations of gases and particles, emitted from automobiles, smokestacks, and other human-made sources, or formed by secondary processes in the atmosphere. **Acid deposition**

occurs following long-range transport of sulfur dioxide gas emitted from coal-fired power plants, conversion of the sulfur dioxide to liquid-phase sulfuric acid, and deposition of sulfuric-acid-related species to the ground by rain or another means. Sulfuric acid harms soils, lakes, and forests and damages structures. Acid deposition also occurs when nitric acid gas, produced chemically from automobile pollutants, dissolves into fog drops, which deposit to ground surfaces or to the lungs of humans. This form of acid deposition is **acid fog.**

**Antarctic ozone depletion** and **global ozone reduction** are caused, to a large extent, by human-made chlorine and bromine compounds that are emitted to the atmosphere and break down only after they have diffused to the upper atmosphere. Ozone reduction increases the intensity of ultraviolet radiation from the sun reaching the ground. Some ultraviolet wavelengths destroy microorganisms on the surface of the earth and cause skin cancer in humans. **Global climate change** is characterized by changes in global temperatures and rainfall patterns due to increases in atmospheric carbon dioxide, methane, nitrous oxide, water vapor, and other gases that absorb infrared radiation. The addition of particles to the atmosphere may offset some of the warming caused by increases in gas concentrations.

Historically, meteorological models have been used to simulate weather, climate, and climate change. Photochemical models have been used to study urban, regional, and global air-pollution emissions, chemistry, aerosol processes, and transport of pollutants. Only recently have meteorological models been merged with photochemical models to tackle these problems together.

One purpose of developing a model is to better understand the physical, chemical, dynamical, and radiative properties of air pollution and meteorology. A second purpose is to improve the model so that it may be used for forecasting. A third purpose is to develop a tool that can be used for policy making. With an accurate model, policy makers can attempt to mitigate pollution problems.

## 1.2. SCALES OF MOTION

Atmospheric problems can be simulated over a variety of spatial scales. **Molecular-scale** motions occur over distances much smaller than 2 mm. Molecular diffusion is a molecular-scale motion. **Microscale** motions occur over distances of 2 mm to 2 km. Eddies, or swirling motions of air, are microscale events. **Mesoscale** motions, such as thunderstorms, occur over distances of 2–2,000 km. The **synoptic scale** covers motions or events on a scale of 500–10,000 km. High- and low-pressure systems and the Antarctic ozone hole occur over the synoptic scale. **Planetary-scale** events are those larger than synoptic-scale events. Global wind systems are planetary-scale motions.

Some phenomena occur on more than one scale. Acid deposition is a mesoscale and synoptic-scale phenomenon. Table 1.1 summarizes atmospheric scales and motions or phenomena occurring on each scale.

**Table 1.1.** Scales of Atmospheric Motion

| Scale Name | Scale Dimension | Examples |
|---|---|---|
| Molecular scale | ≪2 mm | Molecular diffusion, molecular viscosity |
| Microscale | 2 mm–2 km | Eddies, small plumes, car exhaust, cumulus clouds |
| Mesoscale | 2–2,000 km | Gravity waves, thunderstorms, tornados, cloud clusters, local winds, urban air pollution |
| Synoptic scale | 500–10,000 km | High- and low-pressure systems, weather fronts, tropical storms, hurricanes, Antarctic ozone hole |
| Planetary scale | >10,000 km | Global wind systems, Rossby (planetary) waves, stratospheric ozone reduction, global warming |

## 1.3. ATMOSPHERIC PROCESSES

Atmospheric models simulate many processes and feedbacks among them. Figure 1.1 shows a diagram of a model that simulates gas, radiative, aerosol, dynamical, transport, and cloud processes. **Gases** are distinguished from **aerosols** in two ways. First, gases consist of individual atoms or molecules, whereas aerosols consist of aggregates of atoms and/or molecules, bonded together. Thus, aerosols are larger than gases. Second, whereas aerosols are liquids or solids suspended in the air, gases are in their own phase state. Aerosols can contain many components, including liquid water. **Cloud drops** are aerosols on which significant water vapor has condensed or deposited and are larger than most, but not all, aerosols. **Raindrops** are cloud drops that have aggregated together or grown in size. Gas molecules have diameters on the order of $5 \times 10^{-10}$ m, aerosols have diameters that range in size from a few tens of gas molecules to 10 mm, cloud drops have diameters between 10 and 200 μm, and raindrops have diameters between 200 μm and 8 mm. In this text, a **particle** is defined as an aerosol, cloud drop, or raindrop.

An important process that affects gases is chemistry. Gas **kinetic chemistry** is the process by which reactant gases collide with each other and transform to product gases. **Photolysis** is the process by which reactant gases are broken down by sunlight to form products. **Photochemistry** is defined here to encompass kinetic chemistry and photolysis.

Gases are also affected by **gas-to-particle conversion.** Conversion processes include nucleation, condensation/evaporation, dissolution/evaporation, deposition/sublimation, and heterogeneous chemistry. **Nucleation** occurs when gas molecules aggregate and condense in a cluster until it reaches a critical size, at which point it is a small particle. **Condensation** occurs when a gas diffuses to and sticks to the surface of a particle and changes state to a liquid. **Evaporation** occurs when a liquid molecule on the surface of a particle changes state to the gas phase and diffuses away from the surface. **Dissolution** occurs when a gas molecule diffuses to and dissolves into liquid on the surface of a particle. Evaporation, in this case, is the opposite of dissolution. **Gas deposition** is the process by which a gas diffuses

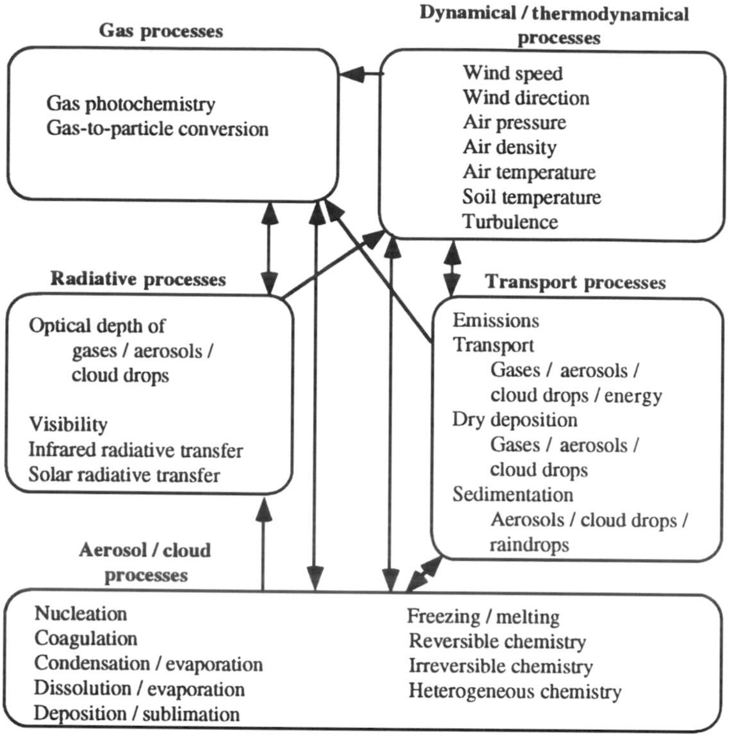


**Figure 1.1.** Diagram of processes simulated in an atmospheric air-pollution model and the interactions among them. Adapted from Jacobson et al. 1996; Jacobson 1997a,b.

to and sticks to the surface of a particle and changes state to a solid. **Sublimation** occurs when a solid molecule on the surface of a particle changes state to the gas phase and diffuses away from the surface. Gases react chemically on the surfaces of particles during **heterogeneous chemistry**.

Gas concentrations are also affected by emissions, transport, and dry deposition. **Emissions** usually originate from near or above the ground. **Transport** may be horizontal or vertical. **Dry deposition** (different from gas deposition) occurs when gases (or particles) impinge upon and stick to a surface. Gases impinge upon a surface when the wind or eddies carry them to the surface. **Molecular diffusion**, which is the movement, collision, and random redirection of molecules, also transports gases to the surface.

Gases influence meteorology through radiative transfer. Gases absorb and scatter **solar radiation** (emitted by the sun) and **infrared radiation** (emitted by the earth, atmosphere, and the sun). Changes in absorption and scattering affect temperatures, which affect pressures. Pressures affect wind speeds and directions.

Aerosol and cloud processes include gas-to-particle conversion, coagulation, freezing/melting, equilibrium chemistry, and aqueous chemistry. **Coagulation** occurs when two particles collide and stick together to form a third, larger particle. During **freezing**, liquid within a particle changes state to a solid. **Melting** is the reverse. **Equilibrium reactions** are reversible chemical reactions that occur between or among liquids, ions, and/or solids within particles. **Aqueous reactions** are irreversible reactions that usually occur among species dissolved in water.

Aerosol and cloud drops are also affected by emissions, transport, dry deposition, and sedimentation. **Sedimentation** is the process by which particles fall from one altitude to another or to the surface due to their weight. This differs from dry deposition, which occurs when particles diffuse to, blow over, or fall on a surface and stick to the surface. Aerosols and cloud drops scatter and absorb solar and infrared radiation. Gases, aerosols, and cloud drops all affect visibility, but aerosols (and clouds) affect visibility more than do gases.

The primary meteorological variables in an atmospheric model are wind speed, wind direction, air temperature, air density, air pressure, and water content. These variables are simulated by solving a set of partial differential equations and parameterized equations, including the **momentum equation**, the **thermodynamic energy equation**, the **continuity equation for air**, the **equation of state**, and the **continuity equation for total water**. The primary radiative variables in a model are heating rates and actinic fluxes. The former are used to calculate changes in temperature, and the latter are used to calculate photolysis rate coefficients. Both variables are calculated with the **radiative-transfer equation**. Pollutant parameters in a model are gas and aerosol concentrations. Changes in concentration are found by solving ordinary differential equations that describe chemistry and physics and partial differential equations that describe transport.

In sum, weather, climate, and air pollution can be modeled by taking into account a fairly well-defined set of physical, chemical, and/or dynamical equations. In this book, the processes shown in Fig. 1.1 are examined. The first discussion centers around atmospheric thermodynamics and dynamics. Descriptions of atmospheric transport, cloud processes, radiation, gas photochemistry, aerosol microphysics and chemistry, deposition, and model application follow.

# 2

# Atmospheric Structure, Composition, and Thermodynamics

THE atmosphere contains a few highly-concentrated gases, such as nitrogen, oxygen, and argon, and many trace gases, among them carbon dioxide, methane, and ozone. All such gases are constituents of air. Important characteristics of air are its pressure, density, and temperature. These parameters vary with altitude, latitude, longitude, and season and are related to each other by the equation of state. Other fundamental equations applicable to the atmosphere are the Clausius–Clapeyron equation and the first law of thermodynamics. The Clausius–Clapeyron equation relates temperature to the quantity of water vapor over a surface at saturation. The first law of thermodynamics relates the temperature change of a gas to energy transfer and the change in work. In this chapter, atmospheric variables and gases are discussed, and basic equations describing atmospheric physics and thermodynamics are introduced.

## 2.1. PRESSURE, DENSITY, AND COMPOSITION

In the earth's atmosphere, air pressure, temperature, and density change with altitude. **Air pressure** is the weight of air above a horizontal plane, divided by the area of the plane. Since the weight of air per unit area above a given altitude is always greater than the weight of air per unit area above any higher altitude, pressure decreases with increasing altitude. In fact, pressure decreases exponentially with altitude. **Standard sea-level surface pressure** is 1013 mb (or $1.013 \times 10^5$ Pa or 760 mm Hg). The sea-level pressure at a given location and time typically varies by +10 to −20 mb from standard sea-level pressure. In a strong low-pressure system, such as at the center of a hurricane, the actual sea-level pressure may be more than 50 mb lower than standard sea-level pressure.

**Air density** is the mass of air per unit volume of air. Since atmospheric mass is concentrated near the surface, air density is highest near the surface. Air density decreases exponentially with altitude. Figure 2.1(a) and (b) show standard profiles of atmospheric pressure and density, respectively. Figure 2.1(a) shows that 50 percent of the atmospheric mass lies between sea level and 5.5 km. About 99.9 percent of the mass lies below about 48 km. The earth's radius is approximately 6370 km. Thus, almost all of the earth's atmosphere lies in a layer thinner than 1 percent of the radius of the earth.

**Table 2.1.** Average Composition of the Lowest 100 km in the Earth's Atmosphere

| Gas | Volume Mixing Ratio (percent) | (ppmv) |
|---|---|---|
| **Fixed Gases** | | |
| Nitrogen ($N_2$) | 78.08 | 780,000 |
| Oxygen ($O_2$) | 20.95 | 209,500 |
| Argon (Ar) | 0.93 | 9,300 |
| Neon (Ne) | 0.0015 | 15 |
| Helium (He) | 0.0005 | 5 |
| Krypton (Kr) | 0.0001 | 1 |
| Xenon (Xe) | 0.000005 | 0.05 |
| **Variable Gases** | | |
| Water vapor ($H_2O$) | 0.00001–4.0 | 0.1–40,000 |
| Carbon dioxide ($CO_2$) | 0.0360 | 360 |
| Methane ($CH_4$) | 0.00017 | 1.7 |
| Ozone ($O_3$) | 0.000003–0.001 | 0.03–10 |

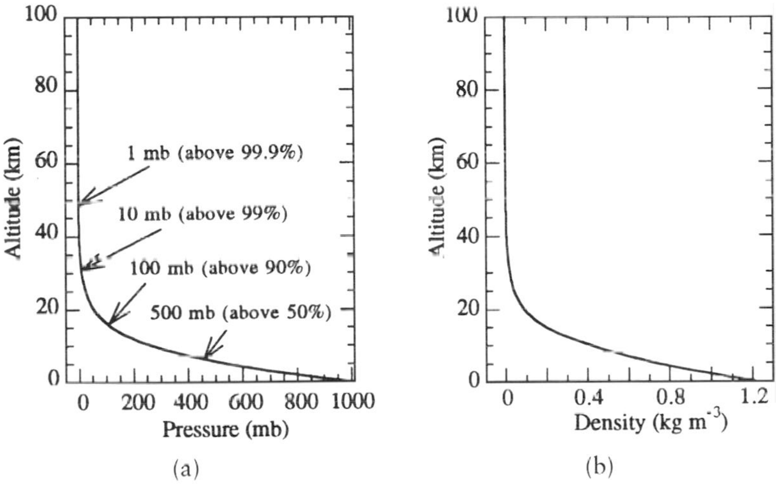


**Figure 2.1.** (a) Pressure and (b) density versus altitude in the earth's lower atmosphere. The pressure diagram shows that 99.9 percent of the atmosphere lies below an altitude of about 48 km (1 mb), and 50 percent lies below about 5.5 km (500 mb).

### 2.1.1. Fixed Gases

Table 2.1 gives the basic composition of the bottom 100 km of the earth's atmosphere, called the **homosphere.** In this region, the primary gases are molecular nitrogen ($N_2$) and molecular oxygen ($O_2$), which together make up over 99 percent of all air molecules. Argon (Ar), a chemically inert gas, makes up most of the remaining 1 percent. Nitrogen, oxygen, and argon are **fixed gases** in that their **volume mixing ratios** (number of molecules of each gas divided by the total number of molecules of dry air) do not change substantially from time to time or place to place. Fixed gases are well mixed in the homosphere. At any given altitude, oxygen makes up about

20.95 percent and nitrogen makes up about 78.08 percent of all non-water gas molecules by volume (23.17 and 75.55 percent, respectively, by mass).

#### 2.1.1.1. *Molecular Nitrogen*

Gas-phase molecular nitrogen is produced biologically in soils. During their growth, denitrifying bacteria reduce nitrate ($NO_3^-$) to $N_2$. This process, called **denitrification**, occurs only in anaerobic (oxygen-depleted) soils. Denitrifying bacteria also reduce nitrate to nitrous oxide gas ($N_2O$), but to a much lesser extent. Nitrate in soils is produced during **nitrification**, which is a two-step process that occurs only in aerobic (oxygen-containing) environments. In the first step, nitrosofying (nitrite-forming) bacteria produce nitrite ($NO_2^-$) from **ammonium** ($NH_4^+$). In the second step, nitrifying (nitrate-forming) bacteria produce nitrate from nitrite.

Ammonium in soils originates from several sources. Bacteria decompose organic compounds in soils to ammonium during **ammonification**. Human-made fertilizers also contain ammonium. Third, during **nitrogen fixation** nitrogen-fixing bacteria, such as *Rhizobium*, *Azotobacter*, and *Beijerinckia*, convert $N_2$ to ammonium. Nitrogen fixation occurs in aerobic environments. Because $N_2$ does not react significantly in the atmosphere and because its removal by nitrogen fixation is slower than its production by denitrification, the concentration of $N_2$ in the atmosphere has built up over time.

#### 2.1.1.2. *Molecular Oxygen*

Gas-phase molecular oxygen is produced by **green-plant photosynthesis**, which occurs when carbon dioxide gas ($CO_2$) reacts with water ($H_2O$) in the presence of sunlight and green-pigmented **chlorophylls**. Chlorophylls exist in plants, trees, blue–green algae, and certain bacteria. They appear green because they absorb red and blue wavelengths of visible light and reflect green wavelengths. Products of photosynthesis reactions are carbohydrates and molecular oxygen gas. One photosynthetic reaction is

$$6CO_2 + 6H_2O + h\nu \rightarrow C_6H_{12}O_6 + 6O_2 \qquad (2.1)$$

where $h\nu$ is a photon of radiation, and $C_6H_{12}O_6$ is glucose. Some bacteria that live in anaerobic environments photosynthesize carbon dioxide with hydrogen sulfide ($H_2S$) instead of with water to produce organic material and elemental sulfur (S). This type of photosynthesis is **anoxygenic photosynthesis** and predates the onset of green-plant photosynthesis.

#### 2.1.1.3. *Argon and Others*

Argon is a noble gas that is colorless and odorless. Like other noble gases, argon is inert and does not react chemically. Other fixed but inert gases present in trace concentrations are neon (Ne), helium (He), krypton (Kr), and xenon (Xe). The mixing ratios of these gases are given in Table 2.1.

### 2.1.2. Variable Gases

Gases in the atmosphere whose volume mixing ratios change significantly from time to time and/or place to place are **variable gases.** Variable gases include water vapor, carbon dioxide, methane ($CH_4$), and ozone ($O_3$), among others.

#### 2.1.2.1. *Water Vapor*

Water vapor enters the air by evaporation from soil, lakes, streams, rivers, and ocean surfaces, sublimation from ice in glaciers, and **transpiration** from plant leaves. It is also produced by gas-phase chemical reactions. Approximately 85 percent of water in the atmosphere originates from ocean surface evaporation. Sinks of water vapor are condensation to the liquid phase, deposition to the ice phase, transfer to the ocean and other surfaces, and gas-phase chemical reaction. The mixing ratio of water vapor in the atmosphere varies with location. When temperatures are low, water vapor readily condenses as a liquid or deposits as ice. Over the North and South Poles, water-vapor mixing ratios are almost zero. When temperatures are high, liquid water readily evaporates and ice readily sublimates to the gas phase. Over equatorial waters, where temperatures are high and evaporation from ocean surfaces readily occurs, the atmosphere contains up to 4 percent or more water vapor by volume. Water vapor is not only a **greenhouse gas** (a gas that readily absorbs infrared radiation), but also a chemical reactant and carrier of latent heat.

#### 2.1.2.2. *Carbon Dioxide*

Carbon dioxide gas is produced by cellular respiration in plants and trees, biological decomposition of dead organic matter, evaporation from the oceans, volcanic outgassing, and fossil-fuel combustion. **Cellular respiration** occurs when oxygen reacts with carbohydrates in the presence of enzymes in living cells. The products of such reactions are $CO_2$, $H_2O$, and energy. The reverse of (2.1) is a cellular respiration reaction. **Biological decomposition** occurs when bacteria and other organisms convert dead organic matter to $CO_2$ and $H_2O$. $CO_2$ is removed from the atmosphere by green-plant photosynthesis and dissolution into ocean water and raindrops.

Figure 2.2 shows how observed $CO_2$ mixing ratios have increased steadily over the last 35 years at a location over the Pacific Ocean. Average global $CO_2$ mixing ratios have increased from approximately 280 parts per million by volume (ppmv) in the mid 1800s to approximately 360 ppmv in the 1990s. The yearly increases are due to increased rates of $CO_2$ emissions from fossil-fuel combustion. The seasonal fluctuation in $CO_2$ mixing ratio is due to photosynthesis and biological decomposition. When annual plants grow in the spring and summer, they remove $CO_2$ from the atmosphere by photosynthesis. When they die in the fall and winter, they return $CO_2$ by biological decomposition.

Like water vapor, carbon dioxide is a greenhouse gas. Unlike water vapor, $CO_2$ does not react chemically in the atmosphere. Its lifetime against chemical destruction is approximately 100–200 years. A more important removal mechanism is its

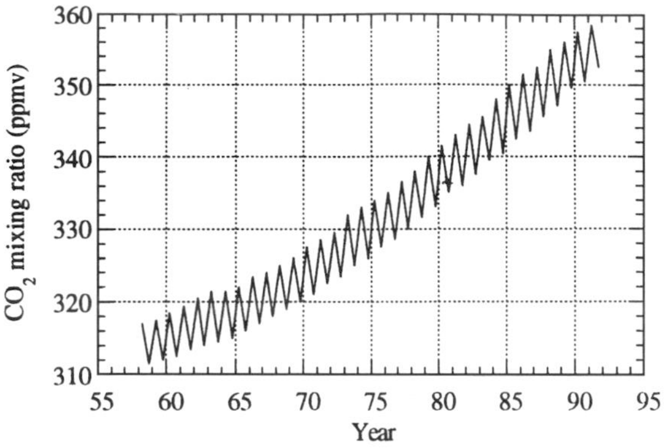


**Figure 2.2.** Yearly and seasonal fluctuations in carbon dioxide mixing ratio since 1958. Data from C. D. Keeling at the Mauna Loa Observatory, Hawaii.

dissolution into ocean water. In the oceans, carbon dioxide dissociates primarily into bicarbonate ($HCO_3^-$). Bicarbonate may be converted to shell material, which eventually sinks to the ocean bottom and fossilizes to form carbonaceous rocks.

### 2.1.2.3. *Methane*

Methane, or natural gas, is a variable gas. Methane is produced in anaerobic environments, where methane-producing bacteria (methanogens) consume organic material and excrete methane. Ripe anaerobic environments include rice paddies, landfills, wetlands, and the digestive tracts of cattle, sheep, and termites. Methane is also produced in the ground from the decomposition of fossilized carbon. This methane often leaks to the atmosphere or is harnessed and used as a source of energy. Additional methane is produced from biomass burning and as a byproduct of atmospheric chemical reactions. Sinks of methane include slow chemical reaction and dry deposition to the ground. Methane's mixing ratio in the lower atmosphere is near 0.17 ppmv, which is an increase from about 0.08 ppmv in the mid-1800s (Ethridge et al. 1992). Its mixing ratio has increased steadily due to increased farming and landfill usage. Methane is important because it is a greenhouse gas that absorbs infrared radiation emitted by the earth 21 times more efficiently, molecule for molecule, than does $CO_2$. Mixing ratios of carbon dioxide however, are much larger than those of methane.

### 2.1.2.4. *Ozone*

Ozone is a trace gas formed by photochemical reaction and not emitted into the atmosphere. In the stratosphere, its production proceeds after photolysis of molecular oxygen. Near the surface of earth, its production proceeds after photolysis of nitrogen dioxide ($NO_2$). Photolysis of molecular oxygen does not occur near the ground, since the wavelengths of radiation required to break apart oxygen are absorbed by molecular oxygen and ozone higher in the atmosphere. In urban regions, ozone production is enhanced by the presence of organic gases. Typical ozone mixing ratios in urban air, rural surface air, and stratospheric air are 0.1, 0.04, and 10 ppmv, respectively. Thus, ozone's mixing ratios are highest in the **stratosphere**, as shown in Fig. 2.4 (Section 2.2.2.1).

The atmosphere contains trace quantities of many other variable gases. These include carbon monoxide (CO), nitrous oxide ($N_2O$), hydrogen ($H_2$), nitric acid ($HNO_3$), ammonia ($NH_3$), sulfuric acid ($H_2SO_4$), nitric oxide (NO), and nitrogen dioxide ($NO_2$). They are discussed in later chapters of this text.

## 2.2. TEMPERATURE STRUCTURE

**Temperature** is proportional to the average kinetic energy of an air molecule. From gas kinetic theory, absolute temperature (K) is obtained from

$$\frac{4}{\pi}k_B T = \frac{1}{2}\bar{M}\bar{v}_a^2 \tag{2.2}$$

where $k_B$ is **Boltzmann's constant** ($1.3807 \times 10^{-23}$ kg m$^2$ s$^{-2}$ K$^{-1}$ molecule$^{-1}$), $\bar{M}$ is the average mass of one air molecule ($4.8096 \times 10^{-26}$ kg molecule$^{-1}$), and $\bar{v}_a$ is the average thermal velocity of an air molecule (m s$^{-1}$). Alternative units of $k_B$ are given in Appendix A. The right side of (2.2) is the average kinetic energy of a molecule.

The bottom 100 km of earth's atmosphere, or homosphere, is the lower atmosphere and is divided into four major regions in which temperature changes with altitude. These are, from bottom to top, the **troposphere, stratosphere, mesosphere,** and **thermosphere**. The troposphere is divided into the **boundary layer**, which is the region from the surface to about 500–3000-m altitude, and the **free troposphere**, which is the rest of the troposphere.

### 2.2.1. Boundary Layer

The **boundary layer** is the portion of the troposphere influenced by the earth's surface and that responds to surface forcings with a time scale of about an hour or less (Stull 1988). The free troposphere is influenced by the boundary layer, but on a longer time scale. The temperature varies significantly in the boundary layer during the day and between day and night. Variations are less significant in the free troposphere. Temperatures in the boundary layer depend on the specific heats of soil and air and on energy transfer processes, such as conduction, radiation, mechanical turbulence, thermal turbulence, and advection. These factors are described below.

#### 2.2.1.1. *Specific Heat*

**Specific heat** is the energy required to increase the temperature of 1 gram of a substance 1 degree Celsius (°C). Soil has a lower specific heat than does water. During the day, the addition of a fixed amount of solar energy increases soil temperatures

**Table 2.2.** Specific Heats and Thermal Conductivities of Four Media

| **Substance** | **Specific Heat ($J\ kg^{-1}\ K^{-1}$)** | **Thermal Conductivity at 298 K ($J\ m^{-1}\ s^{-1}\ K^{-1}$)** |
|---|---|---|
| Dry air at constant pressure | 1004.67 | 0.0256 |
| Liquid water | 4185.5 | 0.6 |
| Clay | 1360 | 0.920 |
| Dry sand | 827 | 0.298 |

more than water temperatures. At night, emissions of a fixed amount of infrared radiation decreases soil temperatures more than water temperatures. Thus, between day and night, soil temperatures vary more than do water temperatures.

Specific heat varies not only between land and water, but also between different soil types, as shown in Table 2.2. The table shows that the specific heat of clay is greater than that of sand, and the specific heat of liquid water is greater than that of clay and sand. If all else is the same, sandy soil heats to a higher temperature than does clay soil, and dry soil heats to a higher temperature than does wet soil during the day. Dry, sandy soils cool to a greater extent than do wet, clayey soils at night.

### 2.2.1.2. *Conduction*

Specific heat is a property of a material that affects its change in temperature. Temperature is also affected by processes that transfer energy within or between materials. One such process is conduction. **Conduction** is the passage of energy from one molecule to the next in a medium (the conductor). The medium, as a whole, experiences no molecular movement. Conduction occurs through soil, air, and particles. Conduction affects ground temperature by transferring energy between the soil surface and bottom molecular layers of the atmosphere, and between the soil surface and molecules of soil just below the surface. The rate of a material's conduction is determined by its **thermal conductivity**, which quantifies the rate of flow of thermal energy through a material in the presence of a temperature gradient. The **thermal conductivity of dry air** ($J\ m^{-1}\ s^{-1}\ K^{-1}$) can be approximated from the empirical equation

$$\kappa_d \approx 0.023807 + 7.1128 \times 10^{-5}(T - 273.16) \tag{2.3}$$

(List 1984), where $T$ is temperature (K). Under atmospheric conditions, the thermal conductivity of dry air is negligibly different from that of air containing water vapor (Pruppacher and Klett 1997). (2.3) is used for dry and moist air throughout this text. Table 2.2 shows that liquid water, clay, and dry sand are more conductive than is dry air. Clay is more conductive and dry sand is less conductive than liquid water.

The vertical flux of energy in the air due to conduction ($W\ m^{-2}$) is approximated with the **conductive heat flux equation,** $H_c = -\kappa_d \Delta T/\Delta z$, where $\Delta T$ (K) is the change in temperature over an incremental height $\Delta z$ (m). Adjacent to the ground, molecules of soil, water, and other surface elements transfer energy by conduction to molecules of air overlying the surface. Since the temperature gradient ($\Delta T/\Delta z$) between the surface and a thin (e.g., 1 mm) layer of air just above the surface is large, the conductive heat flux at the ground is large. Above the ground, temperature gradients are smaller and the conductive heat flux is less important than at the ground.

**Example 2.1**

Find the conductive heat flux through a 1-mm thin layer of air touching the surface if $T = 298$ K and $\Delta T = -12$ K. Do a similar calculation for the free troposphere, where $T = 273$ K and $\Delta T/\Delta z = -6.5$ K km$^{-1}$.

SOLUTION

Near the surface, $\kappa_d = 0.0256$ J m$^{-1}$ s$^{-1}$ K$^{-1}$; thus, the conductive heat flux at the surface is $H_c = 307$ W m$^{-2}$. In the free troposphere, $\kappa_d = 0.0238$ J m$^{-1}$ s$^{-1}$ K$^{-1}$; thus, $H_c = 1.5 \times 10^{-4}$ W m$^{-2}$, which is much smaller than the value at the surface. Heat conduction through the air is important only at the ground surface.

#### 2.2.1.3. *Radiation*

**Radiation** is the transfer of energy by electromagnetic waves. Such waves do not require a medium, such as air, for their transmission. **Solar radiation** is radiation in relatively short wavelength bands emitted by the sun. **Infrared radiation** is radiation in relatively long wavelength bands emitted by the earth, the atmosphere, and other bodies of similar temperatures. The earth's surface receives solar radiation during the daytime only, but its surface and atmosphere emit infrared radiation during the day and night. Radiation is an important energy transfer processes and is discussed in Chapter 10.

#### 2.2.1.4. *Mechanical Turbulence and Forced Convection*

**Convection** is a predominantly vertical motion that results in transport and mixing of atmospheric properties. **Forced convection** is vertical motion produced by mechanical means, such as mechanical turbulence. **Mechanical turbulence** arises when winds blow over objects protruding from a surface, producing swirling motions of air, or eddies. **Turbulence** is the effect of groups of eddies of different sizes. Turbulence mixes energy and other variables vertically and horizontally. Strong winds produce strong eddies and turbulence. Turbulence from wind-generated eddies is mechanical turbulence.

When mechanical turbulence is the dominant process of vertical motion in the boundary layer, the boundary layer is in a state of forced convection. Mechanical turbulence is only one type of forced convection; forced convection also occurs when air rises along a topographical barrier or weather front, or when horizontal winds converge and rise.

### 2.2.1.5. *Thermal Turbulence and Free Convection*

**Free convection** is a predominantly vertical motion produced by buoyancy. The boundary layer is in a state of free convection when thermal turbulence is the dominant process of vertical motion. **Thermal turbulence** occurs when the sun heats the ground differentially. Differential heating occurs because clouds or hills block the sun in some areas but not in others or different areas of a surface lie at different angles. Over a warm, sunlit surface, conduction transfers energy from the ground to molecules of air adjacent to the ground. The warmed air above the ground rises buoyantly and expands. Cool air from nearby is drawn in to replace the rising air. The cool air heats by conduction and rises. The resulting parcel of buoyant air is a **thermal**.

Free convection differs from conduction in that free convection is the mass movement, due to density differences, of molecules containing energy, and conduction is the transfer of energy from molecule to molecule. Free convection occurs most readily over land when the sky is cloud-free and winds are light.

### 2.2.1.6. *Advection and Other Factors*

**Advection** is the horizontal propagation of the mean wind. Horizontal winds advect energy spatially just as they advect gases and particles. Advection is responsible for transferring energy, not only in microscale and mesoscale environments, but also in synoptic and global scale environments.

Other processes that affect temperatures in the boundary layer are surface emissivity, surface albedo, and length of day. **Emissivity** is the ratio of the radiation emitted by an object to the radiation emitted by a perfect emitter. Sand has an emissivity of 0.84–0.91, and clay has an emissivity of 0.9–0.98. The higher its emissivity, the faster a surface cools at night. **Albedo** (or reflectivity) is the ratio of reflected radiation to incident radiation. For dry sand, the albedo varies from 20 to 40 percent. For clay, it varies from 5 to 20 percent. Thus, sand reflects more solar radiation during the day and emits less infrared radiation at night than does clay, counteracting some of the effects of the low specific heat and thermal conductivity of sand.

Temperature variations in the boundary layer are affected by large-scale pressure systems. Within a large-scale high-pressure system, such as the Pacific high, air descends and warms, often on top of cooler surface air. In such cases, an **inversion,** which is an increase in temperature with increasing altitude, is created. An inversion associated with a large-scale pressure system is a **large-scale subsidence inversion.** Temperatures are also affected by length of day. Longer days produce longer periods of surface heating, and longer nights produce longer periods of infrared emission.

### 2.2.1.7. *Boundary-Layer Characteristics*

Figures 2.3(a) and (b) show the temperature variation in the boundary layer over land during the day and night, respectively, under a high pressure system. During the day, the boundary layer is characterized by a surface layer, a convective mixed

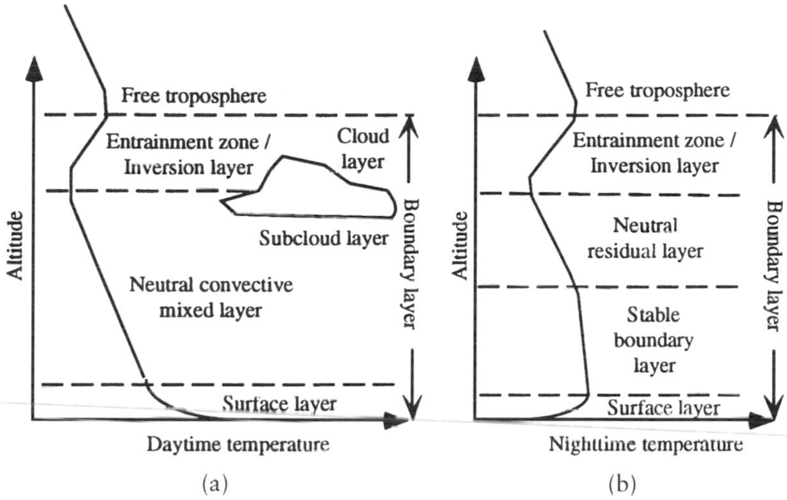


**Figure 2.3.** Variation of temperature with height during the (a) day and (b) night in the atmospheric boundary layer over land under a high-pressure system. Adapted from Stull (1988).

layer, and an entrainment zone (Stull 1988). The **surface layer** is a region of strong wind shear that comprises the bottom 10 percent of the boundary layer. Since the boundary-layer depth ranges from 500–3000 m, the surface layer is about 50–300 m thick.

Over land during the day, temperatures decrease rapidly with altitude in the surface layer but less so in the mixed layer. In the surface layer, the strong temperature gradient is caused by rapid solar heating of the ground. The temperature gradient is usually so strong that the surface layer is **unstably stratified**, and air adjacent to the ground buoyantly rises to the mixed layer. The **mixed layer** is **neutrally stratified**; thus, parcels of air can mix up or down.

When a high-pressure system is present, the top of the mixed layer is affected by a large-scale subsidence inversion, which is **stably stratified.** Thermals originating in the surface or mixed layer cannot easily penetrate through the inversion. Some mixing (entrainment) between the inversion and mixed layer does occur; thus, the inversion layer is an **entrainment zone.**

Other features of the daytime boundary layer are the cloud layer and subcloud layer. A region in which clouds appear in the boundary layer is the **cloud layer,** and the region underneath is the **subcloud layer.** A cloud forms if rising air in a thermal cools sufficiently. An inversion may prevent the thermal from rising past the mixed layer.

During the night, the surface cools radiatively, causing temperatures to increase with height from the ground and creating a surface inversion. Cooling at the top of the surface layer cools the bottom of the mixed layer, increasing the stability of the mixed layer at its base. The portion of the daytime mixed layer that is stable at night is the **stable (nocturnal) boundary layer.** The remaining portion of the mixed layer, which stays neutrally stratified, is the **residual layer.** Because thermals do not form at night, the residual layer does not undergo significant changes, except at its base. At night, the nocturnal boundary layer stabilizes and thickens, eroding the residual-layer base. Above the residual layer, the large-scale subsidence inversion remains.

Over the ocean, the boundary layer is influenced more by large-scale pressure systems than by thermal or mechanical turbulence. Since the temperature of the water does not change significantly during the day, thermal turbulence over the ocean is not so important as over land. Since the ocean surface is relatively smooth, mechanical turbulence is also less than over land. Large-scale high-pressure systems may still cause subsidence inversions to form over the ocean.

### 2.2.2. Free Atmosphere

#### 2.2.2.1. *Troposphere*

The free troposphere and the rest of the lower atmosphere lie above the boundary layer. Figure 2.4 shows a standard profile of the temperature structure of the lower atmosphere, ignoring the boundary layer. The **troposphere**, which is the bottom layer of the lower atmosphere, is the region extending from the surface in which the temperature, on average, decreases with increasing altitude. The average rate of temperature decrease in the free troposphere (above the boundary layer) is about 6.5 K $km^{-1}$. The temperature decreases with increasing altitude in the free troposphere because the pressure decreases with increasing altitude in this region, and air expands with decreasing pressure. For a parcel of air to expand, kinetic energy must be converted to work. The resulting loss of kinetic energy causes the air temperature to decrease. The temperature decreases with increasing altitude up to the **tropopause**. Above the tropopause, the air temperature is constant with increasing altitude (**isothermal**), and then it increases.

Figures 2.5(a) and (b) show global-scale latitude–altitude contour plots of zonally averaged temperatures for a generic January and July, respectively. A **zonally**

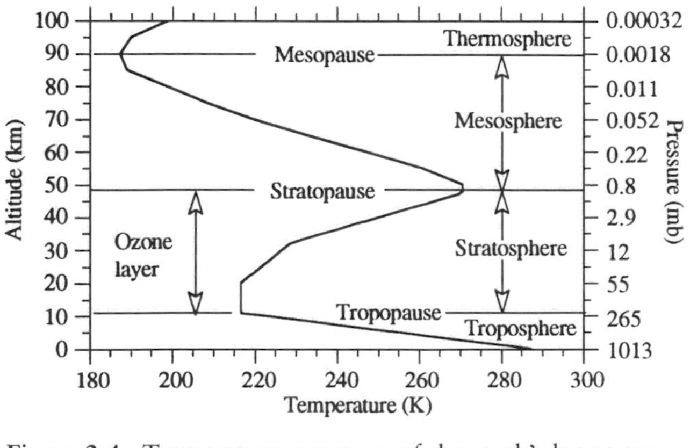


**Figure 2.4.** Temperature structure of the earth's lower atmosphere, ignoring the boundary layer.

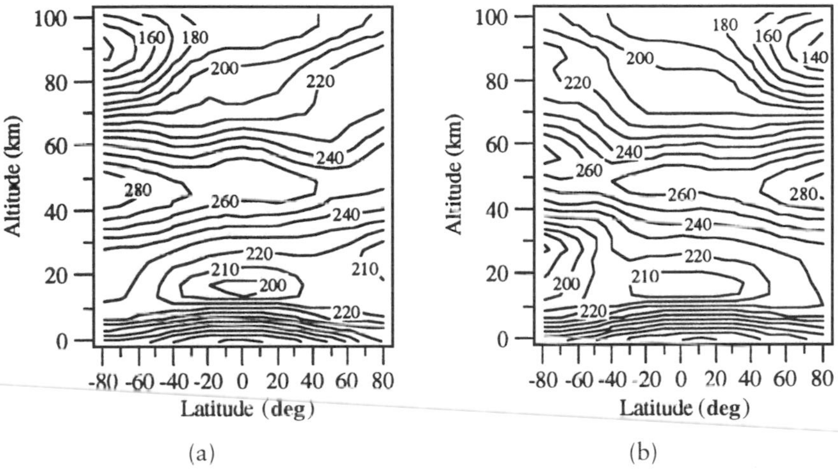


**Figure 2.5.** Zonally and monthly averaged temperatures for (a) January and (b) July. Data for the plots were compiled by Fleming et al. (1988)

**averaged** temperature is found from a set of temperatures, averaged over all longitudes at a given latitude and altitude. The figures indicate that near-surface tropospheric temperatures decrease from the equator to higher latitudes, which is expected, since the earth receives the greatest incident solar radiation near the equator.

The **tropopause** is the upper boundary of the troposphere. Formally, it is defined by the World Meteorological Organization (WMO) as the lowest altitude at which the lapse rate (rate of decrease of temperature with increasing height) decreases to 2 K km$^{-1}$ or less, and at which the lapse rate, averaged between this altitude and any altitude within the next 2 km, does not exceed 2 K km$^{-1}$ (Holten et al. 1995). Above the tropopause base, temperatures are relatively constant before increasing in the stratosphere. Figure 2.5(a) and (b) indicate that tropopuase heights decrease from 15–18 km near the equator to 8 km near the poles. Because temperatures decrease with increasing height in the troposphere, and because tropopause heights are greater over the equator than over the poles, temperatures over the equator decrease to lower values over the equator than over the poles. As a result, minimum tropopause temperatures occur over equatorial regions, as shown in Figs. 2.5(a) and (b). Tropopause heights are highest over the equator because strong vertical motions over the equator raise the base of the ozone layer and force ozone to spread horizontally to higher latitudes. Since ozone is responsible for warming above the tropopause, pushing ozone to greater heights increases the altitude at which warming begins. Near the poles, downward motions push ozone to lower altitudes. The resulting latitudinal gradient of the ozone-layer base decreases tropopause heights from the equator to the poles.

Tropopause temperatures are also affected by water vapor. Water vapor absorbs the earth's infrared radiation, preventing some of it from reaching the tropopause. Water-vapor mixing ratios are much higher over the equator than over the poles. The high mixing ratios over the equator lead to further cooling of the tropopause over the equator.

### 2.2.2.2. *Stratosphere*

The stratosphere is characterized by increasing temperatures with increasing height. $O_3$ and $O_2$ in the stratosphere absorb the sun's ultraviolet radiation and emit infrared radiation, heating the region. Approximately 90 percent of ozone molecules in the atmosphere reside in the stratosphere. Most of the rest reside in the troposphere.

Ozone is produced when molecular oxygen absorbs ultraviolet radiation and photodissociates by the reactions

$$O_2 + h\nu \longrightarrow \cdot\dot{O}(^1D) + \dot{O}\cdot, \qquad \lambda < 175\,\text{nm} \tag{2.4}$$

$$O_2 + h\nu \longrightarrow \dot{O}\cdot + \dot{O}\cdot \qquad 175 < \lambda < 245\,\text{nm} \tag{2.5}$$

where $h\nu$ is a photon of radiation, $O(^1D)$ is excited atomic oxygen, O [$= O\,(^3P)$] is **ground-state atomic oxygen**, the dots identify the valence of an atom (see Chapter 11), and $\lambda$ is a wavelength of radiation affecting the reaction. $O(^1D)$ produced from (2.4) rapidly converts to O via

$$\cdot\dot{O}(^1D) + M \longrightarrow \cdot\dot{O} + M \tag{2.6}$$

where M provides collisional energy for the reaction, but is not created or destroyed by it. Because $N_2$ and $O_2$ are the most abundant gases in the air, M is most likely $N_2$ or $O_2$, although it can be another gas. The most important reaction creating ozone is

$$\boxed{\dot{O}\cdot + O_2 + M \longrightarrow O_3 + M} \tag{2.7}$$

where M, in this case, carries away energy released by the reaction.

Ozone in the stratosphere is destroyed naturally by the photolysis reactions,

$$O_3 + h\nu \longrightarrow O_2 + \cdot\dot{O}(^1D) \qquad \lambda < 310\,\text{nm} \tag{2.8}$$

$$O_3 + h\nu \longrightarrow O_2 + \dot{O}\cdot \qquad \lambda > 310\,\text{nm} \tag{2.9}$$

and the two-body reaction

$$\dot{O}\cdot + O_3 \longrightarrow 2O_2 \tag{2.10}$$

Chapman (1930) postulated that the reactions (2.5), (2.7), (2.9), and (2.10), together with

$$\dot{O}\cdot + \dot{O}\cdot + M \longrightarrow O_2 + M \tag{2.11}$$

describe the natural formation and destruction of ozone in the stratosphere. These reactions make up the **Chapman cycle**, and they simulate the process fairly well. Some Chapman reactions are more important than others. The reactions (2.5), (2.7), and (2.9) affect ozone most significantly. The non-Chapman reaction, (2.8), is also important.

The reactions (2.4)–(2.7) show that ozone production occurs if sufficient oxygen and radiation are present. The oxygen density, like the air density, decreases exponentially with increasing altitude. The solar ultraviolet radiation intensity increases with increasing altitude, since radiation is attenuated through a thinner column at higher altitudes than at lower altitudes. Peak ozone density occurs at about 25–32 km, which is the altitude range in which sufficient radiation encounters sufficient oxygen density. At higher altitudes, the oxygen density is too low to produce peak ozone densities, and at lower altitudes, the radiation is not intense enough to photodissociate enough oxygen to produce peak densities. The exact altitude of the peak also depends on rates of ozone photolysis from the reactions (2.8) and (2.9).

Peak stratospheric temperatures occur at the top of the stratosphere, because this is the altitude at which ozone absorbs the shortest ultraviolet wavelengths reaching the stratosphere (about 0.175 μm). Although the concentration of ozone at the top of the stratosphere is small, each molecule can absorb short wavelengths, increasing the average kinetic energy and temperature of all molecules. In the lower stratosphere, short wavelengths do not penetrate, and temperatures are lower than in the upper stratosphere.

#### 2.2.2.3. *Mesosphere*

Temperatures decrease with increasing altitude in the **mesosphere** for the same reason they do in the free troposphere. Ozone densities are too low for ozone to absorb radiation and affect temperatures in the mesosphere.

#### 2.2.2.4. *Thermosphere*

In the **thermosphere**, temperatures increase with altitude because molecular oxygen and nitrogen absorb very short wavelengths of radiation in this region. Air in the thermosphere would not *feel* hot to the skin, because the thermosphere contains so few gas molecules. But each gas molecule in the thermosphere is highly energized, and the average temperature is high. Because oxygen and nitrogen absorb very short wavelengths of radiation in the thermosphere, such wavelengths do not penetrate to the mesopause.

## 2.3. EQUATION OF STATE

The **equation of state** describes the relationship among pressure, volume, and absolute temperature of a real gas. The **ideal gas law** describes this relationship for an ideal gas. An ideal gas is a gas for which the product of the pressure and volume is proportional to the absolute temperature. A real gas is ideal only to the extent that intermolecular forces are small, which occurs when pressures are low enough or temperatures are high enough for the gas to be sufficiently dilute. Under typical atmospheric temperature and pressure conditions, the ideal gas law gives an error of less than 0.2 percent for dry air and water vapor in comparison with an expanded equation of state (Pruppacher and Klett 1997). Thus, the ideal gas law can reasonably approximate the equation of state.

The **ideal gas law** is expressed as a combination of Boyle's law, Charles' law, and Avogadro's law. In 1661, Robert Boyle (1627–1691) found that doubling the pressure exerted on a gas at constant temperature reduced the volume of the gas by one-half. This relationship is embodied in **Boyle's law**,

$$p \propto \frac{1}{V} \qquad \text{at constant temperature} \tag{2.12}$$

where $p$ is the pressure exerted on the gas (mb), and $V$ is the volume enclosed by the gas ($m^3$ or $cm^3$). Boyle's law describes the compressibility of a gas. When high pressure is exerted on a gas, as in the lower atmosphere, the gas compresses and exerts an equal pressure on its surroundings. When low pressure is exerted on a gas, as in the upper atmosphere, the gas expands and exerts an equal pressure on its surroundings.

In 1787, Jacques Charles (1746–1823) found that increasing the absolute temperature of a gas at constant pressure increased the volume of the gas. This relationship is embodied in **Charles' law**,

$$V \propto T \qquad \text{at constant pressure} \tag{2.13}$$

where $T$ is the temperature of the gas (K). Charles' law states that, at constant pressure, the volume of a gas must decrease when its temperature decreases. Since gases change phase to liquids or solids before 0 K, Charles' law cannot be extrapolated to 0 K.

Amedeo Avogadro (1776–1856) hypothesized that equal volumes of different gases under the same conditions of temperature and pressure contain the same number of molecules. In other words, the volume of a gas is proportional to the number of molecules of gas present and independent of the type of gas. This relationship is embodied in **Avogadro's law**,

$$V \propto n \qquad \text{at constant pressure and temperature} \tag{2.14}$$

where $n$ is the number of gas moles. The number of molecules in a mole is constant for all gases and given by **Avogadro's number**, $A = 6.02252 \times 10^{23}$ molecules mole$^{-1}$.

Combining Boyle's law, Charles' law, and Avogadro's law gives the **ideal gas law** or **simplified equation of state** as

$$p = \frac{nR^*T}{V} = \frac{nA}{V}\left(\frac{R^*}{A}\right)T = Nk_BT \tag{2.15}$$

where $R^*$ is the **universal gas constant** (0.083145 $m^3$ mb mole$^{-1}$ K$^{-1}$ or $8.314 \times 10^4$ $cm^3$ mb mole$^{-1}$ K$^{-1}$), $N = nA/V$ is the number concentration of gas molecules (molecules of gas per cubic meter of cubic centimeter of air), and $k_B = R^*/A$ is **Boltzmann's constant** in units of $1.3807 \times 10^{-25}$ $m^3$ mb K$^{-1}$ or $1.3807 \times 10^{-19}$ $cm^3$ mb K$^{-1}$. Appendix A contains alternative units for $R^*$ and $k_B$.

**Example 2.2**

Calculate the number concentration of air molecules in the atmosphere at standard sea-level pressure and temperature and at a pressure of 1 mb.

SOLUTION

At standard sea level, $p = 1013$ mb and $T = 288$ K. Thus, from (2.15), $N = 2.55 \times 10^{19}$ molecules cm$^{-3}$. From Fig. 2.1(a), $p = 1$ mb occurs at 48 km. At this altitude and pressure, $T = 270$ K, as shown in Fig. 2.4. Under such conditions, $N = 2.68 \times 10^{16}$ molecules cm$^{-3}$.

(2.15) can be used to relate the partial pressure exerted by a gas to its number concentration. In 1803, John Dalton (1766–1844) stated that total atmospheric pressure equals the sum of the partial pressures of the individual gases in a mixture. This is **Dalton's law of partial pressure.** The **partial pressure** exerted by a gas in a mixture is the pressure the gas exerts if it alone occupies the same volume as the mixture. Mathematically, the partial pressure of gas $q$ is

$$p_q = N_q k_B T \tag{2.16}$$

where $N_q$ is the number concentration of the gas (molecules cm$^{-3}$). Total atmospheric pressure is

$$p_a = \sum_q p_q = k_B T \sum_q N_q = N_a k_B T \tag{2.17}$$

where $N_a$ is the number concentration of the air, determined as the sum of the number concentrations of individual gases.

Total atmospheric pressure can be written as $p_a = p_d + p_v$, where $p_d$ is the partial pressure exerted by dry air, and $p_v$ is the partial pressure exerted by water vapor. Similarly, the number concentration of air molecules can be written as $N_a = N_d + N_v$, where $N_d$ is the number concentration of dry air, and $N_v$ is the number concentration of water vapor.

Dry air consists of all gases in the atmosphere, except water vapor. Table 2.1 shows that together, $N_2$, $O_2$, Ar, and $CO_2$ constitute 99.996 percent of dry air by volume. The concentrations of all gases aside from these four can be ignored, without much loss in accuracy, when dry-air pressure is calculated. This assumption is convenient in that the concentrations of most trace gases vary from time to time and place to place.

Partial pressure is related to the mass density and number concentration of dry air through the **equation of state for dry air,**

$$p_d = \frac{n_d R^* T}{V} = \frac{n_d m_d}{V}\left(\frac{R^*}{m_d}\right)T = \rho_d R' T = \frac{n_d A}{V}\left(\frac{R^*}{A}\right)T = N_d k_B T \tag{2.18}$$

where $p_d$ is dry-air partial pressure (mb), $n_d$ is the number of moles of dry air, $m_d$ is the molecular weight of dry air, $\rho_d$ is the mass density of dry air (kg m$^{-3}$ or g cm$^{-3}$), and $R'$ is the gas constant for dry air. The molecular weight of dry air is a volume-weighted average of the molecular weights of $N_2$, $O_2$, Ar, and $CO_2$. The standard value of $m_d$ is 28.966 g mole$^{-1}$. The dry-air mass density, number concentration, and gas constant are, respectively,

$$\rho_d = \frac{n_d m_d}{V} \qquad N_d = \frac{n_d A}{V} \qquad R' = \frac{R^*}{m_d} \tag{2.19}$$

where $R'$ has a value of 2.8704 m$^3$ mb kg$^{-1}$ K$^{-1}$ or 2870.3 cm$^3$ mb g$^{-1}$ K$^{-1}$. Alternative units for $R'$ are given in Appendix A.

**Example 2.3**

When $p_d = 1013$ mb and $T = 288$ K, the density of dry air from (2.18) is $\rho_d = 1.23$ kg m$^{-3}$.

The **equation of state for water vapor** is

$$p_v = \frac{n_v R^* T}{V} = \frac{n_v m_v}{V}\left(\frac{R^*}{m_v}\right)T = \rho_v R_v T = \frac{n_v A}{V}\left(\frac{R^*}{A}\right)T = N_v k_B T \tag{2.20}$$

where $p_v$ is the partial pressure exerted by water vapor (mb), $n_v$ is the number of moles of water vapor, $m_v$ is the molecular weight of water vapor, $\rho_v$ is the mass density of water vapor (kg m$^{-3}$ or g cm$^{-3}$), and $R_v$ is the gas constant for water vapor (4.6140 m$^3$ mb kg$^{-1}$ K$^{-1}$ or 4614.0 cm$^3$ mb g$^{-1}$ K$^{-1}$). Alternative units for $R_v$ are given in Appendix A. The water-vapor mass density, number concentration, and gas constant are, respectively,

$$\rho_v = \frac{n_v m_v}{V} \qquad N_v = \frac{n_v A}{V} \qquad R_v = \frac{R^*}{m_v} \tag{2.21}$$

**Example 2.4.**

When $p_v = 10$ mb and $T = 298$ K, water-vapor density from (2.20) is $\rho_v = 7.27 \times 10^{-3}$ kg m$^{-3}$.

In terms of the dry-air gas constant, the equation of state for water vapor can be rewritten as

$$p_v = \rho_v R_v T = \rho_v \left(\frac{R_v}{R'}\right) R' T = \frac{\rho_v R' T}{\varepsilon} \tag{2.22}$$

where

$$\varepsilon = \frac{R'}{R_v} = \frac{R^*}{m_d}\left(\frac{m_v}{R^*}\right) = \frac{m_v}{m_d} = 0.622 \tag{2.23}$$

The number concentration of a gas (molecules per unit volume of air) is an absolute quantity. The abundance of a gas may also be expressed in terms of a relative quantity, **volume mixing ratio,** defined as the number of gas molecules per molecule of dry air, and expressed for gas $q$ as

$$\chi_q = \frac{N_q}{N_d} = \frac{p_q}{p_d} = \frac{n_q}{n_d} \tag{2.24}$$

where $N_q$, $p_q$, and $n_q$ are the number concentration, partial pressure, and number of moles of gas $q$, respectively. Another relative quantity, mass mixing ratio, is the mass of gas per mass of dry air. The **mass mixing ratio** of gas $q$ (kilograms of gas per kilogram of dry air) is

$$\omega_q = \frac{\rho_q}{\rho_d} = \frac{m_q N_q}{m_d N_d} = \frac{m_q p_q}{m_d p_d} = \frac{m_q n_q}{m_d n_d} = \frac{m_q}{m_d}\chi_q \tag{2.25}$$

where $\rho_q$ is the mass density (kg m$^{-3}$) of gas $q$, and $m_q$ is its molecular weight (g mole$^{-1}$). Volume and mass mixing ratios may be multiplied by $10^6$ and expressed in **parts per million by volume** (ppmv) or **parts per million by mass** (ppmm), respectively.

**Example 2.5.**

Find the mass mixing ratio, number concentration, and partial pressure of ozone if its volume mixing ratio in an urban air parcel is $\chi_q = 0.10$ ppmv. Assume $T = 288$ K.

SOLUTION

The molecular weight of ozone is $m_q = 48.0$ g mole$^{-1}$, and the molecular weight of dry air is $m_d = 28.966$ g mole$^{-1}$. From (2.25), the mass mixing ratio of ozone is $\omega_q = 48.0$ g mole$^{-1}$ $\times 0.10$ ppmv/28.966 g mole$^{-1}$ $= 0.17$ ppmm. From Example 2.2, $N_d = 2.55 \times 10^{19}$ molecules cm$^{-3}$. Thus, from (2.24), the number concentration of ozone is $N_q = 0.10$ ppmv $\times\ 10^{-6} \times 2.55 \times 10^{19}$ molecules cm$^{-3}$ $= 2.55 \times 10^{12}$ molecules cm$^{-3}$. From (2.16), the partial pressure exerted by ozone is $p_q = 0.000101$ mb.

From (2.23) and (2.25), the **mass mixing ratio of water vapor** (kilograms of water vapor per kilogram of dry air) is

$$\omega_v = \frac{\rho_v}{\rho_d} = \frac{m_v p_v}{m_d p_d} = \varepsilon\frac{p_v}{p_d} = \frac{\varepsilon p_v}{p_a - p_v} = \varepsilon\chi_v \tag{2.26}$$

**Example 2.6**

If the partial pressure exerted by water vapor is $p_v = 10$ mb, and the total air pressure is $p_a = 1010$ mb, the mass mixing ratio of water vapor from (2.26) is $\omega_v = 0.00622$ kg kg$^{-1}$.

Water vapor can also be expressed in terms of relative humidity, discussed in Section 2.5, or **specific humidity**. Specific humidity is similar to the mass mixing ratio, except that it expresses the mass of water vapor per unit mass of moist air (dry air plus water vapor). An expression for specific humidity (kilograms of water vapor per kilogram of moist air) is

$$q_v = \frac{\rho_v}{\rho_a} = \frac{\rho_v}{\rho_d + \rho_v} = \frac{\frac{p_v}{R_v T}}{\frac{p_d}{R'T} + \frac{p_v}{R_v T}} = \frac{\frac{R'}{R_v} p_v}{p_d + \frac{R'}{R_v} p_v} = \frac{\varepsilon p_v}{p_d + \varepsilon p_v} \tag{2.27}$$

where $\rho_a = \rho_d + \rho_v$ is the **mass density of moist air**. Specific humidity is related to the mass mixing ratio of water vapor by $q_v = \omega_v \rho_d / \rho_a$.

**Example 2.7.**

If $p_v = 10$ mb and $p_a = 1010$ mb, $p_a = p_d + p_v$ gives $p_d = 1000$ mb. Under such conditions, (2.27) gives the specific humidity as $q_v = 0.00618$ kg kg$^{-1}$.

The **equation of state for moist air** is the sum of the equations of state for dry air and water vapor. Thus,

$$p_a = p_d + p_v = \rho_d R'T + \rho_v R_v T = \rho_a R'T \frac{\rho_d + \rho_v R_v / R'}{\rho_a} \tag{2.28}$$

Substituting $\varepsilon = R'/R_v$, $\rho_a = \rho_d + \rho_v$, and $\omega_v = \rho_v/\rho_d$ into (2.28) yields the equation of state for moist air as

$$p_a = \rho_a R'T \frac{\rho_d + \rho_v/\varepsilon}{\rho_d + \rho_v} = \rho_a R'T \frac{1 + \rho_v/(\rho_d \varepsilon)}{1 + \rho_v/\rho_d} = \rho_a R'T \frac{1 + \omega_v/\varepsilon}{1 + \omega_v} \tag{2.29}$$

This equation can be simplified to

$$p_a = \rho_a R_m T = \rho_a R' T_v \tag{2.30}$$

where

$$R_m = R' \frac{1 + \omega_v/\varepsilon}{1 + \omega_v} = R'(1 + 0.608\omega_v) \tag{2.31}$$

is the **gas constant for moist air** and

$$T_v = T\frac{1+\omega_v/\varepsilon}{1+\omega_v} = T\frac{R_m}{R'} \approx T(1+0.608\omega_v) \quad (2.32)$$

is the **virtual temperature.** This quantity is the temperature of a sample of dry air at the same density and pressure as a sample of moist air at temperature $T$. Since the gas constant for moist air is larger than that for dry air, moist air has a lower density than does dry air at the same temperature and pressure. For the dry-air density to equal the moist-air density at the same pressure, the temperature of the dry air must be higher than that of the moist air by the factor $R_m/R'$. The resulting temperature is the virtual temperature, which is always larger than the actual temperature. In (2.32), the liquid water content is assumed to equal zero.

Equating $R_m = R^*/m_a$ with $R_m$ from (2.31) and noting that $R' = R^*/m_d$ gives the **molecular weight of moist air** as

$$m_a \approx \frac{m_d}{1+0.608\omega_v} \quad (2.33)$$

The molecular weight of moist air is less than that of dry air.

**Example 2.8.**

If $p_d = 1013$ mb, $p_v = 10$ mb, and $T = 298$ K, calculate $\omega_v$, $m_a$, $R_m$, $T_v$, and $\rho_a$.

SOLUTION

From (2.25), $\omega_v = 0.622 \times 10$ mb/1013 mb $= 0.00614$ kg kg$^{-1}$.
From (2.33), $m_a = 28.966/(1. + 0.6 \times 0.00614) = 28.86$ g mole$^{-1}$.
From (2.31), $R_m = 2.8704 \times (1. + 0.6 \times 0.00614) = 2.8809$ m$^3$ mb kg$^{-1}$ K$^{-1}$.
From (2.32), $T_v = 298 \times (1. + 0.6 \times 0.00614) = 299.1$ K.
From (2.30), $\rho_a = p_a/(R_m T) = 1023/(2.8809 \times 298) = 1.19$ kg m$^{-3}$.

## 2.4. CHANGE IN PRESSURE WITH ALTITUDE

The variation of pressure with altitude in the atmosphere can be estimated in one of several ways. The first method discussed is derived from the **hydrostatic equation,**

$$\frac{\partial p_a}{\partial z} = -\rho_a g \quad (2.34)$$

where $g$ is the effective gravitational acceleration, and $z$ is altitude (m). The effective gravitational acceleration, discussed in Chapter 4, varies slightly with latitude and altitude. For convenience, $g$ in (2.34) is set at 9.81 m s$^{-2}$. (2.34) states that, in a hydrostatic atmosphere, the upward gradient in air pressure is balanced by the downward force of gravity multiplied by the air density. Figure 2.1(a) shows that pressure decreases exponentially with increasing altitude, giving rise to an upward pressure gradient. If gravity did not exist, the upward pressure gradient would force air to accelerate to space. The hydrostatic equation does not state that air has no vertical velocity. It states that, in hydrostatic balance, air has no vertical acceleration. Over large horizontal regions, the assumption is often reasonable, since vertical accelerations, when averaged over such regions, are small. Over small regions (<2–3 km in diameter), such as within cumulonimbus clouds, the hydrostatic assumption is not accurate because vertical accelerations are strong.

The pressure at a given altitude can be estimated by approximating the derivative in the hydrostatic equation as

$$\frac{\partial p_a}{\partial z} \approx \frac{p_{a,1} - p_{a,0}}{z_1 - z_0} = -\rho_{a,0} g \tag{2.35}$$

where $p_{a,1}$ is the pressure at altitude $z_1$, and $p_{a,0}$ and $\rho_{a,0}$ are the pressure and density, respectively, at a lower altitude, $z_0$. If the pressure is known at the surface, and either a vertical density or a temperature profile is known, (2.35) can be used to estimate the pressure at each layer above the surface.

**Example 2.9**

If the pressure and density at sea level are $p_{a,0} = 1013$ mb and $\rho_{a,0} = 1.23$ kg m$^{-3}$, respectively, estimate the pressure at 100-m altitude.

SOLUTION

From (2.35), $p_{a,100\,\text{m}} = 1013\ \text{mb} - 1.23\frac{\text{kg}}{\text{m}^3}\left(9.81\frac{\text{m}}{\text{s}^2}\right)(100 - 0\ \text{m})\frac{\text{mb m s}^2}{100\ \text{kg}} =$ 1000.2 mb

Altitude versus pressure can also be determined from a pressure altimeter. A pressure altimeter measures the pressure at an unknown altitude with an **aneroid barometer**. The pressure from the barometer is used to calculate the altitude under standard atmospheric conditions. Applying correction factors to the calculated standard atmosphere altitude gives the actual altitude. Under standard atmospheric conditions, the air is dry and the sea-level pressure, the sea-level temperature, and the free-tropospheric lapse rate are, by definition, $p_{d,s} = 1013.25$ mb, $T_{a,s} = 288$ K, and $\Gamma_s = -\partial T/\partial z = +6.5$ K km$^{-1}$, respectively, (List 1984). $\Gamma_s$ is the average negative change in air temperature with altitude in the free troposphere.

The equation for altitude in a standard atmosphere is derived by substituting $p_d = \rho_d R' T$ into the hydrostatic equation and assuming the air is dry. The result is

$$\frac{\partial p_d}{\partial z} = -\frac{p_d}{R'T} g \tag{2.36}$$

Substituting $T = T_{a,s} - \Gamma_s z$, where $z = 0$ km corresponds to sea level, into (2.36), rearranging, and integrating from $p_{d,s}$ to $p_d$ and 0 to $z$ yields

$$\ln\left(\frac{p_d}{p_{d,s}}\right) = \frac{g}{\Gamma_s R'} \ln\left(\frac{T_{a,s} - \Gamma_s z}{T_{a,s}}\right) \tag{2.37}$$

Rearranging (2.37) gives altitude as a function of pressure in a standard atmosphere as

$$z = \frac{T_{a,s}}{\Gamma_s}\left[1 - \left(\frac{p_d}{p_{d,s}}\right)^{\Gamma_s R'/g}\right] \tag{2.38}$$

Temperature variations with altitude and sea-level pressure in the real atmosphere usually differ from those in the standard atmosphere. Empirical and tabulated expressions correcting for the differences are available (List 1984). Since the corrections are not always accurate, airplanes rarely use pressure altimeters to measure altitude. Instead, they use radar altimeters, which measure altitude with radio waves.

**Example 2.10**

If a pressure altimeter reads $p_d = 850$ mb and the air is dry, find the standard-atmosphere altitude.

SOLUTION

Since $R' = 287.04\ \text{m}^2\,\text{s}^{-2}\,\text{K}^{-1}$, we have $\Gamma_e R'/g = 0.1902$, and $z = 1.45$ km from (2.38).

A third way to estimate pressure versus altitude is with the **scale-height equation.** From the equation of state for moist air, the density is

$$\rho_a = \frac{p_a}{R' T_v} = \frac{m_d}{R^*}\frac{p_a}{T_v} = \frac{p_a}{T_v}\left(\frac{A}{R^*}\right)\frac{m_d}{A} \approx \frac{p_a}{T_v}\left(\frac{1}{k_B}\right)\bar{M} = \frac{p_a \bar{M}}{k_B T_v} \tag{2.39}$$

where $\bar{M} \approx m_d/A$ was previously defined as the **average mass of one air molecule.** Equation (2.39) can be combined with the hydrostatic equation to give

$$\frac{\mathrm{d}p_a}{p_a} = -\frac{\bar{M}g}{k_B T_v}\,\mathrm{d}z = -\frac{\mathrm{d}z}{H} \tag{2.40}$$

where

$$H = \frac{k_B T_v}{\bar{M}g} \tag{2.41}$$

is the **scale height** of the atmosphere at a given virtual temperature. The scale height is the height above a reference height at which pressure decreases to 1/e of its value at the reference height. Since temperature changes with altitude in the atmosphere, scale height also changes with altitude.

**Example 2.11**
If the air is dry, determine the scale height of the atmosphere at $T_v = 298$ K.

SOLUTION

From Appendix A, $\bar{M} = 4.8096 \times 10^{-26}$ kg, $g = 9.81$ m s$^{-2}$, and $k_B = 1.3807 \times 10^{-23}$ kg m$^2$ s$^{-2}$ K$^{-1}$ molecule$^{-1}$. From (2.41), the scale height is $H = 8.72$ km.

Integrating (2.40) at constant temperature gives pressure as a function of altitude as

$$p_a = p_{a,ref}\, e^{-(z-z_{ref})/H} \tag{2.42}$$

where $p_{a,ref}$ is pressure at a reference level, $z = z_{ref}$. If temperature is constant in a layer, the layer is **isothermal**. Temperature is usually not constant with altitude, but in a model, the atmosphere is divided into numerous layers in the vertical, and each layer has an average temperature. The pressure can be estimated in each layer by calculating a scale height for each layer. If the temperature in the middle of the lowest model layer is known and the surface pressure is known, the pressure at the top of the bottom layer can be estimated from (2.42). This pressure is used as the reference pressure for the next layer.

## 2.5. WATER IN THE ATMOSPHERE

Water in the atmosphere appears in three states – gas, liquid, and solid. Sources and sinks of water vapor were discussed in Section 2.1.2.1. Sources of liquid water include sea spray, volcanos, combustion, condensation of water vapor, and melting of ice crystals. Sinks of liquid water include evaporation to the gas phase, freezing to the solid phase, and sedimentation to the surface in the form of aerosols, fog drops, drizzle, and raindrops. Ice in the atmosphere forms from freezing of liquid water and deposition of water vapor. Ice can sublimate to the gas phase, melt to the liquid phase, or sediment to the surface.

### 2.5.1. Types of Energy

When water changes state, it releases or absorbs energy. **Energy** is the capacity of a physical system to do work on matter, which is mass that exists as a solid, liquid, or gas. Energy takes many forms.

**Kinetic energy** is the energy within a body due to its motion and equals one-half the mass of the body multiplied by its velocity squared. The faster a body moves, the greater its kinetic energy. To change the velocity and kinetic energy of a body in motion, mechanical work must be done.

**Potential energy** is the energy of matter that arises due to its position, rather than its motion. Potential energy represents the amount of work that a body can do. A coiled spring, charged battery, and chemical reactant have potential energy. When an object is raised vertically, it possesses gravitational potential energy because it can potentially do work by sinking.

**Internal energy** is the kinetic and/or potential energy of molecules within an object, but does not include the kinetic and/or potential energy of the object as a whole.

**Work** is the energy added to a body by the application of a force that moves the body in the direction of the force.

**Electromagnetic (radiant) energy** is the energy transferred by electromagnetic waves that originate from bodies with temperatures above 0 K.

Heat is not a form of energy. Heat transfer is a term used to describe the energy transfer between two bodies that occurs, for example, when their internal energies (or temperatures) differ. **Heat release (absorption)** occurs when a substance releases (absorbs) energy to (from) the surrounding environment upon a change of state.

### 2.5.2. Latent Heat

During condensation, freezing, and deposition of a substance, energy is released. During evaporation, melting, and sublimation, energy is absorbed. The energy released or absorbed during such processes is **latent heat.** Latent heat absorbed (released) during evaporation (condensation) is **latent heat of evaporation.** It varies with temperature as

$$\frac{dL_e}{dT} = c_{p,V} - c_W \tag{2.43}$$

where $c_{p,V}$ is the specific heat of water vapor at constant pressure, and $c_W$ is the specific heat of liquid water. The latent heat absorbed (released) during sublimation (deposition) is the **latent heat of sublimation.** The latent heat absorbed (released) during melting (freezing) is the **latent heat of melting.** The latent heats of sublimation and melting vary with temperature according to

$$\frac{dL_s}{dT} = c_{p,V} - c_I \qquad \frac{dL_m}{dT} = c_W - c_I \tag{2.44}$$

respectively, where $c_I$ is the specific heat of ice. Figure 2.6 shows the quantity of energy absorbed or released during phase changes of water.

The specific heats of liquid water, water vapor, and ice vary with temperature. Pruppacher and Klett (1997) derived polynomials for the specific heat of liquid water (J kg$^{-1}$ K$^{-1}$) versus temperature from data of Osborne et al. (1939) and

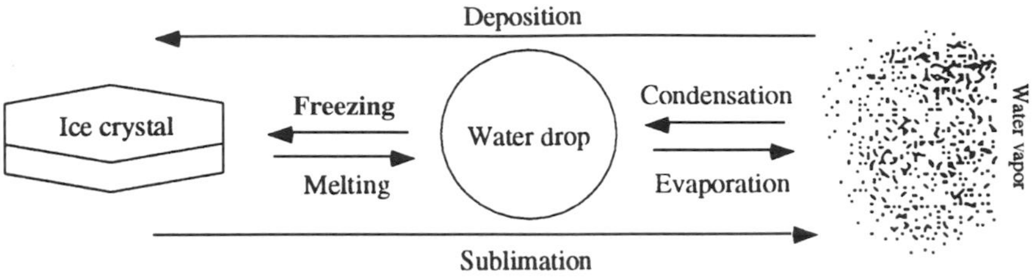


**Figure 2.6.** Phase changes of water. Freezing (melting) at 0°C releases (absorbs) 333.5 J g$^{-1}$, deposition (sublimation) at 0°C releases (absorbs) 2835 J g$^{-1}$, and condensation (evaporation) releases (absorbs) 2510 J g$^{-1}$ at 0°C and 2259 J g$^{-1}$ at 100°C.

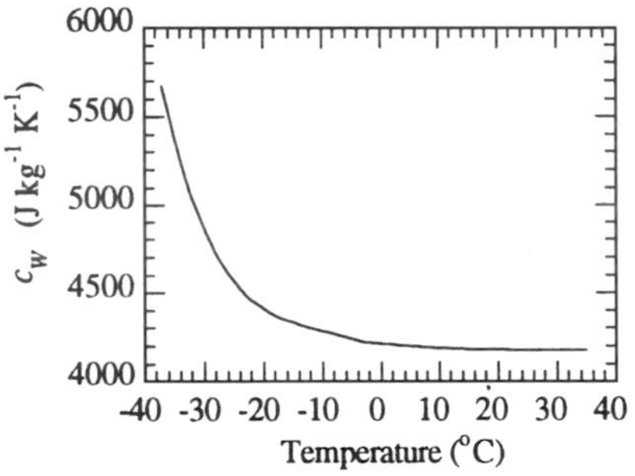


**Figure 2.7.** Variation of specific heat of liquid water with temperature, from (2.45).

Angell et al. (1982). The empirical expressions are

$$c_W = \begin{cases} 4187.9 - 11.319T_c - 0.097215T_c^2 + 0.018317T_c^3 & \\ \quad + 0.0011354T_c^4 & -37 \le T_c < 0 \\ 4175.2 + 0.01297(T_c - 35)^2 & \\ \quad + 1.5899 \times 10^{-5}(T_c - 35)^4 & 0 \le T_c < 35 \end{cases} \tag{2.45}$$

where $T_c$ is temperature in degrees Celsius. Figure 2.7 shows $c_W$ versus temperature from (2.45). The variation of $c_W$ below 0°C is large, but that above 0°C is small.

The specific heat of water vapor at constant pressure also varies slightly with temperature. At 298.15 K, $c_{p,V} \approx 1865.1$ (Lide 1993). At 303.15 K, it is 2 percent larger than at 243.15 K (Rogers and Yau 1989). The specific heat of ice is

$$c_I = 2104.6 + 7.322T_c \qquad -40 \le T_c < 0°\text{C} \tag{2.46}$$

(Pruppacher and Klett 1997; Giauque and Stout 1936; Flubacher et al. 1960).

Because changes in $c_W$ and $c_{p,V}$ are small for temperatures above 0°C, these parameters are often held constant when the variation of the latent heat of evaporation with temperature is calculated. Integrating (2.43) with constant specific heats gives

$$L_e = L_{e,0} - (c_W - c_{p,V})(T - T_0) \tag{2.47}$$

where $L_{e,0}$ is the latent heat of evaporation at temperature $T_0$. An empirical expression for the latent heat of evaporation (J $\text{kg}^{-1}$) is (Bolton 1980; List 1984)

$$L_e \approx 2.501 \times 10^6 - 2370T_c \tag{2.48}$$

> **Example 2.12**
>
> (2.48) predicts that, at 0°C, 2501 J (about 600 cal) is required to evaporate 1 g of liquid water. At 100°C, 2264 J (about 540 cal) is required to evaporate 1 g of liquid water.

An empirical expression for the latent heat of melting (J $\text{kg}^{-1}$), derived from data in List (1984) and valid for temperatures below 0°C, is

$$L_m \approx 3.3358 \times 10^5 + T_c(2030 - 10.46T_c) \tag{2.49}$$

While water at standard pressure always melts at 0°C, it may or may not freeze at temperatures below 0°C. Water that remains liquid below 0°C is **supercooled liquid water**. In the atmosphere, water drops may supercool down to −40°C unless they contain or are intercepted by **ice nuclei**. Ice nuclei include ice particles, minerals, organic material, or pure compounds. Kaolinite, bacteria, and silver iodide (AgI) are ice nuclei. When a supercooled liquid water drop contains or is intercepted by an ice nucleus, the drop may spontaneously freeze and grow by vapor deposition. Small drops are less likely than large drops to freeze because small drops have a small volume and are less likely than large drops to contain ice nuclei. Small drops also have small surface areas and are less likely than large drops to intercept ice nuclei.

> **Example 2.13**
>
> (2.49) predicts that, when 1 g of liquid water freezes at 0°C, 333.5 J (about 80 cal) is released. When 1 g of liquid water freezes at −10°C, 312.2 J (74.6 cal) is released.

The latent heat of sublimation (J $\text{kg}^{-1}$) is the sum of the latent heats of evaporation and melting. Thus,

$$L_s = L_e + L_m \approx 2.83458 \times 10^6 - T_c(340 + 10.46T_c) \tag{2.50}$$

### 2.5.3. Clausius–Clapeyron Equation

The rates of formation and growth of liquid water and ice particles depend on several factors, including the quantity of water vapor available, the temperature, and the size distribution of condensation or deposition nuclei in the air. In this subsection, growth processes are introduced. They are treated more thoroughly in Chapters 15–18.

A parameter that affects liquid-water and ice-particle formation is the **saturation vapor pressure.** The saturation vapor pressure (also called equilibrium or surface vapor pressure) is the partial pressure of a gas at a specific temperature when the gas is in equilibrium with a liquid or solid surface. In a gross sense, it is the maximum amount of vapor the air can hold without the vapor condensing as a liquid or depositing as ice on a surface. When the air is cold, saturation vapor pressures are lower than when it is warm. At low temperatures, liquid-water molecules on particle surfaces have little kinetic energy and cannot easily break free (evaporate) from the surfaces. Thus, in equilibrium, air just above a surface contains few molecules and exerts a low saturation vapor pressure. When the air is warm, liquid-water molecules obtain higher kinetic energies, become more agitated, and break loose from the surface. The resulting evaporation increases the saturation vapor pressure over the surface at equilibrium.

The temperature variation of the saturation vapor pressure of water over a liquid surface ($p_{v,s}$, mb) is approximated with the **Clausius–Clapeyron equation,**

$$\frac{\mathrm{d}p_{v,s}}{\mathrm{d}T} = \frac{\rho_{v,s}}{T} L_e \tag{2.51}$$

where $L_e$ is the latent heat of evaporation of water (J kg$^{-1}$), and $\rho_{v,s}$ is the saturation mass density of water vapor (kg m$^{-3}$). From (2.20), $\rho_{v,s} = p_{v,s}/R_v T$. Combining this expression with (2.51) gives

$$\frac{\mathrm{d}p_{v,s}}{\mathrm{d}T} = \frac{L_e p_{v,s}}{R_v T^2} \tag{2.52}$$

Substituting $L_e$ from (2.48) into (2.52) and rearranging gives

$$\frac{\mathrm{d}p_{v,s}}{p_{v,s}} = \frac{1}{R_v}\left(\frac{A_b}{T^2} - \frac{B_b}{T}\right)\mathrm{d}T \tag{2.53}$$

where $A_b = 3.14839 \times 10^6$ J kg$^{-1}$ and $B_b$ =2370 J kg$^{-1}$ K$^{-1}$. Integrating (2.53) from $p_{v,s,0}$ to $p_{v,s}$ and $T_0$ to $T$, where $p_{v,s,0}$ is a known saturation vapor pressure at $T_0$, gives the **saturation vapor pressure of water over a liquid surface** as

$$p_{v,s} = p_{v,s,0} \exp\left[\frac{A_b}{R_v}\left(\frac{1}{T_0} - \frac{1}{T}\right) + \frac{B_b}{R_v} \ln\left(\frac{T_0}{T}\right)\right] \tag{2.54}$$

At $T = T_0 = 273.15$ K, $p_{v,s,0} = 6.112$ mb. Substituting these values, $R_v = 461.91$ J kg$^{-1}$ K$^{-1}$, $A_b$, and $B_b$ into (2.54) gives

$$p_{v,s} = 6.112 \exp\left[6816\left(\frac{1}{273.15} - \frac{1}{T}\right) + 5.1309 \ln\left(\frac{273.15}{T}\right)\right] \tag{2.55}$$

where $T$ is in kelvin and $p_{v,s}$ is in millibars.

**Example 2.14**

At $T = 253.15$ K ($-20$°C), $p_{v,s} = 1.26$ mb from (2.55).
At $T = 298.15$ K (25°C), $p_{v,s} = 31.60$ mb.

An alternative, empirical parameterization of the saturation vapor pressure of water over a liquid surface is (Bolton 1980)

$$p_{v,s} = 6.112 \exp\left(\frac{17.67T_c}{T_c + 243.5}\right) \quad (2.56)$$

where $T_c$ is in degrees Celsius and $p_{v,s}$ is in millibars. The fit is valid for $-35 < T_c < 35$°C. (2.55) and (2.56) are saturation vapor pressures over flat, dilute liquid surfaces. Saturation vapor pressure are affected by surface curvature, the presence of solutes in water, and the rate of drop heating or cooling. These effects are discussed in Chapter 17.

**Example 2.15**

At $T = 253.15$ K ($-20$°C), the saturation vapor pressure from (2.56) is $p_{v,s} = 1.26$ mb. At $T = 298.15$ K (25°C), $p_{v,s} = 31.67$ mb. A comparison of these results to those from the previous examples shows that (2.55) and (2.56) give similar predictions for $p_{v,s}$.

Figure 2.8(a) shows a plot of saturation vapor pressure of water over a liquid surface versus temperature, obtained from (2.56). The figure shows that saturation vapor pressure increases with increasing temperature. At 0°C, $p_{v,s} = 6.1$ mb, which is equivalent to 0.6 percent of the total sea-level air pressure (1013 mb).

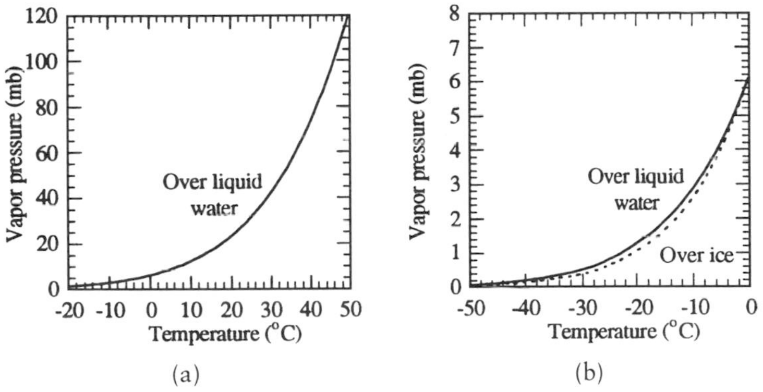


**Figure 2.8.** Saturation vapor pressure over (a) liquid water and (b) liquid water and ice, versus temperature.

At 30°C, $p_{v,s} = 42.5$ mb (4.2 percent of the total sea-level air pressure). Since the partial pressure of water vapor can rarely exceed its saturation vapor pressure, the maximum partial pressure of air varies from <1 percent to ≈4 percent of the total atmospheric pressure. Near the poles, where temperatures are low, saturation vapor pressures and partial pressures approach zero. Near the equator, where temperatures are high, saturation vapor pressures and partial pressures can increase to 4 percent or more of the total air pressure.

The saturation vapor pressure of water over an ice surface is lower than that over a liquid surface at the same subfreezing temperature because ice molecules require more energy to sublimate than liquid molecules require to evaporate at the same temperature. The Clausius–Clapeyron equation for the saturation vapor pressure of water over ice ($p_{v,I}$, mb) is written from (2.52) as

$$\frac{\mathrm{d}p_{v,I}}{\mathrm{d}T} = \frac{L_s p_{v,I}}{R_v T^2} \tag{2.57}$$

Substituting the latent heat of sublimation from (2.50) into (2.57) and integrating gives the **saturation vapor pressure of water over ice as**

$$p_{v,I} = p_{v,I,0} \exp\left[\frac{A_I}{R_v}\left(\frac{1}{T_0} - \frac{1}{T}\right) + \frac{B_I}{R_v}\ln\left(\frac{T_0}{T}\right) + \frac{C_I}{R_v}(T_0 - T)\right] \tag{2.58}$$

where $A_I = 2.14697 \times 10^6$ J kg$^{-1}$, $B_I = -5375$ J kg$^{-1}$ K$^{-1}$, and $C_I = 10.46$ J kg$^{-1}$ K$^{-2}$. At $T_0 = 273.15$ K, the saturation vapor pressure over ice equals that over liquid water ($p_{v,I,0} = 6.112$ mb). Substituting the known parameters into (2.58) results in

$$p_{v,I} = 6.112 \exp\left[4648\left(\frac{1}{273.15} - \frac{1}{T}\right) - 11.64 \ln\left(\frac{273.15}{T}\right) + 0.02265(273.15 - T)\right] \tag{2.59}$$

where $T \leq 273.15$ K and $p_{v,I}$ is in millibars. Figure 2.8(b) shows that, at subfreezing temperatures, $p_{v,I}$ is always less than $p_{v,s}$. Above 273.15 K (0°C), ice surfaces do not exist.

**Example 2.16**

At $T = 253.15$ K (−20°C), (2.59) predicts $p_{v,I} = 1.034$ mb, which is less than $p_{v,s} = 1.26$ mb at the same temperature. Thus, the saturation vapor pressure of water over liquid is greater than that over ice.

### 2.5.4. Condensation and Deposition

Saturation vapor pressures are important for determining whether water vapor condenses as liquid or deposits as ice. When the air temperature is above the

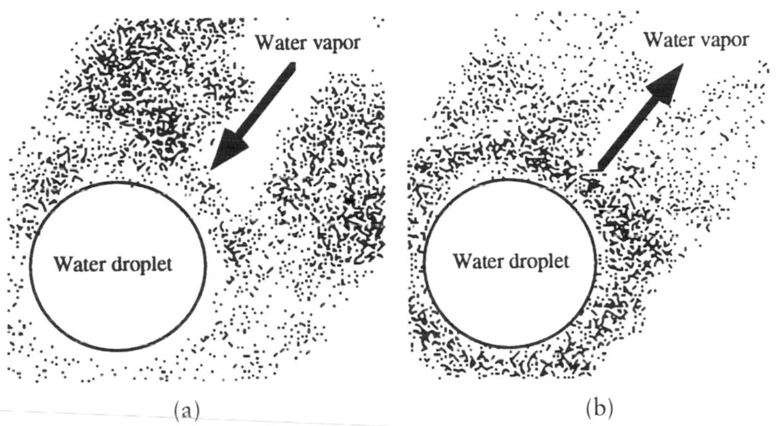


**Figure 2.9.** (a) Condensation occurs when the partial pressure of water vapor exceeds its saturation vapor pressure over liquid water. (b) Evaporation occurs when the saturation vapor pressure exceeds the partial pressure of water vapor. The schematics are not to scale.

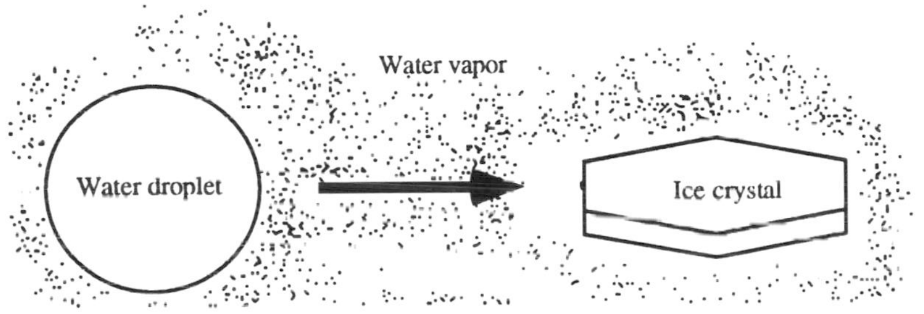


**Figure 2.10.** Bergeron process. When $p_{v,I} < p_v < p_{v,s}$ at a subfreezing temperature liquid evaporates, and the resulting gas diffuses to and deposits onto ice crystals.

freezing temperature of water (273.15 K) and the partial pressure of water is greater than the saturation vapor pressure of water over a liquid surface ($p_v > p_{v,s}$), water vapor generally condenses as a liquid. Liquid water evaporates when $p_v < p_{v,s}$. Figures 2.9(a) and (b) demonstrate growth and evaporation.

If air temperature falls below freezing and $p_v > p_{v,I}$, water vapor deposits as ice. When $p_v < p_{v,I}$, ice sublimates to water vapor. If liquid water and ice coexist in the same parcel of air and $p_{v,I} < p_{v,s} < p_v$, deposition onto ice particles is favored over condensation onto liquid water drops because $p_{v,I} < p_{v,s}$ at the same subfreezing temperature. When $p_{v,I} < p_v < p_{v,s}$, water evaporates from liquid surfaces and deposits onto ice surfaces, as shown in Fig. 2.10. This is the central assumption behind the **Wegener–Bergeron–Findeisen (Bergeron) process** of ice crystal growth in cold clouds. In such clouds, supercooled liquid water cloud droplets coexist with ice crystals. When the ratio of liquid cloud drops to ice crystals is less than 100,000 : 1, each ice crystal receives water from less than 100,000 cloud drops, and ice crystals do not grow large and heavy enough to fall from clouds. When the ratio is greater than 1,000,000 : 1, ice crystals grow large and fall from the cloud before liquid water is depleted. When the ratio is between 100,000 : 1 and 1,000,000 : 1, each ice crystal receives the liquid water from 100,000 to 1,000,000 droplets. The resulting raindrops fall, depleting liquid water in the cloud.

Figure 2.8 (b) shows that the greatest difference between the saturation vapor pressures over liquid water and ice occurs at $T_c = -15°\text{C}$. This is the temperature at which ice crystals grow the fastest. Dendrites, which are snowflake-shaped crystals, form most favorably at this temperature.

The relationship between the partial pressure and saturation vapor pressure of water vapor over a liquid surface is important for determining whether water vapor condenses. This relationship is embodied in the **relative humidity**, defined by the World Meteorological Organization (WMO) as

$$f_r = 100\% \times \frac{\omega_v}{\omega_{v,s}} = 100\% \times \frac{p_v(p_a - p_{v,s})}{p_{v,s}(p_a - p_v)} \approx 100\% \times \frac{p_v}{p_{v,s}} \qquad (2.60)$$

where $\omega_v$ is the mass mixing ratio of water vapor in the air, and $\omega_{v,s}$ is the **saturation mass mixing ratio of water vapor** over a liquid surface. Prior to the WMO definition, the relative humidity was defined exactly as $f_r = 100\% \times p_v/p_{v,s}$.

The mass mixing ratio of water vapor is obtained from (2.26). The saturation mass mixing ratio of water vapor is found in terms of the saturation vapor pressure by substituting $p_{v,s}$ for $p_v$ and $\omega_{v,s}$ for $\omega_v$ in (2.26). The result is

$$\omega_{v,s} = \frac{\varepsilon p_{v,s}}{p_a - p_{v,s}} \approx \frac{\varepsilon p_{v,s}}{p_d} \qquad (2.61)$$

The saturation mass mixing ratio in (2.60), the equation for the relative humidity, is taken with respect to liquid water, even if $T \leq 273.15$ K (List 1984).

**Example 2.17**

If $T = 288$ K and $p_v = 12$ mb, what is the relative humidity?

SOLUTION

From (2.56), $p_{v,s} = 17.04$ mb. From (2.60), $f_r = 100$ percent $\times$ 12 mb/17.04 mb = 70.4 percent.

(2.60) implies that, if the relative humidity exceeds 100 percent and $T >$ 273.15 K, water vapor condenses. If the relative humidity exceeds 100 percent and $T \leq 273.15$ K, water vapor may condense as a liquid or deposit as ice, depending on whether ice nuclei are present.

Another parameter used to predict when bulk condensation occurs is the **dew point**. The dew point ($T_D$) is the temperature to which air must be cooled, at constant water-vapor partial pressure and air pressure, to reach saturation with respect to liquid water. Similarly, the **frost point** is the temperature to which air must be cooled, at constant water-vapor partial pressure and air pressure, to reach saturation with respect to ice. If the air temperature drops below the dew point, the relative humidity increases above 100 percent and condensation occurs. If the dew point is known, the partial pressure of water can be obtained from Fig. 2.8. If the ambient temperature is known, the saturation vapor pressure can be obtained from the figure.

**Example 2.18**

If $T_D = 20°C$ and $T = 30°C$, estimate the partial pressure of water, the saturation vapor pressure of water, and the relative humidity from Fig. 2.8.

SOLUTION

From the figure, $p_v \approx 23.4$ mb and $p_{v,s} \approx 42.5$ mb. Thus, $f_r \approx 55$ percent.

The equation for the dew point can be derived from any equation for saturation vapor pressure by substituting $p_v$ for $p_{v,s}$ and solving for the temperature. Applying this technique to (2.56) and converting the result to absolute temperature gives the dew point as

$$T_D = \frac{4880.357 - 29.66 \ln p_v}{19.48 - \ln p_v} = \frac{4880.357 - 29.66 \ln(\omega_v p_d/\varepsilon)}{19.48 - \ln(\omega_v p_d/\varepsilon)} \qquad (2.62)$$

where $\omega_v = \varepsilon p_v / p_d$ from (2.26), $T$ is in kelvin, and $p_v$ is in millibars.

**Example 2.19**

Calculate the dew point when $p_v = 12$ mb.

SOLUTION

From (2.62), $T_D = 282.8$ K. Thus, if unsaturated air at 288 K cools to 282.8 K when $p_v = 12$ mb, condensation occurs.

When the ambient temperature is close to the dew point ($T \approx T_D$), the relative humidity is high. When ambient and dew-point temperatures are far apart, the relative humidity is low. Figure 2.11(a) and (b) show observed vertical profiles of temperature and dew point at Riverside, California, on the morning and afternoon, respectively, of August 27, 1987. In the morning, near the ground, the dew point and air temperature were close to each other, indicating the air was nearly saturated, the relative humidity was high, and a fog was almost present. Above 950 mb (about

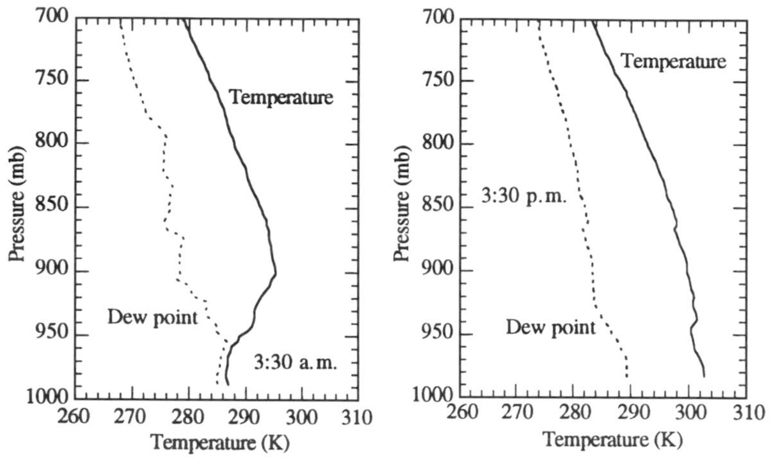


(a) (b)

**Figure 2.11.** Observed vertical profiles of temperature and dew point at (a) 3:30 a.m. and (b) 3:30 p.m. on August 27, 1987 at Riverside, California. The air was nearly saturated near the surface in the morning.

500 m above sea level) in the morning, the dew point and air temperature were far apart, indicating the relative humidity was low and the air was unsaturated. In the afternoon, when air near the surface was hot, the dew point and air temperature were also far apart, and the relative humidity was low.

## 2.6. FIRST LAW OF THERMODYNAMICS

The first law of thermodynamics relates the change in temperature of a gas to energy transfer between the gas and its environment and work done by or on the gas. It is used to derive the **thermodynamic energy equation**, which predicts temperature changes in a model. Changes in temperature affect stability, pressure, density, chemical reaction rates, and other parameters. The first law is also used to predict stability of the atmosphere analytically. The stability determines the extent of mixing of momentum, energy, gases, and particles and the extent of cloud formation and growth.

The **first law of thermodynamics** is

$$dQ^* = dU^* + dW^* \tag{2.63}$$

where $dQ^*$ is the **energy** (J) transferred between an air parcel and its environment, $dU^*$ is the change in **internal energy** (J) of the parcel, and $dW^*$ is the **work** (J) done by or on the parcel. When $dQ^* > 0$, energy is transferred to the parcel from the environment, and the process is **endothermic**. When $dQ^* < 0$, energy is transferred to the environment from the parcel, and the process is **exothermic**. When $dW^* > 0$, work is done by the parcel. When $dW^* < 0$, work is done on the parcel. (2.63) states that if energy is transferred to an air parcel, some of it is used to change the internal energy (and temperature) of the parcel, and the rest is used by the parcel to do work.

The equation also states that changes in the internal energy (temperature) of a parcel result from energy transfer or work. Internal energy changes resulting from energy transfer are **diabatic** while those resulting from work are **adiabatic**. Substituting

$$dQ = \frac{dQ^*}{M_a} \qquad dU = \frac{dU^*}{M_a} \qquad dW = \frac{dW^*}{M_a} \tag{2.64}$$

where $M_a = M_d + M_v$, is the mass of a parcel of air (kg) consisting of dry air mass $M_d$ and water vapor mass $M_v$, into (2.63) gives the first law in terms of energy per unit mass of air (J $kg^{-1}$) as

$$dQ = dU + dW \tag{2.65}$$

Terms in this equation are discussed below.

When a gas expands, work is done by the gas. When air expands, work done by the air is $dW^* = p_a\, dV$, and **work** done per unit mass of air is

$$dW = \frac{dW^*}{M_a} = \frac{p_a\, dV}{M_a} = p_a\, d\alpha_a \tag{2.66}$$

In this equation, $dV$ is the change in volume of the air, and

$$\alpha_a = \frac{V}{M_a} = \frac{1}{\rho_a} \tag{2.67}$$

is the **specific volume of air**. Air expands when it rises to lower pressure. In such cases, work is done by the air, and $dV > 0$. When a parcel of air sinks to higher pressure, the parcel compresses, work is done on the air in the parcel, and $dV < 0$.

**Energy transfer** between a parcel and its environment occurs when $dQ \neq 0$. In the atmosphere, major sources (sinks) of external energy are radiative heating (cooling), condensation (evaporation), deposition (sublimation), and freezing (melting).

The change in **internal energy** of the air is its change in temperature multiplied by the energy required to change its temperature one degree Celsius (one kelvin) without affecting the volume or work done by or on the air. In other words,

$$dU = \left(\frac{\partial Q}{\partial T}\right)_{\alpha_a} dT = c_{v,m}\, dT \tag{2.68}$$

where $(\partial Q/\partial T)_{\alpha_a} = c_{v,m}$ is the **specific heat of moist air at constant volume**. It is the energy required to raise the temperature of 1 g of air 1 K at constant volume and varies with water-vapor mass mixing ratio. An expression for $c_{v,m}$ is derived by noting that, at constant volume,

$$(M_d + M_v)\, dQ = (M_d c_{v,d} + M_v c_{v,V})\, dT \tag{2.69}$$

where $c_{v,d} = 717.63$ J $kg^{-1}$ $K^{-1}$ at 298 K is the specific heat of dry air at constant volume and $c_{v,V} = 1403.2$ J $kg^{-1}$ $K^{-1}$ is the specific heat of water vapor at constant volume. The specific heat of dry air at constant volume decreases by less than 0.2 percent down to 200 K. Dividing (2.69) through by $(M_d + M_v)\, dT$ gives

$$c_{v,m} = \left(\frac{\partial Q}{\partial T}\right)_{\alpha_a} = \frac{M_d c_{v,d} + M_v c_{v,V}}{M_d + M_v} = \frac{c_{v,d} + c_{v,V}\omega_v}{1 + \omega_v} \approx c_{v,d}(1 + 0.96\omega_v) \tag{2.70}$$

Substituting (2.68) and (2.66) into (2.65) gives the **first law of thermodynamics for the atmosphere** as

$$dQ = c_{v,m}\,dT + p_a\,d\alpha_a \tag{2.71}$$

Differentiating $p_a\alpha_a = R_m T$, obtained from $p_a = \rho_a R_m T$ and $\alpha_a = 1/\rho_a$, gives

$$p_a\,d\alpha_a + \alpha_a\,dp_a = R_m\,dT \tag{2.72}$$

Combining (2.72) with (2.71) yields another form of the first law as

$$dQ = c_{p,m}\,dT - \alpha_a\,dp_a \tag{2.73}$$

where

$$c_{p,m} = \left(\frac{dQ}{dT}\right)_{p_a} = \frac{M_d c_{p,d} + M_v c_{p,V}}{M_d + M_v} = \frac{c_{p,d} + c_{p,V}\omega_v}{1+\omega_v} \approx c_{p,d}(1 + 0.859\omega_v) \tag{2.74}$$

is the **specific heat of moist air at constant pressure.** It is derived from (2.73) by noting that $(M_d + M_v)\,dQ = (M_d c_{p,d} + M_v c_{p,V})\,dT$ at constant pressure, where $c_{p,d} = 1004.67$ J kg$^{-1}$ K$^{-1}$ at 298 K is the specific heat of dry air at constant pressure, and $c_{p,V} = 1865.1$ J kg$^{-1}$ K$^{-1}$ is the specific heat of water vapor at constant pressure. Like $c_{v,d}$, $c_{p,d}$ and $c_{p,V}$ vary slightly with temperature. The specific heat of moist air at constant pressure is the energy required to increase the temperature of 1 g of air 1 K without affecting air pressure.

Differentiating (2.71) at constant pressure gives another expression for $c_{p,m}$ as

$$c_{p,m} = \left(\frac{dQ}{dT}\right)_{p_a} = c_{v,m} + p_a\left(\frac{d\alpha_a}{dT}\right)_{p_a} = c_{v,m} + p_a\left(\frac{d}{dT}\frac{R_m T}{p_a}\right)_{p_a} = c_{v,m} + R_m \tag{2.75}$$

When the air is dry, this expression simplifies to $c_{p,d} = c_{v,d} + R'$. (2.75) and the empirical expressions in (2.74) are equivalent, as can be seen by substituting $c_{v,m} \approx c_{v,d}(1 + 0.96\omega_v)$ from (2.70) and $R_m \approx R'(1 + 0.608\omega_v)$ from (2.31) into (2.75).

The first law of thermodynamics can be written in terms of virtual temperature instead of temperature with

$$dQ = c_{p,d}\,dT_v - \alpha_a\,dp_a \tag{2.76}$$

which was derived by substituting $dU = (\partial Q/\partial T_v)_{\alpha_a}\,dT_v = c_{v,d}\,dT_v$ and $dW = p_a\,d\alpha_a = R'\,dT_v - \alpha_a\,dp_a$ (found by differentiating $p_a\alpha_a = R'T_v$) into (2.65), and noting that $c_{p,d} = c_{v,d} + R'$. An advantage of (2.76) is that the specific heat is in terms of dry instead of moist air.

### 2.6.1. Applications of the First Law of Thermodynamics

The first law of thermodynamics can be modified for several special cases. For an **isobaric process** ($\mathrm{d}p_a = 0$), the first law simplifies from (2.73) to

$$\mathrm{d}Q = c_{p,m}\,\mathrm{d}T = \frac{c_{p,m}}{c_{v,m}}\,\mathrm{d}U \tag{2.77}$$

For an **isothermal process** ($\mathrm{d}T = 0$), (2.73) becomes

$$\mathrm{d}Q = -\alpha_a\,\mathrm{d}p_a = p_a\,\mathrm{d}\alpha_a = \mathrm{d}W \tag{2.78}$$

For an **isochoric process** ($\mathrm{d}\alpha_a = 0$), the first law simplifies from (2.71) to

$$\mathrm{d}Q = c_{v,m}\,\mathrm{d}T = \mathrm{d}U \tag{2.79}$$

A fourth case is an **adiabatic process.** Under adiabatic conditions, no energy is transferred to or from a parcel of air ($\mathrm{d}Q = 0$). Instead, the parcel's temperature changes because the parcel expands or contracts as it ascends or descends, respectively. When a parcel rises, it encounters lower pressure and expands. As it expands, kinetic energy is converted to work used to expand the air. Because the kinetic energy is reduced, the air cools. Thus, a rising parcel expands and cools when no energy transfer is considered. An expansion in which a change in temperature is caused by conversion of internal kinetic energy to work is an **adiabatic expansion.** When a parcel sinks, it compresses and warms. When no energy transfer is considered, the compression is an **adiabatic compression.** Under adiabatic conditions, the first law is rewritten from (2.71), (2.73), and (2.76), respectively, to

$$c_{v,m}\,\mathrm{d}T = -p_a\,\mathrm{d}\alpha_a \tag{2.80}$$

$$c_{p,m}\,\mathrm{d}T = \alpha_a\,\mathrm{d}p_a \tag{2.81}$$

$$c_{p,d}\,\mathrm{d}T_v = \alpha_a\,\mathrm{d}p_a \tag{2.82}$$

#### 2.6.1.1. *Dry Adiabatic Lapse Rate*

When an air parcel rises and cools, and no condensation occurs, the parcel's rate of cooling with height is approximately 9.8 K km$^{-1}$, which is the **dry (unsaturated) adiabatic lapse rate.** A lapse rate is the negative of the change in temperature with height. The positive lapse rate indicates that temperature decreases with increasing altitude.

The dry adiabatic lapse rate can be derived in terms of virtual temperature from the hydrostatic equation and the adiabatic form of the first law of thermodynamics. Taking the negative differential of (2.82) with respect to altitude, substituting the hydrostatic equation from (2.34), and substituting $\alpha_a = 1/\rho_a$ gives the dry adiabatic lapse rate as

$$\Gamma_d = -\left(\frac{\partial T_v}{\partial z}\right)_d = -\left(\frac{\alpha_a}{c_{p,d}}\right)\frac{\partial p_a}{\partial z} = \left(\frac{\alpha_a}{c_{p,d}}\right)\rho_a g \tag{2.83}$$
$$= \frac{g}{c_{p,d}} = +9.8\ \mathrm{K\,km^{-1}}$$

where the subscript $d$ indicates that the change is dry (unsaturated) adiabatic. (2.83) states that the virtual temperature of an unsaturated air parcel cools 9.8 K for every kilometer it ascends in the atmosphere under dry adiabatic conditions. The dry adiabatic lapse rate can also be written from (2.81) in terms of temperature as

$$\Gamma_d = -\left(\frac{\partial T}{\partial z}\right)_d = \frac{g}{c_{p,m}} = \frac{g}{c_{p,d}}\left(\frac{1+\omega_v}{1+c_{p,V}\omega_v/c_{p,d}}\right) \tag{2.84}$$

where $c_{p,m}$, obtained from (2.74), varies with moisture content. The advantage of (2.83) is that the right side of the equation does not depend on moisture content.

#### 2.6.1.2. *Potential Temperature*

An important parameter used in atmospheric analysis and modeling is the potential temperature. The **potential temperature** is the temperature an unsaturated air parcel attains if it is brought adiabatically from its altitude down to a pressure of 1000 mb. It is derived by substituting $\alpha_a = R_m T/p_a$ into (2.81) to give

$$\frac{\mathrm{d}T}{T} = \left(\frac{R_m}{c_{p,m}}\right)\frac{\mathrm{d}p_a}{p_a} \tag{2.85}$$

Integrating (2.85) from $T_0$ to $T$ and $p_{a,0}$ to $p_a$ yields **Poisson's equation,**

$$T = T_0\left(\frac{p_a}{p_{a,0}}\right)^{R_m/c_{p,m}} = T_0\left(\frac{p_a}{p_{a,0}}\right)^{R'(1+0.608\omega_v)/c_{p,d}(1+0.859\omega_v)} \approx T_0\left(\frac{p_a}{p_{a,0}}\right)^{\kappa(1-0.251\omega_v)} \tag{2.86}$$

where

$$\kappa = \frac{R'}{c_{p,d}} = \frac{c_{p,d}-c_{v,d}}{c_{p,d}} = 0.286 \tag{2.87}$$

When $p_{a,0} = 1000$ mb, $T_0$ is called the **potential temperature of moist air** ($\theta_{p,m}$), and (2.86) becomes

$$\theta_{p,m} = T\left(\frac{1000\ \text{mb}}{p_a}\right)^{\kappa(1-0.251\omega_v)} \tag{2.88}$$

In the absence of water vapor, $\omega_v = 0$ and $p_a = p_d$. In such a case, (2.88) simplifies to

$$\theta_p = T\left(\frac{1000\ \text{mb}}{p_d}\right)^{\kappa} \tag{2.89}$$

where $\theta_p$ is the **potential temperature of dry air.** Since $\omega_v$ is usually smaller than 0.02 kg kg$^{-1}$, neglecting $\omega_v$ in (2.88) causes an error in $\kappa$ of less than 0.6 percent.

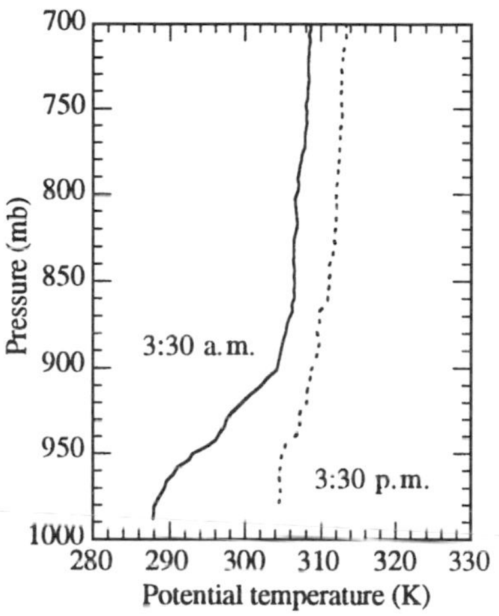


**Figure 2.12.** Observed vertical profiles of potential temperature at 3:30 a.m. and 3:30 p.m. on August 27, 1987 at Riverside, California. Potential temperatures correspond to actual temperatures shown in Fig. 2.11(a) and (b), respectively.

Thus, (2.89) is often used instead of (2.88) for defining potential temperature, even when water vapor is present. Figure 2.12 shows vertical profiles of potential temperature in the morning and afternoon, respectively, at Riverside, California on August 27, 1987.

**Example 2.20**

If the temperature of a dry air parcel at $p_d$ = 800 mb is $T$ = 270 K, then $\theta_p$ = 298.5 K from (2.89).

A parameter related to potential temperature is **potential virtual temperature** ($\theta_v$), which is found by converting all the moisture in a parcel to dry air, then bringing the parcel to a pressure of 1000 mb and determining its temperature. $\theta_v$ is the potential temperature of a sample of moist air as if it were dry and at the same density and pressure as the moist air. It can be derived by substituting $\alpha_a = R'T_v/p_a$ into (2.82), then integrating from $T_0 = \theta_v$ to $T$ and $p_{a,0}$ = 1000 mb to $p_a$. The result is

$$\theta_v = T(1 + 0.608\omega_v)\left(\frac{1000\text{ mb}}{p_a}\right)^{\kappa} = T_v\left(\frac{1000\text{ mb}}{p_a}\right)^{\kappa} \qquad (2.90)$$

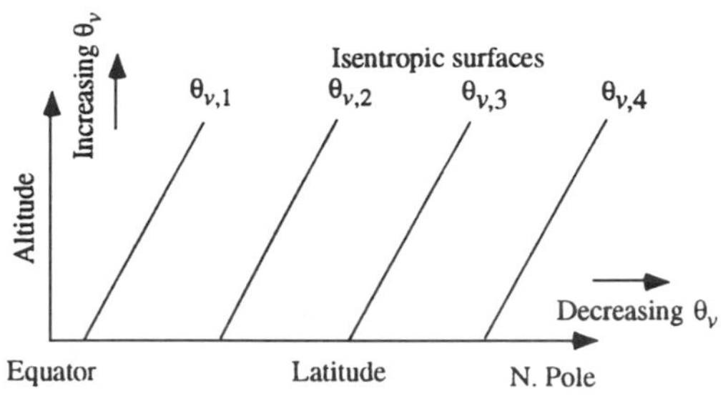


**Figure 2.13.** Isentropic surfaces (surfaces of constant potential virtual temperature) between the equator and the North Pole. Sea-level temperature and, therefore, $\theta_v$ decrease from equator to the pole. Since the free troposphere is stably stratified with respect to unsaturated air, $\theta_v$ increases with altitude, and lines of constant $\theta_v$ slant towards the poles in the vertical.

Potential virtual temperature is conserved if an unsaturated air parcel is displaced adiabatically.

The change in **entropy** of an air parcel per unit mass (J kg$^{-1}$ K$^{-1}$) due to energy transfer between the parcel and its environment is defined as $dS = dQ/T$. During adiabatic expansion and compression, $dQ = 0$ and no change in entropy occurs. Since $dQ = 0$ along surfaces of constant potential virtual temperature, $dS = 0$ along such surfaces as well, and the surface are called **isentropic.** Figure 2.13 illustrates that potential virtual temperatures increase monotonically with height in the Northern-Hemisphere troposphere, causing isentropic surfaces to slant towards the North Pole.

A parameter related to potential virtual temperature is **virtual potential temperature**, which is found by bringing a moist parcel to a pressure of 1000 mb, then converting all moisture in the parcel to dry air and determining the parcel's temperature. It is the virtual temperature of air that has been brought adiabatically to 1000 mb, and it is obtained by applying the virtual temperature correction from (2.32) to the potential temperature of moist air. The result is

$$\theta_{p,v} = \theta_{p,m}(1 + 0.608\omega_v) = T_v\left(\frac{1000\,\text{mb}}{p_a}\right)^{\kappa(1-0.251\omega_v)} \tag{2.91}$$

The difference between potential virtual temperature and virtual potential temperature is small enough that (2.90) can be used for the virtual potential temperature (Brutsaert 1991).

The **Exner function** is defined as $c_{p,d}P$, where

$$P = \left(\frac{p_a}{1000\text{ mb}}\right)^{\kappa} \tag{2.92}$$

Substituting $P$ into (2.89) and (2.90), respectively, gives the parameters

$$T = \theta_p P \qquad T_v = \theta_v P \tag{2.93}$$

### 2.6.2. Stability Criteria for Unsaturated Air

Potential virtual temperature is useful for determining atmospheric stability. The atmosphere is stable when a parcel of air, displaced vertically, returns to its original position. The atmosphere is unstable when a displaced parcel continues in the direction that it is displaced. The atmosphere is neutral when a parcel remains still after being displaced. When the atmosphere is stable near the surface, emissions accumulate and pollution builds up, since parcels of air cannot rise easily. Stability also inhibits clouds of vertical development from forming. When the atmosphere is unstable, pollution emitted near the surface diffuses vertically, decreasing its concentration near the surface. Clouds of vertical development can form in unstable air.

#### 2.6.2.1. *Determining Stability from the Dry Adiabatic Lapse Rate*

When the air is moist but unsaturated, its stability can be determined by comparing the ambient virtual temperature lapse rate with the dry adiabatic lapse rate ($\Gamma_d$). The ambient virtual temperature lapse rate ($\Gamma_v = -\partial T_v/\partial z$) is the negative change of virtual temperature with altitude. For air with no water vapor, it is the negative change of actual temperature with altitude. When temperature decreases with increasing altitude, $\Gamma_v > 0$. In terms of $\Gamma_v$ and $\Gamma_d$, the stability criteria for unsaturated air are

$$\Gamma_v \begin{cases} > \Gamma_d & \text{unsaturated unstable} \\ = \Gamma_d & \text{unsaturated neutral} \\ < \Gamma_d & \text{unsaturated stable} \end{cases} \tag{2.94}$$

Figure 2.14 demonstrates how stability can be determined graphically.

**Example 2.21**

If the change in virtual temperature with altitude is $\partial T_v/\partial z = -15$ K km$^{-1}$, what is the stability class of the atmosphere?

SOLUTION

The ambient virtual temperature lapse rate in this case is $\Gamma_v = +15$ K km$^{-1}$, which is greater than $\Gamma_d = +9.8$ K km$^{-1}$; thus, the atmosphere is unstable with respect to unsaturated air.

(2.94) indicates that virtual temperature can increase or decrease with increasing altitude in stable air but must decrease with increasing altitude in unstable air. In unsaturated air, the atmosphere is stable until the ambient virtual-temperature lapse rate increases to $\Gamma_v = +9.8$ K km$^{-1}$. A **temperature inversion** is an increase in temperature or virtual temperature with increasing altitude. An inversion occurs only in stable air, but the presence of stable air does not necessarily mean that an inversion is present, as illustrated in Example 2.22.

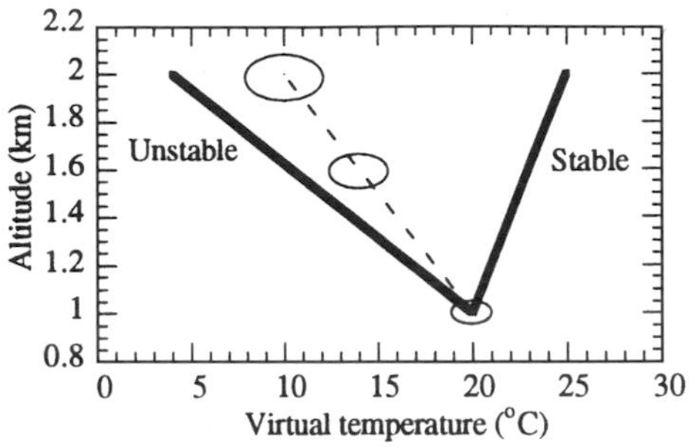


**Figure 2.14.** Demonstration of stability and instability in unsaturated air. When a parcel is displaced vertically, it rises and cools adiabatically (along the dashed line). If the ambient virtual-temperature profile is stable (right thick line), the rising parcel is cooler than the environment and sinks to its original position. If the ambient virtual-temperature profile is unstable (left thick line), the rising parcel is warmer than the air around it and continues to rise. The parcel stops rising only when it encounters air with the same virtual temperature as the parcel. This occurs when the parcel reaches a layer with a new virtual-temperature profile.

**Example 2.22**

When virtual temperature decreases slightly with increasing altitude (e.g., $\Gamma_v = +2.0 \text{ K km}^{-1}$), the atmosphere is stable but an inversion is not present. An inversion occurs when virtual temperature increases with altitude, as demonstrated by the line marked **stable** in Fig. 2.14.

Stability is enhanced by any process that warms air aloft relative to air lower down. At night, surface air stabilizes because the ground and adjacent air cool. During the day, stability is enhanced over land near the sea when cool marine air blows inland and vertically displaces warm air heated by the ground. Stability also increases when warm air blows over a cold surface or when air in a high-pressure system sinks, compresses, and warms on top of cool marine air below. Instability occurs when virtual temperature decreases with increasing height at a rate greater than that of an adiabatic parcel. This occurs when the land heats rapidly during the day or when a cold wind blows over a warm surface.

#### 2.6.2.2. *Determining Stability from Potential Virtual Temperature*

Another method of determining stability in moist but unsaturated air involves potential virtual temperature. The potential-virtual-temperature stability

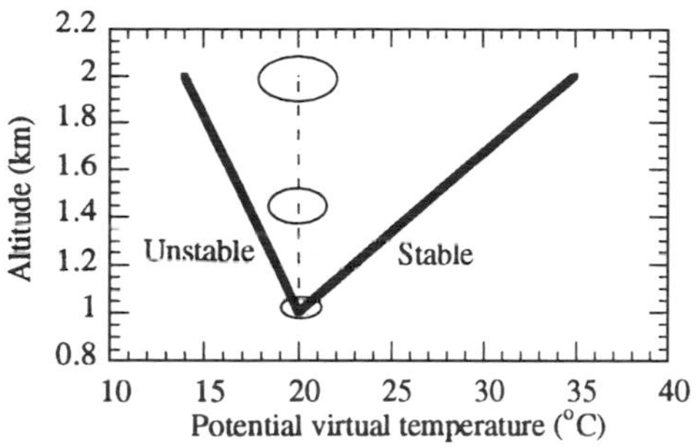


**Figure 2.15.** Demonstration of stability and instability from potential virtual temperature. When an unsaturated parcel is displaced vertically, it rises and cools adiabatically (along the dashed line). If the ambient potential-virtual-temperature profile slopes positively, a rising parcel is always cooler than the environment, the atmosphere is stable, and the parcel sinks back to its original position. If the ambient potential-virtual-temperature profile slopes negatively, a rising parcel is always warmer than the environment, the atmosphere is unstable, and the parcel continues rising.

criteria are

$$\frac{\partial \theta_v}{\partial z} \begin{cases} < 0 & \text{unsaturated unstable} \\ = 0 & \text{unsaturated neutral} \\ > 0 & \text{unsaturated stable} \end{cases} \tag{2.95}$$

Figure 2.15 demonstrates how stability can be determined from a graph of potential virtual temperature versus altitude. Figure 2.12 (a) and (b) showed that potential temperature increased with height in the morning and afternoon at Riverside, indicating the air was stable. Stability near the surface decreased between morning and afternoon, but the air never became unstable.

Potential-virtual-temperature stability criteria are found by differentiating (2.90) as

$$\begin{aligned} \mathrm{d}\theta_v &= \mathrm{d}T_v\left(\frac{1000}{p_a}\right)^{\kappa} + T_v\kappa\left(\frac{1000}{p_a}\right)^{\kappa-1}\left(-\frac{1000}{p_a^2}\right)\mathrm{d}p_a \\ &= \frac{\theta_v}{T_v}\mathrm{d}T_v - \kappa\frac{\theta_v}{p_a}\mathrm{d}p_a \end{aligned} \tag{2.96}$$

Taking the partial derivative of (2.96) with respect to height and substituting $\partial p_a/\partial z = -\rho_a g$ and $\Gamma_v = -\partial T_v/\partial z$ gives

$$\frac{\partial \theta_v}{\partial z} = \frac{\theta_v}{T_v}\frac{\partial T_v}{\partial z} - \kappa\frac{\theta_v}{p_a}\frac{\partial p_a}{\partial z} = -\frac{\theta_v}{T_v}\Gamma_v + \frac{R'}{c_{p,d}}\frac{\theta_v}{p_a}\rho_a g \tag{2.97}$$

Substituting $T_v = p_a/\rho_a R'$ and $\Gamma_d = g/c_{p,d}$ results in

$$\frac{\partial \theta_v}{\partial z} = -\frac{\theta_v}{T_v}\Gamma_v + \frac{\theta_v g}{T_v c_{p,d}} = \frac{\theta_v}{T_v}(\Gamma_d - \Gamma_v) \tag{2.98}$$

(2.94) gave the stability criteria for unsaturated air in terms of $\Gamma_d$ and $\Gamma_v$. Substituting these criteria into (2.98) gives the stability criteria shown in (2.95).

**Example 2.23**

Given $\Gamma_v = +7$ K km$^{-1}$, $p_a = 925$ mb, and $T_v = 290$ K, find $\partial\theta_v/\partial z$.

SOLUTION

From (2.90), $\theta_v = 296.5$ K. Thus, from (2.98), $\partial\theta_v/\partial z = 3.07$ K km$^{-1}$, and the atmosphere is stable with respect to unsaturated air.

### 2.6.2.3. *Determining Stability from the Brunt–Väisälä Frequency*

(2.98) can be rewritten as

$$\frac{\partial \ln \theta_v}{\partial z} = \frac{1}{T_v}(\Gamma_d - \Gamma_v) \tag{2.99}$$

Multiplying through by gravitational acceleration, $g$, gives

$$N_{bv}^2 = g\frac{\partial \ln \theta_v}{\partial z} = \frac{g}{T_v}(\Gamma_d - \Gamma_v) \tag{2.100}$$

where $N_{bv}$ is the **Brunt–Väisälä frequency** (or buoyancy frequency).

The Brunt–Väisälä frequency measures the static stability of the atmosphere. If $\theta_v$ increases with increasing altitude ($\Gamma_d > \Gamma_v$), then $N_{bv}^2 > 0$, and the atmosphere is stable. In such a case, buoyancy acts as a restoring force, causing a perturbed parcel of air to oscillate about its initial altitude with a period $\tau_{bv} = 2\pi/N_{bv}$. During the oscillation, kinetic energy is exchanged with potential energy. The oscillations arising from buoyancy restoration are **gravity waves**, discussed in Chapter 4.

If the potential virtual temperature is constant with increasing altitude ($N_{bv}^2 = 0$), the atmosphere is **neutral**, and displacements occur without resistance from a restoring buoyancy force. If potential virtual temperature decreases with increasing altitude, ($N_{bv}^2 < 0$), the atmosphere is **unstable**, and a parcel's displacement increases exponentially with time. In sum, the stability criteria for unsaturated air can be expressed as

$$N_{bv}^2 \begin{cases} < 0 & \text{unsaturated unstable} \\ = 0 & \text{unsaturated neutral} \\ > 0 & \text{unsaturated stable} \end{cases} \tag{2.101}$$

**Example 2.24**

Given $\Gamma_v = +6.5$ K $km^{-1}$ and $T_v = 288$ K, estimate the Brunt–Väisälä frequency and the period of oscillation of a perturbed parcel of air.

SOLUTION

Since $\Gamma_d = +9.8$ K $km^{-1}$ and $g = 9.81$ m $s^{-2}$, we have $N_{bv} = 0.0106$ $s^{-1}$ from (2.100), and $\tau_{bv} = 593$ s. The atmosphere is statically stable with respect to unsaturated air in this case.

## 2.7. SUMMARY

In this chapter, the structure and composition of the atmosphere were discussed. The bottom 100 km of the atmosphere is divided into four major regions: the troposphere, stratosphere, mesosphere, and thermosphere. The troposphere is divided into a boundary layer and the free troposphere. Characteristics of these regions were discussed. Equations describing basic atmospheric physics, including the equation of state, hydrostatic equation, Clausius–Clapeyron equation, and first law of thermodynamics, were also given. The equation of state and hydrostatic equation were combined to give expressions for altitude as a function of air pressure and vice versa. The Clausius–Clapeyron equation was combined with empirical relationships for latent heat to derive expressions for the saturation vapor pressure of water over liquid and ice. The first law of thermodynamics was used to derive atmospheric stability criteria and potential-virtual-temperature equations. The discussion for unsaturated air is extended to saturated air in Chapter 9. First, equations for simulating atmospheric dynamics are given.

## 2.8. PROBLEMS

**2.1.** If $T = 295$ K at 1 mm above the ground surface and the conductive heat flux is $H_c = 250$ W $m^{-2}$, estimate the temperature at the ground.

**2.2.** If $N_q = 1.5 \times 10^{12}$ molecules $cm^{-3}$ for $O_3$ gas, $T = 285$ K, and $p_d = 980$ mb, find the volume mixing ratio, dry-air mass mixing ratio, and partial pressure of ozone.

**2.3.** If $\omega_q = 1.3$ ppmm for carbon monoxide gas, $T = 285$ K, and $p_d = 980$ mb, find the volume mixing ratio, number concentration, and partial pressure of carbon monoxide.

**2.4.** If $T = 268$ K and $p_d = 700$ mb, respectively, find $p_{v,s}$ in millibars, and find the corresponding mixing ratio of water vapor in percent, ppmm, and ppmv.

**2.5.** Find the mass density of moist air ($\rho_a$) if $T = 283$ K, $f_r = 78$ percent, and $p_d = 850$ mb.

**2.6.** Find the pressure exerted by moist air if $T = 288$ K, $f_r = 82$ percent, and $p_d = 925$ mb.

**2.7.** Find the virtual temperature when $N_a = 2.1 \times 10^{19}$ molecules cm$^{-3}$, $T = 295$ K, and $f_r = 92$ percent.

**2.8.** Find the partial pressure of water vapor if $q_v = 3$ g kg$^{-1}$, $T = 278$ K, and $\rho_d = 0.5$ kg m$^{-3}$.

**2.9.** If $T_v = 281$ K, $p_v = 3$ mb, and $p_a = 972$ mb, find the air temperature.

**2.10.** If the total air pressure, temperature, and relative humidity are $p_a = 945$ mb, $T = 276$ K, and $f_r = 46$ percent, respectively, find $\omega_v$, $m_a$, $R_m$, $T_v$, and $\rho_a$.

**2.11.** If the dry-air pressure, temperature, and water-vapor–dry-air mass mixing ratio are $p_d = 927$ mb, $T = 281$ K, and $\omega_v = 0.005$ kg kg$^{-1}$, find $f_r$, $m_a$, $R_m$, $T_v$, and $\rho_a$.

**2.12.** If the total air pressure, water-vapor volume mixing ratio, and temperature are $p_a = 966$ mb, $\chi_v = 3000$ ppmv, and $T = 284$ K, respectively, find $p_v$, $m_a$, $R_m$, $T_v$, and $\rho_a$.

**2.13.** Find the altitude in a standard atmosphere that a pressure altimeter gives if the pressure measured by the altimeter is $p_d = 770$ mb.

**2.14.** Estimate the scale height of the atmosphere ($H$) and resulting pressure at $z = 200$ m altitude if the air is dry, the pressure at $z = 100$ m is $p_d = 990$ mb, and the average temperature between $z = 100$ m and $z = 200$ m is $T = 284$ K.

**2.15.** If the air is dry, $z = 10$ km, $p_d = 250$ mb, and $T = 218$ K (base of the tropopause), estimate the scale height at $z = 10$ km and pressure at $z = 10.5$ km.

**2.16.** Calculate the saturation vapor pressure of water over liquid and ice if $T_c = -15$°C. Find $\omega_v$ and $p_v$ at this temperature if $f_r = 3$ percent and $p_d = 230$ mb.

**2.17.** Calculate the dew point if $f_r = 54$ percent and $T = 263$ K.

**2.18.** Derive the expression for the water-vapor mass mixing ratio as a function of dew point and pressure from (2.62). If $T_D = 284$ K and $p_d = 1000$ mb, find $\omega_v$.

**2.19.** If $T_D = 279$ K, $T = 281$ K, and $p_d = 930$ mb, calculate $f_r$, $p_v$, $\omega_v$, and $\omega_{v,s}$.

**2.20.** **(a)** Estimate the diabatic energy (dQ) that needs to be added to or removed from a parcel of air to increase its virtual temperature $dT_v = +2$ K when $\rho_a = 1.2$ kg m$^{-3}$ and when the pressure change in the parcel due to adiabatic expansion is $dp_a = -10$ mb.

**(b)** If dQ is removed from part (a) and other conditions stay the same, what is the parcel virtual-temperature change? What type of process is this?

**(c)** If the parcel in part (a) does not rise or expand ($dp_a = 0$), but dQ calculated from part (a) remains, what is the virtual-temperature change of the parcel? What is the name of this type of process?

**(d)** If the parcel in part (a) does not change virtual temperature ($dT_v = 0$), but $dp_a = -10$ mb, what is the new value of dQ? What is the name of this process?

**2.21.** Calculate the potential virtual temperature of dry air when (a) $p_a = 900$ mb and $T = 280$ K; (b) $p_a = 850$ mb and $T = 278$ K. Is the air parcel between 900 and 850 mb stable, unstable, or neutral with respect to unsaturated air?

**2.22.** If $\theta_p = 303$ K at $p_d = 825$ mb, find the air temperature at this pressure.

**2.23.** Calculate the change in potential virtual temperature with altitude ($\partial\theta_v/\partial z$) when the ambient virtual-temperature lapse rate is $\Gamma_v = +6.2$ K km$^{-1}$, the air pressure is $p_a = 875$ mb, and $T_v = 283$ K. Is this air stable, unstable, or neutral with respect to unsaturated air?

**2.24.** If the air is dry and the potential virtual temperature increases at the rate $\partial\theta_v/\partial z = 1$ K km$^{-1}$, calculate the ambient virtual-temperature lapse rate when $p_a = 925$ mb and $T = 288$ K. Is this air stable, unstable, or neutral with respect to unsaturated air?

**2.25.** If the potential virtual temperature increases at the rate $\partial\theta_v/\partial z = 2$ K km$^{-1}$ in dry air at an altitude where $p_a = 945$ mb and $T = 287$ K, estimate $T$ at 100 m above this altitude.

**2.26.** Would liquid water and/or ice particles grow when (a) $p_v = 1$ mb, $T = -30$ K; (b) $p_v = 1.2$ mb, $T = -20$ K; or (c) $p_v = 1$ mb, $T = -16$ K? Use Fig 2.8 (b).

**2.27.** Does potential virtual temperature at sea level increase, decrease, or stay constant (on average) between the equator and North Pole? Why? Does potential virtual temperature increase, decrease, or stay constant with altitude if $\Gamma_v = +6.5$ K km$^{-1}$? Why?

**2.28.** What might Fig. 2.3 (a) and (b) look like under a low-pressure system, if all other conditions were the same?

## 2.9. COMPUTER PROGRAMMING PRACTICE

**2.29.** Write a computer script to calculate air pressure ($p_a$) as a function of altitude from (2.35). Assume $p_a = 1013$ mb and $T = 288$ K at the surface, the temperature decreases at the rate of 6.5 K km$^{-1}$, and the air is dry. Use the program to estimate the pressure from $z = 0$ to 10 km in increments of 100 m. Calculate the density $\rho_a$ with the equation of state at the base of each layer before each pressure calculation for the next layer. Plot the results.

**2.30.** Write a computer script to calculate $p_d$ as a function of altitude from (2.38). Assume $T = 298$ K and $p_d = 1013$ mb at the surface. Use the program to estimate the pressure from $z = 0$ to 10 km in increments of 100 m. Plot the results.

**2.31.** Write a computer script to calculate $p_a$ as a function of altitude from (2.42). Assume $T = 298$ K and $p_a = 1013$ mb at the surface, the air is dry, and the temperature decreases from the surface at 6.5 K km$^{-1}$. Use the program to estimate the pressure from $z = 0$ to 10 km in increments of 100 m, calculating the scale height for each layer. Plot the results.

**2.32.** Write a computer script to calculate the saturation vapor pressure of water over liquid and ice from (2.55) and (2.59), respectively. Use the program to estimate $p_{v,s}$ between $-50°C$ and $+50°C$ and $p_{v,I}$, between $-50°C$ and $0°C$, in increments of $1°C$. Plot the results.

**2.33.** Write a computer script to calculate $p_{v,s}$, $p_{v,I}$, and $T_D$ versus altitude. Assume $T$ is 298 K at $z = 0$ km and decreases 6.5 K km$^{-1}$. Assume also that $f_r = 90$ percent at all altitudes. Use the program to estimate parameters from $z = 0$ to 10 km in increments of 100 m. Plot the results.

# 3

## The Continuity and Thermodynamic Energy Equations

THE continuity and thermodynamic energy equations are important equations for atmospheric modeling. The continuity equation can be written for air, gases, or particles. It is used to simulate changes in concentration or mixing ratio of a variable over time and takes account of transport, external sources, and external sinks of the variable. The thermodynamic energy equation is used to predict changes in virtual or potential virtual temperature with time and takes account of transport, external sources, and external sinks of energy. In this chapter, scalars, vectors, gradient operators, local derivatives, and total derivatives are defined, and continuity and thermodynamic energy equations are derived.

### 3.1. LOCAL AND TOTAL DIFFERENTIATION

#### 3.1.1. Wind Velocity and Speed

Temperature, air pressure, air mass, and other variables defined in terms of magnitude but not direction are **scalars**. Variables that have magnitude and direction are **vectors**. Winds are described by three parameters – velocity, scalar velocity, and speed. **Velocity** is a vector that quantifies the rate at which the position of a body changes over time. Total and horizontal wind velocity vectors are defined in Cartesian (rectangular) horizontal coordinates as

$$\mathbf{v} = \mathbf{i}u + \mathbf{j}v + \mathbf{k}w \qquad \mathbf{v}_h = \mathbf{i}u + \mathbf{j}v \tag{3.1}$$

respectively, where **i**, **j**, and **k** are Cartesian-coordinate west–east, south–north, and vertical unit vectors, respectively, and

$$u = \frac{\mathrm{d}x}{\mathrm{d}t} \qquad v = \frac{\mathrm{d}y}{\mathrm{d}t} \qquad w = \frac{\mathrm{d}z}{\mathrm{d}t} \tag{3.2}$$

are scalar components of velocity (**scalar velocities**) (m s$^{-1}$). Scalar velocities have magnitude only. When applied in (3.1), positive $u$, $v$, and $w$ correspond to winds moving from west to east, south to north, and lower to higher elevation, respectively. The vertical scalar velocity in (3.2) is written in the **altitude vertical coordinate**. In this coordinate, tops and bottom of horizontal layers are defined by surfaces of constant altitude.

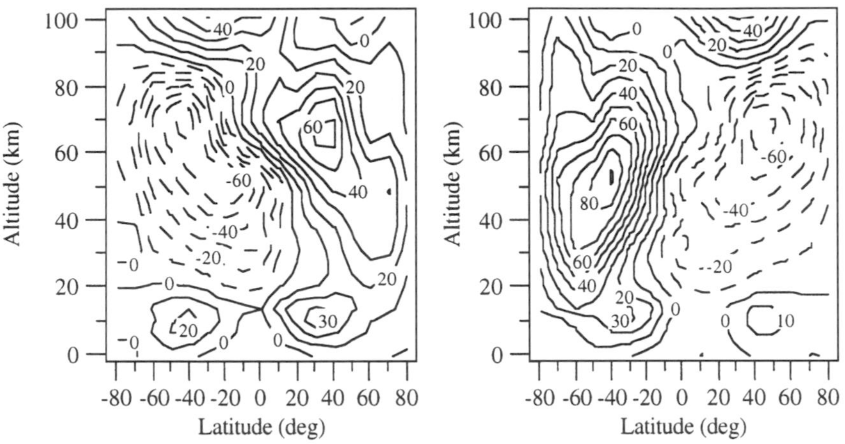


(a) January (b) July

**Figure 3.1.** Zonally averaged west–east scalar velocities (m s$^{-1}$) for (a) January and (b) July. Data for the plots were compiled by Fleming et al. (1988).

The magnitude of the wind is its speed. The total and horizontal wind speeds are, respectively,

$$|\mathbf{v}| = \sqrt{u^2 + v^2 + w^2} \qquad |\mathbf{v}_h| = \sqrt{u^2 + v^2} \tag{3.3}$$

Wind directions are generally named for where winds originate. A westerly wind, southwesterly wind, sea breeze, and mountain breeze originate from the west, the southwest, the sea, and a mountain, respectively. A positive scalar velocity $u$ with no south–north component is a westerly wind. A positive scalar velocity $v$ with no west–east component is a southerly wind.

Velocities vary with latitude, longitude, altitude, and time. Figure 3.1 (a) and (b) show global-scale latitude–altitude contour plots of zonally averaged west–east scalar velocities for a January and a July, respectively. The figures indicate that west–east winds in the upper troposphere almost always originate from the west. The two peaks near 10 km in each figure correspond to **subtropical jet streams**. Near the surface at the equator and poles, winds originate from the east but are weak. Near the surface at midlatitudes (30°–60°) in both hemispheres, winds originate from the west. In the stratosphere, westerly wind speeds increase with height in the winter hemisphere (Northern Hemisphere in January; Southern Hemisphere in July), forming **polar night jets** near 60 km. Easterly wind speeds increase with increasing altitude in the summer hemisphere. Winds above the surface are driven by pressure gradients, and pressure gradients are driven by temperature gradients. Thus, strong winds aloft indicate strong temperature and pressure gradients.

### 3.1.2. Time and Spatial Rates of Change

The time rate of change of a variable, such as concentration, momentum, or temperature, can be determined at a fixed location or in the frame of reference of the variable as it moves. Suppose a plume, carrying a gas with number concentration

$N = N(t, x[t])$ (molecules cm$^{-3}$) travels with the wind from fixed point A in the west to fixed point B in the east. The time rate of change of $N$ anywhere along the plume's trajectory is the **total derivative**, $\mathrm{d}N/\mathrm{d}t$. The total derivative can be expanded with the chain rule in Cartesian coordinates as

$$\frac{\mathrm{d}N}{\mathrm{d}t} = \frac{\partial N}{\partial t}\frac{\mathrm{d}t}{\mathrm{d}t} + \frac{\partial N}{\partial t}\frac{\mathrm{d}x}{\mathrm{d}t} = \frac{\partial N}{\partial t} + u\frac{\partial N}{\partial x} \tag{3.4}$$

where $\partial N/\partial t$ is the time rate of change of concentration at fixed point A (**local derivative**), and $u\,\partial N/\partial x$ is the time rate of change of concentration in the plume that results from a west–east velocity transporting the plume.

The total derivative of a variable is nonzero when processes other than transport affect the variable. In the case of gases, external processes include chemistry and gas-to-particle conversion. If $\mathrm{d}N/\mathrm{d}t = 0$, the concentration of a gas does not change as it travels with the wind. The local derivative of a variable is the difference between the total derivative and rate of change of the variable due to transport. Thus, the local derivative is affected by external processes and transport. If $\partial N/\partial t = 0$, the rate of production (loss) of a variable due to external processes matches the rate of loss (production) of the variable due to transport of a spatial gradient of the variable.

**Example 3.1**

Suppose the time rate of change of concentration of a gas along the path of a hot-air balloon traveling with the wind from east to west at $u = -10$ m s$^{-1}$ is $\mathrm{d}N/\mathrm{d}t = 10^8$ molecules cm$^{-3}$ s$^{-1}$. If the west–east gradient in concentration is $\partial N/\partial x = 10^{10}$ molecules cm$^{-3}$ km$^{-1}$ (concentration increases from west to east), determine the time rate of change of concentration at a fixed point A, which the balloon passes over.

SOLUTION

Since $u\,\partial N/\partial x = -10^8$ molecules cm$^{-3}$ s$^{-1}$, (3.4) predicts $(\partial N/\partial t)_A \approx 2 \times 10^8$ molecules cm$^{-3}$ s$^{-1}$. Thus, transport from the east ($u\,\partial N/\partial x$) accounts for one-half of the production rate of $N$ at point A, and transformations along the trajectory ($\mathrm{d}N/\mathrm{d}t$) account for the other half.

A **Lagrangian** frame of references is a frame of reference that moves relative to a fixed coordinate system. An **Eulerian** frame of reference is a frame of reference fixed relative to a fixed coordinate system. The left side of (3.4) is written in terms of a Lagrangian frame of reference. The right side is written in terms of an Eulerian frame of reference. Generalizing (3.4) to three dimensions gives

$$\frac{\mathrm{d}N}{\mathrm{d}t} = \frac{\partial N}{\partial t} + u\frac{\partial N}{\partial x} + v\frac{\partial N}{\partial y} + w\frac{\partial N}{\partial z} \tag{3.5}$$

### 3.1.3. Gradient Operator

A *gradient operator* (also called a directional derivative, nabla operator, or del operator) is a vector operator of partial derivatives. The **gradient operator in Cartesian–altitude coordinates** is

$$\nabla = \mathbf{i}\frac{\partial}{\partial x} + \mathbf{j}\frac{\partial}{\partial y} + \mathbf{k}\frac{\partial}{\partial z} \tag{3.6}$$

The **dot product** of the velocity vector with the gradient operator is a scalar operator,

$$\mathbf{v} \cdot \nabla = (\mathbf{i}u + \mathbf{j}v + \mathbf{k}w) \cdot \left(\mathbf{i}\frac{\partial}{\partial x} + \mathbf{j}\frac{\partial}{\partial y} + \mathbf{k}\frac{\partial}{\partial z}\right) = u\frac{\partial}{\partial x} + v\frac{\partial}{\partial y} + w\frac{\partial}{\partial z} \tag{3.7}$$

where $\mathbf{i} \cdot \mathbf{i} = 1$, $\mathbf{j} \cdot \mathbf{j} = 1$, and $\mathbf{k} \cdot \mathbf{k} = 1$. Cross terms are zero ($\mathbf{i} \cdot \mathbf{j} = 0$, $\mathbf{i} \cdot \mathbf{k} = 0$, and $\mathbf{j} \cdot \mathbf{k} = 0$), since the unit vectors are orthogonal. The dot product of two vectors is a scalar and symmetric (e.g., $\mathbf{a} \cdot \mathbf{v} = \mathbf{v} \cdot \mathbf{a}$). The dot product of a gradient operator with a vector is a scalar operator but not symmetric ($\nabla \cdot \mathbf{v} \neq \mathbf{v} \cdot \nabla$). Instead,

$$\nabla \cdot \mathbf{v} = \left(\mathbf{i}\frac{\partial}{\partial x} + \mathbf{j}\frac{\partial}{\partial y} + \mathbf{k}\frac{\partial}{\partial z}\right) \cdot (\mathbf{i}u + \mathbf{j}v + \mathbf{k}w) = \frac{\partial u}{\partial x} + \frac{\partial v}{\partial y} + \frac{\partial w}{\partial z} \tag{3.8}$$

which is a scalar **divergence**. When concentration is multiplied by a divergence, the result is the scalar

$$N(\nabla \cdot \mathbf{v}) = N\frac{\partial u}{\partial x} + N\frac{\partial v}{\partial y} + N\frac{\partial w}{\partial z} \tag{3.9}$$

The gradient of a scalar, such as concentration, is a vector. For example,

$$\nabla N = \left(\mathbf{i}\frac{\partial}{\partial x} + \mathbf{j}\frac{\partial}{\partial y} + \mathbf{k}\frac{\partial}{\partial z}\right) N = \mathbf{i}\frac{\partial N}{\partial x} + \mathbf{j}\frac{\partial N}{\partial y} + \mathbf{k}\frac{\partial N}{\partial z} \tag{3.10}$$

Applying the dot product of velocity with the gradient operator to $N$ gives the scalar

$$(\mathbf{v} \cdot \nabla) N = \left(u\frac{\partial}{\partial x} + v\frac{\partial}{\partial y} + w\frac{\partial}{\partial z}\right) N = u\frac{\partial N}{\partial x} + v\frac{\partial N}{\partial y} + w\frac{\partial N}{\partial z} \tag{3.11}$$

Substituting this result into the total-derivative equation (3.5) yields

$$\frac{\mathrm{d}N}{\mathrm{d}t} = \frac{\partial N}{\partial t} + (\mathbf{v} \cdot \nabla) N \tag{3.12}$$

Generalizing (3.12) for any variable gives the **total derivative in Cartesian–altitude coordinates** as

$$\frac{\mathrm{d}}{\mathrm{d}t} = \frac{\partial}{\partial t} + u\frac{\partial}{\partial x} + v\frac{\partial}{\partial y} + w\frac{\partial}{\partial z} = \frac{\partial}{\partial t} + \mathbf{v} \cdot \nabla \tag{3.13}$$

## 3.2. CONTINUITY EQUATIONS

When air circulates in an enclosed volume, and no chemical or physical processes affect it, the mass of the air, summed throughout the volume, is conserved. In an atmospheric model divided into many **grid cells** (or **grid boxes**), the mass of air entering one cell minus the mass leaving the cell equals the final minus the initial mass in the cell. The same is true for other atmospheric variables, such as trace-gas concentrations or energy, when only transport affects these variables.

Figure 3.2 shows a grid cell with dimension $\Delta x$, $\Delta y$, $\Delta z$ (m). The west–east scalar velocities entering and leaving the cell are $u_1$ and $u_2$ (m s$^{-1}$), respectively. Gas concentrations at the west and east boundaries of the cell are $N_1$ and $N_2$ (molecule cm$^{-3}$), respectively. Mass fluxes of gas into the cell and out of the cell are $u_1 N_1$ and $u_2 N_2$ (m [molecule cm$^{-3}$] s$^{-1}$), respectively.

From the information given, the number of molecules entering, leaving, and accumulating in the box during time period $\Delta t$ are $u_1 N_1\ \Delta y\, \Delta z\, \Delta t$, $u_2 N_2\ \Delta y\, \Delta z\, \Delta t$, and

$$\Delta N\, \Delta x\, \Delta y\, \Delta z = u_1 N_1\ \Delta y\, \Delta z\, \Delta t - u_2 N_2\ \Delta y\, \Delta z\, \Delta t \tag{3.14}$$

respectively. Dividing both sides by $\Delta t$ and the box volume ($\Delta x\, \Delta y\, \Delta z$) gives

$$\frac{\Delta N}{\Delta t} = -\left(\frac{u_2 N_2 - u_1 N_1}{\Delta x}\right) \tag{3.15}$$

As $\Delta x \rightarrow 0$ and $\Delta t \rightarrow 0$, this equation approaches

$$\frac{\partial N}{\partial t} = -\frac{\partial (uN)}{\partial x} \tag{3.16}$$

which is the **continuity equation** for a gas affected by velocity in one direction. This

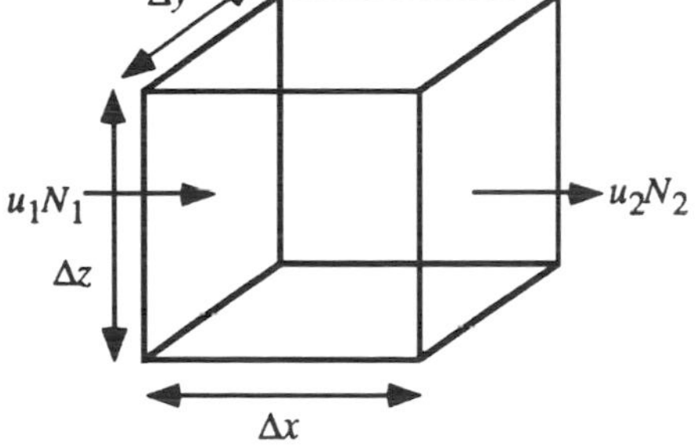


**Figure 3.2.** Example of mass conservation. The number of molecules entering minus the number of molecules leaving the box equals the number of molecules accumulating in the box.

equation expands to three dimensions in Cartesian–altitude coordinates as

$$\frac{\partial N}{\partial t} = -\frac{\partial(uN)}{\partial x} - \frac{\partial(vN)}{\partial y} - \frac{\partial(wN)}{\partial z} = -\nabla \cdot (\mathbf{v}N) \tag{3.17}$$

where $\mathbf{v}N = \mathbf{i}uN + \mathbf{j}vN + \mathbf{k}wN$. A similar equation can be written for air density. (3.17) states that the time rate of change of $N$ at a fixed location equals the negative of the local spatial gradient of the flux of $N$. (3.17) is a **flux divergence form** of the continuity equation, so called because $\nabla \cdot (\mathbf{v}N)$ is a divergence of concentration.

Substituting

$$\nabla \cdot (\mathbf{v}N) = N(\nabla \cdot \mathbf{v}) + (\mathbf{v} \cdot \nabla)N \tag{3.18}$$

into (3.17) and writing a similar equation for air density gives the continuity equations for gas concentration and total air as

$$\frac{\partial N}{\partial t} = -N(\nabla \cdot \mathbf{v}) - (\mathbf{v} \cdot \nabla)N \tag{3.19}$$

$$\frac{\partial \rho_a}{\partial t} = -\rho_a(\nabla \cdot \mathbf{v}) - (\mathbf{v} \cdot \nabla)\rho_a \tag{3.20}$$

respectively. Substituting

$$(\mathbf{v} \cdot \nabla)N = \frac{dN}{dt} - \frac{\partial N}{\partial t} \tag{3.21}$$

from (3.12) and a similar equation for density into (3.19) and (3.20), respectively, gives **velocity divergence forms** of the continuity equations as

$$\frac{dN}{dt} = -N(\nabla \cdot \mathbf{v}) \tag{3.22}$$

$$\frac{d\rho_a}{dt} = -\rho_a(\nabla \cdot \mathbf{v}) \tag{3.23}$$

where $\mathbf{v} \cdot \nabla$ is the divergence of velocity. The equations are also **advective forms** of the continuity equation in that $d/dt$ contains the advection term, $\mathbf{v} \cdot \nabla$. The equations state that the change of a scalar variable over time in a moving parcel equals the scalar variable multiplied by the negative local spatial gradient of velocity.

The gas number concentration $N$ (molecules per cubic centimeter of air) is related to the moist-air mass mixing ratio q (kilograms per kilogram of moist air) of a species with molecular weight $m$ by

$$N = \frac{A\rho_a q}{m} \tag{3.24}$$

Substituting (3.24) into (3.19) and expanding gives

$$q\left(\frac{\partial \rho_a}{\partial t} + \rho_a(\nabla \cdot \mathbf{v}) + (\mathbf{v} \cdot \nabla)\rho_a\right) + \rho_a\frac{\partial q}{\partial t} = -\rho_a(\mathbf{v} \cdot \nabla)q \tag{3.25}$$

Substituting the continuity equation for air from (3.20) into (3.25) gives the **gas continuity equation** in units of the moist-air mass mixing ratio as

$$\frac{\partial q}{\partial t} = -(\mathbf{v} \cdot \nabla)q \tag{3.26}$$

(3.22) and (3.23) assume that air is **compressible**, meaning that total volume of a parcel of air changes over time. Ocean water is considered **incompressible**, meaning that the total volume of a parcel of ocean water does not change over time. Although the total volume remains constant in an incompressible fluid, the density may vary throughout the fluid. If fluid density is constant throughout the volume, the fluid is **homogeneous.** Ocean water is **inhomogeneous** and incompressible. Since the total fluid volume cannot change in an incompressible fluid,

$$\frac{\partial u}{\partial x} + \frac{\partial v}{\partial y} + \frac{\partial w}{\partial z} = 0 \tag{3.27}$$

This is the **continuity equation for an incompressible fluid.** If (3.27) is not satisfied, a net divergence out of or convergence into a fluid volume occurs, causing the volume to expand or contract, respectively. (3.27) states that an **incompressible fluid is nondivergent.** The equation can also be written as $\nabla \cdot \mathbf{v} = 0$. Substituting water density ($\rho_w$) for air density and substituting $\nabla \cdot \mathbf{v} = 0$ into (3.23) gives

$$\frac{d\rho_w}{dt} = 0 \tag{3.28}$$

which states that the density of an incompressible fluid is constant along the motion of the fluid. At a fixed point in the fluid, the density may change. Substituting water density for air density and $\nabla \cdot \mathbf{v} = 0$ into (3.20) gives

$$\frac{\partial \rho_w}{\partial t} = -(\mathbf{v} \cdot \nabla)\rho_w \tag{3.29}$$

which states that the change in water density at a fixed point equals the negative of velocity multiplied by the gradient of density. In sum, the density of an incompressible fluid, such as liquid water, can vary spatially, but the total volume of such a fluid is constant over time. The density and total volume of a compressible fluid can change over time.

## 3.3. EXPANDED CONTINUITY EQUATIONS

(3.17) gave the continuity equation without molecular diffusion or external source and sink terms. A more complete form of the **continuity equation for a gas** is

$$\frac{\partial N}{\partial t} = -\nabla \cdot (\mathbf{v}N) + D\nabla^2 N + \sum_{n=1}^{N_{e,t}} R_n \tag{3.30}$$

(e.g., Reynolds et al. 1973), where $D$ is the **molecular diffusion coefficient** of the gas (m$^2$ s$^{-1}$ or cm$^2$ s$^{-1}$), $R_n$ is the time rate of change of trace-gas concentration due to the $n$th external process (chemistry, emissions, etc.) affecting the gas, and $N_{e,t}$ is the number of external processes affecting the gas. Units of $R$ in (3.30) are molecules cm$^{-3}$ s$^{-1}$. **Molecular diffusion** is the movement of molecules due to their kinetic energy. As molecules move, they collide with other molecules and are redirected along arbitrary paths. A molecular diffusion coefficient quantifies the rate of molecular diffusion, and is defined mathematically in Section 17.3.

The squared gradient in the molecular diffusion term of (3.30) expands to

$$\nabla^2 N = (\nabla \cdot \nabla) N = \left[\left(\mathbf{i}\frac{\partial}{\partial x} + \mathbf{j}\frac{\partial}{\partial y} + \mathbf{k}\frac{\partial}{\partial z}\right) \cdot \left(\mathbf{i}\frac{\partial}{\partial x} + \mathbf{j}\frac{\partial}{\partial y} + \mathbf{k}\frac{\partial}{\partial z}\right)\right] N \tag{3.31}$$

$$= \frac{\partial^2 N}{\partial x^2} + \frac{\partial^2 N}{\partial y^2} + \frac{\partial^2 N}{\partial z^2}$$

Substituting (3.17) and (3.31) into (3.30) gives the continuity equation for a gas as

$$\frac{\partial N}{\partial t} + \frac{\partial (uN)}{\partial x} + \frac{\partial (vN)}{\partial y} + \frac{\partial (wN)}{\partial z} = D\left(\frac{\partial^2 N}{\partial x^2} + \frac{\partial^2 N}{\partial y^2} + \frac{\partial^2 N}{\partial z^2}\right) + \sum_{n=1}^{N_{e,t}} R_n \tag{3.32}$$

### 3.3.1. Time and Grid Volume Averaging

The spatial domain in a model is divided into grid cells of finite size, and time steps are taken to advance species concentrations, velocities, and other variables. In most cases, atmospheric motions occur over spatial scales much smaller than the resolution of the grid cell and over temporal scales smaller than the resolution of the time step. A typical mesoscale model might have horizontal resolution 5 km × 5 km, vertical resolution 100 m, and time resolution 5 s. A global-scale model might have horizontal resolution 400 km × 400 km, vertical resolution 1 km, and time resolution 300 s. Fluctuations in atmospheric motions due to eddies occur on smaller scales in both cases. Eddies range in diameter from less than a millimeter to more than a kilometer and affect wind speeds over small distances and time periods. Variations in wind speeds affect concentrations, temperatures, pressure, and other parameters.

To account for subgrid-scale disturbances, each variable in (3.32) and other model equations can be divided into an average and perturbation component. Such a division is referred to as **Reynolds decomposition**, after Osborne Reynolds. In the case of gas concentration, **actual (precise) concentrations** are decomposed as

$$N = \bar{N} + N' \tag{3.33}$$

where $\bar{N}$ is an **average gas concentration**, and $N'$ is an instantaneous and local **perturbation concentration**. A precise concentration occurs at a given instant and location within a grid cell. An average concentration is obtained by integrating and averaging over a model time step and grid cell volume. Thus,

$$\bar{N} = \frac{1}{\Delta t\,\Delta x\,\Delta y\,\Delta z}\int_t^{t+\Delta t}\left\{\int_x^{x+\Delta x}\left[\int_y^{y+\Delta y}\left(\int_z^{z+\Delta z} N\,\mathrm{d}z\right)\mathrm{d}y\right]\mathrm{d}x\right\}\mathrm{d}t \quad (3.34)$$

(Pielke 1984), where $\Delta t$ is the time step, and $\Delta x$, $\Delta y$, $\Delta z$ are space increments, shown in Fig. 3.2. The average concentrations are averages for one grid cell and time step and differ for each grid cell and time step. Perturbation concentrations are distributed on both sides of the average, so that the spatial and temporal average of all perturbations is zero ($\bar{N'} = 0$), which is the **Reynolds assumption.**

Scalar and vector velocities can be decomposed in a similar manner. Thus,

$$u = \bar{u} + u' \qquad v = \bar{v} + v' \qquad w = \bar{w} + w' \quad (3.35)$$

where $\bar{u}$, $\bar{v}$, and $\bar{w}$ are the time- and volume-averaged scalar velocities, and $u'$, $v'$, and $w'$ are perturbation scalar velocities. The precise velocity vector is

$$\mathbf{v} = \bar{\mathbf{v}} + \mathbf{v}' \quad (3.36)$$

where $\bar{\mathbf{v}} = \mathbf{i}\bar{u} + \mathbf{j}\bar{v} + \mathbf{k}\bar{w}$ is a time- and volume-averaged velocity and $\mathbf{v}' = \mathbf{i}u' + \mathbf{j}v' + \mathbf{k}w'$ is a perturbation velocity. **Advection** is the mean horizontal velocity. Thus, $\bar{u}$ and $\bar{v}$ are components of advection. Figure 3.3 shows an example of precise, mean, and perturbation scalar velocities and trace-gas concentration, respectively.

**Unsteady flow** occurs when **v** varies with time, but not necessarily randomly, at a given location. **Steady flow** occurs when **v** is independent of time. **Turbulent flow** is unpredictable flow in which **v** varies randomly with time at a location. Thus, turbulent flow is unsteady, but unsteady flow is not necessarily turbulent. **Laminar flow** is nonturbulent flow in which **v** may vary, but not randomly, with time at a given location. In laminar flow, fluid particles travel along well-defined streamlines and flood layers flow independent of each other. Laminar flow can be steady or unsteady. Nearly all flows in the atmosphere are turbulent.

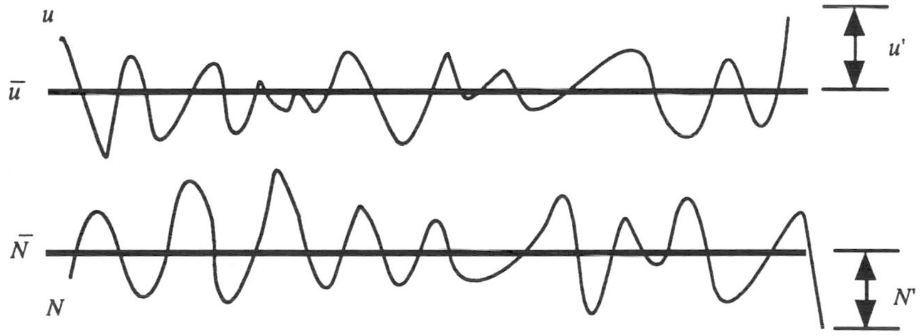


**Figure 3.3.** Precise, mean, and perturbation components of scalar velocity and gas concentration. The precise scalar velocity is denoted by $u$, the precise gas concentration is denoted by $N$, time- and volume-averaged values are denoted by an overbar, and perturbation components are denoted by a prime. Each point on the horizontal axis is a perturbation at a given time and location within a grid cell.

Subgrid-scale effects are estimated by substituting decomposed variables into an equation, then taking a time average and grid volume average of resulting terms. Substituting (3.33) and (3.35) into the species continuity equation from (3.32) and averaging terms over space and time gives

$$\overline{\left(\frac{\partial(\bar{N}+N')}{\partial t}\right)}+\overline{\left(\frac{\partial(\bar{u}+u')(\bar{N}+N')}{\partial x}\right)}+\overline{\left(\frac{\partial(\bar{v}+v')(\bar{N}+N')}{\partial y}\right)}$$
$$+\overline{\left(\frac{\partial(\bar{w}+w')(\bar{N}+N')}{\partial z}\right)}=D\left\{\overline{\left[\frac{\partial^2(\bar{N}+N')}{\partial x^2}\right]}+\overline{\left[\frac{\partial^2(\bar{N}+N')}{\partial y^2}\right]}\right.$$
$$\left.+\overline{\left[\frac{\partial^2(\bar{N}+N')}{\partial z^2}\right]}\right\}+\overline{\sum_{n=1}^{N_{e,t}}R_n} \tag{3.37}$$

Since $\overline{\partial(\bar{N}+N')/\partial t}=\partial(\overline{\bar{N}+N'})/\partial t$, $\overline{\bar{N}+N'}=\bar{\bar{N}}+\bar{N}'$, $\bar{\bar{N}}=\bar{N}$, and $\bar{N}'=0$, the first term in (3.37) simplifies to

$$\overline{\left[\frac{\partial(\bar{N}+N')}{\partial t}\right]}=\frac{\partial(\bar{\bar{N}}+\bar{N}')}{\partial t}=\frac{\partial\bar{N}}{\partial t} \tag{3.38}$$

Since $\overline{u'\bar{N}}=0$, $\overline{\bar{u}N'}=0$, and $\overline{\bar{u}\bar{N}}=\bar{u}\bar{N}$, the second term simplifies to

$$\overline{\left[\frac{\partial(\bar{u}+u')(\bar{N}+N')}{\partial x}\right]}=\frac{\partial(\overline{\bar{u}\bar{N}}+\bar{u}\bar{N}'+\overline{u'\bar{N}}+\overline{u'N'})}{\partial x}=\frac{\partial(\bar{u}\bar{N}+\overline{u'N'})}{\partial x} \tag{3.39}$$

The product $\overline{u'N'}$ represents the west–east transport of $N'$ due to subgrid-scale eddies. When a variable or a flux is divided by the air density, it becomes a **kinematic** variable or flux. $\overline{u'N'}$ (m [molecules cm$^{-3}$] s$^{-1}$) is a **kinematic turbulent flux:** its units are those of concentration flux (kg [molecules cm$^{-3}$] m$^{-2}$ s$^{-1}$) divided by air density (kg m$^{-3}$). The partial derivative $\partial(\overline{u'N'})/\partial x$ (molecules cm$^{-3}$ s$^{-1}$) is a **turbulent flux divergence** term.

**Example 3.2**

Suppose two gas concentrations ($N_1 = 8$ and $N_2 = 4$) and scalar velocities ($u_1 = 3$ and $u_2 = -1$) are measured at different locations within a grid cell at a given time. Estimate $\bar{N}$, $N'_1$, $N'_2$, $\bar{u}$, $u'_1$, $u'_2$, $\overline{u'N'}$, $\bar{u}\bar{N}$, and $\overline{uN}$. Ignore units.

SOLUTION

$$\bar{N}=(N_1+N_2)/2=6 \qquad \bar{u}=(u_1+u_2)/2=1$$
$$N'_1=N_1-\bar{N}=2 \qquad u'_1=u_1-\bar{u}=2$$
$$N'_2=N_2-\bar{N}=-2 \qquad u'_2=u_2-\bar{u}=-2$$
$$\overline{u'N'}=(u'_1N'_1+u'_2N'_2)/2=4\bar{u}\bar{N} \qquad \bar{u}\bar{N}=6$$
$$\overline{uN}=\bar{u}\bar{N}+\overline{u'N'}=(u_1N_1+u_2N_2)/2=10$$

Substituting (3.38), (3.39), and similar terms for other directions into (3.37) gives

$$\frac{\partial \bar{N}}{\partial t} + \frac{\partial(\bar{u}\bar{N})}{\partial x} + \frac{\partial(\bar{v}\bar{N})}{\partial y} + \frac{\partial(\bar{w}\bar{N})}{\partial z} + \frac{\partial \overline{u'N'}}{\partial x} + \frac{\partial \overline{v'N'}}{\partial y} + \frac{\partial \overline{w'N'}}{\partial z} = D\left(\frac{\partial^2 \bar{N}}{\partial x^2} + \frac{\partial^2 \bar{N}}{\partial y^2} + \frac{\partial^2 \bar{N}}{\partial z^2}\right) + \sum_{n=1}^{N_{e,t}} \bar{R}_n \tag{3.40}$$

For motions larger than the molecular scale, the molecular diffusion terms in (3.40) are much smaller than the turbulent flux divergence terms and can be removed. Thus, (3.40) simplifies to

$$\frac{\partial \bar{N}}{\partial t} + \frac{\partial(\bar{u}\bar{N})}{\partial x} + \frac{\partial(\bar{v}\bar{N})}{\partial y} + \frac{\partial(\bar{w}\bar{N})}{\partial z} + \frac{\partial \overline{u'N'}}{\partial x} + \frac{\partial \overline{v'N'}}{\partial y} + \frac{\partial \overline{w'N'}}{\partial z} = \sum_{n=1}^{N_{e,t}} \bar{R}_n \tag{3.41}$$

which compresses to

$$\frac{\partial \bar{N}}{\partial t} + \nabla \cdot (\bar{\mathbf{v}}\bar{N}) + \nabla \cdot (\overline{\mathbf{v}'N'}) = \sum_{n=1}^{N_{e,t}} \bar{R}_n \tag{3.42}$$

An analogous equation for air density gives the **continuity equation** as

$$\frac{\partial \bar{\rho}_a}{\partial t} + \nabla \cdot (\bar{\mathbf{v}}\bar{\rho}_a) + \nabla \cdot (\overline{\mathbf{v}'\rho_a'}) = 0 \tag{3.43}$$

In (3.43), the external source and sink terms for air molecules are neglected because they are small in comparison with the other terms.

If moist-air mass mixing ratio instead of number concentration is used in (3.42), the equation needs to be rederived. (3.26) gave the species continuity equation, without sources or sinks, in units of the moist-air mass mixing ratio. Adding source and sinks yields

$$\frac{\partial q}{\partial t} + \mathbf{v} \cdot \nabla q = \sum_{n=1}^{N_{e,t}} R_n \tag{3.44}$$

where $R_n$ is now in kilograms per kilogram of moist air per second. Multiplying the continuity equation for air from (3.20) by q, multiplying (3.44) by $\rho_a$, adding the results, and compressing gives

$$\frac{\partial(\rho_a q)}{\partial t} + \nabla \cdot (\rho_a \mathbf{v} q) = \rho_a \sum_{n=1}^{N_{e,t}} R_n \tag{3.45}$$

The moist-air mass mixing ratio, velocity, and density can be decomposed as $q = \bar{q} + q'$, $\mathbf{v} = \bar{\mathbf{v}} + \mathbf{v}'$, and $\rho_a = \bar{\rho}_a + \rho_a'$, respectively. Density perturbations in the atmosphere are relatively small; thus, $\rho'_a \ll \bar{\rho}_a$, and $\rho_a \approx \bar{\rho}_a$. Substituting decomposed variable values into all but the $R$ term in (3.45) gives

$$\frac{\partial[\bar{\rho}_a(\bar{q}+q')]}{\partial t} + \nabla \cdot [\bar{\rho}_a(\bar{\mathbf{v}}+\mathbf{v}')(\bar{q}+q')] = \bar{\rho}_a \sum_{n=1}^{N_{e,t}} R_n \tag{3.46}$$

Taking the time and grid volume average of this equation, eliminating zero-value terms, and removing unnecessary overbars results in

$$\bar{\rho}_a\left[\frac{\partial \bar{q}}{\partial t} + (\bar{\mathbf{v}} \cdot \nabla)\bar{q}\right] + \bar{q}\left[\frac{\partial \bar{\rho}_a}{\partial t} + \nabla \cdot (\bar{\mathbf{v}}\bar{\rho}_a)\right] + \nabla \cdot (\bar{\rho}_a\overline{\mathbf{v}'q'}) = \bar{\rho}_a \sum_{n=1}^{N_{e,t}} \bar{R}_n \tag{3.47}$$

(3.47) can be simplified by first noting that, when $\rho'_a \ll \bar{\rho}_a$, (3.43) becomes

$$\frac{\partial \bar{\rho}_a}{\partial t} + \nabla \cdot (\bar{\mathbf{v}}\,\bar{\rho}_a) = 0 \tag{3.48}$$

Substituting this expression into (3.47) and dividing through by $\bar{\rho}_a$ gives

$$\frac{\partial \bar{q}}{\partial t} + (\mathbf{v} \cdot \nabla)\bar{q} + \frac{1}{\bar{\rho}_a}\nabla \cdot (\bar{\rho}_a\overline{\mathbf{v}'q'}) = \sum_{n=1}^{N_{e,t}} \bar{R}_n \tag{3.49}$$

which is the **gas continuity equation** in units of the moist-air mass mixing ratio.

In (3.49), $\overline{u'q'}$, $\overline{v'q'}$, and $\overline{w'q'}$ (m [kg kg$^{-1}$] s$^{-1}$) are **kinematic turbulent fluxes** of mixing ratio. Whereas models calculate spatially and temporally averaged values (e.g., $\bar{N}$, $\bar{q}$, $\bar{\rho}_a$, $\bar{u}$), explicit equations for kinematic turbulent fluxes (e.g., $\overline{u'N'}$) have not been developed. Instead, these quantities are estimated with parameterizations of different complexity. Parameterizations from similarity theory and from $K$-theory are discussed in Chapter 8. Here, a simple overview of $K$-theory is given.

With ***K*-theory** (gradient transport theory), kinematic turbulent fluxes are replaced by the product of a constant and the gradient of the mean value of the fluctuating variable (Calder 1949; Pasquill 1962; Monin and Yaglom 1971; Reynolds et al. 1973; Stull 1988). Kinematic turbulent fluxes of gas concentration are parameterized with

$$\overline{u'N'} = -K_{h,xx}\frac{\partial \bar{N}}{\partial x} \qquad \overline{v'N'} = -K_{h,yy}\frac{\partial \bar{N}}{\partial y} \qquad \overline{w'N'} = -K_{h,zz}\frac{\partial \bar{N}}{\partial z} \tag{3.50}$$

where $K_{h,xx}$, $K_{h,yy}$, and $K_{h,zz}$ (m$^2$ s$^{-1}$ or cm$^2$ s$^{-1}$) are **eddy diffusion coefficients** in the $x$-, $y$-, and $z$-directions, respectively. The subscript $h$ indicates that the **eddy diffusion coefficient for energy** (eddy thermal diffusivity) is used. (3.50) is convenient in that models predict mean quantities. (3.50) uses the eddy diffusion coefficient for energy because the turbulent transport of a gas is similar to that of energy. When turbulent transport of velocity is simulated, an **eddy diffusion coefficient for momentum** (eddy viscosity) term is used. Eddy diffusion coefficients for energy and momentum differ, but not by much. Eddy diffusion coefficients are also called eddy transfer, eddy exchange, turbulent transfer, and gradient transfer coefficients.

Eddy diffusion coefficients are parameterizations of subgrid scale transport of energy and momentum. In the vertical, such transport is caused by **mechanical shear** (*mechanical turbulence*) and/or **buoyancy** (*thermal turbulence*). Horizontal wind shear creates eddies that increase in size when the wind flows over rough surfaces. Buoyancy creates instability, causing eddies to become larger and extend to a greater height. Vertical motions in eddies transfer surface air aloft and air aloft to the surface. Eddy diffusion is a parameterization of the rate of vertical exchange of air, averaged over all eddies. Eddies also exchange air horizontally. Horizontal eddy diffusion is proportional to the rate of horizontal exchange of air, averaged over all horizontal eddies.

Substituting (3.50) into (3.41) gives

$$\frac{\partial \bar{N}}{\partial t} + \frac{\partial(\bar{u}\bar{N})}{\partial x} + \frac{\partial(\bar{v}\bar{N})}{\partial y} + \frac{\partial(\bar{w}\bar{N})}{\partial z} = \frac{\partial}{\partial x}\left(K_{h,xx}\frac{\partial \bar{N}}{\partial x}\right) + \frac{\partial}{\partial y}\left(K_{h,yy}\frac{\partial \bar{N}}{\partial y}\right) + \frac{\partial}{\partial z}\left(K_{h,zz}\frac{\partial \bar{N}}{\partial z}\right) + \sum_{n=1}^{N_{e,t}} \bar{R}_n \tag{3.51}$$

Compressing (3.51) and removing overbars for simplicity gives the **continuity equation for gas number concentration** in Cartesian–altitude coordinates as

$$\frac{\partial N}{\partial t} + \nabla \cdot (\mathbf{v}N) = (\nabla \cdot \mathbf{K}_h \nabla)N + \sum_{n=1}^{N_{e,t}} R_n \tag{3.52}$$

where

$$\mathbf{K}_h = \begin{bmatrix} K_{h,xx} & 0 & 0 \\ 0 & K_{h,yy} & 0 \\ 0 & 0 & K_{h,zz} \end{bmatrix} \tag{3.53}$$

is the **eddy diffusion tensor for energy**. The analogous continuity equation for the gas–moist-air mass mixing ratio is

$$\frac{\partial q}{\partial t} + (\mathbf{v} \cdot \nabla)q = \frac{1}{\rho_a}(\nabla \cdot \rho_a \mathbf{K}_h \nabla)q + \sum_{n=1}^{N_{e,t}} R_n \tag{3.54}$$

The units of $R_n$ differ in the two cases.

### 3.3.2. Continuity Equation for Air

External sources and sinks ($R_n$) are relatively small and can be ignored in the continuity equation for air. For most modeling applications, the turbulent flux divergence term in the equation can also be ignored, since $\rho'_a \ll \bar{\rho}_a$. After removing overbars for convenience and making the above modifications, the continuity equation for air in Cartesian–altitude coordinates reduces from (3.43) to

$$\frac{\partial \rho_a}{\partial t} + \nabla \cdot (\mathbf{v}\rho_a) = 0 \qquad (3.55)$$

### 3.3.3. Gas Continuity Equation

The continuity equations for trace gases and particles include several external source and sink terms. Gases enter the atmosphere from surface and elevated sources through **emissions**. They are removed by water, soil, foliage, roads, buildings, cars, and other surfaces during **dry deposition**. In many cases, gases are swept out of the atmosphere by falling raindrops during **washout**. Gases react chemically with each other and are dissociated by solar radiation during **photochemistry**. Some gases aggregate to form new particles during **homogeneous nucleation** or aggregate on existing particle surfaces during **heterogeneous nucleation**. Once a surface has nucleated, gas molecules may diffuse to and **condense** as a liquid or **deposit** as a solid on the surface. Liquid material may also **evaporate** or solid material may **sublimate** to the gas phase. A gas may also **dissolve** in liquid water on the surface of a particle. Dissolved gases may evaporate. Finally, a gas may react chemically on the surface of a particle during **heterogeneous chemistry**.

A form of the continuity equation for a gas q that accounts for the processes discussed above is

$$\begin{aligned}\frac{\partial N_q}{\partial t} + \nabla \cdot (\mathbf{v}N_q) = (\nabla \cdot \mathbf{K}_h \nabla)N_q &+ R_{emisg} + R_{depg} + R_{washg} \\ &+ R_{chemg} + R_{nucg} + R_{c/eg} + R_{dp/sg} \\ &+ R_{ds/eg} + R_{hrg}\end{aligned} \qquad (3.56)$$

where

$R_{emisg}$ = rate of surface or elevated emissions
$R_{depg}$ = rate of dry deposition to the ground surface
$R_{washg}$ = rate of washout to the ground surface or from one altitude to another
$R_{chemg}$ = rate of photochemical production or loss
$R_{nucg}$ = rate of gas loss due to homogeneous or heterogeneous nucleation
$R_{c/eg}$ = rate of gas loss (production) due to condensation (evaporation)
$R_{dp/sg}$ = rate of gas loss (production) due to depositional growth (sublimation)
$R_{ds/eg}$ = rate of gas loss (production) due to dissolutional growth (evaporation)
$R_{hrg}$ = rate of gas loss (production) due to heterogeneous reactions

All rates are expressed in units of concentration per unit time (e.g., molecules $cm^{-3}$ $s^{-1}$).

### 3.3.4. Particle Continuity Equation

The continuity equation for particles is divided into two subequations. One is for particle number concentration, and the other is for particle volume component concentration. Particles contain anywhere from one to hundreds of components. The volume of each component varies over time due to physical and chemical processes. If the total volume of one particle in a size bin $i$ is denoted by $\upsilon_i$ (cm$^3$/particle), the **volume** of component $q$ within that particle is $\upsilon_{q,i}$. Thus, $\upsilon_{q,i}$ gives information about a component in a single particle of a given size. A variable giving information about that component summed over all particles of the same size is more relevant. Such a parameter is **volume concentration** (cubic centimeters of the component per cubic centimeter of air), defined as

$$v_{q,i} = n_i \upsilon_{q,i} \tag{3.57}$$

where $n_i$ is the **number concentration** of particles of size $i$ (particles per cubic centimeter of air). If two of the three variables in (3.57) are predicted numerically, the third can be found from the equation. Typically, volume concentration and number concentration are predicted numerically. They are found from separate continuity equations, because different external sources and sinks affect the number and volume concentrations.

The continuity equation for the number concentration of particles of size $i$ is

$$\frac{\partial n_i}{\partial t} + \nabla \cdot (\mathbf{v} n_i) = (\nabla \cdot \mathbf{K}_h \nabla) n_i + R_{emisn} + R_{depn} + R_{sedn} + R_{washn} + R_{nucn} + R_{coagn} \tag{3.58}$$

where

$R_{emisn}$ = rate of surface or elevated emissions

$R_{depn}$ = rate of particle dry deposition to the surface

$R_{sedn}$ = rate of sedimentation to surface or between layers

$R_{washn}$ = rate of washout to the surface or from one altitude down to another

$R_{nucn}$ = rate of production of new particles due to homogeneous nucleation

$R_{coagn}$ = rate of coagulation of number concentration

All rates are in units of particles cm$^{-3}$ s$^{-1}$. Sources and sinks that affect particle number concentrations include emissions, dry deposition, sedimentation, washout, homogeneous nucleation, and coagulation. **Sedimentation** occurs when particles fall through the atmosphere due to their mass. Sedimentation by gases is negligible because gas molecules have extremely small masses. **Particle dry deposition** occurs when particles diffuse to or otherwise impact a surface by any transport process. **Particle washout** occurs when rain sweeps particles in its path to lower altitudes or the surface. **Homogenous nucleation** is a source of new particles. **Heterogeneous nucleation** does not produce new particles but allows growth to proceed on existing particles. **Coagulation** occurs when two particles collide and stick to form a single, larger particle.

The continuity equation for the volume concentration of component $q$ in particles of size $i$ is

$$\frac{\partial v_{q,i}}{\partial t} + \nabla \cdot (\mathbf{v} v_{q,i}) = (\nabla \cdot \mathbf{K}_h \nabla) v_{q,i} + R_{emisv} + R_{depv} + R_{sedv} + R_{washv} + R_{nucv} + R_{coagv} + R_{c/ev} + R_{dp/sv} + R_{ds/ev} + R_{eqv} + R_{aqv} + R_{hrv} \quad (3.59)$$

where

$R_{emisv}$ = rate of surface or elevated emissions

$R_{depv}$ = rate of dry deposition to the surface

$R_{sedv}$ = rate of sedimentation to the surface or from one altitude to another

$R_{washv}$ = rate of washout to the surface or from one altitude to another

$R_{nucv}$ = rate of change due to homogeneous or heterogeneous nucleation

$R_{coagv}$ = rate of change due to coagulation

$R_{c/ev}$ = rate of change due to condensational growth (evaporation)

$R_{dp/sv}$ = rate of change due to depositional growth (sublimation)

$R_{ds/ev}$ = rate of change due to dissolutional growth (evaporation)

$R_{eqv}$ = rate of change due to reversible chemical equilibrium reactions

$R_{aqv}$ = rate of change due to irreversible aqueous chemical reactions

$R_{hrv}$ = rate of change due to heterogeneous reactions on particle surfaces

Rates in this equation have units of cubic centimeters of particles per cubic centimeter of air per second. Some processes, such as homogeneous nucleation and coagulation, affect number and volume concentrations. Others, such as heterogeneous nucleation, condensation, deposition, dissolution, heterogeneous reaction, chemical equilibrium, and aqueous chemistry affect volume concentration but not number concentration.

### 3.3.5. Continuity Equation for Gas, Liquid, and Solid Water

Water in the atmosphere appears as a gas, liquid, or solid. In a model, the total water content is estimated as

$$q_T = q_v + \sum_{i=1}^{N_B} (q_{L,i} + q_{I,i}) \quad (3.60)$$

where $N_B$ is the number of particle size categories (bins), $q_v$ is the specific humidity of water vapor (kilograms per kilogram of moist air), $q_{L,i}$ is the moist-air mass mixing ratio of liquid water in a size bin, and $q_{I,i}$ is the moist-air mass mixing ratio of ice in a size bin. Mass mixing ratios are determined from the continuity equations for water vapor, liquid, and ice,

$$\frac{\partial q_v}{\partial t} + (\mathbf{v} \cdot \nabla) q_v = \frac{1}{\rho_a}(\nabla \rho_a \mathbf{K}_h \nabla) q_v + R_{emisV} + R_{depV} + R_{chemV} + R_{nucV} + R_{c/eV} + R_{dp/sV} \tag{3.61}$$

$$\frac{\partial q_{L,i}}{\partial t} + (\mathbf{v} \cdot \nabla) q_{L,i} = \frac{1}{\rho_a}(\nabla \rho_a \mathbf{K}_h \nabla) q_{L,i} + R_{emisL} + R_{depL} + R_{sedL} + R_{nucL} + R_{coagL} + R_{c/eL} + R_{f/mL} \tag{3.62}$$

$$\frac{\partial q_{I,i}}{\partial t} + (\mathbf{v} \cdot \nabla) q_{I,i} = \frac{1}{\rho_a}(\nabla \rho_a \mathbf{K}_h \nabla) q_{I,i} + R_{depI} + R_{sedI} + R_{nucI} + R_{coagI} + R_{f/mI} + R_{dp/sI} \tag{3.63}$$

where

$R_{emis}$ = rate of surface or elevated emissions
$R_{dep}$ = rate of dry deposition to the surface
$R_{sed}$ = rate of sedimentation to the surface or from one altitude to another
$R_{chem}$ = rate of photochemical production or loss
$R_{nuc}$ = rate of liquid or ice nucleation from the gas phase
$R_{coag}$ = rate of liquid or ice production or loss in a size bin due to coagulation
$R_{c/e}$ = rate of change due to condensational growth (evaporation)
$R_{dp/s}$ = rate of change due to depositional growth (sublimation)
$R_{f/m}$ = rate of change due to freezing (melting)

and the units of $R$ are kilograms per kilogram of moist air per second.

Many meteorological models simulate liquid water and ice as **bulk parameters**. In such cases, liquid water and ice are not separated into size categories, and their number concentrations are not tracked. Instead, only the moist-air mass mixing ratios of total liquid water and ice are predicted. Since particles are not size-resolved in a bulk parameterization, some processes, such as coagulation, cannot be treated. When liquid and ice content are treated as bulk parameters, $q_T = q_v + q_L + q_I$, where the subscript $i$ has been dropped because a bulk parameterization has no size resolution.

## 3.4. THERMODYNAMIC ENERGY EQUATION

**Temperatures** in the atmosphere are affected by energy transfer and work. Energy transfer occurs by conduction, mechanical turbulence, thermal turbulence, advection, and radiation, all introduced in Section 2.2. Energy is released to the air during condensation of water vapor, deposition of water vapor, freezing of liquid water, exothermic chemical reactions, and radioactive decay. Energy is removed from the air upon melting of ice, sublimation of ice, and evaporation of liquid water. Energy exchange may also occur upon change of state of substances other than water. Because the quantities of nonwater substances changing state are relatively small, resulting energy exchanges are small.

Work is done by (on) the air during an adiabatic expansion (compression). During an adiabatic expansion, a rising parcel of air expands and cools by converting internal energy, in the form of molecular kinetic energy, to work. Sources and sinks of heat energy that are not adiabatic are **diabatic**. Conduction, radiative cooling and heating, and latent heat release and absorption are diabatic energy sources or sinks.

Energy, like air density and species concentrations, must be conserved. An equation for thermodynamic energy can be derived by combining the first law of thermodynamics with the continuity equation for air. The first law of thermodynamics as expressed in (2.76 ) was $\mathrm{d}Q = c_{p,d}\,\mathrm{d}T_v - \alpha_a\,\mathrm{d}p_a$. Differentiating this equation with respect to time, substituting $\alpha_a = 1/\rho_a$, and rearranging gives the **thermodynamic energy equation** as

$$\frac{\mathrm{d}T_v}{\mathrm{d}t} = \frac{1}{c_{p,d}}\frac{\mathrm{d}Q}{\mathrm{d}t} + \frac{1}{c_{p,d}\rho_a}\frac{\mathrm{d}p_a}{\mathrm{d}t} \tag{3.64}$$

If the thermodynamic energy equation is written in terms of potential virtual temperature, the last term in (3.64) can be eliminated. Differentiating $\theta_v = T_v(1000/p_a)^\kappa$ with respect to time gives

$$\begin{aligned}\frac{\mathrm{d}\theta_v}{\mathrm{d}t} &= \frac{\mathrm{d}T_v}{\mathrm{d}t}\left(\frac{1000}{p_a}\right)^\kappa + T_v\kappa\left(\frac{1000}{p_a}\right)^{\kappa-1}\left(-\frac{1000}{p_a^2}\right)\frac{\mathrm{d}p_a}{\mathrm{d}t} \\ &= \frac{\theta_v}{T_v}\frac{\mathrm{d}T_v}{\mathrm{d}t} - \frac{\kappa\theta_v}{p_a}\frac{\mathrm{d}p_a}{\mathrm{d}t}\end{aligned} \tag{3.65}$$

Substituting (3.65), $\kappa = R'/c_{p,d}$, and $p_a = \rho_a R' T_v$ into (3.64), and expanding the total derivative with (3.13) gives the thermodynamic energy equation in terms of potential virtual temperature as

$$\boxed{\frac{\mathrm{d}\theta_v}{\mathrm{d}t} = \frac{\partial\theta_v}{\partial t} + (\mathbf{v}\cdot\nabla)\theta_v = \frac{\theta_v}{c_{p,d}T_v}\frac{\mathrm{d}Q}{\mathrm{d}t}} \tag{3.66}$$

Multiplying all terms in (3.66) by $c_{p,d}\rho_a$, multiplying all terms in the continuity equation for air from (3.20) by $c_{p,d}\theta_v$, adding the two equations, and compressing gives

$$\frac{\partial(c_{p,d}\rho_a\theta_v)}{\partial t} + \nabla\cdot(\mathbf{v}c_{p,d}\rho_a\theta_v) = \rho_a\frac{\theta_v}{T_v}\frac{\mathrm{d}Q}{\mathrm{d}t} \tag{3.67}$$

Substituting the **energy density** (J m$^{-3}$), defined as $E = c_{p,d}\rho_a\theta_v$, into (3.67) gives the **continuity equation for energy**,

$$\frac{\partial E}{\partial t} + \nabla\cdot(\mathbf{v}E) = \rho_a\frac{\theta_v}{T_v}\frac{\mathrm{d}Q}{\mathrm{d}t} \tag{3.68}$$

This equation is similar to the continuity equations for air or gas concentration. It states that the time rate of change of energy in a box equals the energy flux in minus the energy flux out plus (minus) external sources (sinks). Replacing $N$ with $E$ in Fig. 3.2 yields a diagram of energy fluxes into and out of a hypothetical grid cell.

In a model, eddy diffusion of energy must be taken into account. Quantities in (3.67) can be decomposed as $\mathbf{v} = \bar{\mathbf{v}} + \mathbf{v}'$, $\rho_a = \bar{\rho}_a + \rho'_a$, and $\theta_v = \bar{\theta}_v + \theta'_v$. Since density perturbations are small ($\rho'_a \ll \bar{\rho}_a$), the density is decomposed as $\rho_a \approx \bar{\rho}_a$. Substituting velocity, density, and potential virtual temperature decompositions into (3.67) and taking the time- and grid-volume average of the result yields

$$c_{p,d}\left\{\overline{\frac{\partial[\bar{\rho}_a(\bar{\theta}_v + \theta'_v)]}{\partial t}}\right\} + c_{p,d}\overline{\nabla \cdot [\bar{\rho}_a(\bar{\mathbf{v}}\bar{\theta}_v + \bar{\mathbf{v}}\theta'_v + \mathbf{v}'\bar{\theta}_v + \mathbf{v}'\theta'_v)]} = \bar{\rho}_a\overline{\frac{\theta_v}{T_v}\frac{dQ}{dt}} \tag{3.69}$$

Eliminating zero-value time and spatial derivatives and unnecessary overbars results in

$$\frac{\partial(\bar{\rho}_a\bar{\theta}_v)}{\partial t} + \nabla \cdot (\bar{\rho}_a\bar{\mathbf{v}}\bar{\theta}_v) + \overline{\nabla \cdot (\bar{\rho}_a\mathbf{v}'\theta'_v)} = \frac{\bar{\rho}_a}{c_{p,d}}\overline{\frac{\theta_v}{T_v}\frac{dQ}{dt}} \tag{3.70}$$

which expands to

$$\bar{\rho}_a\left[\frac{\partial\bar{\theta}_v}{\partial t} + (\bar{\mathbf{v}} \cdot \nabla)\bar{\theta}_v\right] + \bar{\theta}_v\left[\frac{\partial\bar{\rho}_a}{\partial t} + \nabla \cdot (\bar{\mathbf{v}}\bar{\rho}_a)\right] + \nabla \cdot (\bar{\rho}_a\overline{\mathbf{v}'\theta'_v}) = \frac{\bar{\rho}_a}{c_{p,d}}\overline{\frac{\theta_v}{T_v}\frac{dQ}{dt}} \tag{3.71}$$

Substituting the continuity equation for air from (3.48) into (3.71) and dividing by $\bar{\rho}_a$ gives the thermodynamic energy equation as

$$\frac{\partial\bar{\theta}_v}{\partial t} + (\bar{\mathbf{v}} \cdot \nabla)\bar{\theta}_v + \frac{1}{\bar{\rho}_a}\nabla \cdot (\bar{\rho}_a\overline{\mathbf{v}'\theta'_v}) = \overline{\frac{\theta_v}{c_{p,d}T_v}\frac{dQ}{dt}} \tag{3.72}$$

The kinematic turbulent sensible-heat fluxes ($\overline{\mathbf{v}'\theta'_v}$) can be parameterized with

$$\overline{u'\theta'_v} = -K_{h,xx}\frac{\partial\bar{\theta}_v}{\partial x} \qquad \overline{v'\theta'_v} = -K_{h,yy}\frac{\partial\bar{\theta}_v}{\partial y} \qquad \overline{w'\theta'_v} = -K_{h,zz}\frac{\partial\bar{\theta}_v}{\partial z} \tag{3.73}$$

where the eddy diffusion coefficients for energy are the same as those used in the continuity equation for a trace species. Substituting $\overline{\mathbf{v}'\theta'_v} = -\mathbf{K}_h\nabla\bar{\theta}_v$ into (3.72) and eliminating overbars for simplicity gives

$$\frac{\partial\theta_v}{\partial t} + (\mathbf{v} \cdot \nabla)\theta_v = \frac{1}{\rho_a}(\nabla \cdot \rho_a\mathbf{K}_h\nabla)\theta_v + \frac{\theta_v}{c_{p,d}T_v}\frac{dQ}{dt} \tag{3.74}$$

The diabatic heating rate consists of the terms

$$\frac{dQ}{dt} = \sum_{n=1}^{N_{e,h}}\frac{dQ_n}{dt} = \frac{dQ_{c/e}}{dt} + \frac{dQ_{f/m}}{dt} + \frac{dQ_{dp/s}}{dt} + \frac{dQ_{solar}}{dt} + \frac{dQ_{ir}}{dt} \tag{3.75}$$

where $N_{e,h}$ is the number of diabatic energy sources and sinks. All Q's are in joules per kilogram. $dQ_{c/e}/dt$ is the rate of energy release (absorption) due to condensation (evaporation), $dQ_{f/m}/dt$ is the rate of energy release (absorption) due to freezing (melting), $dQ_{dp/s}/dt$ is the rate of energy release (absorption) due to deposition (sublimation), $dQ_{solar}/dt$ is the rate of solar heating, and $dQ_{ir}/dt$ is the rate of net infrared heating (cooling). Substituting (3.75) into (3.74) gives the **thermodynamic energy equation** as

$$\frac{\partial \theta_v}{\partial t} + (\mathbf{v} \cdot \nabla)\theta_v = \frac{1}{\rho_a}(\nabla \cdot \rho_a \mathbf{K}_h \nabla)\theta_v + \frac{\theta_v}{c_{p,d}T}\sum_{n=1}^{N_{e,h}} \frac{dQ_n}{dt} \tag{3.76}$$

## 3.5. SUMMARY

In this chapter, local and total derivatives were defined, and the continuity and thermodynamic energy equations were derived. Continuity equations include those for air, trace gases, aerosol number concentration, and aerosol volume concentration. These equations include kinematic turbulent flux terms, which are generally parameterized. A basic parameterization is $K$-theory. Equations described in this chapter are necessary for simulating the transport and transformations of total air, gases, aerosols, and energy. An equation used for predicting wind speed and direction, the momentum equation, is discussed next.

## 3.6. PROBLEMS

**3.1.** Expand the total derivative of the $u$-scalar velocity [e.g., substitute $u$ for $N$ in (3.5)] when the air flow is (a) steady, (b) unsteady.

**3.2.** Explain why (3.17) differs from (3.26).

**3.3.** What is the purpose of Reynolds averaging?

**3.4.** If $u = -5$ m s$^{-1}$ and $v = +5$ m s$^{-1}$, write out the horizontal velocity vector, determine the horizontal wind speed, and name the wind.

**3.5.** Assume that a grid cell has dimension $\Delta x = 5$ km, $\Delta y = 4$ km, and $\Delta z = 0.1$ km and that the west, east, south, north, and lower scalar velocities are $u_1 = +3$, $u_2 = +4$, $v_3 = -3$, $v_4 = +2$, and $w_5 = +0.2$ m s$^{-1}$. If the atmosphere is incompressible, what is $w$ at the top of the cell?

**3.6.** **(a)** A grid cell has dimensions $\Delta x = 5$ km, $\Delta y = 4$ km, and $\Delta z = 0.1$ km. Assume the gas concentration and scalar velocity at the west boundary of the cell are $N_1 = 1 \times 10^{11}$ molecules cm$^{-3}$ and $u_1 = +7$ m s$^{-1}$, respectively, and those at the east boundary of the cell are $N_2 = 5 \times 10^{11}$ molecules cm$^{-3}$ and $u_2 = +8$ m s$^{-1}$, respectively. (i) Assuming external sources and sinks and eddy diffusion are absent, estimate $N$ at the cell center after 60 s if the initial $N$ is an average of the two boundary $N$-values and boundary parameters remain constant. (ii) Calculate the time after the start at which $N$ at the cell center becomes zero.

**(b)** Assume that the gas concentration and scalar velocity at the south boundary of the grid cell in part (a) are $N_3 = 1 \times 10^{11}$ molecules cm$^{-3}$ and $v_3 = -2$ m s$^{-1}$, respectively, and those at the north boundary are $N_4 = 7 \times 10^{11}$ molecules cm$^{-3}$ and $v_4 = +1$ m s$^{-1}$, respectively. Calculate (i) $N$ at the cell center after 60 s and (ii) the time after the start at which $N$ at the center becomes zero. Assume fluxes operate in four directions, and the initial $N$ at the center of the cell is the average of all four grid-boundary $N$-values.

**(c)** Convert the gas number concentrations from part (a) ($N_1$ and $N_2$) to moist-air mass mixing ratios, assuming that the gas is ozone, $T_v = 298$ K, and $p_a = 1013$ mb.

**(d)** Re-solve parts (a)(i) and (a)(ii) with (3.26) using moist-air mass mixing ratios instead of number concentration units. Assume that the west–east velocity for this question is an average of the grid-cell boundary velocities. Convert the mass mixing ratio from the 60-s case back to number concentration units. How does the result differ from that found in part (a)(i)? If it differs, why does it differ?

**3.7.** A grid cell has dimensions $\Delta x = 5$ km, $\Delta y = 4$ km, and $\Delta z = 0.1$ km. Assume that the potential virtual temperature, pressure, and scalar velocity at the west boundary of the grid cell are $\theta_{v,1} = 302$ K, $p_{a,1} = 1004$ mb, and $u_1 = +7$ m s$^{-1}$, respectively, and those at the east boundary of the grid cell are $\theta_{v,2} = 299$ K, $p_{a,2} = 1008$ mb, and $u_2 = +8$ m s$^{-1}$, respectively.

**(a)** Calculate the virtual temperature and air density at the west and east boundaries of the grid cell.

**(b)** Calculate the energy density $E$ at each boundary.

**(c)** Assuming diabatic energy sources and sinks and eddy diffusion are absent, calculate the potential virtual temperature at the center of the grid cell after 10 s.

## 3.7. COMPUTER PROGRAMMING PRACTICE

**3.8.** Assume that the grid-cell size, boundary conditions, and initial $N$ are initially the same as in Problem 3.6 (a). Write a computer script to calculate the final $N$ at the grid cell center after a time step $h$. After each time step, set the east-boundary gas concentration ($N_2$) equal to the final concentration at the center of the cell. Set $h = 3$ s, and run the program for a simulation period of one hour. Plot the grid center concentration versus time.

**3.9.** Assume that grid-cell size, boundary conditions, and initial $\theta_v$ are the same as in Problem 3.7. Write a computer script to calculate the final $\theta_v$ at the cell center after a time step $h$. After each step, set $\theta_{v,2}$ equal to $\theta_v$ at the cell center. Set $h = 3$ s, and run the program for six hours. Plot $\theta_v$ versus time at the cell center.

# 4

# The Momentum Equation in Cartesian and Spherical Coordinates

THE **momentum equation** (equation of motion) describes atmospheric motion. This equation is used to predict wind velocity (speed and direction). The momentum equation and each of its terms are derived in this chapter. These terms include local acceleration, the earth's centripetal acceleration (apparent centrifugal force), the Coriolis acceleration (apparent Coriolis force), the gravitational force, the pressure-gradient force, the viscous force, and the turbulent-flux divergence. The magnitudes of the terms in the equation are compared for different regions of the atmosphere. The equation is derived for Cartesian and spherical horizontal coordinates. Cartesian coordinates are often used when the earth's curvature is effectively small, and spherical coordinates are used when curvature cannot be neglected. The relationships for the geostrophic wind, gradient wind, and surface wind can be derived from the momentum equation. Atmospheric waves are also discussed. Important waves include acoustic, Lamb, gravity, inertia Lamb, inertia gravity, and Rossby waves.

## 4.1. CONVERSION FROM CARTESIAN TO SPHERICAL COORDINATES

Atmospheric modeling equations can be derived for one of several horizontal coordinate systems. Horizontal **Cartesian** (rectangular) coordinates are often used in urban-, plume-, and cloud-scale models. Over short distances, the earth's curvature is not significant, and its surface can be divided into rectangles. Over long distances, curvature prevents the accurate division of the earth's surface into rectangles. In such cases, regions of the earth can still be divided into groups of rectangles, the edges of which overlap other groups of rectangles. This is the idea behind the **universal transverse Mercator** (UTM) coordinate system. In the UTM system, separate rectangular meshes are superimposed over the globe, one piece at a time. UTM coordinate locations are mapped back to spherical coordinate locations with UTM-to-spherical conversion equations (U.S. Department of the Army 1958). Because spherical coordinates are natural to use on a spherical globe, regardless of the scale, the UTM system is not necessary for most modeling applications.

For model simulations on all scales, use of **spherical coordinates** is more natural than use of Cartesian coordinates, since spherical coordinates take into account the earth's curvature. Other horizontal map projections are Mercator, stereographic, and Lambert conformal. A **Mercator projection** is one in which each rhumb line

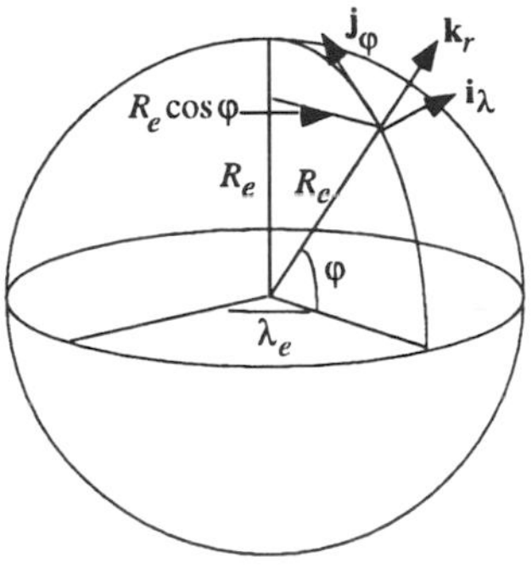


**Figure 4.1.** Spherical coordinate symbols. $R_e$ is the earth's radius, $\varphi$ is latitude, $\lambda_e$ is longitude, and $\mathbf{i}_\lambda$, $\mathbf{j}_\varphi$, and $\mathbf{k}_r$ are west–east, south–north, and vertical unit vectors, respectively.

on a sphere is represented as a straight line. A **rhumb line** is a curve on the surface of a sphere that cuts all **meridians** (south–north lines) at the same angle. Since the angle can be any angle, a rhumb line can spiral to the poles. A **stereographic projection** is one in which points on a sphere correspond exactly to points on an extended plane and in which the North Pole on the sphere corresponds to infinity on the plane. A **Lambert conformal projection** is one in which meridians are represented as straight lines converging towards the nearer pole and parallels (west–east lines) are represented as arc segments of concentric circles. Though these three projections are used in cartography, they are used less frequently in atmospheric modeling.

Figure 4.1 shows the earth as a sphere and the primary components of the spherical coordinate system. Although the earth is an oblate spheroid, slightly bulging at the equator, the difference between the equatorial and polar radii is small enough (21 km) that the earth can be considered a sphere for most applications. The spherical-coordinate unit vectors for the earth are, $\mathbf{i}_\lambda$, $\mathbf{j}_\varphi$, and $\mathbf{k}_r$, which are the west–east, south–north, and vertical unit vectors, respectively. Because the earth's surface is curved, spherical-coordinate unit vectors have a different orientation at each horizontal location. In Cartesian coordinates, **i**, **j**, and **k** were oriented in the same direction everywhere.

In spherical coordinates, west–east and south–north distances are measured in terms of **longitude** ($\lambda_e$) and **latitude** ($\varphi$), respectively. The vertical coordinate for now is the altitude ($z$) coordinate. It will be changed later to the pressure ($p$), sigma-pressure ($\sigma$-$p$), and sigma-altitude ($s$-$z$) coordinates. Conversions between increments of distance in Cartesian coordinates and increments of arc length in spherical coordinates, along the surface of earth, are obtained from the equation for arc length around a circle. In the west–east and south–north directions, these conversions are

$$dx = (R_e \cos\varphi)\, d\lambda_e \qquad dy = R_e\, d\varphi \tag{4.1}$$

respectively, where $R_e \approx 6371$ km is the radius of the earth, $d\lambda_e$ is a west–east arc-length increment (rad), $d\varphi$ is a south–north arc-length increment (rad), and $R_e \cos\varphi$ is the distance from the earth's axis of rotation to the surface of the earth at latitude $\varphi$, as shown in Fig. 4.1. Since $\cos\varphi$ is maximum at the equator (where $\varphi = 0$) and minimum at the poles, $dx$ decreases from equator to pole when $d\lambda_e$ is constant.

**Example 4.1**

If a grid cell has dimensions $d\lambda_e = 5°$ and $d\varphi = 5°$, centered at $\varphi = 30°$N latitude, find $dx$ and $dy$ at the grid cell latitudinal center.

SOLUTION

First, $d\lambda_e = d\varphi = 5° \times \pi/180° = 0.0873$ rad. Substituting these values into (4.1) gives $dx = (6371)(0.866)(0.0873) = 482$ km and $dy = (6371)(0.0873) = 556$ km.

The **local velocity vector** and **local horizontal velocity vector** in spherical coordinates are

$$\mathbf{v} = \mathbf{i}_\lambda u + \mathbf{j}_\varphi v + \mathbf{k}_r w \qquad \mathbf{v}_h = \mathbf{i}_\lambda u + \mathbf{j}_\varphi v \tag{4.2}$$

respectively. When spherical coordinates are used, horizontal scalar velocities can be redefined by substituting (4.1) into (3.2). In sum,

$$u = \frac{dx}{dt} = R_e \cos\varphi \frac{d\lambda_e}{dt} \qquad v = \frac{dy}{dt} = R_e \frac{d\varphi}{dt} \qquad w = \frac{dz}{dt} \tag{4.3}$$

The altitude-coordinate vertical scalar velocity is the same on a spherical grid as on a Cartesian grid. The **gradient operator in spherical-altitude coordinates** is found by substituting (4.1) into (3.6) and replacing Cartesian- with spherical-coordinate unit vectors. The result is

$$\nabla = \mathbf{i}_\lambda \frac{1}{R_e \cos\varphi} \frac{\partial}{\partial \lambda_e} + \mathbf{j}_\varphi \frac{1}{R_e} \frac{\partial}{\partial \varphi} + \mathbf{k}_r \frac{\partial}{\partial z} \tag{4.4}$$

In spherical coordinates, the dot product of the gradient operator with a scalar can be written from (4.4) without modification, but the dot product of the gradient operator with a vector must be rederived because unit vectors change orientation at different locations on a sphere. Expanding $\nabla \cdot \mathbf{v}$ in spherical coordinates gives

$$\nabla \cdot v = \left( \mathbf{i}_\lambda \frac{1}{R_e \cos\varphi} \frac{\partial}{\partial \lambda_e} + \mathbf{j}_\varphi \frac{1}{R_e} \frac{\partial}{\partial \varphi} + \mathbf{k}_r \frac{\partial}{\partial z} \right) \cdot (\mathbf{i}_\lambda u + \mathbf{j}_\varphi v + \mathbf{k}_r w) \tag{4.5}$$

$$= \left( \frac{1}{R_e \cos\varphi} \frac{\partial u}{\partial \lambda_e} + \mathbf{i}_\lambda u \frac{1}{R_e \cos\varphi} \frac{\partial \mathbf{i}_\lambda}{\partial \lambda_e} + \mathbf{i}_\lambda v \frac{1}{R_e \cos\varphi} \frac{\partial \mathbf{j}_\varphi}{\partial \lambda_e} \right.$$

$$\left. + \mathbf{i}_\lambda w \frac{1}{R_e \cos\varphi} \frac{\partial \mathbf{k}_r}{\partial \lambda_e} \right) + \left( \frac{1}{R_e} \frac{\partial v}{\partial \varphi} + \mathbf{j}_\varphi u \frac{1}{R_e} \frac{\partial \mathbf{i}_\lambda}{\partial \varphi} + \mathbf{j}_\varphi v \frac{1}{R_e} \frac{\partial \mathbf{j}_\varphi}{\partial \varphi} \right.$$

$$\left. + \mathbf{j}_\varphi w \frac{1}{R_e} \frac{\partial \mathbf{k}_r}{\partial \varphi} \right) + \left( \frac{\partial w}{\partial z} + \mathbf{k}_r u \frac{\partial \mathbf{i}_\lambda}{\partial z} + \mathbf{k}_r v \frac{\partial \mathbf{j}_\varphi}{\partial z} + \mathbf{k}_r w \frac{\partial \mathbf{k}_r}{\partial z} \right)$$

where some terms were eliminated because $\mathbf{i}_\lambda \cdot \mathbf{j}_\varphi = 0$, $\mathbf{i}_\lambda \cdot \mathbf{k}_r = 0$, and $\mathbf{j}_\varphi \cdot \mathbf{k}_r = 0$. Partial derivatives of the unit vectors in (4.5) can be derived graphically. From Fig. 4.2(a) and the equation for arc length around a circle, we have

$$|\Delta \mathbf{i}_\lambda| = |\mathbf{i}_\lambda| \, \Delta\lambda_e = \Delta\lambda_e \tag{4.6}$$

where $|\Delta \mathbf{i}_\lambda|$ is the magnitude of the change in the west–east unit vector per unit change in longitude, $\Delta\lambda_e$. If $\Delta\lambda_e$ is small, Fig. 4.2(b) indicates that

$$\Delta \mathbf{i}_\lambda = \mathbf{j}_\varphi |\Delta \mathbf{i}_\lambda| \sin\varphi - \mathbf{k}_r |\Delta \mathbf{i}_\lambda| \cos\varphi \tag{4.7}$$

Substituting (4.6) into (4.7), dividing by $\Delta\lambda_e$, and letting $\Delta \mathbf{i}_\lambda \to 0$ and $\Delta\lambda_e \to 0$ gives

$$\frac{\partial \mathbf{i}_\lambda}{\partial \lambda_e} \approx \frac{\mathbf{j}_\varphi \Delta\lambda_e \sin\varphi - \mathbf{k}_r \Delta\lambda_e \cos\varphi}{\Delta\lambda_e} \approx \mathbf{j}_\varphi \sin\varphi - \mathbf{k}_r \cos\varphi \tag{4.8}$$

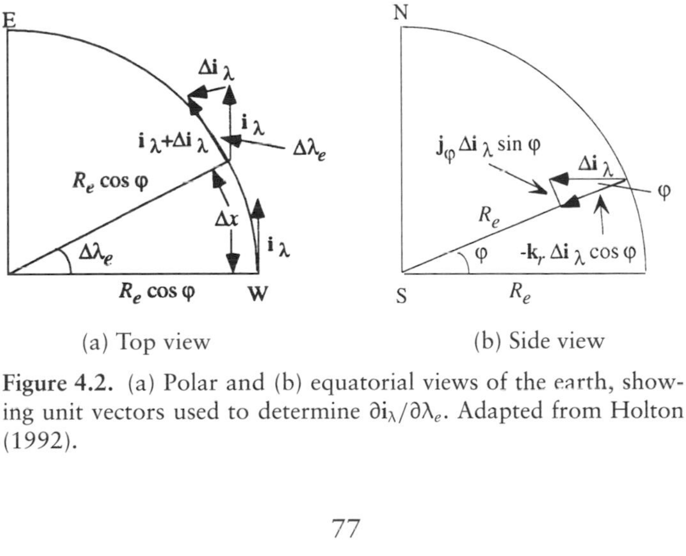


**Figure 4.2.** (a) Polar and (b) equatorial views of the earth, showing unit vectors used to determine $\partial \mathbf{i}_\lambda / \partial \lambda_e$. Adapted from Holton (1992).

Similar derivations for other derivatives yield

$$\frac{\partial \mathbf{i}_\lambda}{\partial \lambda_e} = \mathbf{j}_\varphi \sin\varphi - \mathbf{k}_r \cos\varphi \qquad \frac{\partial \mathbf{i}_\lambda}{\partial \varphi} = 0 \qquad \frac{\partial \mathbf{i}_\lambda}{\partial z} = 0$$
$$\frac{\partial \mathbf{j}_\varphi}{\partial \lambda} = -\mathbf{i}_\lambda \sin\varphi \qquad \frac{\partial \mathbf{j}_\varphi}{\partial \varphi} = -\mathbf{k}_r \qquad \frac{\partial \mathbf{j}_\varphi}{\partial z} = 0$$
$$\frac{\partial \mathbf{k}_r}{\partial \lambda} = \mathbf{i}_\lambda \cos\varphi \qquad \frac{\partial \mathbf{k}_r}{\partial \varphi} = \mathbf{j}_\varphi \qquad \frac{\partial \mathbf{k}_r}{\partial z} = 0 \tag{4.9}$$

Substituting these expressions into (4.5) gives

$$\nabla \cdot \mathbf{v} = \frac{1}{R_e \cos\varphi}\frac{\partial u}{\partial \lambda_e} + \frac{1}{R_e \cos\varphi}\frac{\partial}{\partial \varphi}(v\cos\varphi) + \frac{1}{R_e^2}\frac{\partial}{\partial z}\left(w R_e^2\right) \tag{4.10}$$

which simplifies to

$$\boxed{\nabla \cdot \mathbf{v} = \frac{1}{R_e \cos\varphi}\frac{\partial u}{\partial \lambda_e} + \frac{1}{R_e \cos\varphi}\frac{\partial}{\partial \varphi}(v\cos\varphi) + \frac{\partial w}{\partial z}} \tag{4.11}$$

when $R_e$ is assumed constant. Since the incremental distance $z$ above the earth's surface is much smaller for most modeling applications (<60 km) than the radius of earth (6371 km), the assumption of a constant $R_e$ for use in (4.11) does not induce significant error.

## 4.2. NEWTON'S SECOND LAW OF MOTION

The momentum equation is derived from **Newton's second law of motion,** $F = Ma$, where $F$ is force (N), $M$ is mass (kg), and $a$ is acceleration (m s$^{-2}$). Newton's second law states that the acceleration of a body due to a force is proportional to the force, inversely proportional to the mass of the body, and in the direction of the force. When applied to the atmosphere, the second law can be written in vector form as

$$\mathbf{a}_i = \frac{1}{M_a}\sum \mathbf{F} \tag{4.12}$$

where $\mathbf{a}_i$ is the **total** or **inertial acceleration,** which is the rate of change of velocity of a parcel of air in motion relative to a coordinate system fixed in space (outside the earth–atmosphere system), $M_a$ is the mass of the air parcel, and $\sum \mathbf{F}$ is the sum of the force vectors acting on the parcel. A reference frame at rest, such as one fixed in space, is an **inertial reference frame.** A reference frame that moves, such as one fixed on the rotating earth or on an object in motion relative to the earth, is a **noninertial reference frame.** Passengers traveling in a moving car are in a noninertial frame because the car and the earth are in motion relative to a fixed point in space.

An observer at a fixed point in space is in an inertial reference frame with respect to any body on a rotating earth, even if the body is at rest on the surface of earth.

Inertial acceleration is derived by considering that, to an observer in space, the **absolute velocity** (m $s^{-1}$) of a body in motion near the surface of the earth is

$$\mathbf{v}_A = \mathbf{v} + \boldsymbol{\Omega} \times \mathbf{R}_e \tag{4.13}$$

where $\mathbf{v}$ is the local velocity, defined in (4.2), of the body relative to the earth's surface, $\boldsymbol{\Omega}$ is the angular velocity vector for earth, $\mathbf{R}_e$ is the radius vector for the earth, and $\boldsymbol{\Omega} \times \mathbf{R}_e$ is the rate of change in position of the body due to the earth's rotation. The earth's **angular velocity vector** (rad $s^{-1}$) and **radius vector** (m) are defined as

$$\boldsymbol{\Omega} = \mathbf{j}_\varphi \Omega \cos\varphi + \mathbf{k}_r \Omega \sin\varphi \qquad \mathbf{R}_e = \mathbf{k}_r R_e \tag{4.14}$$

where $\Omega = 2\pi$ rad/86,164 s $= 7.292 \times 10^{-5}$ rad $s^{-1}$ is the magnitude of the angular velocity, and 86,164 is the actual number of seconds that the earth takes to make one revolution around its axis (23 h 56 m 4 s). The angular velocity vector acts perpendicular to the equatorial plane of the earth, as shown in Fig. 4.3. It does not have a west–east component.

Inertial acceleration is defined as

$$\mathbf{a}_i = \frac{d\mathbf{v}_A}{dt} + \boldsymbol{\Omega} \times \mathbf{v}_A \tag{4.15}$$

Substituting (4.13) into (4.15) and noting that $\boldsymbol{\Omega}$ is independent of time gives

$$\mathbf{a}_i = \frac{d\mathbf{v}}{dt} + \boldsymbol{\Omega} \times \frac{d\mathbf{R}_e}{dt} + \boldsymbol{\Omega} \times \mathbf{v} + \boldsymbol{\Omega} \times (\boldsymbol{\Omega} \times \mathbf{R}_e) \tag{4.16}$$

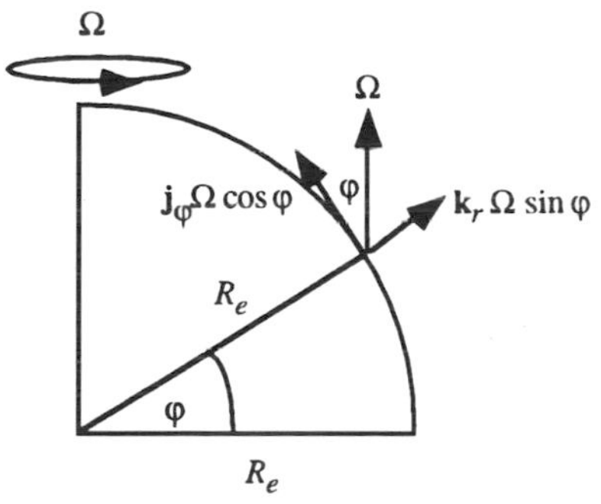


**Figure 4.3.** Components of the earth's angular velocity vector.

The total derivative of $\mathbf{R}_e$ is

$$\frac{\mathrm{d}\mathbf{R}_e}{dt} = R_e \frac{\mathrm{d}\mathbf{k}_r}{dt} = \mathbf{i}_\lambda u + \mathbf{j}_\varphi v \approx \mathbf{v} \tag{4.17}$$

where $\mathrm{d}\mathbf{k}_r/dt$ is derived shortly in (4.28). Substituting (4.17) into (4.16) yields

$$\mathbf{a}_i = \frac{\mathrm{d}\mathbf{v}}{dt} + 2\boldsymbol{\Omega} \times \mathbf{v} + \boldsymbol{\Omega} \times (\boldsymbol{\Omega} \times \mathbf{R}_e) = \mathbf{a}_l + \mathbf{a}_c + \mathbf{a}_r \tag{4.18}$$

where

$$\mathbf{a}_l = \frac{\mathrm{d}\mathbf{v}}{dt} \qquad \mathbf{a}_c = 2\boldsymbol{\Omega} \times \mathbf{v} \qquad \mathbf{a}_r = \boldsymbol{\Omega} \times (\boldsymbol{\Omega} \times \mathbf{R}_e) \tag{4.19}$$

are the local, Coriolis, and earth's centripetal accelerations, respectively. **Local acceleration** is the rate of change of velocity of a parcel of air in motion relative to a coordinate system fixed on earth, **Coriolis acceleration** is the rate of change of velocity of a parcel due to the rotation of a spherical earth underneath the parcel, and the **earth's centripetal acceleration** is the inward-directed rate of change of velocity of a parcel due to its motion around the earth's axis.

While the centripetal and Coriolis effects are treated as accelerations when viewed from an inertial frame of reference, they are treated as apparent forces when viewed from a noninertial frame of reference. An **apparent** (or **inertial**) **force** is a fictitious force that appears to exist when an observation is made in a noninertial frame of reference. When a passenger in a car rounds a curve, the passenger, who is in a noninertial frame of reference, appears to be pulled outward by a local **apparent centrifugal force**, which is equal and opposite to local centripetal acceleration multiplied by mass. In an inertial frame of reference, an outside observer sees the passenger and car exhibiting a constant inward acceleration as the car rounds a curve.

Analogously, in a noninertial frame of reference, such as on earth's surface, an apparent centrifugal force appears to push the earth and atmosphere away from the earth's axis of rotation as the earth rotates. In an inertial frame of reference, such as in space, the earth appears to exhibit a constant inward acceleration during rotation and does not appear to be acted upon by a centrifugal force.

The Coriolis effect can also be viewed from an inertial or a noninertial frame of reference. From a noninertial frame of reference, moving bodies appear to feel the **Coriolis force** pushing them to the right in the Northern Hemisphere and to the left in the Southern Hemisphere. From an inertial frame of reference, such as from space, rotation of the earth underneath a moving body makes the body appear to accelerate towards the right in the Northern Hemisphere or left in the Southern Hemisphere. In sum, the centripetal and Coriolis effects can be treated as either accelerations or apparent forces, depending on the frame of reference considered.

When Reynolds decomposition is applied to the precise local acceleration term in (4.19), the term becomes $\mathbf{a}_l = \bar{\mathbf{a}}_l + \mathbf{a}'_l$, where $\bar{\mathbf{a}}_l$ is a mean local acceleration and $\mathbf{a}'_l$ is a perturbation component, called a **turbulent-flux divergence of momentum.** This term accounts for perturbations to the mean flow of wind, such as those caused by **mechanical shear** (mechanical turbulence), **thermal buoyancy** (thermal turbulence), and **atmospheric waves.** For now, the precise local acceleration term is retained. It will be decomposed later in this section.

The terms on the right side of (4.12) are real forces. Real forces that affect local acceleration of a parcel of air include the force of gravity (**true gravitational force**), the force caused by spatial pressure gradients (**pressure-gradient force**), and the force caused by molecular viscosity of air (**viscous force**). Substituting inertial acceleration terms from (4.18) into (4.12) and expanding the right side gives

$$\mathbf{a}_l + \mathbf{a}_c + \mathbf{a}_r = \frac{1}{M_a}(\mathbf{F}_g^* + \mathbf{F}_p + \mathbf{F}_v) \tag{4.20}$$

where $\mathbf{F}_g^*$ represents true gravitational force, $\mathbf{F}_p$ represents the pressure gradient force, and $\mathbf{F}_v$ represents the force due to air viscosity. Atmospheric models usually require expressions for local acceleration; thus, the momentum equation is written most conveniently for a reference frame fixed on the surface of the earth rather than fixed outside the earth–atmosphere system. In such a case, only local acceleration is treated as an acceleration. The Coriolis acceleration is treated as a Coriolis force per unit mass ($\mathbf{a}_c = \mathbf{F}_c/M_a$), and the earth's centripetal acceleration is treated as an apparent centrifugal (negative centripetal) force per unit mass ($\mathbf{a}_r = -\mathbf{F}_r/M_a$). Combining these terms with (4.20) gives the **momentum equation from a reference frame fixed on earth's surface** as

$$\mathbf{a}_l = \frac{1}{M_a}(\mathbf{F}_r - \mathbf{F}_c + \mathbf{F}_g^* + \mathbf{F}_p + \mathbf{F}_v) \tag{4.21}$$

In the following subsections, terms in (4.21) are discussed.

### 4.2.1. Local Acceleration

The **local acceleration**, or the total derivative of velocity, expands to

$$\mathbf{a}_l = \frac{d\mathbf{v}}{dt} = \frac{\partial \mathbf{v}}{\partial t} + (\mathbf{v} \cdot \nabla)\mathbf{v} \tag{4.22}$$

This equation states that the local acceleration along the motion of a parcel equals the local acceleration at a fixed point plus changes in local acceleration due to gradient fluxes of velocity.

In Cartesian–altitude coordinates, the left side of (4.22) expands to

$$\frac{d\mathbf{v}}{dt} = \frac{d(\mathbf{i}u + \mathbf{j}v + \mathbf{k}w)}{dt} = \mathbf{i}\frac{du}{dt} + \mathbf{j}\frac{dv}{dt} + \mathbf{k}\frac{dw}{dt} \tag{4.23}$$

and the right side expands to

$$\begin{aligned}\frac{\partial \mathbf{v}}{\partial t} + (\mathbf{v} \cdot \nabla)\mathbf{v} &= \left(\frac{\partial}{\partial t} + u\frac{\partial}{\partial x} + v\frac{\partial}{\partial y} + w\frac{\partial}{\partial z}\right)(\mathbf{i}u + \mathbf{j}v + \mathbf{k}w) \\ &= \mathbf{i}\left(\frac{\partial u}{\partial t} + u\frac{\partial u}{\partial x} + v\frac{\partial u}{\partial y} + w\frac{\partial u}{\partial z}\right) + \mathbf{j}\left(\frac{\partial v}{\partial t} + u\frac{\partial v}{\partial x}\right. \\ &\quad \left. + v\frac{\partial v}{\partial y} + w\frac{\partial v}{\partial z}\right) + \mathbf{k}\left(\frac{\partial w}{\partial t} + u\frac{\partial w}{\partial x} + v\frac{\partial w}{\partial y} + w\frac{\partial w}{\partial z}\right)\end{aligned} \tag{4.24}$$

In spherical–altitude coordinates, the left side of (4.22) expands with the chain rule to

$$\begin{aligned}\frac{d\mathbf{v}}{dt} = \frac{d(\mathbf{i}_\lambda u + \mathbf{j}_\varphi v + \mathbf{k}_r w)}{dt} &= \left(\mathbf{i}_\lambda\frac{du}{dt} + u\frac{d\mathbf{i}_\lambda}{dt}\right) \\ &+ \left(\mathbf{j}_\varphi\frac{dv}{dt} + v\frac{d\mathbf{j}_\varphi}{dt}\right)\left(\mathbf{k}_r\frac{dw}{dt} + w\frac{d\mathbf{k}_r}{dt}\right)\end{aligned} \tag{4.25}$$

Time derivatives of the unit vectors are needed to complete this equation. Substituting (4.1) into the total derivative in Cartesian–altitude coordinates from (3.13) gives the **total derivative in spherical–altitude coordinates** as

$$\frac{d}{dt} = \frac{\partial}{\partial t} + u\frac{1}{R_e \cos\varphi}\frac{\partial}{\partial \lambda_e} + v\frac{1}{R_e}\frac{\partial}{\partial \varphi} + w\frac{\partial}{\partial z} \tag{4.26}$$

Applying (4.26) to $\mathbf{i}_\lambda$ yields

$$\frac{d\mathbf{i}_\lambda}{dt} = \frac{\partial \mathbf{i}_\lambda}{\partial t} + u\frac{1}{R_e \cos\varphi}\frac{\partial \mathbf{i}_\lambda}{\partial \lambda_e} + v\frac{1}{R_e}\frac{\partial \mathbf{i}_\lambda}{\partial \varphi} + w\frac{\partial \mathbf{i}_\lambda}{\partial z} \tag{4.27}$$

Since $\mathbf{i}_\lambda$ does not change in time at a given location, $\partial \mathbf{i}_\lambda/\partial t = 0$. Substituting $\partial \mathbf{i}_\lambda/\partial t = 0$ and terms from (4.9) into (4.27) and into like expressions for $d\mathbf{j}_\varphi/dt$ and $d\mathbf{k}_r/dt$ gives

$$\frac{d\mathbf{i}_\lambda}{dt} = \mathbf{j}_\varphi\frac{u\tan\varphi}{R_e} - \mathbf{k}_r\frac{u}{R_e} \qquad \frac{d\mathbf{j}_\varphi}{dt} = -\mathbf{i}_\lambda\frac{u\tan\varphi}{R_e} - \mathbf{k}_r\frac{v}{R_e} \qquad \frac{d\mathbf{k}_r}{dt} = \mathbf{i}_\lambda\frac{u}{R_e} + \mathbf{j}_\varphi\frac{v}{R_e} \tag{4.28}$$

Finally substituting (4.28) into (4.25) results in

$$\begin{aligned}\frac{d\mathbf{v}}{dt} &= \mathbf{i}_\lambda\left(\frac{du}{dt} - \frac{uv\tan\varphi}{R_e} + \frac{uw}{R_e}\right) + \mathbf{j}_\varphi\left(\frac{dv}{dt} + \frac{u^2\tan\varphi}{R_e} + \frac{vw}{R_e}\right) \\ &+ \mathbf{k}_r\left(\frac{dw}{dt} - \frac{u^2}{R_e} - \frac{v^2}{R_e}\right)\end{aligned} \tag{4.29}$$

**Example 4.2**

If $u = 20\ \mathrm{m\ s^{-1}}$, $v = 10\ \mathrm{m\ s^{-1}}$, and $w = 0.01\ \mathrm{m\ s^{-1}}$, and if $du/dt$ scales as $u/(\Delta x/u)$, estimate the value of each term on the right side of (4.29) at $\varphi = 45°$N latitude assuming $\Delta x = 500$ km, $\Delta y = 500$ km, and $\Delta z = 10$ km for large-scale motions.

SOLUTION

From the values given,

$$\frac{du}{dt} \approx 8 \times 10^{-4}\ \mathrm{m\ s^{-2}} \qquad \frac{uv\tan\varphi}{R_e} \approx 3.1 \times 10^{-5}\ \mathrm{m\ s^{-2}}$$

$$\frac{uw}{R_e} \approx 3.1 \times 10^{-8}\ \mathrm{m\ s^{-2}}$$

$$\frac{dv}{dt} \approx 2 \times 10^{-4}\ \mathrm{m\ s^{-2}} \qquad \frac{u^2\tan\varphi}{R_e} \approx 6.3 \times 10^{-5}\ \mathrm{m\ s^{-2}}$$

$$\frac{vw}{R_e} \approx 1.6 \times 10^{-8}\ \mathrm{m\ s^{-2}}$$

$$\frac{dw}{dt} \approx 1 \times 10^{-8}\ \mathrm{m\ s^{-2}} \qquad \frac{u^2}{R_e} \approx 6.3 \times 10^{-5}\ \mathrm{m\ s^{-2}}$$

$$\frac{v^2}{R_e} \approx 1.6 \times 10^{-5}\ \mathrm{m\ s^{-2}}$$

$uw/R_e$ and $vw/R_e$ are small for large and small scale motions. $dw/dt$ is also small for large-scale motions.

Example 4.2 shows that $uw/R_e$ and $vw/R_e$ are small for large-scale motions and can be removed from (4.29). If these terms are removed, $u^2/R_e$ and $v^2/R_e$ must also be removed from the vertical term to avoid a false addition of energy to the system. Fortunately, these latter terms are small in comparison with the gravitational and pressure-gradient forces per unit mass. Implementing the above changes gives the **local acceleration in spherical–altitude coordinates** as

$$\frac{d\mathbf{v}}{dt} = \mathbf{i}_\lambda\left(\frac{du}{dt} - \frac{uv\tan\varphi}{R_e}\right) + \mathbf{j}_\varphi\left(\frac{dv}{dt} + \frac{u^2\tan\varphi}{R_e}\right) + \mathbf{k}_r\frac{dw}{dt} \tag{4.30}$$

Expanding the total derivatives of $\mathbf{v}$, $u$, $v$, and $w$ in (4.30) gives the right side of (4.22) in spherical–altitude coordinates as

$$\begin{aligned}\frac{\partial \mathbf{v}}{\partial t} + (\mathbf{v}\cdot\nabla)\mathbf{v} = {} & \mathbf{i}_\lambda\left(\frac{\partial u}{\partial t} + \frac{u}{R_e\cos\varphi}\frac{\partial u}{\partial \lambda_e} + \frac{v}{R_e}\frac{\partial u}{\partial \varphi} + w\frac{\partial u}{\partial z} - \frac{uv\tan\varphi}{R_e}\right) \\ & + \mathbf{j}_\varphi\left(\frac{\partial v}{\partial t} + \frac{u}{R_e\cos\varphi}\frac{\partial v}{\partial \lambda_e} + \frac{v}{R_e}\frac{\partial v}{\partial \varphi} + w\frac{\partial v}{\partial z} + \frac{u^2\tan\varphi}{R_e}\right) \\ & + \mathbf{k}_r\left(\frac{\partial w}{\partial t} + \frac{u}{R_e\cos\varphi}\frac{\partial w}{\partial \lambda_e} + \frac{v}{R_e}\frac{\partial w}{\partial \varphi} + w\frac{\partial w}{\partial z}\right)\end{aligned} \tag{4.31}$$

In the **horizontal**, local accelerations have magnitude on the order of $10^{-4}$ m s$^{-2}$. These accelerations are less important than Coriolis accelerations or the pressure-gradient force per unit mass, but greater than accelerations due to the viscous force, except adjacent to the ground. In the **vertical**, local accelerations over a large horizontal region have values on the order of $10^{-7}$ m s$^{-2}$ and can be neglected, since gravity and pressure-gradient accelerations are a factor of $10^8$ larger. Over small horizontal distances (<3 km), local accelerations in the vertical are important and cannot be ignored.

### 4.2.2. Coriolis Force

The second term in the momentum equation is the Coriolis force. In a noninertial frame of reference, the Coriolis force appears to push moving bodies to the right in the Northern Hemisphere and to the left in the Southern Hemisphere. In the Northern Hemisphere, it acts 90° to the right of the direction of motion, and in the Southern Hemisphere, it acts 90° to the left of the direction of motion. The Coriolis force is only apparent: no force really acts. Instead, the rotation of a spherical earth below a moving body makes the body accelerate to the right in the Northern Hemisphere or left in the Southern Hemisphere when viewed from an inertial frame of reference, such as from space. The acceleration is zero at the equator, maximum near the poles, and zero for bodies at rest. Moving bodies include winds, ocean currents, airplanes, and baseballs. Figure 4.4 gives an example of Coriolis deflections.

In terms of an apparent force per unit mass, the Coriolis term in (4.19) expands in spherical–altitude coordinates to

$$\frac{\mathbf{F}_c}{M_a} = 2\boldsymbol{\Omega} \times \mathbf{v} = 2\Omega \begin{vmatrix} \mathbf{i}_\lambda & \mathbf{j}_\varphi & \mathbf{k}_r \\ 0 & \cos\varphi & \sin\varphi \\ u & v & w \end{vmatrix}$$
$$= \mathbf{i}_\lambda 2\Omega(w\cos\varphi - v\sin\varphi) + \mathbf{j}_\varphi 2\Omega u \sin\varphi - \mathbf{k}_r 2\Omega u\cos\varphi \quad (4.32)$$

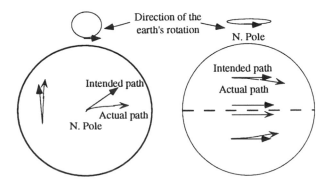

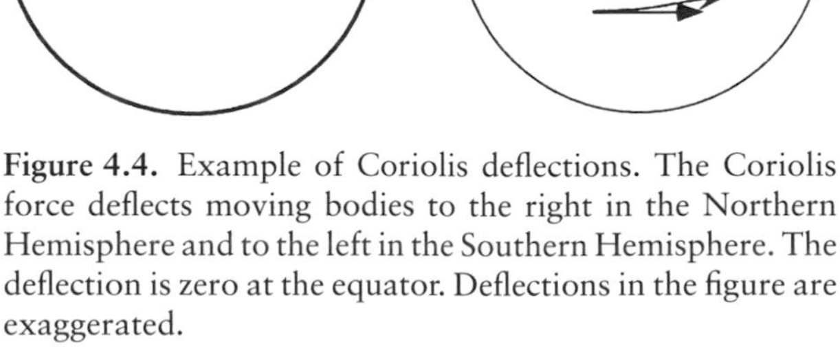


**Figure 4.4.** Example of Coriolis deflections. The Coriolis force deflects moving bodies to the right in the Northern Hemisphere and to the left in the Southern Hemisphere. The deflection is zero at the equator. Deflections in the figure are exaggerated.

If only a **zonal** (west–east) **wind** is considered, (4.32) simplifies to

$$\frac{\mathbf{F}_c}{M_a} = 2\boldsymbol{\Omega} \times \mathbf{v} = \mathbf{j}_\varphi\, 2\Omega u \sin\varphi - \mathbf{k}_r 2\Omega u \cos\varphi \tag{4.33}$$

Example 4.3 uses (4.33) to show that moving bodies are deflected to the right in the Northern Hemisphere.

**Example 4.3**

The fact that the Coriolis effect appears to deflect moving bodies to the right in the Northern Hemisphere can be demonstrated graphically. Consider only local acceleration and the Coriolis force per unit mass in (4.21). In such a case, $\mathbf{a}_l = -\mathbf{F}_c/M_a$. Subsituting (4.33) and $\mathbf{a}_l = \mathrm{d}\mathbf{v}/\mathrm{d}t$ from (4.19) into this expression when a west–east wind is present gives

$$\frac{\mathrm{d}\mathbf{v}}{\mathrm{d}t} = -\mathbf{j}_\varphi\, 2\Omega u \sin\varphi + \mathbf{k}_r\, 2\Omega u \cos\varphi$$

Figure 4.5 shows the directional components on the right side of this equation and the magnitude and direction of the resulting acceleration. The figure shows that the Coriolis effect acts perpendicular to a wind blowing from the west ($+u$), forcing the wind towards the south and vertically. At $\varphi = 0°$N, only the vertical component of the Coriolis effect remains, and the wind is not turned horizontally. At $\varphi = 90°$N, only the horizontal component remains, and the wind is not turned vertically.

Because vertical velocities are much smaller than horizontal velocities, $\mathbf{i}_\lambda 2\Omega w \cos\varphi$ can be removed from (4.32). Because the vertical component of the Coriolis force is smaller than other vertical components in the momentum equation (e.g.,

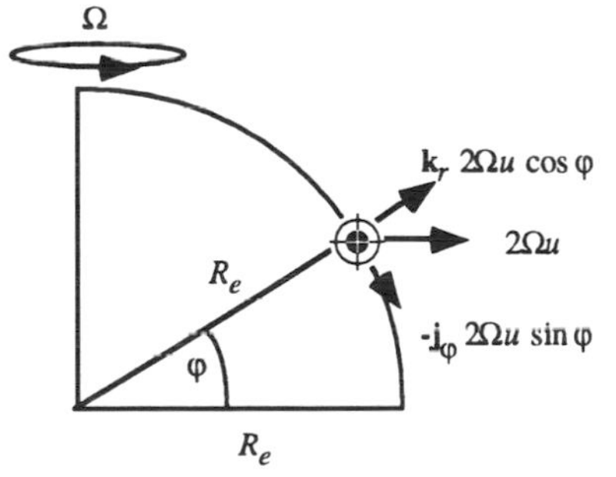


**Figure 4.5.** Coriolis acceleration components that result when the Coriolis force acts on a west-to-east wind traveling around the earth, parallel to the equator, and into the page (denoted by arrow tail). See Example 4.3 for a discussion.

gravity and pressure-gradient terms), $\mathbf{k}_r\, 2\Omega u \cos\varphi$ can also be removed. Making these changes gives the **Coriolis force vector** per unit mass in spherical–altitude coordinate as

$$\frac{\mathbf{F}_c}{M_a} = 2\mathbf{\Omega} \times \mathbf{v} \approx -\mathbf{i}_\lambda\, 2\Omega v \sin\varphi + \mathbf{j}_\varphi\, 2\Omega u \sin\varphi \tag{4.34}$$

Defining the **Coriolis parameter** as

$$f = 2\Omega \sin\varphi \tag{4.35}$$

gives another form of the Coriolis term as

$$\frac{\mathbf{F}_c}{M_a} \approx -\mathbf{i}_\lambda f v + \mathbf{j}_\varphi f u = \begin{vmatrix} \mathbf{i}_\lambda & \mathbf{j}_\varphi & \mathbf{k}_r \\ 0 & 0 & 1 \\ u & v & 0 \end{vmatrix} = f\mathbf{k}_r \times \mathbf{v}_h \tag{4.36}$$

(4.36) can be approximated in Cartesian coordinates by substituting **i**, **j**, and **k** for $\mathbf{i}_\lambda$, $\mathbf{j}_\varphi$, and $\mathbf{k}_r$. The magnitude of (4.36) is $|\mathbf{F}_c|/M_a = f|\mathbf{v}_h| = f\sqrt{u^2 + v^2}$, where $|\mathbf{v}_h|$ is the horizontal wind speed (m s$^{-1}$).

**Example 4.4**

A mean wind speed of $|\mathbf{v}_h| = 10$ m s$^{-1}$ at the North Pole results in a Coriolis acceleration magnitude of about $|\mathbf{F}_c|/M_a = 0.001454$ m s$^{-2}$.

### 4.2.3. Gravitational Force

Gravity is a real force that acts on a parcel of air. **True gravitational force** acts towards the center of the earth. The earth's **apparent centrifugal force**, which acts away from the axis of rotation of the earth, slightly displaces the direction and magnitude of the true gravitational force. The sum of true gravitational and apparent centrifugal force vectors gives an **effective gravitational force vector**, which acts normal to the surface of the earth but not towards its center.

The Earth's apparent centrifugal force and centripetal acceleration arise because the earth rotates. To an observer fixed in space, objects moving with the surface of a rotating earth exhibit an inward centripetal acceleration. The object, itself, feels as if it is being pushed outward during rotation by an apparent centrifugal force. The force is greatest at the equator, where the component of the earth's angular velocity normal to the earth's axis of rotation is greatest, and zero at the poles, where the normal component of the earth's angular velocity is zero. Over time, apparent centrifugal force has caused the earth to bulge at the equator and compress at the poles. The equatorial radius of earth is now about 21 km longer than the polar radius, making the earth an **oblate spheroid**.

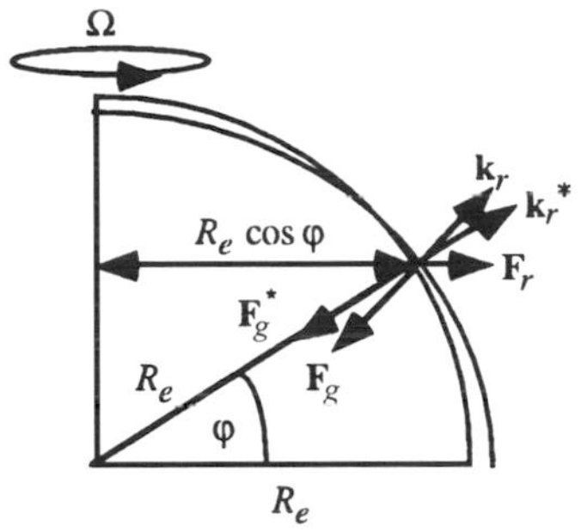


**Figure 4.6.** Gravitational force components for the earth. True gravitational force acts towards the center of the earth, and apparent centrifugal force acts away from its axis of rotation. Effective gravitational force, which is the sum of true gravitational and apparent centrifugal force, acts normal to the true surface but not towards the center of the earth. Apparent centrifugal force (negative centripetal acceleration) has caused the earth to bulge at the equator, as shown in the diagram, making the earth an oblate spheroid. Vectors corresponding to a true sphere are marked with asterisks to distinguish them from those corresponding to the oblate spheroid.

The **true gravitational force vector** per unit mass, which acts towards the center of earth, is

$$\frac{\mathbf{F}_g^*}{M_a} = -\mathbf{k}_r^* g^* \tag{4.37}$$

where $g^*$ is the **true gravitational acceleration.** The vectors $\mathbf{i}_\lambda^*$, $\mathbf{j}_\varphi^*$, and $\mathbf{k}_r^*$ are unit vectors on a true sphere. The vectors $\mathbf{i}_\lambda$, $\mathbf{j}_\varphi$, and $\mathbf{k}_r$ are unit vectors on the earth, which is an oblate spheroid. Figure 4.6 shows the orientation of true gravitational force and vertical unit vectors for earth and a perfect sphere.

True gravitational acceleration is derived from **Newton's law of universal gravitation.** This law gives the gravitational force vector between two bodies as

$$\mathbf{F}_{12,g} = -\mathbf{r}_{21} \frac{G M_1 M_2}{r_{21}^3} \tag{4.38}$$

where $G$ is the **universal gravitational constant** ($6.6720 \times 10^{-11}$ m$^3$ kg$^{-1}$ s$^{-2}$), $M_1$

and $M_2$ are the masses of the two bodies, respectively, $\mathbf{F}_{12,g}$ is the vector force exerted on $M_2$ by $M_1$, $\mathbf{r}_{21}$ is the distance vector pointing from body 2 to body 1, $r_{21}$ is the distance between the centers of the two bodies, and the negative sign indicates that the force acts in a direction opposite to that of the vector $\mathbf{r}_{21}$. The magnitude of the gravitational force vector is $F_g = GM_1M_2/r_{21}^2$.

In the case of the earth, $\mathbf{F}_{12,g} = \mathbf{F}_g^*$, $\mathbf{r}_{21} = \mathbf{R}_e = \mathbf{k}_r^* R_e$, $M_1 = M_a$, $M_2 = M_e$, and $r_{21} = R_e$, where $M_a$ is the mass of a parcel of air, and $M_e$ is the mass of the earth. Substituting these values into (4.38) gives the **true gravitational force vector per unit mass** as

$$\frac{\mathbf{F}_g^*}{M_a} = -\mathbf{k}_r^* \frac{GM_e}{R_e^2} \tag{4.39}$$

Equating (4.39) with (4.37) and taking the magnitude gives

$$\frac{|\mathbf{F}_g^*|}{M_a} = g^* = \frac{GM_e}{R_e^2} \tag{4.40}$$

**Example 4.5**

The mass of the earth is about $M_e = 5.98 \times 10^{24}$ kg, and the mean radius is $R_e = 6.37 \times 10^6$ m. Thus, from (4.40), the true gravitational acceleration has a magnitude of about $g^* = 9.833$ m s$^{-2}$.

For a true sphere, the apparent centrifugal force per unit mass expands to

$$\frac{\mathbf{F}_r}{M_a} = -\mathbf{a}_r = -\boldsymbol{\Omega} \times (\boldsymbol{\Omega} \times \mathbf{R}_e) = -\Omega \begin{vmatrix} \mathbf{i}_\lambda^* & \mathbf{j}_\varphi^* & k_r^* \\ 0 & \cos\varphi & \sin\varphi \\ R_e\cos\varphi & 0 & 0 \end{vmatrix}$$
$$= -\mathbf{j}_\varphi^* R_e\Omega^2 \cos\varphi\sin\varphi + k_r^* R_e\Omega^2\cos^2\varphi \tag{4.41}$$

where

$$\boldsymbol{\Omega} = \mathbf{j}_\varphi^*\Omega\cos\varphi + k_r^*\Omega\sin\varphi \qquad \mathbf{R}_e = k_r^* R_e \tag{4.42}$$

are the angular velocity vector and radius vector, respectively, of the earth as if it were a true sphere, $R_e\cos\varphi$ is the perpendicular distance between the axis of rotation of earth and the surface of earth at latitude $\varphi$, as shown in Fig. 4.6, and

$$\boldsymbol{\Omega} \times \mathbf{R}_e = \Omega \begin{vmatrix} \mathbf{i}^* & \mathbf{j}^* & \mathbf{k}^* \\ 0 & \cos\varphi & \sin\varphi \\ 0 & 0 & R_e \end{vmatrix} = \mathbf{i}^* R_e\Omega\cos\varphi \tag{4.43}$$

The magnitude of (4.41) is $|\mathbf{F}_r|/M_a = R_e\Omega^2\cos\varphi$. Adding (4.41) to (4.37) gives the effective gravitational force vector per unit mass on the earth as

$$\frac{\mathbf{F}_g}{M_a} = \frac{\mathbf{F}_g^*}{M_a} + \frac{\mathbf{F}_r}{M_a} = -\mathbf{j}_\varphi^* R_e\Omega^2\cos\varphi\sin\varphi + \mathbf{k}_r^*(R_e\Omega^2\cos^2\varphi - g^*) = -\mathbf{k}_r g \tag{4.44}$$

where $\mathbf{k}_r$ is the unit vector normal to the oblate spheroid surface of the earth, and

$$g = [(R_e\Omega^2\cos\varphi\sin\varphi)^2 + (g^* - R_e\Omega^2\cos^2\varphi)^2]^{1/2} \tag{4.45}$$

is the magnitude of the gravitational force per unit mass, or **effective gravitational acceleration** (effective gravity). The effective gravity at the surface varies from $g = 9.799$ m s$^{-2}$ at the equator to $g = 9.833$ m s$^{-2}$ at the poles. These values are much larger than accelerations due to the Coriolis effect. The effective gravity at the poles equals the true gravitational acceleration there, since the apparent centrifugal acceleration does not act at the poles. Centripetal acceleration affects true gravitational acceleration by about 0.34% at the equator. The difference between the equatorial and polar radii of earth is about 21 km, or 0.33% of an average earth's radius. Thus, apparent centrifugal force appears to account for the bulging of the earth at its equator.

The **globally averaged effective gravity** at the earth's surface, symbolized by $g_0$, is approximately 9.81 m s$^{-2}$. Since the variation of the effective gravity $g$ with latitude is small, $g$ is usually approximated with $g_0$ in models of the earth's lower atmosphere.

**Example 4.6**

Both $g^*$ and $g$ vary with altitude in the earth's atmosphere. (4.40) predicts that, 100 km above the equator, $g^* \approx 9.531$ m s$^{-2}$, or 3.1 percent lower than its surface value. (4.45) predicts that, at 100 km, $g \approx 9.497$ m s$^{-2}$, also 3.1 percent lower than its surface value. Thus, the variation of gravity with altitude is more significant than the variation of gravity due to centripetal acceleration.

Effective gravity is used in the equation for the geopotential. The **geopotential** is the work done against gravity to raise a unit mass of air from sea level to a given altitude. It is a measure of the gravitational potential energy of air per unit mass. The **geopotential vector** is defined as $\mathbf{\Phi}(z) = \mathbf{k}_r\Phi(z)$, where

$$\Phi(z) = \int_0^z g(z)\,dz \tag{4.46}$$

is the magnitude of the geopotential (m$^2$ s$^{-2}$). In the integral, $z = 0$ corresponds to sea level. The **geopotential height** is defined as

$$Z = \frac{\Phi(z)}{g_0} \tag{4.47}$$

In the lower atmosphere, $g(z) \approx g_0$, and geopotential height approximately equals altitude ($Z \approx z$). Table B.1 of Appendix B shows geopotential heights for different altitudes in the bottom 100 km of the earth's atmosphere. The table shows that geopotential height differs from actual altitude by only 1.55% at 100 km. At 25 km, the difference is 0.39%. The assumptions $Z \approx z$ and $g(z) \approx g_0 = g$ are often made in models of the lowest 100 km of the atmosphere. Under these assumptions, the magnitude and gradient of the geopotential are

$$\Phi(z) \approx gz \qquad \nabla\Phi = \mathbf{k}_r \frac{\partial \Phi}{\partial z} \approx \mathbf{k}_r g \tag{4.48}$$

respectively. Substituting (4.48) into (4.44) gives the **effective gravitational force per unit mass** in spherical coordinates as

$$\frac{\mathbf{F}_g}{M_a} = -\mathbf{k}_r g = -\nabla\Phi \tag{4.49}$$

This equation can be written in Cartesian–altitude coordinates by substituting $\mathbf{k}$ for $\mathbf{k}_r$.

### 4.2.4. Pressure-Gradient Force

The **pressure-gradient force** is a real force that causes air to move from regions of high pressure to regions of low pressure. The force results from pressure differences. Suppose a cubic parcel of air has volume $\Delta x \, \Delta y \, \Delta z$, as shown in Fig. 4.7. Suppose also that air pressures on the right and left sides of the parcel impart the forces

$$F_{p,r} = -\left(p_c + \frac{\partial p}{\partial x}\frac{\Delta x}{2}\right)\Delta y \, \Delta z \qquad F_{p,l} = \left(p_c - \frac{\partial p}{\partial x}\frac{\Delta x}{2}\right)\Delta y \, \Delta z \tag{4.50}$$

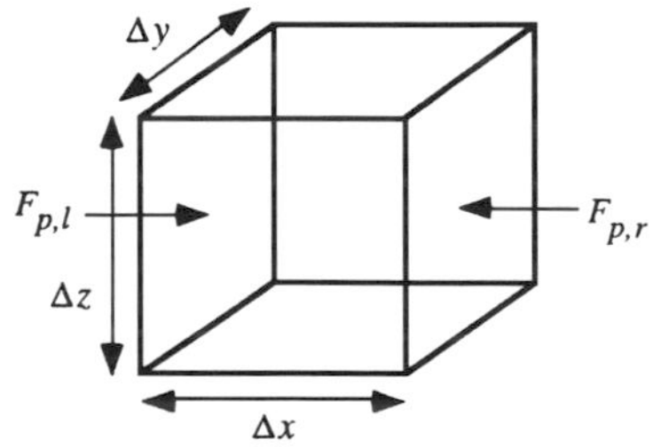


**Figure 4.7.** Example of pressure-gradient forces acting on both sides of a parcel of air. $F_{p,r}$ is the force acting on the right side, and $F_{p,l}$ is the force acting on the left side.

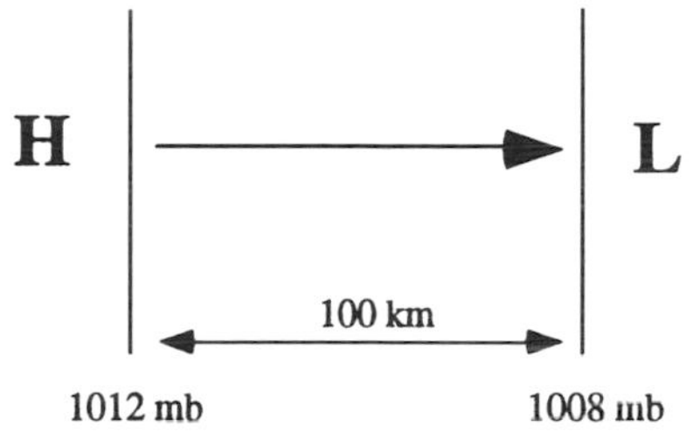


**Figure 4.8.** Example of a pressure gradient. The difference in pressure over a 100-km distance is 4 mb. The letters $H$ and $L$ indicate high and low pressure, respectively. The thick arrow indicates the direction of the pressure-gradient force.

respectively, where $p_c$ is the pressure at the center of the parcel. Dividing the sum of these forces by the mass of the parcel, $M_a = \rho_a \,\Delta x\, \Delta y\, \Delta z$, and allowing $\Delta x$, $\Delta y$, and $\Delta z$ to approach zero gives the pressure-gradient force per unit mass in the $x$-direction as

$$\frac{F_{p,x}}{M_a} = -\frac{1}{\rho_a}\frac{\partial p_a}{\partial x} \tag{4.51}$$

**Example 4.7**

Figure 4.8 shows two **isobars**, or lines of constant pressure, 100 km apart. The pressure difference between the isobars is 4 mb. Assuming $\rho_a = 1.2$ kg m$^{-3}$, the magnitude of the horizontal pressure-gradient force per unit mass is approximately

$$\begin{aligned}\frac{1}{\rho_a}\frac{\partial p_a}{\partial x} &\approx \frac{1}{1.2\,\text{kg m}^{-3}}\left(\frac{1012 - 1008\,\text{mb}}{10^5\,\text{m}}\right)\frac{100\,\text{kg m}^{-1}\text{s}^{-2}}{\text{mb}}\\ &= 0.0033\ \text{m s}^{-2}\end{aligned}$$

which is much smaller than the force per unit mass due to gravity, but on the same scale as the Coriolis force per unit mass.

The **pressure-gradient force per unit mass** can be generalized for three directions in Cartesian–altitude coordinates as

$$\frac{\mathbf{F}_p}{M_a} = -\frac{1}{\rho_a}\nabla p_a = -\frac{1}{\rho_a}\left(\mathbf{i}\frac{\partial p_a}{\partial x} + \mathbf{j}\frac{\partial p_a}{\partial y} + \mathbf{k}\frac{\partial p_a}{\partial z}\right) \tag{4.52}$$

and in spherical–altitude coordinates as

$$\frac{\mathbf{F}_p}{M_a} = -\frac{1}{\rho_a}\nabla p_a = -\frac{1}{\rho_a}\left(\mathbf{i}_\lambda \frac{1}{R_e \cos\varphi}\frac{\partial p_a}{\partial \lambda_e} + \mathbf{j}_\varphi \frac{1}{R_e}\frac{\partial p_a}{\partial \varphi} + \mathbf{k}_r \frac{\partial p_a}{\partial z}\right) \qquad (4.53)$$

**Example 4.8**

In the vertical, the pressure-gradient force per unit mass is much larger than that in the horizontal. Pressures at sea level and 100-m altitude are $p_a \approx 1013$ and 1000 mb, respectively. The average air density at 50 m is $\rho_a \approx 1.2$ kg m$^{-3}$. In this case, the pressure-gradient force per unit mass in the vertical is

$$\frac{1}{\rho_a}\frac{\partial p_a}{\partial z} \approx \frac{1}{1.2\,\text{kg m}^{-3}}\left(\frac{1013 - 1000\,\text{mb}}{100\text{ m}}\right)\frac{100\text{ kg m}^{-1}\text{s}^{-2}}{\text{mb}} = 10.80\text{ m s}^{-2}$$

which is over 3000 times greater than the horizontal pressure-gradient force per unit mass calculated in Example 4.7.

### 4.2.5. Viscous Force

Viscosity is a measure of a fluid's ability to transmit forces laterally to the mean flow. When two layers of air slide over one another at different velocities, they exert a **viscous stress**, or force per unit area, on each other. The force acts parallel to the direction of motion and over the area of the plane between the two layers. The net viscous force over the plane is proportional to the difference between the forces exerted by the two layers. The viscous force is an **internal force** in that it is caused by molecular interactions within a parcel.

The viscous force can be described in terms of wind shear and shearing stress. The change of velocity with height (e.g., $\partial u/\partial z$) is **wind shear.** Wind shear exerts a force per unit area, called **shearing stress,** parallel to the direction of motion and over a plane normal to the direction of shear. If wind shear in the $z$-direction exerts a force in the $x$-direction per unit area of the $x-y$ plane, the resulting shearing stress is

$$\tau_{zx} = \eta_a \frac{\partial u}{\partial z} \qquad (4.54)$$

(N m$^{-2}$ or kg m$^{-1}$ s$^{-2}$), where $\eta_a$ is the dynamic viscosity of air. The **dynamic viscosity of air** is the ratio of shearing stress to shear. An empirical expression for the dynamic viscosity (kg m$^{-1}$ s$^{-1}$) is Sutherland's equation (List 1984),

$$\eta_a \approx 1.8325 \times 10^{-5}\left(\frac{416.16}{T+120}\right)\left(\frac{T}{296.16}\right)^{1.5} \qquad (4.55)$$

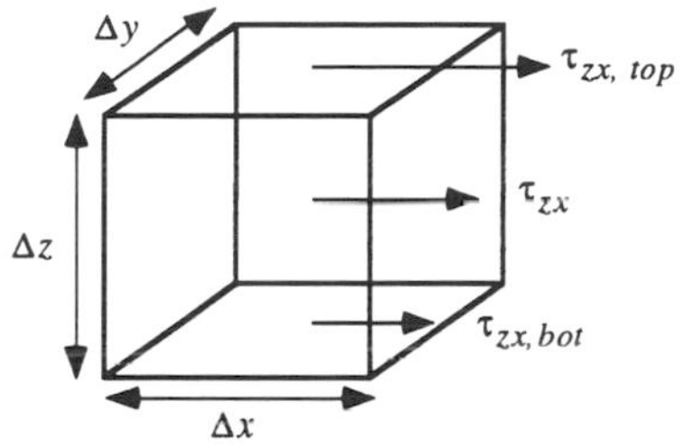


**Figure 4.9.** Example of shearing stress in the $x$-direction on a volume of air.

Shearing stress results when momentum is transported down a gradient of velocity, just as gas molecules are transported by molecular diffusion down a gradient of gas concentration. A cubic parcel of air experiences a shearing stress on its top and bottom, as shown in Fig. 4.9. The net viscous force on the parcel equals the shearing stress on the top minus the shearing stress on the bottom, multiplied by the area over which the stress acts. If $\tau_{zx}$ is the shearing stress in the middle of the parcel, and if $\partial\tau_{zx}/\partial z$ is the vertical gradient of shearing stress, the shearing stresses at the top and bottom, respectively, of the parcel are

$$\tau_{zx,top} = \tau_{zx} + \frac{\partial\tau_{zx}}{\partial z}\frac{\Delta z}{2} \qquad \tau_{zx,bot} = \tau_{zx} - \frac{\partial\tau_{zx}}{\partial z}\frac{\Delta z}{2} \tag{4.56}$$

Subtracting the shearing stress at the bottom from that at the top of the parcel, multiplying by the area, and dividing by the air-parcel mass $M_a = \rho_a \Delta x\,\Delta y\,\Delta z$ gives the net viscous force per unit mass as

$$\frac{F_{v,zx}}{M_a} = (\tau_{zx,top} - \tau_{zx,bot})\frac{\Delta x\,\Delta y}{\rho_a\,\Delta x\,\Delta y\,\Delta z} = \frac{1}{\rho_a}\frac{\partial\tau_{zx}}{\partial z} \tag{4.57}$$

Substituting shearing stress from (4.54) into (4.57), and assuming $\eta_a$ is invariant with altitude, we have

$$\frac{F_{v,zx}}{M_a} = \frac{1}{\rho_a}\frac{\partial}{\partial z}\left(\eta_a\frac{\partial u}{\partial z}\right) \approx \frac{\eta_a}{\rho_a}\frac{\partial^2 u}{\partial z^2} \tag{4.58}$$

This equation predicts that, if wind speed does not change with height or changes linearly with height and $\eta_a$ is constant, the viscous force per unit mass in the $x$-direction due to shear in the $z$-direction is zero. In such cases, shearing stress is constant with height, and no net viscous force occurs. When wind speed changes nonlinearly with height, the shearing stress at the top of a parcel differs from that at the bottom, and the net viscous force is nonzero. Figure 4.10 illustrates these two cases.

Expanding (4.58) gives the **viscous-force vector** per unit mass as

$$\frac{\mathbf{F}_v}{M_a} = \frac{\eta_a}{\rho_a}\nabla^2\mathbf{v} = \nu_a\nabla^2\mathbf{v} \tag{4.59}$$

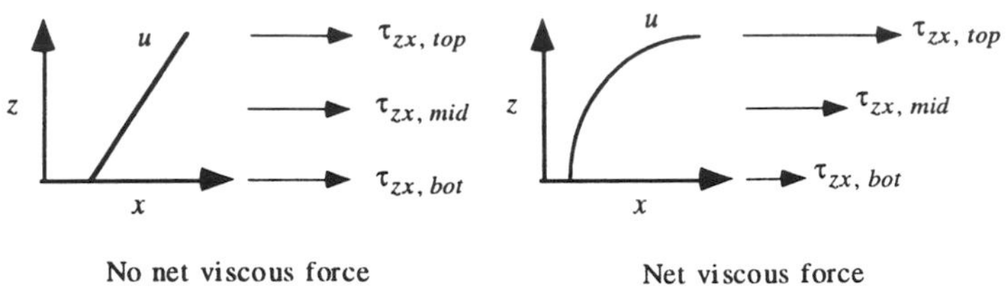


**Figure 4.10.** A linear vertical wind shear results in a constant shearing stress at all heights and no net viscous force. A nonlinear wind shear results in a change of shearing stress with height and a net viscous force.

where $\nu_a = \eta_a/\rho_a$ is the **kinematic viscosity of air** ($m^2\ s^{-1}$). The gradient term can be expanded in Cartesian–altitude coordinates as

$$\nabla^2 \mathbf{v} = (\nabla \cdot \nabla)\mathbf{v} = \mathbf{i}\left(\frac{\partial^2 u}{\partial x^2} + \frac{\partial^2 u}{\partial y^2} + \frac{\partial^2 u}{\partial z^2}\right) + \mathbf{j}\left(\frac{\partial^2 v}{\partial x^2} + \frac{\partial^2 v}{\partial y^2} + \frac{\partial^2 v}{\partial z^2}\right) + \mathbf{k}\left(\frac{\partial^2 w}{\partial x^2} + \frac{\partial^2 w}{\partial y^2} + \frac{\partial^2 w}{\partial z^2}\right) \tag{4.60}$$

Substituting (4.1) and spherical-coordinate unit vectors into (4.60) converts (4.60) to spherical–altitude coordinates. The kinematic viscosity of air is a molecular diffusion coefficient for air, analogous to the molecular diffusion coefficient for a trace gas.

Viscous forces in the atmosphere are small, except adjacent to the ground surface. Near the surface, relatively large, nonlinear velocity variations occur over small distances, increasing the magnitude of the viscous force.

**Example 4.9**

Viscous forces away from the ground surface are small. Suppose $u_1 = 10$ m s$^{-1}$ at altitude $z_1 = 1000$ m, $u_2 = 14$ m s$^{-1}$ at $z_2 = 1250$ m, and $u_3 = 20$ m s$^{-1}$ at $z_3 = 1500$ m. If the average temperature and air density at $z_2$ are $T = 280$ K and $\rho_a = 1.085$ kg m$^{-3}$, respectively, the dynamic viscosity of air is $\eta_a = 1.753 \times 10^{-5}$ kg m$^{-1}$ s$^{-1}$. The resulting viscous force per unit mass in the $x$-direction due to wind shear in the $z$-direction is approximately

$$\frac{F_{v,zx}}{M_a} \approx \frac{\eta_a}{\rho_a} \frac{1}{(z_3 - z_1)/2}\left(\frac{u_3 - u_2}{z_3 - z_2} - \frac{u_2 - u_1}{z_2 - z_1}\right) = 5.17 \times 10^{-10}\ \mathrm{m\ s^{-2}}$$

which is much smaller than the pressure-gradient force per unit mass from Example 4.7.

**Example 4.10**

Viscous forces near the surface are often significant. Suppose $u_1 = 0$ m s$^{-1}$ at altitude $z_1 = 0$ m, $u_2 = 0.4$ m s$^{-1}$ at $z_2 = 0.05$ m, and $u_3 = 1$ m s$^{-1}$ at $z_3 = 0.1$ m. If the average temperature and air density at $z_2$ are $T = 288$ K and $\rho_a = 1.225$ kg m$^{-3}$, the dynamic viscosity of air is $\eta_a = 1.792 \times 10^{-5}$ kg m$^{-1}$ s$^{-1}$. The resulting viscous force per unit mass in the $x$-direction due to wind shear in the $z$-direction is

$$\frac{F_{v,zx}}{M_a} \approx \frac{\eta_a}{\rho_a} \frac{1}{(z_3 - z_1)/2} \left( \frac{u_3 - u_2}{z_3 - z_2} - \frac{u_2 - u_1}{z_2 - z_1} \right) = 1.17 \times 10^{-3} \text{ m s}^{-2}$$

which is comparable in magnitude to the horizontal pressure gradient force per unit mass calculated in Example 4.7.

### 4.2.6. Turbulent-Flux Divergence

At the ground surface, wind speeds are zero. In the **surface layer**, which is a 50–300-m-thick region of the atmosphere adjacent to the surface, wind speeds increase logarithmically with height, creating wind shear. Wind shear produces a shearing stress that enhances motions on a molecular scale and creat eddies. Friction resulting from eddies is **eddy viscosity**, analogous to molecular viscosity that results from shearing stress. Eddies transfer momentum and energy vertically and horizontally. The greater the wind shear, the greater the transfer of momentum and energy. **Turbulence** consists of many eddies of different sizes acting together. Thus, wind shear creates turbulence.

Obstacles at the surface, such as rocks, trees, structures, and mountains, deflect air. The deflected air gives rise to wind shear and eddies downwind of or adjacent to the obstacles. Wind shear and turbulence resulting from obstacles are **mechanical shear** and **mechanical turbulence**, respectively. Eddies created downwind of obstacles are **turbulent wakes**. The sizes of turbulent-wake eddies vary from a few millimeters to the thickness of the boundary layer, which ranges from about 500 to 3000 m above the surface. In most cases, turbulence from surface drag and obstacles does not escape the boundary layer.

When surface air in an eddy rises, it slows the horizontal wind aloft as an obstacle would. When elevated air in an eddy circulates downward, it produces wind gusts near the surface. A **gust** is characterized by the rapid shift of wind speed and direction.

Whereas obstacles create mechanical turbulence, surface heating creates **thermal turbulence**, defined in Chapter 2. Mechanical turbulence is most important in the surface layer, and thermal turbulence is most important in the mixed layer. Thermal turbulence magnifies the effect of mechanical turbulence by enabling eddies to extend to greater heights, increasing their ability to exchange air between the surface and the mixed layer (Stull 1988).

In atmospheric models, the effects of mechanical shear and buoyancy on the acceleration of a parcel of air are modeled as subgrid-scale effects and simulated with turbulent-flux divergence terms for momentum. These terms are derived from precise local acceleration, the continuity equation for air, and Reynolds decomposition. Multiplying the precise acceleration from (4.22) by $\rho_a$, multiplying the continuity equation for air from (3.20) by $\mathbf{v}$, and adding the results yields

$$\rho_a \mathbf{a}_l = \rho_a \left[ \frac{\partial \mathbf{v}}{\partial t} + (\mathbf{v} \cdot \nabla)\mathbf{v} \right] + \mathbf{v} \left[ \frac{\partial \rho_a}{\partial t} + \nabla \cdot (\mathbf{v}\rho_a) \right] \tag{4.61}$$

The variables in this equation are decomposed as $\mathbf{v} = \bar{\mathbf{v}} + \mathbf{v}'$ and $\rho_a = \bar{\rho}_a + \rho'_a$. Because density perturbations are generally small ($\rho'_a \ll \bar{\rho}_a$), the density simplifies to $\rho_a \approx \bar{\rho}_a$. Substituting the decomposed values into (4.61) gives

$$\begin{aligned} \bar{\rho}_a \mathbf{a}_l = \bar{\rho}_a &\left\{ \frac{\partial(\bar{\mathbf{v}} + \mathbf{v}')}{\partial t} + [(\bar{\mathbf{v}} + \mathbf{v}') \cdot \nabla](\bar{\mathbf{v}} + \mathbf{v}') \right\} \\ &+ (\bar{\mathbf{v}} + \mathbf{v}') \left\{ \frac{\partial \bar{\rho}_a}{\partial t} + \nabla \cdot [(\bar{\mathbf{v}} + \mathbf{v}')\bar{\rho}_a] \right\} \end{aligned} \tag{4.62}$$

Taking the time and grid volume average of this equation, eliminating zero terms, and removing unnecessary overbars gives

$$\begin{aligned} \bar{\rho}_a \mathbf{a}_l = \bar{\rho}_a &\left[ \left( \frac{\partial \bar{\mathbf{v}}}{\partial t} + (\bar{\mathbf{v}} \cdot \nabla)\bar{\mathbf{v}} \right) \right] + \bar{\mathbf{v}} \left[ \left( \frac{\partial \bar{\rho}_a}{\partial t} + \nabla \cdot (\bar{\mathbf{v}}\, \bar{\rho}_a) \right) \right] \\ &+ \overline{\bar{\rho}_a (\mathbf{v}' \cdot \nabla)\mathbf{v}' + \mathbf{v}' \nabla \cdot (\mathbf{v}' \bar{\rho}_a)} \end{aligned} \tag{4.63}$$

Substituting the time- and grid-volume-averaged continuity equation for air from (3.48) into (4.63) and dividing through by $\bar{\rho}_a$ yeilds $\mathbf{a}_l = \bar{\mathbf{a}}_l + \mathbf{a}'_l$, where

$$\bar{\mathbf{a}}_l = \frac{\partial \bar{\mathbf{v}}}{\partial t} + (\bar{\mathbf{v}} \cdot \nabla)\bar{\mathbf{v}} \qquad \mathbf{a}'_l = \frac{\mathbf{F}_t}{M_a} = \frac{1}{\bar{\rho}_a} \overline{[\bar{\rho}_a (\mathbf{v}' \cdot \nabla)\mathbf{v}' + \mathbf{v}' \nabla \cdot (\mathbf{v}'\, \bar{\rho}_a)]} \tag{4.64}$$

The second term is treated as a force per unit mass in the momentum equation, where $\mathbf{F}_t$ is a turbulent-flux divergence vector multiplied by air mass. Since $\mathbf{F}_t$ originates on the left side of (4.12), it must be subtracted from the right side when treated as a force. $\mathbf{F}_t$ accounts for mechanical shear, buoyancy, and other eddy effects, such as waves. Expanding the **turbulent-flux divergence** term in (4.64) in Cartesian–altitude coordinates gives

$$\begin{aligned} \frac{\mathbf{F}_t}{M_a} = \mathbf{i} \frac{1}{\rho_a} &\left[ \frac{\partial(\rho_a \overline{u'u'})}{\partial x} + \frac{\partial(\rho_a \overline{v'u'})}{\partial y} + \frac{\partial(\rho_a \overline{w'u'})}{\partial z} \right] \\ + \mathbf{j} \frac{1}{\rho_a} &\left[ \frac{\partial(\rho_a \overline{u'v'})}{\partial x} + \frac{\partial(\rho_a \overline{v'v'})}{\partial y} + \frac{\partial(\rho_a \overline{w'v'})}{\partial z} \right] \\ + \mathbf{k} \frac{1}{\rho_a} &\left[ \frac{\partial(\rho_a \overline{u'w'})}{\partial x} + \frac{\partial(\rho_a \overline{v'w'})}{\partial y} + \frac{\partial(\rho_a \overline{w'w'})}{\partial z} \right] \end{aligned} \tag{4.65}$$

Averages, such as $\overline{u'u'}$ and $\overline{w'v'}$ (m [m s$^{-1}$] s$^{-1}$), are kinematic turbulent fluxes of momentum, since they have units of momentum flux (kg [m s$^{-1}$] m$^{-2}$ s$^{-1}$)

divided by air density (kg $m^{-3}$). Each such average must be parameterized. A parameterization introduced previously and discussed in Chapter 8 is ***K*-theory**. With $K$-theory, vertical kinematic turbulent fluxes of west–east and south–north momentum are approximated with

$$\overline{w'u'} = -K_{m,zx}\frac{\partial \bar{u}}{\partial z} \qquad \overline{w'v'} = -K_{m,zy}\frac{\partial \bar{v}}{\partial z} \tag{4.66}$$

respectively, where the $K_m$'s are **eddy diffusion coefficients for momentum** ($m^2$ $s^{-1}$ or $cm^2$ $s^{-1}$). In all, nine eddy diffusion coefficients for momentum are needed. Only three were required for energy. Since $\overline{u'v'} = \overline{v'u'}$, $\overline{u'w'} = \overline{w'u'}$, and $\overline{v'w'} = \overline{w'v'}$, the number of coefficients for momentum can be reduced to six. Substituting (4.66) and other like terms into (4.65) and dropping overbars for simplicity gives the turbulent-flux divergence as

$$\begin{aligned}\frac{\mathbf{F}_t}{M_a} = &-\mathbf{i}\frac{1}{\rho_a}\left[\frac{\partial}{\partial x}\left(\rho_a K_{m,xx}\frac{\partial u}{\partial x}\right) + \frac{\partial}{\partial y}\left(\rho_a K_{m,yx}\frac{\partial u}{\partial y}\right) + \frac{\partial}{\partial z}\left(\rho_a K_{m,zx}\frac{\partial u}{\partial z}\right)\right] \\ &-\mathbf{j}\frac{1}{\rho_a}\left[\frac{\partial}{\partial x}\left(\rho_a K_{m,xy}\frac{\partial v}{\partial x}\right) + \frac{\partial}{\partial y}\left(\rho_a K_{m,yy}\frac{\partial v}{\partial y}\right) + \frac{\partial}{\partial z}\left(\rho_a K_{m,zy}\frac{\partial v}{\partial z}\right)\right] \\ &-\mathbf{k}\frac{1}{\rho_a}\left[\frac{\partial}{\partial x}\left(\rho_a K_{m,xz}\frac{\partial w}{\partial x}\right) + \frac{\partial}{\partial y}\left(\rho_a K_{m,yz}\frac{\partial w}{\partial y}\right) + \frac{\partial}{\partial z}\left(\rho_a K_{m,zz}\frac{\partial w}{\partial z}\right)\right]\end{aligned} \tag{4.67}$$

(4.67) can be transformed to spherical–altitude coordinates by substituting (4.1) and spherical-coordinate unit vectors into it. Each term in (4.67) is an acceleration in one direction due to transport of momentum normal to that direction. The $zx$ term is an acceleration in the $x$-direction due to a gradient in wind shear and transport of momentum in the $z$-direction. Kinematic turbulent fluxes are analogous to shearing stresses in that both result when momentum is transported down a gradient of velocity. In the case of viscosity, momentum is transported by molecular diffusion. In the case of turbulence, momentum is transported by eddy diffusion. As with the viscous forces per unit mass in (4.59), the turbulent-flux divergence terms in (4.67) equal zero when the $\rho_a K_m$ terms are constant in space and the wind shear either is zero or changes linearly with distance. Usually, the $\rho_a K_m$ terms vary in space.

The eddy diffusion coefficients in (4.67) can be written in tensor form as

$$\mathbf{K}_m = \begin{bmatrix} K_{m,xx} & K_{m,yx} & K_{m,zx} \\ K_{m,xy} & K_{m,yy} & K_{m,zy} \\ K_{m,xz} & K_{m,yz} & K_{m,zz} \end{bmatrix} \tag{4.68}$$

where $K_{m,xy} = K_{m,yx}$, $K_{m,xz} = K_{m,zx}$, and $K_{m,yz} = K_{m,zy}$. In vector and tensor notation, (4.67) simplifies to

$$\frac{\mathbf{F}_t}{M_a} = -\frac{1}{\rho_a}(\nabla \cdot \rho_a \mathbf{K}_m \nabla)\mathbf{v} \tag{4.69}$$

**Example 4.11**

A typical value of $K_m$ in the vertical in the middle of the boundary layer is 50 $m^2$ $s^{-1}$. Assume $u_1 = 10$ m $s^{-1}$ at altitude $z_1 = 300$ m, $u_2 = 12$ m $s^{-1}$ at $z_2 = 350$ m, and $u_3 = 15$ m $s^{-1}$ at $z_3 = 400$ m. If the density and the eddy diffusion coefficient remain constant between 300 and 400 m, the west–east acceleration of wind due to transfer of westerly momentum from 400 to 300 m is approximately

$$\frac{\mathbf{F}_{t,zx}}{M_a} = \frac{1}{\rho_a}\frac{\partial}{\partial z}\left(\rho_a K_{m,zx}\frac{\partial u}{\partial z}\right)$$

$$\approx \frac{K_{m,zx}}{(z_3 - z_1)/2}\left(\frac{u_3 - u_2}{z_3 - z_2} - \frac{u_2 - u_1}{z_2 - z_1}\right) = 0.02\,\mathrm{m\,s^{-2}}$$

**Example 4.12**

In the horizontal, the eddy diffusion coefficient is about $K_m = 2.5 \times 10^3$ $m^2$ $s^{-1}$. Assume $u_1 = 10$ m $s^{-1}$ at location $y_1 = 0$ m, $u_2 = 9$ m $s^{-1}$ at $y_2 = 500$ m, and $u_3 = 7$ m $s^{-1}$ at $y_3 = 1000$ m. If the density and the eddy diffusion coefficient remain constant between $y_1$ and $y_3$, the deceleration of westerly wind in the south due to transfer of momentum to the north is approximately

$$\frac{\mathbf{F}_{t,yx}}{M_a} = \frac{1}{\rho_a}\frac{\partial}{\partial y}\left(\rho_a K_{m,yx}\frac{\partial u}{\partial y}\right)$$

$$\approx \frac{K_{m,yx}}{(y_3 - y_1)/2}\left(\frac{u_3 - u_2}{y_3 - y_2} - \frac{u_2 - u_1}{y_2 - z_1}\right) = -0.01\,\mathrm{m\,s^{-2}}$$

### 4.2.7. Complete Momentum Equation

Table 4.1 summarizes the accelerations and forces per unit mass derived above and gives approximate magnitudes of the terms. Noting that $\mathbf{a}_l = \bar{\mathbf{a}}_l + \mathbf{F}_t/M_a$, removing overbars for simplicity, and substituting terms from Table 4.1 into (4.21) gives the **vector form of the momentum equation** as

$$\frac{d\mathbf{v}}{dt} = -f\mathbf{k} \times \mathbf{v} - \nabla\Phi - \frac{1}{\rho_a}\nabla p_a + \frac{\eta_a}{\rho_a}\nabla^2\mathbf{v} + \frac{1}{\rho_a}(\nabla \cdot \rho_a \mathbf{K}_m \nabla)\mathbf{v} \tag{4.70}$$

Table 4.1 shows that some terms in these equations are unimportant, depending on the scale of motion. Three dimensionless parameters – the Ekman number, Rossby number, and Froude number – are used for scale analysis to estimate the

**Table 4.1.** Terms in the Momentum Equation and Their Horizontal and Vertical Magnitudes

| Term | Acceleration or Force/Mass Expression | Horizontal Accel. (m s$^{-2}$) | Vertical Accel. (m s$^{-2}$) |
|---|---|---|---|
| Local acceleration | $\bar{a}_l = \frac{d\bar{v}}{dt} = \frac{\partial \bar{v}}{\partial t} + (\bar{v} \cdot \nabla)\bar{v}$ | $10^{-4}$ | [a] $10^{-7}$–1 |
| Coriolis force per unit mass | $\frac{F_c}{M_a} = f\mathbf{k} \times \mathbf{v}$ | $10^{-3}$ | 0 |
| Effective gravitational force per unit mass | $\frac{F_g}{M_a} = \frac{F_g^*}{M_a} + \frac{F_r}{M_a} = -\nabla\Phi$ | 0 | 10 |
| Pressure-gradient force per unit mass | $\frac{F_p}{M_a} = -\frac{1}{\rho_a}\nabla p_a$ | $10^{-3}$ | 10 |
| Viscous force per unit mass | $\frac{F_v}{M_a} = \frac{\eta_a}{\rho_a}\nabla^2 \mathbf{v}$ | [b] $10^{-12}$–$10^{-3}$ | [b] $10^{-15}$–$10^{-5}$ |
| Turbulent flux divergence of momentum | $\frac{F_t}{M_a} = -\frac{1}{\rho_a}(\nabla \cdot \rho_a \mathbf{K}_m \nabla)\mathbf{v}$ | [c] 0–1 | [c] 0–1 |

[a] Low value for large-scale motions; high value for small-scale motions (<3 km).
[b] Low value for free atmosphere, high value for air adjacent to the surface.
[c] Low value for no wind shear, high value for large wind shear.

importance of different processes. These parameters are

$$\text{Ek} = \frac{\nu_a u/x^2}{uf} \qquad \text{Ro} = \frac{u^2/x}{uf} \qquad \text{Fr}^2 = \frac{w^2/z}{g} \tag{4.71}$$

respectively. The **Ekman number** gives the ratio of the viscous force to the Coriolis force. Above the ground surface, viscous terms are unimportant relative to Coriolis terms. Thus, the Ekman number is small, and the viscous term is usually removed from the momentum equation. The **Rossby number** gives the ratio of the local acceleration to the Coriolis force per unit mass. Local accelerations are more important than viscous accelerations, but less important than Coriolis accelerations. Thus, the Rossby number is much larger than the Ekman number, but usually less than unity. The **Froude number** gives the ratio of local acceleration to gravitational acceleration in the vertical. Over large horizontal distances (>3 km), vertical accelerations are small in comparison with gravitational accelerations. In such cases, the vertical acceleration term is often removed from the momentum equation, resulting in the hydrostatic assumption.

**Example 4.13**

Over large horizontal regions $\nu_a \approx 10^{-6}$ m$^2$ s$^{-1}$, $u \approx 10$ m s$^{-1}$, $z \approx 10^6$ m, $f \approx 10^{-4}$ s$^{-1}$, $w \approx 0.01$ m s$^{-1}$, and $z \approx 10^4$ m. The Ekman number under these conditions is Ek $= 10^{-14}$, indicating that viscous forces are small. The Rossby number has a value of about Ro $= 0.1$, indicating that local accelerations are an order of magnitude smaller than Coriolis accelerations. The Froude number is Fr $= 3 \times 10^{-5}$. Thus, local accelerations in the vertical are unimportant over large horizontal regions.

Since air viscosity is negligible for most atmospheric scales, it can be ignored in the momentum equation. Removing viscosity from (4.70) and expanding the equation in **Cartesian–altitude coordinates** gives

$$\frac{du}{dt}=\frac{\partial u}{\partial t}+u\frac{\partial u}{\partial x}+v\frac{\partial u}{\partial y}+w\frac{\partial u}{\partial z}=fv-\frac{1}{\rho_a}\frac{\partial p_a}{\partial x}+\frac{1}{\rho_a}\times\left[\frac{\partial}{\partial x}\left(\rho_a K_{m,xx}\frac{\partial u}{\partial x}\right)+\frac{\partial}{\partial y}\left(\rho_a K_{m,yx}\frac{\partial u}{\partial y}\right)+\frac{\partial}{\partial z}\left(\rho_a K_{m,zx}\frac{\partial u}{\partial z}\right)\right] \tag{4.72}$$

$$\frac{dv}{dt}=\frac{\partial v}{\partial t}+u\frac{\partial v}{\partial x}+v\frac{\partial v}{\partial y}+w\frac{\partial v}{\partial z}=-fu-\frac{1}{\rho_a}\frac{\partial p_a}{\partial y}+\frac{1}{\rho_a}\times\left[\frac{\partial}{\partial x}\left(\rho_a K_{m,xy}\frac{\partial v}{\partial x}\right)+\frac{\partial}{\partial y}\left(\rho_a K_{m,yy}\frac{\partial v}{\partial y}\right)+\frac{\partial}{\partial z}\left(\rho_a K_{m,zy}\frac{\partial v}{\partial z}\right)\right] \tag{4.73}$$

$$\frac{dw}{dt}=\frac{\partial w}{\partial t}+u\frac{\partial w}{\partial x}+v\frac{\partial w}{\partial y}+w\frac{\partial w}{\partial z}=-g-\frac{1}{\rho_a}\frac{\partial p_a}{\partial z}+\frac{1}{\rho_a}\times\left[\frac{\partial}{\partial x}\left(\rho_a K_{m,xz}\frac{\partial w}{\partial x}\right)+\frac{\partial}{\partial y}\left(\rho_a K_{m,yz}\frac{\partial w}{\partial y}\right)+\frac{\partial}{\partial z}\left(\rho_a K_{m,zz}\frac{\partial w}{\partial z}\right)\right] \tag{4.74}$$

The momentum equation in spherical–altitude coordinates requires additional terms. When a reference frame fixed on the surface of the earth is used, the spherical–altitude coordinate conversion terms from (4.30), $(uv\tan\varphi)/R_e$ and $(u^2\tan\varphi)/R_e$, should be treated as apparent forces and are moved to the right side of the momentum equation. Implementing this change and substituting (4.1) into (4.72)–(4.74) gives approximate forms of the directional momentum equations in **spherical–altitude coordinates** as

$$\frac{\partial u}{\partial t}+\frac{u}{R_e\cos\varphi}\frac{\partial u}{\partial\lambda_e}+\frac{v}{R_e}\frac{\partial u}{\partial\varphi}+w\frac{\partial u}{\partial z}=\frac{uv\tan\varphi}{R_e}+fv-\frac{1}{\rho_a R_e\cos\varphi}\frac{\partial p_a}{\partial\lambda_e}+\frac{1}{\rho_a}\left[\frac{1}{R_e^2\cos\varphi}\frac{\partial}{\partial\lambda_e}\times\left(\frac{\rho_a K_{m,xx}}{\cos\varphi}\frac{\partial u}{\partial\lambda_e}\right)+\frac{1}{R_e^2}\frac{\partial}{\partial\varphi}\left(\rho_a K_{m,yx}\frac{\partial u}{\partial\varphi}\right)+\frac{\partial}{\partial z}\left(\rho_a K_{m,zx}\frac{\partial u}{\partial z}\right)\right] \tag{4.75}$$

$$\frac{\partial v}{\partial t}+\frac{u}{R_e\cos\varphi}\frac{\partial v}{\partial\lambda_e}+\frac{v}{R_e}\frac{\partial v}{\partial\varphi}+w\frac{\partial v}{\partial z}$$
$$=-\frac{u^2\tan\varphi}{R_e}-fu-\frac{1}{\rho_a R_e}\frac{\partial p_a}{\partial\varphi}+\frac{1}{\rho_a}\left[\frac{1}{R_e^2\cos\varphi}\frac{\partial}{\partial\lambda_e}\right.$$
$$\left.\times\left(\frac{\rho_a K_{m,xy}}{\cos\varphi}\frac{\partial v}{\partial\lambda_e}\right)+\frac{1}{R_e^2}\frac{\partial}{\partial\varphi}\left(\rho_a K_{m,yy}\frac{\partial v}{\partial\varphi}\right)+\frac{\partial}{\partial z}\left(\rho_a K_{m,zy}\frac{\partial v}{\partial z}\right)\right] \tag{4.76}$$

$$\frac{\partial w}{\partial t}+\frac{u}{R_e\cos\varphi}\frac{\partial w}{\partial\lambda_e}+\frac{v}{R_e}\frac{\partial w}{\partial\varphi}+w\frac{\partial w}{\partial z}$$
$$=-g-\frac{1}{\rho_a}\frac{\partial p_a}{\partial z}+\frac{1}{\rho_a}\left[\frac{1}{R_e^2\cos\varphi}\frac{\partial}{\partial\lambda_e}\left(\frac{\rho_a K_{m,xz}}{\cos\varphi}\frac{\partial w}{\partial\lambda_e}\right)\right.$$
$$\left.+\frac{1}{R_e^2}\frac{\partial}{\partial\varphi}\left(\rho_a K_{m,yz}\frac{\partial w}{\partial\varphi}\right)+\frac{\partial}{\partial z}\left(\rho_a K_{m,zz}\frac{\partial w}{\partial z}\right)\right] \tag{4.77}$$

## 4.3. APPLICATIONS OF THE MOMENTUM EQUATION

The continuity equation for air, the species continuity equation, the thermodynamic energy equation, the three momentum equations, and the equation of state are referred to here as the **equations of atmospheric dynamics**. Removing the species continuity equation from this list and substituting the hydrostatic equation in place of the full vertical momentum equation gives the **primitive equations**, which are a basic form of the Eulerian equations of fluid motion. Many atmospheric motions, including the geostrophic wind, surface winds, the gradient wind, surface winds around high- and low-pressure centers, and atmospheric waves, can be understood by looking at simplified forms of the equations of atmospheric dynamics or primitive equations.

### 4.3.1. Geostrophic Wind

The momentum equation can be used to isolate motions, such as the geostrophic (earth-turning) wind. The **geostrophic wind** arises when the Coriolis force exactly balances the pressure-gradient force. Such a balance occurs following a geostrophic adjustment process, described in Fig. 4.19 (Section 4.3.5.3). Many motions in the free troposphere are near-geostrophic, because in the free troposphere, horizontal local accelerations and turbulent accelerations are usually much smaller than Coriolis and pressure-gradient accelerations. Removing local acceleration and turbulence terms from (4.72) and (4.73), and solving for velocities gives the **geostrophic scalar velocities** in Cartesian–altitude coordinates as

$$v_g=\frac{1}{f\rho_a}\frac{\partial p_a}{\partial x}\qquad u_g=-\frac{1}{f\rho_a}\frac{\partial p_a}{\partial y} \tag{4.78}$$

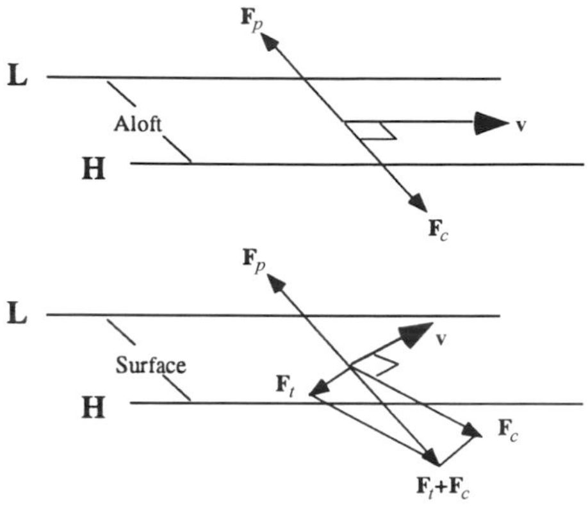


**Figure 4.11.** Force and wind vectors aloft and at the surface in the Northern Hemisphere. The parallel lines are isobars.

In geostrophic equilibrium, the pressure-gradient force is equal and opposite to the Coriolis force, and the geostrophic wind flows 90° to the left of the Coriolis force in the Northern Hemisphere, as shown in the top portion of Fig. 4.11. In the Southern Hemisphere, the geostrophic wind flows 90° to the right of the Coriolis force and 90° to the left of the pressure-gradient force.

**Example 4.14**

If $\varphi = 30°\text{N}$, $\rho_a = 0.76 \text{ kg m}^{-3}$, and the pressure gradient is 4 mb per 150 km in the south-north direction, estimate the west–east scalar geostrophic velocity.

SOLUTION

From (4.35), $f = 7.292 \times 10^{-5}$ rad s$^{-1}$. From (4.78), $u_g = 48.1$ m s$^{-1}$.

In vector form, the geostrophic velocity in Cartesian–altitude coordinates is

$$\mathbf{v}_g = \mathbf{i}u_g + \mathbf{j}v_g = \frac{1}{f\rho_a}\left(-\mathbf{i}\frac{\partial p_a}{\partial y} + \mathbf{j}\frac{\partial p_a}{\partial x}\right) = \frac{1}{f\rho_a}\begin{vmatrix} \mathbf{i} & \mathbf{j} & \mathbf{k} \\ 0 & 0 & 1 \\ \frac{\partial p_a}{\partial x} & \frac{\partial p_a}{\partial y} & 0 \end{vmatrix} \tag{4.79}$$

$$= \frac{1}{f\rho_a}\mathbf{k} \times \nabla_z p_a$$

where

$$\nabla_z = \left(\mathbf{i}\frac{\partial}{\partial x}\right)_z + \left(\mathbf{j}\frac{\partial}{\partial y}\right)_z = \mathbf{i}\frac{\partial}{\partial x} + \mathbf{j}\frac{\partial}{\partial y} \tag{4.80}$$

is the **horizontal gradient operator in Cartesian–altitude coordinates.** The subscript $z$ indicates that the partial derivative is taken along a surface of constant altitude;

thus, $\mathbf{k}\partial/\partial z = 0$. (4.79) indicates that the geostrophic wind flows parallel to lines of constant pressure (isobars).

### 4.3.2. Surface-Layer Winds

In steady state, winds in the surface layer are affected primarily by the pressure-gradient force, Coriolis force, mechanical turbulence, and thermal turbulence. Objects protruding from the surface, such as blades of grass, bushes, trees, structures, hills, and mountains, slow down winds near the surface. Aloft, fewer obstacles exist to resist winds, and their speeds are generally higher than at the surface. In terms of the momentum equation, forces that slow down winds near the surface appear in the turbulent-flux divergence vector $\mathbf{F}_t$.

Near the surface and in steady state, the sum of the Coriolis force and turbulent flux vectors equals the pressure-gradient force vector. Aloft, the Coriolis and pressure-gradient force vectors balance. Figure 4.11 shows forces and resulting winds in the Northern Hemisphere in both cases. On average, friction near the surface shifts winds about 30° counterclockwise in comparison with winds aloft in the Northern Hemisphere, and 30° clockwise in comparison with winds aloft in the Southern Hemisphere. The variation in wind direction may be 45° or more, depending on the roughness of the surface. In both hemispheres, surface winds are tilted towards low pressure.

In the presence of near-surface turbulence, the steady-state horizontal momentum equations simplify in Cartesian–altitude coordinates from (4.72)–(4.73) to

$$-fv = -\frac{1}{\rho_a}\frac{\partial p_a}{\partial x} + \frac{1}{\rho_a}\frac{\partial}{\partial z}\left(\rho_a K_{m,zx}\frac{\partial u}{\partial z}\right) \tag{4.81}$$

$$fu = -\frac{1}{\rho_a}\frac{\partial p_a}{\partial y} + \frac{1}{\rho_a}\frac{\partial}{\partial z}\left(\rho_a K_{m,zy}\frac{\partial v}{\partial z}\right)$$

Figure 4.12(a) and (b) show idealized variations of wind speed with height in the boundary layer during day and night, respectively. During the day, wind speeds

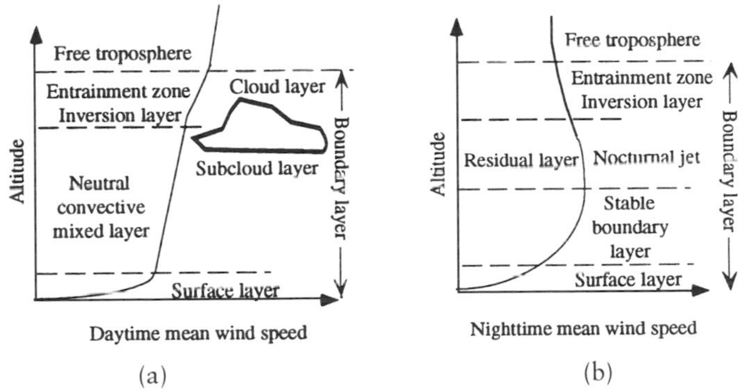


**Figure 4.12.** Variation of wind speed with height during the (a) day and (b) night in the atmospheric boundary layer. Adapted from Stull (1988).

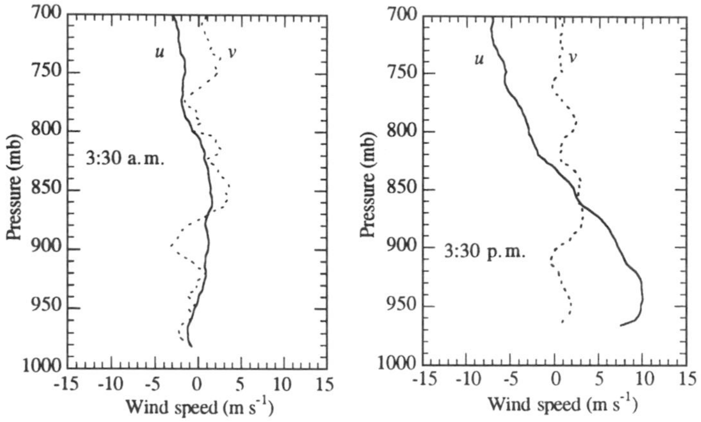


(a) (b)

**Figure 4.13.** Observed variation of $u$ and $v$ scalar velocities with height at (a) 3:30 a.m. and (b) 3:30 p.m. on August 27, 1987, at Riverside, California.

increase logarithmically with height in the surface layer. In the mixed layer, temperature and wind speeds are relatively uniform with height. Above the entrainment zone, wind speeds increase to their geostrophic values.

At night, wind speeds near the surface increase logarithmically with height but are lower than during the day. The reduction of mixing in the stable boundary layer increases wind speeds towards the top of the layer, creating a **nocturnal** or **low-level jet** that is faster than the geostrophic wind. In the residual layer, wind speeds decrease and approach the geostrophic wind speed.

Figure 4.13(a) and (b) show observations of $u$- and $v$-scalar velocities at Riverside, California, during a summer night and day, respectively. Riverside lies between the coast and a mountain range. At night, the $u$- and $v$-winds were small. In the lower boundary layer, during the day, the $u$-winds were strong due to a sea breeze that advected inland. Aloft, flow towards the ocean was strong due to a thermal low inland, which increased pressure and easterly winds aloft inland. The $v$-component of wind was relatively small.

### 4.3.3. The Gradient Wind

When air rotates around a center of low pressure, as in a midlatitude cyclone, tornado, or hurricane, or when it rotates around a center of high pressure, as in an anticyclone, the horizontal momentum equations are more appropriately written in cylindrical coordinates. In Cartesian coordinates the distance variables are $x$, $y$, and $z$, and the unit vectors are **i**, **j**, and **k**. In cylindrical coordinates the distance variables are $R_c$, $\theta$, and $z$, and the unit vectors are $\mathbf{i}_R = \mathbf{i}\cos\theta + \mathbf{j}\sin\theta$, $\mathbf{j}_\theta = -\mathbf{i}\sin\theta + \mathbf{j}\cos\theta$, and **k**. Here $R_c$ is the radius of curvature, or distance from point O to P in Fig. 4.14, and $\theta$ is the angle (rad) measured counterclockwise from the positive $x$-axis. Distances $x$ and $y$ in Cartesian coordinates are related to $R_c$ and

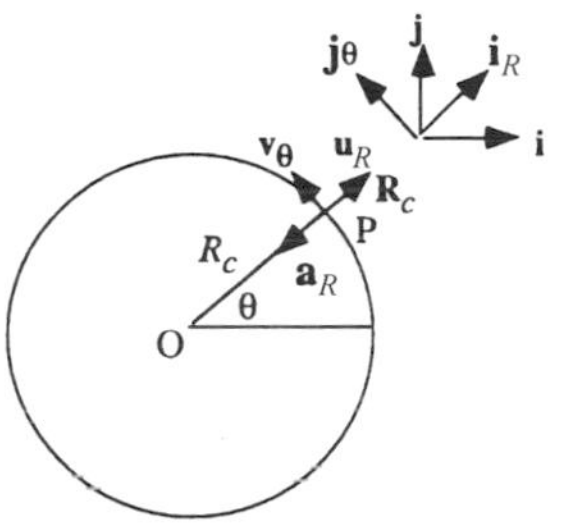


**Figure 4.14.** Cylindrical coordinate components (**i** and **j** are in Cartesian coordinates).

$\theta$ in cylindrical coordinates by $x = R_c \cos\theta$ and $y = R_c \sin\theta$. Thus, $R_c^2 = x^2 + y^2$ and $\theta = \tan^{-1}(y/x)$.

The unit vectors in cylindrical coordinates, $\mathbf{i}_R$ and $\mathbf{j}_\theta$, are directed normal and tangential, respectively, to the circle in Fig. 4.14. The radial vector, directed normal to the circle, is $\mathbf{R}_c = \mathbf{i}_R R_c$. The radial, tangential, and angular velocity vectors in cylindrical coordinates are $\mathbf{u}_R = \mathbf{i}_R u_R$, $\mathbf{v}_\theta = \mathbf{j}_\theta v_\theta$, and $\boldsymbol{\omega}_z = \mathbf{k}\omega_z$, respectively, where $u_R = \mathrm{d}R_c/dt$, $v_\theta = R_c\,\mathrm{d}\theta/dt$, and $\omega_z = \mathrm{d}\theta/dt = v_\theta/R_c$ are the radial, tangential, and angular scalar velocities, respectively.

The tangential and angular velocity vectors in cylindrical coordinates are related to each other by

$$\mathbf{v}_\theta = \boldsymbol{\omega}_z \times \mathbf{R}_c = \begin{vmatrix} \mathbf{i}_R & \mathbf{j}_\theta & \mathbf{k} \\ 0 & 0 & v_\theta/R_c \\ R_c & 0 & 0 \end{vmatrix} = \mathbf{j}_\theta v_\theta \tag{4.82}$$

The local apparent centrifugal force per unit mass, which acts away from the center of curvature, is equal and opposite to local centripetal acceleration ($\mathbf{F}_R/M_a = -\mathbf{a}_R$). The local centripetal acceleration vector is

$$\mathbf{a}_R = \boldsymbol{\omega}_z \times \mathbf{v}_\theta = \begin{vmatrix} \mathbf{i}_R & \mathbf{j}_\theta & \mathbf{k} \\ 0 & 0 & v_\theta/R_c \\ 0 & v_\theta & 0 \end{vmatrix} = -\mathbf{i}_R \frac{v_\theta^2}{R_c^2} = -\mathbf{i}_R R_c a_R = \mathbf{R}_c a_R \tag{4.83}$$

where $a_R = v_\theta^2/R_c$ is the scalar centripetal acceleration.

In cylindrical coordinates and in the absence of eddy diffusion, the horizontal momentum equations transform from (4.72) and (4.73) to

$$\frac{\mathrm{d}u_R}{\mathrm{d}t} = f v_\theta - \frac{1}{\rho_a}\frac{\partial p_a}{\partial R_c} + \frac{v_\theta^2}{R_c} \tag{4.84}$$

$$\frac{\mathrm{d}v_\theta}{\mathrm{d}t} = -f u_R - \frac{u_R v_\theta}{R_c} \tag{4.85}$$

respectively, where $v_\theta^2/R_c$ and $-u_R v_\theta/R_c$ are centripetal accelerations, treated as

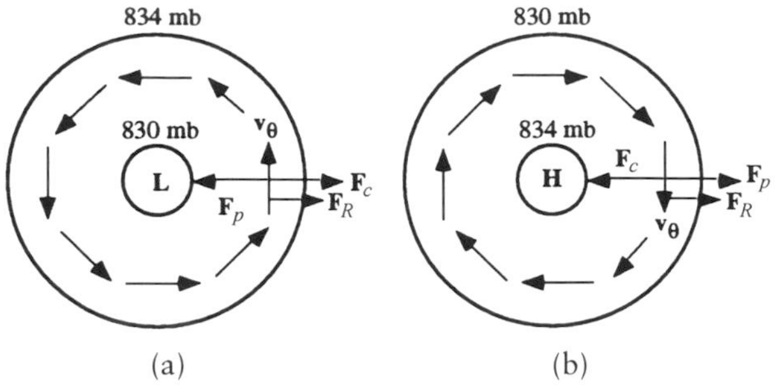


**Figure 4.15.** Gradient winds around a center of (a) low and (b) high pressure, in the Northern Hemisphere, and the forces affecting them.

apparent centrifugal forces per unit mass, that arise from the transformation from Cartesian to cylindrical coordinates. These terms are implicitly included in the Cartesian-coordinate momentum equations.

When air flows around a center of low or high pressure aloft, as in Figs. 4.15(a) or (b), respectively, the primary forces acting on the air are the Coriolis, pressure-gradient, and apparent centrifugal forces. If local acceleration is removed from (4.84), the resulting wind is the **gradient wind**. Assuming $du_R/dt = 0$ and solving (4.84) gives the gradient-wind scalar velocity as

$$v_\theta = -\frac{R_c f}{2} \pm \frac{R_c}{2}\sqrt{f^2 + 4\frac{1}{R_c \rho_a}\frac{\partial p_a}{\partial R_c}} \qquad (4.86)$$

In this solution, the positive square root is correct and results in a $v_\theta$ that is either positive or negative. The negative root is incorrect and results in an unphysical solution.

**Example 4.15**

Suppose the pressure gradient a distance $R_c = 70$ km from the center of a hurricane is $\partial p_a/\partial R_c = 45$ mb per 100 km. Assume $\varphi = 15°$N, $p_a = 850$ mb, and $\rho_a = 1.06$ kg m$^{-3}$. These values give $v_\theta = 52$ m s$^{-1}$ and $v_g = 1123$ m s$^{-1}$. Since $v_\theta < v_g$, the apparent centrifugal force slows down the geostrophic wind.

If the same conditions are used for the high-pressure center case (except the pressure gradient is reversed), the quadratic has an imaginary root, which is an unphysical solution. Reducing $\partial p_a/\partial R_c$ to $-0.1$ mb per 100 km in this example gives $v_\theta = -1.7$ m s$^{-1}$. Thus, the magnitude of pressure gradients and resulting wind speeds around high-pressure centers are smaller than those around comparative low-pressure centers.

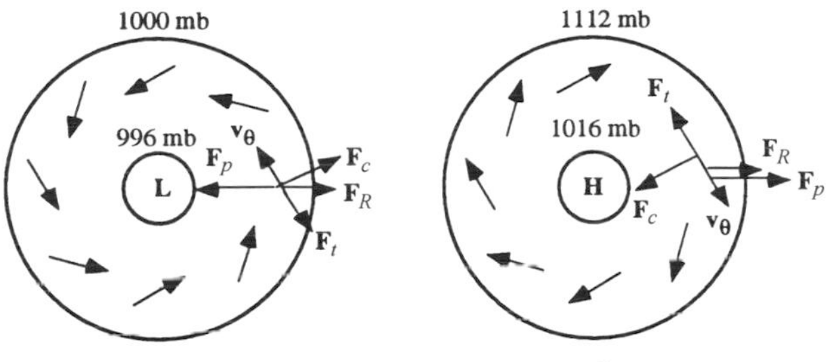


**Figure 4.16.** Surface winds around centers of (a) low and (b) high pressure, in the Northern Hemisphere and the forces affecting them.

### 4.3.4. Surface Winds around Highs and Lows

Near the surface, the deceleration of wind due to drag must be included in the momentum equation when flow around a low- or high-pressure center occurs. Figure 4.16(a) and (b) show forces acting on the wind in such cases. In cylindrical coordinates, the horizontal momentum equations describing flow around these centers are

$$\frac{du_R}{dt} = f v_\theta - \frac{1}{\rho_a}\frac{\partial p_a}{\partial R_c} + \frac{v_\theta^2}{R_c} + \frac{1}{\rho_a}\frac{\partial(\rho_a \overline{w' u'_R})}{\partial z} \tag{4.87}$$

$$\frac{dv_\theta}{dt} = -f u_R - \frac{u_R v_\theta}{R_c} + \frac{1}{\rho_a}\frac{\partial(\rho_a \overline{w' v'_\theta})}{\partial z}$$

where $u'_R$ and $v'_\theta$ are the perturbation components of $u_R$ and $v_\theta$, respectively. The turbulent-flux divergence terms in these equations need to be parameterized.

### 4.3.5. Atmospheric Waves

**Atmospheric waves** are sinusoidal oscillations in pressure and/or velocity that propagate in space and time. Because waves perturb air parcels, which contain energy and gases, waves result in spatial and temporal changes in potential temperature, mixing ratios, and momentum. Many wave types occur in the atmosphere, including acoustic (sound), gravity (buoyancy), and Rossby (planetary) waves. Waves are important to understand from a modeling point of view, because they are often the fastest motion in a model domain and must be filtered from or treated with special numerics in the equations of atmospheric dynamics. Such filtering and special treatment are discussed in Section 5.1. In this section, wave equations are solved analytically from simplified forms of the equations of atmospheric dynamics.

An atmospheric wave is a **pulse** or **group** that consists of individual harmonic (sinusoidal) waves of different (or the same) wavelengths and frequencies, superimposed on each other. Each harmonic wave has its own phase speed $c_\alpha$ (m s$^{-1}$), frequency of oscillation $\nu_\alpha$ (s$^{-1}$), wavenumber in the $x$-, $y$-, and/or $z$-direction ($\tilde{k}$, $\tilde{l}$, and/or $\tilde{m}$) (m$^{-1}$), and wavelength in the $x$-, $y$-, and/or $z$-direction ($\lambda_{\alpha,x}$, $\lambda_{\alpha,y}$, and/or $\lambda_{\alpha,z}$) (m). The **phase speed** (wave speed) is the speed at which a wave travels, the **frequency of oscillation** is the number of wavelengths that pass through a given point per unit time, the **wavenumber** is the number of wavelengths per unit distance, and the **wavelength** is the distance between two crests or troughs in a wave. In the *x*-, *y*-, and *z*-directions, wavenumbers are related to wavelengths by

$$\tilde{k} = \frac{2\pi}{\lambda_{\alpha,x}} \qquad \tilde{l} = \frac{2\pi}{\lambda_{\alpha,y}} \qquad \tilde{m} = \frac{2\pi}{\lambda_{\alpha,z}} \tag{4.88}$$

respectively.

A **dispersion relationship** relates the frequency of oscillation of a harmonic wave to its wavenumbers by

$$\nu_\alpha = c_\alpha\sqrt{\tilde{k}^2 + \tilde{l}^2 + \tilde{m}^2} = c_\alpha|\tilde{\mathbf{K}}| \tag{4.89}$$

where $|\tilde{\mathbf{K}}| = \sqrt{\tilde{k}^2 + \tilde{l}^2 + \tilde{m}^2}$ is the magnitude of the wavenumber vector, $|\tilde{\mathbf{K}}| = \mathbf{i}\tilde{k} + \mathbf{j}\tilde{l} + \mathbf{k}\tilde{m}$. If the phase speeds of individual waves in a pulse are independent of wavenumber (or wavelength), the pulse maintains its shape over time. In such a case, the medium in which the pulse propagates is **nondispersive**. If the phase speeds of individual waves are a function of wavenumber, different harmonic waves travel at different phase speeds, and the pulse changes shape over time. In such a case, the medium in which the pulse propagates is **dispersive**.

The velocity at which the center of the pulse travels is the **group velocity**. In vector form, the group velocity is $\mathbf{c}_g = \mathbf{i}c_{g,x} + \mathbf{j}c_{g,y} + \mathbf{k}c_{g,z}$, where

$$c_{g,x} = \frac{\partial \nu_\alpha}{\partial \tilde{k}} \qquad c_{g,y} = \frac{\partial \nu_\alpha}{\partial \tilde{l}} \qquad c_{g,z} = \frac{\partial \nu_\alpha}{\partial \tilde{m}} \tag{4.90}$$

are **scalar group velocities** in the *x*-, *y*-, and *z*-directions, respectively. Substituting (4.89) into (4.90) gives

$$c_{g,x} = c_\alpha\frac{\tilde{k}}{|\tilde{\mathbf{K}}|} + |\tilde{\mathbf{K}}|\frac{\partial c_\alpha}{\partial \tilde{k}} \qquad c_{g,y} = c_\alpha\frac{\tilde{l}}{|\tilde{\mathbf{K}}|} + |\tilde{\mathbf{K}}|\frac{\partial c_\alpha}{\partial \tilde{l}} \tag{4.91}$$
$$c_{g,z} = c_\alpha\frac{\tilde{m}}{|\tilde{\mathbf{K}}|} + |\tilde{\mathbf{K}}|\frac{\partial c_\alpha}{\partial \tilde{m}}$$

In a **nondispersive medium,** $c_\alpha$ is independent of $\tilde{k}$, $\tilde{l}$, and $\tilde{m}$, and the magnitude of the group velocity vector equals the phase speed of each individual wave in the pulse. Thus, $c_{g,x} = c_\alpha\tilde{k}/|\tilde{\mathbf{K}}|$, $c_{g,y} = c_\alpha\tilde{l}/|\tilde{\mathbf{K}}|$, $c_{g,z} = c_\alpha\tilde{m}\,|\tilde{\mathbf{K}}|$, and $|\mathbf{c}_g| = c_\alpha$.

In a **dispersive medium,** $c_\alpha$ is a function of $\tilde{k}$, $\tilde{l}$, and/or $\tilde{m}$, the magnitude of the group velocity vector differs from the phase speed of each wave, and the shape of the pulse changes over time. Figure 4.17(a) and (b) show examples of wave pulses in nondispersive and dispersive media, respectively.

Waves are generally transverse, longitudinal, or both. A **transverse wave** is a wave in which the disturbance is perpendicular to the direction of propagation of

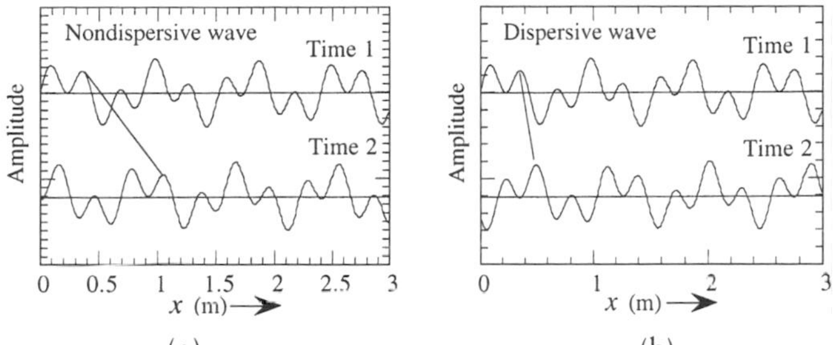


**Figure 4.17.** Wave pulses in (a) nondispersive and (b) dispersive media. In the nondispersive case, $\nu_\alpha = c_s\tilde{k}$ (sound waves) where $c_s$ is the speed of sound (346 m s$^{-1}$ in the example). In the dispersive case, $\nu_\alpha = \sqrt{g\tilde{k}}$ (water waves). For both cases, wave pulses are described by $A = \sin(\tilde{k}_1 x - \nu_{\alpha,1}t) + \sin(\tilde{k}_2 x - \nu_{\alpha,2}t)$, where $\tilde{k}_1 = 0.3$ and $\tilde{k}_2 = 0.8$. Time 1 occurs when $t = 0$ s in both cases, and time 2 occurs when $t = 0.002$ s in the nondispersive case and 0.2 s in the dispersive case. In the nondispersive case, the wave pulse does not change shape over time and the group velocity equals the phase speed of each harmonic component in the pulse. In the dispersive case, the pulse changes shape over time and the group velocity differs from the phase speed of each harmonic component.

the wave. If one end of a spring is fixed to a floor and the other end is lifted vertically, and then oscillated horizontally, a sinusoidal transverse wave propagates down the spring to the floor. Pure gravity waves, light waves, and violin-string waves are transverse waves. A **longitudinal wave** is one in which the disturbance in wave-shape flows parallel to the direction of propagation of the wave. If one end of a spring is fixed to the floor and the other end is lifted vertically and then oscillated vertically, a longitudinal wave propagates to the floor. Pure acoustic waves are longitudinal. Water waves at the top of the ocean surface are transverse and longitudinal. In the following sub-subsections, wave types and their characteristic dispersion relationships are identified and briefly described.

#### 4.3.5.1. *Pure Acoustic Waves*

**Pure acoustic (sound) waves** occur when a vibration causes alternating adiabatic compression and expansion of a compressible fluid, such as air. During compression and expansion, the pressure oscillates, causing acceleration to oscillate. Pure acoustic waves are longitudinal because wave disturbances travel parallel to the direction of wave propagation. Acoustic waves transport not only sound, but also parcels of air containing energy and gases.

The dispersion relationship for pure acoustic waves can be derived in three dimensions by solving the $u$-, $v$-, and $w$-momentum equations (ignoring the Coriolis, gravitational, viscous, and eddy diffusion terms), the continuity equation for air, and the thermodynamic energy equation (ignoring diabatic energy sources and sinks). If the direction of wave propagation is assumed to be along the $x$-axis, the disperson relationship can be simplified by setting $v$- and $w$-terms to zero. In this case, the $u$-momentum, continuity, and thermodynamic energy equations simplify from (4.72), (3.23) and (3.66) to

$$\frac{du}{dt} = -\frac{1}{\rho_a}\frac{\partial p_a}{\partial x} \qquad \frac{d\rho_a}{dt} = -\rho_a\frac{\partial u}{\partial x} \qquad \frac{1}{\theta_v}\frac{d\theta_v}{dt} = \frac{d\ln\theta_v}{dt} = 0 \tag{4.92}$$

respectively. Substituting $\theta_v = T_v(1000/p_a)^\kappa$ and $p_a = \rho_a R' T_v$ into the thermodynamic energy equation gives

$$\frac{d\rho_a}{dt} = \frac{\rho_a}{\gamma}\frac{d\ln p_a}{dt} \tag{4.93}$$

where $\gamma = 1/(1-\kappa) = c_{p,d}/c_{v,d} \approx 1.4$. Substituting (4.93) into the continuity equation of (4.92) gives

$$\frac{d\ln p_a}{dt} = \frac{1}{p_a}\frac{dp_a}{dt} = -\gamma\frac{\partial u}{\partial x} \tag{4.94}$$

Taking the time derivative of (4.94), substituting the momentum equation from (4.92), substituting $p_a = \rho_a R' T_v$, $p_a = \bar{p}_a + p'_a$, $\rho_a = \bar{\rho}_a + \rho'_a$, and $u = \bar{u} + u'$, and eliminating products of perturbation variables yields the **acoustic wave equation,**

$$\frac{d^2 p'_a}{dt^2} = \left(\frac{\partial}{\partial t} + \bar{u}\frac{\partial}{\partial x}\right)^2 p'_a = c_s^2\frac{\partial^2 p'_a}{\partial x^2} \tag{4.95}$$

where $c_s = \pm\sqrt{\gamma R' \bar{T}_v}$ is the **adiabatic speed of sound.**

**Example 4.16**

When $\bar{T}_v = 298$ K, $c_s = 346$ m s$^{-1}$. When $\bar{T}_v = 225$ K, $c_s = 301$ m s$^{-1}$.

A sinusoidal solution to (4.95) is $p'_a = p'_{a,0}\sin(\tilde{k}x - \nu_\alpha t)$, where $x$ is the distance along the direction of phase propagation, $t$ is time, and $p'_{a,0}$ is the amplitude of the oscillation. Substituting this solution into (4.95) gives the **dispersion relation** for pure acoustic waves as

$$\nu_\alpha = (\bar{u} \pm c_s)\tilde{k} \tag{4.96}$$

The phase speed of pure acoustic waves is $c_\alpha = \bar{u} \pm c_s$, found by substituting (4.96) into (4.89). Taking the partial derivative of (4.96) with respect to $\tilde{k}$ gives the scalar group velocity in the $x$-direction as $c_{g,x} = \bar{u} \pm c_s$. The scalar group velocities in the $y$- and $z$-directions are zero in this case. Since the magnitude of the group velocity vector equals the phase speed of all individual waves, acoustic waves are nondispersive.

#### 4.3.5.2. *Gravity, Acoustic-Gravity, and Lamb Waves*

When the atmosphere is stably stratified and a parcel of air is displaced vertically, buoyancy restores the parcel to its equilibrium position in an oscillatory manner. The frequency of oscillation, $N_{bv}$, is the **Brunt–Väisälä** frequency, defined in Chapter 2. The period of oscillation is $\tau_{bv} = 2\pi/N_{bv}$. The wave resulting from this oscillation is a **gravity (buoyancy) wave.** Air-parcel displacements that give rise to gravity waves may result from forced convection, air flow over mountains, wind shear in frontal regions, wind shear associated with jet-stream flow, or perturbations associated with the geostrophic wind. Because gravity waves displace parcels of air, they perturb the potential temperature, gas mixing ratios, vertical velocities, and other parameters associated with the parcel. Gravity-wave motion may increase vertical transfer of ozone between the lower stratosphere and upper troposphere (Lindzen 1981; VanZandt and Fritts 1989). Gravity waves may also cause large periodic fluctuations in ozone at specific locations in the troposphere (Langford et al. 1996).

The dispersion relationship for gravity waves can be isolated from a broader relationship for acoustic-gravity waves. In three dimensions, the **acoustic-gravity-wave dispersion relationship** is found by solving the three momentum equations (retaining gravity in the vertical), the continuity equation for air, and the thermodynamics energy equations. If only the $x$–$z$ plane is considered, a simplified acoustic-gravity wave relationship is obtained by solving

$$\frac{du}{dt} = -\frac{1}{\rho_a}\frac{\partial p_a}{\partial x} \qquad \frac{dw}{dt} = -\frac{1}{\rho_a}\frac{\partial p_a}{\partial z} - g \tag{4.97}$$

$$\frac{d\rho_a}{dt} = -\rho_a\left(\frac{\partial u}{\partial x} + \frac{\partial w}{\partial z}\right) \qquad \frac{d\rho_a}{dt} = \frac{\rho_a}{\gamma}\frac{d\ln p_a}{dt}$$

written from (4.72), (4.74), (3.23), and (4.93) respectively. When $p_a = \bar{p}_a + p'_a$, $\rho_a = \bar{\rho}_a + \rho'_a$, $u = \bar{u} + u'$, and $w = \bar{w} + w'$, (4.97) can be solved analytically for the acoustic-gravity-wave dispersion relationship,

$$\frac{N_{bv}^2}{(\nu_\alpha - \bar{u}\tilde{k})^2}\tilde{k}^2 + \frac{(\nu_\alpha - \bar{u}\tilde{k})^2}{c_s^2} = \tilde{m}^2 + \tilde{k}^2 + \frac{\nu_c^2}{c_s^2} \tag{4.98}$$

where $\nu_c = c_s/2H$ is the **acoustic cutoff frequency** ($s^{-1}$), and $H$ is the scale height of the atmosphere, obtained from (2.41) under the assumption that the atmosphere is isothermal and stably stratified. The acoustic cutoff frequency is a frequency above which acoustic waves may propagate vertically if other conditions are right. Below this frequency, acoustic waves cannot propagate vertically.

**Example 4.17**

For an isothermal layer at $T_v = 298$ K, find the acoustic cutoff and Brunt–Väisälä frequencies.

SOLUTION

From Appendix A, $\bar{M} = 4.8096 \times 10^{-26}$ kg, $g = 9.81$ m s$^{-2}$, and $k_B = 1.3807 \times 10^{-23}$ kg m$^2$ s$^{-2}$ K$^{-1}$ molecule$^{-1}$. From (2.41), the scale height is $H = 8.72$ km. Thus, $c_s = \pm\sqrt{\gamma R' \bar{T}_v} = 346$ m s$^{-1}$, and $\nu_c = c_s/2H = 0.0198$ s$^{-1}$. From (2.100), the Brunt-Väisälä frequency is $N_{bv} = \sqrt{g(\Gamma_d - \Gamma_v)/T_v}$. Substituting $\Gamma_v = 0$ (since the atmosphere is isothermal) and $\Gamma_d = +9.8$ K km$^{-1}$ into this equation gives $N_{bv} = 0.0180$ s$^{-1}$. The acoustic cutoff frequency slightly exceeds the Brunt–Väisälä frequency ($\nu_c > N_{bv}$).

Dispersion relationships for gravity waves, acoustic waves, modifed by the effects of stable, stratification, and mixed waves are isolated from the acoustic–gravity-wave dispersion relationship by considering certain frequency and wavenumber regimes. For several cases, (4.98) simplified to

$$\nu_\alpha = \begin{cases} \bar{u}\tilde{k} + \dfrac{N_{bv}\tilde{k}}{\left(\tilde{k}^2+\tilde{m}^2+v_c^2/c_s^2\right)^{1/2}} & \nu_\alpha^2 \ll v_c^2 & \begin{cases}\text{low-frequency}\\ \text{gravity waves}\end{cases} \\ \bar{u}\tilde{k} + \left(c_s^2\tilde{k}^2 + c_s^2\tilde{m}^2 + v_c^2\right)^{1/2} & \nu_\alpha^2 \gg N_{bv}^2 & \begin{cases}\text{high-frequency}\\ \text{acoustic waves}\end{cases} \\ \bar{u}\tilde{k} + \dfrac{N_{bv}\tilde{k}}{(\tilde{k}^2+\tilde{m}^2)^{1/2}} & \tilde{k} \to \infty & \{\text{mountain waves} \\ (\bar{u}+c_s)\tilde{k} & \begin{cases}\nu_\alpha^2 \ll v_c^2, & \tilde{m}^2 = 0,\ \tilde{k} \to 0 \text{ or} \\ \nu_\alpha^2 \gg N_{bv}^2, & \tilde{m}^2 = 0,\ \tilde{k} \to \infty\end{cases} & \begin{cases}\text{Lamb}\\ \text{waves}\end{cases} \end{cases} \tag{4.99}$$

When $\nu_\alpha^2 \ll \nu_c^2$, (4.98) simplifies to the dispersion relationship for **gravity waves.** The same result is found by assuming the atmosphere is incompressible ($c_s \to \infty$). Figure 4.18 shows that vertically and horizontally propagating, low-frequency gravity waves occur when $\tilde{m}^2 > 0$, $\nu_\alpha - \bar{u}\tilde{k} < c_s\tilde{k}$ as $\tilde{k} \to 0$, and $\nu_\alpha - \bar{u}\,\tilde{k} < N_{bv}$ as $\tilde{k} \to \infty$. Such waves are **internal gravity waves.** An **internal wave** is a wave that exhibits vertical motion trapped beneath a surface. The Brunt–Väisälä frequency demarks the upper frequency for internal gravity waves. Applying (4.89) to the first term on the right side of (4.99) gives the phase speed of internal gravity waves as $c_\alpha = \bar{u} + N_{bv}/\sqrt{\tilde{k}^2 + \tilde{m}^2 + \nu_c^2/c_s^2}$. Applying (4.90) to this phase speed gives scalar group velocities of internal gravity waves as

$$c_{g,x} = \bar{u} + \frac{N_{bv}\left(\tilde{m}^2 + \nu_c^2/c_s^2\right)}{\left(\tilde{k}^2 + \tilde{m}^2 + \nu_c^2/c_s^2\right)^{3/2}} \qquad c_{g,z} = -\frac{N_{bv}\tilde{k}\tilde{m}}{\left(\tilde{k}^2 + \tilde{m}^2 + \nu_c^2/c_s^2\right)^{3/2}} \tag{4.100}$$

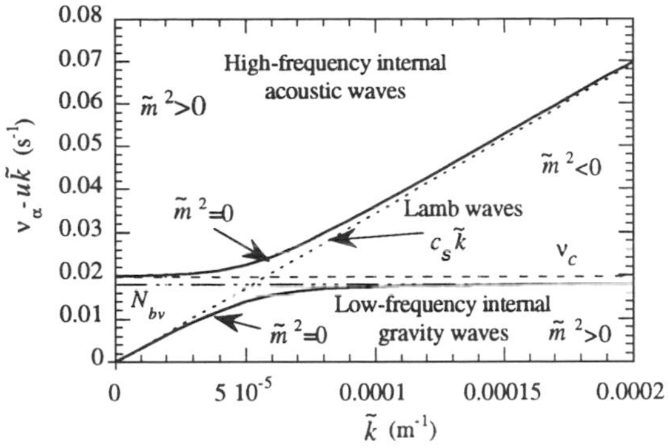


**Figure 4.18.** Vertical wavenumber squared, $\tilde{m}^2$, as a function of $\tilde{k}$ and $\nu_\alpha$ when $H = 8.72$ km, $c_s = 346$ m s$^{-1}$, and $N_{bv} = 0.0180\,\text{s}^{-1}$ (see Example 4.17). The upper region where $\tilde{m}^2 > 0$ represents conditions that give rise to vertically propagating high-frequency acoustic waves. The lower region where $\tilde{m}^2 > 0$ represents conditions that give rise to vertically propagating low-frequency gravity waves. The region in which $\tilde{m}^2 < 0$ represents conditions that give rise to horizontally propagating acoustic–gravity and Lamb waves. Lamb waves connect the point $\tilde{k} = 0$, $\tilde{m}^2 = 0$ to the point $\tilde{k} = \infty$, $\tilde{m}^2 = 0$. The acoustic cutoff frequency slightly exceeds the Brunt–Väisälä frequency.

Internal gravity waves are dispersive, since their phase speed differs from the magnitude of their group velocity vector.

In the region of Fig. 4.18 where $\tilde{m}^2 < 0$, waves are **external acoustic-gravity waves** and propagate only horizontally. An **external wave** is a wave that has its maximum amplitude at the external boundary (free surface) of a fluid. Such waves have no influence in the vertical, since their amplitudes decrease exponentially with distance from the surface.

When $\nu_\alpha^2 \gg N_{bv}^2$, the frequency from (4.98) simplifies to that of acoustic waves, modified by the effects of stable stratification. Figure 4.18 shows that vertically and horizontally propagating acoustic waves occur when $\tilde{m}^2 > 0$, $\nu_\alpha - \bar{u}\tilde{k} > \nu_c$ as $\tilde{k} \to 0$, and $\nu_\alpha - \bar{u}\tilde{k} > c_s\tilde{k}$ as $\tilde{k} \to \infty$. Such waves are **internal acoustic waves.** The dispersion relationship for internal acoustic waves differs from that for pure acoustic waves only in that the latter ignores the effect of buoyancy in a stably stratified atmosphere.

The dotted line in Fig. 4.18 connecting the points between $\tilde{m}^2 = 0$, $\tilde{k} \to 0$ for internal gravity waves and $\tilde{m}^2 = 0$, $\tilde{k} \to \infty$ for internal acoustic waves modified by stratification describes the dispersion relationship for external **Lamb waves** (Lamb 1910). The frequency of a Lamb wave approaches $\nu_\alpha \approx (\bar{u} + c_s)\tilde{k}$, which is the same frequency as that of a pure acoustic wave, as shown in (4.96). Lamb waves are nondispersive and propagate horizontally at the speed of sound. Whereas their energy decreases exponentially with increasing altitude, their amplitude may decrease or increase exponentially with altitude, depending on the value of $\tilde{m}$.

When $\tilde{k} \rightarrow \infty$, we have $\tilde{k} \gg \nu_c/c_s$, and the acoustic–gravity-wave dispersion relationship simplifies from that of gravity waves to that of **mountain lee waves**, as shown in (4.99). Mountain lee waves produce a wavelength on the order of 10 km, which results in a wavenumber of $\tilde{k} = 6.3 \times 10^{-4}\ \mathrm{m}^{-1}$. This is large in comparison with $\nu_c/c_s \approx 5.7 \times 10^{-5}\ \mathrm{m}^{-1}$ from Example 4.17. Since $\tilde{k} > \nu_c/c_s$ for mountain lee waves, $\nu_c/c_s$ can be ignored, giving the simplified dispersion relationship shown in (4.99).

#### 4.3.5.3. *Inertia Gravity Waves and Inertia Lamb Waves*

In addition to pressure and buoyancy oscillations, a third oscillation type is an inertial oscillation. When a parcel of air moving from west to east with the geostrophic wind is perturbed in the south–north direction, the Coriolis force propels the parcel towards or away from its original latitude. In the former case, the atmosphere is **inertially stable** and the oscillation about the initial latitude is an **inertial oscillation**. In the latter case, the atmosphere is **inertially unstable**.

Whether the atmosphere is inertially stable or unstable can be estimated by solving the horizontal momentum equations,

$$\frac{\mathrm{d}u}{\mathrm{d}t} = fv = f\frac{\mathrm{d}y}{\mathrm{d}t} \qquad \frac{\mathrm{d}v}{\mathrm{d}t} = f(u_g - u) \tag{4.101}$$

where only acceleration, the Coriolis force, and the south–north pressure gradient (embodied in $u_g$) are considered. If a parcel of air moving with the geostrophic wind is displaced in the south–north direction a distance $\Delta y$, the $u$-equation in (4.101) can be integrated between $y_0$ and $y_0 + \Delta y$ to give

$$u(y_0 + \Delta y) - u_g(y_0) \approx f\Delta y \tag{4.102}$$

The geostrophic wind at point $y_0$ can be found by taking a first-order Taylor series expansion of the geostrophic wind at $y_0 + \Delta y$. The result is

$$u_g(y_0 + \Delta y) \approx u_g(y_0) + \frac{\partial u_g}{\partial y}\Delta y \tag{4.103}$$

Substituting (4.103) into (4.102) yields

$$u_g(y_0 + \Delta y) - u(y_0 + \Delta y) \approx -\left(f - \frac{\partial u_g}{\partial y}\right)\Delta y \tag{4.104}$$

which substitutes into the $v$-equation of (4.101) to give

$$\boxed{\frac{\mathrm{d}v}{\mathrm{d}t} = -f\left(f - \frac{\partial u_g}{\partial y}\right)\Delta y} \tag{4.105}$$

(4.105) states that, if a parcel of air is displaced a distance $\Delta y$ north of its original latitude in the Northern Hemisphere ($\Delta y > 0$), and if $f - \partial u_g/\partial y > 0$, then $dv/dt < 0$, and the parcel returns southward. If a parcel is displaced a distance $\Delta y$ south of its original latitude ($\Delta y < 0$), and if $f - \partial u_g/\partial y > 0$, then $dv/dt > 0$, and the parcel returns northward. When $f - \partial u_g/\partial y = 0$, a perturbation to the air parcel causes an inertial oscillation about the original latitude. In both cases, a perturbation causes the parcel to accelerate further away from its original latitude. When $f - \partial u_g/\partial y = 0$, a perturbation causes the parcel to move to the new location, but no further. In sum, **inertial stability criteria for the Northern Hemisphere**, where $f > 0$, are

$$f - \frac{\partial u_g}{\partial y} \begin{cases} < 0 & \text{inertially ustable} \\ = 0 & \text{inertially neutral} \\ > 0 & \text{inertially stable} \end{cases} \tag{4.106}$$

which are analogous to the buoyancy-related stability criteria discussed in Chapter 2.

The atmosphere is almost always inertially stable, except near the equator, where the Coriolis parameter is small. When inertial oscillations occur, their frequency $f$ has a value on the order of $10^{-4}$ s$^{-1}$. This is lower than the acoustic cutoff and Brunt–Väisälä frequencies, which have values on the order of $10^{-2}$ s$^{-1}$, as shown in Example 4.17. Inertial oscillations are significant only when perturbations caused by low-frequency waves are considered. Since vertically propagating acoustic waves occur only at frequencies above $\nu_c$, as shown in Fig. 4.18, and since $f$ becomes important only at frequencies below $\nu_c$, inertial oscillations do not affect the dispersion relationship of vertically propagating acoustic waves.

When the frequency of a Lamb wave is below $\nu_c$, it may be affected by inertial oscillations. Under such a condition, the dispersion relationship for **inertia Lamb waves** is rewritten from (4.99) as

$$\nu_\alpha^2 = f^2 + c_s^2 \tilde{k}^2 \tag{4.107}$$

At low frequencies, inertia oscillations affect gravity waves. The dispersion relationship for **inertia gravity waves** is

$$\nu_\alpha^2 = f^2 + \frac{N_{bv}^2 \tilde{k}^2}{\tilde{k}^2 + \tilde{m}^2 + \nu_c^2/c_s^2} \tag{4.108}$$

Inertia gravity waves are driven by inertial and buoyant oscillations and occur when the atmosphere is inertially and buoyantly stable. Inertia Lamb waves are driven by inertial and pressure oscillations.

When the geostrophic velocity in midlatitudes is perturbed, horizontally propagating inertia Lamb waves act to adjust the vertically integrated velocity and surface pressure fields while vertically and horizontally propagating inertia gravity waves act to adjust vertical shear and temperature fields to restore geostrophic balance (Arakawa 1997). The restoration to geostrophic balance is **geostrophic adjustment.** During geostrophic adjustment, the primary factor that determines whether the pressure (mass) field adjusts to the velocity field or the velocity field adjusts to the pressure field is whether the length scale of horizontal motion ($L$) is smaller or larger than the **Rossby radius of deformation.** For atmospheric waves, the length scale is the horizontal wavelength. The Rossby radius of deformation is $\lambda_R = \sqrt{gh_e}/f$, where $h_e$ is the **equivalent depth,** given as

$$h_e = \begin{cases} \frac{c_s^2}{g} & \text{inertia Lamb waves} \\ \frac{N_{bv}^2/g}{\tilde{m}^2+\nu_c^2/c_s^2} & \text{inertia gravity waves} \end{cases} \quad (4.109)$$

A typical value of $\lambda_R$ for inertia Lamb waves is 3,000 km, and typical values for inertia gravity waves are 100–1,000 km.

**Example 4.18**

From the data in Example 4.17, calculate the equivalent depth and the Rossby radius of deformation for inertia gravity waves assuming $\lambda_{\alpha,z}$ = 5 km and $\varphi$ = 30°N latitude.

SOLUTION

From Example 4.17, $c_s$ = 346 m s$^{-1}$, $\nu_c$ = 0.0198 s$^{-1}$, and $N_{bv}$ = 0.0180 s$^{-1}$, since $\tilde{m} = 2\pi/\lambda_{\alpha,z}$ = 0.0013 m$^{-1}$. From (4.109), $h_e$ = 19.5 m. At $\varphi$ = 30°N, we have $f = 7.292 \times 10^{-5}$ s$^{-1}$ and $\lambda_R$ = 190 km.

When $L > \lambda_R$ for inertia Lamb waves, the perturbed vertical mean velocity field adjusts to the pressure field, as roughly demonstrated in two dimensions in Fig. 4.19. If $L < \lambda_R$, the pressure field adjusts to the vertical mean velocity field. If $L = \lambda_R$, a mutual adjustment between the pressure and velocity fields occurs. If $L > \lambda_R$ for inertia gravity waves, the wind shear adjusts to the temperature field. Otherwise the temperature field adjusts to the wind shear or mutual adjustment occurs (Arakawa 1997).

### 4.3.5.4. *Rossby Waves*

**Rossby (planetary) waves** (Rossby et al. 1939) are important for synoptic and planetary-scale meteorology and arise from the change of the Coriolis parameter $f = 2\Omega \sin\varphi$ with latitude. If an air parcel is perturbed in the south–north direction, the change in $f$ with latitude provides a restoring force that gives rise to freely propagating Rossby waves. The presence of topography can initiate forced

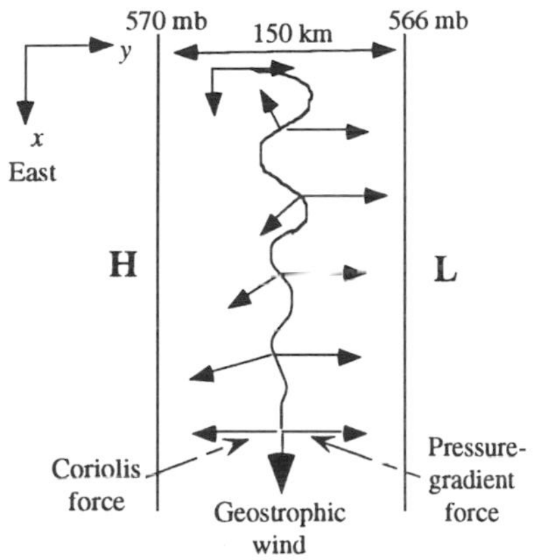


**Figure 4.19.** Geostrophic adjustment of the velocity field to the pressure field at one altitude when $L > \lambda_R$ and when the domain is assumed to be large enough to allow the energy of the oscillations to disperse and decay. In reality, the vertical mean velocity field, not the velocity field at a given altitude, adjusts to the surface pressure. A perturbation to the geostrophic wind causes waves to restore the system to geostrophic balance along an oscillatory path.

topographic Rossby waves. Rossby waves differ from pure inertial oscillations in that the latter do not require a variation of $f$ with latitude.

In an incompressible, frictionless, adiabatic atmosphere of constant depth, absolute vorticity of air is conserved. Rossby waves can be described in terms of the conservation of absolute vorticity. **Vorticity** is the measure of spin around an axis. The faster a body spins, the greater its vorticity. Air spins relative to the surface of earth, and the surface of earth spins relative to its own axis of rotation. The spin of air relative to the surface of earth is the **relative vorticity**, and the spin of earth relative to its own axis is the **earth's vorticity**. Positive vorticity corresponds to counterclockwise spin when the spin is viewed from above the North Pole. Since the earth rotates counterclockwise, the earth's vorticity is positive. For a body at the North Pole, the earth's vorticity is maximum, since the body spins around its own vertical axis. For a body standing at the equator, the earth's vorticity is zero, since the body does not spin around its own vertical axis. The vertical component of the earth's vorticity is the Coriolis parameter $f$. At the equator, North Pole, and South Pole, $f = 0$, $2\Omega$, and $-2\Omega$, respectively.

The **relative vorticity** of air is the vorticity of air relative to the earth. Mathematically, the relative vorticity is the curl of velocity. The relative vorticity vector is

$$\zeta_r = \nabla \times \mathbf{v} = \begin{vmatrix} \mathbf{i} & \mathbf{j} & \mathbf{k} \\ \partial/\partial x & \partial/\partial y & \partial/\partial z \\ u & v & w \end{vmatrix}$$
$$= \left(\frac{\partial w}{\partial y} - \frac{\partial v}{\partial z}\right)\mathbf{i} - \left(\frac{\partial w}{\partial x} - \frac{\partial u}{\partial z}\right)\mathbf{j} + \left(\frac{\partial v}{\partial x} - \frac{\partial u}{\partial y}\right)\mathbf{k} \tag{4.110}$$

The **absolute vorticity** is the sum of the earth's and the relative vorticity. The vertical component of absolute vorticity is $\zeta_{a,z} = f + \zeta_{r,z}$, where $\zeta_{r,z}$ is the vertical component of relative vorticity. In a frictionless, incompressible, and adiabatic atmosphere, the vertical component of absolute vorticity divided by the depth of the atmosphere is constant. In other words,

$$P_v = \frac{f + \zeta_{r,z}}{\Delta z_t} = \frac{f + \frac{\partial v}{\partial x} - \frac{\partial u}{\partial y}}{\Delta z_t} = \text{constant} \tag{4.111}$$

where $P_v$ is a simplified form of **potential vorticity** and $\Delta z_t$ is the depth of the atmosphere from the surface to the tropopause. If the depth of the atmosphere is constant and $v$ is nonzero, a freely propagating Rossby wave arises. As air moves northward, $f$ increases. To conserve potential vorticity, $\zeta_{r,z}$ decreases. For $\zeta_{r,z}$ to decrease, $u$ and $v$ must change. The sinusoidal motion resulting from the continuous changes of $u$, $v$, and $f$ is a freely propagating Rossby wave.

The dispersion relationship for freely propagating Rossby waves in a frictionless, incompressible, and adiabatic atmosphere is found by solving the $u$ and $v$ momentum equations and the continuity equation for air. For this derivation, atmospheric depth is allowed to vary. The constant-depth result is easily extracted from the solution.

If only inertial, Coriolis, and pressure-gradient terms are considered, the horizontal momentum equations simplify from (4.72) and (4.73) to

$$\frac{du}{dt} = fv - \frac{1}{\rho_a}\left(\frac{\partial p_a}{\partial x}\right)_z \qquad \frac{dv}{dt} = -fu - \frac{1}{\rho_a}\left(\frac{\partial p_a}{\partial y}\right)_z \tag{4.112}$$

where the subscript $z$ indicates the partial derivative is taken along a surface of constant altitude. At least three sets of terms can be substituted into these equations.

The first is the **midlatitude beta-plane approximation**, $f = f_0 + \beta(y - y_0)$, which estimates the variation of the Coriolis parameter with latitude. In this equation, $y_0$ is the south–north distance from the equator at which $f = f_0$, $y$ is the south–north distance of interest from the equator, $\beta(y - y_0) \ll f_0$, and

$$\boxed{\beta = \frac{\partial f}{\partial y} = 2\Omega \frac{\partial \varphi}{\partial y} \cos\varphi = \frac{2\Omega}{R_e}\cos\varphi} \tag{4.113}$$

where $\partial y = R_e \partial \varphi$.

The second is a pair of equations relating pressure gradients along surfaces of constant altitude to geopotential gradients along surfaces of constant pressure. The equations, derived in Chapter 5, are

$$\left(\frac{\partial p_a}{\partial x}\right)_z = \rho_a\left(\frac{\partial \Phi}{\partial x}\right)_p \qquad \left(\frac{\partial p_a}{\partial y}\right)_z = \rho_a\left(\frac{\partial \Phi}{\partial y}\right)_p \tag{4.114}$$

where $p$ indicates that a partial derivative is taken along a surface of constant pressure.

The third is a set of equations dividing scalars ($u$, $v$, $\Phi$) into **geostrophic** ($u_g$, $v_g$, $\Phi_g$) and **ageostrophic** ($u_a$, $v_a$, $\Phi_a$) components.

Substituting $f = f_0 + \beta(y - y_0)$, (4.114), $u = u_g + u_a$, $v = v_g + v_a$, and $\Phi = \Phi_g + \Phi_a$ into (4.112) gives

$$\frac{\mathrm{d}(u_g + u_a)}{\mathrm{d}t} = [f_0 + \beta(y - y_0)](v_g + v_a) - \left[\frac{\partial(\Phi_g + \Phi_a)}{\partial x}\right]_p \tag{4.115}$$

$$\frac{\mathrm{d}(v_g + v_a)}{\mathrm{d}t} = -[f_0 + \beta(y - y_0)](u_g + u_a) - \left[\frac{\partial(\Phi_g + \Phi_a)}{\partial y}\right]_p \tag{4.116}$$

These equations can be simplified by combining (4.114) with (4.78) at latitude $y_0$ to give **geostrophic scalar velocities** along surfaces of constant pressure as

$$v_g = \frac{1}{f_0}\left(\frac{\partial \Phi_g}{\partial x}\right)_p \qquad u_g = -\frac{1}{f_0}\left(\frac{\partial \Phi_g}{\partial y}\right)_p \tag{4.117}$$

Substituting these terms into (4.115) and (4.116) and assuming $\mathrm{d}u_a/\mathrm{d}t$, $\mathrm{d}v_a/\mathrm{d}t$, $\beta(y - y_0)u_a$, and $\beta(y - y_0)v_a$ are small gives the **quasigeostrophic horizontal momentum equations,**

$$\frac{\mathrm{d}u_g}{\mathrm{d}t} = f_0 v_a + \beta(y - y_0)v_g - \left(\frac{\partial \Phi_a}{\partial x}\right)_p \tag{4.118}$$

$$\frac{\mathrm{d}v_g}{\mathrm{d}t} = -f_0 u_a - \beta(y - y_0)u_g - \left(\frac{\partial \Phi_a}{\partial y}\right)_p \tag{4.119}$$

Subtracting $\partial/\partial y$ of (4.118) from of $\partial/\partial x$ (4.119) yields

$$\frac{\mathrm{d}}{\mathrm{d}t}\left(\frac{\partial v_g}{\partial x} - \frac{\partial u_g}{\partial y}\right) = -f_0\left(\frac{\partial u_a}{\partial x} + \frac{\partial v_a}{\partial y}\right) - \beta v_g \tag{4.120}$$

where all remaining terms cancel. For example, $\partial u_g/\partial x + \partial v_g/\partial y = 0$, which indicates that the geostrophic wind in nondivergent.

The vertical scalar velocity in the altitude coordinate is

$$w = \frac{\mathrm{d}z}{\mathrm{d}t} = \frac{1}{g}\frac{\mathrm{d}\Phi}{\mathrm{d}t} \tag{4.121}$$

Substituting (4.121), $u = u_g + u_a$, $v = v_g + v_a$, and $\partial u_g/\partial x + \partial v_g/\partial y = 0$ into the incompressible continuity equation from (3.27), and assuming $\mathrm{d}\Phi_a/\mathrm{d}t$ is small gives

$$\frac{1}{g}\frac{\partial}{\partial z}\left(\frac{\mathrm{d}\Phi_g}{\mathrm{d}t}\right) = -\left(\frac{\partial u_a}{\partial x} + \frac{\partial v_a}{\partial y}\right) \tag{4.122}$$

Integrating (4.122) from the surface ($z = 0$) to the mean tropopause height, $z = \Delta z_t$, gives

$$\frac{\mathrm{d}\Phi_g}{\mathrm{d}t} = -g\Delta z_t\left(\frac{\partial u_a}{\partial x} + \frac{\partial v_a}{\partial y}\right) \tag{4.123}$$

When $\mathrm{d}\Phi_g/\mathrm{d}t \neq 0$, the atmosphere has a free surface with average depth $\Delta z_t$. When $\mathrm{d}\Phi_g/\mathrm{d}t = 0$, the atmosphere has a rigid lid with constant depth $\Delta z_t$. Substituting (4.123) into (4.120) gives

$$\frac{\mathrm{d}}{\mathrm{d}t}\left(\zeta_g - \frac{f_0}{g\Delta z_t}\Phi_g\right) = -\beta v_g \tag{4.124}$$

where

$$\zeta_g = \frac{\partial v_g}{\partial x} - \frac{\partial u_g}{\partial y} = \frac{1}{f_0}\left(\frac{\partial^2\Phi_g}{\partial x^2} + \frac{\partial^2\Phi_g}{\partial y^2}\right)_p = \frac{\nabla_p^2\Phi_g}{f_0} \tag{4.125}$$

is the **geostrophic potential vorticity.**

Substituting (4.125) and $v_g$ from (4.117) into (4.124) and expanding the total derivative gives the **quasigeostrophic potential vorticity equation,**

$$\left(\frac{\partial}{\partial t} + \bar{u}\frac{\partial}{\partial x}\right)\left(\nabla_p^2\Phi_g - \frac{f_0^2}{g\Delta z_t}\Phi_g\right) + \beta\left(\frac{\partial\Phi_g}{\partial x}\right)_p = 0 \tag{4.126}$$

where $u$ and $v$ were replaced by $u = \bar{u} + u'$ and $v \approx v'$, respectively, and the equation was Reynolds-averaged. The assumption $v \approx v'$ is reasonable for midlatitudes, where $u$ dominates over $v$. (4.126) has a sinusoidal wave solution, $\Phi_g = \Phi_{g,0}\sin(\tilde{k}x + \tilde{l}y - \nu_\alpha t)$, where $\Phi_{g,0}$ is the amplitude of the wave. Substituting the solution into (4.126) and solving for $\nu_\alpha$ gives the **dispersion relationship for freely-propagating Rossby waves** in a frictionless, incompressible, and adiabatic atmosphere as

$$\nu_\alpha = \left(\bar{u} - \frac{\beta}{\tilde{k}^2 + \tilde{l}^2 + \lambda_R^{-2}}\right)\tilde{k} \tag{4.127}$$

where $\lambda_R = \sqrt{g\Delta z_t}/f_0$ is another form of the **Rossby radius of deformation.** For an atmosphere of constant depth, $d\Phi_g/dt = 0$ in (4.124), and (4.127) simplifies to

$$\nu_\alpha = \left(\bar{u} - \frac{\beta}{\tilde{k}^2 + \tilde{l}^2}\right)\tilde{k} \tag{4.128}$$

Rossby-wave phasc speeds propagate from east to west, as shown in (4.127) and (4.128). Since the mean zonal scalar velocity $\bar{u}$ is from west to east with a magnitude greater than the Rossby-wave phase speed, Rossby waves themselves travel from west to east.

**Example 4.19**

From (4.128), estimate the Rossby-wave phase speed when $\lambda_{\alpha,z} \approx \lambda_{\alpha,y} = 6{,}000$ at $\varphi = 45°$N.

SOLUTION

From (4.88), $\tilde{k} \approx \tilde{l} = 2\pi/\lambda_{\alpha,z} = 1.05 \times 10^{-6}$ m$^{-1}$. From (4.113), $\beta = 1.62 \times 10^{-11}$ m$^{-1}$ s$^{-1}$. From (4.128), $c_\alpha = -\beta/(\tilde{k}^2 + \tilde{l}^2) = -7.3$ m s$^{-1}$. Thus, Rossby-wave phase speed propagates from east to west.

The wave solution for Rossby waves accounts for horizontally-propagating waves in an incompressible atmosphere. Rossby waves propagate vertically as well as horizontally. To account for vertical propagation and air compressibility, the dispersion relationship for Rossby waves can be found using the divergent continuity equation for air (in three dimensions) from (4.97) and the thermodynamic energy equation from (4.97), instead of the nondivergent continuity equation from (3.27). In the new case, the dispersion relationship is the same as (4.127), except the Rossby radius of deformation is $\lambda_R = \sqrt{gh_e}/f_0$, where the equivalent depth is the same as that for inertia gravity waves, given in (4.109). The resulting dispersion relationship for vertically and horizontally propagating Rossby waves in a compressible atmosphere is

$$\nu_\alpha = \bar{u}\tilde{k} - \frac{\beta\tilde{k}}{\tilde{k}^2 + \tilde{l}^2 + \frac{f_0^2}{N_{bv}^2}\left(\tilde{m}^2 + \frac{\nu_c^2}{c_s^2}\right)} \tag{4.129}$$

For Rossby waves that propagate horizontally in a compressible atmosphere, (4.127) and $\lambda_R = \sqrt{gh_e}/f_0$ are used. In this case, $h_e = c_s^2/g$, which is the equivalent depth given in (4.109) for inertia Lamb waves. Rossby waves can be initiated, not only by a perturbation to the geostrophic velocity in the south–north direction, but also by airflow over large topographical barriers, such as the Rocky or Himalayan mountain ranges. (4.111) shows that a change in the depth of the atmosphere forces $f$, $u$, and $v$ to change, initiating **forced topographic Rossby waves.**

## 4.4. SUMMARY

In this chapter, elements of the momentum equation, including the local acceleration, the Coriolis acceleration, the earth's centripetal acceleration, the gravitational force, the pressure-gradient force, the viscous force, and the turbulent flux divergence, were derived. For a coordinate system fixed on the surface of the earth, the Coriolis and the earth's centripetal acceleration are treated as apparent forces. The momentum equation was applied to estimate geostrophic and surface velocities and transformed to cylindrical coordinates to estimate the gradient wind velocity. Atmospheric waves were also discussed. Important waves include acoustic, Lamb, gravity, inertia Lamb, inertia gravity, and Rossby waves.

In the next chapter, the equations of atmospheric dynamics are converted to pressure, sigma-pressure, and sigma-altitude vertical coordinates.

## 4.5. PROBLEMS

**4.1.** Assume a spherical coordinate grid cell, centered at $\varphi = 60°$S, has dimension $d\lambda_e = 2.5°$ and $d\varphi = 2°$. Calculate $dx$ at the north and south boundaries of the cell, and calculate $dy$.

**4.2.** If $u = 10$ m s$^{-1}$, $v = 10$ m s$^{-1}$, $w = 0.01$ m s$^{-1}$, and $du/dt$ scales as $u/(\Delta x/u)$, estimate each term on the right side of (4.29) at $\varphi = 30°$N. Let $\Delta x = 5$ km, $\Delta y = 5$ km, and $\Delta z = 1$ km.

**4.3.** In Cartesian coordinates, show that $(\mathbf{v} \cdot \nabla)\mathbf{v} = \nabla(\mathbf{v}^2/2) + (\nabla \times \mathbf{v}) \times \mathbf{v}$.

**4.4.** Assume a grid cell has dimension $\Delta x = 5$ km, $\Delta y = 4$ km, and $\Delta z = 0.1$ km, and assume the west, east, south, north, lower, and upper boundary scalar velocities are $u_1 = +2$, $u_2 = +3$, $v_3 = +1$, $v_4 = -3$, $w_5 = +0.03$, and $w_6 = +0.04$ m s$^{-1}$, respectively. Estimate the magnitude of the divergence term of the local acceleration $(\mathbf{v} \cdot \nabla \mathbf{v})$.

**4.5.** If $u = 30$ m s$^{-1}$, $v = 10$ m s$^{-1}$, and $\varphi = 45°$N, find the Coriolis-force magnitude per unit mass.

**4.6.** Assume a grid cell has dimension $\Delta x = 500$ km and $\Delta y = 400$ km, centered at $\varphi = 30°$N latitude. Assume potential virtual temperature and air pressure at the west, east, south, and north boundaries are $\theta_{v,1} = 302$ K and $p_{a,1} = 520$ mb, $\theta_{v,2} = 304$ K and $p_{a,2} = 530$ mb, $\theta_{v,3} = 302$ K and $p_{a,3} = 500$ mb, and $\theta_{v,4} = 304$ K and $p_{a,4} = 540$ mb, respectively. Estimate the geostrophic scalar velocities and geostrophic wind speed.

**4.7.** In a region to the east of the center of an intense hurricane at $\varphi = 20°$N latitude, a west-east pressure gradient of 50 mb per 125 km and a gradient wind scalar velocity of $v = 70$ m s$^{-1}$ are observed. What is the distance between the center of the hurricane and the observation? Assume $T_v = 280$ K and $p_a = 930$ mb.

**4.8.** **(a)** Assume $u_1 = 1$ m s$^{-1}$ at $x_1 = 0$ m, $u_2 = 2$ m s$^{-1}$ at $x_2 = 2500$ m, $u_3 = 3$ m s$^{-1}$ at $x_3 = 5000$ m, and $K_{m,xx} = 2.5 \times 10^3$ m$^2$ s$^{-1}$. If density and $K_{m,xx}$ are constant, estimate the change in $u$, due to eddy diffusion alone, at the cell center after one hour.

**(b)** Assume $w_1 = 0.02$ m s$^{-1}$ at location $z_1 = 0$ m, $w_2 = 0.02$ m s$^{-1}$ at $z_2 =$ 50 m, $w_2 = 0.04$ m s$^{-1}$ at $z_3 = 1000$ m, and $K_{m,zz} = 50$ m$^2$ s$^{-1}$. If density and $K_{m,zz}$ are constant, estimate the change in $w$ at the cell center after 100 s, due to eddy diffusion alone.

**4.9.** Calculate the viscous acceleration force per unit mass when the air is dry, $T = 298$ K, $p_a = 995$ mb, and $u = 1.0$ m s$^{-1}$ at 10 cm above the ground and 0.7 m s$^{-1}$ at 5 cm above the ground.

**4.10.** Derive (4.129).

## 4.6. COMPUTER PROGRAMMING PRACTICE

**4.11.** **(a)** Assume $\Delta x = 5$ km for a grid cell, $v = 5$ m s$^{-1}$ at the center of the cell, $\varphi = 35°$N, $T = 285$ K, $\partial u/\partial y = 0$, $\partial u/\partial z = 0$, $p_{a,1} = 1010$ mb and $p_{a,2} = 1006$ mb at the west and east boundaries of the cell, and $u_1 = +3$ m s$^{-1}$ and $u_2 = +4$ m s$^{-1}$ at the west and east boundaries of the cell, respectively. Write a computer script to estimate the change in $u$ over time from (4.72), assuming that viscous and diffusive terms are zero. Assume $\partial u/\partial x$ in the $u\partial u/\partial x$ term is constant throughout the simulation, but $u$ is not. Set the time step to 1 s and run the program for one hour. Plot the results.

**(b)** Change $p_{a,2}$ to (i) 1009 mb and (ii) 1011 mb. Plot the results in both cases. For $p_{a,2} = 1009$ mb, change $u_1$ to $+3.5$ m s$^{-1}$ and plot the results. Comment on the sensitivity of the solution to changes in pressure and velocity.

# 5

# Vertical-Coordinate Conversions

THE equations of atmospheric dynamics are often adapted to vertical coordinates other than the altitude coordinate. Such coordinates include the pressure, sigma-pressure, sigma-altitude, and isentropic coordinates. The first three are discussed in this chapter. A description of the isentropic coordinate, in which layer tops and bottoms are defined as surfaces of constant potential temperature, can be found in Kasahara (1974). A disadvantage of the altitude and pressure coordinates is that both allow model layers to intersect ground topography. The sigma-pressure and sigma-altitude coordinates are terrain-following and do not allow model layers to intersect ground topography. A disadvantage of the pressure and sigma-pressure coordinates is that both assume the atmosphere is in hydrostatic balance. Such an assumption is reasonable for global and most mesoscale models but causes inaccuracy when the horizontal grid resolution falls below 3 km. For smaller horizontal resolutions, the atmosphere is often nonhydrostatic. For nonhydrostatic simulations, the altitude or sigma-altitude coordinate can be used.

## 5.1. ALTITUDE COORDINATE

Basic atmospheric equations were written in Chapters 3 and 4 in Cartesian and spherical horizontal coordinates and the altitude vertical coordinate. In the altitude ($z$) coordinate, layer tops and bottoms are defined as surfaces of constant altitude, and pressure varies in the $x$- and $y$-directions along these surfaces. Altitude is an independent variable, and pressure is a dependent variable. This coordinate causes problems when surface elevations are not uniform. In such cases, planes of constant altitude intercept topography, as shown in Fig. 5.1. Special boundary conditions are needed in this situation.

The altitude coordinate can be used to simulate hydrostatic or nonhydrostatic motions. In both cases, some variables are solved for prognostically while others are solved for diagnostically. A **prognostic equation** is one in which a time derivative is solved. The species continuity equation, the thermodynamic energy equation, and the horizontal momentum equations are prognostic equations for the species mixing ratio, potential virtual temperature, and horizontal scalar velocities, respectively. A **diagnostic equation** is one in which a time derivative is not solved. The equation of state and the hydrostatic equation are diagnostic equations for any of the variables in them.

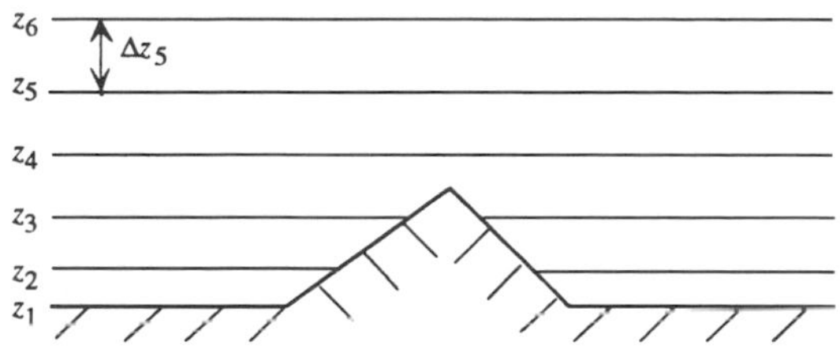


**Figure 5.1.** Heights of altitude coordinate surfaces.

### 5.1.1. Hydrostatic Model Equations

An advantage of the hydrostatic approximation is that it removes vertically propagating acoustic waves as a possible solution to the equations of atmospheric dynamics. Explicit numerical schemes that solve such equations require a time step smaller than the grid spacing divided by the speed of the fastest motion in the grid domain. When acoustic waves are present, they are the fastest motion in the domain, and time steps required to resolve these waves must be short. As shown in (4.99), vertically propagating acoustic waves arise as an analytical solution to the equations of atmospheric dynamics when the equations include the horizontal and vertical momentum equations, continuity equation for air, and thermodynamic energy equation.

To eliminate the need for a short time step, vertically propagating acoustic waves must be removed (filtered out) as a solution to the basic equations. This can be done by removing the total derivative ($dw/dt$) from the vertical momentum equation or the local derivative ($\partial \rho_a/\partial t$) from the continuity equation for air. Removing $dw/dt$ from the vertical momentum equation gives the **hydrostatic equation.** Removing $\partial \rho_a/\partial t$ from the air continuity equation gives the **anelastic continuity equation** (e.g., Ogura and Phillips 1962; Pielke 1984). The hydrostatic assumption filters out vertically propagating acoustic waves in a **hydrostatic model.** The anelastic continuity equation does so in a **nonhydrostatic model.**

Nonhydrostatic models also include those that solve the compressible continuity equation without the anelastic assumption. When an explicit time-differencing scheme is used to solve the resulting **elastic equation**, it requires a short time step to resolve acoustic-wave perturbations. A method to reduce computational time in such cases is to solve terms producing acoustic waves with a shorter time step than other terms (e.g., Klemp and Wilhelmson 1978; Chen 1991; Dudhia 1993). Alternatively, a semiimplicit scheme can be used to integrate terms responsible for waves implicitly (e.g., Tapp and White 1976; Tanguay et al. 1990; Golding 1992).

A hydrostatic model requires the solution to several equations. The potential virtual temperature is predicted from the thermodynamic energy equation given in (3.76), the water-vapor specific humidity is predicted from the species continuity equation given in (3.61), and the horizontal scalar velocities are predicted from the momentum equations given in (4.72), (4.73), (4.75), and/or (4.76). The pressure is extracted diagnostically from the hydrostatic equation given in (2.34), the temperature is calculated from the potential virtual temperature with (2.90), and the density is found as a function of pressure and temperature from the equation of state given in (2.30). Once density is known, $w$ can be extracted diagnostically from the continuity equation for air, given in (3.55).

### 5.1.2. Anelastic Nonhydrostatic Model Equations

In an anelastic nonhydrostatic model, the potential virtual temperature, specific humidity, horizontal scalar velocities, temperature, and density are found from the same equations as in a hydrostatic model, but the pressure and vertical scalar velocity are derived from new equations. The pressure is determined from an equation for nonhydrostatic pressure. This equation is derived by assuming that the pressure at a given location can be decomposed as

$$p_a = \hat{p}_a + p_a'' \tag{5.1}$$

where $\hat{p}_a$ is an average pressure, integrated over a large-scale, hydrostatic environment, and $p_a''$ ($\ll \hat{p}_a$) is the difference between the actual pressure and the large-scale pressure. Other variables are decomposed in a similar manner. In the large-scale environment, the atmosphere is assumed to be in hydrostatic balance. Thus,

$$\frac{1}{\hat{\rho}_a}\frac{\partial \hat{p}_a}{\partial z} = \hat{\alpha}_a\frac{\partial \hat{p}_a}{\partial z} = -g \tag{5.2}$$

At the point of interest, the air is not in hydrostatic balance, and the full vertical momentum equation must be considered. Decomposing gravitational and pressure-gradient terms in the vertical momentum equation and substituting (5.2) into the result gives

$$g + \frac{1}{\rho_a}\frac{\partial p_a}{\partial z} = g + \left(\hat{\alpha}_a + \alpha_a''\right)\frac{\partial}{\partial z}\left(\hat{p}_a + p_a''\right) \approx \hat{\alpha}_a\frac{\partial p_a''}{\partial z} - \frac{\alpha_a''}{\hat{\alpha}_a}g \tag{5.3}$$

where the product of perturbation terms has been removed because it is small. Substituting (5.3) into the vertical momentum equation of (4.74) and ignoring eddy diffusion for now gives

$$\frac{dw}{dt} = \frac{\partial w}{\partial t} + u\frac{\partial w}{\partial x} + v\frac{\partial w}{\partial y} + w\frac{\partial w}{\partial z} = -\hat{\alpha}_a\frac{\partial p_a''}{\partial z} + \frac{\alpha_a''}{\hat{\alpha}_a}g \tag{5.4}$$

Taking the divergence of the sum of (5.4), (4.72), and (4.73), where the last two equations are the horizontal momentum equations, results in

$$\frac{\partial}{\partial t}\nabla\cdot(\mathbf{v}\hat{\rho}_a) + \nabla\cdot[\hat{\rho}_a(\mathbf{v}\cdot\nabla)\mathbf{v}] = -\nabla\cdot(\hat{\rho}_a f\mathbf{k}\times\mathbf{v}) - \nabla_z^2\hat{p}_a - \nabla^2 p_a'' + g\frac{\partial}{\partial z}\left(\frac{\alpha_a''}{\hat{\alpha}_a^2}\right) \tag{5.5}$$

where $\nabla$ and $\nabla_z$ are total and horizontal gradient operators from (3.6) and (4.80), respectively. (5.5) can be simplified by differentiating $p_a\alpha_a = R'T_v$ to obtain $d\alpha_a/\alpha_a = dT_v/T_v - dp_a/p_a$, substituting $T_v = \theta_v(p_a/1000\text{ mb})^\kappa$ from (2.90) into the result, and replacing differential and nondifferential terms with perturbation and large-scale terms, respectively. These steps yield

$$\frac{\alpha_a''}{\hat{\alpha}_a} \approx \frac{\theta_v''}{\hat{\theta}_v} - \frac{c_{v,d}}{c_{p,d}} \frac{p_a''}{\hat{p}_a} \tag{5.6}$$

Another simplification to (5.5) is obtained by removing the local derivative in the continuity equation for air and assuming density in the equation is large-scale density. The result is the **anelastic approximation to the continuity equation,**

$$\nabla \cdot (\mathbf{v}\hat{\rho}_a) = 0 \tag{5.7}$$

Substituting (5.6) and (5.7) into (5.5), adding the eddy diffusion term, and solving for perturbation pressure gives the **diagnostic equation for nonhydrostatic pressure** as

$$\nabla^2 p_a'' - g\frac{c_{v,d}}{c_{p,d}}\frac{\partial}{\partial z}\left(\hat{\rho}_a \frac{p_a''}{\hat{p}_a}\right) = -\nabla \cdot [\hat{\rho}_a(\mathbf{v} \cdot \nabla)\mathbf{v}] - \nabla \cdot [\hat{\rho}_a f \mathbf{k} \times \mathbf{v}]$$
$$- \nabla_z^2 \hat{p}_a + g\frac{\partial}{\partial z}\left(\hat{\rho}_a \frac{\theta_p''}{\hat{\theta}_p}\right) + \nabla \cdot (\nabla \cdot \hat{\rho}_a \mathbf{K}_m \nabla)\mathbf{v} \tag{5.8}$$

(e.g., Pielke 1984). When the vertical scale of circulation is much less than the scale depth of the atmosphere (shallow circulations), the second term on the left side of (5.8) can be removed, and the resulting equation is a **Poisson partial differential equation.** For deep circulations, which many atmospheric circulations are, the full equation is retained. When pressure is calculated from (5.8), vertical velocities can be extracted diagnostically from (5.7), since $u$ and $v$ are known from the horizontal momentum equations and $\hat{\rho}_a$ is available from (5.2). In this type of nonhydrostatic model, the vertical momentum equation in not solved directly. A disadvantage of (5.8) is that its solution is computationally intensive. Equations for an elastic nonhydrostatic model are derived in Section 5.4.

## 5.2. PRESSURE COORDINATE

In the **pressure ($p$ or isobaric) coordinate**, layer tops and bottoms are defined as surfaces of constant pressure. Since altitude is a function of pressure in the $x$- and $y$-directions, pressure is the independent variable and altitude is a dependent variable. The pressure coordinate does not erase problems associated with surface topography. Figure 5.2 shows how surfaces of constant pressure intercept topography in the pressure coordinate. Equations in the pressure coordinate are derived from the hydrostatic assumption. This assumption removes vertically propagating acoustic waves as a solution to the equations of atmospheric dynamics but reduces the accuracy of the solution when horizontal grid cells are less than 3 km in width. In the pressure coordinate, the continuity equation for air requires neither a density nor a time-derivative term. In this section, hydrostatic model equations for the pressure coordinate are derived.

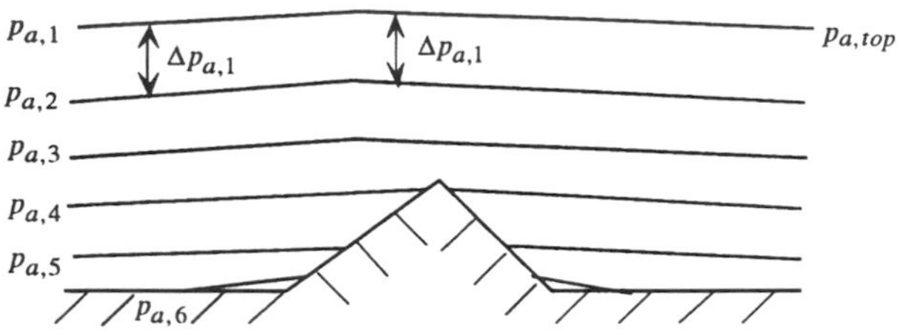


**Figure 5.2.** Heights of pressure coordinate surfaces. Each line is a surface of constant pressure. Pressure thicknesses of each layer are constant throughout the layer.

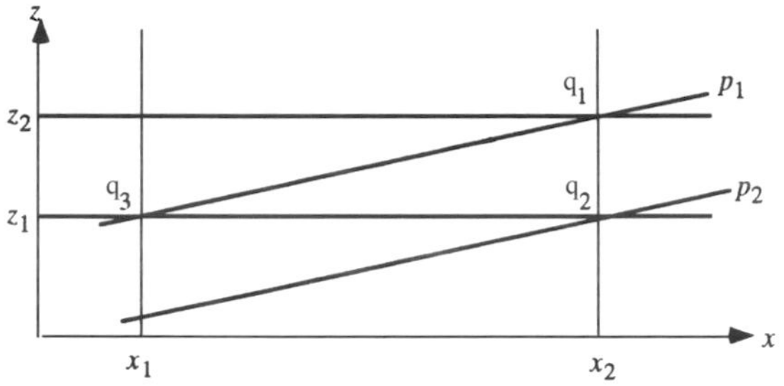


**Figure 5.3.** Intersection of pressure ($p$) with altitude ($z$) surfaces. The moist-air mass mixing ratio (q) varies along each surface of constant pressure.

### 5.2.1. Gradient Conversion from the Altitude to Pressure Coordinate

The gradient conversion from altitude to pressure coordinates can be obtained diagramatically. Figure 5.3 shows two lines of constant pressure (**isobars**) intersecting two lines of constant altitude in the $x$–$z$ plane, where $x$ is a Cartesian coordinate distance. Suppose the moist-air mass mixing ratio q varies along each surface of constant pressure, as shown in the figure. In this case, q varies with distance as

$$\frac{q_2 - q_3}{x_2 - x_1} = \frac{q_1 - q_3}{x_2 - x_1} + \left(\frac{p_2 - p_1}{x_2 - x_1}\right)\left(\frac{q_1 - q_2}{p_1 - p_2}\right) \tag{5.9}$$

This relationship is an exact equivalence.

As $x_2 - x_1 \rightarrow 0$ and $p_1 - p_2 \rightarrow 0$, the differences in (5.9) can be approximated by differentials. From Fig. 5.3, the approximations for each term are

$$\left(\frac{\partial q}{\partial x}\right)_z = \frac{q_2 - q_3}{x_2 - x_1} \qquad \left(\frac{\partial q}{\partial x}\right)_p = \frac{q_1 - q_3}{x_2 - x_1}$$
$$\left(\frac{\partial p_a}{\partial x}\right)_z = \frac{p_2 - p_1}{x_2 - x_1} \qquad \left(\frac{\partial q}{\partial p_a}\right)_x = \frac{q_1 - q_2}{p_1 - p_2} \tag{5.10}$$

Substituting these terms into (5.9) gives

$$\left(\frac{\partial q}{\partial x}\right)_z = \left(\frac{\partial q}{\partial x}\right)_p + \left(\frac{\partial p_a}{\partial x}\right)_z \left(\frac{\partial q}{\partial p_a}\right)_x \tag{5.11}$$

which generalizes for any variable as

$$\left(\frac{\partial}{\partial x}\right)_z = \left(\frac{\partial}{\partial x}\right)_p + \left(\frac{\partial p_a}{\partial x}\right)_z \left(\frac{\partial}{\partial p_a}\right)_x \tag{5.12}$$

A similar equation can be written for the $y$-direction. Combining the $x$- and $y$-equations in vector form yields the **horizontal gradient conversion from Cartesian–altitude to Cartesian–pressure coordinates** as

$$\nabla_z = \nabla_p + \nabla_z(p_a)\frac{\partial}{\partial p_a} \tag{5.13}$$

where $\nabla_z$ is the horizontal gradient operator in Cartesian–altitude coordinates given in (4.80), and

$$\nabla_p = \mathbf{i}\left(\frac{\partial}{\partial x}\right)_p + \mathbf{j}\left(\frac{\partial}{\partial y}\right)_p \tag{5.14}$$

is the **horizontal gradient operator in Cartesian–pressure coordinates.** The gradient operator in Cartesian–pressure coordinates is used to take partial derivatives of a variable along a surface of constant pressure. The vertical derivative with respect to pressure in (5.13) is taken at a fixed location in the $x$–$y$ plane.

In (5.12), any dimension can be substituted for $x$. Substituting time $t$ for $x$ gives the **time-derivative conversion between the altitude and the pressure coordinate** as

$$\left(\frac{\partial}{\partial t}\right)_z = \left(\frac{\partial}{\partial t}\right)_p + \left(\frac{\partial p_a}{\partial t}\right)_z \left(\frac{\partial}{\partial p_a}\right)_t \tag{5.15}$$

(5.13) may be used to obtain the relationship between the variation of geopotential along a surface of constant pressure and the variation of pressure along a surface of constant altitude. From (4.48), the geopotential was defined as $\Phi = gz$. The geopotential is constant along surfaces of constant altitude ($\nabla_z\Phi = 0$). Substituting $\Phi = gz$, $\partial p_a/\partial z = -\rho_a g$, and $\partial\Phi/\partial z = g$ into (5.13) and rearranging gives

$$\nabla_z(p_a) = -\frac{\partial p_a}{\partial \Phi}\nabla_p\Phi = -\frac{\partial p_a}{g\,\partial z}\nabla_p\Phi = \rho_a\nabla_p\Phi \tag{5.16}$$

which relates changes in geopotential along surfaces of constant pressure to changes in pressure along surfaces of constant altitude. This equation expands to

$$\left(\frac{\partial p_a}{\partial x}\right)_z = \rho_a\left(\frac{\partial \Phi}{\partial x}\right)_p \qquad \left(\frac{\partial p_a}{\partial y}\right)_z = \rho_a\left(\frac{\partial \Phi}{\partial y}\right)_p \tag{5.17}$$

### 5.2.2. Continuity Equation for Air in the Pressure Coordinate

(3.20) gave the Cartesian–altitude coordinate continuity equation for air. Dividing (3.20) into horizontal and vertical components gives

$$\left(\frac{\partial \rho_a}{\partial t}\right)_z = -\rho_a\left(\nabla_z \cdot \mathbf{v}_h + \frac{\partial w}{\partial z}\right) - (\mathbf{v}_h \cdot \nabla_z)\rho_a - w\frac{\partial \rho_a}{\partial z} \tag{5.18}$$

where the subscript $z$ indicates that the value is taken along a surface of constant altitude, and $\mathbf{v}_h = \mathbf{i}u + \mathbf{j}v$. Applying (5.13) to $\mathbf{v}_h$ gives

$$\nabla_z \cdot \mathbf{v}_h = \nabla_p \cdot \mathbf{v}_h - \nabla_z(p_a) \cdot \frac{\partial \mathbf{v}_h}{\partial p_a} \tag{5.19}$$

Substituting (5.19) and $dz = -dp_a/\rho_a g$ into (5.18) gives

$$\left(\frac{\partial \rho_a}{\partial t}\right)_z = -\rho_a\left(\nabla_p \cdot \mathbf{v}_h + \nabla_z(p_a) \cdot \frac{\partial \mathbf{v}_h}{\partial p_a}\right) - (\mathbf{v}_h \cdot \nabla_z)\rho_a + \rho_a g\frac{\partial (w\rho_a)}{\partial p_a} \tag{5.20}$$

The **vertical scalar velocity in the pressure coordinate** (mb s$^{-1}$) is defined as

$$\mathrm{w}_p = \frac{dp_a}{dt} = \left(\frac{\partial p_a}{\partial t}\right)_z + (\mathbf{v} \cdot \nabla)p_a = \left(\frac{\partial p_a}{\partial t}\right)_z + (\mathbf{v}_h \cdot \nabla_z)p_a + w\frac{\partial p_a}{\partial z} \tag{5.21}$$

Substituting $\partial p_a/\partial z = -\rho_a g$ into (5.21) yields

$$\mathrm{w}_p = -\left(\rho_a g\frac{\partial z}{\partial t}\right)_z + (\mathbf{v}_h \cdot \nabla_z)p_a - w\rho_a g \tag{5.22}$$

If horizontal and temporal variations in pressure are ignored, the vertical scalar velocity in the pressure coordinate simplifies to $\mathrm{w}_p = -w\rho_a g$. Thus, a positive vertical velocity in the pressure coordinate corresponds to downward motion, or a negative vertical velocity in the altitude coordinate. Horizontal and temporal variations in pressure cannot be excluded when the continuity equation for air is derived. Taking the partial derivative of $\mathrm{w}_p$ from (5.22) with respect to altitude gives

$$\frac{\partial \mathrm{w}_p}{\partial z} = -g\left(\frac{\partial \rho_a}{\partial t}\right)_z + \nabla_z(p_a) \cdot \frac{\partial \mathbf{v}_h}{\partial z} + (\mathbf{v}_h \cdot \nabla_z)\frac{\partial p_a}{\partial z} - g\frac{\partial (w\rho_a)}{\partial z} \tag{5.23}$$

Substituting the hydrostatic equation throughout yields

$$\rho_a\frac{\partial \mathrm{w}_p}{\partial p_a} = \left(\frac{\partial \rho_a}{\partial t}\right)_z + \rho_a\nabla_z(p_a) \cdot \frac{\partial \mathbf{v}_h}{\partial p_a} + (\mathbf{v}_h \cdot \nabla_z)\rho_a - \rho_a g\frac{\partial (w\rho_a)}{\partial p_a} \tag{5.24}$$

Adding (5.20) to (5.24) and compressing gives the **continuity equation for air in Cartesian–pressure coordinates** as

$$\nabla_p \cdot \mathbf{v}_h + \frac{\partial \mathrm{w}_p}{\partial p_a} = 0 \tag{5.25}$$

which expands to

$$\left(\frac{\partial u}{\partial x}+\frac{\partial v}{\partial y}\right)_p+\frac{\partial \mathrm{w}_p}{\partial p_a}=0 \tag{5.26}$$

(5.25) shows that, in the pressure coordinate, the continuity equation for air does not depend on air density, nor does it require a time-derivative term.

**Example 5.1**

Assume a grid cell in the pressure coordinate has dimension $\Delta x = 5$ km, $\Delta y =$ 5 km, and $\Delta p_a = -10$ mb, and that west, east, south, north, and lower boundary scalar velocities are $u_1 = -3$, $u_2 = -1$ m s$^{-1}$, $v_3 = +2$, $v_4 = -2$ m s$^{-1}$, and $\mathrm{w}_{p,5} = +0.02$ mb s$^{-1}$, respectively. Use the continuity equation in the pressure coordinate to estimate the pressure-coordinate vertical scalar velocity at the cell's top.

SOLUTION

Applying the velocities and incremental distances to (5.26) gives

$$\frac{(-1+3)\ \mathrm{m\,s^{-1}}}{5000\ \mathrm{m}}+\frac{(-2-2)\ \mathrm{m\,s^{-1}}}{5000\ \mathrm{m}}+\frac{(\mathrm{w}_{p,6}-0.02)\ \mathrm{mb\,s^{-1}}}{-10\ \mathrm{mb}}=0$$

which has solution $\mathrm{w}_{p,6} = +0.016$ mb s$^{-1}$.

### 5.2.3. Total Derivative in the Pressure Coordinate

The total derivative in the pressure coordinate is used to derive the species continuity equation, the thermodynamic energy equation, and the momentum equations in the pressure coordinate. The total derivative in Cartesian–altitude coordinates was

$$\frac{\mathrm{d}}{\mathrm{d}t}=\left(\frac{\partial}{\partial t}\right)_z+(\mathbf{v}_h\cdot\nabla_z)+w\frac{\partial}{\partial z} \tag{5.27}$$

Substituting time and horizontal-gradient conversions from (5.15) and (5.13), respectively, into (5.27) gives

$$\frac{\mathrm{d}}{\mathrm{d}t}=\left(\frac{\partial}{\partial t}\right)_p+\left(\frac{\partial p_a}{\partial t}\right)_z\frac{\partial}{\partial p_a}+(\mathbf{v}_h\cdot\nabla_p)+[(\mathbf{v}_h\cdot\nabla_z)p_a]\frac{\partial}{\partial p_a}+w\frac{\partial}{\partial z} \tag{5.28}$$

From (5.21), the vertical velocity in the altitude coordinate is related to vertical velocity in the pressure coordinate by

$$w=\frac{\left(\frac{\partial p_a}{\partial t}\right)_z+(\mathbf{v}_h\cdot\nabla_z)p_a-\mathrm{w}_p}{\rho_a g} \tag{5.29}$$

Substituting this equation and the hydrostatic equation into (5.28) and simplifying gives the **total derivative in Cartesian–pressure coordinates** as

$$\frac{\mathrm{d}}{\mathrm{d}t}=\left(\frac{\partial}{\partial t}\right)_p+(\mathbf{v}_h\cdot\nabla_p)+\mathrm{w}_p\frac{\partial}{\partial p_a} \tag{5.30}$$

### 5.2.4. Species Continuity Equation in the Pressure Coordinate

Applying the total derivative in Cartesian–pressure coordinates to (3.54) gives the **species continuity equation in Cartesian–pressure coordinates** as

$$\frac{dq}{dt} = \left(\frac{\partial q}{\partial t}\right)_p + (\mathbf{v}_h \cdot \nabla_p)q + w_p\frac{\partial q}{\partial p_a} = \frac{(\nabla \cdot \rho_a \mathbf{K}_h \nabla)q}{\rho_a} + \sum_{n=1}^{N_{e,t}} R_n \tag{5.31}$$

If the concentration of a species is given as molecules per cubic centimeter of air ($N$), particles per cubic centimeter of air ($n_i$), or cubic centimeters of particle component per cubic centimeter of air ($v_{q,i}$), it can be converted to the moist-air mass mixing ratio for use in (5.31) with

$$q = \frac{Nm}{\rho_a A} \qquad q = \frac{n_i \upsilon_i \rho_p}{\rho_a} \qquad q = \frac{v_{q,i}\rho_p}{\rho_a} \tag{5.32}$$

respectively, where $\upsilon_i$ is the volume of one particle in size bin $i$ (cm$^3$), and $\rho_p$ is the particle mass density (g cm$^{-3}$).

Since $R_n$ does not include spatial derivatives, it does not need to be transformed to the pressure coordinate. The eddy diffusion term in (5.31) includes spatial derivatives. Its transformation is performed by breaking the term into directional components, applying the gradient conversion from (5.13), and applying $dz = -dp_a/\rho_a g$. The result is

$$\begin{aligned}(\nabla \cdot \rho_a \mathbf{K}_h \nabla)q &= \left(\frac{\partial}{\partial x}\right)_z \left[\rho_a K_{h,xx}\left(\frac{\partial q}{\partial x}\right)_z\right] + \left(\frac{\partial}{\partial y}\right)_z \left[\rho_a K_{h,yy}\left(\frac{\partial q}{\partial y}\right)_z\right] \\ &+ \frac{\partial}{\partial z}\left(\rho_a K_{h,zz}\frac{\partial q}{\partial z}\right) = \left[\left(\frac{\partial}{\partial x}\right)_p + \left(\frac{\partial p_a}{\partial x}\right)_z \frac{\partial}{\partial p_a}\right]\left\{\rho_a K_{h,xx}\left[\left(\frac{\partial q}{\partial x}\right)_p\right.\right. \\ &\left.\left.+ \left(\frac{\partial p_a}{\partial x}\right)_z \frac{\partial q}{\partial p_a}\right]\right\} + \left[\left(\frac{\partial}{\partial y}\right)_p + \left(\frac{\partial p_a}{\partial y}\right)_z \frac{\partial}{\partial p_a}\right]\left\{\rho_a K_{h,yy}\left[\left(\frac{\partial q}{\partial y}\right)_p\right.\right. \\ &\left.\left.+ \left(\frac{\partial p_a}{\partial y}\right)_z \frac{\partial q}{\partial p_a}\right]\right\} + \rho_a g^2 \frac{\partial}{\partial p_a}\left(\rho_a^2 K_{h,zz}\frac{\partial q}{\partial p_a}\right)\end{aligned} \tag{5.33}$$

### 5.2.5. Thermodynamic Energy Equation in the Pressure Coordinate

Applying the total derivative in Cartesian–pressure coordinates to (3.76) gives the **thermodynamic energy equation in Cartesian–pressure coordinate** as

$$\left(\frac{\partial \theta_v}{\partial t}\right)_p + (\mathbf{v}_h \cdot \nabla_p)\theta_v + w_p\frac{\partial \theta_v}{\partial p_a} = \frac{(\nabla \cdot \rho_a \mathbf{K}_h \nabla)\theta_v}{\rho_a} + \frac{\theta_v}{c_{p,d} T_v}\sum_{n=1}^{N_{e,h}} \frac{dQ_n}{dt} \tag{5.34}$$

The eddy diffusion terms here are the same as those used in the species continuity equation, except that $\theta_v$ is used here instead of q.

### 5.2.6. Horizontal Momentum Equations in the Pressure Coordinate

Applying the total derivative in Cartesian–pressure coordinates to velocity, neglecting viscous terms, and substituting $\nabla_z(p_a) = \rho_a \nabla_p \Phi$ from (5.15) into (4.70) gives the **horizontal momentum equation in Cartesian–pressure coordinates** as

$$\left(\frac{\partial \mathbf{v}_h}{\partial t}\right)_p + (\mathbf{v}_h \cdot \nabla_p)\mathbf{v}_h + \mathbf{w}_p \frac{\partial \mathbf{v}_h}{\partial p_a} = -f\mathbf{k} \times \mathbf{v}_h - \nabla_p \Phi + \frac{(\nabla \cdot \rho_a \mathbf{K}_m \nabla)\mathbf{v}_h}{\rho_a} \tag{5.35}$$

The eddy diffusion term expands to $(\nabla \cdot \rho_a \mathbf{K}_m \nabla)\mathbf{v}_h = \mathbf{i}(\nabla \cdot \rho_a \mathbf{K}_m \nabla)u + \mathbf{j}(\nabla \cdot \rho_a \mathbf{K}_h \nabla)v$. Applying the gradient conversion from (5.13) to the $u$-term gives

$$\begin{aligned}(\nabla \cdot \rho_a \mathbf{K}_h \nabla)u &= \left[\left(\frac{\partial}{\partial x}\right)_p + \left(\frac{\partial p_a}{\partial x}\right)_z \frac{\partial}{\partial p_a}\right]\left\{\rho_a K_{m,xx}\left[\left(\frac{\partial u}{\partial x}\right)_p \right.\right. \\ &\left.\left. + \left(\frac{\partial p_a}{\partial x}\right)_z \frac{\partial u}{\partial p_a}\right]\right\} + \left[\left(\frac{\partial}{\partial y}\right)_p + \left(\frac{\partial p_a}{\partial y}\right)_z \frac{\partial}{\partial p_a}\right]\left\{\rho_a K_{m,yx}\left[\left(\frac{\partial u}{\partial y}\right)_p \right.\right. \\ &\left.\left. + \left(\frac{\partial p_a}{\partial y}\right)_z \frac{\partial u}{\partial p_a}\right]\right\} + \rho_a g^2 \frac{\partial}{\partial p_a}\left(\rho_a^2 K_{m,zx} \frac{\partial u}{\partial p_a}\right)\end{aligned} \tag{5.36}$$

The equation for the $v$-direction is similar. The three terms on the right side of (5.36) represent accelerations in the $x$-direction due to west–east, south–north, and vertical turbulent transport, respectively, of west–east momentum.

### 5.2.7. Vertical Momentum Equation in the Pressure Coordinate

The vertical momentum equation in the pressure coordinate is the hydrostatic equation, since the conversion from the altitude to the pressure coordinate requires the hydrostatic assumption. Over large horizontal scales (e.g., >3-km resolution), the hydrostatic assumption is reasonable. Over smaller scales and when convective clouds need to be resolved, vertical accelerations are important, and the hydrostatic assumption is not good. In such cases, the altitude or sigma-altitude coordinate is generally used, and a vertical momentum equation that includes a local acceleration term is solved.

When the hydrostatic assumption is used, the vertical momentum equation in the altitude coordinate is $\partial p_a/\partial z = -\rho_a g$. Substituting $g = \partial\Phi/\partial z$ from (4.48), $p_a = \rho_a R' T_v$ from (2.30), and $T_v = \theta_v P$ from (2.93) into the hydrostatic equation gives the **hydrostatic equation in the pressure coordinate** as

$$\frac{\partial \Phi}{\partial p_a} = -\frac{R' T_v}{p_a} = -\frac{R' \theta_v P}{p_a} = -\frac{R' \theta_v}{p_a}\left(\frac{p_a}{1000\,\text{mb}}\right)^{\kappa} \tag{5.37}$$

Substituting $\kappa = R'/c_{p,d}$ into (5.37) and rearranging gives another form of the equation as

$$\mathrm{d}\Phi = -c_{p,d}\theta_v \,\mathrm{d}\left[\left(\frac{p_a}{1000}\right)^{\kappa}\right] = -c_{p,d}\theta_v \,\mathrm{d}P \qquad (5.38)$$

A finite-difference discretization of this equation is given in Chapter 7.

### 5.2.8. Geostrophic Wind in the Pressure Coordinate

Substituting (5.17) into (4.78) gives geostrophic scalar velocities in Cartesian–pressure coordinates as

$$v_g = \frac{1}{f}\left(\frac{\partial\Phi}{\partial x}\right)_p \qquad u_g = -\frac{1}{f}\left(\frac{\partial\Phi}{\partial y}\right)_p \qquad (5.39)$$

The vector form of (5.39) is

$$\mathbf{v}_g = \mathbf{i}u_g + \mathbf{j}v_g = -\mathbf{i}\frac{1}{f}\left(\frac{\partial\Phi}{\partial y}\right)_p + \mathbf{j}\frac{1}{f}\left(\frac{\partial\Phi}{\partial x}\right)_p = \frac{1}{f}(\mathbf{k}\times\nabla_p\Phi) \qquad (5.40)$$

which indicates that the geostrophic wind flows parallel to lines of constant geopotential. Lines of constant geopotential (**contour lines**) on a constant-pressure surface are analogous to lines of constant-pressure (isobars) on a constant-altitude surface. Since the geostrophic wind flows parallel to isobars, it also flows parallel to contour lines. Figure 5.4 shows two west–east contour lines at 5.5 and 5.6 km, respectively, on a surface of constant pressure. The geostrophic wind in this case flows from the west.

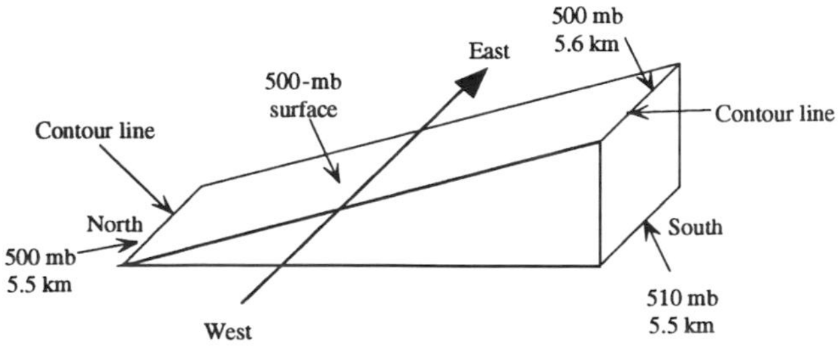


**Figure 5.4.** The geostrophic wind (arrow) flows parallel to contour lines on a constant-pressure surface. In this figure, contour lines at 5.5 and 5.6 km are shown on a 500-mb surface. The resulting geostrophic wind originates from the west.

## 5.3. SIGMA-PRESSURE COORDINATE

A vertical coordinate in which surfaces of the coordinate variable do not intersect ground topography is the **sigma-pressure** ($\sigma$-$p$) **coordinate** (Phillips 1957). In this coordinate, layer tops and bottoms are defined as surfaces of constant $\sigma$, where $\sigma$ equals the difference between the layer bottom and model top pressures divided by the difference between the ground surface and model top pressures. Since the hydrostatic assumption is used to derive equations in the sigma-pressure coordinate, vertically propagating acoustic waves are filtered out as a solution to these equations. Nonhydrostatic flows are usually not simulated in the sigma-pressure coordinate. In this section, hydrostatic model equations for the sigma-pressure coordinate are derived.

### 5.3.1. Definitions

In the sigma-pressure coordinate system, layer tops and bottoms are defined as surfaces of constant $\sigma$, given by

$$\sigma = \frac{p_a - p_{a,top}}{p_{a,surf} - p_{a,top}} = \frac{p_a - p_{a,top}}{\pi_a} \tag{5.41}$$

where $p_a$ is the air pressure at the altitude of interest, $p_{a,surf}$ is the model surface pressure, $p_{a,top}$ is the model top pressure, and $\pi_a = p_{a,surf} - p_{a,top}$ is the pressure difference between the model surface and top (**$\pi$-value**). From (5.41), the pressure at a $\sigma$-level is

$$p_a = p_{a,top} + \sigma\pi_a \tag{5.42}$$

$\sigma$ is a fraction $\leq 1$. At the model top, $\sigma = 0$. At the model surface, $\sigma = 1$. Layer boundaries correspond to $\sigma$-values that are constant in time. The model top pressure, $p_{a,top}$, is constant in space and time along a layer boundary corresponding to $\sigma = 0$. The model surface pressure, $p_{a,surf}$, varies in space and time along the surface corresponding to $\sigma = 1$. Since $p_{a,surf}$, varies, $p_a$, and $\pi_a$ must also vary in space and time. Figure 5.5 shows heights of constant-$\sigma$ surfaces in the sigma-pressure coordinate.

### 5.3.2. Gradient Conversion from the Altitude to the Sigma-Pressure Coordinate

The conversion between pressure in the altitude coordinate and pressure in the sigma-pressure coordinate can be derived from Fig. 5.6. The figure shows the intersection of $\sigma$ with $z$-surfaces and the values of pressure at each intersection point on an $x$–$z$ plane. From the figure, the change in pressure per unit distance is

$$\frac{p_2 - p_3}{x_2 - x_1} = \frac{p_1 - p_3}{x_2 - x_1} + \left(\frac{\sigma_2 - \sigma_1}{x_2 - x_1}\right)\left(\frac{p_1 - p_2}{\sigma_1 - \sigma_2}\right) \tag{5.43}$$

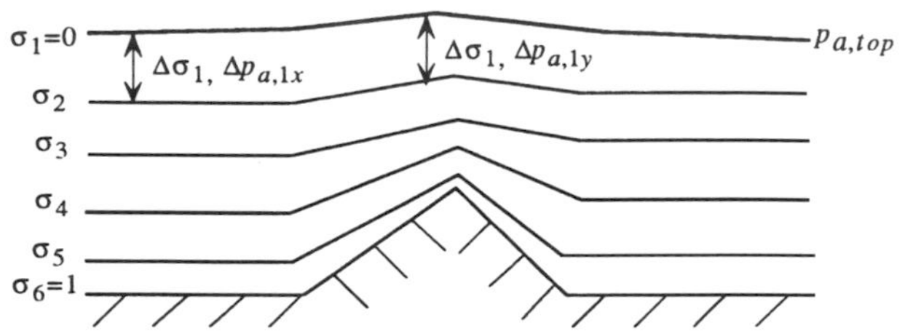


**Figure 5.5.** Heights of sigma-pressure coordinate surfaces. Each layer has the same σ-thickness but a different pressure thickness. Subscripts $x$ and $y$ denote two horizontal locations. Surface pressures vary horizontally.

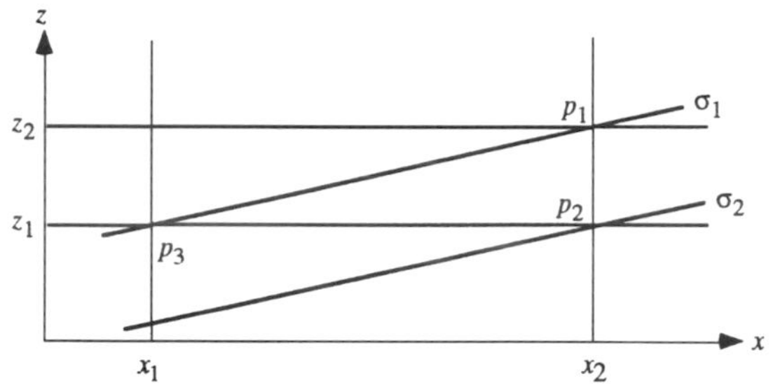


**Figure 5.6.** Intersection of σ-$p$ with $z$ surfaces and values of pressure at each intersection.

which is an exact equivalence. Substituting

$$\left(\frac{\partial p_a}{\partial x}\right)_z = \frac{p_2 - p_3}{x_2 - x_1} \qquad \left(\frac{\partial p_a}{\partial x}\right)_\sigma = \frac{p_1 - p_3}{x_2 - x_1}$$
$$\left(\frac{\partial \sigma}{\partial x}\right)_z = \frac{\sigma_2 - \sigma_1}{x_2 - x_1} \qquad \left(\frac{\partial p_a}{\partial \sigma}\right)_x = \frac{p_1 - p_2}{\sigma_1 - \sigma_2} \tag{5.44}$$

into (5.43) gives

$$\left(\frac{\partial p_a}{\partial x}\right)_z = \left(\frac{\partial p_a}{\partial x}\right)_\sigma + \left(\frac{\partial \sigma}{\partial x}\right)_z \left(\frac{\partial p_a}{\partial \sigma}\right)_x \tag{5.45}$$

A similar expression exists for the $y$-direction. The vector sum of the two equations is

$$\nabla_z(p_a) = \nabla_\sigma(p_a) + (\nabla_z\sigma)\frac{\partial p_a}{\partial \sigma} \tag{5.46}$$

where

$$\nabla_\sigma = \left(\mathbf{i}\frac{\partial}{\partial x}\right)_\sigma + \left(\mathbf{j}\frac{\partial}{\partial y}\right)_\sigma \tag{5.47}$$

is the **horizontal gradient operator in Cartesian–sigma-pressure coordinates.** (5.46) generalizes for any variable to the **gradient conversion from Cartesian–altitude to**

**Cartesian–sigma-pressure coordinates**, given as

$$\nabla_z = \nabla_\sigma + \nabla_z(\sigma)\frac{\partial}{\partial \sigma} \tag{5.48}$$

The gradient of $\sigma$, defined in (5.41), along a surface of constant altitude is

$$\nabla_z \sigma = (p_a - p_{a,top})\nabla_z\left(\frac{1}{\pi_a}\right) + \frac{\nabla_z(p_a)}{\pi_a} = -\frac{\sigma}{\pi_a}\nabla_z(\pi_a) + \frac{\nabla_z(p_a)}{\pi_a} \tag{5.49}$$

Substituting (5.49) into (5.48) yields another form of the gradient conversion,

$$\nabla_z = \nabla_\sigma - \left[\frac{\sigma}{\pi_a}\nabla_z(\pi_a) - \frac{\nabla_z(p_a)}{\pi_a}\right]\frac{\partial}{\partial \sigma} \tag{5.50}$$

### 5.3.3. Gradient Conversion from the Pressure to the Sigma-Pressure Coordinate

The gradient conversion from the pressure to the sigma-pressure coordinate is derived from Fig. 5.7. The figure shows intersections of pressure, altitude, and $\sigma$ surfaces in an $x$–$z$ plane. From the figure, the change in moist-air mass mixing ratio over distance is

$$\frac{q_1 - q_3}{x_2 - x_1} = \frac{q_2 - q_3}{x_2 - x_1} + \left(\frac{\sigma_1 - \sigma_2}{x_2 - x_1}\right)\left(\frac{q_1 - q_2}{\sigma_1 - \sigma_2}\right) \tag{5.51}$$

which is an exact equivalence. Making substitutions similar to those in (5.44) gives

$$\left(\frac{\partial q}{\partial x}\right)_p = \left(\frac{\partial q}{\partial x}\right)_\sigma + \left(\frac{\partial \sigma}{\partial x}\right)_p\left(\frac{\partial q}{\partial \sigma}\right)_x \tag{5.52}$$

A similar equation is written for the $y$-direction. Writing the $x$- and $y$-equations in gradient operator form and generalizing for any variable gives the **gradient**

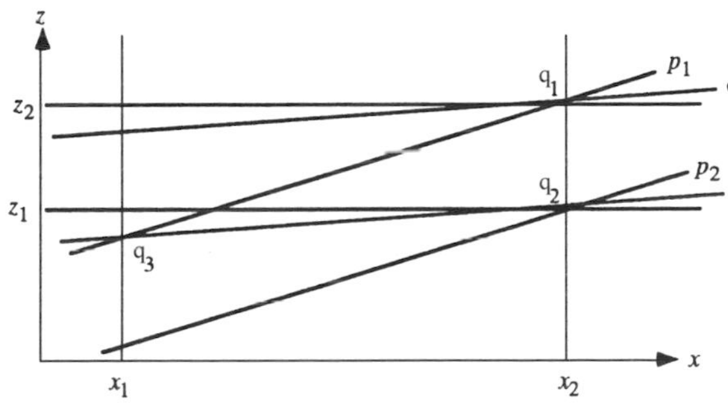


**Figure 5.7.** Intersection of $p$, $z$, and $\sigma$ surfaces in an $x$–$z$ plane. Moist-air mass mixing ratios (q) on constant-pressure surfaces are also shown.

**conversion from Cartesian–pressure to Cartesian–sigma-pressure coordinates** as

$$\nabla_p = \nabla_\sigma + \nabla_p(\sigma)\frac{\partial}{\partial\sigma} \tag{5.53}$$

Taking the gradient of $\sigma$ along a surface of constant pressure and noting that $\nabla_p(p_a) = 0$ and $\nabla_p(p_{top}) = 0$ yields

$$\nabla_p(\sigma) = (p_a - p_{top})\nabla_p\left(\frac{1}{\pi_a}\right) + \frac{\nabla_p(p_a - p_{top})}{\pi_a} = -\frac{\sigma}{\pi_a}\nabla_p(\pi_a) \tag{5.54}$$

Since horizontal gradients of $\pi_a$ are independent of the vertical coordinate system, $\nabla_p(\pi_a) = \nabla_\sigma(\pi_a) = \nabla_z(\pi_a)$. Substituting this expression and (5.54) into (5.53) gives the **gradient conversion from Cartesian–pressure to Cartesian–sigma-pressure coordinates** as

$$\nabla_p = \nabla_\sigma - \frac{\sigma}{\pi_a}\nabla_\sigma(\pi_a)\frac{\partial}{\partial\sigma} \tag{5.55}$$

### 5.3.4. Continuity Equation for Air in the Sigma-Pressure Coordinate

#### 5.3.4.1. *Derivation*

The continuity equation for air in the sigma-pressure coordinate is derived from the continuity equation for air in the pressure coordinate, the hydrostatic equation, and the total derivative in the sigma-pressure coordinate. The partial derivative of pressure with respect to $\sigma$ is $\partial p_a/\partial\sigma = \pi_a$. Substituting this expression and the gradient conversion from (5.55) into the continuity equation for air in the pressure coordinate of (5.25) gives

$$\nabla_\sigma \cdot \mathbf{v}_h - \frac{\sigma}{\pi_a}\nabla_\sigma(\pi_a)\cdot\frac{\partial \mathbf{v}_h}{\partial\sigma} + \frac{1}{\pi_a}\frac{\partial \mathrm{w}_p}{\partial\sigma} = 0 \tag{5.56}$$

An expression for $\partial \mathrm{w}_p/\partial\sigma$ is now needed.

The **vertical scalar velocity in the sigma-pressure coordinate** ($\sigma$ s$^{-1}$) is

$$\dot{\sigma} = \frac{d\sigma}{dt} \tag{5.57}$$

The relationship between $\dot{\sigma}$ and $\mathrm{w}_p$ is found by substituting $p_a = p_{top} + \pi_a\sigma$ and (5.57) into (5.21) to yield

$$\mathrm{w}_p = \frac{dp_a}{dt} = \sigma\frac{d\pi_a}{dt} + \frac{d\sigma}{dt}\pi_a = \sigma\frac{d\pi_a}{dt} + \dot{\sigma}\pi_a \tag{5.58}$$

If $\pi_a$ is assumed constant with time, and horizontal gradients of $\pi_a$ are ignored, (5.22) simplifies to $\mathrm{w}_p = -w\rho_a g$. Substituting this term into (5.58) with constant

$\pi_a$ yields $\dot{\sigma} = -w\rho_a g/\pi_a$. Thus, positive vertical velocities in the sigma-pressure coordinate result in downward motions. Because $\pi_a$ varies in time and space, (5.58) is not simplified.

The **total derivative in Cartesian–sigma-pressure coordinates** is found by replacing $p_a$ with $\sigma$ and $\mathrm{w}_p$ with $\dot{\sigma}$ in (5.30). The result is

$$\frac{\mathrm{d}}{\mathrm{d}t} = \left(\frac{\partial}{\partial t}\right)_\sigma + \mathbf{v}_h \cdot \nabla_\sigma + \dot{\sigma}\frac{\partial}{\partial \sigma} \tag{5.59}$$

Substituting the total derivative of $\pi_a$ into (5.58) gives

$$\mathrm{w}_p = \sigma\left[\left(\frac{\partial \pi_a}{\partial t}\right)_\sigma + (\mathbf{v}_h \cdot \nabla_\sigma)\pi_a\right] + \dot{\sigma}\pi_a \tag{5.60}$$

where $\partial\pi_a/\partial\sigma = 0$. The partial derivative of this equation with respect to $\sigma$ is

$$\frac{\partial \mathrm{w}_p}{\partial \sigma} = \left(\frac{\partial \pi_a}{\partial t}\right)_\sigma + (\mathbf{v}_h \cdot \nabla_\sigma)\pi_a + \sigma\nabla_\sigma(\pi_a)\cdot\frac{\partial \mathbf{v}_h}{\partial \sigma} + \pi_a\frac{\partial \dot{\sigma}}{\partial \sigma} \tag{5.61}$$

Substituting (5.61) into (5.56), canceling terms, and compressing the result yields the **continuity equation for air in Cartesian–sigma-pressure coordinates** as

$$\left(\frac{\partial \pi_a}{\partial t}\right)_\sigma + \nabla_\sigma \cdot (\mathbf{v}_h\pi_a) + \pi_a\frac{\partial \dot{\sigma}}{\partial \sigma} = 0 \tag{5.62}$$

Substituting $\mathbf{v}_h\pi_a$ for $\mathbf{v}$ in the horizontal component of (4.11), substituting the result into (5.62), and multiplying through by $R_e^2 \cos\varphi$ gives the **continuity equation for air in spherical–sigma-pressure coordinates** as

$$R_e^2\cos\varphi\left(\frac{\partial \pi_a}{\partial t}\right)_\sigma + \left[\frac{\partial}{\partial \lambda_e}(u\pi_a R_e) + \frac{\partial}{\partial \varphi}(v\pi_a R_e\cos\varphi)\right]_\sigma + \pi_a R_e^2\cos\varphi\frac{\partial \dot{\sigma}}{\partial \sigma} = 0 \tag{5.63}$$

#### 5.3.4.2. *Pressure and Vertical Velocity from the Continuity Equation for Air*

The time rate of change of air pressure and vertical velocity can be calculated from the continuity equation for air. The integral of (5.62), over all $\sigma$-layers is

$$\int_0^1\left(\frac{\partial \pi_a}{\partial t}\right)_\sigma \mathrm{d}\sigma = -\nabla_\sigma \cdot \int_0^1 (\mathbf{v}_h\pi_a)\,\mathrm{d}\sigma - \pi_a\int_0^0 \mathrm{d}\dot{\sigma} \tag{5.64}$$

The integral limits are $\sigma = 0$ and $\dot{\sigma} = 0$ at the model top and $\sigma = 1$ and $\dot{\sigma} = 0$ at the model surface. **At the ground surface, vertical velocities are always zero,** and at the model top, velocities in the $\sigma$-$p$ coordinate are assumed to be zero to close a boundary condition. In reality, $\dot{\sigma} \neq 0$ at the model top. Integrating (5.64) gives a **prognostic equation for column pressure** in Cartesian–sigma-pressure coordinates as

$$\left(\frac{\partial \pi_a}{\partial t}\right)_\sigma = -\nabla_\sigma \cdot \int_0^1 (\mathbf{v}_h \pi_a)\, d\sigma \tag{5.65}$$

The analogous equation in spherical–sigma-pressure coordinates is

$$R_e^2 \cos\varphi \left(\frac{\partial \pi_a}{\partial t}\right)_\sigma = -\int_0^1 \left[\frac{\partial}{\partial \lambda_e}(u\pi_a R_e) + \frac{\partial}{\partial \varphi}(v\pi_a R_e \cos\varphi)\right]_\sigma d\sigma \tag{5.66}$$

A numerical solution to this equation is shown in Chapter 7.

Once (5.66) has been written, an equation for vertical velocity at any altitude can be derived. Integrating (5.62) with respect to $\sigma$ from the model top to any $\sigma$-level, and rearranging, gives

$$\pi_a \int_0^{\dot{\sigma}} d\dot{\sigma} = -\nabla_\sigma \cdot \int_0^\sigma (\mathbf{v}_h \pi_a)\, d\sigma - \int_0^\sigma \left(\frac{\partial \pi_a}{\partial t}\right)_\sigma d\sigma \tag{5.67}$$

Integrating (5.67) yields

$$\dot{\sigma}\pi_a = -\nabla_\sigma \cdot \int_0^\sigma (\mathbf{v}_h \pi_a)\, d\sigma - \sigma \left(\frac{\partial \pi_a}{\partial t}\right)_\sigma \tag{5.68}$$

which is the vertical velocity at any $\sigma$-level. This equation is solved after $(\partial \pi_a / \partial t)_\sigma$ has been obtained from (5.65). Applying the spherical-coordinate transformation from (4.11) to (5.68) and multiplying through by $R_e^2 \cos\varphi$ gives a **diagnostic equation for vertical velocity at any $\sigma$-level** as

$$\begin{aligned}\dot{\sigma}\pi_a R_e^2 \cos\varphi = &-\int_0^\sigma \left[\frac{\partial}{\partial \lambda_e}(u\pi_a R_e) + \frac{\partial}{\partial \varphi}(v\pi_a R_e \cos\varphi)\right]_\sigma d\sigma \\ &- \sigma R_e^2 \cos\varphi \left(\frac{\partial \pi_a}{\partial t}\right)_\sigma\end{aligned} \tag{5.69}$$

### 5.3.5. Species Continuity Equation in the Sigma-Pressure Coordinate

Applying the total derivative from (5.59) to (3.54) gives the **species continuity equation in Cartesian–sigma-pressure coordinates** as

$$\frac{dq}{dt} = \left(\frac{\partial q}{\partial t}\right)_\sigma + (\mathbf{v}_h \cdot \nabla_\sigma) q + \dot{\sigma}\frac{\partial q}{\partial \sigma} = \frac{(\nabla \cdot \rho_a \mathbf{K}_h \nabla) q}{\rho_a} + \sum_{n=1}^{N_{e,t}} R_n \tag{5.70}$$

The eddy diffusion term expands in Cartesian–sigma-pressure coordinates to

$$
\begin{aligned}
&(\nabla \cdot \rho_a \mathbf{K}_h \nabla)q \qquad (5.71)\\
&= \left[\left(\frac{\partial}{\partial x}\right)_\sigma + \left(\frac{\partial \sigma}{\partial x}\right)_z \frac{\partial}{\partial \sigma}\right]\left\{\rho_a K_{h,xx}\left[\left(\frac{\partial q}{\partial x}\right)_\sigma + \left(\frac{\partial \sigma}{\partial x}\right)_z \frac{\partial q}{\partial \sigma}\right]\right\}\\
&+ \left[\left(\frac{\partial}{\partial y}\right)_\sigma + \left(\frac{\partial \sigma}{\partial y}\right)_z \frac{\partial}{\partial \sigma}\right]\left\{\rho_a K_{h,yy}\left[\left(\frac{\partial q}{\partial y}\right)_\sigma + \left(\frac{\partial \sigma}{\partial y}\right)_z \frac{\partial q}{\partial \sigma}\right]\right\}\\
&+ \frac{\rho_a g^2}{\pi_a^2}\frac{\partial}{\partial \sigma}\left(\rho_a^2 K_{h,zz}\frac{\partial q}{\partial \sigma}\right)
\end{aligned}
$$

The vertical term was found by substituting $\partial p_a/\partial \sigma = \pi_a$ into the vertical term of (5.33).

Multiplying (5.70) by $\pi_a$, multiplying the continuity equation for air from (5.62) by q, and summing the two equations gives the flux-form species continuity equation as

$$
\left[\frac{\partial(\pi_a q)}{\partial t}\right]_\sigma + \nabla_\sigma \cdot (\mathbf{v}_h \pi_a q) + \pi_a \frac{\partial(\dot{\sigma} q)}{\partial \sigma} = \pi_a\left[\frac{(\nabla \cdot \rho_a \mathbf{K}_h \nabla)q}{\rho_a} + \sum_{n=1}^{N_{e,t}} R_n\right] \qquad (5.72)
$$

Applying the spherical coordinate transformation from (4.11) to (5.72) and multiplying through by $R_e^2 \cos\varphi$ gives the **flux form of the species continuity equation in spherical–sigma-pressure coordinates** as

$$
\begin{aligned}
&R_e^2 \cos\varphi \left[\frac{\partial}{\partial t}(\pi_a q)\right]_\sigma + \left[\frac{\partial}{\partial \lambda_e}(u\pi_a q R_e) + \frac{\partial}{\partial \varphi}(v\pi_a q R_e \cos\varphi)\right]_\sigma \qquad (5.73)\\
&+ \pi_a R_e^2 \cos\varphi \frac{\partial}{\partial \sigma}(\dot{\sigma} q) = \pi_a R_e^2 \cos\varphi \left[\frac{(\nabla \cdot \rho_a \mathbf{K}_h \nabla)q}{\rho_a} + \sum_{n=1}^{N_{e,t}} R_n\right]
\end{aligned}
$$

### 5.3.6. Thermodynamic Equation in the Sigma-Pressure Coordinate

Applying the total derivative from (5.59) to (3.76) gives the **thermodynamic energy equation in Cartesian–sigma-pressure coordinates** as

$$
\left(\frac{\partial \theta_v}{\partial t}\right)_\sigma + (\mathbf{v}_h \cdot \nabla_\sigma)\theta_v + \dot{\sigma}\frac{\partial \theta_v}{\partial \sigma} = \frac{(\nabla \cdot \rho_a \mathbf{K}_h \nabla)\theta_v}{\rho_a} + \frac{\theta_v}{c_{p,d} T_v}\sum_{n=1}^{N_{e,h}} \frac{dQ_n}{dt} \qquad (5.74)
$$

Multiplying (5.74) by $\pi_a$, multiplying (5.62) by $\theta_v$, adding the two equations, and compressing terms gives the flux form of the thermodynamic energy equation as

$$
\begin{aligned}
&\left[\frac{\partial(\pi_a \theta_v)}{\partial t}\right]_\sigma + \nabla_\sigma \cdot (\mathbf{v}_h \pi_a \theta_v) + \pi_a \frac{\partial(\dot{\sigma}\theta_v)}{\partial \sigma} \qquad (5.75)\\
&= \pi_a\left[\frac{(\nabla \cdot \rho_a \mathbf{K}_h \nabla)\theta_v}{\rho_a} + \frac{\theta_v}{c_{p,d} T_v}\sum_{n=1}^{N_{e,h}} \frac{dQ_n}{dt}\right]
\end{aligned}
$$

The eddy diffusion term in this equation is the same as that in (5.71) except $\theta_v$ is used instead of q in the present case. Applying the spherical coordinate transformation from (4.11) to (5.75) and multiplying through by $R_e^2 \cos\varphi$ gives the **flux form of the thermodynamic energy equation in spherical–sigma-pressure coordinates** as

$$R_e^2 \cos\varphi\left[\frac{\partial}{\partial t}(\pi_a\theta_v)\right]_\sigma + \left[\frac{\partial}{\partial\lambda_e}(u\pi_a\theta_v R_e) + \frac{\partial}{\partial\varphi}(v\pi_a\theta_v R_e\cos\varphi)\right] + \pi_a R_e^2\cos\varphi\frac{\partial}{\partial\sigma}(\dot{\sigma}\theta_v) = \pi_a R_e^2\cos\varphi\left[\frac{(\nabla\cdot\rho_a\mathbf{K}_h\nabla)\theta_v}{\rho_a} + \frac{\theta_v}{c_{p,d}T_v}\sum_{n=1}^{N_{e,h}}\frac{dQ_n}{dt}\right] \tag{5.76}$$

### 5.3.7. Horizontal Momentum Equations in the Sigma-Pressure Coordinate

Applying the total derivative from (5.59) to the horizontal momentum equation in Cartesian–altitude coordinates from (4.70) gives

$$\left(\frac{\partial\mathbf{v}_h}{\partial t}\right)_\sigma + (\mathbf{v}_h\cdot\nabla_\sigma)\mathbf{v}_h + \dot{\sigma}\frac{\partial\mathbf{v}_h}{\partial\sigma} + f\mathbf{k}\times\mathbf{v}_h = -\frac{1}{\rho_a}\nabla_z(p_a) + \frac{(\nabla\cdot\rho_a\mathbf{K}_m\nabla)\mathbf{v}_h}{\rho_a} \tag{5.77}$$

The pressure-gradient term in (5.77) can be converted to the sigma-pressure coordinate by combining (5.16) with (5.55). The result is

$$\frac{1}{\rho_a}\nabla_z(p_a) = \nabla_p\Phi = \nabla_\sigma\Phi - \frac{\sigma}{\pi_a}\nabla_\sigma(\pi_a)\frac{\partial\Phi}{\partial\sigma} \tag{5.78}$$

Substituting (5.78) into (5.77) gives the **horizontal momentum equation in Cartesian–sigma-pressure coordinates** as

$$\left(\frac{\partial\mathbf{v}_h}{\partial t}\right)_\sigma + (\mathbf{v}_h\cdot\nabla_\sigma)\mathbf{v}_h + \dot{\sigma}\frac{\partial\mathbf{v}_h}{\partial\sigma} + f\mathbf{k}\times\mathbf{v}_h = -\nabla_\sigma\Phi + \frac{\sigma}{\pi_a}\nabla_\sigma(\pi_a)\frac{\partial\Phi}{\partial\sigma} + \frac{(\nabla\cdot\rho_a\mathbf{K}_m\nabla)\mathbf{v}_h}{\rho_a} \tag{5.79}$$

### 5.3.8. Coupling Horizontal and Vertical Momentum Equation

The hydrostatic equation is used in lieu of a prognostic vertical momentum equation in the sigma-pressure coordinate because the conversion of altitude to sigma-pressure coordinate requires the hydrostatic assumption. Substituting $p_a = \rho_a R' T_v$ and $\partial p_a/\partial\sigma = \pi_a$ into (5.37) gives the **hydrostatic equation in the sigma-pressure coordinate** as

$$\frac{\partial\Phi}{\partial\sigma} = -\frac{\pi_a R' T_v}{p_a} = -\frac{\pi_a}{\rho_a} = -\alpha_a\pi_a \tag{5.80}$$

Substituting (5.80) into (5.79) gives the horizontal momentum equation as

$$\left(\frac{\partial \mathbf{v}_h}{\partial t}\right)_\sigma + (\mathbf{v}_h \cdot \nabla_\sigma)\mathbf{v}_h + \dot{\sigma}\frac{\partial \mathbf{v}_h}{\partial \sigma} \tag{5.81}$$
$$= -f\,\mathbf{k} \times \mathbf{v}_h - \nabla_\sigma \Phi - \sigma\alpha_a \nabla_\sigma(\pi_a) + \frac{(\nabla \cdot \rho_a \mathbf{K}_m \nabla)\mathbf{v}_h}{\rho_a}$$

The term $\alpha_a$ should be modified to make the finite differencing of (5.81) consistent with that of geopotential in Chapter 7. Combining $\alpha_a = R' T_v / p_a$ with $R' = \kappa c_{p,d}$, $T_v = \theta_v P$, $\partial p_a / \partial \sigma = \pi_a$, and $\partial P / \partial p_a = \kappa P / p_a$ gives

$$\alpha_a = \frac{R' T_v}{p_a} = \frac{\kappa c_{p,d} \theta_v P}{p_a} = c_{p,d}\theta_v \frac{\partial P}{\partial p_a} = \frac{c_{p,d}\theta_v}{\pi_a}\frac{\partial P}{\partial \sigma} \tag{5.82}$$

Substituting (5.82) into (5.81), multiplying the result by $\pi_a$, multiplying (5.62) by $\mathbf{v}_h$, and summing the two equations gives the flux form of the horizontal momentum equation in Cartesian–sigma-pressure coordinates as

$$\left[\frac{\partial(\mathbf{v}_h \pi_a)}{\partial t}\right]_\sigma + \mathbf{v}_h\, \nabla_\sigma \cdot (\mathbf{v}_h \pi_a) + \pi_a(\mathbf{v}_h \cdot \nabla_\sigma)\mathbf{v}_h + \pi_a \frac{\partial}{\partial \sigma}(\dot{\sigma}\mathbf{v}_h) \tag{5.83}$$
$$= -\pi_a f\,\mathbf{k} \times \mathbf{v}_h - \pi_a \nabla_\sigma \Phi - \sigma c_{p,d}\theta_v \frac{\partial P}{\partial \sigma} \nabla_\sigma(\pi_a) + \pi_a \frac{(\nabla \cdot \rho_a \mathbf{K}_m \nabla)\mathbf{v}_h}{\rho_a}$$

The advection terms in this equation expand to

$$\mathbf{v}_h \nabla_\sigma \cdot (\mathbf{v}_h \pi_a) = \mathbf{i}u\left[\frac{\partial(u\pi_a)}{\partial x} + \frac{\partial(v\pi_a)}{\partial y}\right] + \mathbf{j}v\left[\frac{\partial(u\pi_a)}{\partial x} + \frac{\partial(v\pi_a)}{\partial y}\right] \tag{5.84}$$

$$\pi_a(\mathbf{v}_h \cdot \nabla_\sigma)\mathbf{v}_h = \mathbf{i}\pi_a\left(u\frac{\partial u}{\partial x} + v\frac{\partial u}{\partial y}\right) + \mathbf{j}\pi_a\left(u\frac{\partial v}{\partial x} + v\frac{\partial v}{\partial y}\right) \tag{5.85}$$

Substituting these terms into (5.83), applying the transformation from (4.30), applying (4.1), and multiplying through by $R_e^2 \cos\varphi$ gives the **flux forms of the horizontal momentum equations in spherical–sigma-pressure coordinates** as

$$R_e^2 \cos\varphi \left[\frac{\partial}{\partial t}(\pi_a u)\right]_\sigma + \left[\frac{\partial}{\partial \lambda_e}(\pi_a u^2 R_e) + \frac{\partial}{\partial \varphi}(\pi_a u v R_e \cos\varphi)\right]_\sigma \tag{5.86}$$
$$+ \pi_a R_e^2 \cos\varphi \frac{\partial}{\partial \sigma}(\dot{\sigma} u) = \pi_a u v R_e \sin\varphi + \pi_a f v R_e^2 \cos\varphi$$
$$- R_e\left(\pi_a \frac{\partial \Phi}{\partial \lambda_e} + \sigma c_{p,d}\theta_v \frac{\partial P}{\partial \sigma}\frac{\partial \pi_a}{\partial \lambda_e}\right)_\sigma + R_e^2 \cos\varphi \frac{\pi_a}{\rho_a}(\nabla \cdot \rho_a \mathbf{K}_m \nabla)u$$

$$R_e^2 \cos\varphi \left[\frac{\partial}{\partial t}(\pi_a v)\right]_\sigma + \left[\frac{\partial}{\partial \lambda_e}(\pi_a u v R_e) + \frac{\partial}{\partial \varphi}(v^2 \pi_a R_e \cos\varphi)\right]_\sigma + \pi_a R_e^2 \cos\varphi \frac{\partial}{\partial \sigma}(\dot{\sigma} v) = -\pi_a u^2 R_e \sin\varphi - \pi_a f u R_e^2 \cos\varphi - R_e \cos\varphi \left[\pi_a \frac{\partial \Phi}{\partial \varphi} + \sigma c_{p,d} \theta_v \frac{\partial P}{\partial \sigma} \frac{\partial \pi_a}{\partial \varphi}\right]_\sigma + R_e^2 \cos\varphi \frac{\pi_a}{\rho_a} (\nabla \cdot \rho_a \mathbf{K}_m \nabla) v \tag{5.87}$$

The west–east eddy diffusion term expands in Cartesian–sigma-pressure coordinates to

$$(\nabla \cdot \rho_a \mathbf{K}_h \nabla) u = \left[\left(\frac{\partial}{\partial x}\right)_\sigma + \left(\frac{\partial \sigma}{\partial x}\right)_z \frac{\partial}{\partial \sigma}\right] \left\{\rho_a K_{m,xx} \left[\left(\frac{\partial u}{\partial x}\right)_\sigma + \left(\frac{\partial \sigma}{\partial x}\right)_z \frac{\partial u}{\partial \sigma}\right]\right\} + \left[\left(\frac{\partial}{\partial y}\right)_\sigma + \left(\frac{\partial \sigma}{\partial y}\right)_z \frac{\partial}{\partial \sigma}\right] \left\{\rho_a K_{m,yx} \times \left[\left(\frac{\partial u}{\partial y}\right)_\sigma + \left(\frac{\partial \sigma}{\partial y}\right)_z \frac{\partial u}{\partial \sigma}\right]\right\} + \frac{\rho_a g^2}{\pi_a^2} \frac{\partial}{\partial \sigma}\left(\rho_a^2 K_{m,zx} \frac{\partial u}{\partial \sigma}\right) \tag{5.88}$$

The expression for $(\nabla \cdot \rho_a \mathbf{K}_h \nabla) v$ is similar to (5.88), except $v$, $K_{m,xy}$, $K_{m,yy}$, and $K_{m,zy}$ are used in the new term instead of $u$, $K_{m,xx}$, $K_{m,yx}$, and $K_{m,zx}$, respectively.

## 5.4. SIGMA-ALTITUDE COORDINATE

The **sigma-altitude** (*s-z*) **coordinate** is defined such that layer tops and bottoms are surfaces of constant $s$, where $s$ equals the difference between the model top and the bottom altitude of a layer divided by the difference between the model top and the ground surface altitude (e.g., Kasahara 1974). This coordinate is used to simulate nonhydrostatic or hydrostatic flows. In the sigma-altitude coordinate, layer thicknesses do not change. In the sigma-pressure coordinate, layer thicknesses change continuously. An advantage of the sigma-altitude and sigma-pressure coordinates in comparison with the altitude and pressure coordinates is that the former do not permit model layers to intercept ground topography, whereas the latter do. In this section elastic nonhydrostatic model equations for the sigma-altitude coordinate are derived. Conversion to elastic and hydrostatic equations is also discussed.

### 5.4.1. Definitions

Layer tops and bottoms in the sigma-altitude coordinate are defined as surfaces of constant $s$. In this case, $s$ is defined as

$$s = \frac{z_{top} - z}{z_{top} - z_{surf}} = \frac{z_{top} - z}{Z_t} \tag{5.89}$$

where $z$ is the altitude of interest, $z_{surf}$ is the ground surface altitude, $z_{top}$ is the model top altitude, $Z_t = z_{top} - z_{surf}$ is the altitude difference between the model surface and top, and $s$ is a fraction $\le 1$. At the model top, $s = 0$. At the model surface, $s = 1$. From (5.89), the altitude at a given $s$-level is

$$z = z_{top} - Z_t s \tag{5.90}$$

Figure 5.5, which shows heights of constant sigma-pressure surfaces, can also be used to describe heights of constant-sigma-altitude surfaces by replacing σ's with $s$'s. In the $s$-$z$ coordinate, layer tops and bottoms correspond to $s$-values that are constant in time. Altitudes, including $z$, $z_{surf}$, and $z_{top}$, vary in space, but not in time, along surfaces of constant $s$. The pressure ($p_a$) varies in space and time along surfaces of constant $s$.

### 5.4.2. Gradient Conversion from Altitude to Sigma-Altitude Coordinate

The horizontal gradient conversion of a variable from Cartesian–altitude to Cartesian–sigma-altitude coordinates is obtained in a manner similar to that for the conversion from Cartesian–altitude to Cartesian–pressure coordinates. The result is

$$\nabla_z = \nabla_s + \nabla_z(s)\frac{\partial}{\partial s} \tag{5.91}$$

Taking the horizontal gradient of (5.89) and noting that $\nabla_z(z) = \nabla_z(z_{top}) = 0$ gives

$$\nabla_z(s) = -\frac{z_{top} - z}{Z_t^2}\nabla_z(Z_t) = -\frac{s}{Z_t}\nabla_z(Z_t) \tag{5.92}$$

Substituting this result into (5.91) yields another form of the gradient conversion as

$$\nabla_z = \nabla_s - \frac{s}{Z_t}\nabla_z(Z_t)\frac{\partial}{\partial s} \tag{5.93}$$

The time-derivative conversion from the altitude to sigma-altitude coordinate is

$$\left(\frac{\partial}{\partial t}\right)_z = \left(\frac{\partial}{\partial t}\right)_s \tag{5.94}$$

which was derived by substituting $s$ for $p$ in (5.15) and noting that $(\partial s/\partial t)_z = 0$.

The **vertical scalar velocity in the sigma-altitude coordinate** ($s$ s$^{-1}$) is

$$\dot{s} = \frac{\mathrm{d}s}{\mathrm{d}t} = (\mathbf{v}_h \cdot \nabla_z)s + w\frac{\partial s}{\partial z} = (\mathbf{v}_h \cdot \nabla_z)s - \frac{w}{Z_t} \tag{5.95}$$

which has a zero local time derivative, since $(\partial s/\partial t)_z = 0$. (5.95) was derived by noting that $\partial s/\partial z = -1/Z_t$, obtained by differentiating (5.89). A positive vertical velocity in the sigma-altitude coordinate corresponds to downward motion, just as in the sigma-pressure and pressure coordinates. Substituting $\dot{s}$ for $w_p$ and $s$ for $p$ in (5.30) gives the **total derivative in Cartesian–sigma-altitude coordinates** as

$$\frac{d}{dt} = \left(\frac{\partial}{\partial t}\right)_s + (\mathbf{v}_h \cdot \nabla_s) + \dot{s}\frac{\partial}{\partial s} \tag{5.96}$$

### 5.4.3. Continuity Equation for Air in the Sigma-Altitude Coordinate

The continuity equation for air in Cartesian–sigma-altitude coordinates is derived from the continuity equation for air in Cartesian–altitude coordinates. Splitting (3.20) into horizontal and vertical components and applying (5.91) to $\mathbf{v}_h$ and $\rho_a$ gives

$$\begin{aligned}\left(\frac{\partial \rho_a}{\partial t}\right)_s &= -\rho_a\left[\nabla_s \cdot \mathbf{v}_h + \nabla_z(s)\frac{\partial \mathbf{v}_h}{\partial s} + \frac{\partial w}{\partial z}\right] \\ &\quad - \mathbf{v}_h \cdot \left[\nabla_s(\rho_a) + \nabla_z(s)\frac{\partial \rho_a}{\partial s}\right] - w\frac{\partial \rho_a}{\partial z}\end{aligned} \tag{5.97}$$

Rewriting (5.95) as $w = Z_t[\mathbf{v}_h \cdot \nabla_z(s) - \dot{s}]$ and differentiating gives

$$\frac{\partial w}{\partial z} = Z_t\left[\nabla_z(s)\frac{\partial \mathbf{v}_h}{\partial z} + (\mathbf{v}_h \cdot \nabla_z)\frac{\partial s}{\partial z} - \frac{\partial \dot{s}}{\partial z}\right] \tag{5.98}$$

Substituting $\partial s/\partial z = -1/Z_t$ into (5.98) yields

$$\frac{\partial w}{\partial z} = \frac{\partial \dot{s}}{\partial s} - \nabla_z(s)\frac{\partial \mathbf{v}_h}{\partial s} = \frac{1}{Z_t}(\mathbf{v}_h \cdot \nabla_z)Z_t \tag{5.99}$$

Substituting $w = Z_t[\mathbf{v}_h \cdot \nabla_z(s) - \dot{s}]$, (5.99), and $\partial s/\partial z = -1/Z_t$ into (5.97) gives

$$\left(\frac{\partial \rho_a}{\partial t}\right)_s = -\rho_a\left[\nabla_s \cdot \mathbf{v}_h + \frac{\partial \dot{s}}{\partial s} + \frac{1}{Z_t}(\mathbf{v}_h \cdot \nabla_z)Z_t\right] - (\mathbf{v}_h \cdot \nabla_s)\rho_a - \dot{s}\frac{\partial \rho_a}{\partial s} \tag{5.100}$$

Substituting $\nabla_z(Z_t) = \nabla_s(Z_t)$ into (5.100) gives the **elastic nonhydrostatic continuity equation for air in Cartesian–sigma-altitude coordinates** as

$$\begin{aligned}\left(\frac{\partial \rho_a}{\partial t}\right)_s &= -\frac{1}{Z_t}\nabla_s \cdot (\mathbf{v}_h \rho_a Z_t) - \frac{\partial}{\partial s}(\dot{s}\rho_a) \\ &= -\frac{1}{Z_t}\left[\frac{\partial(u\rho_a Z_t)}{\partial x} + \frac{\partial(v\rho_a Z_t)}{\partial y}\right]_s - \frac{\partial(\dot{s}\rho_a)}{\partial s}\end{aligned} \tag{5.101}$$

When (5.101) is used to predict air density, vertically propagating acoustic waves are one solution to the equations of atmospheric dynamics. To eliminate these waves and enhance numerical stability, the **anelastic approximation** is made to (5.101) in the same manner as described in Section 5.1.

If hydrostatic balance is assumed, vertically propagating acoustic waves are also eliminated as a possible solution to the basic equations. Substituting $\partial s/\partial z = -1/Z_t$ into (2.34) gives the **hydrostatic equation in the sigma-altitude coordinate** as

$$\rho'_a = -\frac{1}{g}\frac{\partial p'_a}{\partial z} = \frac{1}{Z_t g}\frac{\partial p'_a}{\partial s} \tag{5.102}$$

where primes have been added to indicate that density and pressure are in hydrostatic balance. Substituting (5.102) into (5.101) gives the **hydrostatic continuity equation in Cartesian–sigma-altitude coordinates** as

$$\frac{\partial}{\partial t}\left(\frac{\partial p'_a}{\partial s}\right) = -\nabla_s \cdot \left(\mathbf{v}_h \frac{\partial p'_a}{\partial s}\right) - \frac{\partial}{\partial s}\left(\dot{s}\frac{\partial p'_a}{\partial s}\right) \tag{5.103}$$

### 5.4.4. Species Continuity Equation in the Sigma-Altitude Coordinate

The **species continuity equation in Cartesian–sigma-altitude coordinates** is obtained by applying the total derivative from (5.96) to (3.54). The result is

$$\left(\frac{dq}{dt}\right)_s = \left(\frac{\partial q}{\partial t}\right)_s + (\mathbf{v}_h \cdot \nabla_s)q + \dot{s}\frac{\partial q}{\partial s} = \frac{(\nabla \cdot \rho_a \mathbf{K}_h \nabla)q}{\rho_a} + \sum_{n-1}^{N_{e,t}} R_n \tag{5.104}$$

The eddy diffusion term in (5.104) expands in Cartesian–sigma-altitude coordinates to

$$\begin{aligned}(\nabla \cdot \rho_a \mathbf{K}_h \nabla)q = &\left[\left(\frac{\partial}{\partial x}\right)_s + \left(\frac{\partial s}{\partial x}\right)_z \frac{\partial}{\partial s}\right]\left\{\rho_a K_{h,xx}\left[\left(\frac{\partial q}{\partial x}\right)_s \right.\right.\\ &\left.\left. + \left(\frac{\partial s}{\partial x}\right)_z \frac{\partial q}{\partial s}\right]\right\} + \left[\left(\frac{\partial}{\partial y}\right)_s + \left(\frac{\partial s}{\partial y}\right)_z \frac{\partial}{\partial s}\right]\left\{\rho_a K_{h,yy}\right.\\ &\left. \times \left[\left(\frac{\partial q}{\partial y}\right)_s + \left(\frac{\partial s}{\partial y}\right)_z \frac{\partial q}{\partial s}\right]\right\} + \frac{1}{Z_t^2}\frac{\partial}{\partial s}\left(\rho_a K_{h,zz}\frac{\partial q}{\partial s}\right)\end{aligned} \tag{5.105}$$

where $\partial s/\partial z = -1/Z_t$ was used to obtain the vertical term.

### 5.4.5. Thermodynamic Equation in the Sigma-Altitude Coordinate

The **thermodynamic energy equation in Cartesian–sigma-altitude coordinates** is obtained by applying (5.96) to (3.76). The result is

$$\left(\frac{\partial \theta_v}{\partial t}\right)_s + (\mathbf{v}_h \cdot \nabla_s)\theta_v + \dot{s}\frac{\partial \theta_v}{\partial s} = \frac{(\nabla \cdot \rho_a \mathbf{K}_h \nabla)\theta_v}{\rho_a} + \frac{\theta_v}{c_{p,d} T_v}\sum_{n=1}^{N_{e,h}} \frac{dQ_n}{dt} \tag{5.106}$$

Eddy diffusion is treated as in (5.105), except $\theta_v$ is used instead of q in the new term.

### 5.4.6. Horizontal Momentum Equations in the Sigma-Altitude Coordinate

Applying (5.96) to (4.70) gives the **horizontal momentum equation in Cartesian–sigma-altitude coordinates** as

$$\left(\frac{\partial \mathbf{v}_h}{\partial t}\right)_s + (\mathbf{v}_h \cdot \nabla_s)\mathbf{v}_h + \dot{s}\frac{\partial \mathbf{v}_h}{\partial s} + f\mathbf{k}\times\mathbf{v}_h = -\frac{1}{\rho_a}\nabla_z(p_a) + \frac{(\nabla_z \cdot \rho_a \mathbf{K}_m \nabla_z)\mathbf{v}_h}{\rho_a} \tag{5.107}$$

Applying the coordinate conversion from (5.93) to pressure gives

$$\nabla_z(p_a) = \nabla_s(p_a) - \frac{s}{Z_t}\nabla_z(Z_t)\frac{\partial p_a}{\partial s} \tag{5.108}$$

Substituting (5.108) into (5.107) gives the horizontal momentum equation as

$$\left(\frac{\partial \mathbf{v}_h}{\partial t}\right)_s + (\mathbf{v}_h \cdot \nabla_s)\mathbf{v}_h + \dot{s}\frac{\partial \mathbf{v}_h}{\partial s} = -f\mathbf{k}\times\mathbf{v}_h - \frac{1}{\rho_a}\left[\nabla_s(p_a) - \frac{s}{Z_t}\nabla_z(Z_t)\frac{\partial p_a}{\partial s} - (\nabla \cdot \rho_a \mathbf{K}_m \nabla)\mathbf{v}_h\right] \tag{5.109}$$

(5.109) expands in the $x$- and $y$-directions to

$$\left(\frac{\partial u}{\partial t} + u\frac{\partial u}{\partial x} + v\frac{\partial u}{\partial y}\right)_s + \dot{s}\frac{\partial u}{\partial s} = fv - \frac{1}{\rho_a}\left[\left(\frac{\partial p_a}{\partial x}\right)_s - \frac{s}{Z_t}\left(\frac{\partial Z_t}{\partial x}\right)_z\frac{\partial p_a}{\partial s} - (\nabla \cdot \rho_a \mathbf{K}_m \nabla)u\right] \tag{5.110}$$

$$\left(\frac{\partial v}{\partial t} + u\frac{\partial v}{\partial x} + v\frac{\partial v}{\partial y}\right)_s + \dot{s}\frac{\partial v}{\partial s} = -fu - \frac{1}{\rho_a}\left[\left(\frac{\partial p_a}{\partial y}\right)_s - \frac{s}{Z_t}\left(\frac{\partial Z_t}{\partial y}\right)_z\frac{\partial p_a}{\partial s} - (\nabla \cdot \rho_a \mathbf{K}_m \nabla)v\right] \tag{5.111}$$

respectively. The eddy diffusion term in the $x$-direction is

$$(\nabla \cdot \rho_a \mathbf{K}_m \nabla)u = \left[\left(\frac{\partial}{\partial x}\right)_s + \left(\frac{\partial s}{\partial x}\right)_z\frac{\partial}{\partial s}\right]\left\{\rho_a K_{m,xx}\left[\left(\frac{\partial u}{\partial x}\right)_s + \left(\frac{\partial s}{\partial x}\right)_z\frac{\partial u}{\partial s}\right]\right\} + \left[\left(\frac{\partial}{\partial y}\right)_s + \left(\frac{\partial s}{\partial y}\right)_z\frac{\partial}{\partial s}\right]\left\{\rho_a K_{m,yx}\left[\left(\frac{\partial u}{\partial y}\right)_s + \left(\frac{\partial \sigma}{\partial y}\right)_z\frac{\partial u}{\partial s}\right]\right\} + \frac{1}{Z_t^2}\frac{\partial}{\partial s}\left(\rho_a K_{m,zx}\frac{\partial u}{\partial s}\right) \tag{5.112}$$

The expression for $(\nabla \cdot \rho_a K_m \nabla)v$ is similar.

### 5.4.7. Vertical Momentum Equation in the Sigma-Altitude Coordinate

Substituting $\partial s/\partial z = -1/Z_t$ into the vertical momentum equation in Cartesian–altitude coordinates from (4.74) and expanding the total derivative of $w$ with (5.96) gives

$$\left(\frac{\partial w}{\partial t} + u\frac{\partial w}{\partial x} + v\frac{\partial w}{\partial y}\right)_s + \dot{s}\frac{\partial w}{\partial s} = -g + \frac{1}{Z_t\rho_a}\frac{\partial p_a}{\partial s} + \frac{(\nabla\cdot\rho_a\mathbf{K}_m\nabla)w}{\rho_a} \tag{5.113}$$

Substituting $w = Z_t[\mathbf{v}_h\cdot\nabla_z(s) - \dot{s}]$ from (5.95) into (5.113) gives the **vertical momentum equation in Cartesian–sigma-altitude coordinates** as

$$\begin{aligned}\left[\left(\frac{\partial}{\partial t}\right)_s + u\left(\frac{\partial}{\partial x}\right)_s + v\left(\frac{\partial}{\partial y}\right)_s + \dot{s}\frac{\partial}{\partial s}\right]\Bigg[Z_t u\left(\frac{\partial s}{\partial x}\right)_z \\ + Z_t v\left(\frac{\partial s}{\partial y}\right)_z - Z_t\dot{s}\Bigg] = -g + \frac{1}{Z_t\rho_a}\frac{\partial p_a}{\partial s} + \frac{1}{\rho_a}(\nabla\cdot\rho_a\mathbf{K}_m\nabla) \\ \times\left[Z_t u\left(\frac{\partial s}{\partial x}\right)_z + Z_t v\left(\frac{\partial s}{\partial y}\right)_z - Z_t\dot{s}\right]\end{aligned} \tag{5.114}$$

$(\nabla\cdot\rho_a\mathbf{K}_m\nabla)(Z_t\dot{s})$ is similar to $(\nabla\cdot\rho_a\mathbf{K}_m\nabla)u$, except that $Z_t\dot{s}$, $K_{m,xz}$, $K_{m,yz}$, and $K_{m,zz}$ are used in the new term instead of $u$, $K_{m,xx}$, $K_{m,yx}$, and $K_{m,zx}$.

### 5.4.8. Basic Equations in Spherical–Sigma-Altitude Coordinates

The elastic nonhydrostatic sigma-altitude equations derived above can be converted from Cartesian to spherical horizontal coordinates with (4.1), (4.11), and (4.30). The resulting elastic continuity equation for air from (5.101), species continuity equation from (5.104), thermodynamic energy equation from (5.106), and momentum equations from (5.110), (5.111), and (5.114) are, respectively,

$$\begin{aligned}R_e^2\cos\varphi\left(\frac{\partial\rho_a}{\partial t}\right)_s = -\frac{1}{Z_t}\Bigg[\frac{\partial}{\partial\lambda_e}(u\rho_a Z_t R_e) \\ + \frac{\partial}{\partial\varphi}(v\rho_a Z_t R_e\cos\varphi)\Bigg]_s - R_e^2\cos\varphi\frac{\partial}{\partial s}(\dot{s}\rho_a)\end{aligned} \tag{5.115}$$

$$\left(\frac{\partial q}{\partial t} + \frac{u}{R_e\cos\varphi}\frac{\partial q}{\partial\lambda_e} + \frac{v}{R_e}\frac{\partial q}{\partial\varphi}\right)_s + \dot{s}\frac{\partial q}{\partial s} = \frac{(\nabla\cdot\rho_a\mathbf{K}_h\nabla)q}{\rho_a} + \sum_{n=1}^{N_{e,t}} R_n \tag{5.116}$$

$$\begin{aligned}\left(\frac{\partial\theta_v}{\partial t} + \frac{u}{R_e\cos\varphi}\frac{\partial\theta_v}{\partial\lambda_e} + \frac{v}{R_e}\frac{\partial\theta_v}{\partial\varphi}\right)_s + \dot{s}\frac{\partial\theta_v}{\partial s} \\ = \frac{(\nabla\cdot\rho_a\mathbf{K}_h\nabla)\theta_v}{\rho_a} + \frac{\theta_v}{c_{p,d}T_v}\sum_{n=1}^{N_{e,h}}\frac{dQ_n}{dt}\end{aligned} \tag{5.117}$$

$$\left(\frac{\partial u}{\partial t}+\frac{u}{R_e\cos\varphi}\frac{\partial u}{\partial\lambda_e}+\frac{v}{R_e}\frac{\partial u}{\partial\varphi}\right)_s+\dot{s}\frac{\partial u}{\partial s}$$
$$=\frac{uv\tan\varphi}{R_e}+fv-\frac{1}{\rho_a R_e\cos\varphi}\left[\left(\frac{\partial p_a}{\partial\lambda_e}\right)_s-\frac{s}{Z_t}\left(\frac{\partial Z_t}{\partial\lambda_e}\right)_z\frac{\partial p_a}{\partial s}\right]$$
$$+\frac{(\nabla\cdot\rho_a\mathbf{K}_m\nabla)u}{\rho_a} \tag{5.118}$$

$$\left(\frac{\partial v}{\partial t}+\frac{u}{R_e\cos\varphi}\frac{\partial v}{\partial\lambda_e}+\frac{v}{R_e}\frac{\partial v}{\partial\varphi}\right)_s+\dot{s}\frac{\partial v}{\partial s}=-\frac{u^2\tan\varphi}{R_e}$$
$$-fu-\frac{1}{R_e\rho_a}\left[\left(\frac{\partial p_a}{\partial\varphi}\right)_s-\frac{s}{Z_t}\left(\frac{\partial Z_t}{\partial\varphi}\right)_z\frac{\partial p_a}{\partial s}\right]+\frac{(\nabla\cdot\rho_a\mathbf{K}_m\nabla)v}{\rho_a} \tag{5.119}$$

$$\left[\left(\frac{\partial}{\partial t}\right)_s+\frac{u}{R_e\cos\varphi}\left(\frac{\partial}{\partial\lambda_e}\right)_s+\frac{v}{R_e}\left(\frac{\partial}{\partial\varphi}\right)_s+\dot{s}\frac{\partial}{\partial s}\right]$$
$$\times\left[\frac{Z_t u}{R_e\cos\varphi}\left(\frac{\partial s}{\partial\lambda_e}\right)_z+\frac{Z_t v}{R_e}\left(\frac{\partial s}{\partial\varphi}\right)_z-Z_t\dot{s}\right]=-g+\frac{1}{Z_t\rho_a}\frac{\partial p_a}{\partial s}$$
$$+\frac{1}{\rho_a}(\nabla\cdot\rho_a\mathbf{K}_m\nabla)\left[\frac{Z_t u}{R_e\cos\varphi}\left(\frac{\partial s}{\partial\lambda_e}\right)_z+\frac{Z_t v}{R_e}\left(\frac{\partial s}{\partial\varphi}\right)_z-Z_t\dot{s}\right] \tag{5.120}$$

As in Section 5.1 for the altitude coordinate, the elastic equations above can be converted to anelastic form by deriving a diagnostic equation for nonhydrostatic pressure. In such a case the vertical momentum equation from (5.114) is not solved, since the vertical velocity is determined diagnostically from the anelastic continuity equation, discussed following (5.101). In this scenario, the potential virtual temperature is still found from the thermodynamic energy equation, the horizontal scalar velocities are still found from the horizontal momentum equations, and the specific humidity is still found from the species continuity equation.

## 5.5. SUMMARY

The equations of atmospheric dynamics include the continuity equation for air, the species continuity equation, the thermodynamic energy equation, the horizontal momentum equation, and the vertical momentum equation. In this chapter, the vertical coordinate in the equations was transformed from the altitude to the pressure, sigma-pressure, and sigma-altitude coordinates. With the altitude (and pressure) coordinate, model layers may intersect surface topography. With the sigma-altitude and sigma-pressure coordinates, model layers are terrain-following and cannot intersect topography. When the pressure or sigma-pressure coordinates are used, the atmosphere is assumed to be in hydrostatic balance. This assumption is reasonable when large horizontal scales are simulated. For small scales, the altitude or sigma-altitude coordinate should be used so that nonhydrostatic motions can be simulated.

## 5.6. PROBLEMS

**5.1.** **(a)** Assume $\Delta x = 5$ km, $\Delta y = 5$ km, and $\Delta p_a = -10$ mb for a grid cell in the pressure coordinate and that the west, east, south, north, and lower boundary scalar velocities are $u_1 = -2$, $u_2 = +1$ m s$^{-1}$, $v_3 = +1$, $v_4 = -2$ m s$^{-1}$, and $w_5 = +0.03$ m s$^{-1}$, respectively. Convert $w_5$ to the pressure coordinate with $w_p = -w\rho_a g$, assuming $T = 284$ K and $p_a = 980$ mb. Use the pressure-coordinate continuity equation for air to estimate $w_p$ at the top of the cell.

**(b)** For the same cell as in part (a), assume the west, east, south, north, lower, and upper boundary mass mixing ratios of a gas are 0.004, 0.005, 0.003, 0.004, 0.0045, and 0.0055 kg kg$^{-1}$, respectively. Estimate the change in q at the center of the cell after 500 s, ignoring diffusion and external sources/sinks.

**5.2.** Assume a horizontal grid cell has dimension $\Delta x = 500$ km and $\Delta y = 400$ km, centered at $\varphi = 30°$N. Assume that the cell is on a surface of constant pressure at $p_a = 500$ mb and that the altitudes of the west, east, south, and north boundary are 5.5, 5.4, 5.6, and 5.3 km, respectively. Calculate geostrophic scalar velocities ($u_g$ and $v_g$) and the geostrophic wind speed.

**5.3.** Assume that a cell in the $\sigma$-$p$ coordinate has dimension $\Delta x = 4$ km, $\Delta y = 5$ km, and $\Delta\sigma = 0.05$. The west, east, south, and north boundary $u$- and $\pi_a$-values are $u_1 = -2$ and $\pi_{a,1} = 748$ mb, $u_2 = +1$ m s$^{-1}$ and $\pi_{a,2} = 752$ mb, $v_3 = -1$ and $\pi_{a,3} = 749$ mb, and $v_4 = -2$ m s$^{-1}$ and $\pi_{a,4} = 753$ mb, respectively. Assume $p_{a,top} = 250$ mb, $\sigma = 0.9$ at the cell bottom, the grid cell center $\pi_a$ value is an average of the four boundary values, $T_v = 298$ K, the lower boundary vertical scalar velocity in the altitude coordinate is $w_5 = +0.02$ m s$^{-1}$, and the air is dry.

**(a)** Convert vertical scalar velocity from the altitude to the sigma-pressure coordinate with $\dot{\sigma} = -w\rho_a g/\pi_a$. Use the continuity equation for air in the sigma-pressure coordinate to estimate the sigma-pressure coordinate vertical scalar velocity at the top of the cell, assuming $\partial\pi_a/\partial t = 0$.

**(b)** Assume that the west, east, south, north, lower, and upper boundary values of $\theta_v$ are 299, 297, 304, 301, 300, and 302 K, respectively, and that no eddy diffusion or external sources/sinks exist. Estimate the value of $\theta_v$ at the center of the grid cell after 200 s.

**5.4.** Assume that a horizontal grid cell has dimension $\Delta x = 5$ km and $\Delta y = 4$ km, and that $\theta_v$ and $p_a$ on the west, east, south, and north boundary of the cell are $\theta_{v,1} = 298$ K and $p_{a,1} = 1010$ mb, $\theta_{v,2} = 304$ K and $p_{a,2} = 1004$ mb, $\theta_{v,3} = 302$ K and $p_{a,3} = 1000$ mb, and $\theta_{v,4} = 301$ K and $p_{a,4} = 1006$ mb, respectively. What is the change in the $u$- and $v$-component scalar velocities after 10 min due to the pressure gradient force alone? Assume the air is dry.

**5.5.** Derive the continuity equation for air in Cartesian–sigma-pressure coordinates from the continuity equation for air in Cartesian–altitude coordinates.

**5.6.** Derive the horizontal momentum equation in Cartesian–pressure coordinates from that in Cartesian–sigma-pressure coordinates. Ignore conversion of the eddy diffusion term.

## 5.7. COMPUTER PROGRAMMING PRACTICE

**5.7.** Write a computer script to set up a model grid over the globe. Assume the grid stretches from $\varphi$ centered at $-88°$S to $+88°$N, where $d\varphi = 4°$ and from $\lambda_e$ centered at $-177.5°$W to $+177.5°$E, where $d\lambda_e = 5°$. Calculate $dx$ and $dy$ at the top and bottom of every grid-cell, and print the values to a table.

# 6

# Numerical Solutions to Partial Differential Equations

ATMOSPHERIC models simulate physical processes described by ordinary and partial differential equations. Gas and aqueous chemistry and gas-to-particle conversion processes are simulated with ordinary differential equations, and transport processes are simulated with partial differential equations. In this chapter, ordinary and partial differential equations are defined, and numerical methods of solving partial differential equations are discussed. Methods of solving partial differential equations include finite-difference and series expansion methods. In both cases, partial derivatives are approximated with analogs, and the analogs are solved numerically. Below, finite-difference analogs and their numerical solutions are given for the advection-diffusion equation, which is a unidirectional form of the species continuity equation. Solutions that assume constant and variable grid spacing and eddy diffusion coefficients are shown. A special case of the finite-difference method is the semi-Lagrangian method. Two series expansion methods, the finite-element and pseudospectral methods, are discussed.

## 6.1. ORDINARY AND PARTIAL DIFFERENTIAL EQUATIONS

An **ordinary differential equation** (ODE) is an equation with one independent variable, such as time, and a **partial differential equation** (PDE) is an equation with more than one independent variable, such as time and space. ODEs and PDEs are classified by their order and degree. The **order** is the highest derivative rank of the equation, and the **degree** is the highest polynomial exponent of the highest derivative. A **homogeneous** differential equation is an equation that does not contain a term involving the independent variable. A **linear** differential equation is one in which the dependent variable and its derivatives do not appear in second-degree or higher terms and in which the dependent variable is not multiplied by other derivatives of itself.

Table 6.1 shows ordinary and partial differential equations of varying orders and degrees. In the table, (a), (b), (d), (e), and (f) are homogeneous, and the remaining equations are inhomogeneous. Equations (b) and (e) are linear, and the rest are nonlinear. Chemical equations are first-order, first-degree, homogeneous ODEs, such as (a) and (b). These equations are either linear or nonlinear. The species continuity equation and the thermodynamic energy equation are first-order, first-degree, homogeneous, linear PDEs, such as (e). The momentum equation is a first-order, first-degree, homogeneous, nonlinear PDE, such as (f).

**Table 6.1.** Examples of the Orders and Degrees of Ordinary and Partial Differential Equations

| Order, Degree | Ordinary Differential Equations | Partial Differential Equations |
|---|---|---|
| First-order, first-degree | (a) $\frac{dN}{dt} = 16 - 4N^2$ | (e) $\frac{\partial N}{\partial t} + \frac{\partial (uN)}{\partial x} = 0$ |
| First-order, first-degree | (b) $\frac{dN}{dt} = 3AB - 4NC$ | (f) $\frac{\partial u}{\partial t} + u\frac{\partial u}{\partial x} + v\frac{\partial u}{\partial y} = 0$ |
| Second-order, first-degree | (c) $\frac{d^2N}{dt^2} + \frac{dN}{dt} + 5t = 0$ | (g) $\frac{\partial^2 N}{\partial t^2} + \frac{\partial^2 N}{\partial x^2} = 3t^2 + x$ |
| Second-order, second-degree | (d) $\left(\frac{d^2N}{dt^2}\right)^2 + \frac{dN}{dt} + 4 = 0$ | (h) $\left(\frac{\partial^2 N}{\partial t^2}\right)^2 + \frac{\partial N}{\partial x} = t - x$ |

The variable $t$ is time, $x$ is west–east distance, $y$ is south–north distance, $N$, $A$, $B$ and $C$ are concentrations, and $u$ is west–east scalar velocity.

Boundary conditions for ODEs and PDEs must be specified. When conditions are known at one end of a domain but not the other, an **initial-value problem** arises. If the concentrations ($N$) are known at time $t = 0$, if time is the independent variable, and if concentration is the dependent variable, the set of ODEs is an initial-value problem. When conditions are known at both ends of a domain, a **boundary-value problem** arises. If time and west–east direction ($x$) are independent variables, if the concentration is the dependent variable, and if the concentrations are known everywhere at $t = 0$ and at both ends of the spatial domain at all times, the set of PDEs is an initial-value problem with respect to time and a boundary-value problem with respect to space.

## 6.2. OPERATOR-SPLITTING

Each major process in an atmospheric model is generally solved separately from each other process. Suppose a model takes account of dynamics, transport, and gas chemistry. Each of these processes may be solved sequentially during a common time interval, with a unique numerical scheme that takes a unique number of time steps. A **time step** is an increment in time for a given process. A **time interval** is the period during which several time steps of a process are solved without interference by other processes. Suppose the time step for dynamics is 6 s, the transport time step is 300 s, the chemistry time step is variable, and the time interval common to all processes is 300 s. During the time interval, 50 dynamics time steps are taken, followed by 1 transport time step, followed by a variable number of chemistry time steps. After the dynamics steps, average predicted wind speeds are used as inputs into the transport calculations. During the transport step, gases are moved around the grid. Final concentrations from the transport step are used as initial values for the chemistry steps. Final values from the chemistry steps are used as initial values for dynamics steps in the next time interval. Figure 6.1 illustrates this example.

The isolation of individual processes during a time interval is called **time or operator-splitting.** Operator-splitting is used because computers today cannot solve

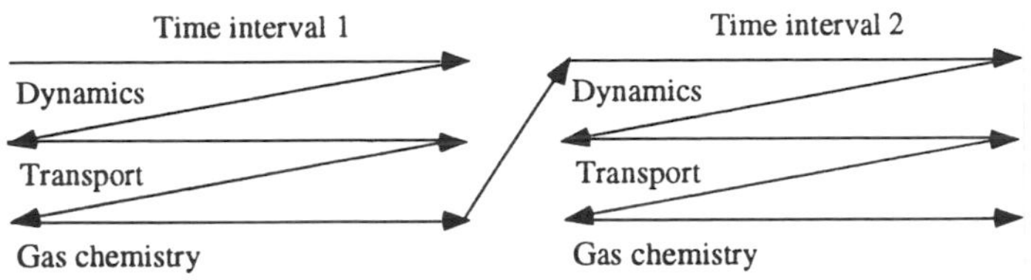


**Figure 6.1.** Example of operator-splitting scheme. During the first time interval, dynamics, transport, and gas chemistry are solved sequentially. Values determined from the end of a time interval after one processes are used to initialize values at the beginning of the same time interval for another process. Values from the end of the last process in one time interval are used at the beginning of the first process in the next time interval.

all model ODEs and PDEs simultaneously in three dimensions. Yanenko (1971) discusses the theoretical basis behind operator-splitting with respect to certain mathematical equations.

## 6.3. ADVECTION–DIFFUSION EQUATIONS

First-order, first-degree, homogeneous, linear or nonlinear partial differential equations solved in atmospheric models include the species continuity equation, the thermodynamic energy equation, and the directional momentum equations. In Chapter 7, a method of solving these equations together is given. In this chapter, methods of solving **advection–diffusion equations**, which are operator split forms of the species continuity equation, are discussed. Advection–diffusion equations are derived by considering that the four-dimensional ($t$, $x$, $y$, $z$) species continuity equation can be divided into three two-dimensional partial differential equations ([$t$, $x$], [$t$, $y$], and [$t$, $z$]) and a single one-dimensional ($t$) ordinary differential equation. The sequential solution to the four operator-split equations approximates the solution to the original four-dimensional equation. This method of operator-splitting a mathematical equation is called the **locally one-dimensional** (LOD) **procedure** or the **method of fractional steps** (e.g., Yanenko 1971; Mitchell 1969) and has been used widely in atmospheric models (e.g., Reynolds et al. 1973; Carmichael et al. 1986; Toon et al. 1988).

From the four-dimensional species continuity equation given in (3.52), the west–east, south–north, and vertical **unidirectional advection–diffusion equations** can be written in number concentration units as

$$\frac{\partial N}{\partial t} + \frac{\partial (uN)}{\partial x} - \frac{\partial}{\partial x}\left(K_{h,xx}\frac{\partial N}{\partial x}\right) = 0 \tag{6.1}$$

$$\frac{\partial N}{\partial t} + \frac{\partial (vN)}{\partial y} - \frac{\partial}{\partial y}\left(K_{h,yy}\frac{\partial N}{\partial y}\right) = 0 \tag{6.2}$$

$$\frac{\partial N}{\partial t} + \frac{\partial (wN)}{\partial z} - \frac{\partial}{\partial z}\left(K_{h,zz}\frac{\partial N}{\partial z}\right) = 0 \tag{6.3}$$

respectively. The solution order of these equations may be reversed each time interval to improve accuracy (e.g., Yanenko 1971). Thus, if the equations are solved in the order (6.1), (6.2), (6.3) during one time interval, they may be solved in the order (6.3), (6.2), (6.1) during the next time interval. Fractional-step schemes associated with order reversal are called **alternating-direction schemes**.

The remaining terms in the species continuity equation are external source/sink terms. These terms are operator-split from the advection–diffusion terms as a single ordinary differential equation,

$$\frac{\partial N}{\partial t} = \sum_{n=1}^{N_{e,t}} R_n \tag{6.4}$$

or may be split further into several ODEs. (6.4) can be solved before or after (6.1)–(6.3) are solved.

In moist-air mass-mixing-ratio units, the operator-split west–east advection–diffusion equation can be written from (3.54) as

$$\frac{\partial \text{q}}{\partial t} + u\frac{\partial \text{q}}{\partial x} - \frac{1}{\rho_a}\frac{\partial}{\partial x}\left(\rho_a K_{h,xx}\frac{\partial \text{q}}{\partial x}\right) = 0 \tag{6.5}$$

Analogous equations can be written for the south–north and vertical directions and for external sources/sinks. In the following sections, finite-difference and series expansion methods of approximating derivatives and of solving advection–diffusion equations are discussed.

## 6.4. FINITE-DIFFERENCE APPROXIMATIONS

Approximate solutions to partial differential equations, such as advection–diffusion equations, can be found with finite-difference or series expansion methods. The purpose of using an approximation is to reduce the solution space for each continuous differential function from an infinite to a finite number of spatial or temporal nodes in order to speed up computation of the differential equation.

A **finite-difference** approximation involves the replacement of each continuous differential operator (d) with a discrete difference analog ($\Delta$). This analog is an approximation written in terms of a finite number of values of the variable being operated on at each temporal or spatial node. If the west–east scalar velocity is a continuous function in space at a given time, values of the velocity can be mapped from the function to a discretized west–east grid, as shown in Fig. 6.2. The grid consists of several **grid cells** (also called **grid boxes, grid points,** or **nodes**) placed any distance apart.

Finite-difference approximations of a variable are often made with respect to time or space. Table 6.1 (e) and (f) are partial differential equations commonly simulated in atmospheric models. In (e), finite-difference analogs are required for $\partial N/\partial t$ and $\partial(uN)/\partial x$. In (f), finite-difference analogs are required for $\partial u/\partial t$, $\partial u/\partial x$, and $\partial u/\partial y$.

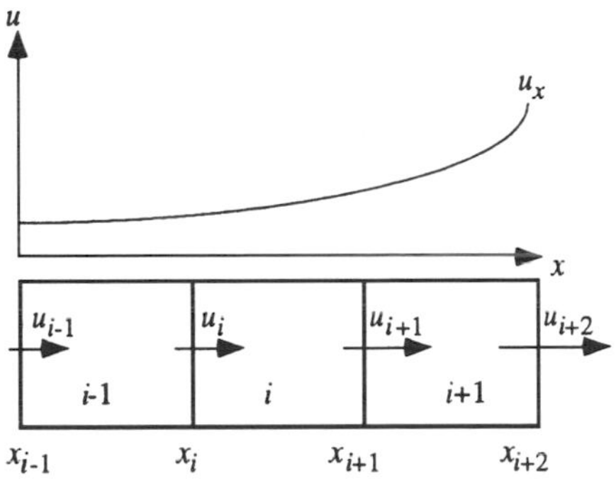


**Figure 6.2.** Discretization of a continuous west–east scalar velocity $u_x$. The west–east grid is broken into discrete cells, and $u$-values are mapped from the continuous function to the edge of each cell. The arrows in the cells represent magnitudes of the wind speed. Distances along the $x$-axis are also mapped to the cells.

### 6.4.1. Consistency, Convergence, and Stability

The ability of a numerical solution to replicate the exact solution to a partial differential equation depends on several factors. A finite-difference analog in space or time must **converge** to its differential expression when the analog is reduced towards zero. If $\Delta N/\Delta x$ is a finite-difference analog of $\partial N/\partial x$, the condition

$$\frac{\partial N}{\partial x} = \lim_{\Delta x \to 0} \left\| \frac{\Delta N}{\Delta x} \right\| \tag{6.6}$$

must be satisfied for the approximation to be useful.

The finite-difference analog $\Delta N/\Delta x$ in (6.6) is obtained from a Taylor series expansion. In the expansion, high-order terms are neglected to reduce the computational burden of the approximation. The difference between the full Taylor series expansion and the truncated approximation is the **truncation error**. A finite-difference approximation of a derivative is **consistent** if the truncation error of the approximation approaches zero as $\Delta x$ (or $\Delta t$) approaches zero. Consistency occurs when

$$\lim_{\Delta x \to 0} \left\| \text{T.E.}\left(\frac{\Delta N}{\Delta x}\right) \right\| = 0 \tag{6.7}$$

where T.E. is the truncation error of the approximation $\Delta N/\Delta x$.

If a finite-difference approximation is consistent, the rate at which its truncation error approaches zero depends on the order of approximation. The **order of approximation** is the order of the lowest-order term in the Taylor series expansion neglected in the approximation. The higher the order of approximation, the faster the truncation error converges towards zero upon an increase in spatial (or temporal) resolution. Thus, with the same $\Delta x$, a high-order approximation is more accurate than a low-order approximation. For the same truncation error, a low-order approximation requires a smaller $\Delta x$ than does a high-order approximation. In sum, a high-order approximation with a large $\Delta x$ can have the same truncation error as a low-order approximation with a small $\Delta x$. Because a high-order approximation includes more terms, it requires more computations than does a low-order approximation with the same $\Delta x$.

Obtaining high order with respect to one variable, such as space, is useful only if the order of the other variable, such as time, is also high. Otherwise, low accuracy in the time derivative swamps the high accuracy in the space derivative. An optimal finite-difference solution has similar order in space and time.

While individual finite-difference analogs must converge towards exact differentials, the overall numerical solution to a PDE must converge to an exact solution when spatial and temporal differences decrease towards zero. If $N_{e,x,t}$ is an exact solution, and $N_{f,x,t}$ is a finite-difference approximation of a PDE, **overall convergence** occurs when

$$\lim_{\Delta x, \Delta t \to 0} \| N_{e,x,t} - N_{f,x,t} \| = 0 \tag{6.8}$$

If a numerical solution is **nonconvergent,** it is not useful.

For a numerical method to be successful, it must be stable. **Stability** occurs if the absolute-value difference between the numerical and exact solutions does not grow over time. Thus,

$$\lim_{t \to \infty} \| N_{e,x,t} - N_{f,x,t} \| \le C \tag{6.9}$$

where $C$ is a constant. Stability often depends on the time step size used. If a numerical solution is stable for any time step smaller than a specified value, the solution is **conditionally stable**. If a solution is stable, regardless of the time step, it is **unconditionally stable**. If a solution is unstable, regardless of the time step, it is **unconditionally unstable**.

Other problems arising from finite-difference and other solutions to partial differential equations are **numerical diffusion** (artificial spreading of peak values across several grid cells) and **numerical dispersion** (waves appearing ahead of and behind peak values). These problems can usually be mitigated by increasing the resolution of the spatial grid (e.g. decreasing $\Delta x$), decreasing the time step, or increasing the order of approximation of the finite-difference analog.

### 6.4.2. Low-Order Approximations of Derivatives

A finite-difference approximation of a differential, such as $\partial u/\partial x$, involves the replacement of individual differential expressions, such as d$u$ or d$x$, with finite-difference analogs, such as $\Delta u$ or $\Delta x$, respectively. Suppose $\partial u/\partial x$ is discretized over a west–east grid, as shown in Fig. 6.2, where all grid cells are rectangular. Each cell is denoted by an index number $i$, and the distance from the western edge of the entire grid to the western edge of cell $i$ is $x_i$.

On the grid layout just defined, the differential scalar velocity d$u$ at point $x_i$ can be estimated as $\Delta u_i = u_{i+1} - u_{i-1}$, $\Delta u_i = u_{i+1} - u_i$, or $\Delta u_i = u_i - u_{i-1}$, which

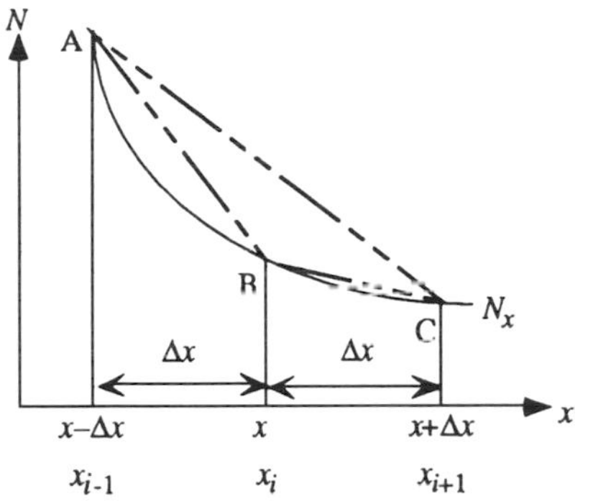


**Figure 6.3.** Derivative approximations at a point on a continuous function. The derivative at point B is approximated with chords AC, BC, or AB, which give the slope of the tangent at point B for the central-, forward-, and backward-difference approximations, respectively.

are the **central-**, **forward-**, and **backward-difference** approximations, respectively. The corresponding discretizations of d$x$ are $\Delta x_i = x_{i+1} - x_{i-1}$, $\Delta x_i = x_{i+1} - x_i$, and $\Delta x_i = x_i - x_{i-1}$, respectively. In the **central-difference case**, the slope of the tangent at point $x_i$ in Fig. 6.2 is approximately

$$\frac{\partial u}{\partial x} \approx \frac{\Delta u_i}{\Delta x_i} = \frac{u_{i+1} - u_{i-1}}{x_{i+1} - x_{i-1}} \tag{6.10}$$

Similar equations can be written for the forward- and backward-difference cases.

The approximations just discussed can be derived from a Taylor series expansion. If gas concentration is a continuous function of west–east distance, as shown in Fig. 6.3, the values of $N$ at points $x + \Delta x$ and $x - \Delta x$, respectively, are determined from **Taylor's theorem** as

$$N_{x+\Delta x} = N_x + \Delta x \frac{\partial N_x}{\partial x} + \frac{1}{2}\Delta x^2 \frac{\partial^2 N_x}{\partial x^2} + \frac{1}{6}\Delta x^3 \frac{\partial^3 N_x}{\partial x^3} + \frac{1}{24}\Delta x^4 \frac{\partial^4 N_x}{\partial x^4} + \cdots \tag{6.11}$$

$$N_{x-\Delta x} = N_x - \Delta x \frac{\partial N_x}{\partial x} + \frac{1}{2}\Delta x^2 \frac{\partial^2 N_x}{\partial x^2} - \frac{1}{6}\Delta x^3 \frac{\partial^3 N_x}{\partial x^3} + \frac{1}{24}\Delta x^4 \frac{\partial^4 N_x}{\partial x^4} - \cdots \tag{6.12}$$

If grid spacing is uniform ($\Delta x$ is constant), the sum of (6.11) and (6.12) is

$$N_{x+\Delta x} + N_{x-\Delta x} = 2N_x + \Delta x^2 \frac{\partial^2 N_x}{\partial x^2} + \frac{1}{12}\Delta x^4 \frac{\partial^4 N_x}{\partial x^4} + \cdots \tag{6.13}$$

Rearranging (6.13) gives

$$\frac{\partial^2 N_x}{\partial x^2} = \frac{N_{x+\Delta x} - 2N_x + N_{x-\Delta x}}{\Delta x^2} + O(\Delta x^2) \tag{6.14}$$

where

$$O(\Delta x^2) = -\frac{1}{12}\,\Delta x^2\,\frac{\partial^4 N_x}{\partial x^4} - \cdots \tag{6.15}$$

includes all terms of order $\Delta x^2$ and higher. If $O(\Delta x^2)$ is small, (6.14) simplifies to

$$\frac{\partial^2 N_x}{\partial x^2} \approx \frac{N_{x+\Delta x} - 2N_x + N_{x-\Delta x}}{\Delta x^2} \tag{6.16}$$

where $O(\Delta x^2)$ is now the **truncation error.** (6.16) is a **second-order central difference approximation** of $\partial^2 N_x/\partial x^2$. The equation is second-order because the lowest-order exponent in the truncation error is two. It is a central-difference approximation because it relies on equally weighted values of $N$ on each side of node $x$.

Subtracting (6.12) from (6.11) gives

$$N_{x+\Delta x} - N_{x-\Delta x} = 2\,\Delta x\,\frac{\partial N_x}{\partial x} + \frac{1}{3}\,\Delta x^3\,\frac{\partial^3 N_x}{\partial x^3} + \cdots \tag{6.17}$$

Rearranging this equation results in

$$\frac{\partial N_x}{\partial x} = \frac{N_{x+\Delta x} - N_{x-\Delta x}}{2\Delta x} + O(\Delta x^2) \tag{6.18}$$

where

$$O(\Delta x^2) = -\frac{1}{6}\,\Delta x^2\,\frac{\partial^3 N_x}{\partial x^3} - \cdots \tag{6.19}$$

includes all terms of order $\Delta x^2$ and higher. If $O(\Delta x^2)$ is small, (6.18) simplifies to

$$\frac{\partial N_x}{\partial x} \approx \frac{N_{x+\Delta x} - N_{x-\Delta x}}{2\Delta x} = \frac{N_{i+1} - N_{i-1}}{2\Delta x} \tag{6.20}$$

where $i + 1$ and $i - 1$ are surrogates for $x + \Delta x$ and $x - \Delta x$, respectively. This equation is a **second-order central-difference approximation of the first derivative of** $N_x$. (6.20) gives the slope of the tangent (represented by chord AC) of $N_x$ at point B in Fig. 6.3.

Another approximation of the first derivative of $N_x$ is obtained from the first two terms of (6.11). Rearranging these terms gives

$$\frac{\partial N_x}{\partial x} \approx \frac{N_{x+\Delta x} - N_x}{\Delta x} = \frac{N_{i+1} - N_i}{\Delta x} \tag{6.21}$$

The truncated portion of the approximation includes terms first-order and higher [$O(\Delta x)$]; thus, (6.21) is a **first-order forward-difference approximation of the first derivative** of $N_x$. The slope of this derivative is represented by chord BC in Fig. 6.3.

Rearranging the first two terms of (6.12) gives

$$\frac{\partial N_x}{\partial x} \approx \frac{N_x - N_{x-\Delta x}}{\Delta x} = \frac{N_i - N_{i-1}}{\Delta x} \tag{6.22}$$

which is the **first-order backward-difference approximation of the first derivative** of $N_x$, represented by chord AB in Fig. 6.3.

If time, not space, is the independent variable, the second-order central-, first-order forward-, and first-order backward-difference approximations of $\partial N_t/\partial t$ are

$$\frac{\partial N_t}{\partial t} \approx \frac{N_{t+h} - N_{t-h}}{2h} \qquad \frac{\partial N_t}{\partial t} \approx \frac{N_{t+h} - N_t}{h} \qquad \frac{\partial N_t}{\partial t} \approx \frac{N_t - N_{t-h}}{h} \tag{6.23}$$

respectively, where $h = \Delta t$ is the **time-step size,** $t$ is the current time, $t + h$ is one time step forward, and $t - h$ is one time step backward. These equations are derived in the same manner as (6.20), (6.21), and (6.22), respectively.

### 6.4.3. Arbitrary-Order Approximations of Derivatives

Finite-difference approximations of arbitrary order can be obtained systematically (e.g., Celia and Gray 1992). The approximation of $\partial^m N/\partial x^m$, which is the $m$th **derivative** of $N$, can be obtained by expanding the derivative across $q$ **discrete nodes** in the $x$-direction. If the independent variable is time, the derivative can be expanded along $q$ time steps. The minimum number of nodes allowed in the expansion is $m + 1$. In general, the maximum **order of approximation** of a finite-difference solution is $q - m$, although it may be smaller or larger for individual cases. For instance, when $m$ is even and the grid spacing is constant, the order of approximation can be increased to $q - m + 1$.

Figure 6.4 shows the arbitrary grid spacing for the derivation to come. The point at which the derivative is taken does not need to correspond to a node point, although in the figure the derivative is assumed to be taken at node point $x_3$. The distance between two nodes is $\Delta x_i = x_{i+1} - x_i$, where $i$ varies from 1 to $q - 1$.

The finite difference solution to the $m$th derivative across $q$ nodes is approximately

$$\frac{\partial^m N}{\partial x^m} \approx \sum_{i=1}^{q} \gamma_i N_i = \gamma_1 N_1 + \gamma_2 N_2 + \cdots + \gamma_q N_q \tag{6.24}$$

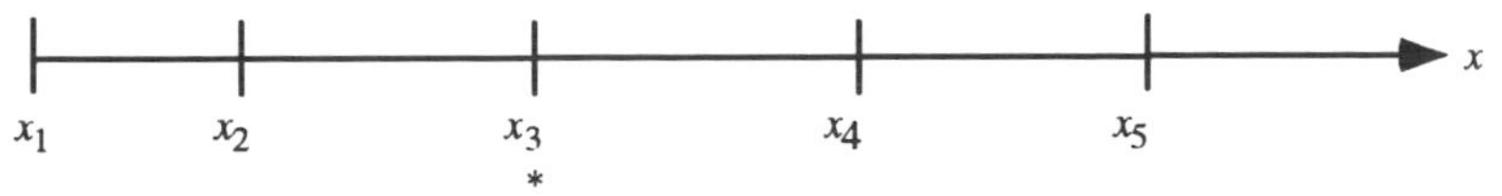


**Figure 6.4.** Grid spacing of an arbitrary-spaced grid where $q = 5$. The derivative is taken at point $x_3$, marked $*$.

where the $\gamma_j$'s are constants to be determined. A Taylor series expansion of $N$ at node $i$ across the point at which the derivative is taken $(*)$ is

$$N_i = N_* + (x_i - x_*)\frac{\partial N_*}{\partial x} + \frac{1}{2}(x_i - x_*)^2\frac{\partial^2 N_*}{\partial x^2} + \frac{1}{6}(x_i - x_*)^3\frac{\partial^3 N_*}{\partial x^3} + \cdots \quad (6.25)$$

Combining (6.24) with (6.25) and gathering terms gives

$$\frac{\partial^m N}{\partial x^m} \approx \sum_{i=1}^{q}\gamma_i N_i = \sum_{i=1}^{q}\gamma_i N_* + \sum_{i=1}^{q}\gamma_i(x_i - x_*)\frac{\partial N_*}{\partial x} + \sum_{i=1}^{q}\gamma_i\frac{1}{2}(x_i - x_*)^2\frac{\partial^2 N_*}{\partial x^2} + \cdots \quad (6.26)$$

This equation can be rewritten as

$$\sum_{i=1}^{q}\gamma_i N_i = B_0 N_* + B_1\frac{\partial N_*}{\partial x} + B_2\frac{\partial^2 N_*}{\partial x^2} + \cdots \quad (6.27)$$

where

$$B_n = \sum_{i=1}^{q}\gamma_i\frac{1}{n!}(x_i - x_*)^n \qquad \text{for} \quad n = 0, \ldots, q-1 \quad (6.28)$$

(6.28) represents a matrix of $q$ equations and unknowns. Multiplying (6.28) by $n!$ gives

$$\begin{bmatrix} 1 & 1 & 1 & \cdots & 1 \\ (x_1 - x_*) & (x_2 - x_*) & (x_3 - x_*) & \cdots & (x_q - x_*) \\ (x_1 - x_*)^2 & (x_2 - x_*)^2 & (x_3 - x_*)^2 & \cdots & (x_q - x_*)^2 \\ \vdots & \vdots & \vdots & & \vdots \\ (x_1 - x_*)^q & (x_2 - x_*)^q & (x_3 - x_*)^q & \cdots & (x_q - x_*)^q \end{bmatrix} \begin{bmatrix} \gamma_1 \\ \gamma_2 \\ \gamma_3 \\ \vdots \\ \gamma_q \end{bmatrix} = \begin{bmatrix} 0!B_0 \\ 1!B_1 \\ 2!B_2 \\ \vdots \\ (q-1)!B_{q-1} \end{bmatrix} \quad (6.29)$$

The highest-order derivative is found when $B_n = 1$ for $n = m$ and $B_n = 0$ for all other $n$.

The **first-order backward-difference approximation** of $\partial N/\partial x$ $(m = 1)$ is found from (6.29) by discretizing $\partial N/\partial x$ across two equally spaced grid cells $(q = 2)$, setting $B_1 = 1$, and setting $B_n = 0$ for all other $n$. The resulting matrix is

$$\begin{bmatrix} 1 & 1 \\ -\Delta x & 0 \end{bmatrix} \begin{bmatrix} \gamma_{i-1} \\ \gamma_i \end{bmatrix} = \begin{bmatrix} 0 \\ 1 \end{bmatrix} \quad (6.30)$$

where the subscript $i-1$ indicates one node to the left of $i$. The matrix has solution $\gamma_{i-1} = -1/\Delta x$ and $\gamma_i = 1/\Delta x$. Substituting these coefficients into

$$\frac{\partial N}{\partial x} \approx \gamma_1 N_1 + \gamma_2 N_2 = \gamma_{i-1} N_{i-1} + \gamma_i N_i \quad (6.31)$$

from (6.24) gives the approximation shown in (6.22) and Table 6.2 (a).

**Table 6.2.** Finite-Difference Approximations of $\partial N/\partial x$ and $\partial^2 N/\partial x^2$

| Order | $m$ | $q$ | Approximation |
|---|---|---|---|
| (a) 1st order backward | 1 | 2 | $\frac{\partial N}{\partial x} \approx \frac{N_i - N_{i-1}}{\Delta x}$ |
| (b) 1st order forward | 1 | 2 | $\frac{\partial N}{\partial x} \approx \frac{N_{i+1} - N_i}{\Delta x}$ |
| (c) 2nd order central | 1 | 3 | $\frac{\partial N}{\partial x} \approx \frac{N_{i+1} - N_{i-1}}{2\,\Delta x}$ |
| (d) 2nd order backward | 1 | 3 | $\frac{\partial N}{\partial x} \approx \frac{N_{i-2} - 4N_{i-1} + 3N_i}{2\,\Delta x}$ |
| (e) 2nd order forward | 1 | 3 | $\frac{\partial N}{\partial x} \approx \frac{-3N_i + 4N_{i+1} - N_{i+2}}{2\,\Delta x}$ |
| (f) 3rd order backward | 1 | 4 | $\frac{\partial N}{\partial x} \approx \frac{N_{i-2} - 6N_{i-1} + 3N_i + 2N_{i+1}}{6\,\Delta x}$ |
| (g) 3rd order forward | 1 | 4 | $\frac{\partial N}{\partial x} \approx \frac{-2N_{i-1} - 3N_i + 6N_{i+1} - N_{i+2}}{6\,\Delta x}$ |
| (h) 4th order central | 1 | 5 | $\frac{\partial N}{\partial x} \approx \frac{N_{i-2} - 8N_{i-1} + 8N_{i+1} - N_{i+2}}{12\,\Delta x}$ |
| (i) 4th order backward (I) | 1 | 5 | $\frac{\partial N}{\partial x} \approx \frac{-N_{i-3} + 6N_{i-2} - 18N_{i-1} + 10N_i + 3N_{i+1}}{12\,\Delta x}$ |
| (j) 4th order forward (I) | 1 | 5 | $\frac{\partial N}{\partial x} \approx \frac{-3N_{i-1} - 10N_i + 18N_{i+1} - 6N_{i+2} + N_{i+3}}{12\,\Delta x}$ |
| (k) 4th order backward (II) | 1 | 5 | $\frac{\partial N}{\partial x} \approx \frac{-3N_{i-4} + 16N_{i-3} - 36N_{i-2} + 48N_{i-1} - 25N_i}{12\,\Delta x}$ |
| (l) 4th order forward (II) | 1 | 5 | $\frac{\partial N}{\partial x} \approx \frac{25N_i - 48N_{i+1} + 36N_{i+2} - 16N_{i+3} + 3N_{i+4}}{12\,\Delta x}$ |
| (m) 2nd order central | 2 | 3 | $\frac{\partial^2 N}{\partial x^2} \approx \frac{N_{i+1} - 2N_i + N_{i-1}}{\Delta x^2}$ |
| (n) 4th order central | 2 | 5 | $\frac{\partial^2 N}{\partial x^2} \approx \frac{-N_{i-2} + 16N_{i-1} - 30N_i + 16N_{i+1} - N_{i+2}}{12\,\Delta x^2}$ |

**Second-order central- and backward-difference approximations** of $\partial N/\partial x$ ($m = 1$) are found by discretizing $\partial N/\partial x$ across three nodes ($q = 3$). The resulting matrices are

$$\begin{bmatrix} 1 & 1 & 1 \\ -\Delta x & 0 & \Delta x \\ (-\Delta x)^2 & 0 & (\Delta x)^2 \end{bmatrix} \begin{bmatrix} \gamma_{i-1} \\ \gamma_i \\ \gamma_{i+1} \end{bmatrix} = \begin{bmatrix} 0 \\ 1 \\ 0 \end{bmatrix} \tag{6.32}$$

$$\begin{bmatrix} 1 & 1 & 1 \\ -2\,\Delta x & -\Delta x & 0 \\ (-2\,\Delta x)^2 & (-\Delta x)^2 & 0 \end{bmatrix} \begin{bmatrix} \gamma_{i-2} \\ \gamma_{i-1} \\ \gamma_i \end{bmatrix} = \begin{bmatrix} 0 \\ 1 \\ 0 \end{bmatrix}$$

respectively. Substituting solutions to these matrices into (6.24) gives the approximation shown in Table 6.2 (c) and (d), respectively. The **second-order forward-difference approximation** of $\partial N/\partial x$ is found by discretizing around the first column in (6.32). The result is shown Table 6.2 (e). Forward- and backward-difference discretizations are negatively symmetric to each other.

**Third-order backward- and forward-difference approximations** of $\partial N/\partial x$ are found in a similar manner. The results are shown in Table 6.2 (f) and (g), respectively, where the discretizations are around four cells.

A **fourth-order central-difference approximation** of $\partial N/\partial x$ is found from

$$\begin{bmatrix} 1 & 1 & 1 & 1 & 1 \\ -2\,\Delta x & -\Delta x & 0 & \Delta x & 2\,\Delta x \\ (-2\,\Delta x)^2 & (-\Delta x)^2 & 0 & (\Delta x)^2 & (2\,\Delta x)^2 \\ (-2\,\Delta x)^3 & (-\Delta x)^3 & 0 & (\Delta x)^3 & (2\,\Delta x)^3 \\ (-2\,\Delta x)^4 & (-\Delta x)^4 & 0 & (\Delta x)^4 & (2\,\Delta x)^4 \end{bmatrix} \begin{bmatrix} \gamma_{i-2} \\ \gamma_{i-1} \\ \gamma_i \\ \gamma_{i+1} \\ \gamma_{i+2} \end{bmatrix} = \begin{bmatrix} 0 \\ 1 \\ 0 \\ 0 \\ 0 \end{bmatrix} \tag{6.33}$$

The approximation resulting from this matrix is shown in Table 6.2 (h). One **fourth-order backward-difference approximation** of $\partial N/\partial x$ is obtained by solving (6.33) after discretizing around the fourth instead of the third column. The result is shown in Table 6.2 (i). The corresponding **fourth-order forward-difference approximation** is shown in Table 6.2 (j). Another fourth-order backward-difference approximation of $\partial N/\partial x$ is obtained by solving (6.33) after discretizing around the fifth column. The result appears in Table 6.2 (k). The corresponding fourth-order forward difference approximation appears in Table 6.2 (l). A **fourth-order central-difference approximation** of $\partial^2 N/\partial x^2$ is obtained by solving (6.33), but setting $B_2 = 1$ and $B_n = 0$ for all other $n$. The solution is shown in Table 6.2 (n).

### 6.4.4. Solutions to the Advection–Diffusion Equation of Low Order in Time

The finite-difference approximations discussed above can be readily applied to the advection–diffusion equations, given in (6.1)–(6.3). In the following sub-subsections, approximations of low order in time are discussed with respect to the west–east advection–diffusion equation.

#### 6.4.4.1. *Courant–Friedrichs–Lewy Stability Criterion*

If the eddy diffusion term is removed, (6.1) simplifies to the west–east **advection equation.** For certain explicit finite-difference approximations of the advection equation, stability is maintained when

$$h < \Delta x_{\min}/|u_{\max}| \tag{6.34}$$

where $u_{\max}$ is the maximum west–east scalar velocity, and $\Delta_{x_{\min}}$ is the minimum west–east grid-cell length in the domain. (6.34) is the **Courant–Friedrichs–Lewy (CFL) stability criterion** (Courant, et al. 1928). If the maximum scalar velocity is $u_{\max} = 20$ m s$^{-1}$, and the minimum grid-cell length is $\Delta x_{\min} = 5$ km, (6.34) predicts that the maximum time step for maintaining stability is $h = 250$ s. To maintain stability, a parcel of air is not allowed to travel across a grid cell during a single time step.

For equations more general than the advection–diffusion equation, $u_{\max}$ should be replaced by $c_{\max}$, where $c_{\max}$ is the maximum speed of propagation in the domain. In the case of the primitive equations, the maximum speed of propagation corresponds to the speed of horizontally propagating acoustic waves.

If advection is ignored, (6.1) simplifies to the west–east **diffusion equation**. A stability criterion for this equation, analogous to (6.34), is $h < \Delta x_{\min}^2 / K_{\max}$, where $K_{\max}$ is the largest eddy diffusion coefficient in the domain. For a typical vertical eddy diffusion coefficient of 50 $m^2\ s^{-1}$, this stability criterion predicts that the time step for eddy diffusion through an altitude of 100 m needs to be less than 200 s. For a typical horizontal eddy diffusion coefficient of 2,500 $m^2\ s^{-1}$, the time step for eddy diffusion across a 5-km cell must be less than 10,000 s. Explicit solutions to the diffusion equation are more likely to become unstable in the vertical than in the horizontal direction in an atmospheric model.

#### 6.4.4.2. *Forward Euler Scheme*

A basic time-discretization scheme used to solve the west–east advection–diffusion equation is the forward Euler scheme. If $u$ varies and $K_{h,xx}$ is constant along $x$, and if $u$ and $K$ are constant during a time step, the time, advection, and diffusion derivatives in (6.1) can be discretized with (6.23), (6.20), and (6.16), respectively, to yield

$$\frac{N_{i,t} - N_{i,t-h}}{h} + \frac{(uN)_{i+1,t-h} - (uN)_{i-1,t-h}}{2\Delta x} - K\frac{N_{i+1,t-h} - 2N_{i,t-h} + N_{i-1,t-h}}{\Delta x^2} = 0 \tag{6.35}$$

where $K = K_{h,xx}$. The time and spatial derivatives in this equation are first- and second-order approximations, respectively. The equation is called the **forward-in-time, centered-in-space** (FTCS) approximation because the time derivative uses information from one previous time step, and the advection terms are central-difference expressions. Since all terms in (6.35), except the final concentration, are evaluated at time $t-h$, the FTCS approximation is **explicit.** In an explicit equation, final terms (time $t$) are calculated explicitly from known values (e.g., from values at times $t-h$, $t-2h$, etc.). When a time derivative is first-order and spatial derivatives are determined explicitly, as in the example above, the time scheme is a **forward Euler** one. For all values of $u$, this equation is unconditionally unstable for $K = 0$ and for large values of $K$ and conditionally stable for small values of $K$, except when $K = 0$ (Mesinger and Arakawa 1976).

Because values on the right side of (6.35) are known, the equation can be solved immediately for $i = 1, \ldots, I$ (where $I$ is the number of west–east nodes). If the grid contains lateral boundaries, the solution depends on the terms $(uN)_{0,t-h}$, $(uN)_{I+1,t-h}$, $(KN)_{0,t-h}$, and $(KN)_{I+1,t-h}$, which lie beyond the boundaries. Outside boundary values may be set equal to values just inside the boundary (e.g., at nodes $i = 1$ and $i = I$) from the previous time step. When the grid wraps around on itself, node $i = 0$ is also node $i = I$, and node $i = 1$ is also node $i = I + 1$, and outside boundary values are not needed.

### 6.4.4.3. *Implicit Scheme*

(6.35) can be solved with an **implicit** time-stepping scheme, which is a scheme that evaluates all terms in an equation at time $t$. In implicit form, (6.35) becomes

$$\frac{N_{i,t}-N_{i,t-h}}{h}+\frac{(uN)_{i+1,t}-(uN)_{i-1,t}}{2\Delta x}-K\frac{N_{i+1,t}-2N_{i,t}+N_{i-1,t}}{\Delta x^2} = 0 \tag{6.36}$$

The solution to (6.36) along $i = 1, \ldots, I$ is obtained by rearranging the equation as

$$A_i N_{i-1,t}+B_i N_{i,t}+D_i N_{i+1,t}=N_{i,t-h} \tag{6.37}$$

where

$$A_i=-h\left(\frac{u}{2\,\Delta x}+\frac{K}{\Delta x^2}\right)_{i-1} \qquad B_i=1+h\left(\frac{2K}{\Delta x^2}\right)_i \qquad D_i=h\left(\frac{u}{2\,\Delta x}-\frac{K}{\Delta x^2}\right)_{i+1} \tag{6.38}$$

For a domain with **boundaries at both ends,** the matrix arising from (6.37) is

$$\begin{bmatrix} B_1 & D_1 & 0 & 0 & \cdots & 0 & 0 & 0 \\ A_2 & B_2 & D_2 & 0 & \cdots & 0 & 0 & 0 \\ 0 & A_3 & B_3 & D_3 & \cdots & 0 & 0 & 0 \\ 0 & 0 & A_4 & B_4 & \cdots & 0 & 0 & 0 \\ \vdots & \vdots & \vdots & \vdots & & \vdots & \vdots & \vdots \\ 0 & 0 & 0 & 0 & \cdots & B_{I-2} & D_{I-2} & 0 \\ 0 & 0 & 0 & 0 & \cdots & A_{I-1} & B_{I-1} & D_{I-1} \\ 0 & 0 & 0 & 0 & \cdots & 0 & A_I & B_I \end{bmatrix} \begin{bmatrix} N_{1,t} \\ N_{2,t} \\ N_{3,t} \\ N_{4,t} \\ \vdots \\ N_{I-2,t} \\ N_{I-1,t} \\ N_{I,t} \end{bmatrix} = \begin{bmatrix} N_{1,t-h} \\ N_{2,t-h} \\ N_{3,t-h} \\ N_{4,t-h} \\ \vdots \\ N_{I-2,t-h} \\ N_{I-1,t-h} \\ N_{I,t-h} \end{bmatrix} - \begin{bmatrix} A_1 N_{0,t} \\ 0 \\ 0 \\ 0 \\ \vdots \\ 0 \\ 0 \\ D_I N_{I+1,t} \end{bmatrix} \tag{6.39}$$

where $A_1 N_{0,t}$ and $D_I N_{I+1,t}$ are outside boundary values. Outside values are assumed to be known in advance although they carry the subscript $t$. (6.39) is a **tridiagonal matrix,** which is solved by matrix decomposition and backsubstitution in the order:

*Decomposition:*

$$\gamma_1=-\frac{D_1}{B_1} \qquad \gamma_i=-\frac{D_i}{B_i+A_i\gamma_{i-1}} \quad \text{for } i=2,\ldots,I \tag{6.40}$$

$$\alpha_1=\frac{R_1}{B_1} \qquad \alpha_i=\frac{R_i-A_i\alpha_{i-1}}{B_i+A_i\gamma_{i-1}} \quad \text{for } i=2,\ldots,I$$

*Backsubstitution:*

$$N_{I,t}=\alpha_I \qquad N_{i,t}=\alpha_i+\gamma_i N_{i+1,t} \quad \text{for } i=I-1,\ldots,1,-1 \tag{6.41}$$

The solution to (6.39) is mass-conserving and unconditionally stable for all values of $u$ and $K$, but it is numerically diffusive. To obtain the solution, (6.36) was converted from a partial differential equation (dependent on time and space) to an ordinary differential equation (dependent on time only). The linear ODE was then solved by matrix decomposition and backsubstitution.

For a grid that wraps around on itself (no lateral boundaries), the matrix from (6.39) becomes

$$\begin{bmatrix} B_1 & D_1 & 0 & 0 & \cdots & 0 & 0 & A_1 \\ A_2 & B_2 & D_2 & 0 & \cdots & 0 & 0 & 0 \\ 0 & A_3 & B_3 & D_3 & \cdots & 0 & 0 & 0 \\ 0 & 0 & A_4 & B_4 & \cdots & 0 & 0 & 0 \\ \vdots & \vdots & \vdots & \vdots & & \vdots & \vdots & \vdots \\ 0 & 0 & 0 & 0 & \cdots & B_{I-2} & D_{I-2} & 0 \\ 0 & 0 & 0 & 0 & \cdots & A_{I-1} & B_{I-1} & D_{I-1} \\ D_I & 0 & 0 & 0 & \cdots & 0 & A_I & B_I \end{bmatrix} \begin{bmatrix} N_{1,t} \\ N_{2,t} \\ N_{3,t} \\ N_{4,t} \\ \vdots \\ N_{I-2,t} \\ N_{I-1,t} \\ N_{I,t} \end{bmatrix} = \begin{bmatrix} N_{1,t-h} \\ N_{2,t-h} \\ N_{3,t-h} \\ N_{4,t-h} \\ \vdots \\ N_{I-2,t-h} \\ N_{I-1,t-h} \\ N_{I,t-h} \end{bmatrix} \tag{6.42}$$

where $D_I = D_0$ and $A_1 = A_{I+1}$. The solution is obtained by solving (6.40), followed by

$$\begin{aligned} \chi_1 &= -\frac{A_1}{B_1} & \chi_i &= -\frac{A_i \chi_{i-1}}{B_i + A_i \gamma_{i-1}}, & &\text{for } i = 2, \ldots, I \\ \psi_I &= 1 & \psi_i &= \gamma_i \psi_{i+1} + \chi_i, & &\text{for } i = I-1, \ldots, 1, -1 \\ \beta_I &= 0 & \beta_i &= \gamma_i \beta_{i+1} + \alpha_i & &\text{for } i = I-1, \ldots, 1, \ 1 \end{aligned} \tag{6.43}$$

$$N_{I,t} = \frac{\alpha_I - \frac{\beta_1 D_I}{B_I + A_I \gamma_{I-1}}}{1 + \frac{D_I \psi_1 + A_I \chi_I}{B_I + A_I \gamma_{I-1}}}$$

$$N_{i,t} = \alpha_i + \gamma_i N_{i+1,t} + \chi_i N_I, t \qquad \text{for } i = I-1, \ldots, 1, -1$$

This solution is mass-conserving but does not require outside boundary information.

#### 6.4.4.4. *Crank–Nicolson Scheme*

The implicit approximation just described was first-order in time and second-order in space. The order of approximation in time can be improved to second order with the **Crank–Nicolson** (trapezoidal) **scheme** (Crank and Nicolson 1947). This scheme is **semiimplicit** in that some terms are evaluated at time $t$ and others are evaluated at time $t - h$. With the Crank–Nicolson scheme, spatial derivatives are weighted 50 percent between each the initial and final times. Rewriting (6.35) gives

$$\begin{aligned} &\frac{N_{i,t} - N_{i,t-h}}{h} + \left[\mu_c \frac{(uN)_{i+1,t} - (uN)_{i-1,t}}{2\,\Delta x}\right. \\ &\left. + (1-\mu_c)\frac{(uN)_{i+1,t-h} - (uN)_{i-1,t-h}}{2\,\Delta x}\right] - K\left[\mu_c \frac{N_{i+1,t} - 2N_{i,t} + N_{i-1,t}}{\Delta x^2}\right. \\ &\left. + (1-\mu_c)\frac{N_{i+1,t-h} - 2N_{i,t-h} + N_{i-1,t-h}}{\Delta x^2}\right] = 0 \end{aligned} \tag{6.44}$$

where $\mu_c$ is the **Crank–Nicolson parameter**. When a finite-difference equation is written in terms of $\mu_c$, the equation is in **Crank–Nicolson form**. When $\mu_c = 0.5$, (6.44) reduces to the Crank–Nicolson scheme, which is second-order in time and absolutely stable for all values of $u$ and $K$. When $\mu_c = 0$, (6.44) reduces to (6.35), the forward Euler scheme, and when $\mu_c = 1$, (6.44) reduces to (6.36), the implicit scheme. (6.44) can be rewritten as

$$A_i N_{i-1,t} + B_i N_{i,t} + D_i N_{i+1,t} = E_i N_{i-1,t-h} + F_i N_{i,t-h} + G_i N_{i+1,t-h} \quad (6.45)$$

where

$$A_i = -\mu_c h\left(\frac{u}{2\,\Delta x} + \frac{K}{\Delta x^2}\right)_{i-1} \qquad B_i = 1 + \mu_c h\left(\frac{2K}{\Delta x^2}\right)_i \quad (6.46)$$

$$D_i = \mu_c h\left(\frac{u}{2\,\Delta x} - \frac{K}{\Delta x^2}\right)_{i+1} \qquad E_i = (1-\mu_c)h\left(\frac{u}{2\,\Delta x} + \frac{K}{\Delta x^2}\right)_{i-1}$$

$$F_i = 1 - (1-\mu_c)h\left(\frac{2K}{\Delta x^2}\right)_i \qquad G_i = -(1-\mu_c)h\left(\frac{u}{2\,\Delta x} - \frac{K}{\Delta x^2}\right)_{i+1}$$

With lateral boundaries, the matrix arising from (6.45) is the same as (6.39), except that the right side of (6.39) is replaced with

$$= \begin{bmatrix} F_1 & G_1 & 0 & 0 & \cdots & 0 & 0 & 0 \\ E_2 & F_2 & G_2 & 0 & \cdots & 0 & 0 & 0 \\ 0 & E_3 & F_3 & G_3 & \cdots & 0 & 0 & 0 \\ 0 & 0 & E_4 & F_4 & \cdots & 0 & 0 & 0 \\ \vdots & \vdots & \vdots & \vdots & & \vdots & \vdots & \vdots \\ 0 & 0 & 0 & 0 & \cdots & F_{I-2} & G_{I-2} & 0 \\ 0 & 0 & 0 & 0 & \cdots & E_{I-1} & F_{I-1} & G_{I-1} \\ 0 & 0 & 0 & 0 & \cdots & 0 & E_I & F_I \end{bmatrix} \begin{bmatrix} N_{1,t-h} \\ N_{2,t-h} \\ N_{3,t-h} \\ N_{4,t-h} \\ \vdots \\ N_{I-2,t-h} \\ N_{I-1,t-h} \\ N_{I,t-h} \end{bmatrix} + \begin{bmatrix} E_1 N_{0,t-h} - A_1 N_{0,t} \\ 0 \\ 0 \\ 0 \\ \vdots \\ 0 \\ 0 \\ G_I N_{I+1,t-h} - D_I N_{I+1,t} \end{bmatrix} \quad (6.47)$$

(6.39) and (6.47) are solved with (6.40) and (6.41). For a grid that wraps around on itself, (6.39) and (6.47) are solved after the rightmost column in (6.47) is removed, after $A_1$ and $D_I$ are placed in the top right and bottom left corners, respectively, of (6.39), and after $E_1$ and $G_I$ are placed in the top right and bottom left corners, respectively, of (6.47).

#### 6.4.4.5. *Leapfrog Scheme*

Another scheme that increases the order of approximation in time to second-order is the **leapfrog scheme.** This scheme uses information from two previous time steps to predict information for a third. More specifically, spatial derivatives from time $t - h$ are used to evaluate differences between times $t$ and $t - 2h$. The leapfrog solution to the west–east advection–diffusion equation is

$$\frac{N_{i,t} - N_{i,t-2h}}{2h} + \frac{(uN)_{i+1,t-h} - (uN)_{i-1,t-h}}{2\,\Delta x} - K\frac{N_{i+1,t-h} - 2N_{i,t-h} + N_{i-1,t-h}}{\Delta x^2} = 0 \quad (6.48)$$

where values from time $t - h$ and $t - 2h$ are determined from previous time steps.

The discretization above is second-order in time and space. When used alone, the leapfrog scheme is unconditionally unstable for all nonzero values of $K$. When $K = 0$, the leapfrog scheme is conditionally stable for linear equations. For nonlinear equations, the scheme destabilizes over time. To suppress such instability, computations from another scheme must be inserted every few leapfrog steps (Mesinger and Arakawa 1976). A scheme used to stabilize the leapfrog scheme is the Matsuno scheme.

#### 6.4.4.6. *Matsuno Scheme*

The **Matsuno scheme** (Matsuno 1966) is an explicit time-stepping scheme commonly used to stabilize and initialize the leapfrog scheme. With the Matsuno scheme, time derivatives are predicted with a forward-difference approximation. The predicted values are substituted into the spatial derivatives to estimate final values. The prediction and correction steps are

$$\frac{N_{i,est} - N_{i,t-h}}{h} + \frac{(uN)_{i+1,t-h} - (uN)_{i-1,t-h}}{2\,\Delta x} - K\frac{N_{i+1,t-h} - 2N_{i,t-h} + N_{i-1,t-h}}{\Delta x^2} = 0 \tag{6.49}$$

$$\frac{N_{i,t} - N_{i,t-h}}{h} + \frac{(uN)_{i+1,est} - (uN)_{i-1,est}}{2\,\Delta x} - K\frac{N_{i+1,est} - 2N_{i,est} + N_{i-1,est}}{\Delta x^2} = 0 \tag{6.50}$$

respectively. Although the Matsuno scheme requires twice as many computations as either the forward Euler or the leapfrog scheme per time step, the Matsuno scheme is still a first-order approximation in time. It is conditionally stable for all values of $u$ when $K$ is zero or small but absolutely unstable for large values of $K$ (Mesinger and Arakawa 1976). When combined with the leapfrog scheme to solve the equations of atmospheric dynamics, Matsuno steps are usually taken every 5–15 leapfrog steps.

#### 6.4.4.7. *Heun Scheme*

With the **Heun scheme**, time derivatives are estimated with a forward-difference approximation that uses initial values in the spatial derivative. Final time derivatives are determined with an average spatial derivative. The average is taken as one-half the spatial derivative determined from initial values plus one-half the spatial derivative determined from estimated values. With respect to the advection–diffusion equation, the Heun scheme involves solving (6.49) followed by

$$\frac{N_{i,t} - N_{i,t-h}}{h} + \frac{1}{2}\left(\frac{(uN)_{i+1,est} - (uN)_{i-1,est}}{2\,\Delta x}\right) - \frac{K}{2}\left(\frac{N_{i+1,est} - 2N_{i,est} + N_{i-1,est}}{\Delta x^2}\right) + \frac{1}{2}\left(\frac{(uN)_{i+1,t-h} - (uN)_{i-1,t-h}}{2\,\Delta x}\right) - \frac{K}{2}\left(\frac{N_{i+1,t-h} - 2N_{i,t-h} + N_{i-1,t-h}}{\Delta x^2}\right) = 0 \tag{6.51}$$

The Heun scheme is a second-order approximation in time. For all values of $u$, this scheme is unconditionally unstable for $K = 0$ and for large values of $K$ and conditionally stable for small values of $K$, except when $K = 0$ (Mesinger and Arakawa 1976).

#### 6.4.4.8. *Adams–Bashforth Scheme*

The last time-differencing scheme discussed is a simplified version of the **Adams–Bashforth scheme.** Like the leapfrog scheme, this scheme is explicit, uses three time levels, and is a second-order approximation in time. The scheme discretizes the advection–diffusion equation as

$$\begin{aligned} &\frac{N_{i,t} - N_{i,t-h}}{h} + \frac{3}{2}\left(\frac{(uN)_{i+1,t-h} - (uN)_{i-1,t-h}}{2\Delta x}\right) \\ &- \frac{3}{2}K\left(\frac{N_{i+1,t-h} - 2N_{i,t-h} + N_{i-1,t-h}}{\Delta x^2}\right) \\ &- \frac{1}{2}\left(\frac{(uN)_{i+1,t-2h} - (uN)_{i-1,t-2h}}{2\Delta x}\right) \\ &+ \frac{1}{2}K\left(\frac{N_{i+1,t-2h} - 2N_{i,t-2h} + N_{i-1,t-2h}}{\Delta x^2}\right) = 0 \end{aligned} \tag{6.52}$$

where the $t - h$ time level is favored over the $t - 2h$ time level. For all values of $u$, the Adams–Bashforth scheme is unconditionally unstable for $K = 0$ and for large values of $K$ and conditionally stable for small values of $K$, except when $K = 0$ (Mesinger and Arakawa 1976). This scheme is useful for short integration periods when a small time step is taken.

### 6.4.5. Fourth Order in Space Solution to the Advection–Diffusion Equation

The fully explicit, fully implicit, and Crank–Nicolson time-difference schemes can be extended to higher order in space by using finite-difference approximations derived earlier. Substituting fourth-order expansions of advection and diffusion terms from Table 6.2(h) and (n), respectively, into (6.1) gives a fully implicit, unconditionally stable form of the equation as

$$\begin{aligned} &\frac{N_{i,t} - N_{i,t-h}}{h} + \frac{(uN)_{i-2,t} - 8(uN)_{i-1,t} + 8(uN)_{i+1,t} - (uN)_{i+2,t}}{12\,\Delta x} \\ &- K\frac{-N_{i-2,t} + 16N_{i-1,t} - 30N_{i,t} + 16N_{i+1,t} - N_{i+2,t}}{12\,\Delta x^2} = 0 \end{aligned} \tag{6.53}$$

This equation can also be written in banded matrix form with five terms on both sides and solved with a banded-matrix method (e.g., Press et al. 1992). Such methods may combine matrix decomposition and/or backsubstitution with a sparse-matrix technique to reduce the number of computations.

### 6.4.6. Variable Grid Spacing and Eddy Diffusion Coefficients

The previous solutions to the advection–diffusion equation were obtained by assuming constant grid spacing and eddy diffusion coefficients. If grid spacing and diffusion coefficients vary in space, the second-order discretization of the eddy diffusion term is

$$\frac{\partial}{\partial x}\left(K\frac{\partial N}{\partial x}\right) = \frac{\frac{K_{i+1/2}(N_{i+1}-N_i)}{x_{i+1}-x_i} - \frac{K_{i-1/2}(N_i-N_{i-1})}{x_i-x_{i-1}}}{x_{i+1/2}-x_{i-1/2}} \tag{6.54}$$

where $K_{i+1/2} = 0.5(K_i + K_{i+1})$, $K_{i-1/2} = 0.5(K_{i-1} + K_i)$, $x_{i+1/2} = 0.5(x_i + x_{i+1})$, and $x_{i-1/2} = 0.5(x_{i-1} + x_i)$. Another form of (6.54) is

$$\frac{\partial}{\partial x}\left(K\frac{\partial N}{\partial x}\right) = \beta_{K,i-1}N_{i-1} + \beta_{K,i}N_i + \beta_{K,i+1}N_{i+1} \tag{6.55}$$

where

$$\beta_{K,i-1} = \frac{2K_{i-1/2}}{(x_i - x_{i-1})(x_{i+1} - x_{i-1})} \tag{6.56}$$

$$\beta_{K,i} = \frac{-2\left[(x_{i+1} - x_i)K_{i-1/2} + (x_i - x_{i-1})K_{i+1/2}\right]}{(x_{i+1} - x_i)(x_i - x_{i-1})(x_{i+1} - x_{i-1})}$$

$$\beta_{K,i+1} = \frac{2K_{i+1/2}}{(x_{i+1} - x_i)(x_{i+1} - x_{i-1})}$$

If grid spacing and wind speed vary in space, the second-order central-difference form of the advection term in (6.1) can be approximated as

$$\frac{\partial(uN)}{\partial x} = \gamma_{a,i-1}(uN)_{i-1} + \gamma_{a,i}(uN)_i + \gamma_{a,i+1}(uN)_{i+1} \tag{6.57}$$

The coefficients for this equation are

$$\gamma_{a,i-1} = \frac{(x_{i+1} - x_i) - 2(x_i - x_{i-1})}{(x_i - x_{i-1})(x_{i+1} - x_{i-1})} \tag{6.58}$$

$$\gamma_{a,i} = \frac{(x_{i+1} - x_i) - (x_i - x_{i-1})}{(x_{i+1} - x_i)(x_i - x_{i-1})}$$

$$\gamma_{a,i+1} = \frac{x_i - x_{i-1}}{(x_{i+1} - x_i)(x_{i+1} - x_{i-1})}$$

which were derived from

$$\begin{bmatrix} 1 & 1 & 1 \\ -(x_i - x_{i-1}) & 0 & (x_{i+1} - x_i) \\ (x_i - x_{i-1})^2 & 0 & (x_{i+1} - x_i)^2 \end{bmatrix} \begin{bmatrix} \gamma_{a,i-1} \\ \gamma_{a,i} \\ \gamma_{a,i+1} \end{bmatrix} = \begin{bmatrix} 0 \\ 1 \\ 0 \end{bmatrix} \tag{6.59}$$

Applying (6.55) and (6.57) to the advection–diffusion equation in (6.1) gives a solution of second order in space that allows for variable grid spacing, wind speeds, and eddy diffusion coefficients. In Crank–Nicolson form, the resulting advection–diffusion equation is

$$\begin{aligned}\frac{N_{i,t} - N_{i,t+h}}{h} &= \mu_c\{[(\gamma_a u - \beta_K)N]_{i-1} + [(\gamma_a u - \beta_K)N]_i \\ &+ [(\gamma_a u - \beta_K)N]_{i+1}\}_t + (1 - \mu_c)\{[(\gamma_a u - \beta_K)N]_{i-1} \\ &+ [(\gamma_a u - \beta_K)N]_i + [(\gamma_a u - \beta_K)N]_{i+1}\}_{t-h}\end{aligned} \tag{6.60}$$

This equation can be written in tridiagonal matrix form and solved.

### 6.4.7. Finite-Differencing in Two Directions

Combining west–east and south–north directions gives the horizontal advection–diffusion equation as

$$\frac{\partial N}{\partial t} + \frac{\partial(uN)}{\partial x} + \frac{\partial(vN)}{\partial y} - \frac{\partial}{\partial x}\left(K_{h,xx}\frac{\partial N}{\partial x}\right) - \frac{\partial}{\partial y}\left(K_{h,yy}\frac{\partial N}{\partial y}\right) = 0 \tag{6.61}$$

Finite-differencing of this equation is similar to finite-differencing of the unidirectional west–east equation. When constant grid spacing and eddy diffusion coefficients are assumed, the implicit expansion of (6.61) to second order in space is

$$\begin{aligned}\frac{N_{i,j,t} - N_{i,j,t-h}}{h} &+ \left[\frac{(uN)_{i+1,j} - (uN)_{i-1,j}}{2\,\Delta x}\right. \\ &+ \left.\frac{(vN)_{i,j+1} - (vN)_{i,j-1}}{2\,\Delta y}\right]_t - \left(K_{h,xx}\frac{N_{i-1,j} - 2N_{i,j} + N_{i+1,j}}{\Delta x^2}\right. \\ &+ \left.K_{h,yy}\frac{N_{i,j-1} - 2N_{i,j} + N_{i,j+1}}{\Delta y^2}\right)_t = 0\end{aligned} \tag{6.62}$$

where $j$ is the grid index in the north–south direction. This equation is linear for all $i$ and $j$ and can be solved implicitly in matrix form, just as with the one-dimensional case. Although the matrix is not banded, it may be solved for by decomposition and backsubstitution. (6.62) can also be extended to three dimensions, put in Crank–Nicolson form, and solved with variable grid spacing and eddy diffusion coefficients.

The primary disadvantage of solving (6.62) implicitly is that the matrix order quickly becomes large. For a $100 \times 100$ horizontal grid, the required matrix is $10{,}000 \times 10{,}000$. The advantage of solving the equation implicitly is that the solution is unconditionally stable for all values of $u$, $v$, $K_{h,xx}$, and $K_{h,yy}$. If (6.62) is solved explicitly, it is unconditionally unstable for all values of $u$ and $v$ when

$K = 0$ or $K$ is large and conditionally stable for all values of $u$ and $v$ for other values of $K$.

### 6.4.8. The Semi-Lagrangian Method

A special case of the finite-difference method is the **semi-Lagrangian method** (e.g., Pepper et al. 1979; Robert 1982; Staniforth and Cote 1991; Makar and Karpik 1996). With this method, the value of a variable at a specific location and time is obtained by tracing back where the air parcel containing the variable came from during the last time step. Suppose points A and B lie on a straight line in the $x$-direction. If the wind speed at point B is 5 m $s^{-1}$ and the model time step is $h = 300$ s, the concentration of water vapor at point B and time $t$ is estimated as the concentration at point A at time $t-h$, where point A is $\Delta x = 5 \text{ m s}^{-1} \times 300 \text{ s} = 1500$ m to the west of point B. During the time interval, the wind advects the scalar from point A to B. Whereas location B is a grid node in a numerical model, point A is a random location that must be determined by examining the west–east scalar velocity and time step. Once point A has been found, values of the variable at point A must be interpolated from values at nearby fixed grid nodes.

In sum, with the semi-Lagrangian method, the concentration of a gas at time $t$ and location $x$ is estimated as

$$N_{x,t} = N_{x-uh,t-h} \tag{6.63}$$

where $N$ at location $x-uh$ and time $t-h$ is interpolated from values of $N$ at nearby node points. Mixing ratios, potential temperatures, $u$-velocities, and $v$-velocities can be estimated in a similar manner. (6.63) can be written for one, two, or three directions.

Several methods exist to interpolate a variable to point A in the west–east case. One method is to estimate the slope of the variable between two nodes surrounding point A and interpolate linearly between the two nodes. A more complex method is to fit a polynomial through three or more adjacent nodes surrounding point A and to calculate the value at point A from the polynomial. When the polynomial is a cubic, the method is the **cubic spline method** (Price and MacPherson 1973; Purnell 1976).

An advantage of semi-Lagrangian schemes is that they can be subjected to long time steps without concern for stability. Advection of tracers, such as water vapor and potential temperature, is sometimes more accurate with a semi-Lagrangian scheme than with a pure finite-difference scheme. A disadvantage of semi-Lagrangian schemes is that, unless proper steps are taken, the mass of a trace species is not conserved during advection. Mass loss affects results over long periods. Transported mass can be conserved during a time step if the fitted spatial concentration curve, used for interpolation, is normalized so that the integral of mass under the curve equals the total mass in the system at the beginning of the time step.

## 6.5. SERIES EXPANSION METHODS

With a finite-difference method, each differential in a PDE is replaced with a difference analog written in terms of a finite number of values along a temporal or spatial direction. With a **series expansion method**, a dependent variable (e.g., $u$, $v$, $w$, $N$) in a PDE is replaced with a finite series that approximates its value. If the PDE arising from the west–east advection equation at node $i$ is

$$\frac{\partial N_i}{\partial t} + \frac{\partial (uN)_i}{\partial x} = 0 \tag{6.64}$$

the series-expansion approximation of the number concentration at node $i$ is

$$\mathbf{N}_i = \sum_j N_j \mathrm{e}_j \tag{6.65}$$

which is a **trial function**. For now, $u$ is assumed constant. The set of $j$ nodes over which the trial function is approximated is a **trial space**. The trial function is a linear combination of the true value of $N_j$ multiplied by a **basis function** $\mathrm{e}_j$ at each node. Basis functions are chosen so that $\mathbf{N}_i \approx N_i$. Because $\mathbf{N}_i$ is a finite series approximation of $N_i$, the two are not exactly equal, and the difference between them is a **residual**, $R_i$. The residual is thus the difference between the approximation and the exact function.

A series expansion method that uses a local basis function is a **finite-element method**. A series expansion method that uses a global basis function orthogonal to the residual is a **spectral method**. A common finite-element method is the **Galerkin finite-element method**. With this method, the local basis function is also orthogonal to the residual. The basis functions of other finite-element methods may or may not be orthogonal to the residual. Below, the Galerkin finite-element method is discussed, and a type of spectral method is briefly described.

### 6.5.1. Finite-Element Method

With the Galerkin finite-element method, basis functions are treated like weight functions in that they weight a residual at each of several nodes along a spatial grid (**test space**). The sum, over the test space, of the residuals multiplied by the weights is zero (e.g., Pepper et al. 1979; Celia and Gray 1992). In other words,

$$\int_x R_i \mathrm{e}_i \, \mathrm{d}x = 0 \tag{6.66}$$

In this equation, the local basis function is orthogonal to the residual. With the Galerkin method, the weight function $\mathrm{e}_i$ in (6.66) is the same as the basis function $\mathrm{e}_j$ in (6.65). The test space in the latter case also equals the trial space in the former case. For other methods, $\mathrm{e}_i$ and $\mathrm{e}_j$ may differ, and the test space may differ from the trial space.

The residual of (6.64) is

$$R_i = \left[\frac{\partial \mathbf{N}_i}{\partial t} + \frac{\partial (u\mathbf{N})_i}{\partial x}\right] - \left[\frac{\partial N_i}{\partial t} + \frac{\partial (uN)_i}{\partial x}\right] = \left[\frac{\partial \mathbf{N}_i}{\partial t} + \frac{\partial (u\mathbf{N})_i}{\partial x}\right] - 0 \quad (6.67)$$

which is the difference between the approximated and the exact form of the equation. Substituting (6.67) and (6.65) into (6.66) gives

$$\int_x \left[\frac{\partial \mathbf{N}_i}{\partial t} + \frac{\partial (u\mathbf{N}_i)}{\partial x}\right] \mathrm{e}_i \, dx \quad (6.68)$$

$$= \int_x \left[\frac{\partial}{\partial t}\left(\sum_j \mathrm{N}_j \mathrm{e}_j\right) + u\frac{\partial}{\partial x}\left(\sum_j \mathrm{N}_j \mathrm{e}_j\right)\right] \mathrm{e}_i \, dx$$

$$= \sum_j \left(\frac{\partial \mathrm{N}_j}{\partial t} \int \mathrm{e}_j \mathrm{e}_i \, dx\right) + u \sum_j \left(\mathrm{N}_j \int \frac{d\mathrm{e}_j}{dx} \mathrm{e}_i \, dx\right) = 0$$

Expanding the basis function in (6.68) over three nodes ($j = i-1, \ldots, i+1$), centered at $i$, and taking a first-order forward-difference approximation in time of the result gives

$$\frac{\mathrm{N}_{i-1,t} - \mathrm{N}_{i-1,t-h}}{h} \int_{x_{i-1}}^{x_i} \mathrm{e}_{i-1}\mathrm{e}_i \, dx + \frac{\mathrm{N}_{i,t} - \mathrm{N}_{i,t-h}}{h} \int_{x_{i-1}}^{x_{i+1}} \mathrm{e}_i \mathrm{e}_i \, dx \quad (6.69)$$

$$+ \frac{\mathrm{N}_{i+1,t} - \mathrm{N}_{i+1,t-h}}{h} \int_{x_i}^{x_{i+1}} \mathrm{e}_{i+1}\mathrm{e}_i \, dx + u\left(\mathrm{N}_{i-1,t} \int_{x_{i-1}}^{x_i} \frac{d\mathrm{e}_{i-1}}{dx} \mathrm{e}_i \, dx\right.$$

$$\left. + \mathrm{N}_{i,t} \int_{x_{i-1}}^{x_{i+1}} \frac{d\mathrm{e}_i}{dx} \mathrm{e}_i \, dx + \mathrm{N}_{i+1,t} \int_{x_i}^{x_{i+1}} \frac{d\mathrm{e}_{i+1}}{dx} \mathrm{e}_i \, dx\right) = 0$$

One set of basis functions is the **chapeau (hat) functions,**

$$\mathrm{e}_i = \begin{cases} \frac{x - x_{i-1}}{x_i - x_{i-1}} & x_{i-1} \le x \le x_i \\ \frac{x_{i+1} - x}{x_{i+1} - x_i} & x_i \le x \le x_{i+1} \\ 0 & \text{all other cases} \end{cases} \quad (6.70)$$

These weightings favor the center cell ($i$) and decrease to zero at $i-1$ and $i+1$, giving them the appearance of a peaked hat. The functions are the same for $\mathrm{e}_j$. When chapeau functions are used, the integral in the first term of (6.69) simplifies to

$$\int_{x_{i-1}}^{x_i} \mathrm{e}_{i-1}\mathrm{e}_i \, dx = \int_{x_{i-1}}^{x_i} \left(\frac{x_i - x}{x_i - x_{i-1}}\right)\left(\frac{x - x_{i-1}}{x_i - x_{i-1}}\right) dx = \frac{x - x_{i-1}}{6} \quad (6.71)$$

After other terms have been integrated, (6.69) becomes

$$\frac{(\mathrm{N}_{i-1,t} - \mathrm{N}_{i-1,t-h})\Delta x_i + (\mathrm{N}_{i,t} - \mathrm{N}_{i,t-h})\,2(\Delta x_{i+1} + \Delta x_i) + (\mathrm{N}_{i+1,t} - \mathrm{N}_{i+1,t-h})\,\Delta x_{i+1}}{6h} \quad (6.72)$$

$$+ u\frac{\mathrm{N}_{i+1,t} - \mathrm{N}_{i-1,t}}{2} = 0$$

where $\Delta x_i = x_i - x_{i-1}$ and $\Delta x_{i+1} = x_{i+1} - x_i$.

(6.69) assumes $u$ is constant. When $u$ varies, its trial function is $\mathbf{U}_i = \sum_j u_j \mathrm{e}_j$. A similar equation applies to the eddy diffusion coefficient. Toon et al. (1988) and Pepper et al. (1979) show a Galerkin method with chapeau functions as finite-elements and a Crank–Nicolson time-stepping scheme. The method uses a fourth-order approximation in space and a second-order approximation in time and solves the unidirectional advection–diffusion equation given in (6.1). The method allows variable scalar velocities, grid spacing, and eddy diffusion coefficients. The solution is found by setting up $i = 2, \ldots, I-1$ equations of the form,

$$\begin{aligned}&\frac{(N_{i-1,t} - N_{i-1,t-h})\,\Delta x_i + (N_{i,t} - N_{i,t-h})\,2(\Delta x_{i+1} + \Delta x_i) + (N_{i+1,t} - N_{i+1,t-h})\,\Delta x_{i+1}}{h}\\ &\quad + \mu_c(\gamma_i N_{i+1} + \beta_i N_i - \alpha_i N_{i-1})_t\\ &\quad + (1-\mu_c)(\gamma_i N_{i+1} + \beta_i N_i - \alpha_i N_{i-1})_{t-h} = 0\end{aligned} \tag{6.73}$$

where $\mu_c = 1, 0.5$, or 0 implies an implicit, Crank–Nicolson, or forward Euler solution, respectively, and

$$\begin{aligned}\alpha_i &= (u_i + 2u_{i-1}) + \frac{3(\rho_{a,i}K_i + \rho_{a,i-1}K_{i-1})}{\rho_{a,i-1}\,\Delta x_i}\\ \beta_i &= (-u_{i-1} + u_{i+1})\\ &\quad + \frac{3[(\rho_{a,i-1}K_{i-1} + \rho_{a,i}K_i)\Delta x_{i+1} + (\rho_{a,i}K_i + \rho_{a,i+1}K_{i+1})\Delta x_i]}{\rho_{a,i}\,\Delta x_i\,\Delta x_{i+1}}\\ \gamma_i &= (u_i + 2u_{i+1}) - \frac{3(\rho_{a,i}K_i + \rho_{a,i+1}K_{i+1})}{\rho_{a,i+1}\,\Delta x_{i+1}}\end{aligned} \tag{6.74}$$

Velocities, eddy diffusion coefficients, and densities in (6.74) are located at grid-cell boundaries, as shown in Fig. 6.5. (6.73) can be written in tridiagonal form as

$$A_i N_{i-1,t} + B_i N_{i,t} + D_i N_{i+1,t} = E_i N_{i-1,t-h} + F_i N_{i,t-h} + G_i N_{i+1,t-h} \tag{6.75}$$

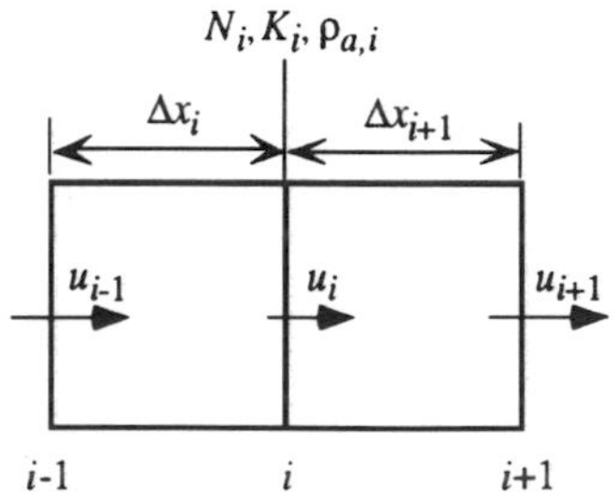


**Figure 6.5.** Locations of variables along a west–east grid for a Galerkin finite-element scheme.

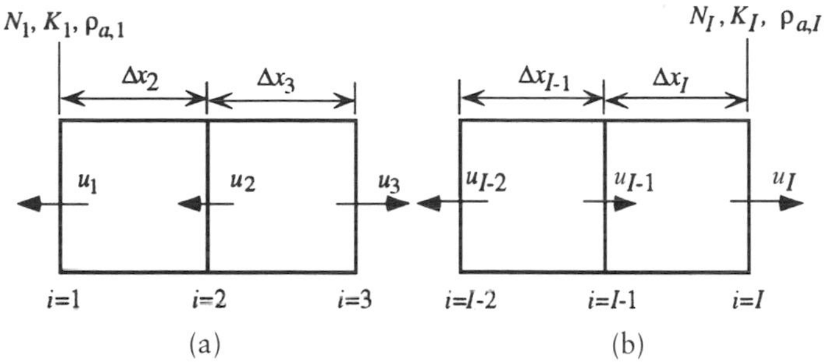


**Figure 6.6.** Location of grid variables at (a) low and (b) high boundaries.

where

$$
\begin{aligned}
A_i &= \Delta x_i - h\mu_c\alpha_i & E_i &= \Delta x_i + (1-\mu_c)\alpha_i \\
B_i &= 2(\Delta x_i + \Delta x_{i+1}) + h\mu_c\beta_i & F_i &= 2(\Delta x_i + \Delta x_{i+1}) - h(1-\mu_c)\beta_i \\
D_i &= \Delta x_{i+1} + h\mu_c\gamma_i & G_i &= \Delta x_{i+1} - h(1-\mu_c)\gamma_i
\end{aligned}
\tag{6.76}
$$

Like (6.45), (6.75) may be solved with a tridiagonal matrix technique.

Figure 6.6 shows locations of variables for boundaries on a nonglobal grid. The coefficients for **outflow from the high boundary** are

$$
\begin{aligned}
A_I &= \Delta x_I - h\mu_c\alpha_I & E_I &= \Delta x_I + h(1-\mu_c)\alpha_I \\
B_I &= 2\Delta x_I + h\mu_c\beta_I & F_I &= 2\Delta x_I - h(1-\mu_c)\beta_I \\
D_I &= 0 & G_I &= 0
\end{aligned}
\tag{6.77}
$$

where

$$
\alpha_I = u_I + 2u_{I-1} + \frac{3(\rho_{a,I}K_I + \rho_{a,I-1}K_{I-1})}{\rho_{a,I-1}\Delta x_I}
\tag{6.78}
$$

$$
\beta_I = 4u_I - u_{I-1} + \frac{9\rho_{a,I}K_I + 3\rho_{a,I-1}K_{I-1}}{\rho_{a,I}\Delta x_I}
$$

The coefficients for **outflow from the low boundary** are

$$
\begin{aligned}
A_1 &= 0 & E_1 &= 0 \\
B_1 &= 2\,\Delta x_1 + h\mu_c\beta_1 & F_1 &= 2\,\Delta x_2 - h(1-\mu_c)\beta_1 \\
D_1 &= \Delta x_2 + h\mu_c\gamma_1 & G_1 &= \Delta x_2 - h(1-\mu_c)\gamma_1
\end{aligned}
\tag{6.79}
$$

where

$$
\begin{aligned}
\beta_1 &= -4u_1 + u_2 + \frac{9\rho_{a,1}K_1 + 3\rho_{a,2}K_2}{\rho_{a,1}\,\Delta x_2} \\
\gamma_1 &= u_1 + 2u_2 - \frac{3(\rho_{a,1}K_1 + \rho_{a,2}K_2)}{\rho_{a,2}\,\Delta x_2}
\end{aligned}
\tag{6.80}
$$

**Inflow** boundary equations are obtained by extending (6.75) one node beyond the boundary and estimating concentrations and wind speeds in this virtual node.

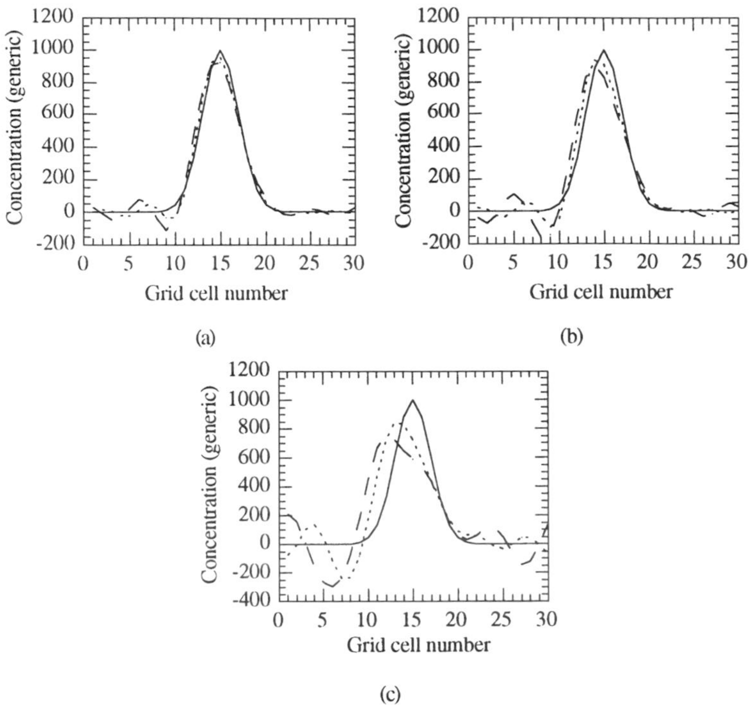


**Figure 6.7.** Preservation of a Gaussian peak during finite-element transport over a grid that circles around on itself when (a) $uh/\Delta x = 0.02$, (b) $uh/\Delta x = 0.25$, (c) $uh/\Delta x = 0.6$. Solid lines are initial values, short-dashed lines are values after four revolutions, and long-dashed lines are values after eight revolutions.

Virtual-node values may be set to values just inside the boundary or extrapolated from two or three nodes inside the boundary. Another option is to set virtual-node values equal to an average of values just inside the boundary from the current and the previous time step.

(6.75) can also be solved over a global domain. Figure 6.7 (a)–(c) show results from the finite-element method described above when a Gaussian plume was advected over a grid that circles around on itself. Grid spacing and wind speeds were uniform for the test. Each figure shows a result when a different value of $uh/\Delta x$ was used. The figures indicate that $uh/\Delta x$ should be 0.25 or less to minimize **numerical diffusion**, which is the artificial spreading of the peak, and **numerical dispersion**, which are oscillations upwind and downwind of the peak. If grid-spacing is 5 km and the average wind speed is 10 m s$^{-1}$, the time step in this case should be no larger than $0.25 \times 5000$ m/10 m s$^{-1}$ = 125 s.

### 6.5.2. Pseudospectral Method

Like the finite-element method, the **pseudospectral method** involves the replacement of the spatial differential operator by a finite series of basis functions. In the case of the finite-element method, the basis functions are local functions. In the case of the pseudospectral method, the basis functions are a finite series of orthogonal functions. The difference between a spectral and a pseudospectral method is that, with the former, time and space derivatives are both approximated with a finite series. With the latter, spatial derivatives are approximated with a finite series, but time derivatives are approximated with an explicit Taylor series expansion or another method.

If wind speed in the $x$-direction is constant, the west–east advection equation for gas number concentration is

$$\frac{\partial N}{\partial t} + u\frac{\partial N}{\partial x} = 0 \tag{6.81}$$

The pseudospectral solution to this equation can be found by representing $N(x, t)$ over the interval $0 \le x \le L$ by the **Fourier series**

$$N(x, t) = \sum_{k=0}^{\infty} a_k(t)e^{ik2\pi x/L} \tag{6.82}$$

(Orszag 1971; Wengle and Seinfeld 1978; Hack 1992), where $k$ is the wavenumber and $a_k(t)$ are **complex Fourier coefficients.** At $t = 0$, $N$ is a known function of $x$. Values of $a_k(0)$ are found by integrating both sides of (6.82) from $0 \le x \le L$. The result is

$$a_k(0) = \frac{1}{L}\int_0^L N(x, 0)e^{-ik2\pi x/L}\,dx \tag{6.83}$$

For practical application, the infinite series in (6.82) is truncated to a finite number of wavenumbers, $K$, giving

$$N(x, t) = \sum_{k=0}^{K} a_k(t)e^{ik2\pi x/L} \tag{6.84}$$

The larger the value of $K$, the more accurate the estimate of $N$.

A pseudospectral solution to (6.81) can be found by taking a second-order central-difference approximation of (6.84) with respect to time and the partial derivative of (6.84) with respect to space. The resulting expressions are

$$\frac{\partial N}{\partial t} \approx \frac{1}{2h}\left(\sum_{k=0}^{K} a_{k,t}e^{ik2\pi x/L} - \sum_{k=0}^{K} a_{k,t-2h}e^{ik2\pi x/L}\right) \tag{6.85}$$

$$\frac{\partial N}{\partial x} = \sum_{k=0}^{K} \frac{ik2\pi a_{k,t-h}}{L}e^{ik2\pi x/L} \tag{6.86}$$

respectively. Substituting these into (6.81) yields

$$\frac{1}{2h}\sum_{k=0}^{K}(a_{k,t} - a_{k,t-2h})e^{ik2\pi x/L} = -u\sum_{k=0}^{K}\frac{ik2\pi a_{k,t-h}}{L}e^{ik2\pi x/L} \tag{6.87}$$

which can be separated into $K$ equations of the form

$$\frac{a_{k,t} - a_{k,t-2h}}{2h} = -\frac{uik2\pi a_{k,t-h}}{L} \tag{6.88}$$

(6.88) is explicit and can be solved immediately, since values of $a_k$ at time $t-h$ and $t-2h$ are known from previous time steps. Fourier coefficients for the first time step ($t = h$) are found by taking a forward- instead of central-difference approximation of (6.84). Once Fourier coefficients have been determined from (6.88), they are substituted back into (6.84) to give an estimate of $N$ at time $t$ for any value of $x$.

An advantage of a pseudospectral scheme over a finite-difference approximation is that only $K$ equations need to be solved in the pseudospectral scheme. In a finite-difference scheme, $I$ finite-difference equations need to be solved per time step, where $I$ is the number of grid cells in one direction. Usually, $I > K$. While the pseudospectral solution to the linear advection problem is easy to implement, it is not readily applied to nonlinear problems, such as when $u$ varies in space or when $u$ is a prognostic variable. In such cases, a separate basis function for $u$ is required. The multiplication of two finite series, such as one for $u$ and one for $N$, results in additional terms, slowing the pseudospectral solution. One way to avoid the multiplication of spectral-basis-function products is with the **spectral transform method** (Eliasen et al. 1970; Orszag 1970).

For global modeling, the basis functions used are the **spherical harmonics**. These functions are a combination of sine and cosine functions along the zonal (west–east) direction and Legendre functions along the meridianal (south–north) direction on a sphere. They are computationally fast in comparison with some other basis functions. Spectral and pseudospectral methods are discussed in more detail in Orszag (1970), Washington and Parkinson (1986), Holton (1992), Hack (1992), and Krishnamurti et al. (1998). Pseudospectral techniques are commonly used to discretize horizontal advection terms in global models.

## 6.6. ADVECTION SCHEMES USED IN AIR-QUALITY MODELS

In the previous sections, several schemes used for solving unidirectional advection–diffusion equations were given. Some, such as the Galerkin finite-element method with chapeau basis functions and a Crank–Nicolson time-stepping scheme, have been used in models in which hundreds to thousands of gases and particles were transported. A related Galerkin method that uses a slightly different time-discretization scheme is the forward Euler Taylor–Galerkin method (Donea 1984). A Galerkin method that uses a chapeau basis function but a modified weighting function is the Petrov–Galerkin method (Hughes and Brooks 1979).

Other schemes used in air-quality models consist, in part, of combinations of methods discussed previously. The schemes of Bott (1989) and Tremback et al. (1987) are explicit, mass-conservative, positive-definite backward-difference schemes in which concentrations in the advection–diffusion equation are expanded with polynomials that are a function of distance. Each polynomial is derived from concentrations in several adjacent cells. The accurate space derivative (ASD) scheme of Gazdag (1973) is a pseudospectral scheme in which time and spatial derivatives are approximated with truncated Taylor series expansions and truncated Fourier series expansions, respectively. In the trajectory grid scheme (Chock et al. 1996), advection is solved with a fully Lagrangian method and eddy diffusion is solved with an Eulerian diffusion scheme. The scheme of Smolarkiewicz (1983) is iterative and uses a backward-difference approximation to preserve the positive-definiteness of the initial condition and an antidiffusion velocity to reduce numerical diffusion. Many of the schemes described above have been compared with each other when used alone (e.g., Chock 1991) or coupled with chemistry (e.g., Chock and Winkler 1994; Dabdub and Seinfeld 1994).

## 6.7. SUMMARY

In this chapter, methods of solving partial differential equations, such as advection–diffusion equations, were discussed. The methods include finite-difference and series expansion methods. A finite-difference approximation involves the replacement of a continuous differential operator with a discrete difference analog along a predetermined number of spatial or temporal nodes. The order of approximation increases with the number of nodes along which the differential is discretized. Forward-, backward-, and central-difference approximations of the first derivative of a variable were derived for a variety of orders under the assumption of constant grid spacing. Finite-difference approximations of the advection–diffusion equation were also discussed for cases of nonuniform grid spacing and variable eddy diffusion coefficients. The semi-Lagrangian method, which is in the family of finite-difference methods, was also discussed, and series expansion methods were described. These latter methods, which include finite-element and pseudospectral methods, involve the replacement of a dependent variable with a finite series that approximates the variable.

## 6.8. PROBLEMS

**6.1.** If an implicit approximation such as (6.36) is unconditionally stable, why is it less accurate, for the same grid spacing and time step, than (6.44) when $\mu_c = 0.5$?

**6.2.** What advantage does the finite-element scheme of fourth-order in space and second-order in time shown in this chapter have over a finite-difference scheme of the same orders?

**6.3.** Find a sixth-order central-difference approximation of $\partial N/\partial x$. Assume constant grid spacing.

**6.4.** Find a fifth-order backward-difference approximation of $\partial N/\partial x$. Assume constant grid spacing.

## 6.9. COMPUTER PROGRAMMING PRACTICE

**6.5.** Set up a west–east grid that wraps around on itself with 100 uniformly spaced grid cells. Assume the last grid cell to the east meets the first grid cell to the west. Set $\Delta x = 5$ km in each cell and $u = 5$ m s$^{-1}$ at each cell west–east boundary. Initialize the gas concentration on the grid with the Gaussian distribution $N_i = N_0 e^{-(i-50)^2/8}$, where $N_0 = 10^{12}$ molecules cm$^{-3}$ is a peak concentration and $i$ corresponds to a cell boundary. Solve (6.36), assuming the eddy diffusion coefficient is zero, with the matrix given in (6.42). Use a time step of 5 s, and solve until 10 revolutions around the grid have been completed. Plot concentration versus grid cell at the end of each revolution. Discuss the numerical diffusion of the peak over time.

# 7

# Finite-Differencing the Equations of Atmospheric Dynamics

Many models have been developed to solve the equations of atmospheric dynamics over regional and global domains. Such models usually discretize time and vertical spatial derivatives with finite-difference schemes. Horizontal advection terms are discretized with finite-difference, spectral, finite-element, or semi-Lagrangian schemes. In this chapter, a finite-difference solution to the equations of atmospheric dynamics in spherical–sigma-pressure coordinates is discussed. The equations solved include the continuity equation for air, species continuity equation, thermodynamic energy equation, horizontal momentum equations, and hydrostatic equation. The solution scheme originated from Arakawa and Lamb (1977) and Arakawa and Suarez (1983). Although more recent and advanced versions of the scheme have been developed, and although other numerical solutions exist, the version presented here has been chosen because it is readily coded and adaptable to regional- and global-scale models. For the global scale, boundary conditions at the poles are more difficult to treat than for the regional scale. Global-scale boundary conditions are discussed in the above-mentioned references.

## 7.1. VERTICAL MODEL GRID

In this section, atmospheric dynamical equations are discretized over the sigma-pressure vertical coordinate and spherical horizontal coordinates. Figure 7.1 shows the location of model variables in the vertical when the sigma-pressure coordinate is used. Vertical scalar velocities are located at the top and bottom of each layer. Horizontal scalar velocities and other variables (e.g., gas and particle mixing ratios and potential virtual temperature) are located at the vertical midpoint of the layer. This type of grid is called the **Lorenz grid**, after Lorenz (1960). A similar grid that stores potential virtual temperature at layer top and bottom boundaries instead of centers is called the **Charney–Phillips grid**, after Charney and Phillips (1953). Both grids are described in more detail in Arakawa and Konor (1995). In Fig. 7.1, layer center indices increase from 1, at the top of the model, to $N_L$, at the bottom, and layer boundary indices increase from $\frac{1}{2}$, at the top, to $N_L + \frac{1}{2}$ at the bottom.

Sigma-pressure coordinate vertical scalar velocities at the bottom of the model are set to zero, since the model bottom is assumed to be a material surface, namely,

| | |
|---|---|
| Model top boundary | $\dot{\sigma}_{1/2}=0,\quad \sigma_{1/2}=0,\quad p_{a,top}$ |
| - - - - - - | $q_1,\quad \theta_{v,1},\quad u_1,\quad v_1,\quad p_{a,1}$ |
| ——— | $\dot{\sigma}_{1+1/2},\quad \sigma_{1+1/2},\quad p_{a,1+1/2}$ |
| - - - - - - | $q_2,\quad \theta_{v,2},\quad u_2,\quad v_2,\quad p_{a,2}$ |
| ——— | $\dot{\sigma}_{k-1/2},\quad \sigma_{k-1/2},\quad p_{a,k-1/2}$ |
| - - - - - - | $q_k,\quad \theta_{v,k},\quad u_k,\quad v_k,\quad p_{a,k}$ |
| ——— | $\dot{\sigma}_{k+1/2},\quad \sigma_{k+1/2},\quad p_{a,k+1/2}$ |
| - - - - - - | $q_{k+1},\quad \theta_{v,k+1},\quad u_{k+1},\quad v_{k+1},\quad p_{a,k+1}$ |
| ——— | $\dot{\sigma}_{N_L-1/2},\quad \sigma_{N_L-1/2},\quad p_{a,N_L-1/2}$ |
| - - - - - - | $q_{N_L},\quad \theta_{v,N_L},\quad u_{N_L},\quad v_{N_L},\quad p_{a,N_L}$ |
| Model bottom boundary | $\dot{\sigma}_{N_L+1/2}=0,\quad \sigma_{N_L+1/2}=1,\quad p_{a,surf}$ |

**Figure 7.1.** Location of variables in the vertical direction of a sigma-pressure coordinate Lorenz grid. Integer subscripts correspond to layer centers, and fraction subscripts correspond to layer boundaries.

the ground surface. Setting the model bottom vertical scalar velocities to zero also helps to filter **Lamb waves**, which are horizontally propagating acoustic waves that give rise to pressure, velocity, and density perturbations. Since the amplitude of Lamb-wave oscillations is usually greatest at the lowest model level, setting the vertical velocity to zero at that level decreases perturbations due to the wave.

At the model top, vertical scalar velocities are also set to zero for lack of a better assumption. Zero velocities at the top are not realistic, unless the model top is the top of the atmosphere. An alternative to setting top vertical scalar velocities to zero is to estimate them, which is also an error-prone process. In either case, errors can be reduced by raising the model top to a higher altitude. This may require adding more model layers, increasing computational requirements.

In the sigma-pressure coordinate, the **$\sigma$-thickness** of a layer is the difference in $\sigma$-values between the top and bottom of the layer. Thus,

$$\Delta\sigma_k = \sigma_{k+1/2} - \sigma_{k-1/2} \qquad \text{for} \quad k = 1, \ldots, N_L \tag{7.1}$$

where the $\sigma$-value at the top and bottom of each layer is predetermined and constant throughout the model domain and during the simulation. $\sigma$ at the top of the model usually equals zero ($\sigma_{1/2} = \sigma_{top} = 0$), and $\sigma$ at the ground surface usually equals unity ($\sigma_{N_L+1/2} = \sigma_{surf} = 1$) although these limits are arbitrary. $\sigma$-values at the boundaries of other layers vary, as described below.

When a model extends from the surface to the top of the stratosphere, the choice of $\sigma$-values is important for maintaining numerical stability. In the stratosphere, small pressure changes correspond to large altitude changes. Near the surface, small pressure changes correspond to small altitude changes. Thus, the $\sigma$- (and pressure-) thickness of a layer near the model top should not be too large and that near the ground should not be too small.

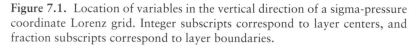

One way to select $\sigma$-values is to set up a test column and assume that the altitude thickness of all layers in the test column is the same. If the base of the test column is at sea level, the altitude at the bottom of each test column layer is

$$z_{k+1/2,test} = z_{top,test}\left(1 - \frac{k}{N_L}\right) \qquad \text{for} \quad k = 0, \ldots, N_L \tag{7.2}$$

where $z_{top,test}$ is the test-column top altitude (m) and $k$ is the layer index. In the sigma-pressure coordinate, the model top pressure, $p_{a,top}$, is assumed to be known and constant throughout the model domain. The test-column top pressure, $p_{a,top,test}$, equals the model top pressure. Table B.1 in Appendix B gives $z_{top,test}$ for any $p_{a,top,test}$. Once the altitude of each test-layer boundary is found from (7.2), Table B.1 can be used to determine each test-layer boundary pressure, $p_{a,k+1/2,test}$. From the layer boundary pressure, the σ-value at the boundary is

$$\sigma_{k+1/2} = \frac{p_{a,k+1/2,test} - p_{a,top,test}}{p_{a,N_L+1/2,test} - p_{a,top,test}} \qquad \text{for} \quad k = 1, \ldots, N_L \tag{7.3}$$

The σ-value at the boundary of each real model layer can then be set equal to the σ-value at the boundary of each test layer.

Once σ-values have been determined for each model layer boundary, they are used to calculate the pressure there. The pressure at the boundary of a model layer is

$$p_{a,k+1/2} = p_{a,top} + \sigma_{k+1/2}\pi_a \tag{7.4}$$

where $\pi_a = p_{a,surf} - p_{a,top}$ is the pressure thickness of each model column. Although $p_{a,top}$ is the same in each model column, $p_{a,surf}$ differs for each column; thus, $\pi_a$ and $p_{a,k+1/2}$ also differ for each column and grid cell, respectively.

The vertical midpoint of a layer can be chosen as the center of height, the center of mass, or the pressure at which the mass-weighted mean of a variable is located. At the mass center of a layer,

$$p_{a,k} = p_{a,k-1/2} + 0.5(p_{a,k+1/2} - p_{a,k-1/2}) \tag{7.5}$$

Alternatively, if potential virtual temperature increases monotonically with height, as shown in Fig. 2.13, the vertical midpoint can be defined as the height where the mass-weighted mean of $P$, the potential temperature factor defined in (2.92), is located (Arakawa and Suarez 1983). The mass-weighed mean of $P$ in layer $k$ is

$$\begin{aligned} P_k &= \frac{1}{p_{a,k+1/2} - p_{a,k-1/2}} \int_{p_{a,k-1/2}}^{p_{a,k+1/2}} P\,\mathrm{d}p_a \\ &= \frac{1}{1+\kappa}\left(\frac{P_{k+1/2}p_{a,k+1/2} - P_{k-1/2}p_{a,k-1/2}}{p_{a,k+1/2} - p_{a,k-1/2}}\right) \end{aligned} \tag{7.6}$$

where values of $p_a$ and $P$ at layer boundaries are $p_{a,k+1/2}$ and

$$P_{k+1/2} = (p_{a,k+1/2}/1000\ \text{mb})^{\kappa} \tag{7.7}$$

respectively. Once $P_k$ has been found from (7.6), the pressure corresponding to $P_k$ is

$$p_{a,k} = (1000\ \text{mb}) P_k^{1/\kappa} \tag{7.8}$$

In the model discussed here, potential virtual temperatures are predicted at layer vertical midpoint, but values at layer tops and bottoms are needed to calculate vertical energy fluxes. If (7.8) is used to locate the vertical pressure-midpoint of a layer, a consistent formulation for $\theta_v$ at a layer boundary is

$$\theta_{v,k+1/2} = \frac{(P_{k+1/2} - P_k)\theta_{v,k} + (P_{k+1} - P_{k+1/2})\theta_{v,k+1}}{P_{k+1} - P_k} \tag{7.9}$$

**Example 7.1**

Given the following pressure and potential virtual temperature profiles, calculate the pressure at the midpoints, with respect to mass and to the mass integral of $P$, of layers $k$ and $k+1$ and potential virtual temperature at the layer boundary, $k+\frac{1}{2}$. Assume the air is dry.

| | |
|---|---|
| ———————— | $p_{a,k-1/2} = 700$ mb |
| – – – – – – – – – – | $\theta_{v,k} = 308$ K |
| ———————— | $p_{a,k+1/2} = 750$ mb |
| – – – – – – – – – – | $\theta_{v,k+1} = 303$ K |
| ———————— | $p_{a,k+3/2} = 800$ mb |

SOLUTION

Substituting values above into (7.5) gives $p_{a,k} = 725$ mb and $p_{a,k+1} = 775$ mb. From (7.7), $P_{k-1/2} = 0.9030214$, $P_{k+1/2} = 0.9210167$, and $P_{k+3/2} = 0.9381747$. Substituting these into (7.6) gives $P_k = 0.9120929$ and $P_{k+1} = 0.9296616$. From (7.8), $p_{a,k}=$ and 724.897 mb and $p_{a,k+1} = 774.904$ mb. In sum, pressure at the midpoint with respect to the mass integral of $P$ is slightly lower than pressure at the mass center of a layer. From (7.9), $\theta_{v,k+1/2} = 305.54$ K.

## 7.2. THE CONTINUITY EQUATION FOR AIR

In the sigma-pressure coordinate, the continuity equation for air is used prognostically to estimate changes in total column pressure and diagnostically to estimate vertical velocities at the top and bottom of each grid cell. From (5.66), the prognostic equation for the change in column pressure was

$$R_e^2 \cos\varphi \left(\frac{\partial \pi_a}{\partial t}\right)_\sigma = -\int_0^1 \left[\frac{\partial}{\partial \lambda_e}(u\pi_a R_e) + \frac{\partial}{\partial \varphi}(v\pi_a R_e \cos\varphi)\right]_\sigma \mathrm{d}\sigma \qquad (7.10)$$

Replacing d, d$t$, d$\lambda_e$, d$\varphi$, and d$\sigma$ with $\Delta$, $h$, $\Delta\lambda_e$, $\Delta\varphi$, and $\Delta\sigma$, respectively, and multiplying through by $\Delta\lambda_e\,\Delta\varphi$, gives a first-order in time and second-order in space approximation for the column presssure as

$$\left(R_e^2 \cos\varphi\,\Delta\lambda_e\,\Delta\varphi\right)_{i,j}\left(\frac{\pi_{a,t}-\pi_{a,t-h}}{h}\right)_{i,j} \qquad (7.11)$$

$$= -\sum_{k=1}^{N_L}\left[\frac{(u\pi_a R_e\,\Delta\varphi\,\Delta\lambda_e\,\Delta\sigma)_{i+1/2,j} - (u\pi_a R_e\,\Delta\varphi\,\Delta\lambda_e\,\Delta\sigma)_{i-1/2,j}}{\Delta\lambda_e}\right]_{k,t-h}$$

$$-\sum_{k=1}^{N_L}\left[\frac{(v\pi_a R_e \cos\varphi\,\Delta\varphi\,\Delta\lambda_e\,\Delta\sigma)_{i,j+1/2} - (v\pi_a R_e \cos\varphi\,\Delta\varphi\,\Delta\lambda_e\,\Delta\sigma)_{i,j-1/2}}{\Delta\varphi}\right]_{k,t-h}$$

The horizontal finite-difference grid chosen for discretizing this equation is the **Arakawa C-grid** (Arakawa and Lamb 1977), shown in Fig. 7.2. In this grid, column pressures ($\pi_a$) are evaluated at $\pi_a$-points, which are bounded to the west and east, respectively, by $u$-velocity points and to the south and north, respectively, by $v$-velocity points. The left sides of the second and third numerators in (7.11) are fluxes entering the solid box shown in Fig. 7.2, and the right sides are fluxes leaving the box.

The grid shown in Fig. 7.2 is a **limited-area grid**, which is a grid with lateral boundaries. $U$-points lie on western and eastern boundaries, and $v$-points lie on

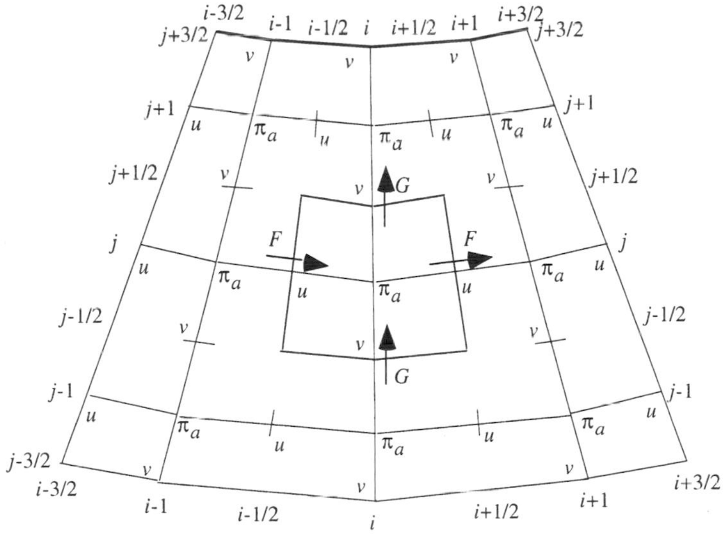


**Figure 7.2.** The Arakawa C-grid. The grid is centered around $\pi_a$ for finite-differencing of the continuity equation. The figure shows exact locations of $u$, $v$, and $\pi_a$ relative to lateral boundaries.

southern and northern boundaries. A **global grid** circles around on itself in the west–east direction and converges to a singularity at the poles in the south–north direction. In such cases, the poles must be treated carefully; otherwise mass converges at the poles and computational instabilities arise. One method of treating the poles on a global grid is to set $v$-velocities to zero in cells at the southernmost and northernmost latitudes. As a result, winds advect from west to east or east to west around the poles ($u \neq 0$), avoiding the singularity.

(7.11) can be solved to give a **prognostic equation for the column pressure** as

$$\pi_{a,i,j,t} = \pi_{a,i,j,t-h} - \frac{h}{\left(R_e^2 \cos\varphi\, \Delta\lambda_e\, \Delta\varphi\right)_{i,j}} \times \sum_{k=1}^{N_L} \left[\left(F_{i+1/2,j} - F_{i-1/2,j} + G_{i,j+1/2} - G_{i,j-1/2}\right)_{k,t-h} \Delta\sigma_k\right] \tag{7.12}$$

where $F$ and $G$ are fluxes defined at $u$- and $v$-points, respectively, as shown in Fig. 7.2. At interior points in the domain, $F$ and $G$ are

$$F_{i+1/2,j,k,t-h} = \left[\frac{\pi_{a,i,j} + \pi_{a,i+1,j}}{2} (u R_e\, \Delta\varphi)_{i+1/2,j,k}\right]_{t-h} \tag{7.13}$$

$$G_{i,j+1/2,k,t-h} = \left[\frac{\pi_{a,i,j} + \pi_{a,i,j+1}}{2} (v R_e \cos\varphi\, \Delta\lambda_e)_{i,j+1/2,k}\right]_{t-h} \tag{7.14}$$

where the division by two indicates that the $\pi_a$ values at $i$, $j$ points have been averaged to obtain approximate $\pi_a$-values at $i + 1/2$, $j$ and $i$, $j + 1/2$ points. At eastern and northern lateral boundaries, $F$ and $G$ are

$$F_{I+1/2,j,k,t-h} = [\pi_{a,I,j} (u R_e\, \Delta\varphi)_{i+1/2,j,k}]_{t-h} \tag{7.15}$$

$$G_{i,J+1/2,k,t-h} = [\pi_{a,i,J} (v R_e \cos\varphi\, \Delta\lambda_e)_{i,J+1/2,k}]_{t-h} \tag{7.16}$$

respectively, where $I + 1/2$ and $J + 1/2$ are the easternmost and northernmost $u$ and $v$ points, respectively (e.g., points $i+3/2$ and $j+3/2$, respectively, in Fig. 7.2). Similar equations can be written for the western and southern boundaries. Once new column pressures have been calculated from (7.12), they are used in (7.4) to obtain new pressures at the top and bottom of each grid cell. Layer vertical midpoint pressures are then obtained from (7.6)–(7.8).

After new column pressures have been determined, vertical velocities at the top and bottom of each grid are calculated diagnostically from (5.69), given as

$$\dot{\sigma}\pi_a R_e^2 \cos\varphi = -\int_0^{\sigma} \left[\frac{\partial}{\partial\lambda_e}(u\pi_a R_e) + \frac{\partial}{\partial\varphi}(v\pi_a R_e \cos\varphi)\right]_{\sigma} d\sigma - \sigma R_e^2 \cos\varphi \left(\frac{\partial\pi_a}{\partial t}\right)_{\sigma} \tag{7.17}$$

Replacing differential and integral operators in this equation with finite-difference analogs and multiplying through by $\Delta\lambda_e\, \Delta\varphi$ yields the **vertical scalar velocity** at the bottom of a layer as

$$
\begin{aligned}
&\left(\dot{\sigma}\pi_a R_e^2 \cos\varphi\, \Delta\lambda_e\, \Delta\varphi\right)_{i,j,k+1/2,t} \\
&= -\sum_{l=1}^{k}\left[\frac{(u\pi_a R_e\, \Delta\lambda_e\, \Delta\varphi\, \Delta\sigma)_{i-1/2,j} - (u\pi_a R_e\, \Delta\lambda_e\, \Delta\varphi\, \Delta\sigma)_{i+1/2,j}}{\Delta\lambda_e}\right]_{l,t-h} \\
&\quad -\sum_{l=1}^{k}\left[\frac{(v\pi_a R_e \cos\varphi\Delta\,\lambda_e\, \Delta\varphi\, \Delta\sigma)_{i,j-1/2} - (v\pi_a R_e \cos\varphi\, \Delta\lambda_e\, \Delta\varphi\, \Delta\sigma)_{i,j+1/2}}{\Delta\,\varphi}\right]_{l,t-h} \\
&\quad -\sigma_{k+1/2}\left(R_e^2 \cos\varphi\, \Delta\lambda_e\, \Delta\varphi\right)_{i,j}\left(\frac{\pi_{a,t} - \pi_{a,t-h}}{h}\right)_{i,j}
\end{aligned}
\tag{7.18}
$$

Substituting fluxes from (7.13)–(7.16) into (7.18) gives the vertical velocity as

$$
\begin{aligned}
\dot{\sigma}_{i,j,k+1/2,t} = &-\frac{1}{\left(\pi_a R_e^2 \cos\varphi\Delta\lambda_e\Delta\varphi\right)_{i,j,t}} \\
&\times \sum_{l=1}^{k}\left[(F_{i+1/2,j} - F_{i-1/2,j} + G_{i,j+1/2} - G_{i,j-1/2})_{l,t-h}\Delta\sigma_l\right] \\
&- \sigma_{k+1/2}\left(\frac{\pi_{a,t} - \pi_{a,t-h}}{h\pi_{a,t}}\right)_{i,j}
\end{aligned}
\tag{7.19}
$$

Because $\dot{\sigma}_{1/2} = \dot{\sigma}_{top} = 0$, (7.19) is solved in the order, $k = 1, \ldots, N_L - 1$. At $k = N_L$, we have $\dot{\sigma}_{N_L+1/2} = 0$ (the vertical scalar velocity at the lowest boundary of the model equals zero). This characteristic is demonstrated by combining (7.12) and (7.19) when $k = N_L$ in the latter equation.

## 7.3. THE SPECIES CONTINUITY EQUATION

Transport of each gas and aerosol component in a model is simulated with the species continuity equation. This equation can be solved as three one-dimensional equations, a two-dimensional and one-dimensional equation, or one three-dimensional equation. In Chapter 6, techniques for solving the one- and two-dimensional equations were given. Here, an explicit solution to the three dimensional equation is shown.

The flux form of the **species continuity equation** in spherical-sigma-pressure coordinates was given in (5.73) as

$$
\begin{aligned}
&R_e^2 \cos\varphi\left[\frac{\partial}{\partial t}(\pi_a q)\right]_\sigma + \left[\frac{\partial}{\partial\lambda_e}(u\pi_a q R_e) + \frac{\partial}{\partial\varphi}(v\pi_a q R_e \cos\varphi)\right]_\sigma \\
&+ \pi_a R_e^2 \cos\varphi\frac{\partial}{\partial\sigma}(\dot{\sigma}q) = \pi_a R_e^2 \cos\varphi\left[\frac{(\nabla \cdot \rho_a \mathbf{K}_h \nabla)q}{\rho_a} + \sum_{n=1}^{N_{e,t}} R_n\right]
\end{aligned}
\tag{7.20}
$$

Replacing differential operators with finite-difference analogs and multiplying through by $\Delta\lambda_e \Delta\varphi$ gives

$$\left(R_e^2 \cos\varphi\, \Delta\lambda_e\, \Delta\varphi\right)_{i,j} \left(\frac{\pi_{a,t} q_t - \pi_{a,t-h} q_{t-h}}{h}\right)_{i,j,k} \tag{7.21}$$

$$+ \frac{(u\pi_a q R_e\, \Delta\lambda_e\, \Delta\varphi)_{i+1/2,j,k,t-h} - (u\pi_a q R_e\, \Delta\lambda_e\, \Delta\varphi)_{i-1/2,j,k,t-h}}{\Delta\lambda_e}$$

$$+ \frac{(v\pi_a q R_e \cos\varphi\, \Delta\lambda_e\, \Delta\varphi)_{i,j+1/2,k,t-h} - (v\pi_a q R_e \cos\varphi \Delta\lambda_e\, \Delta\varphi)_{i,j-1/2,k,t-h}}{\Delta\varphi}$$

$$+ \left[\pi_{a,t} R_e^2 \cos\varphi\, \Delta\lambda_e\, \Delta\varphi \frac{(\dot{\sigma}_t q_{t-h})_{k+1/2} - (\dot{\sigma}_t q_{t-h})_{k-1/2}}{\Delta\sigma_k}\right]_{i,j}$$

$$= \left\{\pi_a R_e^2 \cos\varphi\, \Delta\lambda_e\, \Delta\varphi \left[\frac{(\nabla_z \cdot \rho_a \mathbf{K}_h \nabla_z) q}{\rho_a} + \sum_{n=1}^{N_{e,t}} R_n\right]\right\}_{i,j,k,t-h}$$

where the eddy diffusion term is not differenced here for simplicity. In this equation, $\pi_a \dot{\sigma}$ at time $t$, calculated from (7.19), is required. Substituting (7.13)–(7.16) into (7.21) and interpolting q's to $u$ and $v$ points where necessary gives a **prognostic form of the species continuity equation** as

$$q_{i,j,k,t} = \frac{(\pi_a q)_{i,j,k,t-h}}{\pi_{a,i,j,t}} + \frac{h}{\left(\pi_{a,t} R_e^2 \cos\varphi\, \Delta\lambda_e\, \Delta\varphi\right)_{i,j}} \tag{7.22}$$

$$\times \left\{ \left( F_{i-1/2,j} \frac{q_{i-1,j} + q_{i,j}}{2} - F_{i+1/2,j} \frac{q_{i,j} + q_{i+1,j}}{2} \right.\right.$$

$$\left. + G_{i,j-1/2} \frac{q_{i,j-1} + q_{i,j}}{2} - G_{i,j+1/2} \frac{q_{i,j} + q_{i,j+1}}{2} \right)_{k,t-h}$$

$$+ \left[\pi_{a,t} R_e^2 \cos\varphi\, \Delta\lambda_e\, \Delta\varphi \frac{(\dot{\sigma}_t q_{t-h})_{k-1/2} - (\dot{\sigma}_t q_{t-h})_{k+1/2}}{\Delta\sigma_k}\right]_{i,j}$$

$$\left. + \left[\pi_a R_e^2 \cos\varphi\, \Delta\lambda_e\, \Delta\varphi \left(\frac{(\nabla_z \cdot \rho_a \mathbf{K}_h \nabla_z) q}{\rho_a} + \sum_{n=1}^{N_{e,t}} R_n\right)\right]_{i,j,k,t-h} \right\}$$

where the fluxes are the same as those shown in Fig. 7.2. In this equation, q values outside lateral boundaries [e.g., at $(0,j)$, $(I+1,j)$ $(i,0)$, and $(i,J+1)$ points] can be specified, set to nearest q values inside the boundary from the current time step, or set to nearest q values inside the boundary from the previous time step. Arakawa (1982) and Lu (1994) discuss a more detailed treatment of lateral boundary conditions in which fluxes of scalars at the boundaries are derived from a mass-conservation relationship.

Mixing ratios for (7.22) at the top and bottom of a layer can be interpolated from layer midpoint values with

$$q_{i,j,k-1/2} = \frac{\ln q_{i,j,k-1} - \ln q_{i,j,k}}{(1/q_{i,j,k}) - (1/q_{i,j,k-1})} \qquad (7.23)$$

$$q_{i,j,k+1/2} = \frac{\ln q_{i,j,k} - \ln q_{i,j,k+1}}{(1/q_{i,j,k+1}) - (1/q_{i,j,k})}$$

respectively. Mixing ratios at the surface and model top are not needed, since $\dot{\sigma} = 0$ at those locations.

## 7.4. THE THERMODYNAMIC ENERGY EQUATION

The finite-difference form of the thermodynamic energy equation is similar to that of the species continuity equation. From (5.76), the flux form of the **thermodynamic energy equation in spherical–sigma-pressure coordinates** was

$$R_e^2 \cos\varphi \left[\frac{\partial}{\partial t}(\pi_a \theta_v)\right]_\sigma + \left[\frac{\partial}{\partial \lambda_e}(u\pi_a\theta_v R_e) + \frac{\partial}{\partial \varphi}(v\pi_a\theta_v R_e \cos\varphi)\right] + \pi_a R_e^2 \cos\varphi \frac{\partial}{\partial \sigma}(\dot{\sigma}\theta_v) = \pi_a R_e^2 \cos\varphi \left[\frac{(\nabla \cdot \rho_a \mathbf{K}_h \nabla)\theta_v}{\rho_a} + \frac{\theta_v}{c_{p,d} T_v}\sum_{n=1}^{N_{e,h}} \frac{dQ_n}{dt}\right] \qquad (7.24)$$

Replacing differential operators with finite-difference analogs, multiplying through by $\Delta\lambda_e \Delta\varphi$, taking the finite-difference of each term in (7.24), and substituting fluxes from (7.13)–(7.16) gives potential virtual temperature as

$$\begin{aligned}
\theta_{v,i,j,k,t} &= \frac{(\pi_a\theta_v)_{i,j,k,t-h}}{\pi_{a,i,j,t}} + \frac{h}{\left(\pi_{a,t} R_e^2 \cos\varphi\, \Delta\lambda_e\, \Delta\varphi\right)_{i,j}} \\
&\times \Bigg\{ \Bigg( F_{i-1/2,j}\frac{\theta_{v,i-1,j} + \theta_{v,i,j}}{2} - F_{i+1/2,j}\frac{\theta_{v,i,j} + \theta_{v,i+1,j}}{2} + G_{i,j-1/2} \\
&\times \frac{\theta_{v,i,j-1} + \theta_{v,i,j}}{2} - G_{i,j+1/2}\frac{\theta_{v,i,j} + \theta_{v,i,j+1}}{2}\Bigg)_{k,t-h} + \Bigg[\pi_{a,t} R_e^2 \cos\varphi \Delta\lambda_e \\
&\times \frac{(\dot{\sigma}_t\theta_{v,t-h})_{k-1/2} - (\dot{\sigma}_t\theta_{v,t-h})_{k+1/2}}{\Delta\sigma_k}\Bigg]_{i,j} + \Bigg(\pi_a R_e^2 \cos\varphi \Delta\lambda_e \Delta\varphi \\
&\times \left[\frac{(\nabla_z \cdot \rho_a \mathbf{K}_h \nabla_z)\theta_v}{\rho_a} + \frac{\theta_v}{c_{p,d} T_v}\sum_{n=1}^{N_{e,h}} \frac{dQ_n}{dt}\right]\Bigg)_{i,j,k,t-h} \Bigg\}
\end{aligned} \qquad (7.25)$$

where the eddy diffusion term is not differenced for simplicity. $\theta_v$ values outside lateral boundaries [e.g., at $(0, j)$, $(I + 1, j)(i, 0)$, and $(i, J + 1)$ points] can be specified, set to nearest $\theta_v$ values inside the boundary from the current time step, or set to nearest $\theta_v$ values inside the boundary from the previous time step. $\theta_{v,i,j,k-1/2}$ and $\theta_{v,i,j,k+1/2}$ are potential virtual temperatures at the top and bottom of a layer, respectively, found from (7.9). $\theta_v$ at the surface and model top are not needed, since $\dot{\sigma} = 0$ at those locations.

## 7.5. THE HORIZONTAL MOMENTUM EQUATIONS

The horizontal momentum equation is differenced with a more accurate scheme than was the previous equations. Following Arakawa and Lamb (1977), a second-order scheme that conserves kinetic energy from inertial processes, integrated over the domain, and enstrophy during advection by the nondivergent part of the horizontal velocity, is shown. **Enstrophy** is one-half the mean squared vorticity. Conserving kinetic energy and enstrophy is important for reducing nonlinear computational instabilities. The kinetic energy constraint reduces error by preventing computational cascades of kinetic energy into small-scale motions (Arakawa and Lamb 1977). Conservation of kinetic energy and enstrophy requires strict adherence to the horizontal interpolations presented below and permits solutions to remain stable for long integration periods.

The **flux form of the momentum equation in the *u*-direction** from (5.86) was

$$R_e^2 \cos\varphi \left[\frac{\partial}{\partial t}(\pi_a u)\right]_\sigma + \left[\frac{\partial}{\partial \lambda_e}(\pi_a u^2 R_e) + \frac{\partial}{\partial \varphi}(\pi_a u v R_e \cos\varphi)\right]_\sigma \tag{7.26}$$

$$+ \pi_a R_e^2 \cos\varphi \frac{\partial}{\partial \sigma}(\dot{\sigma} u) = \pi_a u v R_e \sin\varphi + \pi_a f v R_e^2 \cos\varphi$$

$$- R_e \left(\pi_a \frac{\partial \Phi}{\partial \lambda_e} + \sigma c_{p,d} \theta_v \frac{\partial P}{\partial \sigma}\frac{\partial \pi_a}{\partial \lambda_e}\right)_\sigma + R_e^2 \cos\varphi \frac{\pi_a}{\rho_a}(\nabla \cdot \rho_a \mathbf{K}_m \nabla) u$$

The finite-difference form of this equation is obtained by assuming fluxes enter or leave a grid cell through each of eight points, as shown in Fig. 7.3. This differs from Fig. 7.2, where fluxes entered or left through each of four points. The locations of variables and index values in Fig. 7.3 are the same as those in Fig. 7.2.

Substituting differential operators for difference operators in (7.26), multiplying through by $\Delta\lambda_e \Delta\varphi$, and interpolating variables in a manner consistent with energy and enstrophy conservation gives the momentum equation in the *u*-direction as

*Time difference term*

$$u_{i+1/2,j,k,t} = \frac{(\pi_{a,t-h}\Delta A)_{i+1/2,j}}{(\pi_{a,t}\Delta A)_{i+1/2,j}} u_{i+1/2,j,k,t-h} + \frac{h}{(\pi_{a,t}\Delta A)_{i+1/2,j}} \times \Bigg\{ \tag{7.27}$$

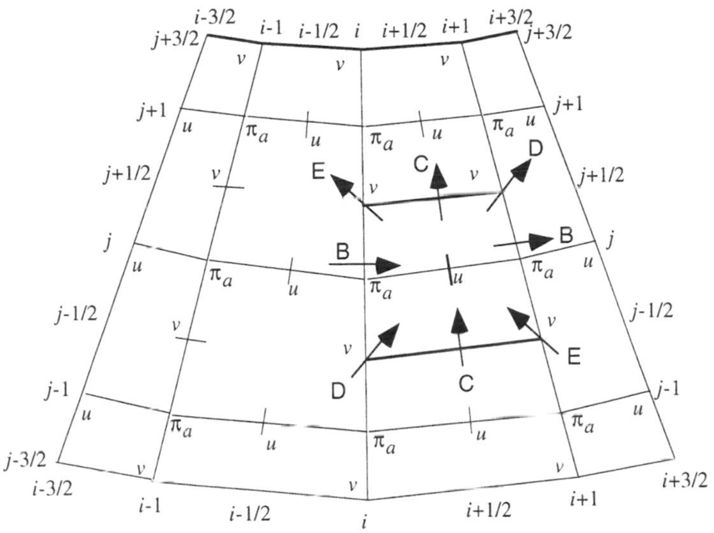


**Figure 7.3.** Location of fluxes for finite-differencing the momentum equation in the $u$-direction. The grid is the same as that in Fig. 7.2. The figure shows exact locations of $u$, $v$, and $\pi_a$ relative to lateral boundaries.

*Horizontal advection terms*

$$\left( \mathrm{B}_{i,j} \frac{u_{i-1/2,j} + u_{i+1/2,j}}{2} - \mathrm{B}_{i+1,j} \frac{u_{i+1/2,j} + u_{i+3/2,j}}{2} \right. \tag{7.28}$$

$$+ \mathrm{C}_{i+1/2,j-1/2} \frac{u_{i+1/2,j-1} + u_{i+1/2,j}}{2} - \mathrm{C}_{i+1/2,j+1/2} \frac{u_{i+1/2,j} + u_{i+1/2,j+1}}{2}$$

$$+ \mathrm{D}_{i,j-1/2} \frac{u_{i-1/2,j-1} + u_{i+1/2,j}}{2} - \mathrm{D}_{i+1,j+1/2} \frac{u_{i+1/2,j} + u_{i+3/2,j+1}}{2}$$

$$\left. + \mathrm{E}_{i+1,j-1/2} \frac{u_{i+3/2,j-1} + u_{i+1/2,j}}{2} - \mathrm{E}_{i,j+1/2} \frac{u_{i+1/2,j} + u_{i-1/2,j+1}}{2} \right)_{k,t-h}$$

*Vertical transport of horizontal momentum*

$$+ \frac{1}{\Delta\sigma_k} (\pi_{a,t} \Delta A \dot{\sigma}_{k-1/2,t} u_{k-1/2,t-h} - \pi_{a,t} \Delta A \dot{\sigma}_{k+1/2,t} u_{k+1/2,t-h})_{1+1/2,j} \tag{7.29}$$

*Coriolis and spherical grid conversion terms*

$$+ \frac{R_e (\Delta\lambda_e \, \Delta\varphi)_{i+1/2,j}}{2} \left[ \pi_{a,i,j} \frac{v_{i,j-1/2} + v_{i,j+1/2}}{2} \left( f_j R_e \cos\varphi_j \right. \right. \tag{7.30}$$

$$\left. + \frac{u_{i-1/2,j} + u_{i+1/2,j}}{2} \sin\varphi_j \right) + \pi_{a,i+1,j} \frac{v_{i+1,j-1/2} + v_{i+1,j+1/2}}{2}$$

$$\left. \times \left( f_j R_e \cos\varphi_j + \frac{u_{i+1/2,j} + u_{i+3/2,j}}{2} \sin\varphi_j \right) \right]_{k,t-h}$$

*Pressure gradient terms*

$$-R_e\Delta\,\varphi_{i+1/2,j}\left[(\Phi_{i+1,j,k}-\Phi_{i,j,k})\frac{\pi_{a,i,j}+\pi_{a,i+1,j}}{2}+(\pi_{a,i+1,j}-\pi_{a,i,j})\right.$$
$$\times\frac{c_{p,d}}{2}\left(\left[\theta_{v,k}\frac{\sigma_{k+1/2}(P_{k+1/2}-P_k)+\sigma_{k-1/2}(P_k-P_{k-1/2})}{\Delta\sigma_k}\right]_{i,j}\right.$$
$$\left.\left.+\left[\theta_{v,k}\frac{\sigma_{k+1/2}(P_{k+1/2}-P_k)+\sigma_{k-1/2}(P_k-P_{k-1/2})}{\Delta\sigma_k}\right]_{i+1,j}\right)\right]_{t-h} \tag{7.31}$$

*Eddy diffusion term*

$$\left.+\left[\left(\pi_{a,t-h}\Delta\mathrm{A}\right)_{i+1/2,j}\left[\frac{(\nabla_z\cdot\rho_a\mathbf{K}_m\nabla_z)u}{\rho_a}\right]_{i+1/2,j,k,t-h}\right]\right\} \tag{7.32}$$

All $\dot{\sigma}$ and some $\pi_a$ values are evaluated at time $t$. All other terms are evaluated at time $t-h$. The interpolation of the second pressure gradient term in (7.31) is differenced in a manner consistent with the differencing of geopotential, discussed shortly.

The equations above require additional interpolations for certain terms. First, column pressure multiplied by grid cell area at a $u$-point is interpolated with

$$(\pi_a\Delta\mathrm{A})_{i+1/2,j}=\tfrac{1}{8}\{(\pi_a\,\Delta\mathrm{A})_{i,j+1}+(\pi_a\,\Delta\mathrm{A})_{i+1,j+1}+2[(\pi_a\,\Delta\mathrm{A})_{i,j}+(\pi_a\,\Delta\mathrm{A})_{i+1,j}]+(\pi_a\,\Delta\mathrm{A})_{i,j-1}+(\pi_a\,\Delta\mathrm{A})_{i+1,j-1}\} \tag{7.33}$$

where

$$\Delta\mathrm{A}=R_e^2\cos\varphi\Delta\lambda\Delta\varphi \tag{7.34}$$

is the horizontal area of a grid cell. (7.33) relies on column pressures (at times $t$ or $(t-h)$ and grid areas at $\pi_a$ points. An interpolation similar to (7.33), required for (7.29), is

$$(\pi_{a,t}\,\Delta\mathrm{A}\dot{\sigma}_{k-1/2,t})_{i+1/2,j}=\tfrac{1}{8}\{(\pi_{a,t}\,\Delta\mathrm{A}\dot{\sigma}_{k-1/2,t})_{i,j+1}+(\pi_{a,t}\,\Delta\mathrm{A}\dot{\sigma}_{k-1/2,t})_{i+1,j+1}$$
$$+2[(\pi_{a,t}\,\Delta\mathrm{A}\dot{\sigma}_{k-1/2,t})_{i,j}+(\pi_{a,t}\,\Delta\mathrm{A}\dot{\sigma}_{k-1/2,t})_{i+1,j}]$$
$$+(\pi_{a,t}\,\Delta\mathrm{A}\dot{\sigma}_{k-1/2,t})_{i,j-1}+(\pi_{a,t}\,\Delta\mathrm{A}\dot{\sigma}_{k-1/2,t})_{i+1,j-1}\} \tag{7.35}$$

All values at time $t$ on the right side of this equation are known from (7.12) and (7.19) at the time the equation is evaluated.

A third set of interpolation, required for the fluxes in (7.28) shown in Fig. 7.3, is

$$\mathrm{B}_{i,j}=\tfrac{1}{12}[F_{i-1/2,j-1}+F_{i+1/2,j-1}+2(F_{i-1/2,j}+F_{i+1/2,j})+F_{i-1/2,j+1}+F_{i+1/2,j+1}] \tag{7.36}$$

$$\mathrm{C}_{i+1/2,j-1/2}=\tfrac{1}{12}[G_{i,j-3/2}+G_{i+1,j-3/2}+2(G_{i,j-1/2}+G_{i+1,j-1/2})+G_{i,j+1/2}+G_{i+1,j+1/2}] \tag{7.37}$$

$$\mathrm{D}_{i,j+1/2} = \frac{1}{24}(G_{i,j-1/2} + 2G_{i,j+1/2} + G_{i,j+3/2} + F_{i-1/2,j} + F_{i-1/2,j+1} + F_{i+1/2,j} + F_{i+1/2,j+1}) \tag{7.38}$$

$$\mathrm{E}_{i,j+1/2} = \frac{1}{24}(G_{i,j-1/2} + 2G_{i,j+1/2} + G_{i,j+3/2} - F_{i-1/2,j} - F_{i-1/2,j+1} - F_{i+1/2,j} - F_{i+1/2,j+1}) \tag{7.39}$$

where $F$ and $G$ are the fluxes defined in (7.13) and (7.14), respectively. Finally, $u$-values at the bottom of a layer, required for (7.29), are interpolated with

$$u_{i+1/2,j,k+1/2,t-h} = \frac{\Delta\sigma_{k+1}u_{i+1/2,j,k,t-h} + \Delta\sigma_k u_{i+1/2,j,k+1,t-h}}{\Delta\sigma_k + \Delta\sigma_{k+1}} \tag{7.40}$$

Boundary conditions for some terms in the momentum equation must be treated carefully. At the horizontal boundaries, a virtual row may be added to obtain variable values just outside the boundaries. If $i + 3/2$ in Fig. 7.3 is the eastern boundary, values of $\pi_a$, $\Phi$, and $\dot{\sigma}$ are needed at $\pi_a$ points ($i+2$ points) just outside the boundary. Values of $u$ at $u$ points ($i + 5/2$ points) are also needed.

In the pressure gradient term of the momentum equation, $\Phi_{t-h}$ and $\pi_{a,t-h}$ values outside a boundary should be set to $\Phi_{t-2h}$ and $\pi_{a,t-2h}$ values just inside the boundary (e.g., Lu 1994). In such a case, (7.31) is replaced with

*Pressure gradient terms*

$$-R_e\Delta\varphi_{I+1/2,j}\Bigg[(\Phi_{I,j,k,t-2h} - \Phi_{I,j,k,t-h})\pi_{a,I,j,t-h} + (\pi_{a,I,j,t-2h} - \pi_{a,I,j,t-h}) \times c_{p,d}\left(\theta_{v,k}\frac{\sigma_{k+1/2}(P_{k+1/2} - P_k) + \sigma_{k-1/2}(P_k - P_{k-1/2})}{\Delta\sigma_k}\right)_{I,j,t-h}\Bigg] \tag{7.41}$$

This boundary condition dampens the effect of gravity waves, which rapidly create instabilities when the boundary condition is absent.

For simplicity, outside boundary values of $\dot{\sigma}$ and $u$ can be set to nearest inside boundary values from the current time step or previous time step. Arakawa (1982) and Lu (1994) contain a more rigorous treatment of these variables at the boundaries. In that treatment, momentum fluxes at the boundaries are derived so as to conserve mass at the boundaries.

The momentum equation in the south–north direction is differenced in a manner similar to that in the west–east direction. Figure 7.4 shows the locations of fluxes for finite differencing this equation. The finite-difference form of the ***v*-momentum equation**, given in (5.87), is

*Time difference term*

$$v_{i,j+1/2,k,t} = \frac{(\pi_{a,t-h}\Delta A)_{i,j+1/2}}{(\pi_{a,t}\Delta A)_{i,j+1/2}}v_{i,j+1/2,k,t-h} + \frac{h}{(\pi_{a,t}\Delta A)_{i,j+1/2}} \times \Bigg\{ \tag{7.42}$$

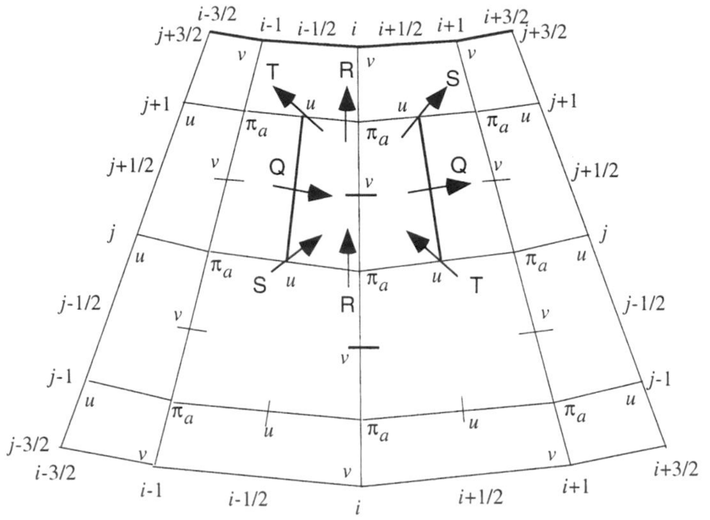


**Figure 7.4.** Location of fluxes for the finite-differencing of the momentum equation in the $v$-direction. The grid is the same as that in Fig. 7.2. The figure shows exact locations of $u$, $v$, and $\pi_a$ relative to lateral boundaries.

*Horizontal advection terms*

$$\left(\mathrm{Q}_{i-1/2,j+1/2}\frac{v_{i-1,j+1/2}+v_{i,j+1/2}}{2}-\mathrm{Q}_{i+1/2,j+1/2}\frac{v_{i,j+1/2}+v_{i+1,j+1/2}}{2}\right. \tag{7.43}$$

$$+\mathrm{R}_{i,j}\frac{v_{i,j-1/2}+v_{i,j+1/2}}{2}-\mathrm{R}_{i,j+1}\frac{v_{i,j+1/2}+v_{i,j+3/2}}{2}$$

$$+\mathrm{S}_{i-1/2,j}\frac{v_{i-1,j-1/2}+v_{i,j+1/2}}{2}-\mathrm{S}_{i+1,j+1}\frac{v_{i,j+1/2}+v_{i+1,j+3/2}}{2}$$

$$\left.+\mathrm{T}_{i+1/2,j}\frac{v_{i+1,j-1/2}+v_{i,j+1/2}}{2}-\mathrm{T}_{i-1/2,j+1}\frac{v_{i,j+1/2}+v_{i-1,j+3/2}}{2}\right)_{k,t-h}$$

*Vertical transport of horizontal momentum*

$$+\frac{1}{\Delta\sigma_k}(\pi_{a,t}\,\Delta\mathrm{A}\dot{\sigma}_{k-1/2,t}u_{k-1/2,t-h}-\pi_{a,t}\,\Delta\mathrm{A}\dot{\sigma}_{k+1/2,t}u_{k+1/2,t-h})_{i,j+1/2} \tag{7.44}$$

*Coriolis and spherical grid conversion terms*

$$-\frac{R_e(\Delta\lambda_e\,\Delta\varphi)_{i,j+1/2}}{2}\times\left[\pi_{a,i,j}\frac{u_{i-1/2,j}+u_{i+1/2,j}}{2}\left(f_j R_e\cos\varphi_j\right.\right. \tag{7.45}$$

$$\left.+\frac{u_{i-1/2,j}+u_{i+1/2,j}}{2}\sin\varphi_j\right)+\pi_{a,i,j+1}\frac{u_{i-1/2,j+1}+u_{i+1/2,j+1}}{2}$$

$$\left.\times\left(f_{j+1}R_e\cos\varphi_{j+1}+\frac{u_{i-1/2,j+1}+u_{i+1/2,j+1}}{2}\sin\varphi_{j+1}\right)\right]_{k,t-h}$$

*Pressure gradient terms*

$$-R_e(\cos\varphi\Delta\lambda_e)_{i,j+1/2}\left[(\Phi_{i,j+1,k}-\Phi_{i,j,k})\frac{\pi_{a,i,j}+\pi_{a,i,j+1}}{2}+(\pi_{a,i,j+1}-\pi_{a,i,j})\right. \quad (7.46)$$

$$\times\frac{c_{p,d}}{2}\left(\left[\theta_{v,k}\frac{\sigma_{k+1/2}(P_{k+1/2}-P_k)+\sigma_{k-1/2}(P_k-P_{k-1/2})}{\Delta\sigma_k}\right]_{i,j}\right.$$

$$\left.\left.+\left[\theta_{v,k}\frac{\sigma_{k+1/2}(P_{k+1/2}-P_k)+\sigma_{k-1/2}(P_k-P_{k-1/2})}{\Delta\sigma_k}\right]_{i,j+1}\right)\right]_{t-h}$$

*Eddy diffusion term*

$$+\left[\left(\pi_{a,t-h}\Delta A\right)_{i,j+1/2}\left[\frac{(\nabla_z\cdot\rho_a\mathbf{K}_m\nabla_z)u}{\rho_a}\right]_{i,j+1/2,k,t-h}\right]\Bigg\} \quad (7.47)$$

In the equations above, column pressure multiplied by the area of a grid cell at a $v$-point is interpolated with

$$(\pi_a\Delta A)_{i,j+1/2}=\tfrac{1}{8}\{(\pi_a\Delta A)_{i+1,j}+(\pi_a\Delta A)_{i+1,j+1}+2[(\pi_a\Delta A)_{i,j} \quad (7.48)$$

$$+(\pi_a\Delta A)_{i,j+1}]+(\pi_a\Delta A)_{i-1,j}+(\pi_a\Delta A)_{i-1,j+1}\}$$

Similarly,

$$(\pi_{a,t}\Delta A\dot{\sigma}_{k-1/2,t})_{i,j+1/2}=\tfrac{1}{8}\{(\pi_{a,t}\Delta A\dot{\sigma}_{k-1/2,t})_{i+1,j}+(\pi_{a,t}\Delta A\dot{\sigma}_{k-1/2,t})_{i+1,j+1} \quad (7.49)$$

$$+2[(\pi_{a,t}\Delta A\dot{\sigma}_{k-1/2,t})_{i,j}+(\pi_{a,t}\Delta A\dot{\sigma}_{k-1/2,t})_{i,j+1}]$$

$$+(\pi_{a,t}\Delta A\dot{\sigma}_{k-1/2,t})_{i-1,j}+(\pi_{a,t}\Delta A\dot{\sigma}_{k-1/2,t})_{i-1,j+1}\}$$

Fluxes in (7.43) are found with

$$Q_{i-1/2,j+1/2}=\tfrac{1}{12}[F_{i-3/2,j}+F_{i-3/2,j+1}+2(F_{i-1/2,j} \quad (7.50)$$

$$+F_{i-1/2,j+1})+F_{i+1/2,j}+F_{i+1/2,j+1}]$$

$$R_{i,j}=\tfrac{1}{12}[G_{i-1,j-1/2}+G_{i-1,j+1/2}+2(G_{i,j-1/2} \quad (7.51)$$

$$+G_{i,j+1/2})+G_{i+1,j-1/2}+G_{i+1,j+1/2}]$$

$$S_{i+1/2,j}=\tfrac{1}{24}(G_{i,j-1/2}+G_{i,j+1/2}+G_{i+1,j-1/2}+G_{i+1,j+1/2} \quad (7.52)$$

$$+F_{i-1/2,j}+2F_{i+1/2,j}+F_{i+3/2,j})$$

$$T_{i+1/2,j}=\tfrac{1}{24}(G_{i,j-1/2}+G_{i,j+1/2}+G_{i+1,j-1/2}+G_{i+1,j+1/2} \quad (7.53)$$

$$-F_{i-1/2,j}-2F_{i+1/2,j}-F_{i+3/2,j})$$

The south–north scalar velocity at the bottom of a layer is interpolated with

$$v_{i,j+1/2,k+1/2,t-h} = \frac{\Delta\sigma_{k+1} v_{i,j+1/2,k,t-h} + \Delta\sigma_k v_{i,j+1/2,k+1,t-h}}{\Delta\sigma_k + \Delta\sigma_{k+1}} \tag{7.54}$$

Boundary conditions for the $v$-equation are similar to those for the $u$-equation. For instance, pressure-gradient boundary conditions in the $v$-equation are

*Pressure gradient terms*

$$\begin{aligned} &-R_e(\cos\varphi\Delta\lambda_e)_{i,J+1/2} \\ &\quad\times\Bigg\{(\Phi_{i,J,k,t-2h} - \Phi_{i,J,k,t-h})\pi_{a,i,J,t-h} + (\pi_{a,i,J,t-2h} - \pi_{a,i,J,t-h}) \\ &\quad\times c_{p,d}\left[\theta_{v,k}\frac{\sigma_{k+1/2}(P_{k+1/2} - P_k) + \sigma_{k-1/2}(P_k - P_{k-1/2})}{\Delta\sigma_k}\right]_{i,J,t-h}\Bigg\} \end{aligned} \tag{7.55}$$

## 7.6. THE HYDROSTATIC EQUATION

When a grid cell is larger than 3 km on a side, the model atmosphere is generally in hydrostatic equilibrium, and the geopotential used in the horizontal momentum equations can be calculated diagnostically with the hydrostatic equation. From (5.38), the hydrostatic equation was $d\Phi = -c_{p,d}\theta_v dP$. A finite-difference form of this equation giving the geopotential at the vertical midpoint of the bottom layer at time $t-h$ is (Arakawa and Suarez 1983)

$$\Phi_{i,j,N_L,t-h} = \Phi_{i,j,N_L+1/2} - c_{p,d}\left[\theta_{v,N_L}(P_{N_L} - P_{N_L+1/2})\right]_{i,j,t-h} \tag{7.56}$$

where $P_{N_L+1/2}$ and $P_{N_L}$ are found from (7.7) and (7.6), respectively. The geopotential at the surface, $\Phi_{i,j,N_L+1/2}$, is the topographical surface altitude multiplied by gravitational acceleration. The geopotential at the bottom and vertical center of each layer above the surface are found with

$$\Phi_{i,j,k+1/2,t-h} = \Phi_{i,j,k+1,t-h} - c_{p,d}[\theta_{v,k+1}(P_{k+1/2} - P_{k+1})]_{i,j,t-h} \tag{7.57}$$

$$\Phi_{i,j,k,t-h} = \Phi_{i,j,k+1/2,t-h} - c_{p,d}[\theta_{v,k}(P_k - P_{k+1/2})]_{i,j,t-h} \tag{7.58}$$

respectively. The geopotential is solved for from the model surface to top. The boundary conditions for the geopotential were discussed in Section 7.5.

## 7.7. ORDER OF CALCULATIONS

The equations in the preceding sections should be solved during a time step in the following order: column pressure from (7.12), vertical scalar velocity from (7.19), species moist-air mass mixing ratios from (7.22), potential virtual temperature from (7.25), geopotential from (7.56)–(7.58), west–east scalar velocities from (7.27)–(7.32), and south–north scalar velocities from (7.42)–(7.47). Quantities in the equations are used explicitly (from time $t - h$), except for all $\dot{\sigma}$-values and specified $\pi_a$-values. Since most equations rely on $\dot{\sigma}$ and some $\pi_a$-values from time $t$, these parameters must be determined first and second, respectively,

## 7.8. TIME-STEPPING SCHEMES

In this chapter, prognostic equations were written as explicit, first-order (forward Euler) approximations in time. Since the forward Euler is unconditionally unstable with respect to scalar velocities in the absence of eddy diffusion, as discussed in Chapter 6, the forward Euler causes solutions to break down quickly. To increase stability and length of simulation, a better time-stepping scheme is needed. Stability and accuracy can be improved by using a Matsuno time-stepping scheme alone or a Matsuno scheme combined with a leapfrog scheme.

The Matsuno scheme is first-order in time but conditionally stable. This scheme, alone, can be used to solve the equations in this chapter. During a Matsuno time step, variable values at time $t$ are estimated from values at time $t - h$. If the time derivative of an advected species with units of moist-air mixing ratio is

$$\frac{\partial q}{\partial t} = f(q) \tag{7.59}$$

then the explicit estimated value of q is $q_{est} = q_{t-h} + hf(q_{t-h})$. The estimate is used to calculate the spatial derivative in a second equation that also relies on information from one time step backward. The final value of q from the second equation is $q_t = q_{t-h} + hf(q_{est})$.

The time-derivative approximation can be improved to second order by using the **leapfrog scheme**. Since the leapfrog scheme destabilizes over time when used alone to solve nonlinear equations, the **Matsuno scheme** or another scheme must be inserted every 5–15 leapfrog steps to maintain stability. With the leapfrog scheme, time derivatives leap over a time step in the spatial derivative terms. The mixing ratio at the end of a leapfrog time step is $q_{t+h} = q_{t-h} + 2hf(q_t)$, which requires information from two time steps backward. The Matsuno scheme is used not only to stabilize the leapfrog scheme but also to initialize a new simulation or time interval. Figure 7.5 shows a sequence of dynamics calculations when Matsuno

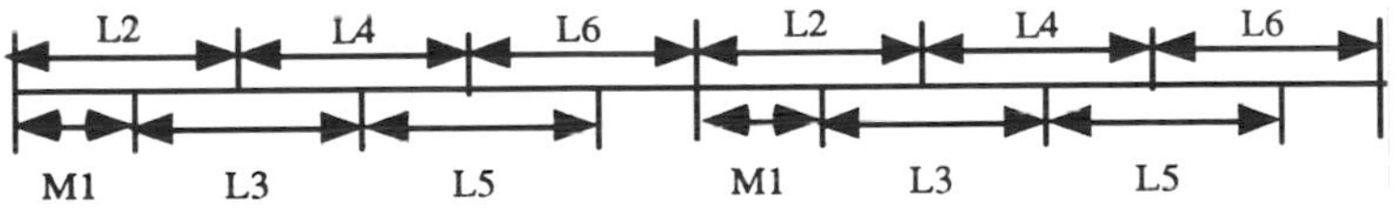


**Figure 7.5.** Example of a dynamical time-stepping scheme that uses a Matsuno (M) and a leapfrog (L) scheme. Two interruptions with a Matsuno step are shown – one to start the simulation and one after several leapfrog steps. The number next to each letter indicates the order in which the step is taken during a sequence.

and leapfrog schemes are used together. In Chapter 6, a **time step** was defined as an increment in time for a given algorithm, and a **time interval** was defined as the period during which several time steps of a process are completed without interference by another process. In Fig. 7.5, M1, L2, L3, ... , L6 are time steps, and each set of M1, ... , L6 is a time interval.

Dynamical, chemical, and physical processes often require different time step sizes in a model. In a global model, typical time steps for physics, chemistry, and dynamics are 1–10 min, $\ll$1 s–30 min, and $<$1 min–1 h, respectively. In a regional model, typical time steps for the three processes are 5–15 s, $\ll$1 s–15 min, and $<$1–30 min, respectively.

## 7.9. SUMMARY

In this chapter, finite-difference solutions to the equations of atmospheric dynamics, including the continuity equation for air, the species continuity equation, the thermodynamic energy equation, the horizontal momentum equations, and the hydrostatic equation, were given. The thermodynamic energy equation is used to calculate the potential virtual temperature, and the horizontal momentum equations are used to obtain horizontal scalar velocities. When the atmosphere is in hydrostatic equilibrium, the continuity equation for air is used to solve for vertical velocity and column pressure, and the hydrostatic equation is used to solve for geopotential. A common time-stepping scheme in dynamical models is the combination of a Matsuno and leapfrog scheme.

## 7.10. PROBLEMS

**7.1.** For the same time step and grid spacing, should the model accuracy increase or decrease when Matsuno but no leapfrog steps are taken versus when Matsuno and leapfrog steps are taken? Why? How is the computer time affected in each case?

**7.2.** **(a)** Assume a grid cell, centered at $\varphi = 45°$N, has dimensions $\Delta\lambda_e = 5°$, $\Delta\varphi = 4°$, and $\Delta\sigma = 0.05$. Calculate q at the center of the cell after one 5-min time step if $u_1 = +3$ m s$^{-1}$, $\pi_{a,1} = 745$ mb, $q_1 = 0.0062$ kg kg$^{-1}$; $u_2 = +1$ m s$^{-1}$, $\pi_{a,2} = 752$ mb, $q_2 = 0.0047$ kg kg$^{-1}$; $v_3 = +2$ m s$^{-1}$, $\pi_{a,3} = 754$ mb, $q_3 = 0.0064$ kg kg$^{-1}$; $v_4 = +1$ m s$^{-1}$, $\pi_{a,4} = 746$ mb, $q_4 = 0.0054$ kg kg$^{-1}$, $\dot{\sigma}_5 = -4.58 \times 10^{-6}\sigma$ s$^{-1}$, $q_5 = 0.005$ kg kg$^{-1}$; and $\dot{\sigma}_6 = -6.11 \times 10^{-6}\sigma$ s$^{-1}$, $q_6 = 0.006$ kg kg$^{-1}$, where subscripts 1, ... , 6 denote the west-, east-, south-, north-, lower-, and upper-boundaries, respectively. Average the four $\pi_a$-values and six q-values to obtain center values. Neglect eddy diffusion and external sources and sinks. Assume the air is dry and the $\pi_a$-values stay constant for the time step.

**(b)** For the conditions of part (a), calculate $\theta_v$ at the center of the cell after 30 minutes if the west, east, south, north, lower, and upper grid-cell boundary values are $\theta_v = 297$, 298, 300, 299, 298.3, and 298.7 K, respectively. Average the six boundary values to obtain an initial center value of $\theta_v$.

## 7.11. COMPUTER PROGRAMMING PRACTICE

**7.3.** Set up a 20-layer (21-boundary) sigma-pressure-coordinate vertical model grid between $p_{a,top} = 250$ mb and $p_{a,surf} = 1000$ mb. Assume each layer has equal sigma-thickness, and assume $T = 288$ at the surface. Assume also that temperature decreases with altitude at 6.5 K km$^{-1}$. Calculate pressure at the center of each layer with (7.8). Plot pressure versus altitude.

## 7.12. MODELING PROJECT

The purpose of this project is to develop a basic mesoscale model that solves the equations of atmospheric dynamics, except for the water-vapor continuity equation. Diabatic energy sources and sinks and eddy diffusion are ignored. Winds are driven primarily by pressure gradients.

(a) Choose a southwest-corner latitude and longitude anywhere on earth away from the poles.

(b) Set a model top pressure everywhere of 250 mb (near the tropopause).

(c) Set NLAT, NLONG, and NVERT, which denote the numbers of latitudinal, longitudinal, and vertical grid centers in the model. Note that 1, ... , NLAT, 1, ... , NLONG, and 1, ... , NVERT increase from south to north, west to east, and top to bottom, respectively. NLAT+1, NLONG + 1, and NVERT + 1 equal the numbers of boundaries in the respective direction. Allow parameters to be variable, but set them initially to NLAT = 40, NLONG = 50, and NVERT = 15.

(d) Use spherical horizontal coordinates and the sigma-pressure vertical coordinate. Select $\Delta\lambda_e = 0.05°$ and $\Delta\varphi = 0.05°$. Select values of $\Delta\sigma$ for each layer using the method discussed in this chapter. Ensure that the bottom layer is at least 150 m thick (1 mb $\approx$ 10 m).

(e) Select a surface pressure for the horizontal centers of the grid cells at each of the four corners of the model. Interpolate surface pressure horizontally to the center of each interior grid cell with a $1/r^2$ interpolation. The interpolated pressure at any interior location, $i$, $j$ is,

$$p_{a,i,j} = \frac{r_s^{-2}\, p_{a,1,1} + r_t^{-2}\, p_{a,\mathrm{NLONG},1} + r_u^{-2}\, p_{a,1,\mathrm{NLAT}} + r_v^{-2}\, p_{a,\mathrm{NLONG},\mathrm{NLAT}}}{r_s^{-2} + r_t^{-2} + r_u^{-2} + r_v^{-2}}$$

where $r_s$, $r_t$, $r_u$, and $r_v$ are distances from the four respective corner centers to point $i$, $j$.

(f) Use (7.4) to calculate pressure at the horizontal midpoint/vertical boundary of each cell. Use (7.8) to calculate pressure at the horizontal midpoint/vertical midpoint of each cell from horizontal midpoint/vertical boundary values.

(g) Initialize temperatures at the horizontal midpoint/vertical midpoint of each cell in one corner column of the model. Estimate temperatures from Table B.1 of Appendix B given the pressures from part (f).

(h) Set the temperature at the horizontal midpoint/vertical midpoint of each cell in all other columns equal to that in the initialized column. Calculate potential virtual temperature at the horizontal midpoint/vertical midpoint of each cell. Assume the air is dry. Interpolate potential virtual temperature to the horizontal midpoint/vertical boundary of each cell with (7.9). No values are needed for the model top or bottom boundaries ($k = 1/2$ or $N_L + 1/2$).

(i) Use the equation of state to calculate the density at the horizontal midpoint/vertical midpoint of each cell.

(j) Use (7.12) to solve for the column pressure at the horizontal midpoint of each column in the model.

(k) Use (7.19) to solve for vertical velocities at the horizontal midpoint/vertical boundary of each cell.

(l) Use (7.25) to solve for the potential virtual temperature at the horizontal midpoint/vertical midpoint of each cell. Ignore diabatic sources and sinks and ignore eddy diffusion.

(m) Solve for the geopotential in each cell with (7.56)–(7.58). Assume the surface geopotential is zero.

(n) Use (7.27)–(7.32) and (7.42)–(7.47) to solve for $u$ and $v$, at locations shown in Figs. 7.3 and 7.4, respectively. Ignore eddy diffusion. Assume $u$, $v$, and $\dot{\sigma}$ are initially zero everywhere.

(o) Use a Matsuno or Matsuno-plus-leapfrog time-stepping scheme instead of the forward Euler scheme to advance all prognostic equations for this problem. Recalculate all diagnostic variables during each Matsuno and/or leapfrog step.

(p) Debug, and run simulations. Change conditions by changing surface pressures at the corners.

(q) Test the effect of adding topography by changing the geopotential at the surface.

(r) Plot horizontal and/or vertical fields of temperature, pressure, and velocities.

# 8

# Boundary-Layer Processes

THE boundary layer is more difficult to treat in a numerical model than the free troposphere, because turbulence is stronger and surface processes have a greater influence in the boundary layer. Parameters important for modeling the boundary layer include turbulent fluxes, eddy diffusion coefficients, and ground temperatures. Parameterizations that describe boundary-layer and soil processes are discussed in this chapter. Expressions for surface-layer kinematic turbulent fluxes of momentum, energy, and moisture are given in terms of bulk aerodynamic formulae and Monin–Obukhov similarity theory. Eddy diffusion coefficients for momentum and energy are derived. These coefficients depend on mechanical shear and buoyancy. During the day, mechanical shear is important in the surface layer, and buoyancy is important in the mixed layer. During the night, shear and buoyancy also affect turbulence in the mixed layer. Buoyancy decreases but does not always disappear at night in the mixed layer. Vertical profiles of wind speed, potential virtual temperature, and water-vapor specific humidity are derived for the surface layer. Prognostic equations for ground temperature and soil moisture are also given.

## 8.1. TURBULENT FLUXES OF MOMENTUM

Turbulence mixes momentum, energy, moisture, and trace gases vertically and horizontally. The degree of turbulence can be quantified with a turbulent flux term. In the case of horizontal momentum, the vertical turbulent flux (kg m$^{-1}$ s$^{-2}$) is a function of $\overline{w'u'}$ and $\overline{w'v'}$ (m$^2$ s$^{-1}$), which are mean kinematic vertical turbulent fluxes, introduced in Section 4.1.6.

Mean kinematic turbulent fluxes of momentum are negatively proportional to **Reynolds stresses**. A **stress** is a force per unit area that causes a body to deform. A Reynolds stress, which arises when a fluid undergoes turbulent motion, causes a parcel of air to deform. Suppose an air parcel fluctuates randomly in time. Precise scalar velocities, $w$ and $u$, in the parcel have mean components $\bar{w}$ and $\bar{u}$ and eddy components $w'$ and $u'$, respectively. A $w'$-velocity has the effect of mixing $u'$-velocities in the $z$-direction. The vertical mixing of the $u'$-velocity exerts a force in the $x$-direction over an area normal to the $z$-direction. The effect of the force per unit area, or stress, is to induce a drag on $\bar{u}$ and to cause the air parcel to deform, as shown in Fig. 8.1.

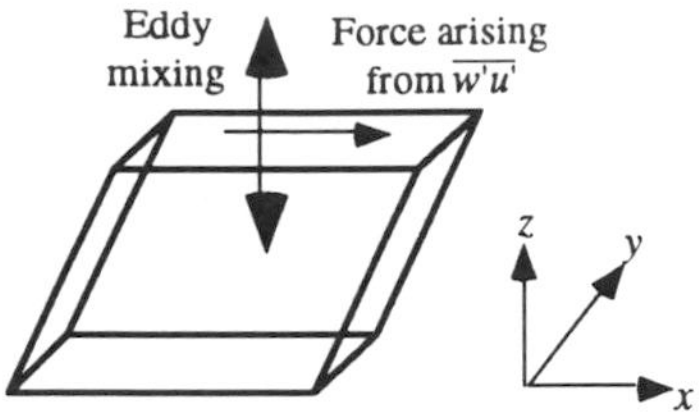


**Figure 8.1.** Deformation of a cubic air parcel caused by a vertical kinematic momentum flux $\overline{w'u'}$, as described in the text. Adapted from Stull (1988).

In the example above, the scalar component of Reynolds stress (e.g., kg m$^{-1}$ s$^{-2}$ or N m$^{-2}$) in the $x$-direction along a plane normal to the $z$-direction is

$$\tau_{zx} = -\rho_a \overline{w'u'} \tag{8.1}$$

This stress results from the vertical transport of a west–east gradient of momentum. Since $w'$ mixes $u'$ in the same way that $u'$ mixes $w'$, we have $\overline{u'w'} = \overline{w'u'}$, $\tau_{zx} = \tau_{xz}$, and $\tau_{xz} = -\rho_a \overline{u'w'}$. The Reynolds stress in the $y$-direction along a plane normal to the $z$-direction is $\tau_{zy} = -\rho_a \overline{w'v'}$. Both $\tau_{zx}$ and $\tau_{zy}$ have the effect of transporting gradients of horizontal momentum in the vertical direction. Combining the two gives the magnitude of the **vertical turbulent flux of horizontal momentum** as

$$|\tau_z| = \rho_a [(\overline{w'u'})^2 + (\overline{w'v'})^2]^{1/2} \tag{8.2}$$

**Example 8.1**

Reynolds stresses are usually much larger than viscous stresses. From (4.54), the shearing stress in the $x$-direction over the $x$–$y$ plane due to wind shear in the $z$-direction was $\tau_{zx} = \eta_a\, \partial u/\partial z$. If $T = 288$ K and $u = 5$ m s$^{-1}$ at $z = 10$ m, then $\eta_a = 1.792 \times 10^{-5}$ kg m$^{-1}$ s$^{-1}$ and $\partial u/\partial z = 0.5$ s$^{-1}$. The resulting shearing stress due to viscosity is $\tau_{zx} = 8.96 \times 10^{-6}$ kg m$^{-1}$ s$^{-2}$.

A typical value of $\overline{w'u'}$ is 0.4 m$^2$ s$^{-2}$. For a surface air density of $\rho_a = 1.25$ kg m$^{-3}$, (8.1) predicts a Reynolds stress due to turbulence of $\tau_{zx} = 0.5$ kg m$^{-1}$ s$^{-2}$.

## 8.2. TURBULENT FLUXES OF ENERGY AND WATER VAPOR

Energy in the surface layer is transferred vertically by turbulence arising from mechanical shear and buoyancy. The **vertical turbulent sensible-heat flux** (W m$^{-2}$) due to shear and buoyancy is

$$H_f = \rho_a c_{p,d} (\overline{w'\theta'_v})_s \tag{8.3}$$

where $\theta'_v$ is the turbulent fluctuation of potential virtual temperature (K), and the subscript $s$ indicates that values are taken in the surface layer. The product $(\overline{w'\theta'_v})_s$ (m K s$^{-1}$) is the **vertical kinematic turbulent sensible-heat flux** in the surface layer and can be calculated from observed quantities of $\theta'_v$ and $w'$. In a model $(\overline{w'\theta'_v})_s$ is parameterized.

Water vapor is also transferred vertically from the surface by mechanical shear and buoyancy. The **vertical turbulent flux of water vapor** (kg $m^{-2}$ $s^{-1}$) is

$$E_f = \rho_a(\overline{w'q'_v})_s \tag{8.4}$$

where $q'_v$ is the turbulent fluctuation of water-vapor specific humidity (kg $kg^{-1}$), and $(\overline{w'q'_v})_s$ (m kg $s^{-1}$ $kg^{-1}$) is the **vertical kinematic moisture flux** in the surface layer, which can be obtained from observations or parametrized.

## 8.3. FRICTION VELOCITY

The magnitude of the Reynolds stress is used to calculate $u_*$, a scaling variable called the **friction velocity**. The friction velocity gives a measure of the vertical kinematic turbulent flux of horizontal momentum in the surface layer. When horizontal winds flow over roughness elements protruding from a surface, drag slows wind speeds near the surface relative to those aloft, creating vertical wind shear. Wind shear produces eddies that exchange momentum, energy, gases, and aerosols vertically. The greater the height that roughness elements protrude from a surface, and the greater the horizontal wind speed, the greater the resulting wind shear and **mechanical turbulence**. The greater the mechanical turbulence, the greater the friction velocity, and the faster momentum, energy, and pollutants from aloft mix with surface air. Typical roughness elements at the surface include rocks, trees, buildings, grass, and sand.

The friction velocity,

$$u_* = \left[(\overline{w'u'})_s^2 + (\overline{w'v'})_s^2\right]^{1/4} = (|\tau_z|/\rho_a)_s^{1/2} \tag{8.5}$$

can be parameterized or found from field experiments in which $u'$, $v'$, and $w'$ are measured.

**Example 8.2**

If two measurements of west–east and vertical scalar wind velocity yield $u_1 = 10$ m $s^{-1}$, $w_1 = 0.1$ m $s^{-1}$, and $u_2 = 6$ m $s^{-1}$, $w_2 = 0.2$ m $s^{-1}$, respectively, estimate $u_*$. Assume the south–north scalar velocity is zero.

SOLUTION

$$\bar{u} = (u_1 + u_2)/2 = 8\,\mathrm{m\,s^{-1}} \qquad \bar{w} = (w_1 + w_2)/2 = 0.15\,\mathrm{m\,s^{-1}}$$
$$u'_1 = u_1 - \bar{u} = 2\,\mathrm{m\,s^{-1}} \qquad w'_1 = w_1 - \bar{w} = -0.05\,\mathrm{m\,s^{-1}}$$
$$u'_2 = u_2 - \bar{u} = -2\,\mathrm{m\,s^{-1}} \qquad w'_2 = w_2 - \bar{w} = 0.05\,\mathrm{m\,s^{-1}}$$

Therefore,

$$\overline{w'u'} = (w'_1u'_1 + w'_2u'_2)/2 = -0.1\,\mathrm{m^2\,s^{-2}} \qquad u_* = (\overline{w'u'})_s^{1/2} = 0.32\,\mathrm{m\,s^{-1}}$$

## 8.4. SURFACE ROUGHNESS LENGTHS

Three variables used frequently in boundary-layer parameterizations are surface roughness lengths for momentum, energy, and moisture. The **surface roughness length for momentum** ($z_{0,m}$), or **aerodynamic roughness,** is the height above a surface at which the logarithmic profile of wind speed versus altitude extrapolates to zero wind speed. It gives a measure of vertical turbulence that occurs when a horizontal wind flows over a rough surface. The greater $z_{0,m}$, the greater the turbulence that arises when wind passes over a roughness element. For a perfectly smooth surface, the roughness length is zero, and mechanical turbulence is minimized. For other surfaces, it is sometimes approximated as 1/30 the height of the average roughness element protruding from the surface. For surfaces with sparsely placed roughness elements $z_{0,m}$ is the height above the base of the roughness elements (Brutsaert 1991). For densely placed roughness elements of average height $h_c$, $z_{0,m}$ is the height above a displacement height ($d_c$) that lies between 0 and $h_c$, as discussed in Section 8.6.8.

Figure 8.2 describes a method of estimating $z_{0,m}$ from observed wind speed profiles. Parameterizations of $z_{0,m}$ from field data have also been developed. For smooth surfaces, such as over a smooth ocean with low wind speeds, one parameterization is

$$z_{0,m} \approx 0.11\nu_a/u_* \tag{8.6}$$

(e.g., Hinze 1975; Garratt 1992). Over a rough ocean with high wind speeds,

$$z_{0,m} = \alpha_c u_*^2/g \tag{8.7}$$

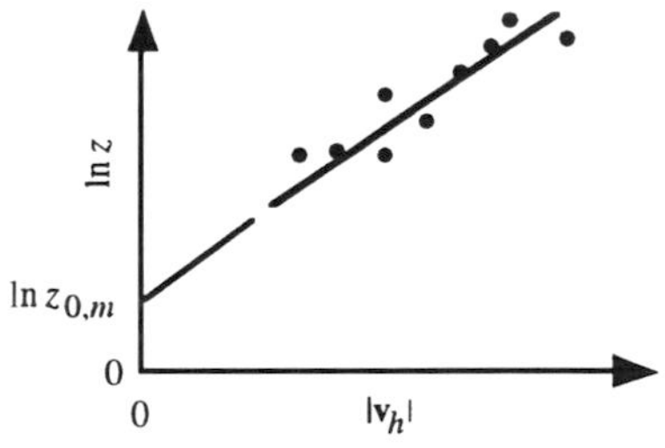


**Figure 8.2.** The surface roughness length for momentum is calculated by taking wind speed measurements at several heights at a given location when the wind is strong. The data are plotted as the natural log of height versus wind speed. Wind speeds from higher altitudes on the plot are extrapolated until a zero wind speed intersects the altitude axis. The altitude at which this occurs is the surface roughness length for momentum. Points in the diagram are observations.

**Table 8.1.** Aerodynamic Roughnesses ($z_{0,m}$), Canopy Heights ($h_c$), and Displacement Heights ($d_c$) for Several Surfaces

| Surface Type | $z_{0,m}$(m) | $h_c$(m) | $d_c$(m) | Reference |
|---|---|---|---|---|
| Smooth sea | 0.00001 | | | Garratt (1992), (8.6) |
| Rough sea | 0.000015–0.0015 | | | Garratt (1992), (8.7) |
| Ice | 0.00001 | | | Oke (1978) |
| Snow | 0.00005–0.0001 | | | Oke (1978) |
| Level desert | 0.0003 | | | Sehmel (1980) |
| Short grass | 0.03–0.01 | 0.02–0.1 | | Oke (1978) |
| Long grass | 0.04–0.1 | 0.25–1.0 | | Oke (1978) |
| Savannah | 0.4 | 8 | 4.8 | Garratt (1992) |
| Agricultural crops | 0.04–0.2 | 0.4–2 | 0.27–1.3 | Oke (1978) |
| Orchard | 0.5–1.0 | 5–10 | 3.3–6.7 | Oke (1978) |
| Coniferous forest | 0.28–3.9 | 10.4–27.5 | 6.3–25.3 | Jarvis et al. (1976) |
| Tropical forest | 2.2 | 35 | 29.8 | Shuttleworth (1989) |
| 2,500-m$^2$ lot with building 8 m high and 160-m$^2$ silhouette | 0.26 | 8 | | (8.8) |
| 25,000-m$^2$ lot with building 80 m high and 3200-m$^2$ silhouette | 5.1 | 80 | | (8.8) |

which is the **Charnock relation** (Charnock 1955), where $\alpha_c \approx 0.016$ is the **Charnock constant** (e.g., Garratt 1992). Over urban areas containing structures,

$$z_{0,m} \approx 0.5 h_0 S_0 / A_0 \tag{8.8}$$

(Lettau 1969; Petersen 1997), where 0.5 is the average drag coefficient over the structures, $h_0$ (m) is the average structure height, $A_0$ (m$^2$) is the area over which $z_{0,m}$ is estimated, and $S_0$ (m$^2$) is the silhouette area, measured in a vertical plane normal to the mean wind, of all structures in surface area $A_0$. Table 8.1 gives $z_{0,m}$ for several surfaces.

**Example 8.3**

If $T = 288$ K, $p_d = 1013$ mb, and $u_* = 0.165$ m s$^{-1}$, estimate $z_{0,m}$ over a smooth ocean. Find $z_{0,m}$ over a rough sea when $u_* = 0.5$ m s$^{-1}$.

SOLUTION

From (4.55), $\eta_a = 0.0000179$ kg m$^{-1}$ s$^{-2}$; from (2.18), $\rho_d = 1.225$ kg m$^{-3}$; and from $\nu_a = \eta_a/\rho_a$, $\nu_a = 0.0000146$ m$^2$ s$^{-2}$. Thus, from (8.6), $z_{0,m} \approx 1 \times 10^{-5}$ m over a smooth sea. From (8.7), $z_{0,m} \approx 0.0004$ m over a rough sea.

The surface roughness length for momentum characterizes the ability of surface elements to absorb momentum. Surface roughness lengths for energy and water vapor characterize the ability of elements protruding from the surface to absorb energy and moisture, relative to their ability to absorb momentum (Kaimal and Finnigan 1994). Both lengths are integration constants, used to derive vertical profiles of potential virtual temperature and specific humidity, respectively, in the surface layer. Brutsaert (1991) gives several parameterizations of $z_{0,h}$ and $z_{0,v}$ based on measurements over different terrain. More general expressions for the **energy and moisture roughness lengths** are

$$z_{0,h} = \frac{D_h}{ku_*} \qquad z_{0,v} = \frac{D_v}{ku_*} \tag{8.9}$$

respectively (Garratt and Hicks 1973), where $D_h$ is the **molecular thermal diffusion coefficient** (molecular thermal diffusivity) (m$^2$ s$^{-1}$ or cm$^2$ s$^{-1}$), $D_v$ is the **molecular diffusion coefficient of water vapor** (m$^2$ s$^{-1}$ or cm$^2$ s$^{-1}$), and $k$ is the **von Kármán constant**. The von Kármán constant has a value of between 0.35 and 0.43 (e.g., Hogstrom 1988). A value of 0.40 is used here. Expressions for $D_h$ (Pruppacher and Klett 1997) and $D_v$ (Chapman and Cowling 1970; Davis 1983) are

$$D_h = \frac{\kappa_d}{\rho_a c_{p,d}} \qquad D_v = \frac{3}{8Ad_v^2\rho_a}\sqrt{\frac{R^*Tm_a}{2\pi}\left(\frac{m_v+m_a}{m_v}\right)} \tag{8.10}$$

respectively, where $\kappa_d$ is the thermal conductivity of dry air (J m$^{-1}$ s$^{-1}$ K$^{-1}$), $A$ is Avogadro's number, and $d_v$ is the diameter of a water-vapor molecule, typically near $4.5 \times 10^{-10}$ m (4.5 Å). For very smooth surfaces, $z_{0,m}$ is slightly less than $z_{0,h}$ and $z_{0,v}$. For rough surfaces, $z_{0,m}$ is up to $10^5$ times larger than $z_{0,h}$ and $z_{0,v}$ (Brutsaert 1991). A typical ratio is about 100 (Garratt and Hicks 1973). For rough surfaces, surface elements are more efficient at absorbing momentum than at absorbing energy or water vapor.

**Example 8.4**

Assuming $u_* = 0.4$ m s$^{-1}$, $P_d = 1013$ mb, and the air is relatively dry, calculate the surface roughness lengths for energy and water vapor when $T = 298$ K.

SOLUTION

At $T = 298$ K, $\kappa_d = 0.0256$ J m$^{-1}$ s$^{-1}$ K$^{-1}$ from (2.3) and $\rho_a \approx \rho_d = 1.18$ kg m$^{-3}$ from (2.18). Thus, $D_h = 2.16 \times 10^{-5}$ m$^2$ s$^{-1}$ and $D_v = 1.4 \times 10^{-5}$ m$^2$ s$^{-1}$ from (8.10). Substituting these values into (8.9) gives $z_{0,h} = 0.000135$ m and $z_{0,v} = 0.0000875$ m, respectively.

## 8.5. BULK AERODYNAMIC EQUATIONS FOR EDDY DIFFUSION

Prognostic equations for kinematic turbulent fluxes of momentum, sensible heat, and specific humidity contain more unknowns than equations available to solve for them. When a new equation is written for an unknown, the equation creates more unknowns, giving rise to the problem of closure. For any set of equations describing kinematic turbulent fluxes, the set of prognostic or diagnostic equations is not closed because at least some terms in the equations must be parameterized as a function of known or derived variables. The **order of closure** is the highest order of a prognostic equation retained in a parameterization. If a kinematic turbulent flux equation is prognostic to first order, and all higher-order terms are parameterized, it is a first-order closure equation. Stull (1988) contains more details about the closure problem.

In the surface layer, kinematic turbulent fluxes of momentum, sensible heat, and specific humidity are often estimated with bulk aerodynamic formulae or Monin–Obukhov similarity theory. Both methods are zero-order closure techniques in that the resulting equations are fully parameterized and have no prognostic parts. Gradient transport theory ($K$-theory) is a first-order closure technique. $K$-theory assumes that the kinematic turbulent flux of a quantity is negatively proportional to a constant multiplied by the gradient of the mean value of the quantity. When bulk aerodynamic formulae or scaling parameters from similarity theory are used to calculate eddy diffusion coefficients in a $K$-theory expression for kinematic turbulent flux, the result is also a first-order closure equation. In the next sections, zero- and first-order closure techniques are described. Higher-order closure techniques are discussed in Mellor and Yamada (1982).

### 8.5.1. Bulk Aerodynamic Kinematic Fluxes of Momentum, Sensible Heat, and Moisture

**Bulk aerodynamic formulae** are equations for surface-layer vertical kinematic turbulent fluxes that assume a constant drag coefficient. The constant drag coefficient is most applicable when the boundary layer is well mixed, the surface layer is thin, and velocities and potential virtual temperatures do not change significantly with height above the surface layer. In the absence of wind shear above the thin surface layer, the vertical turbulent transfer of horizontal momentum is affected primarily by skin drag, form drag, and wave drag. **Skin drag** is the near-surface drag that results from molecular diffusion of momentum across the surface–air interface. **Form drag** is the near-surface drag resulting from turbulence and vertical momentum transfer that occurs when winds hit large obstacles, such as rocks or trees. **Wave drag** is the near-surface drag that results from vertical transfer of momentum by gravity waves, which propagate vertically and horizontally, as discussed in Chapter 4.

Expressions for bulk aerodynamic kinematic turbulent **momentum fluxes** are

$$
\begin{aligned}
(\overline{w'u'})_s &= -C_D|\bar{\mathbf{v}}_h(z_r)|[\bar{u}(z_r) - \bar{u}(z_{0,m})] \\
(\overline{w'v'})_s &= -C_D|\bar{\mathbf{v}}_h(z_r)|[\bar{v}(z_r) - \bar{v}(z_{0,m})]
\end{aligned}
\tag{8.11}
$$

where $\bar{u}(z_{0,m}) = \bar{v}(z_{0,m}) = 0$ are the mean scalar velocities at height $z_{0,m}$, $z_r$ is a **reference height** above the surface (usually 10 m), $C_D$ is the dimensionless **coefficient**

**of drag** determined at the reference height, and $|\bar{\mathbf{v}}_h(z_r)|$ is the mean horizontal wind speed at the reference height. The coefficient of drag, which accounts for skin, form, and wave drag, ranges from 0.001 for smooth surfaces to 0.02 for rough surfaces. Over the ocean a typical value is 0.0015. In tropical storms and haricanes, $C_D$ varies with wind speed over the ocean, since increased wind speeds increase ocean surface roughness (e.g., Krishamurti et al. 1998).

The bulk aerodynamic vertical kinematic turbulent **sensible-heat flux** is

$$(\overline{w'\theta'_v})_s = -C_H|\bar{v}_h(z_r)|[\bar{\theta}_v(z_r) - \bar{\theta}(z_{0,h})] \tag{8.12}$$

where $C_H$ is the **bulk heat-transfer coefficient** or **Stanton number** (unitless), and $\bar{\theta}_v(z_{0,h})$ is the mean potential virtual temperature at height $z_{0,h}$. In a model, $\bar{\theta}_v$ at the ground surface is often used instead of $\bar{\theta}_v(z_{0,h})$ for convenience. $\bar{\theta}_v$ at the ground surface can be estimated with a soil model, as discussed in Section 8.8. $C_H$ depends on the molecular thermal diffusivity across the surface–air interface (skin drag). Its value ranges from 0.001 for smooth surfaces to 0.02 for rough surfaces. $C_H$ differs from $C_D$ because momentum is affected by skin, form, and wave drag while energy is affected primarily by skin drag.

A bulk aerodynamic term for vertical kinematic turbulent **water-vapor flux** is

$$(\overline{w'q'_v})_s = -C_E|\bar{\mathbf{v}}_h(z_r)|[\bar{q}_v(z_r) - \bar{q}_v(z_{0,h})] \tag{8.13}$$

where $C_E$ (unitless) is the **bulk transfer coefficient for water vapor** and $\bar{q}_v(z_{0,v})$ is the mean specific humidity at height $z_{0,v}$. In a model, the specific humidity in the top molecular soil or water surface layers is often used instead of $\bar{q}_v(z_{0,v})$. $C_E$ is usually set equal to $C_H$. Vertical kinematic turbulent fluxes of gases other than water vapor are written in a manner similar to (8.13).

Substituting (8.12) and (8.13) into (8.3) and (8.4), respectively, gives the vertical turbulent sensible-heat and water vapor fluxes as

$$H_f \approx \rho_a c_{p,d} C_H|\bar{\mathbf{v}}_h(z_r)|[\bar{\theta}_v(z_{0,h}) - \bar{\theta}_v(z_r)] \tag{8.14}$$

$$E_f \approx \rho_a C_E|\bar{v}_h(z_r)|[\bar{q}_v(z_{0,v}) - \bar{q}_v(z_r)]$$

respectively. (8.14) states that energy is transferred upward when $\bar{\theta}_v$ at $z_{0,h}$ is greater than that at $z_r$. Moisture is transferred upward when $\bar{q}_v$ at $z_{0,v}$ is greater than that at $z_r$.

### 8.5.2. Bulk Aerodynamic Eddy Diffusion Coefficients

The vertical eddy diffusion coefficient for momentum in the surface layer can be found from the bulk aerodynamic equations by combining (4.66), (8.11), $\partial\bar{u}/\partial z \approx \bar{u}(z_r)/(z_r - z_{0,m})$, and $\partial\bar{v}/\partial z \approx \bar{v}(z_r)/(z_r - z_{0,m})$. The result is

$$K_{m,zx} = -\frac{(\overline{w'u'})_s}{\partial\bar{u}/\partial z} = C_D|\bar{\mathbf{v}}_h(z_r)|(z_r - z_{0,m}) \approx -\frac{(\overline{w'v'})_s}{\partial\bar{v}/\partial z} = K_{m,zy} \tag{8.15}$$

Combining (3.50), (8.12), and $\partial\theta_v/\partial z \approx [\bar{\theta}_v(z_r) - \bar{\theta}_v(z_{0,h})]/(z_r - z_{0,h})$ gives the vertical eddy diffusion coefficient for energy in the surface layer as

$$K_{h,zz} = -\frac{(\overline{w'\theta_v'})_s}{\partial\bar{\theta}_v/\partial z} \approx C_H|\bar{\mathbf{v}}_h(z_r)|(z_r - z_{0,h}) \tag{8.16}$$

A similar expression may be written for water vapor.

**Example 8.5**

If $C_D = 0.02$ (rough surface), $u = 10$ m s$^{-1}$, and $v = 0$ m s$^{-1}$ at $z = 10$ m, estimate $K_{m,zx}$.

SOLUTION

Assuming $z_{0,m}$ is negligible relative to $z$, we have $K_{m,zx} = 2$ m$^2$ s$^{-1}$ from (8.15).

## 8.6. MONIN–OBUKHOV SIMILARITY THEORY

Eddy diffusion in the surface layer is affected primarily by mechanical shear (mechanical turbulence), but also by buoyancy (thermal turbulence). In the presence of strong shear above the surface, bulk aerodynamic formulae are not useful. A better method of parameterizing kinematic fluxes near the surface is with Monin–Obukhov similarity theory. **Similarity theory** is a method by which variables are first combined into dimensionless groups. Experiments are then conducted to obtain values for each variable in the dimensionless group. The dimensionless group, as a whole, is then fitted, as a function of some parameter, with an empirical equation. The experiment is repeated. Usually, equations obtained from later experiments are similar to those from the first experiment. Hence, this method of obtaining an empirical equation for the dimensionless group is called similarity theory, and the relationship between the empirical equation and the dimensionless group is a **similarity relationship**. When similarity theory is applied to the surface layer, it is usually called **Monin–Obukhov similarity theory** or **surface-layer similarity theory** (Monin and Obukhov 1954; Stull 1988).

### 8.6.1. Friction Velocity

One similarity relationship is that for the **dimensionless wind shear**,

$$\frac{\phi_m}{k} = \frac{z}{u_*}\frac{\partial|\bar{\mathbf{v}}_h|}{\partial z} \tag{8.17}$$

The right side of this equation, as a whole, is dimensionless. Individual factors, such as the wind shear ($\partial|\bar{\mathbf{v}}_h|/\partial z$) and $u_*$, are found from field experiments. Wind shear is measured directly, and $u_*$ is found from (8.5), the terms of which are measured. The parameter $\phi_m/k$ is determined as a function of $z/L$ by substituting measurements of $\partial|\bar{\mathbf{v}}_h|/\partial z$ and $u_*$ into (8.17) for different values of $z/L$ and fitting curves to the resulting data. $L$ is the **Monin–Obukhov length** (m), discussed shortly, and $z/L$ is a dimensionless group. The von Kármán constant $k$ is found by substituting measurements of $\partial|\bar{v}_h|/\partial z$ and $u_*$ into (8.17) under neutral conditions, when $\phi_m = 1$, then solving for $k$. Businger et al. (1971) derived $\phi_m$ from field data when $k = 0.35$ as

$$\phi_m = \begin{cases} 1 + \beta_m \frac{z}{L} & \frac{z}{L} > 0 \quad \text{stable} \\ \left(1 - \gamma_m \frac{z}{L}\right)^{-1/4} & \frac{z}{L} < 0 \quad \text{unstable} \\ 1 & \frac{z}{L} = 0 \quad \text{netural} \end{cases} \tag{8.18}$$

where $\beta_m = 4.7$ and $\gamma_m = 15.0$. When $k = 0.4$, the values $\beta_m = 6.0$ and $\gamma_m = 19.3$, should be used instead to obtain the same values of $\phi_m$ as when $k = 0.35$ (Hogstrom 1988). (8.18) was derived for the range, $|z/L| < 2$, but has been used successfully beyond the range under unstable conditions (San Jose et al. 1985). Other similarity relationships for $\phi_m$ include those by Dyer (1974) and Dyer and Bradley (1982), among others.

Integrating both sides of (8.17) between $z_{0,m}$ and $z_r$, solving for $u_*$, and noting that $|\mathbf{v}_h(z_{0,m})| = 0$ gives

$$u_* = \frac{k|\bar{\mathbf{v}}_h(z_r)|}{\int_{z_{0,m}}^{z_r} \phi_m \frac{dz}{z}} \tag{8.19}$$

where

$$\int_{z_{0,m}}^{z_r} \phi_m \frac{dz}{z} \tag{8.20}$$

$$= \begin{cases} \ln \frac{z_r}{z_{0,m}} + \frac{\beta_m}{L}(z_r - z_{0,m}) & \frac{z}{L} > 0 \quad \text{stable} \\ \ln \frac{\left(1-\gamma_m \frac{z_r}{L}\right)^{1/4} - 1}{\left(1-\gamma_m \frac{z_r}{L}\right)^{1/4} + 1} - \ln \frac{\left(1-\gamma_m \frac{z_{0,m}}{L}\right)^{1/4} - 1}{\left(1-\gamma_m \frac{z_{0,m}}{L}\right)^{1/4} + 1} + 2\tan^{-1} & \\ \quad \times \left(1 - \gamma_m \frac{z_r}{L}\right)^{1/4} - 2\tan^{-1}\left(1 - \gamma_m \frac{z_{0,m}}{L}\right)^{1/4} & \frac{z}{L} < 0 \quad \text{unstable} \\ \ln \frac{z_r}{z_{0,m}} & \frac{z}{L} = 0 \quad \text{neutral} \end{cases}$$

The integral in (8.20) exceeds zero and increases with increasing $z/L$. Thus, for the same wind speed, $u_*$ from (8.19) is larger in unstable air ($z/L < 0$) than in stable air ($z/L > 0$). Increasing values of $z_{0,m}$ also increase $u_*$ by decreasing the integral in (8.20).

### 8.6.2. Monin–Obukhov Length

The **Monin–Obukhov length** ($L$) is a height proportional to the height above the surface at which buoyant production of turbulence first dominates mechanical (shear) production of turbulence. Mathematically,

$$L = -\frac{u_*^3 \bar{\theta}_v}{kg(\overline{w'\theta'_v})_s} = \frac{u_*^2 \bar{\theta}_v}{kg\theta_*} \tag{8.21}$$

where $\theta_*$ is a **potential temperature scale** (K), discussed shortly, and the second expression is derived by substituting the similarity-theory approximation

$$(\overline{w'\theta'_v})_s \approx -u_*\theta_* \tag{8.22}$$

into the first expression. (8.21) may be solved with parameterized values of $u_*$ and $\theta_*$.

$\theta_*$ is proportional to $\bar{\theta}_v(z_r) - \bar{\theta}_v(z_{0,h})$, the vertical difference in potential virtual temperature. The greater $\bar{\theta}_v$ at $z_{0,h}$ in comparison with its value at $z_r$, the more negative the change in $\bar{\theta}_v$ with increasing height, and the greater the instability of the surface layer. In such cases, $L$ is negative but has a small magnitude, since it is inversely proportional to $\theta_*$. When $L$ is negative with a small magnitude, $z/L$ is negative with a large magnitude. Such values of $z/L$ correspond to large instability due to buoyancy. Positive values of $z/L$ correspond to increasing $\bar{\theta}_v$ with altitude and stable stratification.

### 8.6.3. Potential Temperature Scale

An expression for the potential temperature scale, $\theta_*$, can be obtained from a similarity relationship for the **dimensionless potential temperature gradient,**

$$\frac{\phi_h}{k} \approx \frac{z}{\theta_*}\frac{\partial \bar{\theta}_v}{\partial z} \tag{8.23}$$

where $\partial\bar{\theta}_v/\partial z$ is the change in mean potential virtual temperature with height. Businger et al. (1971) performed experiments to find $\phi_h$ for different stability regimes when $\bar{\theta}_p$ was used instead of $\bar{\theta}_v$ and $k = 0.35$. The resulting parameterization was

$$\phi_h = \begin{cases} \mathrm{Pr}_t + \beta_h \frac{z}{L} & \frac{z}{L} > 0 \quad \text{stable} \\ \mathrm{Pr}_t\left(1 - \gamma_h \frac{z}{L}\right)^{-1/2} & \frac{z}{L} < 0 \quad \text{unstable} \\ \mathrm{Pr}_t & \frac{z}{L} = 0 \quad \text{neutral} \end{cases} \tag{8.24}$$

where $\beta_h = 4.7$, $\gamma_h = 9.0$, and $\mathrm{Pr}_t = K_{m,zx}/K_{h,zz}$ is the **turbulent Prandtl number**, which approximates the ratio of the eddy diffusion coefficient for momentum to that for energy. For $k = 0.35$, Businger et al. estimated $\mathrm{Pr}_t \approx 0.74$. Hogstrom (1988) noted that, when $k = 0.4$, Businger et al.'s constants should be modified to $\beta_h = 7.8, \gamma_h = 11.6$, and $\mathrm{Pr}_t \approx 0.95$ to obtain the same relationship as when $k = 0.35$. Integrating both sides of (8.23) between $z_{0,h}$ and $z_r$ and solving for

$\theta_*$ gives

$$\theta_* = \frac{k[\bar{\theta}_v(z_r) - \bar{\theta}_v(z_{0,h})]}{\int_{z_{0,h}}^{z_r} \phi_h \frac{dz}{z}} \tag{8.25}$$

where

$$\int_{z_{0,h}}^{z_r} \phi_h \frac{dz}{z} = \begin{cases} \mathrm{Pr}_t \ln \frac{z_r}{z_{0,h}} + \frac{\beta_h}{L}(z_r - z_{0,h}) & \frac{z}{L} > 0 \quad \text{stable} \\ \mathrm{Pr}_t \left[ \ln \frac{\left(1-\gamma_h \frac{z_r}{L}\right)^{1/2} - 1}{\left(1-\gamma_h \frac{z_r}{L}\right)^{1/2} + 1} - \ln \frac{\left(1-\gamma_h \frac{z_{0,h}}{L}\right)^{1/2} - 1}{\left(1-\gamma_h \frac{z_{0,h}}{L}\right)^{1/2} + 1} \right] & \frac{z}{L} < 0 \quad \text{unstable} \\ \mathrm{Pr}_t \ln \frac{z_r}{z_{0,h}} & \frac{z}{L} = 0 \quad \text{neutral} \end{cases} \tag{8.26}$$

In sum, $u_*$ can be determined from (8.19) and (8.20) if $L$ is known. If $L$ is not known, $L$, $u_*$, and $\theta_*$ must be determined simultaneously by solving (8.19), (8.21), and (8.25).

### 8.6.4. Noniterative Parameterization for Momentum and Energy Scales

A noniterative method of determining $u_*$, $\theta_*$, and $L$ is with the parameterization of Louis (1979). The first step is to calculate the **bulk Richardson number,**

$$\mathrm{Ri}_b = \frac{g[\bar{\theta}_v(z_r) - \bar{\theta}_v(z_{0,h})](z_r - z_{0,m})^2}{\bar{\theta}_v(z_{0,h})[\bar{u}(z_r)^2 + \bar{v}(z_r)^2](z_r - z_{0,h})} \tag{8.27}$$

which quantifies the ratio of buoyancy to mechanical shear. The second step is to calculate $u_*$ and $\theta_*$ as

$$u_* \approx \frac{k|\bar{\mathbf{v}}_h(z_r)|}{\ln(z_r/z_{0,m})}\sqrt{G_m} \qquad \theta_* \approx \frac{k^2|\bar{\mathbf{v}}_h(z_r)|[\theta_v(z_r) - \theta_v(z_{0,h})]}{u_* \,\mathrm{Pr}_t \ln^2(z_r/z_{0,m})} G_h \tag{8.28}$$

where

$$\begin{aligned} G_m &= 1 - \frac{9.4\mathrm{Ri}_b}{1 + \frac{70k^2(|\mathrm{Ri}_b| z_r/z_{0,m})^{0.5}}{\ln^2(z_r/z_{0,m})}} && \mathrm{Ri}_b \le 0 \\ G_h &= 1 - \frac{9.4\mathrm{Ri}_b}{1 + \frac{50k^2(|\mathrm{Ri}_b| z_r/z_{0,m})^{0.5}}{\ln^2(z_r/z_{0,m})}} && \mathrm{Ri}_b \le 0 \\ G_m, G_h &= \frac{1}{(1 + 4.7\mathrm{Ri}_b)^2} && \mathrm{Ri}_b > 0 \end{aligned} \tag{8.29}$$

The surface roughness length for momentum is used for momentum and energy terms in (8.29) to maintain consistency. From values of $u_*$ and $\theta_*$, $L$ can be determined diagnostically with (8.21).

**Example 8.6**

Given

| | | |
|---|---|---|
| $z_{0,m} = 0.01\,\text{m}$ | $z_{0,h} = 0.0001\,\text{m}$ | $z_r = 10\,\text{m}$ |
| $\bar{u}(z_r) = 10\,\text{m s}^{-1}$ | $\bar{v}(z_r) = 5\ \text{m s}^{-1}$ | $k = 0.4$ |
| $\theta_v(z_r) = 285\,\text{K}$ | $\bar{\theta}(z_{0,h}) = 288\,\text{K}$ | $\text{Pr}_t = 0.95$ |

calculate $u_*$, $\theta_*$, and $L$.

SOLUTION

From (8.27) and (8.28),

| | | |
|---|---|---|
| $\lvert\bar{\mathbf{v}}_h(z_r)\rvert = 11.18\,\text{m s}^{-1}$ | $\text{Ri}_b = -8.15 \times 10^{-3}$ | $G_m = 1.046$ |
| $G_h = 1.052$ | $u_* = 0.662\,\text{m s}^{-1}$ | $\theta_* = -0.188\,\text{K}$ |
| $L = -169\,\text{m}$ | | |

### 8.6.5. Gradient Richardson Number

The bulk Richardson number is used for practical application in meteorological modeling. It is derived from the **gradient Richardson number,**

$$\text{Ri}_g = \frac{\dfrac{g}{\bar{\theta}_v}\dfrac{\partial\bar{\theta}_v}{\partial z}}{\left(\dfrac{\partial\bar{u}}{\partial z}\right)^2 + \left(\dfrac{\partial\bar{v}}{\partial z}\right)^2} \tag{8.30}$$

The gradient and bulk Richardson numbers give the ratio of turbulence due to shear relative to that due to buoyancy. When $\text{Ri}_g$, $\text{Ri}_b < 0$, the potential virtual temperature decreases with increasing altitude, and the atmosphere is buoyantly unstable and turbulent. When $\text{Ri}_g$, $\text{Ri}_b$ are small and negative, wind shear is large in comparison with buoyancy, and turbulence due to mechanical shear dominates over turbulence due to buoyancy. When $\text{Ri}_g$, $\text{Ri}_b$ are large and negative, turbulence due to free convection dominates turbulence due to forced convection. When $\text{Ri}_g$, $\text{Ri}_b > 0$, the potential virtual temperature exceeds zero, the atmosphere is buoyantly stable, and turbulence due to free convection does not occur. When $\text{Ri}_g$, $\text{Ri}_b$ are small and positive, wind shear is large in comparison with buoyant stability, and turbulence due to forced convection occurs. When $\text{Ri}_g$, $\text{Ri}_b$ are large and positive, wind shear is low in comparison with buoyant stability, and turbulence due to forced or free convection does not occur. Instead, air flow is laminar. Table 8.2 summarizes the flow regimes obtained from different Richardson numbers.

When $\text{Ri}_b$ and $\text{Ri}_g$ are large and positive and decrease to less than a **critical Richardson number** ($\text{Ri}_c$) of 0.25, laminar flow becomes turbulent and wind shear increases. When $\text{Ri}_b$ and $\text{Ri}_g$ are small and positive and increase to above the **termination Richardson number** ($\text{Ri}_T$) of 1.0, turbulent flow becomes laminar.

**Table 8.2.** Characteristics of Vertical Flow of Air for Different Values of $Ri_b$ and $Ri_g$

| $Ri_b$ or $Ri_g$ | Type of Flow | Level of Turbulence Due to Buoyancy | Level of Turbulence Due to Shear |
|---|---|---|---|
| Large, negative | Turbulent | Large | Small |
| Small, negative | Turbulent | Small | Large |
| Small, positive | Turbulent | None (weakly stable) | Large |
| Large, positive | Laminar | None (strongly stable) | Small |

The level of turbulence due to buoyancy is relative to that due to shear, and vice versa.

### 8.6.6. Vertical Fluxes of Energy and Moisture

The turbulent sensible-heat and moisture fluxes can be derived from similarity theory. From (8.22), $(\overline{w'\theta_v'})_s = -u_*\theta_*$. A similar equation for the vertical kinematic turbulent water vapor flux in the surface layer is $(\overline{w'q_v'})_s = -u_*q_*$, where $q_*$ is the water-vapor scale (kg kg$^{-1}$). Substituting these expressions into (8.3) and (8.4), respectively, gives the surface-layer **vertical turbulent sensible-heat and moisture fluxes** as

$$H_f \approx -\rho_a c_{p,d} u_* \theta_* \qquad E_f \approx -\rho_a u_* q_* \tag{8.31}$$

Like $\theta_*$, $q_*$ is a found from a similarity relationship,

$$\frac{\phi_q}{k} = \frac{z}{q_*}\frac{\partial \bar{q}_v}{\partial z} \tag{8.32}$$

where $\phi_q \approx \phi_h$ is a dimensionless specific humidity gradient. Substituting $\phi_h \approx \phi_q$ into (8.32), integrating the equation between $z_{0,v}$ and $z_r$, and solving for $q_*$ gives

$$q_* = \frac{k[\bar{q}_v(z_r) - \bar{q}_v(z_{0,v})]}{\int_{z_{0,v}}^{z_r} \phi_h \frac{dz}{z}} \tag{8.33}$$

This integral of $\phi_h$ is found from (8.26), but with $z_{0,v}$ substituted for $z_{0,h}$.

### 8.6.7. Eddy Diffusion Coefficients in the Surface Layer

Eddy diffusion coefficients for the surface layer can be obtained from similarity theory. Substituting (8.11) into (8.5) gives the relationship between friction velocity and the coefficient of drag as

$$u_* = \bar{v}_h(z_r)\sqrt{C_D} \tag{8.34}$$

**Example 8.7**

If $C_D = 0.001$ and $|\bar{v}_h(z_r)| = 10$ m s$^{-1}$, then $u_* = 0.32$ m s$^{-1}$ from (8.34). If $C_D = 0.02$ and $|\bar{v}_h(z_r)| = 10$ m s$^{-1}$, then $u_* = 1.41$ m s$^{-1}$. These are typical values of $u_*$ over smooth and rough surfaces, respectively, when the surface layer is well mixed.

Substituting $C_D = [u_*/\bar{\mathbf{v}}_h(z_r)]^2$ from (8.34) into (8.11) gives the **vertical kinematic turbulent fluxes of momentum from similarity theory** as

$$(\overline{w'u'})_s = -\frac{u_*^2}{|\bar{\mathbf{v}}_h(z_r)|}\bar{u}(z_r) \qquad (\overline{w'v'})_s = -\frac{u_*^2}{|\bar{\mathbf{v}}_h(z_r)|}\bar{v}(z_r) \tag{8.35}$$

Substituting (8.35), $\partial\bar{u}/\partial z \approx \bar{u}(z_r)/(z_r - z_{0,m})$, and $\partial\bar{v}/\partial z \approx \bar{v}(z_r)/(z_r - z_{0,m})$ into (8.15) yields

$$K_{m,zx} = \frac{u_*^2}{|\bar{\mathbf{v}}_h(z_r)|}\frac{\bar{u}(z_r)}{\partial\bar{u}/\partial z} \approx \frac{u_*^2}{|\bar{\mathbf{v}}_h(z_r)|}(z_r - z_{0,m}) \approx \frac{u_*^2}{|\bar{\mathbf{v}}_h(z_r)|}\frac{\bar{v}(z_r)}{\partial\bar{v}/\partial z} = K_{m,zy} \tag{8.36}$$

Further substituting $\partial|\bar{\mathbf{v}}_h|/\partial z \approx |\bar{\mathbf{v}}_h|/(z_r - z_{0,m})$ and (8.17) into (8.36) gives the surface-layer eddy diffusion coefficient for momentum from similarity theory under stable, unstable, or neutral conditions as

$$K_{m,zx} = \frac{kzu_*}{\phi_m} = K_{m,zy} \tag{8.37}$$

where $kz = l$ is the **mixing length** near the surface. The mixing length is the average distance an eddy travels before it exchanges momentum with surrounding eddies. Near the surface, mixing is limited by the ground. Under neutral conditions, $\phi_m = 1$, and (8.37) simplifies to $K_{m,zx} = kzu_* = K_{m,zy}$.

Expressions for the **vertical eddy diffusion coefficient for energy** from similarity theory are obtained by substituting (8.22) and (8.23) into (8.16). The result is

$$K_{h,zz} = \frac{u_*\theta_*}{\partial\bar{\theta}_v/\partial z} = \frac{kzu_*}{\phi_h} \tag{8.38}$$

The eddy diffusion coefficients for water vapor and other species are set equal to those for energy.

Combining (8.37) and (8.38) with $\mathrm{Pr}_t = K_{m,zx}/K_{h,zz}$ gives the similarity-theory expression, $\mathrm{Pr}_t = \phi_h/\phi_m$. Another parameterization for $\mathrm{Pr}_t$ was developed by Tjernstrom (1993), who estimated $\mathrm{Pr}_t \approx (1 + 4.47\,\mathrm{Ri}_g)^{1/2}$ from turbulence observations in the boundary layer. In this equation, $\mathrm{Ri}_g$ varies from 0.01 to 10.

**Example 8.8**

Given the conditions in Example 8.6, find $K_{h,zz}$, $K_{m,zx}$, $K_{m,zy}$, and $K_{m,zx}/K_{h,zz}$.

SOLUTION

Substituting $u_*$, $\theta_*$, $\partial\bar{u}/\partial z$, $\partial\bar{v}/\partial z$, $|\bar{\mathbf{v}}_h(z_r)|$, and $\partial\bar{\theta}_v/\partial z$ into (8.36) and (8.38) gives

$$K_{h,zz} = 0.41\ \mathrm{m^2\ s^{-1}} \qquad K_{m,zx} = 0.39\ \mathrm{m^2\ s^{-1}}$$
$$K_{m,zy} = 0.39\ \mathrm{m^2\ s^{-1}} \qquad K_{m,zx}/K_{h,zz} = 0.95$$

Thus, the Louis equations predict $\mathrm{Pr}_t = K_m/K_h$ consistently with the value of $\mathrm{Pr}_t$ used in (8.28).

### 8.6.8. Vertical Profiles of Wind Speed, Potential Virtual Temperature, and Moisture

Similarity theory can be used to derive profiles of vertical wind speed, potential virtual temperature, and specific humidity in the surface layer. From (8.17) and (8.23), the vertical slopes of wind speed and potential virtual temperature in the surface layer are

$$\frac{\partial|\bar{\mathbf{v}}_h(z)|}{\partial z} = \frac{u_*}{kz}\phi_m = \frac{u_*}{kz}[1-(1-\phi_m)] \qquad \frac{\partial\bar{\theta}_v}{\partial z} = \frac{\theta_*}{kz}\phi_h = \frac{\theta_*}{kz}[1-(1-\phi_h)] \tag{8.39}$$

respectively. Integrating the first equation between $z_{0,m}$ and $z$ and the second equation between $z_{0,h}$ and $z$ gives wind speed and potential virtual temperature versus altitude as

$$|\bar{\mathbf{v}}_h(z)| = \frac{u_*}{k}\left[\ln\left(\frac{z}{z_{0,m}}\right) - \psi_m\right] \tag{8.40}$$

$$\bar{\theta}_v(z) = \bar{\theta}_v(z_{0,h}) + \mathrm{Pr}_t\frac{\theta_*}{k}\left[\ln\left(\frac{z}{z_{0,h}}\right) - \psi_h\right] \tag{8.41}$$

respectively, where

$$\psi_m = \int_{z_{0,m}}^{z}(1-\phi_m)\frac{\mathrm{d}z}{z} \qquad \psi_h = \int_{z_{0,h}}^{z}(1-\phi_h)\frac{\mathrm{d}z}{z} \tag{8.42}$$

are **influence functions for momentum and energy**. Integrating (8.42) with values of $\phi_m$ and $\phi_h$ from (8.18) and (8.24), respectively, gives

$$\psi_m = \begin{cases} -\frac{\beta_m}{L}(z - z_{0,m}) & \frac{z}{L} > 0 \quad \text{stable} \\ \ln\frac{[1+\phi_m(z)^{-2}][1+\phi_m(z)^{-1}]^2}{[1+\phi_m(z_{0,m})^{-2}][1+\phi_m(z_{0,m})^{-1}]^2} & \\ \quad -2\tan^{-1}[\phi_m(z)]^{-1} + 2\tan^{-1}[\phi_m(z_{0,m})]^{-1} & \frac{z}{L} < 0 \quad \text{unstable} \\ 0 & \frac{z}{L} = 0 \quad \text{neutral} \end{cases} \tag{8.43}$$

$$\psi_h = \begin{cases} -\frac{1}{\mathrm{Pr}_t}\frac{\beta_h}{L}(z - z_{0,h}) & \frac{z}{L} > 0 \quad \text{stable} \\ 2\ln\frac{1+\phi_h(z)^{-1}}{1+\phi_h(z_{0,h})^{-1}} & \frac{z}{L} < 0 \quad \text{unstable} \\ 0 & \frac{z}{L} = 0 \quad \text{neutral} \end{cases} \tag{8.44}$$

The influence function for momentum accounts for the difference between a logarithmic wind speed profile and an actual profile under stable and unstable conditions. The influence function for energy is analogous to that for momentum.

Under neutral conditions, $\phi_m = 1$, and (8.40) reduces to a standard **logarithmic wind profile** for a neutrally stratified surface layer,

$$|\bar{\mathbf{v}}_h(z)| = \frac{u_*}{k}\ln\frac{z}{z_{0,m}} \tag{8.45}$$

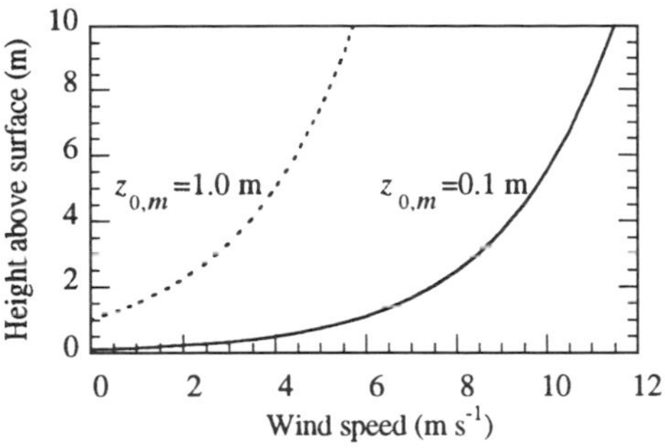


**Figure 8.3.** Logarithmic wind profiles in the surface layer from (8.45) when $u_* = 1$ m s$^{-1}$.

This equation states that the wind speed at $z_{0,m}$ is zero but increases logarithmically with altitude. Figure 8.3 shows two examples of logarithmic wind profiles.

The **vertical profile of specific humidity in the surface layer** is derived in the same way as that for potential virtual temperature. The result is

$$\bar{q}_v(z) = \bar{q}_v(z_{0,v}) + \mathrm{Pr}_t \frac{q_*}{k}\left[\ln\left(\frac{z}{z_{0,v}}\right) - \psi_h\right] \tag{8.46}$$

where the energy integral from (8.44) is used, but with $z_{0,v}$ instead of $z_{0,h}$. In (8.41) and (8.46), values of the potential virtual temperature and specific humidity at the ground surface are often substituted for values at the surface roughness length.

In a canopy, such as in a field of crops, an orchard, or a forest, the canopy top affects velocity, potential virtual temperature, and water vapor more than does the ground surface. In such cases, $z_{0,m}$, $z_{0,h}$, and $z_{0,v}$ are displaced a vertical distance $d_c$ above the surface. This height is the **displacement height**, which usually lies within 70 to 80 percent of the **canopy height** $h_c$ (Kaimal and Finnigan 1994). When a displacement height exists, $z_{0,m}$, $z_{0,h}$, and $z_{0,v}$ are defined as heights above the displacement height. The mean wind speed extrapolates to zero at the height $d_c + z_{0,m}$. Wind speed, potential virtual temperature, and specific humidity profiles in a canopy are redefined as

$$|\bar{\mathbf{v}}_h(z)| = \frac{u_*}{k}\left[\ln\left(\frac{z-d_c}{z_{0,m}}\right) - \psi_m\left(\frac{z-d_c}{L}\right)\right] \tag{8.47}$$

$$\bar{\theta}_v(z) = \bar{\theta}_v(d_c + z_{0,h}) + \mathrm{Pr}_t \frac{\theta_*}{k}\left[\ln\left(\frac{z-d_c}{z_{0,h}}\right) - \psi_h\left(\frac{z-d_c}{L}\right)\right] \tag{8.48}$$

$$\bar{q}_v(z) = \bar{q}_v(d_c + z_{0,v}) + \mathrm{Pr}_t \frac{q_*}{k}\left[\ln\left(\frac{z-d_c}{z_{0,v}}\right) - \psi_h\left(\frac{z-d_c}{L}\right)\right] \tag{8.49}$$

respectively. Figure 8.4 shows the relationship among $d_c$, $h_c$, and $z_{0,m}$ and discusses how to calculate $d_c$. Table 8.1 gives values of $h_c$ and $d_c$ for some surfaces.

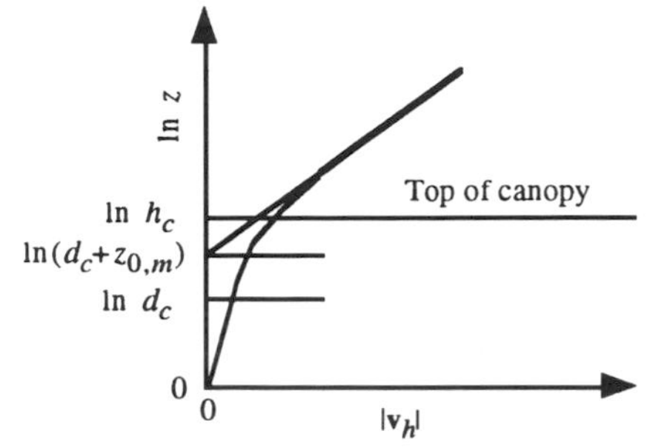


**Figure 8.4.** Relationship among $d_c$, $h_c$, and $z_{0,m}$. The displacement height is found by plotting wind speed over a canopy versus ln $z$. The plotted wind speed is extrapolated to zero to obtain $d_c + z_{0,m}$. Different values of $d_c$ are substituted into $d_c + z_{0,m}$ to estimate $z_{0,m}$. Both $d_c$ and $z_{0,m}$ are substituted into (8.47) until the predicted curve of wind speed with height matches the logarithmic curve shown in the diagram. This method works best when winds aloft are strong.

## 8.7. EDDY DIFFUSION ABOVE THE SURFACE LAYER

Similarity-theory expressions for eddy diffusion coefficients are used for the surface layer. Above the surface layer, turbulent transport of momentum and energy is parameterized with a high-order turbulence closure scheme or a hybrid model. A summary of high-order closure schemes is discussed in Mellor and Yamada (1982).

In a hybrid model, vertical turbulent transport is treated differently for stable and weakly unstable conditions versus for strongly unstable conditions. For stable and weakly unstable conditions, turbulent transport is simulated with $K$-theory using an **eddy diffusion coefficient for momentum** that depends on mechanical shear and buoyancy. One such diffusion coefficient is

$$K_{m,zx} \approx K_{m,zy} \approx \left(\frac{kz}{1 + kz/\lambda_m}\right)^2 \sqrt{\left(\frac{\Delta \bar{u}}{\Delta z}\right)^2 + \left(\frac{\Delta \bar{v}}{\Delta z}\right)^2} \frac{\text{Ri}_c - \text{Ri}_b}{\text{Ri}_c} \qquad (8.50)$$

(e.g., Blackadar 1976; Stull 1988), where $z$ is in meters. The term, $kz/(1+kz/\lambda_m)$ is an expression for mixing length that simplifies to $kz$ near the surface (small $z$) and to $\lambda_m = 70$–$200$ m in the free atmosphere (large $z$). Near the surface, turbulence and mixing length are limited by the ground. In the free atmosphere, turbulence is limited by a maximum mixing length. (8.50) is valid for $\text{Ri}_b < \text{Ri}_c \approx 0.25$. When $\text{Ri}_b \geq \text{Ri}_c$, the atmosphere is strongly stable, (8.50) predicts no turbulence will occur (a negative value $K_{m,zx}$), and the diffusion coefficient is set to a minimum positive value.

When $Ri_c > Ri_b > 0$, turbulence is due to mechanical shear, and the atmosphere is weakly stable. When $Ri_b$ is small and negative, turbulence is due to shear and weak buoyancy, and the atmosphere is weakly unstable. Under such conditions, (8.50) is valid.

In a jet, wind speeds increase and then decrease with altitude, minimizing wind shear in the jet's peak. In such a case, and in the absence of buoyancy, (8.50) predicts a small $K_{m,zx}$. Some observations of turbulence near a jet's maximum speed indicate that turbulence peaks near the maximum (Lenschow et al. 1988; Tjernstrom 1993), while others observations indicate that turbulence is lowest near the jet maximum (e.g., Mahrt et al. 1979; Lenschow et al. 1988; Tjernstrom 1993). In cases when the turbulence peaks near the jet maximum, (8.50) may not properly predict $K_{m,zx}$ (Shir and Bornstein 1976).

The eddy diffusion coefficient for energy is related to that for momentum by $K_{m,zx} \approx K_{m,zy} \approx \mathrm{Pr}_t\ K_{h,zz}$. Eddy diffusion coefficients for trace gases or particles are usually set equal to those for energy.

When $Ri_b$ is large and negative, the atmosphere is strongly unstable, and free convection occurs. Since (8.50) captures effects of small eddies but not large eddies that arise during free convection (Stull 1988), (8.50) is not used to simulate free convection in a hybrid model. Instead, vertical turbulent transport under strong, unstable conditions is simulated with a **free convective plume scheme** (Blackadar 1978; Zhang and Anthes 1982; Lu 1994). A free convective plume scheme assumes buoyant plumes from the surface rise and mix with every level in the boundary layer, exchanging momentum, energy, moisture, and gases. The result of mixing is to evenly distribute these parameters throughout the boundary layer in a short time. A plume model differs from a $K$-theory parameterization in that the latter mixes material between two adjacent layers at a time, and the former mixes all layers simultaneously.

## 8.8. GROUND SURFACE TEMPERATURE AND SOIL MOISTURE

Predictions of ground surface temperature and soil moisture (liquid-water content) are necessary to calculate the transport of energy and moisture, respectively, from the ground surface to boundary layer. Such predictions are important, since ground temperatures and moisture influence mixing heights, wind speeds, and pollutant concentrations. Low soil moisture increases ground temperatures, increasing thermal turbulence, increasing mixing heights, decreasing pollutant concentrations, and increasing wind speeds (Jacobson 1998c). Low soil moisture contents increase near-surface wind speeds by enhancing turbulence transport of momentum from aloft to the surface. Downward momentum transfer can occur, since wind speeds aloft are generally faster than near the surface. Variations in soil moisture over inhomogeneous topography also affect pressure gradients and corresponding wind speeds. Over a two-day period, perturbations to soil moisture over a mesoscale region can affect vertical temperature profiles up to 600-millibars altitude (Jacobson 1998c). In this section, factors that affect surface temperatures and moisture contents are discussed, and equations describing soil energy and liquid-water transport are described.

### 8.8.1. Factors Affecting Soil Temperature

Ground surface temperatures are affected by soil specific heat, soil moisture content, soil thermal conductivity, solar and infrared radiative fluxes, turbulent sensible heat fluxes, vegetation cover, transpiration, and precipitation. Dry soil contains solid soil and air (voids). Specific heats vary with soil type, as shown in Table 2.2. Since the specific heat of dry air is lower than that of solid soil (Table 2.2), the average specific heat of soil plus air is less than that of soil alone. When liquid water is added to soil, it replaces air. Since the specific heat of liquid water is much larger than that of air, the specific heat of a soil-water-air mixture is greater than that of a soil-air mixture of the same volume. Thus, a wet, sandy soil heats up less during the day and cools down less during the night than does a dry, sandy soil when only specific heat is considered.

Soil moisture affects ground temperatures not only by affecting the specific heat of a soil-water-air mixture, but also by affecting evaporation rates. The lesser the liquid-water content of soil, the lesser the rate of evaporation of water to the air (or greater the rate of condensation of water vapor to the ground), the lesser the latent heat flux to the air, and the lesser the cooling (greater the warming) of the ground due to latent heat absorption (release).

Conduction between surface soil and molecules below the surface is another factor that affects ground temperatures. During the night, the ground surface cools radiatively, creating a temperature gradient in the top soil layers, and forcing energy stored below the surface to conduct upward to replenish the lost radiative energy. The greater the thermal conductivity of the soil, the faster the energy transfer occurs. Table 2.2 shows that clay is more conductive than sand. At night, the replenishment of energy to a clay surface from the subsoil is faster than for a sandy surface. During the day, conduction of absorbed radiation from the surface to the subsoil is faster for clay than for sand.

Several remaining factors affect soil moisture and ground temperatures. Solar radiation heats the ground during the day. Infrared emissions cool the ground during the day and the night. Vegetation cover reduces the solar radiation reaching the ground. Transpiration removes liquid water from deep soil covered by vegetation. The water is drawn from the soil through roots and xylem (water-conducting tissue within plants) to the leaves. Water vapor escapes through plant leaf stomata (pores). Finally, a major source of liquid water in soil is precipitation.

### 8.8.2. Modeling Surface Temperature and Moisture

Soil models predict ground surface temperatures by dividing the soil near the surface into multiple layers. In this subsection, soil energy and moisture transfer equations are shown, and equations describing changes in temperature and soil moisture at the ground surface are given.

At the ground, conductive, radiative, sensible, and evaporative energy fluxes affect soil temperatures. Below the surface, conduction is the most important factor affecting soil temperatures. Temperature changes due to conduction in a homogeneous soil below the ground are estimated with the **heat conduction equation,**

$$\frac{\partial T_s}{\partial t} = \frac{1}{\rho_g c_G} \frac{\partial}{\partial z}\left(\kappa_s \frac{\partial T_s}{\partial z}\right) \tag{8.51}$$

where $T_s$ is the soil temperature, $\kappa_s$ is the thermal conductivity of the soil-water-air mixture (J m$^{-1}$ s$^{-1}$ K$^{-1}$), $\rho_g$ is the density of the mixture (kg m$^{-3}$), $c_G$ is the specific heat of the mixture (J kg$^{-1}$ K$^{-1}$), and $\kappa_s \partial T_s / \partial z$ is the conductive heat flux through the soil-water-air mixture (J m$^{-2}$ s$^{-1}$). The **thermal conductivity** (J m$^{-1}$ s$^{-1}$ K$^{-1}$) of a soil-water-air mixture may be approximated with

$$\kappa_s = \max\left(418 e^{-\log_{10}|\psi_p| - 2.7}, 0.172\right) \tag{8.52}$$

(McCumber and Pielke 1981; Al Nakshabandi and Konhke 1965), where $\psi_p$ is the moisture potential (cm) of soil tension, and "max" indicates the larger of the two values. The **moisture potential** is the potential energy required to extract water from capillary and adhesive forces in the soil. Clapp and Hornberger (1978) parameterized the moisture potential as

$$\psi_p = \psi_{p,s}\left(\frac{w_{g,s}}{w_g}\right)^b \tag{8.53}$$

where $\psi_{p,s}$ is the moisture potential when the soil is saturated (cm), $w_g$ is the **volumetric water content of the soil (soil moisture)** in cubic meters of liquid water per cubic meter of soil-water-air mixture, $w_{g,s}$ is the maximum volumetric water content that a given soil type can hold (m$^3$ m$^{-3}$), and $b$ is a coefficient required to fit (8.53) to data. Values of $\psi_{p,s}$, $w_{g,s}$, and $b$ are given in Table 8.3 for different soil

**Table 8.3.** Soil Parameters for 11 Soil Types

| Soil Type | $b$ | $w_{g,s}$ (m$^3$ m$^{-3}$) | $\psi_{p,s}$ (cm) | $K_{g,s}$ (m s$^{-1}$) | $\rho_s c_S$ (J m$^{-3}$ K$^{-1}$) |
|---|---|---|---|---|---|
| Sand | 4.05 | 0.395 | −12.1 | 1.76(−4) | 1.47(6) |
| Loamy sand | 4.38 | 0.410 | −9.0 | 1.56(−4) | 1.41(6) |
| Sandy loam | 4.90 | 0.435 | −21.8 | 3.41(−5) | 1.34(6) |
| Silt loam | 5.30 | 0.485 | −78.6 | 7.20(−6) | 1.27(6) |
| Loam | 5.39 | 0.451 | −47.8 | 7.00(−6) | 1.21(6) |
| Sandy clay loam | 7.12 | 0.420 | −29.9 | 6.30(−6) | 1.18(6) |
| Silty clay loam | 7.75 | 0.477 | −35.6 | 1.70(−6) | 1.32(6) |
| Clay loam | 8.52 | 0.476 | −63.0 | 2.50(−6) | 1.23(6) |
| Sandy clay | 10.40 | 0.426 | −15.3 | 2.20(−6) | 1.18(6) |
| Silty clay | 10.40 | 0.492 | −49.0 | 1.00(−6) | 1.15(6) |
| Clay | 11.40 | 0.482 | −40.5 | 1.30(−6) | 1.09(6) |

Adapted from Clapp and Hornberger (1978), Pielke (1984), Noilhan and Planton (1989), and Mahlouf and Noilan (1996). 1.00(6) means $1.00 \times 10^6$.

types. An alternative to the Clapp and Hornberger moisture potential equation is one developed by van Genuchten (1980) that has been widely used in soil-physics studies (Cuenca et al. 1996).

The density multiplied by the specific heat of a soil-water-air mixture is

$$\rho_g c_G = (1 - w_{g,s})\rho_s c_S + w_g \rho_w c_W \tag{8.54}$$

where $\rho_s$ is the density of solid soil (kg m$^{-3}$), $c_S$ is the specific heat of solid soil (J kg$^{-1}$ K$^{-1}$), $\rho_w$ is the density of liquid water (1000 kg m$^{-3}$), $c_W$ is the specific heat of liquid water (J kg$^{-1}$ K$^{-1}$), $1 - w_{g,s}$ is the volumetric content of solid soil, $w_g$ is the volumetric content for water in the soil, and $w_{g,s} - w_g$ is the volumetric air content in the soil-water-air mixture. In (8.54), the specific heat and mass density of air are neglected, since they are assumed to be much smaller than those of soil and water. Values of $\rho_s c_S$ are given in Table 8.3 for different soil types.

The time rate of change of volumetric water content of soil below the surface can be approximated with (McCumber and Pielke 1981)

$$\frac{\partial w_g}{\partial t} = \frac{\partial}{\partial z}\left[K_g\left(\frac{\partial \psi_p}{\partial z} + 1\right)\right] = \frac{\partial}{\partial z}\left(D_g \frac{\partial w_g}{\partial z} + K_g\right) \tag{8.55}$$

where $K_g$ is the coefficient of permeability of liquid water through soil (hydraulic conductivity), $D_g$ is the diffusion coefficient for water in soil, and $K_g \partial(\psi_p + z)/\partial z$ is the kinematic flux of liquid water through the soil (m [m$^3$ m$^{-3}$] s$^{-1}$). Liquid water affects soil temperatures through the thermal conductivity and specific heat in (8.51).

The hydraulic conductivity (m s$^{-1}$), which accounts for gravity drainage through a viscous soil, is affected by water viscosity and the shapes and sizes of voids between soil particles. Clapp and Hornberger (1978) parameterized it as

$$K_g = K_{g,s}\left(\frac{w_g}{w_{g,s}}\right)^{2b+3} \tag{8.56}$$

where $K_{g,s}$ is the hydraulic conductivity at saturation (m s$^{-1}$), and values of $b$ and $w_{g,s}$ are shown in Table 8.3. An alternative parameterization is given by van Genuchten (1980). The **diffusion coefficient for water in soil** (m$^2$ s$^{-1}$) is

$$D_g = K_g \frac{\partial \psi_p}{\partial w_g} = -\frac{b K_{g,s} \psi_{p,s}}{w_g}\left(\frac{w_g}{w_{g,s}}\right)^{b+3} = -\frac{b K_{g,s} \psi_{p,s}}{w_{g,s}}\left(\frac{w_g}{w_{g,s}}\right)^{b+2} \tag{8.57}$$

where the second expression was obtained by substituting (8.53) and (8.56) into the first.

At the ground surface, (8.51) and (8.55) can be modified to

$$\frac{\partial T_s}{\partial t} = \frac{1}{\rho_g c_G}\frac{\partial}{\partial z}\left(\kappa_s \frac{\partial T_s}{\partial z} + F_{n,g} - H_f - L_e E_f\right) \tag{8.58}$$

$$\frac{\partial w_g}{\partial t} = \frac{\partial}{\partial z}\left(D_g \frac{\partial w_g}{\partial z} + K_g + \frac{E_f - P_g}{\rho_w}\right) \tag{8.59}$$

respectively, where $\kappa_s \partial T_s/\partial z$ (W m$^{-2}$) is the conductive heat flux between the ground surface and the layer of soil just below the surface, $F_{n,g}$ (W m$^{-2}$) is the solar plus infrared radiative flux absorbed by the surface minus the infrared flux emitted by the surface (positive $F_{n,g}$ is down), $H_f$ (W m$^{-2}$) is the vertical turbulent sensible-heat flux at the surface (positive is up), $E_f$ (kg m$^{-2}$ s$^{-1}$) is the evaporation rate at the surface (positive is up), $L_e$ (J kg$^{-1}$) is the latent heat of evaporation, and $P_g$ (kg m$^{-2}$ s$^{-1}$) is the net flux of liquid water reaching the soil surface (precipitation minus runoff) (positive is down). The product $L_e E_f$ is the net latent heat flux between the ground surface and the atmosphere. $F_{n,g}$ is determined from a radiative-transfer calculation (e.g., Chapter 10).

$H_f$ and $E_f$ can be estimated from similarity theory with (8.31) or from bulk aerodynamic equations with (8.14). $H_f$ depends on the potential virtual temperature of air at height $z_{0,h}$, and $E_f$ depends on the specific humidity of air at height $z_{0,v}$. Two methods of calculating the specific humidity at height $z_{0,h}$ are

$$\bar{q}_v(z_{0,v}) = \alpha_g q_{v,s}(T_g) \tag{8.60}$$

$$\bar{q}_v(z_{0,v}) = \beta_g q_{v,s}(T_g) + (1 - \beta_g)\bar{q}_v(z_r) \tag{8.61}$$

where $q_{v,s}(T_g) = \varepsilon p_s/(p_d + \varepsilon p_s)$ is the saturation specific humidity at the ground temperature, and $\alpha_g$ and $\beta_g$ are wetness functions. The function $\alpha_g$ is like a relative humidity adjacent to the water in soil pores. Philip (1957) estimated $\alpha_g \approx \exp[\psi_p g/R_v T_g]$, but Wetzel and Chang (1987), Avissar and Mahrer (1988), Kondo et al. (1990), Lee and Pielke (1992), and Mihailovic et al. (1995) pointed out that this equation is incorrect. These papers and Mahfouf and Noilhan (1991) give formulations for $\beta_g$ and alternative formulations for $\alpha_g$.

Together, (8.51), (8.55), (8.58), and (8.59), or variations thereof, can be solved numerically among several soil layers (e.g., McCumber and Pielke 1981). An alternative to a multilayer soil model is a two-compartment soil model, described by Noilhan and Planton (1989) and Mahfouf and Noilhan (1996). This model is a generalization of the **force-restore method**, which involves forcing the temperature and liquid water content in the top centimeter of soil over a short time and restoring these variables with deep-soil values over a longer time (Bhumralkar 1975; Blackadar 1976; Deardorff 1977).

## 8.9. SUMMARY

In this chapter, boundary-layer and soil processes were discussed, and expressions for kinematic turbulent fluxes of momentum, sensible heat, and moisture were derived. In the surface layer, such fluxes were described in terms of bulk aerodynamic formulae and Monin–Obukhov similarity theory. Some similarity-theory parameters are friction velocity, potential temperature scale, and Monin–Obukhov length. For turbulent transfer of momentum and energy above the surface layer, eddy diffusion may be parameterized with a high-order closure technique or with a hybrid model that accounts for different stability regimes. Equations describing soil temperatures and moisture contents were also given. Ground temperatures depend on the specific heat, thermal conductivity, and moisture content of soil and fluxes of radiative, sensible, and latent heat between the soil and air.

## 8.10. PROBLEMS

**8.1.** If conditions at the surface are $z_r = 10$ m, $z_{0,m} = 0.05$ m, $z_{0,h} = 0.0005$ m, $\bar{u}(z_r) = 8$ m s$^{-1}$, $\bar{v}(z_r) = 2$ m s$^{-1}$, $\bar{T}(z_r) = 285$ K, $\bar{T}(z_{0,h}) = 286$ K, $p_a(z_r) = 1004$ mb, and $p_a(z_{0,h}) = 1005$ mb, calculate $u_*$, $\theta_*$, $L$, $\mathrm{Ri}_b$, and $K_{m,zx}$ at the reference height. Assume dry air.

**8.2.** Assume $z_{0,m} = 0.01$ m, $u_* = 1$ m s$^{-1}$, $\bar{u}(z_r) = 10$ m s$^{-1}$, and $\bar{v}(z_r) = 5$ m s$^{-1}$ at $z_r = 10$ m. Calculate the eddy diffusion coefficient for momentum, $K_{m,zx}$.

**8.3.** Assuming $u_* = 0.1$ m s$^{-1}$, $p_a = 998$ mb, and the air is dry, calculate the surface roughness lengths for energy and water vapor when $\bar{T} = 288$ K.

**8.4.** Compare vertical turbulent sensible-heat fluxes over the ocean from bulk aerodynamic formulae and Monin–Obukhov similarity theory. Assume the same conditions as in Problem 8.1, except assume $z_{0,m} = 0.00001$ m. Assume $\bar{\theta}_v$ at the roughness length for energy equals that at the roughness length for momentum. Discuss differences in results.

**8.5.** Using conditions from Problems 8.1 and 8.4, calculate the eddy diffusion coefficient for energy with bulk aerodynamic formulae and similarity theory. Discuss differences in results.

**8.6.** Assume all conditions are the same as in Problem 8.1, except that the relative humidity at the reference height is now 85 percent, and the specific humidity at the roughness length for moisture equals the saturation specific humidity at the roughness length for energy. Estimate the vertical turbulent flux of moisture from similarity theory.

## 8.11. COMPUTER PROGRAMMING PRACTICE

**8.7.** If $z_r = 10$ m, $\bar{u}(z_r) = 3$ m s$^{-1}$, $\bar{v}(z_r) = 16$ m s$^{-1}$, $\bar{T}(z_r) = 293$ K, $\bar{T}(z_{0,h}) = 292.5$ K, $p_a(z_r) = 1002$ mb, and $p_a(z_{0,h}) = 1003$ mb, write a script to calculate $u_*$, $\theta_*$, $L$, $\mathrm{Ri}_b$, and $K_{m,zz}$ as a function of $z_{0,m}$ when the air is dry. Calculate values for $0.00001\text{m} < z_{0,m} < 5$ m. Assume $z_{0,h} \approx z_{0,m}/100$. Plot the results for each variable.

**8.8.** Write a computer script to calculate the change of wind speed and potential virtual temperature with height in the surface layer from similarity theory. Assume the same conditions as in problem 8.1. Plot the results from the ground surface up to 10 m altitude. Change $\bar{T}(z_r)$ to 286 K, and replot the results. Discuss the differences between the two cases.

# 9

# Cloud Thermodynamics and Dynamics

CLOUDS affect the atmosphere in several ways. They absorb and reflect radiation, modify air temperatures, pressures, and wind speeds, produce precipitation, mix air and gases rapidly in the vertical, remove gases and particles from the air, and alter photolysis coefficients. In this chapter, cloud formation is described, and the thermodynamics of moist air is discussed. The pseudoadiabatic lapse rate and thermodynamic energy equation for a cloud are derived. The thermodynamic energy equation includes an entrainment term, which is also derived. A method of estimating the base height and vertical extent of a cloud is given, and the vertical momentum equation in a cloud is derived. Cumulus parameterizations for simulating effects of subgrid scale clouds in a model are also described. Further discussions on clouds can be found in Cotton and Anthes (1989), Rogers and Yau (1989), and Houze (1993).

## 9.1. FOG AND CLOUD TYPES AND FORMATION MECHANISMS

Clouds are a type of hydrometeor. A **hydrometeor** is an ensemble of liquid or solid water particles suspended in or falling through the air. In this section, cloud types and their formation mechanisms are discussed.

### 9.1.1. Cloud Classification

Clouds form primarily in the troposphere. In the tropics, the highest clouds reach near 18-km altitude. In midlatitudes and at the poles, tropospheric clouds reach a maximum altitude of near 13 and 8 km, respectively. Ice-crystal and nitric acid ice clouds can form in the polar stratosphere. These clouds are discussed in Chapter 12. For purposes of cloud classification, the troposphere is divided into three altitude ranges or **étages**, in which clouds of different types (genera) form most frequently. These étages encompass high, middle, and low altitude ranges, as shown in Table 9.1.

In 1802, Jean Baptiste Lamark (1744–1829) proposed a cloud classification scheme, but the cloud types he suggested were not generally accepted. In 1803, Luke Howard proposed an alternative identification scheme that used Latin roots in the cloud names. Sheetlike clouds were called **stratus** (Latin for *layer*). Puffy

**Table 9.1.** Altitude Range of Cloud Étages

| | Range (km) | | |
|---|---|---|---|
| **Étage** | **Polar Regions** | **Temperate Regions** | **Tropical Regions** |
| High | 3–8 | 5–13 | 6–18 |
| Middle | 2–4 | 2–7 | 2–8 |
| Low | 0–2 | 0–2 | 0–2 |

Source: WMO (1975).

clouds were called **cumulus** (*heap*), featherlike clouds were called **cirrus** (*curl of hair*), and rain clouds were called **nimbus** (*violent rain*).

Today, clouds are divided into 10 primary categories, or **genera**. These include cirrus, cirrocumulus, cirrostratus, altocumulus, altostratus, nimbostratus, stratocumulus, stratus, cumulus, and cumulonimbus. A cloud can belong to one genus only. Table 9.2 lists and describes the 10 cloud genera and identifies the étage(s) in which each genus is (are) most commonly observed.

A **fog** is a cloud touching the ground, but a fog is not classified as a cloud. Instead, a fog is a separate type of hydrometeor and is defined as a suspension of liquid water drops that causes visibility to be reduced to less than 1 km (WMO 1975). If visibility is greater than 1 km, the fog is called a **mist**. If the fog contains ice crystals instead of liquid water, it is an **ice fog**.

Other hydrometeors suspended in the air include rain, supercooled rain, drizzle, supercooled drizzle, snow, snow grains, snow pellets, diamond dust, hail, small hail, ice pellets, drifting snow, blowing snow, and spray. **Rain** is precipitation of water drops falling from a cloud. **Supercooled rain** is liquid rain at a temperature below 0°C. **Drizzle** is precipitation of water drops with diameter smaller than 0.5 mm. **Supercooled drizzle** is liquid drizzle at a temperature below 0°C. **Snow** is precipitation from a cloud of single or agglomerated ice crystals. **Snow grains** are opaque ice crystals of diameter less than 1 mm. **Snow pellets** are rounded white and opaque ice particles with diameters up to 5 mm. **Diamond dust** is precipitation of small ice crystals from the clear sky. **Hail** is precipitation of clumpy, spheroidal ice particles 5–50 mm in diameter. **Small hail** is precipitation of translucent, spherical ice particles near 5 mm in diameter. **Ice pellets** are transparent ice particles, either spheroidal or irregular, with diameter less than 5 mm in diameter.

Hydrometeors affixed to a surface include fog deposits, dew, frozen dew, hoar frost, rime, glaze, and freezing rain. **Fog deposits** are fog drops fixed to a surface when the surface temperature exceeds 0°C. **Dew** is liquid water on a surface produced from condensation of water vapor. **Frozen dew** forms from dew if the temperature drops below 0°C. **Hoar frost** or **frost** is ice produced by sublimation of water vapor on a surface. **Rime** is ice produced by the freezing of supercooled fog droplets on a surface when the surface temperature is below 0°C. **Glaze** is transparent ice that forms when supercooled drizzle or rain freezes on contact with

**Table 9.2.** Cloud Classification

| Genera | Description | Étage |
|---|---|---|
| Cirrus (Ci) | Detached with white, delicate filaments, whitish patches, or narrow bands. Fibrous appearance and/or silky sheen. | High |
| Cirrocumulus (Cc) | Thin, white patch, sheet, or layer cloud, made of small elements with grains or ripples, merged or separate, but regularly arranged. | High |
| Cirrostratus (Cs) | Transparent, whitish cloud veil. Fibrous or smooth, totally or partly covering the sky. Produces a halo. | High |
| Altocumulus (Ac) | White and/or gray patch, sheet, or layer, generally with shading. Contains rounded masses, sometimes fibrous, which may or may not be merged. | Middle |
| Altostratus (As) | Grayish sheet or layer, fibrous or uniform, totally or partly covers the sky. Has parts thin enough to reveal the sun vaguely. No halo. | Middle<br>High |
| Nimbostratus (Ns) | Gray, often dark. Contains rain that almost always reaches the ground. Thick enough to block the sun. Low, ragged clouds appear below the cloud. | Low<br>Middle<br>High |
| Stratocumulus (Sc) | Gray and/or whitish, patch, sheet, or layer. Almost always has dark parts and rolls, which may or may not be merged. | Low |
| Stratus (St) | Gray cloud layer with uniform base that may produce drizzle, ice, or snow. When the sun is visible through the cloud, its outline is clear. Stratus does not produce a halo, except at low temperatures. | Low |
| Cumulus (Cu) | Detached clouds, usually dense, with sharp outlines. Vertically developed as a rising mound, dome, or tower, of which the upper part often looks like a cauliflower. The sunlit portion appears white. Its base is dark and horizontal. | Low<br>Middle |
| Cumulonimbus (Cb) | Heavy, dense cloud with great vertical extent, in the form of a mountain or tower. Part of upper cloud is smooth, fibrous, or striated, and almost always flattened, spreading out like anvil. Under its dark base, precipitation occurs. | Low<br>Middle<br>High |

Source: WMO (1995).

a surface. **Freezing rain** occurs when nonsupercooled drizzle or rain freezes on contact with a surface.

### 9.1.2. Cloud Formation

Clouds form by one of several mechanisms. Clouds that form by surface heating and free convection are **convective clouds**. When the ground surface is exposed to intense sunlight, air just above the surface heats by conduction. Elevated air is not affected immediately. The resulting unstable lapse rate causes buoyancy and lifting (free convection). The altitude at which lifting by free convection starts is the **level of free convection** (LFC). As the parcel rises, it expands and cools dry adiabatically. If the buoyant parcel contains water vapor, and if the parcel temperature cools to the **isentropic condensation temperature** (ICT), the vapor condenses. At the altitude of initial condensation, which is the **lifting condensation level** (LCL), a cloud base forms.

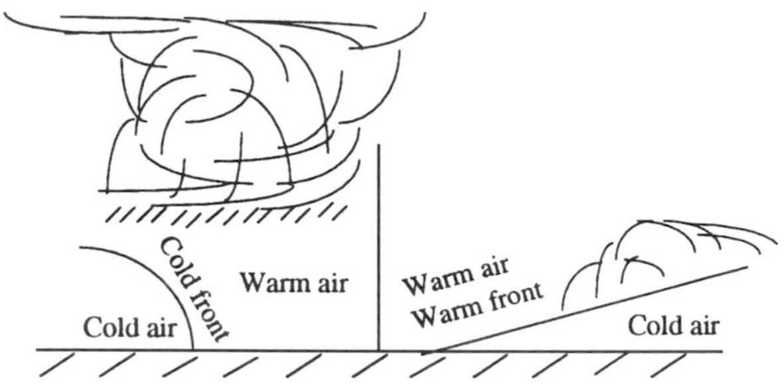


**Figure 9.1.** Formation of clouds along a cold and a warm front. The cold air behind a cold front forces warm air vertically. The warm air behind a warm front slides up over cold air ahead of the front. In both cases, the rising air cools and may result in cloud formation.

**Orographic clouds** appear when a horizontal wind encounters a topographic barrier, such as a mountain, which forces the wind to shift slantwise or vertically (orographic uplifting). As the air rises, it expands and cools adiabatically. If the parcel rises to the LCL, a cloud forms. Clouds can also form when surface air converges, as when winds converge horizontally around a center of a low pressure. Convergence forces air to rise vertically (forced convection). Air that rises above the LCL during orographic uplifting or forced convection may reach a level of free convection (LFC) and become unstable and rise buoyantly on its own.

Clouds also form along weather fronts, as shown in Fig. 9.1. Along a **cold front**, cold, dense air pushes warm, moist air vertically, causing the warm air to expand and cool until condensation occurs. Along a **warm front**, warm, moist air slides over cold air and expands and cools until a cloud forms. The types of clouds that form along a cold front differ from those that form along a warm front. Ahead of a cold front, cumulonimbus, altocumulus, cirrostratus, and cirrus are the most prevalent cloud types. In front of a warm front, stratocumulus, stratus, nimbostratus, altostratus, cirrostratus, and cirrus are most likely to form.

The type of cloud that results from a given formation process depends on the height of the lifting condensation level, the stability of the atmosphere above the LCL, and the rate of mixing of surrounding air into the cloud. When the LCL is below 2 km, the resulting clouds are stratus, stratocumulus, nimbostratus, cumulus, or cumulonimbus. When the LCL is in the middle étage, the resulting clouds are altostratus, altocumulus, cumulus, or cumulonimbus. Cirrus clouds often form as the top portion of a cumulonimbus cloud dissipates or as a pocket of moist air enters a cold region.

Atmospheric stability above the LCL affects the cloud type. If a cloud base is below 2 km, and the atmosphere is stable just above the LCL, the cloud will probably be a cumulus humilis (cumulus cloud of slight vertical extent), stratus, or stratocumulus, since it cannot rise significantly. If the atmosphere is stable at a higher altitude, but below the tropopause, a cumulus congestus cloud (cumulus cloud of great vertical extent) may form. If the atmosphere is stable near the tropopause, cumulonimbus clouds may form.

**Entrainment** is the mixing of cool, dry air into a warm, moist cloud. If the air around a cloud is dry, air entraining into the cloud will evaporate water, causing the cloud to cool near the edges. Such cooling reduces the maximum height of the cloud and accelerates downdrafts along its edges. Since the cloud type depends on the maximum height of the cloud, entrainment affects cloud type.

Some cloud types form from interactions between other cloud types and the environment. Cirrocumulus, altocumulus, and stratocumulus may form when the top of a cirrostratus, altostratus, or stratus cloud cools radiatively and the bottom warms radiatively. Resulting buoyant motions within the cloud produce pockets of air that rise to form cirro- alto-, or stratocumulus clouds.

### 9.1.3. Fog Formation

Fogs form in one of several ways. A **radiation fog** forms when air near the ground cools radiatively during the night and early morning to below the dew point. An **advection fog** forms when warm, moist air moves over a colder surface and cools to below the dew point. Such fogs often occur in coastal regions. An **upslope fog** forms when warm, moist air flows up a topographical barrier and cools adiabatically to below the dew point. The formation of this type of fog is similar to the formation of an orographic cloud.

An **evaporation fog** forms when water evaporates in warm, moist air, then mixes with cool, dry air, in which the moisture recondenses. Two types of evaporation fogs are steam fogs and frontal fogs. A **steam fog** occurs when warm surface water evaporates and then rises into cooler air before recondensing and giving the appearance of rising steam. A **frontal fog** occurs when water from warm raindrops evaporates as the drops fall through a cold air mass. The moisture recondenses in the cold air to form a fog. These conditions often exist ahead of an approaching front. After a fog has formed, it typically burns off from the top and bottom. Solar energy evaporates drops at the fog top, and infrared energy emitted from the ground evaporates drops at the fog base. When the fog no longer touches the ground, it is a stratus cloud. Stratus clouds often form from the dissipation of a fog.

## 9.2. MOIST- AND PSEUDOADIABATIC PROCESSES

In this section, atmospheric stability in the presence of moist air is discussed. If water vapor in a cloud condenses, it releases latent heat, affecting the temperature profile. Since latent-heat release at a given altitude increases instability below that altitude, condensation increases cloud buoyancy, increasing the clouds potential to rise further.

A parcel of air rising dry adiabatically cools at the rate of $-9.8$ K km$^{-1}$. If condensation occurs as the parcel rises, latent heat is released, offsetting the dry-adiabatic cooling by up to $+4$ K km$^{-1}$. When the released latent heat is absorbed by dry air, water vapor, and liquid water, the process is **moist adiabatic**. When it is assumed to be absorbed by only dry air and water vapor, the process is **pseudoadiabatic.** Moist- and pseudoadiabatic ascent occur only within a cloud, since, by definition, a cloud forms when the air is supersaturated and water condenses (or deposits).

### 9.2.1. Pseudoadiabatic Lapse Rate

Here, we consider absorption of latent heat by only dry air and water vapor and derive the pseudoadiabatic lapse-rate equation in terms of virtual temperature. During a pseudoadiabatic ascent, the latent heat energy per unit mass of air (J $kg^{-1}$) released during condensation is

$$\mathrm{d}Q = -L_e\,\mathrm{d}\omega_{v,s} \tag{9.1}$$

where $L_e$ is the latent heat of evaporation (J $kg^{-1}$) from (2.48), and $\omega_{v,s} = \varepsilon p_{v,s}/p_d$ is the mass mixing ratio of water vapor at saturation over a liquid surface (kilograms of vapor per kilogram of dry air) from (2.61). Combining (9.1) with the first law of thermodynamics in terms of virtual temperature from (2.76) yields

$$-L_e\,\mathrm{d}\omega_{v,s} = c_{p,d}\,\mathrm{d}T_v - \alpha_a\,\mathrm{d}p_a \tag{9.2}$$

Substituting $\alpha_a = 1/\rho_a = R'T_v/p_a$ from (2.30) into (9.2) and rearranging, gives

$$\mathrm{d}T_v = \frac{R'T_v}{c_{p,d}p_a}\,\mathrm{d}p_a - \frac{L_e}{c_{p,d}}\,\mathrm{d}\omega_{v,s} \tag{9.3}$$

Differentiating (9.3) with respect to altitude, and combining the result with $\partial p_a/\partial z = -\rho_a g$ and $p_a = \rho_a R'T_v$ yields

$$\left(\frac{\partial T_v}{\partial z}\right)_w = \frac{R'T_v}{c_{p,d}p_a}\frac{\partial p_a}{\partial z} - \frac{L_e}{c_{p,d}}\frac{\partial \omega_{v,s}}{\partial z} = -\frac{g}{c_{p,d}} - \frac{L_e}{c_{p,d}}\frac{\partial \omega_{v,s}}{\partial z} \tag{9.4}$$

where subscript $w$ signifies pseudoadiabatic. (9.4) simplifies to (2.83), the equation for the dry adiabatic lapse rate, when $\partial\omega_{v,s}/\partial z = 0$. Differentiating $\omega_{v,s} = \varepsilon p_{v,s}/p_d$ with respect to altitude, then substituting $\mathrm{d}p_{v,s} = L_e p_{v,s}\,\mathrm{d}T/R_v T^2$ (Clausius–Clapeyron equation), $\mathrm{d}T = R'\,\mathrm{d}T_v/R_m = T\mathrm{d}T_v/T_v$, $\omega_{v,s} = \varepsilon p_{v,s}/p_d$, $R' = \varepsilon R_v$, and $\partial p_d/\partial z = -\ p_d g/R'T$ yields

$$\frac{\partial \omega_{v,s}}{\partial z} = \frac{\varepsilon}{p_d}\left(\frac{\partial p_{v,s}}{\partial z} - \frac{p_{v,s}}{p_d}\frac{\partial p_d}{\partial z}\right) = \frac{L_e\varepsilon\omega_{v,s}}{R'TT_v}\frac{\partial T_v}{\partial z} + \frac{\omega_{v,s}g}{R'T} \tag{9.5}$$

Substituting (9.5) and $\Gamma_d = g/c_{p,d}$ into (9.4) and rearranging gives the **pseudoadiabatic virtual temperature change with altitude** as

$$\left(\frac{\partial T_v}{\partial z}\right)_w = -\Gamma_w = -\Gamma_d\left(1 + \frac{L_e\omega_{v,s}}{R'T}\right)\bigg/\left(1 + \frac{L_e^2\varepsilon\omega_{v,s}}{R'c_{p,d}TT_v}\right) \tag{9.6}$$

where $\Gamma_w$ is the **pseudoadiabatic lapse rate** (K km$^{-1}$).

**Example 9.1**
Find $\Gamma_w$ if $p_d = 950$ mb and (a) $T = 283$ K; (b) $T = 293$ K.

SOLUTION

(a) From (2.56), $p_{v,s} = 12.27$ mb; from (2.61), $\omega_{v,s} = 0.00803$ kg kg$^{-1}$; from (2.32), $T_v = 284.4$ K; from (2.48), $L_e = 2.4761 \times 10^6$ J kg$^{-1}$; and from (9.6) $\Gamma_w = 5.26$ K km$^{-1}$. (b) When $T = 293$ K, $p_{v,s} = 23.37$ mb, $\omega_{v,s} = 0.0153$ kg kg$^{-1}$, $T_v = 295.7$ K, $L_e = 2.4522 \times 10^6$ J kg$^{-1}$, and $\Gamma_w = 4.26$ K km$^{-1}$. Thus, the pseudoadiabatic lapse rate changes with temperature and pressure.

### 9.2.2. Stability Criteria

If a parcel of air is dry or moist but unsaturated, the dry adiabatic lapse rate ($\Gamma_d$) is used to determine stability. The stability criteria for dry or unsaturated air were given in (2.94). If condensation occurs in a parcel, the pseudoadiabatic lapse rate may be used to estimate stability. Combining the stability criteria for unsaturated and saturated air gives

$$\begin{cases} \Gamma_v > \Gamma_d & \text{absolutely unstable} \\ \Gamma_v = \Gamma_d & \text{unsaturated neutral} \\ \Gamma_d > \Gamma_v > \Gamma_w & \text{conditionally unstable} \\ \Gamma_v = \Gamma_w & \text{saturated neutral} \\ \Gamma_v < \Gamma_w & \text{absolutely stable} \end{cases} \tag{9.7}$$

If, for example, $\Gamma_w = +6.0$ K km$^{-1}$ and $\Gamma_v = +8.0$ K km$^{-1}$, the atmosphere is conditionally unstable, meaning the atmosphere is stable if a parcel is unsaturated but unstable if condensation occurs in the parcel. Figure 9.2 shows stability criteria for unsaturated and saturated air, and Fig. 9.3 shows how stability is determined in atmospheric layers when each layer has a different ambient virtual temperature profile.

A second method of estimating stability in the presence of unsaturated or saturated air is with equivalent potential temperature. The **equivalent potential temperature** is the potential temperature a parcel of air would have if all its water vapor were condensed and the resulting latent heat were released and used to heat the parcel. It can be found by lifting a parcel from 1000 mb to low pressure until all water vapor has condensed, then lowering the parcel dry adiabatically back to 1000 mb, as shown in Fig. 9.4. In reality, not all water vapor in a parcel condenses because the saturation mass mixing ratio of water slightly exceeds zero for all atmospheric temperatures. At high altitudes, the additional condensation from an incremental decrease in temperature results in negligible latent-heat release, and the pseudoadiabatic lapse rate approaches the dry adiabatic rate. The parcel can be brought back dry-adiabatically from this altitude to 1000 mb dry to determine

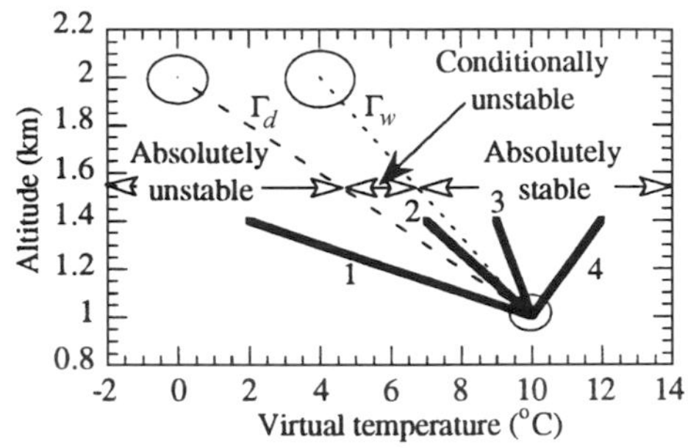


**Figure 9.2.** Stability criteria for unsaturated and saturated air. If air is saturated, the virtual-temperature profile is compared with the pseudoadiabatic profile to determine stability. Virtual-temperature profiles 3 and 4 are stable and 1 and 2 are unstable with respect to saturated air. Profiles 2, 3, and 4 are stable and 1 is unstable with respect to unsaturated air. A rising or sinking air parcel follows the $\Gamma_d$-line when the air is unsaturated and the $\Gamma_w$-line when the air is saturated.

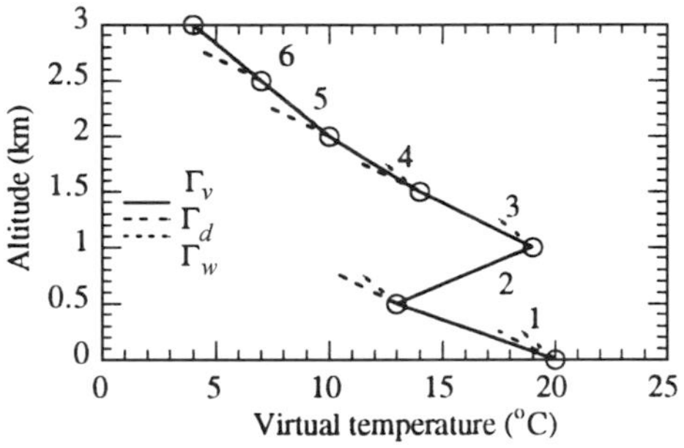


**Figure 9.3.** Determination of stability in multiple layers of air. Layer 1 is absolutely unstable, layer 2 is absolutely stable and an inversion, layer 3 is dry neutral, layer 4 is conditionally unstable, and layers 5 and 6 are moist neutral. Stability is determined by comparing the slope of the ambient virtual temperature lapse rate profile with the slope of the dry and pseudoadiabatic lapse-rate temperature profiles.

the equivalent potential temperature. If the air is initially saturated, as it is above the LCL in Fig. 9.4, the equivalent potential temperature is

$$\theta_{p,e} \approx \theta_p \exp\left(\frac{L_e}{c_{p,d}T}\omega_{v,s}\right) \tag{9.8}$$

where $T$ is the actual initial temperature of the saturated parcel (K), and $\omega_{v,s}$ is the saturation mass mixing ratio of water vapor at that temperature (kg kg$^{-1}$). If the air

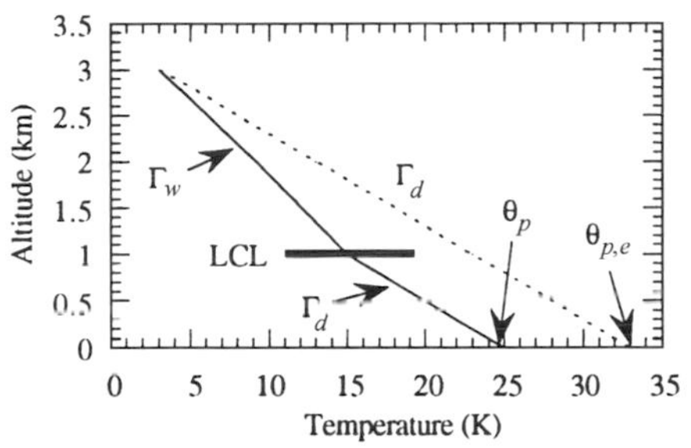


**Figure 9.4.** Schematic showing the relationship between potential temperature and equivalent potential temperature. Suppose a parcel with $\theta_p = 298.15$ K (25°C) rises to the LCL dry-adiabatically and then to 3 km pseudoadiabatically, at which point it has lost its water vapor. If the parcel descends dry-adiabatically back to the surface, its final potential temperature is $\theta_{p,e} = 306.15$ K (33°C).

is initially unsaturated, as it is below the LCL, equivalent potential temperature is

$$\theta_{p,e} \approx \theta_p \exp\left(\frac{L_e}{c_{p,d} T_{\text{LCL}}} \omega_v\right) \tag{9.9}$$

where $T_{\text{LCL}}$ is the temperature (K) of the parcel if it were lifted dry-adiabatically to the LCL, and $\omega_v$ is the initial, unsaturated mass mixing ratio of water vapor in the parcel.

Stability in unsaturated or saturated air can be found from equivalent potential temperature by defining a variable, $\hat{\theta}_{p,e}$ that equals $\theta_{p,e}$ from (9.8) for both saturated and unsaturated air. For unsaturated air, the temperature in (9.8) is the initial temperature of the environment, not $T_{\text{LCL}}$. In saturated air, the temperature is the saturated-parcel temperature. In both cases, $\omega_{v,s}$ is the saturation mixing ratio at the temperature used. Conceptually, $\hat{\theta}_{p,e}$ is the value of $\theta_{p,e}$ in a hypothetically saturated parcel at the temperature of the parcel. The stability criteria in terms of $\hat{\theta}_{p,e}$ are

$$\frac{\partial \hat{\theta}_{p,e}}{\partial z} \begin{cases} < 0 & \text{saturated unstable} \\ = 0 & \text{saturated neutral} \\ > 0 & \text{saturated stable} \end{cases} \tag{9.10}$$

## 9.3. CLOUD DEVELOPMENT BY FREE CONVECTION

A fog or cloud forms when the air temperature drops below the dew point. Figure 9.5 shows an example of how the environmental, dry adiabatic, and pseudoadiabatic lapse rates can be applied together with the dew point to estimate the extent of cloud development. If air at the surface is moist but unsaturated, and the environment is unstable with respect to dry air, surface air rises in the presence of

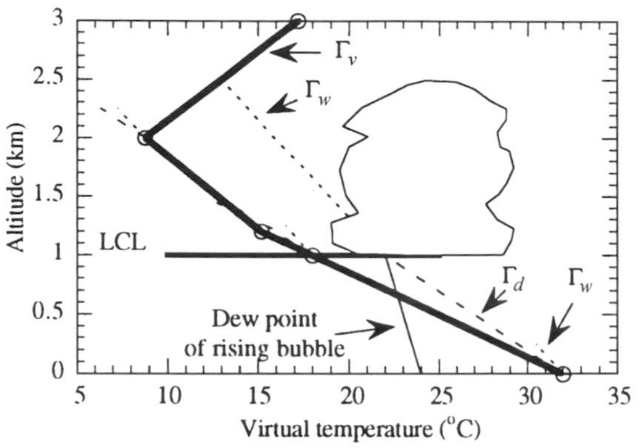


**Figure 9.5.** Simplified model of cumulus-cloud formation. An unsaturated surface parcel of air at 32°C virtual temperature in an unstable environment is perturbed and starts rising. As it rises, it cools at the dry adiabatic lapse rate ($\Gamma_d$). The dew point in the parcel decreases at a slower rate. At the lifting condensation level (LCL), the dew point equals the parcel virtual temperature, and a cloud begins to form. Because the parcel virtual temperature is still greater than that of the environment, the parcel continues to rise but cools at the pseudoadiabatic lapse rate ($\Gamma_w$). When the parcel virtual temperature equals that of the environment (cloud top), the cloud parcel stops rising.

external forcing (e.g., from forced convection). As the parcel rises, the dry-air pressure ($p_d$) in the parcel decreases because pressure decreases with increasing altitude. If the water-vapor mixing ratio ($\omega_v$) is constant in the parcel, $p_v$ must decrease with increasing altitude to satisfy $\omega_v = \varepsilon p_v / p_d$. Because $p_v$ decreases, the dew point must also decrease, since it is an increasing function of $p_v$, as seen from (2.62). The decrease in dew point with increasing altitude is relatively modest (e.g., around 2 K km$^{-1}$).

The virtual temperature in an unsaturated rising parcel decreases at the dry adiabatic lapse rate. When the dew point and the parcel virtual temperature meet, the air is saturated, condensation usually occurs, and a cloud usually forms, as shown in Fig. 9.5. The altitude of cloud formation is the LCL. The virtual temperature at saturation is the isentropic condensation temperature (ICT), which usually occurs at the LCL. The ICT is more formally defined as the virtual temperature at which saturation is reached when unsaturated air is cooled adiabatically at a constant mass mixing ratio of water vapor. The ICT (K) can be approximated from

$$T_{v,L} \approx \frac{4880.357 - 29.66 \ln\left[\frac{\omega_v p_{a,0}}{\varepsilon}\left(\frac{T_{v,L}}{T_{v,0}}\right)^{1/\kappa}\right]}{19.48 - \ln\left[\frac{\omega_v p_{a,0}}{\varepsilon}\left(\frac{T_{v,L}}{T_{v,0}}\right)^{1/\kappa}\right]} \tag{9.11}$$

where $p_{a,0}$ is the surface air pressure (not necessarily 1000 mb), $T_{v,0}$ is the surface virtual temperature (K), and $\omega_v$ is assumed to be constant between the surface and LCL. (9.11) was found by substituting $T_{v,L}$ for $T_D$ in (2.62), combining the result with $p_{a,L} = p_{a,0}(T_{v,L}/T_{v,0})^{1/\kappa}$, derived from (2.90), and assuming $p_{a,L} \approx p_{d,L}$ at the LCL. The solution to (9.11) must be found iteratively.

Condensation releases latent heat, giving a parcel buoyancy. As a parcel rises above the LCL, as in Fig. 9.5, its virtual temperature and dew point decrease pseudoadiabatically. So long as the virtual temperature of the cloudy air exceeds that of the ambient air, the parcel remains buoyant and continues to rise and expand. When the cloud and ambient virtual temperatures meet, the cloud stops rising. As shown in Fig. 9.5, the ambient virtual-temperature profile, which was unstable below the cloud base and within part of the cloud, must become stable for the cloud to stop rising.

## 9.4. ENTRAINMENT

Entrainment, which is the mixing of relatively cool, dry air from outside a cloud with warm, moist air inside the cloud, causes evaporation and cooling at the sides of a cloud, forcing downdrafts at its edges and slowing its rise. If ambient air is dry, entrainment can inhibit cloud vertical growth. Entrainment may also occur at the top or base of a cloud. **Detrainment** is the opposite of entrainment, except that during detrainment, cloud mass mixes with and is removed by the air around the cloud.

A simple model of the effects of entrainment on a cumulus cloud was developed by Stommel (1947). This model assumes that entrained air continuously enters a rising cloud from its sides and instantaneously mixes uniformly throughout the cloud. In reality, entrainment is not entirely lateral, instantaneous, or continuous (Houze 1993). Stommel's model is a useful tool for one-dimensional studies of clouds and can be used to derive a simplified entrainment term in the thermodynamic energy equation.

The model assumes that entrainment affects cloud temperatures in two ways. First, it forces the cloud to expend energy heating cool, entrained air to the cloud virtual temperature. This energy loss (J) from the cloud is

$$\mathrm{d}Q_1^* = -c_{p,d}(T_v - \hat{T}_v)\,\mathrm{d}M_c \tag{9.12}$$

where $T_v$ is the cloud virtual temperature, $\hat{T}_v$ is the ambient virtual temperature, and $\mathrm{d}M_c$ is the mass of ambient dry air plus water vapor entrained in the cloud. Second, entrainment forces the cloud to expend energy evaporating liquid water to maintain saturation of dry, entrained air. This energy loss (J) is

$$\mathrm{d}Q_2^* = -L_e(\omega_{v,s} - \hat{\omega}_v)\,\mathrm{d}M_c \tag{9.13}$$

where $\omega_{v,s}$ is the saturation mass mixing ratio of water vapor in the cloud (kg kg$^{-1}$), and $\hat{\omega}_v$ is the mass mixing ratio of water vapor outside the cloud.

The entrainment region also gains latent heat energy when rising water vapor condenses. The energy gained (J) from latent-heat release is

$$\mathrm{d}Q_3^* = -M_c L_e\, \mathrm{d}\omega_{v,s} \tag{9.14}$$

where $M_c$ is the total mass of air within an entrainment region of the cloud, which consists of dry air, water vapor, and liquid water.

The sum of the three sources and sinks of energy is $\mathrm{d}Q^* = \mathrm{d}Q_1^* + \mathrm{d}Q_2^* + \mathrm{d}Q_3^*$. Substituting (9.12)–(9.14) into this equation gives the total change of energy in the entrainment region of a cloud as

$$\mathrm{d}Q^* = -c_{p,d}(T_v - \hat{T}_v)\,\mathrm{d}M_c - L_e(\omega_{v,s} - \hat{\omega}_v)\,\mathrm{d}M_c - M_c L_e\,\mathrm{d}\omega_{v,s} \tag{9.15}$$

From (2.76) and (2.64), the first law of thermodynamics for this problem requires

$$\mathrm{d}Q^* = M_c(c_{p,d}\,\mathrm{d}T_v - \alpha_a\,\mathrm{d}p_a) \tag{9.16}$$

Subtracting (9.16) from (9.15) and rearranging gives

$$c_{p,d}\,\mathrm{d}T_v - \alpha_a\,\mathrm{d}p_a = -[c_{p,d}(T_v - \hat{T}_v) + L_e(\omega_{v,s} - \hat{\omega}_v)]\frac{\mathrm{d}M_c}{M_c} - L_e\,\mathrm{d}\omega_{v,s} \tag{9.17}$$

Dividing (9.17) by $c_{p,d}T_v$ and substituting $\alpha_a = R'T_v/p_a$ results in

$$\frac{\mathrm{d}T_v}{T_v} - \frac{R'}{c_{p,d}}\frac{\mathrm{d}p_a}{p_a} = -\left[\frac{T_v - \hat{T}_v}{T_v} + \frac{L_e(\omega_{v,s} - \omega_v')}{c_{p,d}T_v}\right]\frac{\mathrm{d}M_c}{M_c} - \frac{L_e\,\mathrm{d}\omega_{v,s}}{c_{p,d}T_v} \tag{9.18}$$

Differentiating (9.18) with respect to height and then substituting $\partial p_a/\partial z = -\rho_a g$ and $p_a = \rho_a R'T_v$ gives the virtual temperature change with altitude in an entraining cloud as

$$\frac{\partial T_v}{\partial z} = -\frac{g}{c_{p,d}} - \left[(T_v - \hat{T}_v) + \frac{L_e}{c_{p,d}}(\omega_{v,s} - \hat{\omega}_v)\right]\frac{1}{M_c}\frac{\partial M_c}{\partial z} - \frac{L_e}{c_{p,d}}\frac{\partial \omega_{v,s}}{\partial z} \tag{9.19}$$

When no entrainment occurs ($\mathrm{d}M_c = 0$), (9.19) simplifies to (9.4).

Rearranging (2.97) as

$$\frac{\partial T_v}{\partial z} = \frac{T_v}{\theta_v}\frac{\partial \theta_v}{\partial z} + \frac{R'T_v}{c_{p,d}p_a}\frac{\partial p_a}{\partial z} = \frac{T_v}{\theta_v}\frac{\partial \theta_v}{\partial z} - \frac{g}{c_{p,d}} \tag{9.20}$$

and substituting the result into (9.19) gives the change in potential virtual temperature with altitude in an entrainment region of a cloud as

$$\boxed{\frac{\partial \theta_v}{\partial z} = -\frac{\theta_v}{T_v}\left[(T_v - \hat{T}_v) + \frac{L_e}{c_{p,d}}(\omega_{v,s} - \hat{\omega}_v)\right]\frac{1}{M_c}\frac{\partial M_c}{\partial z} - \frac{\theta_v}{T_v}\frac{L_e}{c_{p,d}}\frac{\partial \omega_{v,s}}{\mathrm{d}z}} \tag{9.21}$$

Multiplying through by d$z$ and dividing through by d$t$ gives the time rate of change of potential virtual temperature in a cloud as

$$\frac{d\theta_v}{dt} = -\frac{\theta_v}{T_v}\left[(T_v - \hat{T}_v) + \frac{L_e}{c_{p,d}}(\omega_{v,s} - \hat{\omega}_v)\right]E - \frac{\theta_v L_e}{c_{p,d} T_v}\frac{d\omega_{v,s}}{dt} \tag{9.22}$$

where $E = (1/M_c)\,dM_c/dt$ is the **entrainment rate** ($s^{-1}$) of outside air into the cloud. If a thermal is assumed to be a spherical bubble with radius $r_t \approx 0.2z_c$ (m), where $z_c$ is the altitude (m) of the center of the thermal above its starting point, then the rate of entrainment can be approximated as (Houze 1993)

$$E = \frac{1}{M_c}\frac{dM_c}{dt} \approx \frac{3}{4\pi r_t^3}\frac{d}{dt}\left(\frac{4\pi r_t^3}{3}\right) \tag{9.23}$$

(3.76) gave the thermodynamic energy equation with diabatic source and sink terms. One such term is energy release due to condensation, $dQ_{c/e} = -L_e\,d\omega_{v,s}$. Two other terms are energy release due to freezing of liquid water and deposition of vapor to ice. These terms can be quantified as $dQ_{f/m} = -L_m\,d\omega_L$ and $dQ_{dp/s} = -L_s\,d\omega_{v,I}$, respectively, where $L_m$ is the latent heat of melting from (2.49), $d\omega_L$ is the change in mass mixing ratio (kg $kg^{-1}$) of liquid water upon freezing, $L_s$ is the latent heat of sublimation from (2.50), $\omega_{v,I} \approx \varepsilon p_{v,I}/p_d$ is the **saturation mass mixing ratio of water vapor over an ice surface**, which is analogous to (2.61), and $d\omega_{v,I}$ is the change in saturation mass mixing ratio upon sublimation.

Adding these and remaining terms from (3.76) to (9.22) gives the **thermodynamic energy equation in a cloud** as

$$\frac{d\theta_v}{dt} = -\frac{\theta_v}{T_v}\left[(T_v - \hat{T}_v) + \frac{L_e}{c_{p,d}}(\omega_{v,s} - \hat{\omega}_v)\right]E + \frac{1}{\rho_a}(\nabla\cdot\rho_a\mathbf{K}_h\nabla)\theta_v + \frac{\theta_v}{c_{p,d}T_v}\left(-L_e\frac{d\omega_{v,s}}{dt} - L_m\frac{d\omega_L}{dt} - L_s\frac{d\omega_{v,I}}{dt} + \frac{dQ_{solar}}{dt} + \frac{dQ_{ir}}{dt}\right) \tag{9.24}$$

Water categories in a cloud include water vapor, liquid water, drizzle, rainwater, cloud ice, snow, graupel, and hail. Conversion from one form of water to another results in a release or gain of energy. The latent-heat terms in (9.24) take account of these processes in a general sense. In a numerical model, each conversion process is often parameterized. Houze (1993), Fowler et al. (1996), and Pruppacher and Klett (1997) give parameterizations for these processes.

## 9.5. VERTICAL MOMENTUM EQUATION IN A CLOUD

In a cloud, vertical velocities are affected by local acceleration, gravity, pressure gradients, and turbulence. From (4.74), the vertical momentum equation in Cartesian–altitude coordinates was

$$\frac{dw}{dt} = -g - \frac{1}{\rho_a}\frac{\partial p_a}{\partial z} + \frac{1}{\rho_a}(\nabla\cdot\rho_a\mathbf{K}_m\nabla)w \tag{9.25}$$

If $\hat{p}_a$ and $\hat{\rho}_a$ are the pressure and density, respectively, of the air outside a cloud, and if the ambient air is in hydrostatic balance, $\partial \hat{p}_a / \partial z = -\hat{\rho}_a g$. Adding this equation to (9.25) gives

$$\frac{\mathrm{d}w}{\mathrm{d}t} = -g\frac{\rho_a - \hat{\rho}_a}{\rho_a} - \frac{1}{\rho_a}\frac{\partial(p_a - \hat{p}_a)}{\partial z} + \frac{1}{\rho_a}(\nabla \cdot \rho_a \mathbf{K}_m \nabla)w \tag{9.26}$$

where $p_a - \hat{p}_a$ and $\rho_a - \hat{\rho}_a$ are the deviations of cloud pressure and density from ambient pressure and density, respectively.

The **buoyancy factor** is defined as

$$B = -\frac{\rho_a - \hat{\rho}_a}{\rho_a} = -\frac{p_a \hat{T}_v - \hat{p}_a T_v}{p_a \hat{T}_v} \tag{9.27}$$

$$= -\frac{\hat{T}_v - T_v}{\hat{T}_v} + \left(\frac{T_v}{\hat{T}_v}\right)\frac{\hat{p}_a - p_a}{p_a} \approx -\frac{\hat{\theta}_v - \theta_v}{\hat{\theta}_v}$$

where $T_v$, $\theta_v$, and $\rho_a = p_a / R' T_v$ are the virtual temperature, potential virtual temperature, and density, respectively, of cloudy air, and $\hat{T}_v$, $\hat{\theta}_v$, and $\hat{\rho}_a = \hat{p}_a / R' \hat{T}_v$ are the analogous variables for ambient air. The approximation on the right of (9.27) was obtained by assuming $(\hat{p}_a - p_a)/p_a$ is small. Virtual potential temperature can replace potential virtual temperature in (9.27), but the difference is small. If a parcel contains liquid water, as it does above the lifting condensation level, condensate (condensed water) adds a downward force to the parcel. Condensate in the surrounding air also adds a downward force to the surrounding air. To allow for condensate, the buoyancy factor can be adjusted to

$$B = -\frac{\rho_a - \hat{\rho}_a}{\rho_a} = -\frac{\hat{\theta}_v(1 + \omega_L) - \theta_v(1 + \hat{\omega}_L)}{\hat{\theta}_v} \approx \frac{\theta_v - \hat{\theta}_v}{\hat{\theta}_v} - \omega_L \tag{9.28}$$

where $\omega_L$ and $\hat{\omega}_L$ are the mass mixing ratios of liquid water in the parcel and ambient air, respectively (kilograms per kilogram of dry air). The second expression assumes the liquid-water content of the ambient air is small compared to that in a cloud.

Substituting (9.28) into (9.26) gives the vertical momentum equation as

$$\frac{\mathrm{d}w}{\mathrm{d}t} = g\left(\frac{\theta_v - \hat{\theta}_v}{\hat{\theta}_v} - \omega_L\right) - \frac{1}{\rho_a}\frac{\partial(p_a - \hat{p}_a)}{\partial z} + \frac{1}{\rho_a}(\nabla \cdot \rho_a \mathbf{K}_m \nabla)w \tag{9.29}$$

From (2.34), (4.48), and (5.38),

$$\frac{1}{\rho_a}\frac{\partial p_a}{\partial z} = -g = -\frac{\partial \Phi}{\partial z} = c_{p,d}\theta_v \frac{\partial P}{\partial z} \tag{9.30}$$

Substituting (9.30) into (9.29) for cloudy and ambient air gives the **vertical**

**momentum equation in a cloud** as

$$\frac{dw}{dt} = g\left(\frac{\theta_v - \hat{\theta}_v}{\hat{\theta}_v} - \omega_L\right) - c_{p,d}\theta_v \frac{\partial(P - \hat{P})}{\partial z} + \frac{1}{\rho_a}(\nabla \cdot \rho_a \mathbf{K}_m \nabla) w \quad (9.31)$$

This equation is similar in several respects to (5.4), the vertical momentum equation derived for nonhydrostatic motions. As with (5.4), (9.31) can be solved directly or used to derive a diagnostic equation for nonhydrostatic pressure. In the latter case, the vertical scalar velocity may be determined diagnostically from the anelastic continuity equation for air, given in (5.7).

If the pressure perturbation and the eddy diffusion term are ignored, (9.31) becomes

$$\frac{dw}{dt} = \frac{dw}{dz}\frac{dz}{dt} = \frac{dw}{dz}w = g\left(\frac{\theta_v - \hat{\theta}_v}{\hat{\theta}_v} - \omega_L\right) = gB \quad (9.32)$$

where $w = dz/dt$. Rearranging (9.32) gives $w\,dw = gB\,dz$. Integrating this equation from a reference height $z_a$, where the vertical velocity is $w_a$, to height $z$ yields a simplified expression for the vertical velocity in a cloud,

$$w^2 = w_a^2 + 2g\int_{z_a}^{z}\left(\frac{\theta_v - \hat{\theta}_v}{\hat{\theta}_v} - \omega_L\right)dz = w_a^2 + 2g\int_{z_a}^{z} B\,dz \quad (9.33)$$

In this expression, the vertical velocity is the integral of buoyancy between the base of the cloud and the altitude of interest.

## 9.6. CONVECTIVE AVAILABLE POTENTIAL ENERGY

(9.33) can be modified to give an expression for the **convective available potential energy** (CAPE), which describes the growth potential of a cloud. The CAPE determines the buoyant stability of the atmosphere and correlates positively with growth of and rainfall production in cumulus clouds (Zawadski et al. 1981). CAPE (m$^2$ s$^{-2}$) is defined as

$$\mathrm{CAPE} = g\int_{z_{\mathrm{LFC}}}^{z_{\mathrm{LNB}}} B\,dz \approx g\int_{z_{\mathrm{LFC}}}^{z_{\mathrm{LNB}}}\left(\frac{\theta_v - \hat{\theta}_v}{\hat{\theta}_v}\right)dz \quad (9.34)$$

where $z_{\mathrm{LFC}}$ is the **level of free convection** (LFC), $z_{\mathrm{LNB}}$ is the **level of neutral buoyancy** (LNB), $\theta_v$ is the potential virtual temperature of a rising parcel of air, and $\hat{\theta}_v$ is the potential virtual temperature of the environment. The LFC is the altitude at which a parcel of rising air first becomes warmer than the environment, it may be below or above the LCL. The LNB is the altitude near the cloud top at which environmental and cloud temperatures equalize and the cloud is no longer buoyant. In Fig. 9.5, the CAPE is the integrated difference of $(\theta_v - \hat{\theta}_v)/\hat{\theta}_v$ between 0 and 2.5 km.

**Example 9.2**

Estimate CAPE and $w$ for a 10-km-thick cumulonimbus cloud over the ocean and over land, if $\theta_v - \hat{\theta}_v \approx 1.5$ K over the ocean, $\theta_v - \hat{\theta}_v \approx 8$ K over land, and the average ambient virtual temperature between 0 and 10 km is $\theta_v = 288$ K in both cases.

SOLUTION

From (9.34), CAPE $\approx 511$ m$^2$ s$^{-2}$ for the ocean case and 2725 m$^2$ s$^{-2}$ for the land case. If $\omega_L$ is ignored, (9.33) gives $w = 32$ m s$^{-1}$ for the ocean case and $w = 74$ m s$^{-1}$ for the land case. These values are higher than observed maximums over the ocean and land, which are 10 and 50 m s$^{-1}$, respectively, in thunderstorm clouds.

## 9.7. CUMULUS PARAMETERIZATIONS

Vertical cloud motions vary over horizontal scales of tens to hundreds of meters. When a model's horizontal resolution is smaller than this, the vertical momentum equation can be used to reproduce cloud structure. The vertical momentum equation can also be used to reproduce much of the structure but not the details of a squall-line convective system for grid resolution of up to 4 km (Weisman et al. 1997). Many mesoscale and global models have horizontal resolutions of 4–50 km and 100–600 km, respectively. In both types of models, cloud motions are subgrid-scale phenomena and must be parameterized.

Techniques have been developed to estimate the effects of subgrid-scale cumulus clouds on the model-scale environment. These techniques are called **cumulus parameterizations** and require input variables from the model-scale environment. Important model-scale variables used to predict subgrid effects are horizontal and vertical wind speeds, potential temperatures, and total water mixing ratios. Cumulus parameterizations use these variables to adjust potential temperature, total water, and momentum fields and to predict precipitation rates. The effects of a cumulus parameterization on the model-scale environment are **feedback effects**.

Cumulus parameterizations include moist convective adjustment schemes (e.g., Manabe et al. 1965; Miyakoda et al. 1969; Krishnamurti and Moxim 1971; Kurihara 1973), Kuo schemes (Kuo 1965, 1974; Anthes 1977; Krishnamurti et al. 1980; Molinari 1982), the Arakawa–Schubert scheme (Arakawa and Schubert 1974; Lord and Arakawa 1980; Kao and Ogura 1987; Moorthi and Suarez 1992; Ding and Randall 1997; Cheng and Arakawa 1997), and other schemes (Ooyoma 1971; Kreitzberg and Perkey 1976; Fritsch and Chappell 1980; Betts 1986; Betts and Miller 1986; Frank and Cohen 1987; Tiedke 1989; Emanuel 1991; Grell 1993; Hack 1994). Cotton and Anthes (1989) discuss several schemes in detail. Three are briefly described below.

**Moist convective adjustment schemes** are the most basic cumulus parameterizations. In these schemes, the model-scale vertical temperature profile is adjusted to a critical, stable profile when the relative humidity exceeds a specified value and the temperature profile is unstable with respect to moist air. During adjustment, the temperature profile is adjusted to the pseudoadiabatic rate, the large-scale relative humidity is unchanged, condensed water vapor precipitates, and total moist enthalpy is conserved.

In **Kuo schemes**, rainfall from cumulus convection is assumed to occur following model-scale convergence of moisture. Part of the moisture condenses, releasing latent heat and increasing rainfall. The rest is used to increase the relative humidity of the environment. Cloud dynamics and microphysics are not computed in Kuo schemes, cloud types are not classified, and the altitudes of cloud bases and tops cannot be found.

In the **Arakawa–Schubert scheme** the model-scale environment is divided into a cloud layer, where clouds form, and a subcloud mixed layer. The sum of individual clouds in a model column make up a cloud ensemble. The cloud ensemble is assumed to occupy a large horizontal area, although the area must be much smaller than the area of a grid cell. Each cloud in an ensemble has its own fractional entrainment rate, vertical mass flux across the cloud base, and cloud top height. The fractional entrainment rate is defined as the entrainment rate per unit height divided by the vertical mass flux. The ensemble is divided into subensembles, which consist of clouds with similar fractional entrainment rates. Equations are derived for each subensemble and summed over all subensembles to obtain the net effect of the ensemble on the model-scale environment.

Cloud ensembles affect the model-scale environment in two ways. First, when saturated air containing liquid water detrains (escapes) from cloud tops and evaporates, it cools the model-scale environment. Evaporation increases the water-vapor content in the environment. Rates of detrainment differ for different cloud types and cloud-top heights. Second, cumulus convection, which occurs when clouds grow vertically, induces subsidence between clouds. During subsidence, the model-scale temperature increases and relative humidity decreases. An advantage of the Arakawa-Schubert scheme is its sophisticated treatment of subgrid-scale clouds. Disadvantages of the original scheme were the omission of convective downdrafts and the assumption that all clouds started in the lowest model layer. Convective downdrafts were recently added (Cheng and Arakawa 1997). Some versions of the model now permit cloud bases to appear at any altitude (Ding and Randall 1998).

## 9.8. SUMMARY

In this chapter, cloud and fog classification, formation, development, and modeling were discussed. Clouds form by free convection, orographic uplifting, forced convection, and lifting along frontal boundaries. Fogs form by radiational cooling, advection of warm moist air over a cool surface, orographic uplifting, and evaporation of warm water into cold air. When air rises in a cloud, it expands and cools pseudoadiabatically. Condensation adds energy and buoyancy to the cloud, and entrainment of outside air causes cooling at the edges, enhancing downdrafts in the cloud. The thermodynamic energy equation for a cloud takes account of convection, latent-heat release, entrainment, and radiative effects. In a cloud, the hydrostatic approximation is not valid; thus, a vertical momentum equation with an inertial acceleration term is used instead.

## 9.9. PROBLEMS

**9.1.** Calculate the pseudoadiabatic lapse rate when $p_d = 900$ mb and $T = 273.15$ K.

**9.2.** Calculate the dew-point lapse rate in a parcel of air that rises from $p_d = 1000$ mb to $p_d = 900$ mb. Assume $\omega_v = 0.003$ kg kg$^{-1}$ does not change in the parcel, and the average $T_v = 279$ K.

## 9.10. COMPUTER PROGRAMMING PRACTICE

**9.3.** Write a computer script to calculate the pseudoadiabatic lapse rate between $z = 0$ and 10 km. Assume $T = 288$ K at the surface and decreases 6.5 K km$^{-1}$. Assume $p_d = 1013.25$ mb at the surface, and use (2.37) to estimate $p_d$ at each subsequent altitude. Use the program to estimate the pseudoadiabatic lapse rate at 100-m increments.

**9.4.** Assume that, at the surface, $T = 285$ K, $T_D = 278$ K, $p_d = 998$ mb, and $\partial T/\partial z = -11$ K km$^{-1}$. Write a computer script to estimate $p_a = p_d + p_v$ and $T$ at the lifting condensation level. [Hint: First estimate $\omega_v$ at the surface from (2.62) and assume it stays constant in the rising parcel of air. Estimate the decrease in $p_a$ with altitude from (2.35). Use the mass mixing ratio and $p_d$ to estimate $T_D$ at each altitude.]

# 10

# Radiative Energy Transfer

RADIATION in the earth's atmosphere affects temperatures, gas concentrations, visibility, colors, and biological organisms. Red sunsets, blue skies, white clouds, rainbows, green trees, and brown smog are caused by interactions of visible light with gases, particles, cloud drops, or objects. Ultraviolet, visible, and infrared radiation all play a role in the atmosphere. In this chapter, radiation laws, optical properties of gases and particles, light processes, and the radiative transfer equation are discussed. The radiative transfer equation is used to determine rates of heating, cooling, and molecular photolysis. Important radiative laws discussed here include Planck's law, Wien's law, and the Stefan–Boltzmann law. Equations quantifying the level of attenuation and redirection of radiation by gases, particles, and cloud drops are also described.

## 10.1. ENERGY TRANSFER PROCESSES

**Radiation** is the emission or propagation of energy in the form of a photon, or electromagnetic wave. A **photon** is a particle or quantum of electromagnetic energy that has no mass, no electric charge, and an indefinite lifetime. An **electromagnetic wave** is a disturbance traveling through a medium, such as air or space, that transfers energy from one object to another without permanently displacing the medium itself. Electromagnetic waves may be considered as dual transverse waves in that they consist of an electric wave and a magnetic wave in phase with and at right angles to each other and to the direction of propagation. Radiation is emitted by all bodies in the universe that have a temperature above absolute zero (0 K). Once emitted, radiation passes through space or air to another body. Upon reaching the second body, the radiation can be reflected, scattered, absorbed, refracted, dispersed, or transmitted. Each of these processes is discussed in this chapter.

When a body, such as air, emits more radiation than it absorbs, its temperature decreases. When a body absorbs more radiation than it emits, its temperature increases. Other processes that affect air temperatures include advection, forced convection, turbulence, and latent-heat exchange. The thermodynamic energy equation in (3.76) takes these processes into account.

Figure 10.1 shows the relative importance of several energy transfer processes on the atmospheric energy budget. Solar radiation drives atmospheric, cloud, and surface heating. When the earth's surface heats up, it emits infrared radiation, much of which is absorbed by greenhouse gases and clouds and some of which escapes

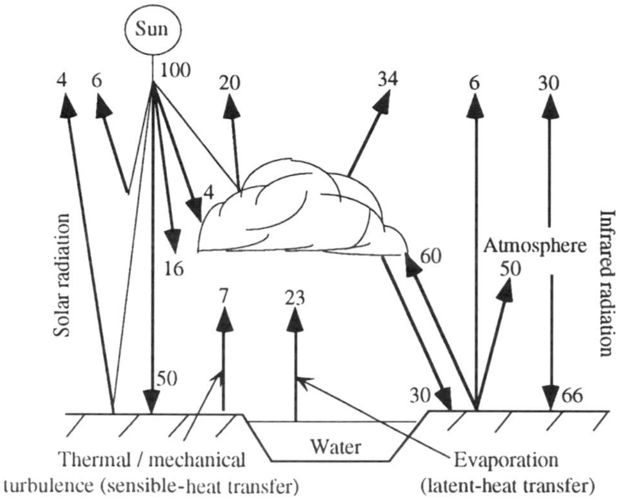


**Figure 10.1.** Energy balance for earth–atmosphere system. Values are unitless relative quantities pf energy. The sum of sources minus sinks for clouds, the atmosphere, or the earth equals zero. For example, clouds are in radiative balance, since they absorb and emit 64 units of radiation.

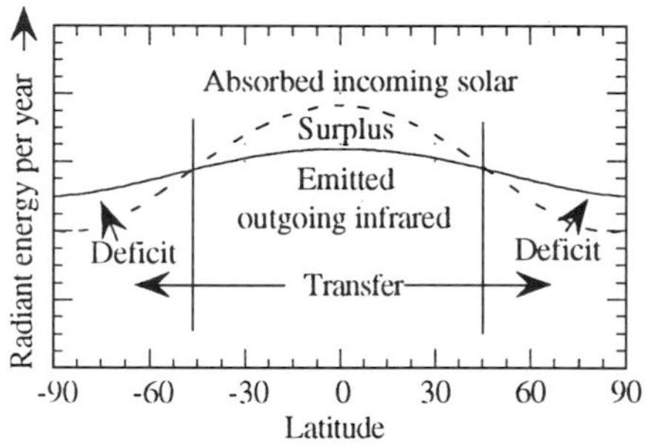


**Figure 10.2.** Schematic showing that a globally and yearly averaged energy surplus at the equator and deficit at the poles is compensated for by energy transfer from the equator towards the poles.

to space. After absorbing, the atmosphere and clouds reemit infrared radiation in all directions. Turbulence transfers surface energy to the troposphere. Evaporation transfers latent heat to the troposphere, where it may be released by condensation.

The **transport of latent heat** is an important process by which solar energy absorbed near the equator is transferred poleward. Figure 10.2 shows that, at the equator, on average, the surface absorbs more solar radiation than it emits infrared radiation, causing an energy surplus. At the poles, the reverse is true, and an energy deficit occurs. In the absence of energy transfer between the equator and poles, temperatures near the poles would continuously decrease, and those near the equator would continuously increase. On earth, energy is continuously supplied from the equator to the poles by three processes. One is poleward transport of energy by advection, the second is poleward transport of energy by ocean currents, and the third is the poleward advection of latent heat. When liquid water evaporates near the equator, latent heat is stored in water vapor. As the vapor advects poleward, it often condenses, releasing latent heat.

## 10.2. ELECTROMAGNETIC SPECTRUM

Whether radiation is considered an electromagnetic wave or photon, it travels at the speed of light. If radiation propagates as a wave, its **wavelength** is

$$\lambda = \frac{c}{\nu} = \frac{1}{\tilde{\nu}} \tag{10.1}$$

where $\nu$ is the wave's frequency of oscillation ($s^{-1}$), $c$ is the speed of light (e.g., $m\ s^{-1}$) and $\tilde{\nu}$ is the wavenumber (e.g., $m^{-1}$), which is the number of wavelengths per unit distance. Units for wavelength are micrometers ($1\ \mu m = 10^{-6}$ m), nanometers ($1\ nm = 10^{-9}$ m), centimeters (cm), or meters (m). In a vacuum, the speed of light is $c = 2.9979 \times 10^{8}\ m\ s^{-1}$.

In 1900, Max Planck theorized that the total radiative energy (J) emitted or absorbed by a substance was

$$E_t = nh\nu = n\frac{hc}{\lambda} \tag{10.2}$$

where n is an integer, called a **quantum number**, and $h = 6.6256 \times 10^{-34}$ J s is **Planck's constant.** (10.2) states that substances do not emit or absorb radiative energy continuously, but in packets or quanta. Although Planck believed that light was emitted or absorbed discontinuously, he held that it traveled through space as an electromagnetic wave. In 1905, Albert Einstein postulated that light energy traveled through space in concentrated bundles, and the energy of each bundle was

$$E_p = h\nu = \frac{hc}{\lambda} \tag{10.3}$$

In 1926, Gilbert Lewis termed these bundles **photons.** (10.3) represents the energy emitted or absorbed per photon (J $photon^{-1}$). The quantity of energy emitted or absorbed depends on the wavelength (or frequency) of radiation. The greater the frequency, the shorter the wavelength, and the greater the energy of a quantum. The electromagnetic and photon theories of light are interrelated by (10.2). In 1918, Planck won a Nobel prize in physics for his discovery of quanta. In 1921, Einstein won the Nobel prize in physics for his work on the particle theory of light.

**Example 10.1**

Find the energy emitted per photon, the frequency, and the wavenumber of a $\lambda = 0.5$-$\mu$m and $\lambda = 10.0$-$\mu$m wavelength of energy.

SOLUTION

When $\lambda = 0.5$ $\mu$m, $E = 3.97 \times 10^{-19}$ J photon$^{-1}$, $\nu = 5.996 \times 10^{14}$ s$^{-1}$, and $\tilde{\nu} = 2$ $\mu$m$^{-1}$. When $\lambda = 10.0$ $\mu$m, $E = 1.98 \times 10^{-20}$ J photon$^{-1}$, $\nu = 2.998 \times 10^{13}$ s$^{-1}$, and $\tilde{\nu} = 0.1$ $\mu$m$^{-1}$. Shorter wavelengths generate more energy per photon than do longer wavelengths.

From (10.2), Planck derived an equation relating the intensity of radiant emissions from a perfectly emitting substance to the absolute temperature of the substance and the wavelength of emissions. A perfectly emitting substance is one that, in thermodynamic equilibrium, emits all radiation that it absorbs. If its emissions were less than its absorption, a body would be out of equilibrium, and its temperature would increase.

**Absorption** occurs when electromagnetic energy enters a substance and is converted to internal energy. A **blackbody** is a substance that absorbs all radiation incident upon it. No incident radiation is reflected by a blackbody. No substances are true blackbodies. Some that are close include elemental carbon, platinum black, and black gold (e.g., Siegel and Howell 1992). (The term blackbody was coined because good absorbers of visible radiation generally appear black. But good absorbers of infrared radiation are not necessarily black. One such absorber is white oil-based paint.)

Blackbodies are useful to study because they not only absorb, but also emit the maximum possible intensity of radiant energy at a given wavelength and temperature. Planck determined this intensity as

$$B_{\lambda,T} = \frac{2hc^2}{\lambda^5\left[\exp\left(\frac{hc}{\lambda k_B T}\right) - 1\right]} \tag{10.4}$$

now called **Planck's law,** where $B_{\lambda,T}$ is **radiant intensity** or **radiance** (W m$^{-2}$ $\mu$m$^{-1}$ sr$^{-1}$), and $k_B$ is Boltzmann's constant ($1.38 \times 10^{-23}$ J K$^{-1}$ = W s K$^{-1}$). Radiance is the energy emitted per unit area per unit time per unit wavelength per incremental solid angle (units of steradians). Since the radiance in (10.4) is defined for individual wavelengths, it is called a **spectral radiance.** Incremental solid angle is defined shortly.

(10.4) applies to any blackbody with a temperature above absolute zero (0 K). Real substances are not perfect emitters. Instead, they emit a fraction of the radiance that a blackbody emits. The radiance (W m$^{-2}$ $\mu$m$^{-1}$ sr$^{-1}$) actually emitted by any

**Table 10.1.** Emissivities of Different Surface Types for a Typical Infrared Wavelength

| Surface Type | Emissivity (Fraction) | Surface Type | Emissivity (Fraction) |
|---|---|---|---|
| Liquid water | 1.0[a] | Soil | 0.9–0.98[b] |
| Fresh snow | 0.99[b] | Grass | 0.9–0.95[b] |
| Old snow | 0.82[b] | Desert | 0.84–0.91[b] |
| Liquid water clouds | 0.25–1.0[c] | Forest | 0.95–0.97[b] |
| Cirrus clouds | 0.1–0.9[c] | Concrete | 0.71–0.9[b] |
| Ice | 0.96[d] | Urban | 0.85–0.87[a] |

[a]Seaman et al. (1989). [b]Oke (1978). [c]Liou (1992). [d]Sellers (1965).

substance is approximately

$$e_\lambda = \varepsilon_\lambda B_{\lambda,T} \tag{10.5}$$

where $\varepsilon_\lambda$ (unitless) is the **emissivity** of the substance. Emissivity is the fraction ($\leq 1$) of $B_{\lambda,T}$ actually emitted. Emissivities depend on wavelength. Table 10.1 gives emissivities for some surface types in the infrared part of the electromagnetic spectrum.

**Absorptivity** ($a_\lambda$) is the fraction ($\leq 1$) of incident radiation that a substance actually absorbs. **Kirchoff's law** states that, in thermodynamic equilibrium, the emissivity and absorptivity of a substance equal each other ($a_\lambda = \varepsilon_\lambda$ ). Thus, the efficiency at which a substance absorbs radiation equals that at which it emits radiation, and a perfect emitter of radiation ($\varepsilon_\lambda = 1$) is also a perfect absorber of radiation ($a_\lambda = 1$) and a blackbody. Suppose a blackbody object is placed in a vacuum enclosed by blackbody walls. Over time, the temperatures of the object and walls equalize. If Kirchoff's law did not hold at this point, a net heat transfer would occur between the object and the walls even though the object and wall temperatures were the same. Such a heat transfer violates the second law of thermodynamics. Thus, the absorptivity and emissivity of a substance must equal each other in equilibrium.

An **incremental solid angle,** $d\Omega_a$, used in the definition of radiance, is an incremental surface area on a unit sphere. A unit sphere is a sphere with radius normalized to unity. The equation for incremental solid angle is

$$d\Omega_a = \frac{dA_s}{r_s^2} \tag{10.6}$$

where $dA_s$ is an incremental surface area, and $r_s$ is the radius of a true sphere. Incremental solid angle has units of steradians (sr), which is analogous to units of radian for a circle.

The incremental surface area in (10.6) can be found from Fig. 10.3. In the figure, a line is drawn from the center of the sphere to the center of an incremental area $dA_s$, which is a distance $r_s$. from the sphere's center. The line is directed at a **zenith angle** $\theta$ from the surface normal (where the surface is on the $x$–$y$ plane). The line is also located at an **azimuth angle** $\phi$, directed counterclockwise from the positive

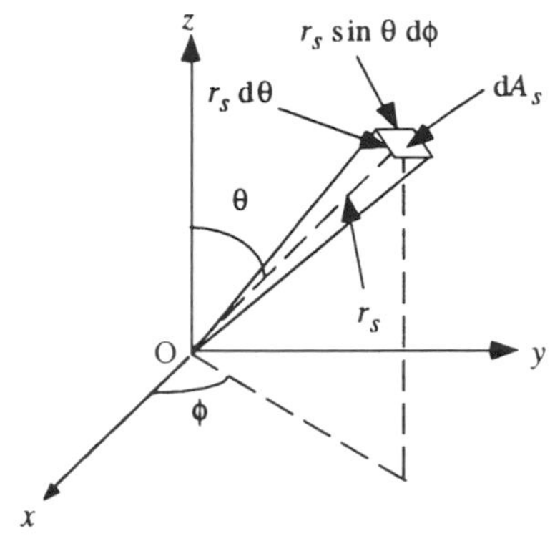


**Figure 10.3.** Radiance, emitted from point (O) on a horizontal plane, passes through an incremental area $dA_s$ at a distance $r_s$ from the point of emissions. The angle between the *z*-axis and the angle of emissions is the zenith angle (θ), and the horizontal angle between a reference axis (*x*-axis) and the line of emissions is the azimuth angle (ϕ). The size of the incremental surface area is exaggerated.

*x*-axis to a horizontal line dropped from the line. From the geometry shown, the incremental surface area is

$$dA_s = (r_s\, d\theta)(r_s \sin\theta\, d\phi) = r_s^2 \sin\theta\, d\theta\, d\phi \tag{10.7}$$

where $d\theta$ and $d\phi$ are incremental zenith and azimuth angles, respectively. Substituting (10.7) into (10.6) gives the **incremental solid angle** as

$$d\Omega_a = \sin\theta\, d\theta\, d\phi \tag{10.8}$$

Integrating $d\Omega_a$ over all possible solid angles around the center of a sphere gives the solid angle around a sphere as

$$\Omega_a = \int_{\Omega_a} d\Omega_a = \int_0^{2\pi}\int_0^{\pi} \sin\theta\, d\theta\, d\phi = 4\pi \text{ steradians} \tag{10.9}$$

The solid angle around the center of the base of a hemisphere is $2\pi$ steradians.

Planck's law gives the spectral radiance emitted by a blackbody, $B_{\lambda,T}$, in W m$^{-2}$ μm$^{-1}$ sr$^{-1}$. In terms of Fig. 10.3, imagine that the *x*–*y* plane is the surface of an object. The radiance emitted from point O can travel in any direction above the *x*–*y* plane. If it travels through area $dA_s$, it passes through an incremental solid angle, $d\Omega_a$, which is through a cone originating at point O. $B_{\lambda,T}$ is the spectral radiance emitted by a blackbody at its surface. Radiance changes as it travels through space. $I_\lambda$ is defined here as the spectral radiance at a given point in space, regardless of the source of the radiance.

In atmospheric models, two quantities related to spectral radiance are commonly calculated. These are spectral actinic flux and spectral irradiance (e.g., Madronich 1987). **Spectral actinic flux** (also called spherical intensity or actinic irradiance) is the integral of spectral radiance over all solid angles of a sphere. It is used for calculating photolysis coefficients of gases in the atmosphere. Since gas molecules can absorb radiation, regardless of the direction the radiation originates from, radiance is integrated over a sphere to obtain actinic flux. Spectral actinic flux can be calculated with a **photoelectric detector** (e.g., de Arellana et al. 1994) or calculated numerically in a model.

Spectral actinic flux is determined by integrating incremental actinic flux over all solid angles around the center of a sphere. The **incremental actinic flux** is

$$\mathrm{d}E_\lambda = I_\lambda \,\mathrm{d}\Omega_a \tag{10.10}$$

where $E_\lambda$ is in units of W m$^{-2}$ μm$^{-1}$ but is typically converted to units of photons cm$^{-2}$ s$^{-1}$ μm$^{-1}$ for photolysis calculations (Section 10.8.4). The integral of incremental actinic flux over a sphere is

$$E_\lambda = \int_{\Omega_a} \mathrm{d}E_\lambda = \int_{\Omega_a} I_\lambda \,\mathrm{d}\Omega_a = \int_0^{2\pi}\int_0^{\pi} I_\lambda \sin\theta \,\mathrm{d}\theta \,\mathrm{d}\phi \tag{10.11}$$

Radiance usually propagates with equal intensity in all directions; thus, $I_\lambda$ is independent of direction. In such cases, the radiance is called **isotropic**, and (10.11) simplifies to

$$E_\lambda = I_\lambda \int_0^{2\pi}\int_0^{\pi} \sin\theta \,\mathrm{d}\theta \,\mathrm{d}\phi = 4\pi I_\lambda \tag{10.12}$$

(10.12) states that the isotropic spectral actinic flux equals $4\pi$ multiplied by the spectral radiance, where the units of $4\pi$ are steradians.

**Spectral irradiance** (also called the net flux or energy flux) is the radiant energy propagating from all directions across a flat surface per unit surface area, time, and wavelength. In terms of Fig. 10.3, it is the vertical component of radiance, integrated over the hemisphere above the $x$–$y$ plane. For a zenith angle of $\theta = 0°$, the irradiance impinging on the $x$–$y$ plane is maximum. For a zenith angle of $\theta = 90°$, the irradiance impinging on the $x$–$y$ plane is zero. Irradiance is often measured with a flat-plate radiometer, a spectral pyranometer, or a spectroradiometer.

Spectral irradiance is calculated by integrating the component of radiance normal to the $x$–$y$ plane over all solid angles of the hemisphere above the $x$–$y$ plane. If $I_\lambda$ is the spectral radiance passing through point O in Fig. 10.3, $I_\lambda \cos\theta$ is the component of radiance normal to the $x$–$y$ plane. Multiplying this quantity by incremental solid angle gives the **incremental spectral irradiance** normal to the $x$–$y$ plane as

$$\mathrm{d}F_\lambda = I_\lambda \cos\theta \,\mathrm{d}\Omega_a \tag{10.13}$$

where $F_\lambda$ is in W m$^{-2}$ μm$^{-1}$.

Spectral irradiance is found by integrating (10.13) over the hemisphere above the $x$–$y$ plane in Fig. 10.3. Thus

$$F_\lambda = \int_{\Omega_a} dF_\lambda = \int_{\Omega_a} I_\lambda \cos\theta \, d\Omega_a = \int_0^{2\pi}\int_0^{\pi/2} I_\lambda \cos\theta \sin\theta \, d\theta \, d\phi \qquad (10.14)$$

For isotropic emissions, (10.14) simplifies to

$$F_\lambda = I_\lambda \int_0^{2\pi}\int_0^{\pi/2} \cos\theta \sin\theta \, d\theta \, d\phi = \pi I_\lambda \qquad (10.15)$$

Thus, the isotropic spectral irradiance is $\pi$ multiplied by the isotropic radiance where units of $\pi$ are steradians. At the surface of a blackbody, the **spectral irradiant emissions** are

$$F_\lambda = \pi I_\lambda = \pi B_{\lambda,T} \qquad (10.16)$$

Figure 10.4 shows irradiant emissions from a blackbody as a function of wavelength and temperature, obtained from (10.16) and (10.4). The figure shows that hotter substances emit much more energy than do cooler substances.

In the earth's atmosphere, the sources of irradiant energy are the sun, the surface of the earth, and the atmosphere itself. The center of the sun has a temperature of about $1.5 \times 10^6$ K and emits primarily X rays, as shown in Fig. 10.4. Most of these waves are absorbed during their random walk towards the outside of the sun. Solar radiation that reaches earth originates mainly from the visible surface of the sun, called the **photosphere**, which has an effective temperature of near 6000 K (closer to 5800 K). This temperature results in the **solar spectrum** of radiation shown in Fig. 10.5. The solar spectrum falls into the ultraviolet, visible, and infrared wavelength regions. The **ultraviolet region** consists of wavelengths shorter than

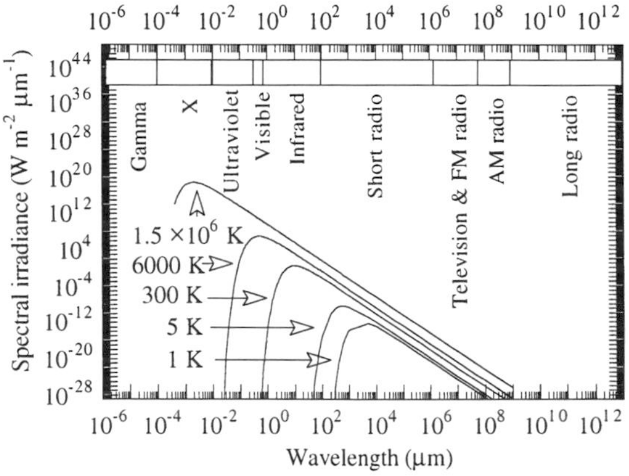


**Figure 10.4.** Spectral irradiance as a function of wavelength and temperature resulting from emissions by a blackbody. Data from (10.16) and (10.4).

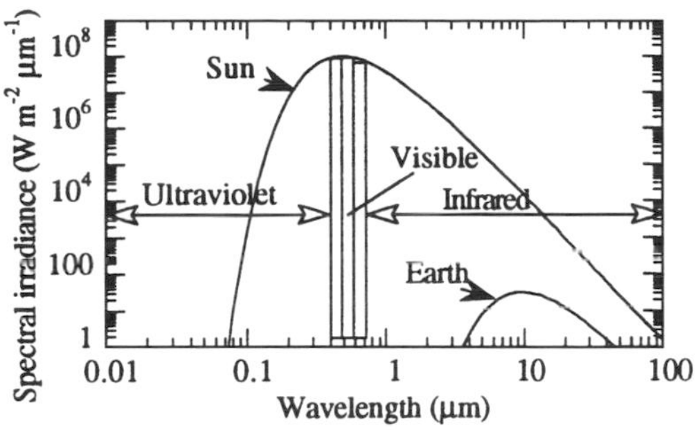


**Figure 10.5.** Irradiance emissions versus wavelength for the sun and the earth when both are considered perfect emitters. Data were obtained from (10.16) and (10.4).

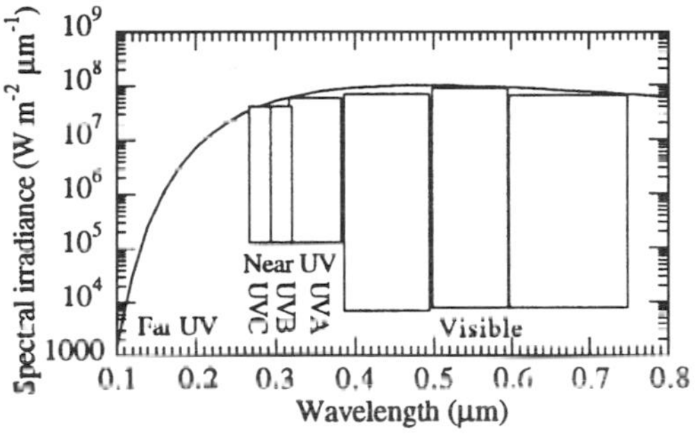


**Figure 10.6.** Ultraviolet and visible portions of the solar spectrum. Data from (10.16) and (10.4).

about 0.38 µm, the **infrared region** consists of wavelengths longer than about 0.75 µm, and the **visible region** consists of wavelengths between 0.38 and 0.75 µm. The infrared spectrum is divided into the **near-infrared** (0.75–4 µm) and the **far-infrared** (>4 µm) spectra. Most infrared radiation from the sun that reaches the earth is near-infrared radiation.

Figure 10.5 shows the irradiant emissions spectrum of the earth. The effective temperature at the surface of earth is approximately 288 K; thus, almost all emissions from the earth are in the far-infrared part of the electromagnetic spectrum. Even in the stratosphere, where temperatures can drop below 200 K, and in hot deserts, where temperatures exceed 315 K, emissions are in the far infrared. Thus, with respect to studies of the earth's atmosphere, incoming wavelengths of interest are ultraviolet, visible, and near infrared, and outgoing wavelengths of interest are far infrared.

Figure 10.6 shows the ultraviolet and visible portions of the solar spectrum. The ultraviolet (UV) is broken into **near UV** and **far UV**. Near UV consists of 0.25–0.38-µm wavelengths, and far UV consists of 0.01–0.25-µm wavelengths. Near UV is further divided into **UVA** (0.32–0.38 µm), **UVB** (0.29–0.32 µm), and **UVC** (0.25–0.29 μm) radiation. All solar radiation below 0.28 μm is absorbed by earth's atmosphere above the troposphere. Nitrogen gas ($N_2$) absorbs wavelengths less than 0.1 μm in the thermosphere, and oxygen absorbs wavelengths less than 0.245 μm in the thermosphere, mesosphere, and stratosphere. Ozone in the stratosphere and troposphere absorbs significantly in the 0.17–0.35-μm wavelength region. Some UVA, UVB, and UVC radiation longer than 0.28 μm, passes through the stratosphere and troposphere to the ground.

UVA radiation (tanning radiation) is relatively harmless to most humans. UVB radiation can cause sunburn. UVC radiation can cause severe damage to many forms of life when received in high dosages. Since ozone in the stratosphere absorbs most UVB and UVC radiation, ozone reduction in the stratosphere would allow more UVB and UVC radiation to reach the surface, causing severe damage to exposed life forms.

The visible spectrum contains the colors of the rainbow. For convenience, visible light is divided into blue (0.38–0.5 μm), green (0.5–0.6 μm), and red (0.6–0.75 μm).

The peak radiation wavelength emitted at a given temperature can be found from **Wien's displacement law** (Wien's law). This law is derived by differentiating (10.4) with respect to wavelength at a constant $T$ and setting the derivative equal to zero. The result is

$$\lambda_p(\mu\mathrm{m}) \approx \frac{2897}{T(K)} \qquad (10.17)$$

where $\lambda_p$ is the peak wavelength of emissions from a blackbody. The law states that the hotter a body, the shorter the peak wavelength of radiation emitted. Figure 10.7 shows how peak wavelengths of radiation at different temperatures can be connected by a line on a log–log scale, where the line is described by (10.17).

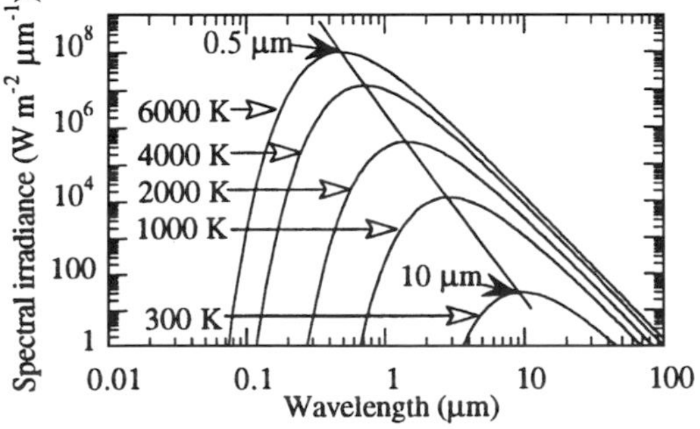


**Figure 10.7.** Radiation spectrum as a function of temperature and wavelength. Data from (10.16) and (10.4). The line through the peaks was obtained from Wien's law.

**Example 10.2**

In the sun's photosphere, the peak wavelength of emissions is near $\lambda_p = 2897/5800 = 0.5\ \mu\text{m}$, which is in the visible part of the solar spectrum. At the earth's surface, the peak wavelength of emissions is $\lambda_p = 2897/288 = 10.1$ μm, which is in the infrared part of the spectrum. Both peak wavelengths are shown in Fig. 10.7.

Integrating the spectral irradiance in (10.16) over all wavelengths gives the **Stefan–Boltzmann law,**

$$F_b = \pi \int_0^\infty B_{\lambda,T}\,d\lambda = \sigma_B T^4 \tag{10.18}$$

where $F_b$ is total irradiant energy emitted by a blackbody at a given temperature and $\sigma_B = 2k_B^4\pi^4/15h^3c^2 = 5.67 \times 10^{-8}$ W m$^{-2}$ K$^{-4}$ is the Stefan–Boltzmann constant. (10.18) yields the area under the curves shown in Figs. 10.4–10.7.

**Example 10.3**

For an effective temperature of $T = 5800$ K, the irradiance emitted by the photosphere is $F_b = 64\times10^6$ W m$^{-2}$. For an effective temperature of $T = 288$ K, the irradiance emitted by the earth's surface is $F_b = 390$ W m$^{-2}$.

## 10.3. LIGHT PROCESSES

Absorption is one process that affects electromagnetic radiation in the earth's atmosphere. Other processes include reflection, refraction, scattering, and diffraction. **Reflection** occurs when a wave or photon of radiation is absorbed by an object and reemitted with an angle of reflection equal to the angle of incidence. No energy is lost during absorption and reemission. Figure 10.8 shows an example of reflection.

**Albedo** (or **reflectivity**) is the fraction of incident sunlight reflected at a given wavelength. Table 10.2 gives mean albedos in the visible spectrum for several surface types. The table shows that the albedo of the earth and atmosphere together (**planetary albedo**) is about 30 percent. Two-thirds of earth's surface is covered with water, which has an albedo of 5–20 percent (typical value of 8 percent), depending largely on the angle of the sun. Soils and forests also have low albedos. Much of the earth-atmosphere reflectivity is due to clouds and ice, which have high albedos.

The albedos of grassland and other non-snow surfaces in the UVB sprectrum are much less than those in the visible spectrum. The albedo of snow surfaces in

**Table 10.2.** Albedos in the Visible Spectrum for Several Surface Types

| Surface Type | Albedo (Fraction) | Surface Type | Albedo (Fraction) |
|---|---|---|---|
| Earth and atmosphere | 0.3[a] | Soil | 0.05–0.2[b] |
| Liquid water | 0.05–0.2[b] | Grass | 0.16–0.26[c] |
| Fresh snow | 0.75–0.95[d] | Desert | 0.20–0.40[b] |
| Old snow | 0.4–0.7[d] | Forest | 0.10–0.25[b] |
| Thick clouds | 0.3–0.9[b] | Asphalt | 0.05–0.2[c] |
| Thin clouds | 0.2–0.7[b] | Concrete | 0.1–0.35[c] |
| Sea ice | 0.25–0.4[b] | Urban | 0.1–0.27[c] |

[a]Liou (1992). [b]Hartmann (1994). [c]Oke (1978). [d]Sellers (1965).

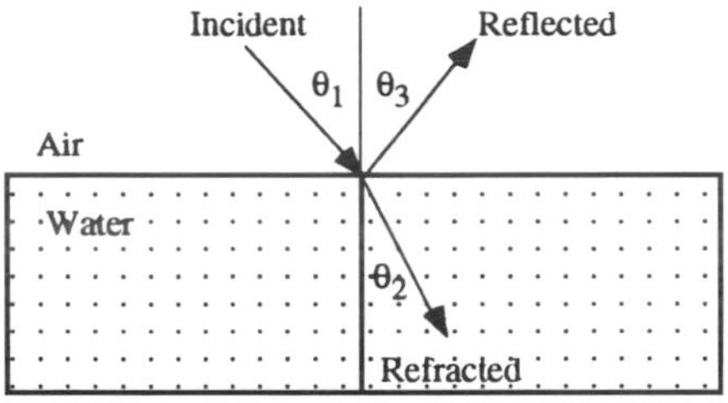


**Figure 10.8.** Examples of reflection and refraction. During reflection, $\theta_1 = \theta_3$. During refraction, the angles of incidence and refraction are related by Snell's law.

the UVB spectrum slightly exceeds that in the visible spectrum (Blumthaler and Ambach 1988). Harvey et al. (1977) found the UVA-UVB albedo to be 3.0 percent for water and 0.8 percent for grassland.

**Refraction** occurs when a wave or photon travels through a medium of one density and enters a medium of another density. In such a case, the speed of the wave changes, changing the angle of the incident wave relative to a surface normal, as shown in Fig. 10.8. If a wave travels from a medium of one density to a medium of a higher density, it bends (refracts) towards the surface normal. The angle of refraction is related to the angle of incidence by **Snell's law**,

$$\frac{n_2}{n_1} = \frac{\sin\theta_1}{\sin\theta_2} \tag{10.19}$$

In this equation, $n$ is the real index of refraction (unitless), $\theta$ is the angle of incidence or refraction, and subscripts 1 and 2 refer to incident and refracted light, respectively. The **real index of refraction** is the ratio of the speed of light in a vacuum ($c$) to that in a different medium ($c_1$). Thus,

$$n_1 = c/c_1 \tag{10.20}$$

Since light cannot travel faster than its speed in a vacuum, the real index of refraction of a medium other than a vacuum must exceed unity. The index of refraction is

**Table 10.3.** Real Indices of Refraction of Air and Liquid Water versus Wavelength

| Wavelength (μm) | $n_{air}$ | $n_{water}$ | Wavelength (μm) | $n_{air}$ | $n_{water}$ |
|---|---|---|---|---|---|
| 0.2 | 1.000324 | 1.396 | 1.0 | 1.000274 | 1.327 |
| 0.3 | 1.000292 | 1.349 | 4.0 | 1.000273 | 1.351 |
| 0.4 | 1.000283 | 1.339 | 7.0 | 1.000273 | 1.317 |
| 0.5 | 1.000279 | 1.335 | 10.0 | 1.000273 | 1.218 |
| 0.6 | 1.000277 | 1.332 | 20.0 | 1.000273 | 1.480 |
| 0.7 | 1.000276 | 1.331 | | | |

Data for air were obtained from (10.21), and data for liquid water were obtained from Hale and Querry (1973).

wavelength-dependent. The real index of refraction of air has been approximated empirically as a function of wavelength with

$$n_{a,\lambda} - 1 = 10^{-8}\left(8342.13 + \frac{2,406,030}{130 - \lambda^{-2}} + \frac{15,997}{38.9 - \lambda^{-2}}\right) \quad (10.21)$$

(Edlen 1966) where λ is in micrometers. Real indices of refraction of air and liquid water are given in Table 10.3 for several wavelengths.

**Example 10.4**

Find the angle of refraction in water if light of wavelength $\lambda = 0.5$ μm enters water from air at an incident angle of $\theta_1$ is 45°. Also find the speed of light in air and water at this wavelength.

SOLUTION

At $\lambda = 0.5$ μm, $n_{air} = 1.000279$ and $n_{water} = 1.335$ from Table 10.3. Thus, from (10.19), the angle of refraction in water is $\theta_2 = 32°$. From (10.20), the speeds of light in air and water at $\lambda = 0.5$ μm are $2.9971 \times 10^8$ m s$^{-1}$ and $2.2456 \times 10^8$ m s$^{-1}$, respectively.

Refraction plays a role in several optical phenomena. **Rainbows** appear when incident light is refracted, dispersed, reflected, then refracted again, as shown in Fig. 10.9. **Dispersion** is the separation of white light into colors. During dispersion, the shortest wavelengths of visible light (blue light) are bent the most. As a beam of visible light (mixture of many wavelengths) enters a raindrop, all wavelengths bend towards the surface normal due to refraction. Blue light bends the most due to dispersion. When the light hits the back of the drop, much of it reflects. If reflected light reaches another surface of the drop, it may leave the drop and refract away from the surface normal. The angle of the blue and red wavelengths that reach a viewer's eye are 40° and 42°, respectively, in relation to the incident beam. Only

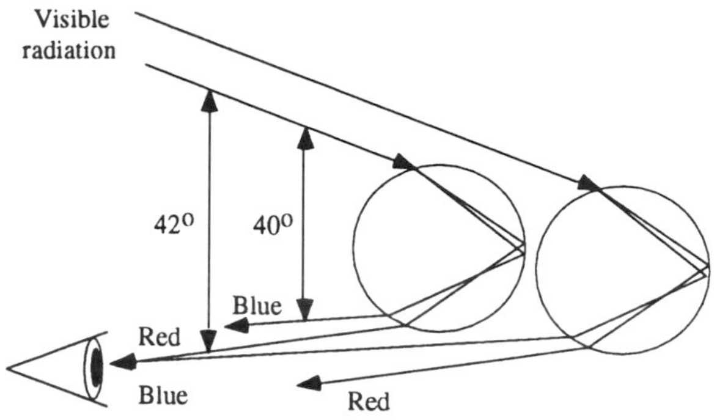


**Figure 10.9.** Geometry of a primary rainbow.

one wavelength from each raindrop impinges upon a viewer's eye. Thus, a rainbow appears when waves from many raindrops hit the eye. The geometry of a rainbow is such that the sun must be at the viewer's back for the viewer to see a rainbow. As shown in Fig. 10.9, blue (or violet) appears on the bottom of a primary rainbow. A **secondary rainbow** can occur if a second reflection occurs inside each drop. Blue light appears on top of red light in the secondary rainbow.

**Diffraction** is a process by which the direction of propagation of a wave changes when the wave encounters an obstruction. In terms of visible wavelengths, it is the bending of light as it passes an obstruction. In the atmosphere, a wavelength in any part of the electromagnetic spectrum can diffract as it passes close enough to the surface of an aerosol, cloud drop, or raindrop.

Diffraction can be explained in terms of **Huygens' principle**, which states that each point of an advancing wavefront may be considered the source of a new series of secondary waves. A **wavefront** is a surface of constant phase in a wave's motion. If a stone is dropped in a tank of water, waves move out horizontally in all directions, and wavefronts are seen as concentric circles around the stone. If a point source emits waves in three dimensions, wavefronts are concentric spherical surfaces. When a wavefront encounters an obstacle, such as in Fig. 10.10, waves appear to bend (diffract) around the obstacle because a series of secondary concentric waves is emitted at the surface of the obstacle (and at other points along the primary wavefronts). New waves do not appear in the backward direction, because the intensity of the secondary wavelet depends on angle and is zero in the backward direction, as Gustav Kirchoff (1824–1887) demonstrated.

When some waves bend more than others, constructive or destructive interference may occur. **Constructive interference** results when the crests or troughs of two waves meet. In the visible spectrum, this results in a band of bright light. **Destructive interference** results when a crest meets a trough. In the visible spectrum, this results in a band of darkness. Constructive and destructive interference cause bands of visible light to appear past an obstruction. In the atmosphere, a **corona** appears around the sun or moon when thin clouds containing spherical liquid water drops are present. The corona, caused by diffraction of sunlight or moonlight as it passes by the surfaces of cloud drops, appears as a set of prismatic colored rings

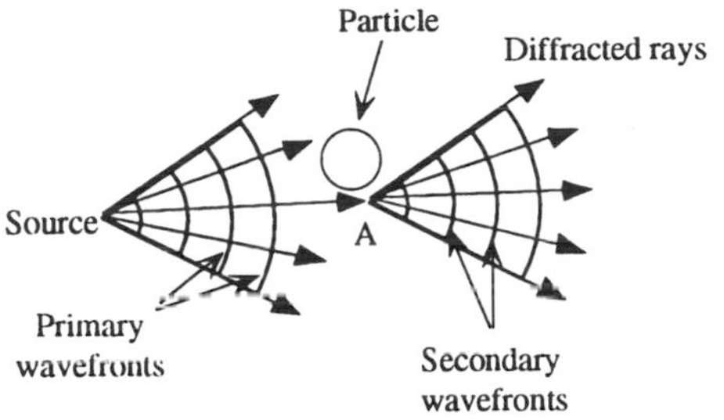


**Figure 10.10.** Diffraction around a particle. Any point along a wavefront may be taken as the source of a new series of secondary waves. Rays emitted from point A appear to cause waves from the original source to bend around the particle.

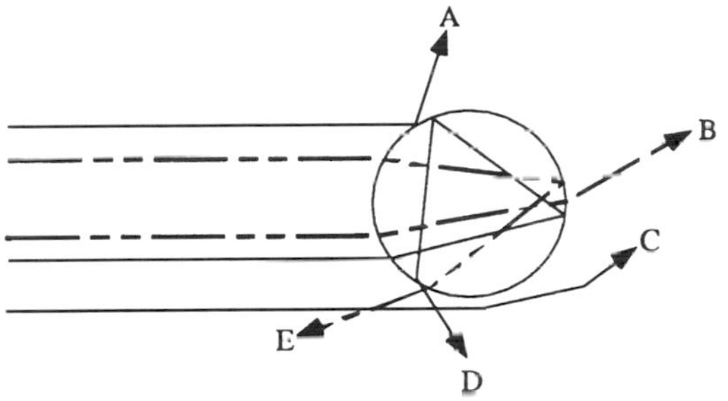


**Figure 10.11.** Radiative scattering by a sphere. Ray A is reflected; B is refracted twice; C is diffracted; D is refracted, reflected twice, then refracted; and E is refracted, reflected once, and refracted. Rays A, B, C, and D scatter in the forward or sideward direction. E scatters in the backward direction.

around the sun or moon. Sometimes, the corona around the moon appears white with alternating bands of light and dark.

When an electromagnetic wave enters a spherical particle, such as a cloud drop, it can reflect off the particle, diffract around the surface of the particle, or enter the particle then refract. When a wave refracts in the particle, it can refract out of the particle, reflect internally and then refract out, or reflect twice internally and then refract out. **Particle scattering** is the sum of the five processes discussed above, each of which is shown in Fig. 10.11. The processes that affect particle scattering the most are diffraction and double refraction, identified by rays C and B, respectively. Thus, particles scatter light primarily in the forward direction. They also scatter some light to the side and in the backward direction. **Backscattered** light results primarily from a single internal reflection (ray E in Fig. 10.11). Figure 10.12 shows the primary processes by which a cloud drop might scatter radiation in the forward and backward directions.

Whether one or more internal reflections occurs within a particle depends on the angle at which a wave within the particle strikes a surface and on the indices

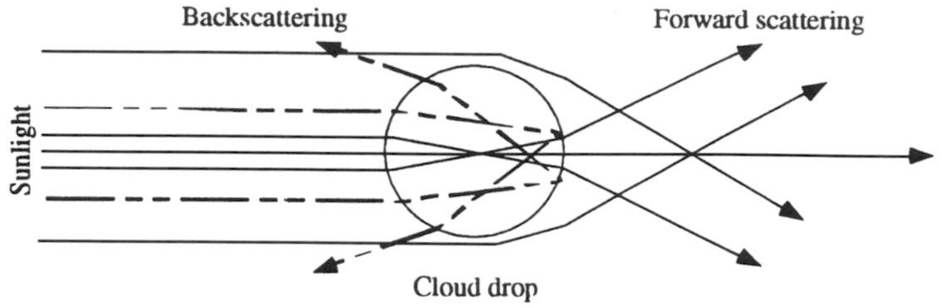


**Figure 10.12.** Forward scattering and backscattering by a cloud drop. Forward scattering is due primarily to diffraction and double refraction. Backscattering is due primarily to a single internal reflection.

of refraction of the particle and air. Suppose light travels from water to air instead of from air to water in Fig. 10.8. In the figure, **total internal reflection** within the water occurs only if $\theta_1$ equals or exceeds 90°. From (10.19), $\theta_1 = 90°$ when

$$\theta_{2,c} = \sin^{-1}\left(\frac{n_1}{n_2}\sin 90°\right) \tag{10.22}$$

which is the **critical angle.** When $\theta_2 < \theta_{2,c}$, some light reflects internally, but most refracts out of the water to the air. When $\theta_2 \geq \theta_{2,c}$, total internal reflection occurs within the water. Total internal reflection can occur only when light originates from the medium of lower index of refraction.

**Example 10.5**

Find the critical angle of light at $\lambda = 0.5$ μm in a liquid water drop suspended in air.

SOLUTION

From Table 10.3, the indices of refraction of water and air are 1.335 and 1.000279, respectively. Substituting these values into (10.22) gives $\theta_{2,c} = 48.53°$. Thus, total internal reflection within the drop occurs when $\theta_2 \geq 48.53°$.

Whereas raindrops, cloud drops, and aerosols scatter mostly in the forward direction, gases scatter equally in the forward and backward directions. Visible radiation does not diffract readily around gas molecules, because visible electromagnetic wavelengths are significantly larger than the diameters of gas molecules.

Cloud drops and large particles often scatter all wavelengths of visible light, but gas molecules selectively scatter the shortest (blue) wavelengths of such light. When a gas molecule intercepts a blue-light wavelength, the molecule is equally likely to scatter the wave in the forward direction as in the backward direction. Because gas molecules are close together, the scattered light is quickly intercepted by another

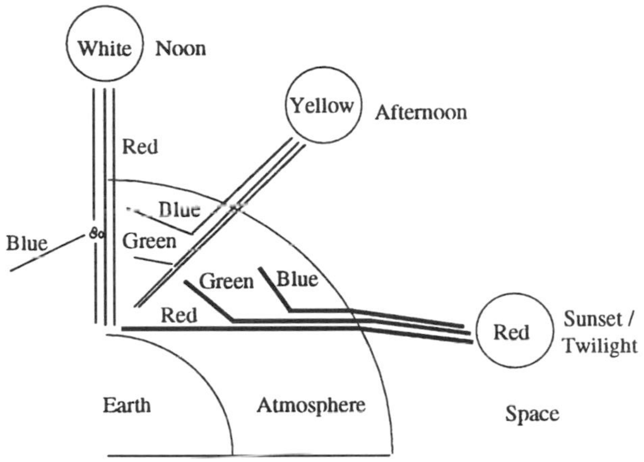


**Figure 10.13.** Colors of the sun. At noon, the sun appears white because red, green, and some blue light transmit to a viewer's eye. In the afternoon, sunlight traverses a longer path in the atmosphere removing more blue. At sunset, most green is removed from the line of sight, leaving a red sun. After sunset, the sky appears red due to refraction between space and the earth's atmosphere.

molecule, and additional scattering occurs. Blue wavelengths are scattered in many directions, and when an observer looks away from the sun, the sky appears blue.

The sun, when observed directly, appears white at noon, yellow in the afternoon, and red at night, as shown in Fig. 10.13. At noon, the sun's radiation traverses a short distance through the atmosphere. Most visible light is transmitted, although some blue light is scattered by air molecules. In the afternoon, the sun traverses a longer path in the atmosphere. Most blue and some green wavelengths are scattered, and all red and some green wavelengths are transmitted, making the sun look yellow. Although green wavelengths are less likely than blue wavelengths to be scattered by a single gas molecule, the number of gas molecules along the line of sight is so large that the probability of green light being scattered is sizable. At sunset, visible light from the sun traverses a long distance through the atmosphere, and all blue and green and some red wavelengths are scattered. Only some red light transmits, and the sun appears red at sunset. The horizon appears red even after the sun sets because light from the sun refracts as it enters earth's atmosphere, as shown in Fig. 10.13.

## 10.4. ABSORPTION AND SCATTERING BY GASES AND PARTICLES

Absorption removes radiation from an incident beam, and scattering redirects the radiation. In both cases, radiation in the beam is attenuated, which reduces the amount of radiation received past the point of absorption or scattering. Certain particles (aerosols and cloud drops) and gases absorb and/or scatter radiation at

**Table 10.4.** Wavelengths of Absorption in the Solar Spectrum by Several Atmospheric Gases

| Gas | Absorption Wavelengths (μm) | Gas | Absorption Wavelengths (μm) |
|---|---|---|---|
| $N_2$ | <0.1 | $N_2O_5$ | <0.38 |
| $O_2$ | <0.245 | $HNO_3$ | <0.33 |
| $O_3$ | 0.17–0.35, 0.45–0.75 | $HO_2NO_2$ | <0.33 |
| $CO_2$ | <0.21 | HCHO | 0.25–0.36 |
| $H_2O$ | <0.21 | $CH_3CHO$ | <0.345 |
| $H_2O_2$ | <0.35 | $CH_3CO_3NO_2$ | <0.3 |
| $NO_2$ | <0.71 | HCl | <0.22 |
| $N_2O$ | <0.24 | $CFCL_3$ | <0.23 |
| $NO_3$ | 0.41–0.67 | $CF_2CL_2$ | <0.23 |
| HONO | <0.4 | $CH_3Cl$ | <0.22 |

specific wavelengths. In polluted air, visible light is attenuated primarily by particle scattering. Visible-light particle absorption is important only when elemental carbon is present, and gas absorption is important only when nitrogen dioxide concentrations are high. In the free troposphere, visible-light gas scattering is usually more important than other light-attenuation processes, except when clouds are present. In such cases, particle scattering is most important. For each region and wavelength of light, gas absorption, gas scattering, particle absorption, and particle scattering have their own significance, as discussed next.

### 10.4.1. Gas Absorption

Gases selectively absorb radiation in different portions of the electromagnetic spectrum. In this subsection, gas absorption in the solar and infrared spectra are discussed, the gas absorption extinction coefficient is defined, and effects of gas absorption on visibility are analyzed.

#### 10.4.1.1. *Gas Absorption in the Solar Spectrum*

Certain gases that absorb solar (ultraviolet, visible, and near-infrared) radiation and dissociate are shown in Table 10.4. Of the gases listed, only ozone ($O_3$), nitrogen dioxide ($NO_2$), and nitrate ($NO_3$) absorb in the visible spectrum. The rest absorb in the ultraviolet spectrum. Absorption by ozone is weak in the visible spectrum, and concentrations of the nitrate radical are relatively low, except when sunlight is absent. Thus, in polluted air, nitrogen dioxide is the only gas that absorbs enough radiation to affect visibility. Nitrogen dioxide appears yellow, brown, or red because it absorbs blue light preferentially and green light to a lesser extent.

When a gas absorbs radiation, it often breaks into smaller molecules or atoms during **photolysis**. Table 10.4 shows that ozone absorbs wavelengths below 0.35 and above 0.45 μm. Below 0.31 μm, ozone photolysis produces $O_2 + O(^1D)$. These wavelength bands of ozone photolysis are the **Hartley bands**. Between 0.31 and 0.35 μm, ozone photolysis produces $O_2+O(^3P)$. These wavelength bands of ozone photolysis are the **Huggins bands.** Between 0.45 and 0.75 μm, called the **Chappuis bands,** ozone also photolyzes to $O_2 + O(^3P)$. Oxygen ($O_2$) photolyzes below 0.175 μm to produce $O(^1D) + O(^3P)$. This wavelength region of $O_2$ photolysis is the **Schumann–Runge** system. Photolysis at wavelengths between 0.175 and 0.245 μm produces $2O(^3P)$. This region of $O_2$ photolysis is the **Herzberg continuum.**

Of the gases shown in Table 10.4, **nitrogen** and **oxygen** have the highest mixing ratios in the atmosphere (78 and 21 percent, respectively). The mixing ratio of **ozone** reaches its peak in the stratosphere (up to 10 ppmv or more), and is much lower near the surface (40 ppbv in the free troposphere and 0.1 ppmv or more in polluted air). As discussed in Section 2.2.2, the presence of $N_2$, $O_2$, and $O_3$ above the troposphere prevents ultraviolet radiation of wavelengths <0.28 μm from reaching the troposphere.

Although most gases in Table 10.4 absorb UV radiation, their mixing ratios are usually too low to significantly attenuate ultraviolet radiation. **Water-vapor** mixing ratios range from 0–4000 ppmv near the surface to 0–4 ppmv in the stratosphere. Stratospheric mixing ratios of water vapor are much lower than those of oxygen, and oxygen absorbs many of the same wavelengths as does water vapor. Thus, water vapor has little effect on UV light attenuation in the stratosphere. Since wavelengths less than 0.28 μm do not reach the surface, and water vapor absorbs wavelengths shorter than 0.21 μm, water vapor has no effect on visible or UV light attenuation in the troposphere. **Carbon dioxide** absorbs wavelengths shorter than 0.21 μm, which do not reach the surface. Thus, $CO_2$ does not attenuate UV radiation in the troposphere. Since its mixing ratio is only 360 ppmv, $CO_2$ is not an important UV absorber in the stratosphere.

#### **10.4.1.2.** *Gas Absorption in the Infrared Spectrum – the Greenhouse Effect*

Table 10.5 lists selected gases in the atmosphere and their peak absorption bands in the near and far infrared spectra. The gases listed are **greenhouse gases:** they are relatively transparent to incoming solar radiation but opaque to certain wavelengths of infrared radiation. Their presence permits a net warming of the earth's atmosphere like a glass enclosure permits warming of its interior. Since solar radiation can penetrate glass, and certain wavelengths of outgoing infrared radiation cannot, air inside a glass enclosure (such as a glass house) warms during the day so long as mass (such as plant mass) is present in the enclosure to absorb solar radiation and reemit infrared radiation. The surface of the earth, like the plants, absorbs solar radiation and reemits infrared radiation. Greenhouse gases, like glass, are transparent to solar radiation but absorb infrared radiation.

The **natural greenhouse effect** is the warming of the earth's atmosphere due to the presence of background greenhouse gases. The natural greenhouse effect is responsible for about 33 K of the earth's average near-surface air temperature of 288 K. Without the natural greenhouse effect, earth's average near-surface temperature would be about 255 K, which is too cold for most life on the planet. Thus, the presence of natural greenhouse gases is beneficial. Human emissions of certain

**Table 10.5.** Peak Absorption Wavelengths (μm) of Some Greenhouse Gases

| $H_2O$ | $CO_2$ | $O_3$ | $N_2O$ | $CH_4$ | $CF_2Cl_2$ | $CFCl_3$ | $CH_3Cl$ |
|---|---|---|---|---|---|---|---|
| 0.72 | 1.4 | 9.6 | 4.5 | 3.31 | 8.6 | 11.8 | 11.8 |
| 0.82 | 1.6 | 14.27 | 7.78 | 3.43 | 9.13 | 9.2 | |
| 0.94 | 2.0 | | 17.0 | 6.55 | | | |
| 1.1 | 2.7 | | | 7.65 | | | |
| 1.38 | 4.3 | | | | | | |
| 1.87 | 4.8 | | | | | | |
| 2.7 | 5.2 | | | | | | |
| 3.2 | 15.0 | | | | | | |
| 6.25 | | | | | | | |
| >12.0 | | | | | | | |

*Source:* Liou (1992).

greenhouse gases have increased the concentrations of such gases, causing global warming. **Global warming** is the increase in earth's temperature above that due to the natural greenhouse effect.

The most important greenhouse gas is water vapor, which accounts for approximately 90 percent of the 33-K temperature increase due to natural greenhouse warming. Carbon dioxide ($CO_2$) is the second most important and abundant greenhouse gas. Water vapor absorbs wavelengths below 0.21 μm and above 0.72 μm. Carbon dioxide absorbs wavelengths below 0.21 μm and above 1.4 μm. Both gases are transparent to visible radiation. Most infrared absorption by water vapor and carbon dioxide occurs outside the wavelength region, 8–12 μm, called the **atmospheric window**. In the atmospheric window, gases are relatively transparent to the earth's infrared radiation, allowing the radiation to escape to space. Ozone, several chlorofluorocarbons ($CF_2Cl_2$ and $CFCl_3$), and methyl chloride ($CH_3Cl$) absorb radiation within the atmospheric window. Nitrous oxide ($N_2O$) and methane ($CH_4$) absorb at the edges of the window. Increases in concentration of these gases enhance global warming by strengthening infrared absorption at wavelengths that usually escape to space.

Each gas in Table 10.5 has a different potential to absorb and increase temperatures. A $CH_4$ molecule is approximately 21 times more efficient at absorbing radiation than is a $CO_2$ molecule. $N_2O$ and $CFCl_3$ are 206 and 12,400 times more efficient at absorbing than is $CO_2$ (IPCC, 1990). The increase in concentrations of greenhouse gases other than $CO_2$ can affect global climate significantly.

#### 10.4.1.3. *Gas Absorption Extinction Coefficient*

A parameter that quantifies absorption by gases is the gas absorption extinction coefficient $\sigma_{a,g,\lambda}$ which has units of $cm^{-1}$, $m^{-1}$, or $km^{-1}$ and is a function of wavelength. An **extinction coefficient** is a parameter that measures the loss of electromagnetic radiation due to a specific process, per unit distance and may be determined as the product of an effective cross section and a number concentration. In the case of gas absorption, the extinction coefficient through a uniformly mixed

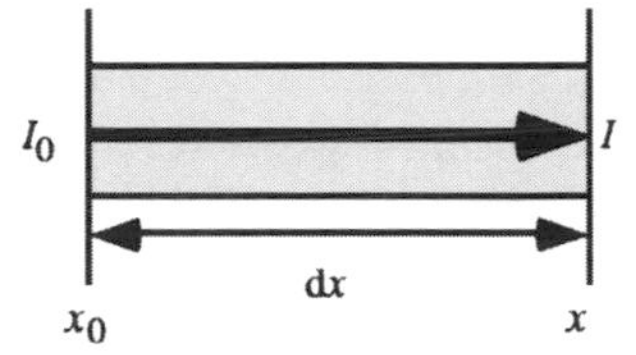


**Figure 10.14.** Attenuation of incident radiance $I_0$ due to absorption in a column of gas.

gas $q$ is $\sigma_{a,g,q,\lambda,T} = N_q b_{a,g,q,\lambda,T}$, where $N_q$ is the number concentration of the gas (molecules cm$^{-3}$) and $b_{a,g,q,\lambda,T}$ is the absorption cross section of the gas (cm$^2$) at wavelength $\lambda$ and temperature $T$.

Suppose incident radiance $I_{0,\lambda}$ travels a distance d$x$ through the uniformly mixed absorbing gas, as shown in Fig. 10.14. The reduction in radiance with distance through the gas is

$$\frac{dI_\lambda}{dx} = -N_q b_{a,g,q,\lambda,T} I_\lambda = -\sigma_{a,g,q,\lambda,T} I_\lambda \tag{10.23}$$

Integrating (10.23) from $I_{0,\lambda}$ to $I_\lambda$ and $x_0$ to $x$ gives

$$I_\lambda = I_{0,\lambda} e^{-N_q b_{a,g,q,\lambda,T}(x-x_0)} = I_{0,\lambda} e^{-\sigma_{a,g,q,\lambda,T}(x-x_0)} \tag{10.24}$$

(10.24) states that incident radiation is reduced from $I_{0,\lambda}$ to $I_\lambda$ over distance d$x$ by gas absorption. The absorption extinction coefficient quantifies the rate of absorption per unit distance. The **absorption cross section** is an effective cross section of the gas that results in radiance reduction by absorption. It can be extracted from (10.24) if the attenuation of radiance from $I_{0,\lambda}$ to $I_\lambda$ through a column of gas with known **path length** $N_q(x - x_0)$ (molecules cm$^{-2}$) is measured experimentally. Absorption cross section data as a function of wavelength and temperature can be found for many gases in DeMore et al. (1997) and Atkinson et al. (1997).

For radiative transfer calculations, the extinction coefficient due to gas absorption, summed over all absorbing gases, is required. This quantity is

$$\sigma_{a,g,\lambda} = \sum_{q=1}^{N_{ag}} N_q b_{a,g,q,\lambda,T} = \sum_{q=1}^{N_{ag}} \sigma_{a,g,q,\lambda,T} \tag{10.25}$$

where $N_{ag}$ is the number of absorbing gases and the subscript $T$ was dropped in $\sigma_{a,g,\lambda}$ for convenience. In the infrared, the gas absorption extinction coefficient may be written in terms of a **mass absorption coefficient** $k_{a,g,q,\lambda,T}$ (cm$^2$ g$^{-1}$),

$$\sigma_{a,g,\lambda} = \sum_{q=1}^{N_{ag}} \rho_q k_{a,g,q,\lambda,T} \tag{10.26}$$

**Table 10.6.** Extinction Coefficients ($\sigma_{a,g}$) and Meteorological Ranges ($x_{a,g}$) Due to $NO_2$ Absorption at Selected Wavelength Intervals ($\lambda \pm 0.005$ μm) and Concentrations

| | | 0.01 | | 0.25 ppmv $NO_2$ | | |
|---|---|---|---|---|---|---|
| $\lambda$ (μm) | $b$ ($10^{-19}$ cm$^2$) | $\sigma_{a,g}$ ($10^8$ cm$^{-1}$) | $x_{a,g}$ (km) | $\sigma_{a,g}$ ($10^{-8}$ cm$^{-1}$) | $x_{a,g}$ (km) | $x_{s,g}$ (km) |
| 0.42 | 5.39 | 13.2 | 296 | 330 | 11.8 | 112 |
| 0.45 | 4.65 | 11.4 | 343 | 285 | 13.7 | 148 |
| 0.50 | 2.48 | 6.10 | 641 | 153 | 25.6 | 227 |
| 0.55 | 0.999 | 2.46 | 1,590 | 61.5 | 63.6 | 334 |
| 0.60 | 0.292 | 0.72 | 5,430 | 18.0 | 217 | 481 |
| 0.65 | 0.121 | 0.30 | 13,000 | 7.5 | 520 | 664 |

Absorption-cross-section data ($b$) for $NO_2$ are from Schneider et al. (1987). Also shown is the meteorological range due to Rayleigh scattering only ($x_{s,g}$). $T = 298$ K and $p_a = 1$ atm.

where $\rho_q$ is the mass density (g cm$^{-3}$) of gas $q$ in air. Mass absorption coefficients are cross sections, averaged over a wavelength increment, per unit mass of absorbing gas.

#### 10.4.1.4. *Gas Absorption Effects on Visibility*

The furthest distance a typical eye can see may be estimated with the **Koschmieder equation,** $x = 3.912/\sigma_{ext,\lambda}$, derived in Section 10.5. This equation gives the visibility (or **meteorological range**) as a function of the total extinction coefficient at a given wavelength ($\sigma_{ext,\lambda}$). The total extinction includes extinction due to gas absorption, gas scattering, particle absorption, and particle scattering. Table 10.6 shows extinction coefficients and meteorological ranges due to $NO_2$ absorption, alone.

Table 10.6 shows that $NO_2$ absorbs more strongly at shorter (blue) than at longer (green or red) wavelengths. At low concentrations (0.01 ppmv), the effect of $NO_2$ absorption on visibility is less than that of gas scattering at all wavelengths. At typical polluted-air concentrations (0.1–0.25 ppmv), $NO_2$ reduces visibility significantly for wavelengths less than 0.50 μm and moderately for wavelengths between 0.50 and 0.60 μm.

Studies have found that most effects of $NO_2$ on visibility are limited to times when its concentration peaks. Results from a project studying Denver's **brown cloud** showed that $NO_2$ accounted for about 6.6 percent of total extinction averaged over all sampling periods, and 37 percent of extinction during periods of maximum $NO_2$ concentration. Scattering and absorption by particles caused most remaining extinction (Groblicki *et al.* 1981).

Figure 10.15 shows extinction coefficients due to nitrogen dioxide and ozone absorption at different mixing ratios. The figure indicates that nitrogen dioxide affects extinction (and therefore radiative transfer and visibility) only at high mixing ratios and at wavelengths below about 0.5 μm. In polluted air, such as in Los Angeles, $NO_2$ mixing ratios typically range from 0.01 to 0.1 ppmv and peak near

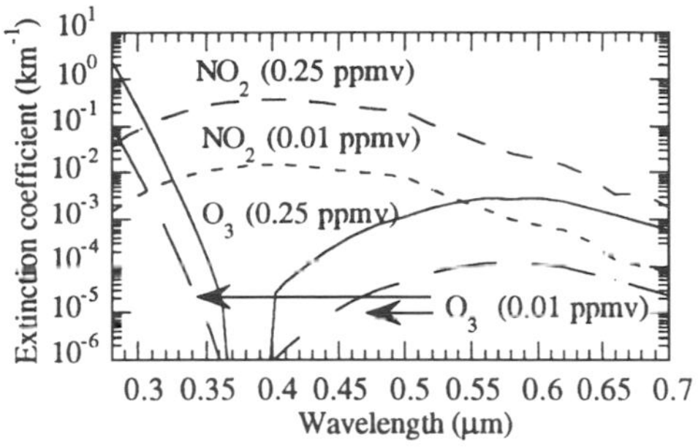


**Figure 10.15.** Extinction coefficients due to $NO_2$ and $O_3$ absorption when $T = 298$ K and $p_a = 1013$ mb. Cross-section data for $NO_2$ are from DeMore et al. (1997) and interpolated from Schneider et al. (1987). Data for ozone are from Atkinson et al. (1997).

0.15 ppmv during the morning. A typical value is 0.05 ppmv, which results in an extinction coefficient of about 0.07 km$^{-1}$ at 0.4 μm and 0.01 km$^{-1}$ at 0.55 μm. The visibility (meteorological range) is about 390 km with this latter value. Ozone has a larger effect on extinction than does nitrogen dioxide at wavelengths below about 0.32 μm. Ozone mixing ratios in polluted air usually peak between 0.05 and 0.25 ppmv. Nevertheless, the cumulative effect of ozone, nitrogen dioxide, and other gases on extinction is small in comparison with the effects of scattering and absorption by particles.

### 10.4.2. Gas Scattering

The only important gas-scattering process in the atmosphere is **Rayleigh scattering,** which is the scattering of radiation by gas molecules (primarily $N_2$ and $O_2$). Rayleigh scattering gives the sky its blue color. It causes dark objects a few kilometers away to appear behind a blue haze of scattered light, and bright objects more than 30 km away to appear reddened (Waggoner et al. 1981).

A Rayleigh scatterer has molecular radius $r$ much smaller than the wavelength $\lambda$ of interest ($2\pi r/\lambda \ll 1$). The radius of a gas molecule is sufficiently small in relation to visible light wavelengths to meet this criterion. Since oxygen and nitrogen are the most abundant gases in the atmosphere, they are the most abundant Rayleigh scatterers.

The extinction coefficient (cm$^{-1}$) due to Rayleigh scattering is

$$\sigma_{s,g,\lambda} = N_a b_{s,g,\lambda} \tag{10.27}$$

where $N_a$ is the number concentration of air molecules (molecules cm$^{-3}$) at a given altitude, $b_{s,g,\lambda}$ and is the **scattering cross section** of a typical air molecule. This cross section can be estimated with

$$b_{s,g,\lambda} = \frac{8\pi^3\left(n_{a,\lambda}^2 - 1\right)^2}{3\lambda^4 N_{a,0}^2} f(\delta_*) \approx \frac{32\pi^3(n_{a,\lambda} - 1)^2}{3\lambda^4 N_{a,0}^2} f(\delta_*) \qquad (10.28)$$

where $n_{a,\lambda}$ is the real index of refraction of air, obtained from (10.21), $N_{a,0} = 2.55 \times 10^{19}$ cm$^{-3}$ is the number concentration of air molecules at standard temperature and pressure (288 K and 1 atm), and $f(\delta_*)$ is the **anisotropic correction term**. This term accounts for deviations of Rayleigh scattering from perfect isotropic scattering and can be estimated with

$$f(\delta_*) = \frac{6 + 3\delta_*}{6 - 7\delta_*} \approx 1.05 \qquad (10.29)$$

where $\delta_* \approx 0.0279$ (Young 1980). $f(\delta_*)$ is a function of wavelength and varies from 1.08 at $\lambda = 0.2$ μm to 1.047 at $\lambda = 1.0$ μm (Liou 1992). The second expression in (10.28) was found by assuming $n_{a,\lambda}^2 - 1 \approx 2(n_{a,\lambda} - 1)$, which is possible because $n_{a,\lambda} \approx 1$.

Since air density decreases exponentially with increasing altitude, extinction due to Rayleigh scattering decreases exponentially with increasing altitude according to (10.27). (10.28) shows that the Rayleigh scattering cross section is inversely proportional to the fourth power of wavelength. Gases scatter short wavelengths more effectively than they do long wavelengths. Table 10.6 shows that, at short visible wavelengths (e.g., 0.42 μm), the meteorological range due to Rayleigh scattering is much smaller than that at longer wavelengths (e.g., 0.65 μm). The larger scattering cross section at short wavelengths accounts for the reduction in meteorological range at those wavelengths.

**Example 10.6**

For $\lambda = 0.5$ μm, $p_a = 1$ atm (sea level), and $T = 288$ K, $\sigma_{s,g,\lambda} = 1.72 \times 10^{-7}$ cm$^{-1}$ from (10.27). This extinction coefficient results in a meteorological range of $x = 3.912/\sigma_{s,g,\lambda} = 227$ km.

For $\lambda = 0.55$ μm, the extinction coefficient decreases to $\sigma_{s,g,\lambda} = 1.17 \times 10^{-7}$ cm$^{-1}$ and the meteorological range increases to $x = 334$ km. Waggoner et al. (1981) reported a total extinction at Bryce Canyon, Utah corresponding to a visual range of within a few percent of 400 km for a wavelength of 0.55 μm.

### 10.4.3. Particle Absorption and Scattering

Particulate scattering is the most important solar radiation attenuation process in polluted air, followed, in order, by particulate absorption, gas absorption, and gas scattering. All particulate species scatter solar and infrared radiation. All particle species absorb infrared radiation, but few absorb solar radiation. For several species, the ability to absorb solar radiation increases from the visible to ultraviolet spectra.

The strongest particulate absorber of solar radiation is **elemental carbon** (Horvath 1995), which is an amorphous solid emitted during combustion. Other absorbers include hematite ($Fe_2O_3$) and aluminum oxide (alumina – $Al_2O_3$). Hematite is found in soil particles, especially near iron ore mining areas. Some forms of aluminum oxide are found in soil particles and others are found in combustion particles.

Although near UV radiation (0.25–0.38 μm) makes up only 5 percent of total solar radiation, near UV wavelengths are responsible for most gas photolysis. Certain organic and inorganic compounds are strong or moderate absorbers of UV but weak absorbers of visible radiation. Particulate organics with at least one near-UV absorption peak include certain nitrated aromatics, benzaldehydes, benzoic acids, aromatic polycarboxylic acids, phenols, and polycyclic aromatic hydrocarbons (Jacobson 1998b). The strong absorptivity of nitrated aromatics occurs because substitution of a nitrate group onto a benzene ring shifts the peak absorption 0.057 μm towards a longer wavelength (e.g., Dean 1992). The strong absorptivity of benzaldehydes and benzoic acids results because the addition of an aldehyde or an acid group to a benzene ring shifts the peak absorption 0.046 or 0.0255 μm, respectively, towards a longer wavelength (Dean 1992). Particulate inorganics with a near-UV peak include the nitrate ion (0.302 μm), ammonium nitrate (0.308 μm), and sodium nitrate (0.297 μm) (Cleaver et al. 1963, Sommer 1989).

Most particulate components are weak absorbers of solar radiation. Silicon dioxide (silica – $SiO_2$), which is a white, colorless, crystalline compound found in quartz, sand, and other minerals, is a weak but nonnegligible absorber of solar radiation (e.g., Krekov 1993). Sodium chloride (NaCl) and ammonium sulfate [$(NH_4)_2SO_4$] are also weak solar absorbers. Sulfuric acid ($H_2SO_4$) and water are negligible absorbers of solar radiation.

Soil particles, which contain $SiO_2$, $Al_2O_3$, $Fe_2O_3$, $CaCO_3$, $MgCO_3$, and other substances, are moderate absorbers of solar radiation. Their absorptivity increases from the visible to the ultraviolet spectrum (e.g., Gillette et al. 1993; Sokolik et al. 1993). Most emitted soil particles range between 1 and 100 μm in diameter. Those larger than 10 mm in diameter fall out of the atmosphere after a few minutes to a few days, depending on their size and height of emission. During dust storms, mass concentrations of large soil particles increase, increasing light absorption and decreasing visibility significantly.

The extent to which a substance absorbs radiation can be quantified by its **imaginary index of refraction.** If scattering is ignored, the attenuation, due to absorption, of radiant energy as it propagates through a particle is approximately

$$\frac{dI}{dx} = -\frac{4\pi\kappa}{\lambda} I \tag{10.30}$$

where $\kappa$ is the imaginary index of refraction. Although $\kappa$ is a function of wavelength, the wavelength subscript in (10.30) has been dropped. $4\pi\kappa/\lambda$ is an absorption extinction coefficient for a particle. Figure 10.16 shows a possible path of radiant

**Table 10.7.** Real and Imaginary Indices of Refraction for Some Substances at $\lambda = 0.50$ and 10.0 μm

| | 0.5 μm | | 10 μm | |
|---|---|---|---|---|
| | **Real ($n_\lambda$)** | **Imaginary ($\kappa_\lambda$)** | **Real ($n_\lambda$)** | **Imaginary ($\kappa_\lambda$)** |
| $H_2O$(aq)[a] | 1.34 | $1.0 \times 10^{-9}$ | 1.22 | $5.0 \times 10^{-2}$ |
| Elemental C(s)[b] | 1.82 | $7.4 \times 10^{-1}$ | 2.40 | $1.0 \times 10^{0}$ |
| Organic C(s)[b] | 1.45 | $1.0 \times 10^{-3}$ | 1.77 | $1.2 \times 10^{-1}$ |
| $H_2SO_4$(aq)[b] | 1.43 | $1.0 \times 10^{-8}$ | 1.89 | $4.6 \times 10^{-1}$ |
| $(NH_4)_2SO_4$(s)[b] | 1.52 | $5.0 \times 10^{-4}$ | 2.15 | $2.0 \times 10^{-2}$ |
| NaCl(s)[b] | 1.45 | $1.5 \times 10^{-4}$ | 1.53 | $5.3 \times 10^{-2}$ |

[a]Hale and Querry (1973). [b]Krekov (1993).

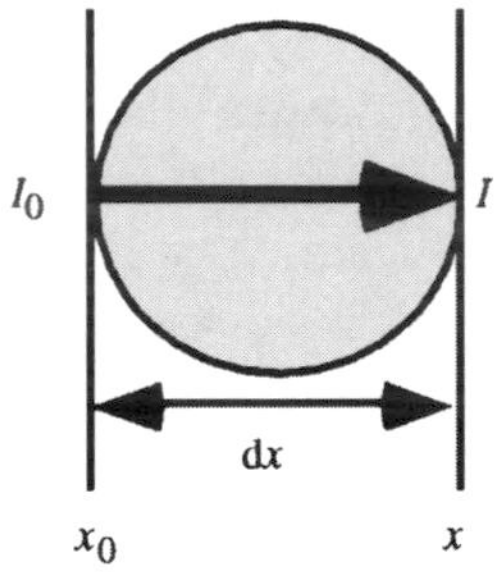


**Figure 10.16.** Attenuation of incident radiance $I_0$ due to absorption in a particle.

energy through a particle. The attenuation of this beam is described by (10.30), which is similar to (10.23). Integrating (10.30) from $I_0$ to $I$ and $x_0$ to $x$ gives

$$I = I_0 e^{-4\pi\kappa(x - x_0)/\lambda} \tag{10.31}$$

Imaginary and real indices of refraction are combined in the **complex index of refraction,**

$$m_\lambda = n_\lambda - i\kappa_\lambda \tag{10.32}$$

Optical properties of a substance can be described in terms of this complex index. Table 10.7 gives the real ($n_\lambda$) and imaginary ($\kappa_\lambda$) indices of refraction for some substances at $\lambda = 0.50$ and 11.5 μm. Elemental carbon has the largest imaginary index of refraction in the visible spectrum among the substances shown.

Figure 10.17 shows the imaginary index of refraction of liquid nitrobenzene as a function of wavelength. Although liquid nitrobenzene has little presence in atmospheric particles, the shape of its index-of-refraction curve is important because it suggests that particulate organics with peak absorption wavelengths below 0.3 μm absorb moderately from 0.3 to 0.4 μm as well, thereby affecting photolysis coefficients. Most organic materials in atmospheric particles have peak absorption wavelengths below 0.3 μm.

**Table 10.8.** Light Transmission through Elemental Carbon or Liquid Water Particles at $\lambda = 0.50$ μm

| Substance | κ | Particle Diameter (μm) | Transmission ($I/I_0$) |
|---|---|---|---|
| Elemental carbon | 0.74 | 0.1 | 0.16 |
| | | 1.0 | $8.0 \times 10^{-9}$ |
| | | 10.0 | 0 |
| Water | $10^{-9}$ | 0.1 | 0.999999997 |
| | | 1.0 | 0.99999997 |
| | | 10.0 | 0.9999997 |

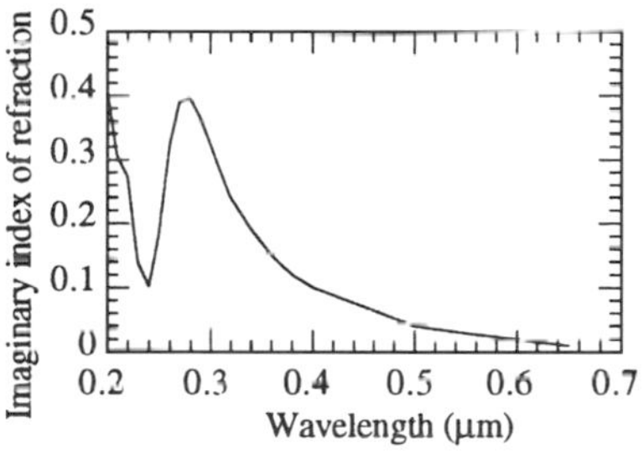


**Figure 10.17.** Imaginary index of refraction of liquid nitrobenzene versus wavelength. Data from Foster (1992).

Table 10.8 shows the extent of transmission of light through particles made of elemental carbon and water at $\lambda = 0.50$ μm. These calculations were obtained from (10.31). The **transmission**, defined as $I/I_0$, is the fraction of incident radiation passing through a substance. The table shows that a 0.1-μm-diameter elemental carbon particle transmits 16 percent of incident 0.50-μm radiation, and a 0.1-μm water drop transmits nearly 100 percent of the radiation if only absorption is considered. A 10-μm liquid water drop absorbs a negligible amount of incident 0.50-μm radiation. Thus, elemental carbon is a strong absorber of visible radiation, and liquid water is not.

Particle absorption and scattering extinction coefficients can be separated into aerosol and cloud drop coefficients, since the optical properties of aerosols differ from those of cloud drops. Extinction coefficients due to aerosol absorption and scattering can be modeled as

$$\sigma_{a,a,\lambda} = \sum_{i=1}^{N_B} n_i b_{a,a,i,\lambda} \qquad \sigma_{s,a,\lambda} = \sum_{i=1}^{N_B} n_i b_{s,a,i,\lambda} \tag{10.33}$$

respectively, where $n_i$ is the number concentration of aerosols of a given size (particles cm$^{-3}$), $N_B$ is the number of aerosol sizes, $b_{a,a,i,\lambda}$ is the effective **absorption cross section of a single aerosol**, and $b_{s,a,i,\lambda}$ is the effective **scattering cross section of a single aerosol.** The equations for extinction due to cloud-drop absorption and scattering are identical, except that $\sigma_{a,c,\lambda}$, $b_{a,c,i,\lambda}$, $\sigma_{s,c,\lambda}$, $b_{s,c,i,\lambda}$ replace $\sigma_{a,a,\lambda}$, $b_{a,a,i,\lambda}$, $\sigma_{s,a,\lambda}$, $b_{s,a,i,\lambda}$, respectively, when cloud drops are considered

For spherical aerosols, the effective absorption and scattering cross sections are

$$b_{a,a,i,\lambda} = \pi r_i^2 Q_{a,i,\lambda} \qquad b_{s,a,i,\lambda} = \pi r_i^2 Q_{s,i,\lambda} \tag{10.34}$$

respectively, where $\pi r_i^2$ is the actual aerosol cross-sectional area, $Q_{a,i,\lambda}$ (unitless) is the single-particle absorption efficiency, and $Q_{s,i,\lambda}$ is the single-particle scattering efficiency. The absorption and scattering efficiencies are functions of the complex index of refraction $m_\lambda$ and of the **size parameter**

$$\alpha_{i,\lambda} = 2\pi r_i/\lambda \tag{10.35}$$

The former quantity depends on wavelength and particle composition, and the latter depends on particle size and wavelength. Absorption cross sections for cloud drops are calculated in the same manner as in (10.34).

The **single-particle scattering efficiency** is the ratio of the effective scattering cross section of a particle to its actual cross section. The scattering efficiency can exceed unity, since a portion of the radiation diffracting around a particle can be intercepted and scattered by the particle. Scattering efficiencies above unity account for this additional scattering. The greater the absorption by a particle, the lesser the scattering efficiency, since absorption hinders scattering. The scattering efficiency is highest when the wavelength of light is close to the particle radius.

The **single-particle absorption efficiency** is the ratio of the effective absorption cross section of a particle to its actual cross section. The absorption efficiency can exceed unity, since a portion of the radiation diffracting around a particle can be intercepted and absorbed by the particle. Absorption efficiencies above unity account for this additional absorption. The larger the imaginary index of refraction of a particle, the greater its absorption efficiency. The real index of refraction affects the angle that radiation bends upon entering a particle, affecting the distance it travels and the cumulative absorption that occurs before it exits.

Single-particle absorption and scattering efficiencies vary with particle size and radiation wavelength. When a particle's diameter is much smaller than the wavelength of the light ($r_i \ll 0.1\lambda$, or $\alpha_{i,\lambda} < 0.1$), the particle is in the **Rayleigh regime** and is called a **Tyndall absorber** or **scatterer**. Such particles have an absorption efficiency of

$$Q_{a,i,\lambda} = -4\frac{2\pi r_i}{\lambda}\,\mathrm{Im}\left(\left|\frac{m_\lambda^2 - 1}{m_\lambda^2 + 2}\right|^2\right) \approx \frac{2\pi r_i}{\lambda}\left[\frac{24 n_\lambda \kappa_\lambda}{\left(n_\lambda^2 + \kappa_\lambda^2\right)^2 + 4\left(n_\lambda^2 - \kappa_\lambda^2 + 1\right)}\right] \tag{10.36}$$

As $\kappa_\lambda \to 0$, (10.36) approaches

$$Q_{a,i,\lambda} = \frac{2\pi r_i}{\lambda}\left[\frac{24 n_\lambda \kappa_\lambda}{\left(n_\lambda^2 + 2\right)^2}\right] \tag{10.37}$$

which is a linear function of $\kappa_\lambda$. When $r_i \ll \lambda$, the absorption efficiency is small. Thus, small particles are relatively inefficient absorbers of radiation.

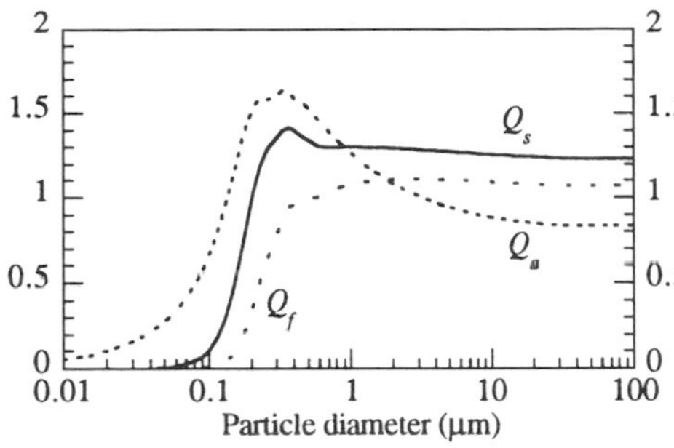


**Figure 10.18.** Single-particle absorption ($Q_a$), total scattering ($Q_s$), and forward scattering ($Q_f$) efficiencies of elemental carbon (soot) particles of different sizes at $\lambda = 0.50$ μm ($n_\lambda = 1.94$, $\kappa_\lambda = 0.66$). The efficiency for soot at any other wavelength $\lambda_1$ in the visible or UV spectrum and diameter $d_1$ is the efficiency in the figure located at diameter $d = 0.5$ μm $\times\ d_1/\lambda_1$.

The **Tyndall scattering efficiency** is

$$Q_{s,i,\lambda} = \frac{8}{3}\left(\frac{2\pi r_i}{\lambda}\right)^4 \left|\frac{m_\lambda^2 - 1}{m_\lambda^2 + 2}\right|^2 \qquad (10.38)$$

(10.38) indicates that small particles are inefficient scatterers of radiation. (10.38) and (10.36) show that the scattering and absorption efficiencies of small particles are proportional to $(r_i/\lambda)^4$ and $r_i/\lambda$, respectively. At a given wavelength, the scattering efficiency of an absorbing particle (large $\kappa$) is much less than the absorption efficiency. If a particle is nonabsorbing, its Tyndall scattering efficiency, although small, exceeds its absorption efficiency. In the atmosphere, particles that contain elemental carbon absorb visible light. Figures 10.18 and 10.19 show the absorption and scattering efficiencies, including Tyndall efficiencies, of elemental carbon and liquid water particles, respectively.

**Example 10.7**

At $\lambda = 0.5$ μm and $r_i = 0.01$ μm, find the single-particle scattering and absorption efficiencies of liquid water.

SOLUTION

From Table 10.7, $n_\lambda = 1.34$ and $\kappa_\lambda = 1.0\times10^{-9}$ at $\lambda = 0.5$ μm. From (10.38) and (10.37), respectively, $Q_{s,i,\lambda} = 2.92\times10^{-5}$ and $Q_{a,i,\lambda} = 2.8\times10^{-10}$. Since liquid water is nonabsorbing, its single-particle absorption efficiency is expected to be small in comparison with its single-particle scattering efficiency.

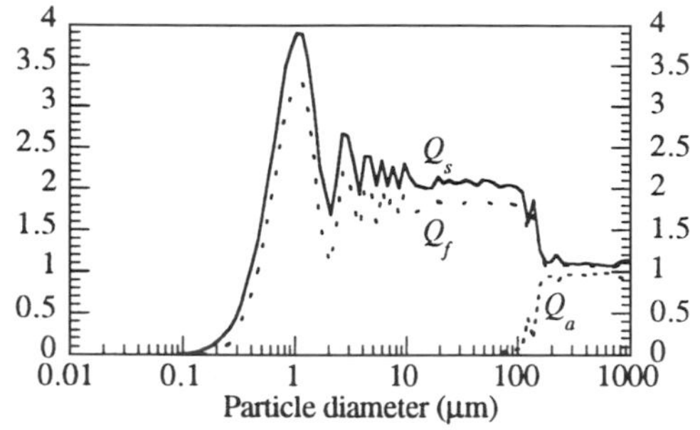


**Figure 10.19.** Single-particle absorption ($Q_a$), total scattering ($Q_s$), and forward scattering ($Q_f$) efficiencies of liquid water drops of different sizes at $\lambda = 0.50$ μm ($n_\lambda = 1.34$, $\kappa_\lambda = 0$). The efficiency for water at any other wavelength $\lambda_1$ in the visible or UV spectrum and diameter $d_1$ is the efficiency in the figure located at diameter $d = 0.5$ μm $\times\, d_1/\lambda_1$.

When a particle has diameter approximately equal to the wavelength of light ($r_i \sim \lambda$ or $0.1 < \alpha_{i,\lambda} < 100$), it is in the **Mie regime.** In this regime, absorption and scattering efficiencies are approximated with Mie's solution to Maxwell's equations (van de Hulst 1957; Kerker 1969). The **scattering efficiency of a particle in the Mie regime** is determined by

$$Q_{s,i,\lambda} = \frac{2}{\alpha_{i,\lambda}} \sum_{k=1}^{\infty} (2k+1)\left(|a_k|^2 + |b_k|^2\right) \tag{10.39}$$

where $a_k$ and $b_k$ are complex functions. The single-particle absorption efficiency of a particle in the Mie regime is estimated from $Q_{a,i,\lambda} = Q_{e,i,\lambda} - Q_{s,i,\lambda}$, where

$$Q_{e,i,\lambda} = \frac{2}{\alpha_{i,\lambda}} \sum_{k=1}^{\infty} (2k+1)\,\mathrm{Re}(a_k + b_k) \tag{10.40}$$

is the single-particle extinction efficiency in the Mie regime. A solution to these equations can be found in Toon and Ackerman (1981). Figures 10.18 and 10.19 show $Q_{a,i,\lambda}$ and $Q_{s,i,\lambda}$ for elemental carbon and liquid water, respectively, as a function of particle size at $\lambda = 0.50$ μm. The figures also show the **forward scattering efficiency** $Q_{f,i,\lambda}$, which is the efficiency with which a particle scatters light in the forward direction. Figure 10.18 shows that elemental carbon particles 0.2–0.4 μm in diameter are most efficient at absorbing visible radiation. The absorption efficiency decreases to a constant value at larger sizes and to a lower, Tyndall-regime solution at smaller sizes.

Automobiles emit particles containing elemental carbon. Such particles generally have diameter $\leq 0.1$ μm. Absorption efficiencies for such particles are smaller than those for larger particles, as shown in Fig. 10.18, but the number concentration of small particles far exceeds that of large particles. Thus, small elemental-carbon-containing particles affect the absorption extinction $\sigma_{ap,\lambda}$ to a greater extent than do large ones.

Figure 10.19 shows that particles 0.3–2.0 μm in diameter scatter more efficiently than do smaller or larger particles. The intermediate size range is part of the **accumulation mode**, discussed in Chapter 14. For particles smaller than 0.3 μm, the scattering efficiency decreases rapidly. For particles larger than 2.0 μm, the number concentration decreases rapidly. Because the accumulation mode contains sufficient particle number concentration and scattering efficiency, it almost always determines the extent of particle extinction (Waggoner et al. 1981). In many urban regions, 20–50 percent of the accumulation-mode mass is sulfate. Thus, sulfate is correlated to particle scattering more closely than is any other particulate species, aside from liquid water.

When a particle has diameter much larger than the wavelength of light ($\alpha_{i,\lambda} > 100$ or $d_i > 16$ μm in Fig. 10.19), the particle is in the **geometrical regime**. Such particles reflect, refract, and diffract light significantly. Figure 10.19 shows that their scattering efficiencies are lower than those of accumulation-mode particles. Although large particles, such as fog drops, cloud drops, dust, and sand, have relatively small scattering efficiencies, they obscure visibility because their cross-sectional areas are large.

As $\alpha_{i,\lambda}$ increases from the Mie to the geometric regime, the scattering efficiency approaches a constant in a sinusoidal fashion, as shown in Fig. 10.19. As $\alpha_{i,\lambda} \rightarrow \infty$, the scattering efficiency drops and then approaches another constant, given by Chylek (1977) as

$$\lim_{\alpha_{i,\lambda}\rightarrow\infty} Q_{s,i,\lambda} = 1 + \left|\frac{m_\lambda - 1}{m_\lambda + 2}\right| \tag{10.41}$$

---

**Example 10.8**

At $\lambda = 0.5$ μm, calculate the scattering efficiency of liquid water as $\alpha_{i,\lambda} \rightarrow \infty$.

SOLUTION

From Table 10.7, $n_\lambda = 1.34$ at $\lambda = 0.5$ μm for liquid water. From (10.41), $Q_{s,i,\lambda} = 1.1$. Figure 10.19 shows that, as $d \rightarrow \infty$ ($\alpha_{i,\lambda} \rightarrow \infty$), the scattering efficiency approaches $Q_{s,i,\lambda} = 1.1$ as well, which is consistent with (10.41).

---

Deirmendjian (1969) showed that, as $\alpha_{i,\lambda} \rightarrow \infty$, a particle's absorption efficiency converges to its scattering efficiency regardless of how weak the imaginary index of refraction is. Thus, **large particles absorb radiation, regardless of their imaginary index of refraction.** Figure 10.19 shows how the absorption efficiency of water increases to unity for particles larger than $d_i > 100$ μm.

In the previous discussion, particles were assumed to be uniformly mixed. In reality, each particle has a unique composition. One way to account for the different indices of refraction in a particle is to volume-average the indices. The resulting

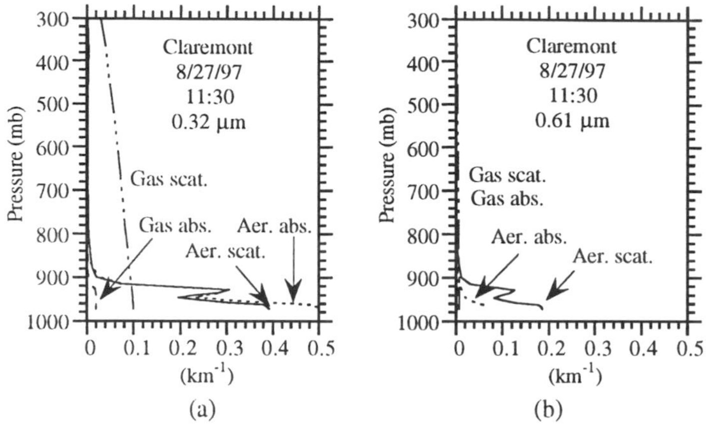


**Figure 10.20.** Predicted profiles from Claremont, California, of extinction coefficients (km$^{-1}$) due to gas scattering, aerosol scattering, gas absorption, and aerosol absorption at (a) $\lambda = 0.32$ and (b) 0.61 μm. The peaks at 920 mb indicate elevated aerosol layers. From Jacobson (1998b).

**effective real index of refraction** of a particle of a given size can be calculated as

$$n_\lambda = \sum_{q=1}^{N_V} \left( \frac{v_{q,i}}{v_i} n_{\lambda,q} \right) \tag{10.42}$$

where $N_V$ is the number of volume components in a particle, $v_i$ is the volume concentration of the particles (cubic centimeters of particles per cubic centimeter of air), and $v_{q,i}$ is the volume concentration of component $q$ (cubic centimeters of component $q$ per cubic centimeter air). A similar equation can be written for $\kappa_\lambda$.

Figure 10.20 (a) and (b) show vertical profiles of extinction coefficients over a polluted airshed, predicted from a model simulation. Both figures indicate that extinction coefficients increased from the free troposphere (altitudes above 900 mb) to the boundary layer. The first figure indicates that, at $\lambda = 0.32$ μm (ultraviolet light), extinction due to aerosol scattering in the boundary layer was less than that due to aerosol absorption. At $\lambda = 0.61$ μm (visible light), the reverse was true because the aerosol absorptivity decreased from the ultraviolet to visible spectra. Above the boundary layer, gas scattering dominated aerosol scattering at 0.32 and 0.61 μm.

## 10.5. VISIBILITY

A result of aerosol buildup is visibility degradation. Although unnatural visibility degradation, itself, causes no adverse health effects, it usually indicates the presence of pollutants, which are often harmful. Several terms describe maximum visibility. The **meteorological range** is the distance from an ideal dark object at which the object has a 0.02 liminal contrast ratio against a white background. The liminal or threshold **contrast ratio** is the lowest visually perceptible brightness contrast

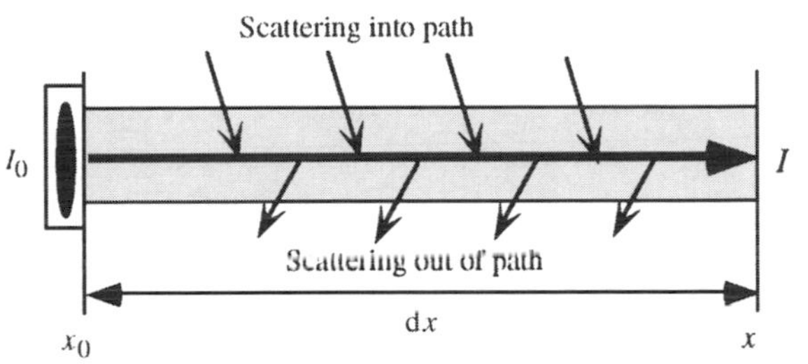


**Figure 10.21.** Determination of meteorological range. The intensity $I_0 = 0$ from a dark object at point $x_0 = 0$ increases over the distance d$x$ due to the scattering of background light into its path and decreases due to absorption along its path and scattering out of its path. At point $x$, the intensity of the radiation originating from the direction of the object has increased to the background intensity.

a person can see. It varies from individual to individual. Although Koschmieder (1924) selected a value of 0.02, Middleton (1952) tested 1000 people and found a threshold-contrast range of between 0.01 and 0.20 with the mode of the sample between 0.02 and 0.03. Campbell and Maffel (1974) found a liminal contrast of 0.003 in laboratory studies of monocular vision. Nevertheless, 0.02 has become an accepted liminal contrast value for meteorological-range measurements.

Another measure of visibility, **visual range**, is the actual distance at which a person can discern an ideal dark object against the horizon sky. **Prevailing visibility** is the greatest visual range a person can see anywhere along a 180° horizontal arc.

Meteorological range, typically used for visibility calculations, is derived in the following manner. Suppose a perfectly absorbing dark object lies against a white background at a point $x_0$, as shown in Fig. 10.21. Since the object is perfectly absorbing, it reflects and emits no visible radiation; thus, its visible radiance at $x_0$ is zero ($I_0 = 0$). As a viewer backs away from the object, background white light scatters into the field of view, increasing the background light that the viewer sees. Some of this light is scattered out of the field of view or absorbed along the field of view by gases and particles. At some distance away from the object, so much background light has entered the path between the viewer and the object that the viewer can barely discern the object against the background light.

The **change in object intensity** along the path described can be approximated by

$$\frac{dI}{dx} = \sigma_b I_B - \sigma_{ext} I \qquad (10.43)$$

where all wavelength subscripts have been removed. In this equation, $I_B$ is the background intensity (or radiance) of light, which is constant along the path, $\sigma_b I_B$ is the constant allowing for the scattering of background light into the line of vision at all points along the path, and $\sigma_{ext}$ is the total extinction coefficient along the path, which is the sum of extinctions due to scattering and absorption by gases,

**Table 10.9.** Meteorological Ranges Due to Gas Scattering $(s, g)$, Gas Absorption $(a, g)$, Aerosol Scattering $(s, a)$, Aerosol Absorption $(a, a)$, and Total Extinction at a 0.55-μm Wavelength on a Clean and a Polluted Day in Los Angeles

| | **Meteorological Range (km)** | | | | |
|---|---|---|---|---|---|
| **Day** | $x_{s,g,\lambda}$ | $x_{a,g,\lambda}$ | $x_{s,a,\lambda}$ | $x_{a,a,\lambda}$ | $x_\lambda$ |
| Clean (4/7/83) | 352 | 326 | 151 | 421 | 67.1 |
| Polluted (8/25/83) | 366 | 130 | 9.59 | 49.7 | 7.42 |

Extinction coefficients from Larson et al. (1984). The meteorological range due to total extinction is $x_\lambda = 3.912/(\sigma_{s,g,\lambda} + \sigma_{a,g,\lambda} + \sigma_{s,a,\lambda} + \sigma_{a,a,\lambda})$.

particles, and cloud drops. The product $\sigma_{ext} I$ is the total attenuation of radiation along the path due to scattering and absorption.

The **change in background intensity** with distance is

$$\frac{dI_B}{dx} = \sigma_b I_B - \sigma_{ext} I_B = 0 \tag{10.44}$$

Because the background intensity does not change along the path, (10.44) is set to zero, and $\sigma_b = \sigma_{ext}$. Substituting this result into (10.43) and rearranging gives

$$\frac{dI}{I_B - I} = \sigma_{ext}\, dx \tag{10.45}$$

Integrating (10.45) from $I_0(= 0)$ to $I$ and $x_0(= 0)$ to $x$ yields the **contrast ratio**

$$C_{ratio} = \frac{I_B - I}{I_B} = e^{-\sigma_{ext} x} \tag{10.46}$$

which is the relative ratio of intensity between an object and a background. When $C_{ratio} = 0.02$ at $\lambda = 0.55$ μm, the resulting distance is the meteorological range (**Koschmieder equation**),

$$x = \frac{3.912}{\sigma_{ext}} \tag{10.47}$$

In polluted tropospheric air, the only important gas-phase visible-light attenuation processes are Rayleigh scattering and absorption by nitrogen dioxide (Waggoner et al. 1981). Scattering by particles – particularly those containing sulfate, organic carbon, and nitrate – causes between 60 and 95 percent of visibility reduction. Absorption by soot particles causes between 5 and 40 percent of reduction (Cass 1979; Tang et al. 1981; Waggoner et al. 1981).

Larson et al. (1984) studied visibility in Los Angeles on two days in 1983: one day clear, and the second, polluted. They calculated gas and particle scattering and absorption coefficients, which are shown converted to meteorological ranges in Table 10.9.

Table 10.9 shows that particle scattering at 0.55 μm dominated light extinction on the polluted and on the clear day. Even on a clear day in Los Angeles, the only impediment to visibility was particle scattering. On the clear day, gas absorption by $NO_2$ had an effect comparable to that of Rayleigh scattering. On the polluted day, its effect was only slightly more significant. Absorption by soot had a significant effect on the polluted day.

## 10.6. OPTICAL DEPTH

A purpose of simulating atmospheric radiation is to estimate the spectral irradiance ($F_\lambda$) that reaches a given layer of the atmosphere. The first step in such a calculation is to calculate the **total spectral extinction coefficient** at each wavelength as

$$\sigma_\lambda = \sigma_{s,g,\lambda} + \sigma_{a,g,\lambda} + \sigma_{s,a,\lambda} + \sigma_{a,a,\lambda} + \sigma_{s,c,\lambda} + \sigma_{a,c,\lambda} \tag{10.48}$$

where the individual coefficients are for scattering by gases, absorption by gases, scattering by aerosols, absorption by aerosols, scattering by cloud drops, and absorption by cloud drops, respectively. The next step is to define the vertical component of the incremental distance along a beam of interest through which radiation travels. When the beam of interest is the solar beam, this component is

$$dz = \cos\theta_s \, dS_b = \mu_s \, dS_b \tag{10.49}$$

where $dS_b$ is the **incremental distance along the beam**, $\theta_s$ is the **solar zenith angle**, and

$$\mu_s = \cos\theta_s \tag{10.50}$$

The solar zenith angle is the angle between the surface normal (a line directed from the center of the earth that extends vertically above the surface) and the direction of the sun. Figure 10.22 shows the relationship among $dz$, $dS_b$, and $\theta_s$.

An **optical depth** (unitless) is found by integrating an extinction coefficient over an incremental distance. For radiative transfer calculations, the distance of interest is in the vertical direction. In such cases, optical depth quantifies scattering and absorbing that occurs between the top of the atmosphere and a given altitude. The optical depth increases from zero at the top of the atmosphere to a maximum at the ground surface. The **incremental optical depth** is

$$d\tau_\lambda = -\sigma_\lambda \, dz = -\sigma_\lambda \mu_s \, dS_b \tag{10.51}$$

which increases in the opposite direction to the incremental altitude. Integrating (10.51) from the top of the atmosphere ($z = S_b = \infty$) to any altitude $z$, which corresponds to a location $S_b$ along the beam of interest, gives the optical depth as

$$\tau_\lambda = \int_\infty^z \sigma_\lambda \, dz = \int_\infty^{S_b} \sigma_\lambda \mu_s \, dS_b \tag{10.52}$$

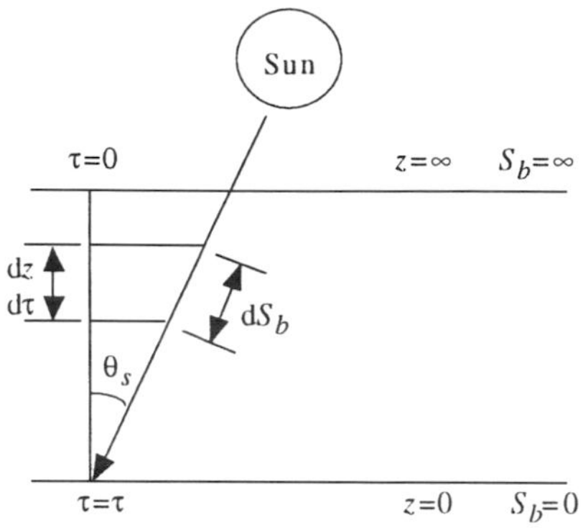


**Figure 10.22.** Relationship among incremental optical depth ($d\tau$), incremental altitude ($dz$), solar zenith angle ($\theta_s$), and incremental distance along the beam ($dS_b$).

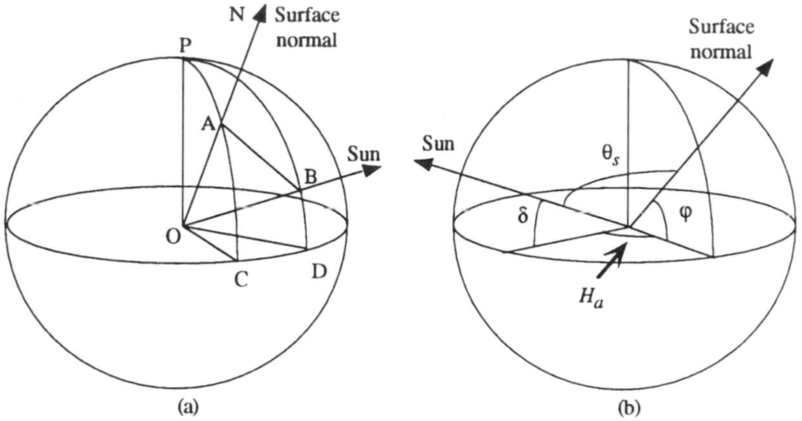


**Figure 10.23.** (a) Geometry for zenith angle calculations on a sphere. The ray OAN is the surface normal above the point of interest. The ray OB points to the sun, which is incident directly over point B. Point B is the subsolar point. Angle AOB is the solar zenith angle ($\theta_s$). Angle BOD is the solar declination ($\delta$), angle AOC is the latitude ($\varphi$) of the surface normal, angle CAB is the solar azimuth angle ($\phi_s$), and angles COD = CPD = APB are hour angles ($H_a$). (b) Geometry for a different solar zenith angle.

## 10.7. SOLAR ZENITH ANGLE

The **solar zenith angle** is estimated from

$$\cos\theta_s = \sin\varphi\sin\delta + \cos\varphi\cos\delta\cos H_a \tag{10.53}$$

where $\varphi$ is the latitude, $\delta$ is the solar declination angle, and $H_a$ is the local hour angle of the sun, as shown in Fig. 10.23(a) and (b). The **declination angle** is the angle between the equator and the north or south latitude of the **subsolar point,** which is the point at which the sun is directly overhead. The **local hour angle** is the angle, measured westward, between the longitude (meridian) of the subsolar point and the longitude of the location of interest.

(10.53) is obtained from Fig. 10.23 (a) by applying the law of cosines to triangle APB. Since arcangle AOP = $90^\circ - \varphi$, arcangle BOP = $90^\circ - \delta$ , and the distances OP, OA, and OB are known, the arc lengths AP and BP can be determined from the law of sines. With these distances and the fact that angle APB = $H_a$, the arc length AB can be obtained from the law of cosines. The solar zenith angle (AOB) can be determined from the law of sines, since the distances AB, OA, and OB are known (e.g., Hartman 1994).

The **solar declination angle** is found from

$$\delta = \sin^{-1}(\sin \varepsilon_{ob} \sin \lambda_{ec}) \tag{10.54}$$

where $\varepsilon_{ob}$ is the obliquity of the ecliptic and $\lambda_{ec}$ is the ecliptic longitude of the earth. The **ecliptic** is the mean plane of the earth's orbit around the sun. It is fixed in space relative to the earth, and the earth rotates through it. It cuts through the Tropic of Capricorn on one side of the earth, the equator in the middle and the Tropic of Cancer on the other side of the earth. The **obliquity of the ecliptic** is the angle between the plane of the earth's equator and the plane of the ecliptic, approximated as

$$\varepsilon_{ob} = 23^\circ.439 - 0^\circ.0000004 N_{JD} \tag{10.55}$$

(NAO 1993), where

$$N_{JD} = \begin{cases} -4384.5 + (Y - 1988) \times 365 + D_J & 1985 \le Y < 1989 \\ -4018.5 + (Y - 1989) \times 365 + D_J & 1989 \le Y < 1993 \\ -2557.5 + (Y - 1993) \times 365 + D_J & 1993 \le Y < 1997 \\ \quad 1096.5 + (Y - 1997) \times 365 + D_J & 1997 \le Y < 2001 \\ 364.5 - (Y - 2001) \times 365 + D_J & 2001 \le Y < 2005 \end{cases} \tag{10.56}$$

is the number of days from the beginning of Julian year 2000. In (10.56), $Y$ is the current year, and $D_J$ is the **Julian day of the year**, which varies from 1 on January 1 to 365 (for nonleap years) or 366 (for leap years) on December 31. Leap years occur every year evenly divisible by 4. The **ecliptic longitude** of the earth is approximately

$$\lambda_{ec} = L_M + 1^\circ.915 \sin g_M + 0^\circ.020 \sin 2g_M \tag{10.57}$$

where $L_M$ is the mean longitude of the sun and $g_M$ is the mean anomaly. An **anomaly** is the angular measurement of a planet, in its orbit, from its perihelion, which is the point at which the planet, in orbit around the sun, is closest to the sun. The **mean longitude of the sun** and the **mean anomaly** are approximated, respectively, as

$$L_M = 280^\circ.460 + 0^\circ.9856474 N_{JD} \qquad g_M = 357^\circ.528 + 0^\circ.9856003 N_{JD} \tag{10.58}$$

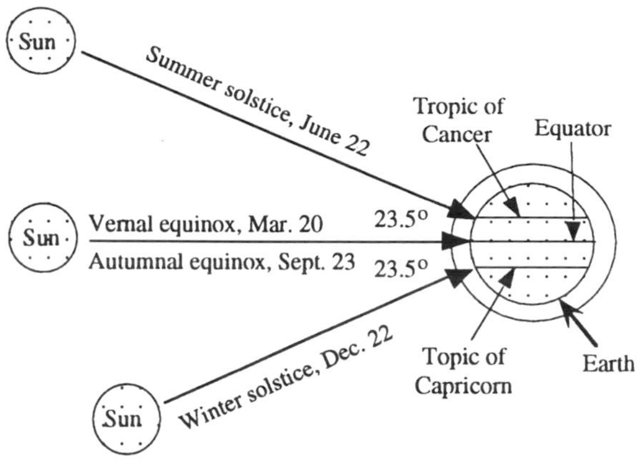


**Figure 10.24.** Solar declination angles during solstices and equinoxes of the four times shown, the earth–sun distance is greatest at the summer solstice.

The **local hour angle** (in radians) in (10.53) is

$$H_a = \frac{2\pi t_s}{86{,}400} \tag{10.59}$$

where $t_s$ is the number of seconds past local noon, and 86,400 is the number of seconds in a day. At noon, when the sun is highest, the local hour angle is zero, and (10.53) simplifies to $\cos\theta_s = \sin\varphi\sin\delta + \cos\varphi\cos\delta$. When the sun is over the equator, the declination angle and latitude are zero, and (10.53) simplifies to $\cos\theta_s = \cos H_a$. Figure 10.24 shows that the sun reaches its maximum declination ($\pm 23.5°$) at the summer and winter solstices and its minimum declination (0°) at the vernal and autumnal equinoxes.

**Example 10.9**

Calculate the solar zenith angle at 1:00 p.m. PST, on February 27, 1994 at a latitude of $\varphi = 35°$ N.

SOLUTION

February 27 corresponds to Julian day $D_J = 58$. From (10.56), $N_{JD} = -2134.5$. From (10.58), $L_m = -1823.40°$ and $g_m = -1746.23°$. Thus, from (10.57), (10.55), and (10.54), $\lambda_{ec} = -1821.87°$, $\varepsilon_{ob} = 23.4399°$, and $\delta = -8.52°$, respectively. (10.59) gives $H_a = 15.0°$. Thus, from (10.53),

$$\cos\theta_s = \sin 35° \sin(-8.52°) + \cos 35° \cos(-8.52°)\cos 15.0° \qquad \theta_s = 45.8°$$

## 10.8. THE RADIATIVE TRANSFER EQUATION

The **radiative transfer equation** gives the change in radiance and/or irradiance along a beam of electromagnetic energy at a point in the atmosphere. Radiances are used to calculate actinic fluxes, which are used in photolysis coefficient equations. Irradiances are used to calculate heating rates, which are used in temperature prediction equations. The processes that affect radiation along a beam are scattering of radiation out of the beam, absorption of radiation along the beam, multiple scattering of indirect, diffuse radiation into the beam, single scattering of direct, solar radiation into the beam, and emission of infrared radiation into the beam. **Single scattering** occurs when a photon of radiation is redirected into a beam after it collides with a particle or gas molecule, as shown in Fig. 10.25. **Multiple scattering** occurs when a photon enters a beam after colliding sequentially with several particles or gas molecules, each of which redirects the photon. Solar radiation that has not yet been scattered is **direct** radiation. Radiation, either solar or infrared, that has been scattered is **diffuse** radiation.

The change in spectral radiance over the distance $dS_b$ along a beam is

$$dI_\lambda = -dI_{so,\lambda} - dI_{ao,\lambda} + dI_{si,\lambda} + dI_{Si,\lambda} + dI_{ei,\lambda} \tag{10.60}$$

where

$$dI_{so,\lambda} = I_\lambda \sigma_{s,\lambda}\, dS_b \tag{10.61}$$

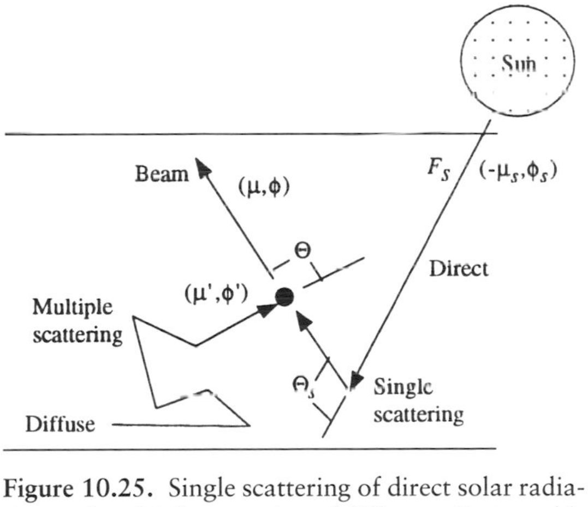


**Figure 10.25.** Single scattering of direct solar radiation and multiple scattering of diffuse radiation adds to the intensity along a beam of orientation μ, ϕ. The parameter $\mu = \cos\theta$ is always positive, but when a ray is directed upward, $+\mu$ is used, and when a ray is directed downward, $-\mu$ is used. Adapted from Liou (1992).

represents **scattering of radiation out of the beam,**

$$\mathrm{d}I_{ao,\lambda} = I_\lambda \sigma_{a,\lambda}\,\mathrm{d}S_b \tag{10.62}$$

represents **absorption of radiation along the beam,**

$$\mathrm{d}I_{si,\lambda} = \left[\sum_k \left(\frac{\sigma_{s,k,\lambda}}{4\pi}\int_0^{2\pi}\int_{-1}^{1} I_{\lambda,\mu',\phi'} P_{s,k,\lambda,\mu,\mu',\phi,\phi'}\,\mathrm{d}\mu'\,\mathrm{d}\phi'\right)\right]\mathrm{d}S_b \tag{10.63}$$

represents **multiple scattering of diffuse radiation into the beam,**

$$\mathrm{d}I_{Si,\lambda} = \left[\sum_k \left(\frac{\sigma_{s,k,\lambda}}{4\pi} P_{s,k,\lambda,\mu,-\mu_s,\phi,\phi_s}\right)\right] F_{s,\lambda} e^{-\tau_\lambda/\mu_s}\,\mathrm{d}S_b \tag{10.64}$$

represents **single scattering of direct solar radiation into the beam,** and

$$\mathrm{d}I_{ei,\lambda} = \sigma_{a,\lambda} B_{\lambda,T}\,\mathrm{d}S_b \tag{10.65}$$

represents **emission of infrared radiation into the beam.** In (10.63) and (10.64), the summations are over all scattering processes ($k = g$ for gases, $a$ for aerosols, and $c$ for cloud drops), and the $P_s$ factors are scattering phase functions, to be defined shortly. In (10.61), (10.62), and (10.65), the factors

$$\sigma_{s,\lambda} = \sigma_{s,g,\lambda} + \sigma_{s,a,\lambda} + \sigma_{s,c,\lambda} \qquad \sigma_{a,\lambda} = \sigma_{a,g,\lambda} + \sigma_{a,a,\lambda} + \sigma_{a,c,\lambda} \tag{10.66}$$

are extinction coefficients due to total scattering and absorption, respectively. The extinction coefficient due to scattering plus absorption is

$$\sigma_\lambda = \sigma_{s,\lambda} + \sigma_{a,\lambda} \tag{10.67}$$

In the solar spectrum, single scattering of solar radiation is more important than radiative emission. In the infrared spectrum, the reverse is true. Thus, (10.64) is used for solar wavelengths and (10.65) is used for infrared wavelengths.

### 10.8.1. Phase Function and Asymmetry Factor

In (10.63), $P_{s,k,\lambda,\mu,\mu',\phi,\phi'}$ is the **scattering phase function,** which gives the angular distribution of scattered energy as a function of direction. It relates how diffuse radiation which has direction $\mu'$, $\phi'$, is redirected by gases or particles towards the beam of interest, which has direction $\mu$, $\phi$, as shown in Fig. 10.25. In the figure, $\mu' = \cos\theta'$ and $\mu = \cos\theta$, where $\theta'$ and $\theta$ are the zenith angles of the diffuse radiation and the beam of interest, respectively. Similarly, $\phi'$ and $\phi$ are the azimuth angles of the diffuse radiation and the beam of interest, respectively. The integral in (10.63) is over all possible angles of incoming multiple-scattered radiation. Scattering phase functions vary with wavelength and differ for gases, aerosols, and cloud drops.

In (10.64), $P_{s,k,\lambda,\mu,-\mu_s,\phi,\phi_s}$ is the scattering phase function for direct radiation. The function relates how direct solar radiation, with direction $-\mu_s (= -\cos\theta_s)$, $\phi_s$, is redirected by gases or particles to μ, ϕ, as shown in Fig. 10.25. This phase function is not integrated over all solid angles, since single-scattered radiation originates from one angle.

The scattering phase function is defined so that

$$\frac{1}{4\pi}\int_{4\pi} P_{s,k,\lambda}(\Theta)\, d\Omega_a = 1 \tag{10.68}$$

In this equation, Θ is the angle between directions μ′, ϕ′ and μ, ϕ, as shown in Fig. 10.25. Thus, $P_{s,k,\lambda,\mu,\mu',\phi,\phi'} = P_{s,k,\lambda}(\Theta)$. Here, $\Theta_s$ is the angle between the solar beam $(-\mu_s, \phi_s)$ and the beam of interest (μ, ϕ). Thus, $P_{s,k,\lambda,\mu,-\mu_s,\phi,\phi_s} = P_{s,k,\lambda}(\Theta_s)$.

Substituting $d\Omega_a = \sin\Theta\, d\Theta\, d\phi$ from (10.8) into (10.68) gives

$$\frac{1}{4\pi}\int_0^{2\pi}\int_0^{\pi} P_{s,k,\lambda}(\Theta)\sin\Theta\, d\Theta\, d\phi = 1 \tag{10.69}$$

The integral limits are defined so that the integration is over a full sphere. For **isotropic scattering,** the phase function is

$$P_{s,k,\lambda}(\Theta) = 1 \tag{10.70}$$

which satisfies (10.68). For **Rayleigh (gas) scattering,** the phase function is

$$P_{s,k,\lambda}(\Theta) = \tfrac{3}{4}(1+\cos^2\Theta) \tag{10.71}$$

which also satisfies (10.68). Figure 10.26(a) and (b) show scattering phase functions for isotropic and Rayleigh scattering, respectively, versus scattering angle. The phase function for isotropic scattering projects equally in all directions, and that for Rayleigh scattering is symmetric, but projects mostly in the forward and backward directions.

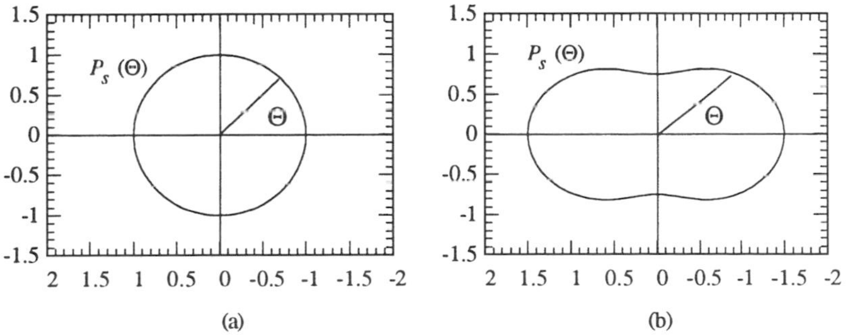


**Figure 10.26.** Scattering phase function distribution in polar coordinates for (a) isotropic and (b) Rayleigh scattering. The diagrams were generated from (10.70) and (10.71), respectively.

The **asymmetry factor**, or first moment of the phase function, is

$$g_{a,k,\lambda} = \frac{1}{4\pi}\int_{4\pi} P_{s,k,\lambda}(\Theta)\cos\Theta\, d\Omega_a \tag{10.72}$$

which gives the relative direction of scattering by particles or gases. In general,

$$g_{a,k,\lambda}\begin{cases} > 0 & \text{forward (Mie) scattering} \\ = 0 & \text{isotropic or Rayleigh scattering} \\ < 0 & \text{backward scattering} \end{cases} \tag{10.73}$$

The asymmetry factor approaches +1 for scattering strongly peaked in the forward direction and −1 for scattering strongly peaked in the backward direction. Expanding (10.72) with $d\Omega_a = \sin\Theta\, d\Theta\, d\phi$ yields

$$g_{a,k,\lambda} = \frac{1}{4\pi}\int_0^{2\pi}\int_0^{\pi} P_{s,k,\lambda}(\Theta)\cos\Theta\sin\Theta\, d\Theta\, d\phi \tag{10.74}$$

For **isotropic scattering**, where $P_{s,k,\lambda}(\Theta) = 1$, the asymmetry factor simplifies to

$$g_{a,k,\lambda} = \frac{1}{4\pi}\int_0^{2\pi}\int_0^{\pi}\cos\Theta\sin\Theta\, d\Theta\, d\phi = -\frac{1}{2}\int_1^{-1}\mu\, d\mu = 0 \tag{10.75}$$

where $\mu = \cos\Theta$. The zero asymmetry factor for isotropic scattering is expected, since isotropic scattering distributes radiation equally in all directions. Substituting the phase function for **Rayleigh scattering** into (10.74) gives

$$\begin{aligned} g_{a,k,\lambda} &= \frac{1}{4\pi}\int_0^{2\pi}\int_0^{\pi}\frac{3}{4}(1+\cos^2\Theta)\cos\Theta\sin\Theta\, d\Theta\, d\phi \\ &= -\frac{3}{8}\int_0^{2\pi}\int_1^{-1}(\mu+\mu^3)\, d\mu\, d\phi = 0 \end{aligned} \tag{10.76}$$

The Rayleigh scattering intensity is evenly distributed between forward and backward directions, but Rayleigh scattering is not isotropic, since the radiation is not scattered equally in all directions.

An approximation to the phase function for Mie scattering by aerosols and cloud drops is the **Henyey–Greenstein function** (Henyey and Greenstein 1941), which gives the phase function as

$$P_{s,k,\lambda}(\Theta) = \frac{1 - g_{a,k,\lambda}^2}{\left(1 + g_{a,k,\lambda}^2 - 2g_{a,k,\lambda}\cos\Theta\right)^{3/2}} \tag{10.77}$$

This equation is valid primarily for scattering that is not strongly peaked in the forward direction (Liou 1992) and requires knowledge of the asymmetry factor in advance.

The asymmetry factor for aerosols or cloud drops can be found with

$$g_{a,k,\lambda} = Q_{f,i,\lambda}/Q_{s,i,\lambda} \tag{10.78}$$

where $Q_{f,i,\lambda}$ is the single-particle forward scattering efficiency and $Q_{s,i,\lambda}$ is the single-particle total scattering efficiency, both of which can be obtained from a Mie-scattering algorithm. Curves of both parameters were shown in Figs. 10.18 and 10.19 for two substances. These figures indicate that the asymmetry factors for elemental carbon and water particles are between 0 and 1.

### 10.8.2. Incident Solar Radiation

In (10.64), $F_{s,\lambda}$ is the incident solar radiation at the top of the atmosphere in a wavelength interval centered at $\lambda$. The radiation intensity at the top of the atmosphere depends on the solar luminosity and the earth–sun distance. The **solar luminosity** ($L_p$) is the total energy per unit time emitted by the sun's photosphere and is approximately $3.9 \times 10^{26}$ W. The irradiance, or luminosity emitted per unit area (W m$^{-2}$) at the photosphere, is approximately

$$F_p = \frac{L_p}{4\pi R_p^2} = \sigma_B T_p^4 \tag{10.79}$$

where $R_p = 6.96 \times 10^8$ m is the radius of the sun, from its center to the photosphere, $T_p$ is the temperature of the sun at the photosphere, and (10.79) is a summation over all wavelength intervals.

**Example 10.10**

For a photosphere temperature of $T_p = 5796$ K, $F_p \approx 6.4 \times 10^7$ W m$^{-2}$ from (10.79).

Irradiance decreases proportionally to the inverse square distance from its source. The yearly mean solar irradiance reaching the top of earth's atmosphere, or **solar constant,** is related to $F_p$ by

$$\bar{F}_s = \left(\frac{R_p}{\bar{R}_{es}}\right)^2 F_p = \left(\frac{R_p}{\bar{R}_{es}}\right)^2 \sigma_B T_p^4 \tag{10.80}$$

where $\bar{R}_{es} = 1.5 \times 10^{11}$ m is the mean distance from the center of the sun to the earth. (10.80) gives $\bar{F}_s \approx 1379$ W m$^{-2}$, which is close to measurements of $\bar{F}_s = 1365$ W m$^{-2}$. The value of $\bar{F}_s$ varies by $\pm 1$ W m$^{-2}$ over each 11-year sunspot cycle.

The actual total irradiance at the top of the atmosphere at a given time is

$$F_s = \left(\frac{R_{es}}{\bar{R}_{es}}\right)^2 \bar{F}_s \tag{10.81}$$

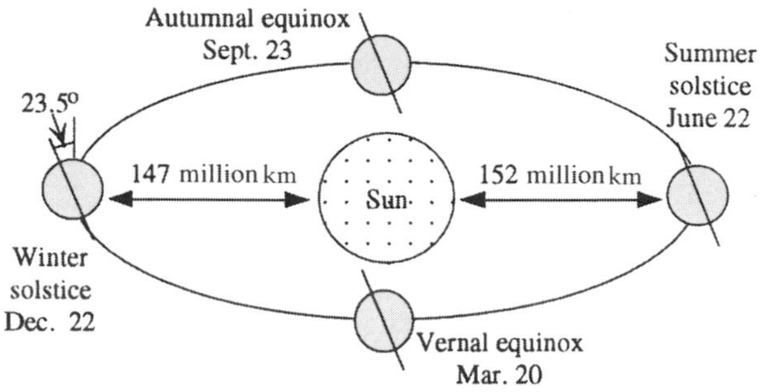


**Figure 10.27.** Relationship between the sun and earth during the solstices and equinoxes.

where $R_{es}$ is the actual distance between the sun and the earth on a given day, and

$$\left(\frac{R_{es}}{\bar{R}_{es}}\right)^2 \approx 1.00011 + 0.034221\cos\theta_J + 0.00128\sin\theta_J + 0.000719\cos 2\theta_J + 0.000077\sin 2\theta_J \quad (10.82)$$

(Spencer 1971). In (10.82), $\theta_J = 2\pi D_J / D_Y$, where $D_Y$ is the number of days in a year (365 for nonleap years and 366 for leap years), and $D_J$ is the Julian day of the year. Figure 10.27 shows that the earth is further from the sun in the Northern Hemisphere summer than in the winter by about 3.4 percent.

**Example 10.11**

On December 22, the total irradiance at the top of earth's atmosphere is $F_s = 1365 \times 1.034 = 1411$ W m$^{-2}$, and on June 22, the irradiance is $F_s = 1365 \times 0.967 = 1321$ W m$^{-2}$. Thus, the irradiance varies by 90 W m$^{-2}$ (6.6 percent) between December and June.

In a model, the actual total solar irradiance reaching the top of earth's atmosphere ($F_s$) (W m$^{-2}$) is related to the actual solar irradiance in each wavelength interval ($F_{s,\lambda}$) (W m$^{-2}$ μm$^{-1}$), the mean solar irradiance in each wavelength interval ($\bar{F}_{s,\lambda}$) (W m$^{-2}$ μm$^{-1}$), and the solar constant $\bar{F}_s$ (W m$^{-2}$) by

$$F_s = \sum_\lambda (F_{s,\lambda}\Delta\lambda) = \left(\frac{R_{es}}{\bar{R}_{es}}\right)^2 \sum_\lambda (\bar{F}_{s,\lambda}\Delta\lambda) = \left(\frac{R_{es}}{\bar{R}_{es}}\right)^2 \bar{F}_s \quad (10.83)$$

where $\Delta\lambda$ is a wavelength interval (μm). Values for $\bar{F}_{s,\lambda}$ are given in Table B.2 of Appendix B.

The **solar constant** $\bar{F}_s$ can be used to determine earth's equilibrium temperature in the absence of the greenhouse effect. The total energy (W) absorbed by the earth–atmosphere system is about

$$E_{\text{in}} = \bar{F}_s(1 - A_{e,0})(\pi R_e^2) \tag{10.84}$$

where $R_e$ is the earth's radius, $\pi R_e^2$ is the earth's cross-sectional area (effective area that solar irradiance impinges upon), and $A_{e,0}$ is the globally and wavelength-averaged earth–atmosphere albedo ($\approx$0.30). The outgoing energy from the earth is

$$E_{\text{out}} = \varepsilon_{e,0}\sigma_B T_e^4 (4\pi R_e^2) \tag{10.85}$$

where $T_e$ is the temperature of the earth's surface, and $\varepsilon_{e,0}$ is earth's globally and wavelength-averaged emissivity. The actual average emissivity is 0.96–0.98, but if the earth is considered a blackbody, $\varepsilon_{e,0} = 1$. Equating incoming and outgoing energy gives

$$T_e = \left[\frac{\bar{F}_s(1 - A_{e,0})}{4\varepsilon_{e,0}\sigma_B}\right]^{1/4} \tag{10.86}$$

which is the equilibrium temperature of the earth in the absence of a greenhouse effect.

**Example 10.12**

For $F_s = 1365$ W m$^{-2}$, $A_{e,0} = 0.3$, and $\varepsilon_{e,0} = 1$, (10.86) predicts $T_e = 254.8$ K. The actual average surface temperature on earth is about 288 K, and the difference is due primarily to absorption by greenhouse gases.

### 10.8.3. A Solution to the Radiative Transfer Equation

(10.60)–(10.65) can be combined to give the radiative transfer equation as

$$\begin{aligned}\frac{dI_{\lambda,\mu,\phi}}{dS_b} &= -I_{\lambda,\mu,\phi}(\sigma_{s,\lambda} + \sigma_{a,\lambda}) \\ &\quad + \sum_k \left(\frac{\sigma_{s,k,\lambda}}{4\pi}\int_0^{2\pi}\int_{-1}^{1} I_{\lambda,\mu',\phi'} P_{s,k,\lambda,\mu,\mu',\phi,\phi'}\, d\mu'\, d\phi'\right) \\ &\quad + F_{s,\lambda}e^{-\tau_\lambda/\mu_s}\sum_k \left(\frac{\sigma_{s,k,\lambda}}{4\pi} P_{s,k,\lambda,\mu,-\mu_s,\phi,\phi_s}\right) + \sigma_{a,\lambda}B_{\lambda,T}\end{aligned} \tag{10.87}$$

The fraction of total extinction due to scattering is the **single-scattering albedo,**

$$\omega_{s,\lambda} = \frac{\sigma_{s,\lambda}}{\sigma_\lambda} = \frac{\sigma_{s,g,\lambda} + \sigma_{s,a,\lambda} + \sigma_{s,c,\lambda}}{\sigma_{s,g,\lambda} + \sigma_{a,g,\lambda} + \sigma_{s,a,\lambda} + \sigma_{a,a,\lambda} + \sigma_{s,c,\lambda} + \sigma_{a,c,\lambda}} \tag{10.88}$$

Substituting (10.88), $\sigma_\lambda = \sigma_{s,\lambda} + \sigma_{a,\lambda}$ from (10.67) and $d\tau_\lambda = -\sigma_\lambda \mu_s \, dS_b$ from (10.51) into (10.87) gives the radiative transfer equation as

$$\mu \frac{dI_{\lambda,\mu,\phi}}{d\tau_\lambda} = I_{\lambda,\mu,\phi} - J^{diffuse}_{\lambda,\mu,\phi} - J^{direct}_{\lambda,\mu,\phi} - J^{emis}_{\lambda,\mu,\phi} \tag{10.89}$$

where

$$J^{diffuse}_{\lambda,\mu,\phi} = \frac{1}{4\pi} \sum_k \left( \frac{\sigma_{s,k,\lambda}}{\sigma_\lambda} \int_0^{2\pi} \int_{-1}^{1} I_{\lambda,\mu',\phi'} P_{s,k,\lambda,\mu,\mu',\phi,\phi'} \, d\mu' \, d\phi' \right) \tag{10.90}$$

$$J^{direct}_{\lambda,\mu,\phi} = \frac{1}{4\pi} F_{s,\lambda} e^{-\tau_\lambda/\mu_s} \sum_k \left[ \frac{\sigma_{s,k,\lambda}}{\sigma_\lambda} P_{s,k,\lambda,\mu,-\mu_s,\phi,\phi_s} \right] \tag{10.91}$$

$$J^{emis}_{\lambda,\mu,\phi} = (1 - \omega_{s,\lambda}) B_{\lambda,T} \tag{10.92}$$

The equation must be solved numerically, except when simplifying assumptions are made. An analytical solution can be derived when absorption is considered but scattering and emissions are neglected. In this case, (10.89) simplifies for the upward and downward directions as

$$\mu \frac{dI_{\lambda,\mu,\phi}}{d\tau_{a,\lambda}} = I_{\lambda,\mu,\phi} \qquad -\mu \frac{dI_{\lambda,-\mu,\phi}}{d\tau_{a,\lambda}} = I_{\lambda,-\mu,\phi} \tag{10.93}$$

respectively, where $+\mu$ is used in the upward case, $-\mu$ is used in the downward case, and $\tau_{a,\lambda}$ is the optical depth due to absorption only. Integrating the **upward** equation from a lower optical depth $\tau_{a,\lambda,b}$ to the optical depth of interest, $\tau_{a,\lambda}$, gives

$$I_{\lambda,\mu,\phi}(\tau_{a,\lambda}) = I_{\lambda,\mu,\phi}(\tau_{a,\lambda,b}) e^{(\tau_{a,\lambda} - \tau_{a,\lambda,b})/\mu} \tag{10.94}$$

Integrating the **downward** equation from an upper optical depth $\tau_{a,\lambda,t}$ to the optical depth of interest, $\tau_{a,\lambda}$, gives

$$I_{\lambda,-\mu,\phi}(\tau_{a,\lambda}) = I_{\lambda,-\mu,\phi}(\tau_{a,\lambda,t}) e^{-(\tau_{a,\lambda} - \tau_{a,\lambda,t})/\mu} \tag{10.95}$$

(10.94) and (10.95) describe **Beer's law**, which states that the absorption of radiation increases exponentially with the optical depth of the absorbing species.

A second analytical solution can be derived when absorption and infrared emissions are considered, but scattering is neglected. In this case, (10.89) simplifies for the upward and downward directions as

$$\mu \frac{dI_{\lambda,\mu,\phi}}{d\tau_{a,\lambda}} = I_{\lambda,\mu,\phi} - B_{\lambda,T} \qquad -\mu \frac{dI_{\lambda,-\mu,\phi}}{d\tau_{a,\lambda}} = I_{\lambda,-\mu,\phi} - B_{\lambda,T} \tag{10.96}$$

respectively, which are **Schwartzchild's equations**. Integrating (10.96) gives upward and downward radiances as

$$I_{\lambda,\mu,\phi}(\tau_{a,\lambda}) = I_{\lambda,\mu,\phi}(\tau_{a,\lambda,b})e^{(\tau_{a,\lambda} - \tau_{a,\lambda,b})/\mu} - \frac{1}{\mu}\int_{\tau_{a,\lambda,b}}^{\tau_{a,\lambda}} \left[B_{\lambda,T(\tau'_{a,\lambda})}e^{(\tau_{a,\lambda} - \tau'_{a,\lambda})/\mu}\right] d\tau'_{a,\lambda} \tag{10.97}$$

$$I_{\lambda,-\mu,\phi}(\tau_{a,\lambda}) = I_{\lambda,-\mu,\phi}(\tau_{a,\lambda,t})e^{-(\tau_{a,\lambda} - \tau_{a,\lambda,t})/\mu} + \frac{1}{\mu}\int_{\tau_{a,\lambda,t}}^{\tau_{a,\lambda}} \left[B_{\lambda,T(\tau'_{a,\lambda})}e^{-(\tau_{a,\lambda} - \tau'_{a,\lambda})/\mu}\right] d\tau'_{a,\lambda} \tag{10.98}$$

respectively. In each equation, the Planck function, which varies with temperature, must be integrated between the two optical depths.

When scattering is included in the radiative transfer equation, analytical solutions become difficult to obtain and numerical solutions are needed. One numerical solution is found with the **two-stream method**. With this method, radiance is divided into an upward (+) and downward (−) component, each of which is approximated with a forward and a backward scattering term in the diffuse phase function integral of (10.90). One approximation to the integral is

$$\frac{1}{4\pi}\int_0^{2\pi}\int_{-1}^{1} I_{\lambda,\mu',\phi'} P_{s,k,\lambda,\mu,\mu',\phi,\phi'}\, d\mu' d\phi' \approx \begin{cases} \frac{(1+g_{a,k,\lambda})}{2} I_+ + \frac{(1-g_{a,k,\lambda})}{2} I_- & \text{upward} \\ \frac{(1+g_{a,k,\lambda})}{2} I_- + \frac{(1-g_{a,k,\lambda})}{2} I_+ & \text{downward} \end{cases} \tag{10.99}$$

where $(1 + g_{a,k,\lambda})/2$ and $(1 - g_{a,k,\lambda})/2$ are integrated fractions of the forward- and backward-scattered energy, respectively. Wavelength subscripts have been omitted on $I_+$ and $I_-$. (10.99) is the **two-point quadrature** approximation to the phase-function integral (Liou 1974, 1992; Meador and Weaver 1980). Substituting (10.99) into (10.90) gives

$$\frac{1}{4\pi}\sum_k \left[\frac{\sigma_{s,k,\lambda}}{\sigma_\lambda}\int_0^{2\pi}\int_{-1}^{1} I_{\lambda,\mu',\phi'} P_{s,k,\lambda,\mu,\mu',\phi,\phi'}\, d\mu'\, d\phi'\right] \approx \begin{cases} \omega_{s,\lambda}(1 - b_\lambda) I_+ + \omega_{s,\lambda} b_\lambda I_- \\ \omega_{s,\lambda}(1 - b_\lambda) I_- + \omega_{s,\lambda} b_\lambda I_+ \end{cases} \tag{10.100}$$

where $\omega_{s,\lambda}$ is the single-scattering albedo from (10.88),

$$1 - b_\lambda = \frac{1 + g_{a,\lambda}}{2} \qquad b_\lambda = \frac{1 - g_{a,\lambda}}{2} \tag{10.101}$$

are effective integrated fractions of forward- and backward-scattered energy, and

$$g_{a,\lambda} = \frac{\sigma_{s,a,\lambda} g_{a,a,\lambda} + \sigma_{s,c,\lambda} g_{a,c,\lambda}}{\sigma_{s,g,\lambda} + \sigma_{s,a,\lambda} + \sigma_{s,c,\lambda}} \tag{10.102}$$

is an **effective asymmetry factor.**

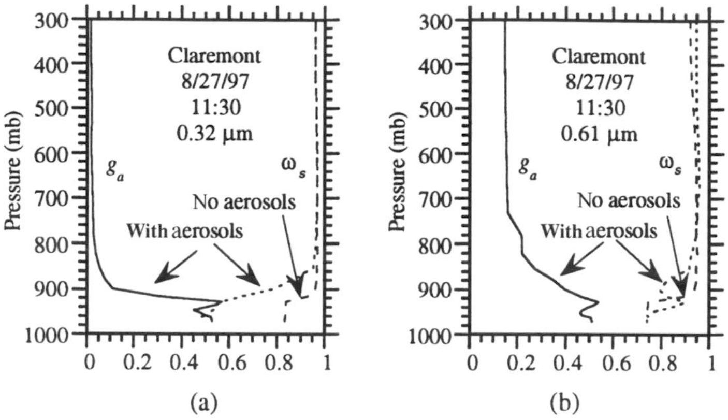


**Figure 10.28.** Predicted vertical profiles from Claremont, California, of the effective asymmetry factor and the single-scattering albedo at (a) 0.32 and (b) 0.61 μm when aerosols were and were not included in the simulations. The overall asymmetry factor was zero when aerosols were absent. The peaks at 920 mb indicate elevated aerosol layers. From Jacobson (1998b).

The effective asymmetry factor is the weighed sum of asymmetry factors for gases ($g_{a,g,\lambda}$), aerosols ($g_{a,a,\lambda}$), and cloud drops ($g_{a,c,\lambda}$). Asymmetry factors for gases are zero. Figure 10.28 (a) and (b) show vertical profiles of effective asymmetry factors and single-scattering albedos over an urban airshed from two model simulations. In one simulation, aerosols were assumed to be present, and in the other, they were ignored. Clouds did not appear during the simulation period. The figures indicate that effective asymmetry factors increased from the free troposphere to the boundary layer when aerosols were present. Above the boundary layer, the effective asymmetry factors were small because Rayleigh scattering, which has a zero asymmetry factor, dominated scattering extinction (see Fig. 10.20). Within the boundary layer, effective asymmetry factors increased because aerosol-scattering extinction dominated Rayleigh-scattering extinction.

The single-scattering albedo curves in Fig. 10.28 (a) and (b) indicate that, above the boundary layer, ultraviolet extinction was due almost entirely to gas scattering when aerosols were absent, and to gas and particle scattering when aerosols were present. Within the boundary layer, aerosol absorption played a larger role than above the boundary layer.

The phase function for the single scattering of solar radiation can be estimated with the Eddington approximation of the solar phase function (Eddington 1916)

$$P_{s,k,\lambda}(\Theta_s) \approx 1 \pm 3g_{a,\lambda}\mu_1\mu_s \tag{10.103}$$

where $\mu_1$ is the diffusivity factor, set to $\mu_1 = 1/\sqrt{3}$ when the quadrature approximation for diffuse radiation is used (Liou 1974).

With the parameters above, the **spectral radiance** from (10.89) can be written for solar wavelengths (where the emission term is neglected) in terms of an upward and a downward component (Liou 1992) as

$$\mu_1 \frac{dI_+}{d\tau} = I_+ - \omega_s(1-b)I_+ - \omega_s b I_- - \frac{\omega_s}{4\pi}(1 - 3g_a\mu_1\mu_s)F_s e^{-\tau/\mu_s} \quad (10.104)$$

$$-\mu_1 \frac{dI_-}{d\tau} = I_- - \omega_s(1-b)I_- - \omega_s b I_+ - \frac{\omega_s}{4\pi}(1 + 3g_a\mu_1\mu_s)F_s e^{-\tau/\mu_s} \quad (10.105)$$

respectively, where $I_-$ is the **downward radiance**, $I_+$ is the **upward radiance**, $+\mu_1$ is used for upward radiance, $-\mu_1$ is used for downward radiance, and all wavelength subscripts have been omitted. In (10.104), $-\mu_1$ is used for direct solar radiation because solar radiation is downward relative to $I_+$. In (10.105), $+\mu_1$ is used for direct solar radiation because solar radiation is in the same direction as $I_-$.

(10.104) and (10.105) can be written in terms of **spectral irradiance** with the conversions $F_+ = 2\pi\mu_1 I_+$ and $F_- = 2\pi\mu_1 I_-$, and generalized for the two-point quadrature or any other two-stream approximation. The resulting equations are (Meador and Weaver 1980; Toon et al. 1989b; Liou 1992)

$$\frac{dF_+}{d\tau} = \gamma_1 F_+ - \gamma_2 F_- - \gamma_3 \omega_s F_s e^{-\tau/\mu_s} \quad (10.106)$$
$$\frac{dF_-}{d\tau} = -\gamma_1 F_- + \gamma_2 F_+ + (1-\gamma_3)\omega_s F_s e^{-\tau/\mu_s}$$

where the $\gamma$'s, defined in Table 10.10, are coefficients derived from the two point quadrature and Eddington approximations of the diffuse phase-function integral. The Eddington approximation of the diffuse phase function integral is discussed in Irvine (1968, 1975), Kawata and Irvine (1970), and Meador and Weaver (1980). Replacing the solar term in (10.106) with an emission term gives an equation for infrared irradiance as

$$\frac{dF_+}{d\tau} = \gamma_1 F_+ - \gamma_2 F_- - 2\pi(1-\omega_s)B_T \quad (10.107)$$
$$\frac{dF_-}{d\tau} = -\gamma_1 F_- + \gamma_2 F_+ + 2\pi(1-\omega_s)B_T$$

(10.104)–(10.107) may be solved in a model after a vertical grid is defined, the equations are discretized, and the boundary conditions are defined. Figure 7.1 shows a model atmosphere in which layer bottoms and tops are defined by indices, $k \pm 0.5$. For this grid, upward and downword irradiances at the bottom of a layer $k$ are $F_{+,k+1/2}$ and $F_{-,k+1/2}$ respectively, where, $k = 0, \ldots, N_L$. The derivatives in (10.104)–(10.107) can be discretized over this grid with a finite-difference expansion of any order. A discretization requires boundary conditions at the ground and top of the atmosphere. At the ground, boundary conditions for irradiance at a given wavelength are

$$F_{+,N_L+1/2} = A_e F_{-,N_L+1/2} + \begin{cases} A_e \mu_s F_s e^{-\tau_{N_L+1/2}/\mu_s} & \text{solar} \\ \varepsilon\pi B_T & \text{infrared} \end{cases} \quad (10.108)$$

**Table 10.10.** Coefficients for the Two-Stream Method for Two Approximations of the Diffuse Phase Function

| **Approximation** | $\gamma_1$ | $\gamma_2$ | $\gamma_3$ |
|---|---|---|---|
| Quadrature | $\frac{1-\omega_s(1+g_a)/2}{\mu_1}$ | $\frac{\omega_s(1-g_a)}{2\mu_1}$ | $\frac{1-3g_a\mu_1\mu_s}{2}$ |
| Eddington | $\frac{7-\omega_s(4+3g_a)}{4}$ | $-\frac{1-\omega_s(4-3g_a)}{4}$ | $\frac{2-3g_a\mu_s}{4}$ |

*Sources:* Liou (1992), Meador and Weaver (1980), Toon et al. (1989a).

where $A_e$ is the albedo and $\varepsilon$ is the emissivity. (10.108) states that upward irradiance at the surface equals reflected downward diffuse irradiance plus reflected direct solar or emitted infrared irradiance. At the top of the atmosphere, the boundary condition is $F_{-,1/2} = 0$. The boundary conditions for radiance are similar. Terms from the discretization of (10.104)–(10.107) can be placed in a matrix. If a second-order central difference discretization is used, the matrix is tridiagonal and can be solved noniteratively to obtain the upward and downward fluxes at the boundary of each layer (e.g., Toon et al. 1989a).

When the atmosphere absorbs significantly, the two-stream and Eddington approximations result in an underprediction of forward scattering because the expansion of the phase function is too simple to obtain the strong peak in the scattering efficiency. As a partial remedy, the effective asymmetry factor, single-scattering albedo, and optical depth can be adjusted with the **delta functions**

$$g'_a = \frac{g_a}{1+g_a} \qquad \omega'_s = \frac{(1-g_a^2)\omega_s}{1-\omega_s g_a^2} \qquad \tau' = (1-\omega_s g_s^2)\tau \tag{10.109}$$

respectively (Hanson 1969; Potter 1970; Joseph 1976; Wiscombe 1977; Liou 1992), which replace terms in (10.89)–(10.92) and in Table 10.10.

### 10.8.4. Heating and Photolysis Rates

Spectral irradiances are used to estimate changes in air temperature due to radiative heating or cooling. If only radiative effects are considered, the local time rate of change in temperature of a layer is found from the net flux divergence equation,

$$\left(\frac{\partial T}{\partial t}\right)_r = \frac{1}{c_{p,m}}\left(\frac{dQ_{solar}}{dt} + \frac{dQ_{ir}}{dt}\right) = \frac{1}{c_{p,m}\rho_a}\frac{\partial F_n}{\partial z} \tag{10.110}$$

where $F_n = \int_0^\infty (F_{-,\lambda} - F_{+,\lambda})\, d\lambda$ is the net downward minus upward radiative flux (W $m^{-2}$), summed over all wavelengths, and $Q_{solar}$ and $Q_{solar}$ are solar and infrared **radiative heating rates** (J $kg^{-1}$), respectively, from the thermodynamic energy equation of (3.75). $F_n$ is the same as $F_{n,g}$ from (8.58), except that $F_{n,g}$ applies only to the ground. Positive values of $F_n$ and $F_{n,g}$ correspond to net downward radiation. The partial derivative of $F_n$ with respect to altitude can be discretized over layer $k$ with

$$\frac{\partial F_{n,k}}{\partial z} \approx \frac{\sum_\lambda[(F_{-,\lambda,k-1/2} - F_{+,\lambda,k-1/2}) - (F_{-,\lambda,k+1/2} - F_{+,\lambda,k+1/2})]}{z_{k-1/2} - z_{k+1/2}} \tag{10.111}$$

where $k$ corresponds to the center of a layer, $k - 1/2$ corresponds to the top of the layer, and $k + 1/2$ corresponds to the bottom of the layer. The resulting change in temperature in the layer due to diabatic radiative heating is

$$\Delta T_k \approx \frac{1}{c_{p,m}\rho_a}\frac{\partial F_n}{\partial z}h \tag{10.112}$$

where $h$ is the time step.

**Photolysis rate coefficients** are found by solving (10.104) and (10.105) for the solar spectral radiance. The spectral radiance is multiplied by $4\pi$ steradians to obtain the spectral actinic flux, as defined by (10.12). The photolysis rate coefficient ($s^{-1}$) of a species $q$ producing products $p$ at the bottom of layer $k$ (denoted by subscript $k + 1/2$) is

$$J_{q,p,k+1/2} = \int_0^\infty 4\pi I_{p,\lambda,k+1/2} b_{a,g,q,\lambda,T} Y_{q,p,\lambda,T}\, d\lambda \tag{10.113}$$

where $b_{ag,q,\lambda,T}$, previously defined, is the temperature- and wavelength-dependent absorption cross section of gas $q$ ($cm^2$ molecules$^{-1}$), $Y_{q,p,\lambda,T}$ is the temperature- and wavelength-dependent quantum yield of $q$ producing products $p$ (molecules photon$^{-1}$), and

$$I_{p,\lambda,k+1/2} = (I_{-,\lambda,k+1/2} - I_{+,\lambda,k+1/2})10^{-8}\frac{\lambda}{hc} \tag{10.114}$$

is the spectral radiance at the bottom of layer $k$ (photons $cm^{-2}$ $s^{-1}$ $\mu m$ $sr^{-1}$). The units of $I_{+/-}$, $\lambda$, $c$, and h (Planck's constant) are W $m^{-2}$ $\mu m$ $sr^{-1}$, $\mu m$, cm $s^{-1}$, and J $s^{-1}$, respectively. The **quantum yield** is the fractional number ($\leq 1$) of molecules of a specific product formed per photon of radiation absorbed at a given wavelength. A photolysis reaction may produce different sets of products. Each set of products has its own set of wavelength-dependent quantum yields, as discussed in Chapter 13.

**Example 10.14**

Convert $I_\lambda = 12$ W $m^{-2}$ in the wavelength region, $0.495\ \mu m < \lambda < 0.505\ \mu m$, to photons $cm^{-2}$ $s^{-1}$.

SOLUTION

The mean wavelength is $\lambda = 0.5\ \mu m$. Thus, $I_{p,\lambda} = I_\lambda 10^{-8}\ \lambda/hc = 3.02 \times 10^{15}$ photons $cm^{-2}$ $s^{-1}$.

## 10.9. SUMMARY

In this chapter, radiation laws, radiation processes, optical properties, and the radiative transfer equation were described. The laws include Planck's law, Wien's law, and the Stefan–Boltzmann law. The radiation processes include reflection, refraction, scattering, absorption, diffraction, and transmission. Gases and particles attenuate radiation by absorbing and scattering. All gases scatter ultraviolet and short visible wavelengths. Some gases selectively absorb ultraviolet, visible, and infrared wavelengths. All particle components scatter visible light, but only a few absorb such radiation. The radiative transfer equation determines the change in radiance and irradiance along a beam by allowing for scattering out of the beam, absorption along the beam, multiple scattering of diffuse radiation into the beam, single scattering of direct solar radiation into the beam, and emission of infrared radiation into the beam. Changes in radiance and irradiance are used to calculate photolysis and heating rates, respectively.

## 10.10. PROBLEMS

**10.1.** Calculate the radiance and irradiance from the Planck function at $T =$ 273 K and (a) $\lambda = 0.4$ μm, (b) $\lambda = 1.0$ μm, and (c) $\lambda = 15$ μm.

**10.2.** Calculate the equilibrium surface temperature of the earth if its emissivity and albedo were 70 percent and 25 percent, respectively. Does this represent an increase or decrease with respect to the equilibrium temperature (when the emissivity and albedo are 100 and 30 percent, respectively)?

**10.3.** Calculate the transmission of light through a 0.5-μm-diameter particle made of ammonium sulfate at a wavelength of $\lambda = 0.5$ μm (use the data from Table 10.7).

**10.4.** Calculate the extinction coefficient, meteorological range, and optical depth in a 1-km region of the atmosphere resulting from nitrogen dioxide absorption at a wavelength of $\lambda = 0.55$ μm when the volume mixing ratio of $NO_2$ is $\chi = 0.05$ ppmv, $T = 288$ K, and $p_d = 980$ mb.

**10.5.** Find the meteorological ranges at $\lambda = 0.53$ μm under the following conditions:

**(a)** In a rain shower with 1-mm-diameter water drops and a mass loading of 1 g m$^{-3}$.

**(b)** In a fog with 10-μm-diameter water drops and a mass loading of 1 g m$^{-3}$.

**(c)** In a haze with 0.5-μm-diameter particles, 40 percent ammonium sulfate by volume, 60 percent liquid water by volume, and a total mass loading of 50 μg m$^{-3}$.

**(d)** Behind a diesel exhaust, with 50 μg m$^{-3}$ of 0.5-μm-diameter soot particles.

In all cases, assume particles are spherical, neglect Rayleigh scattering, and neglect gas absorption. In the case of soot and raindrops, assume particle scattering and absorption occur. In the other cases, assume only particle scattering occurs. In the case of haze, calculate the scattering efficiency by weighting the volume fractions of each component by the corresponding scattering efficiency. Use Figs. 10.18 and 10.19 to determine the efficiencies. Assume the scattering efficiency of ammonium sulfate is the same as that of water. Assume the mass densities of soot, ammonium sulfate, and liquid water are 1.25, 1.77, and 1.0 g cm$^{-3}$, respectively.

## 10.11. COMPUTER PROGRAMMING PRACTICE

**10.6.** Write a script to calculate radiance and irradiance from the Planck function versus temperature and wavelength. Use the program to calculate values between $\lambda = 0.01$ μm and $\lambda = 100$ mm for $T = 6000$ K and $T = 300$ K. Plot the results.

**10.7.** Divide the atmosphere from $z = 0$ to 10 km into 100 vertical layers. Assume $T = 288$ K and $p_d = 1013$ mb at the surface, and assume the temperature decreases 6.5 K $\text{km}^{-1}$. Use (2.35) to estimate the air pressure in each layer. Calculate the extinction coefficient and optical depth in each layer due to Rayleigh scattering at $\lambda = 0.4$ μm. Estimate the cumulative optical depth at the surface.

# 11

# Gas-Phase Species, Chemical Reactions, and Reaction Rates

THE atmosphere contains numerous gases that undergo chemical reactions. Because many chemical pathways are initiated by sunlight, atmospheric reactions are collectively called **photochemical reactions**. Lightning and molecular collisions also initiate reactions.

Photochemistry is responsible for the transformation of gases in all regions of the atmosphere. It converts nitrogen oxide and reactive organic gases, emitted during fuel combustion, to ozone, peroxyacetyl nitrate (PAN), and a host of other products. It produces and destroys ozone in the free troposphere and stratosphere. Photochemistry also converts dimethyl sulfide (DMS) to sulfur dioxide and sulfuric acid over the oceans. In this chapter, chemical species, structures, reactions, rate coefficients, rates, and lifetimes are discussed. In Chapter 12, photochemical reactions important in different regions of the atmosphere are described. In Chapter 13, numerical methods of solving chemical ordinary differential equations arising from such reactions are given.

## 11.1. ATMOSPHERIC GASES AND THEIR MOLECULAR STRUCTURES

Gases consist of neutral or charged single atoms or molecules. Elements that make up most gases are hydrogen (H), carbon (C), nitrogen (N), oxygen (O), fluorine (F), sulfur (S), chlorine (Cl), and bromine (Br). Particles contain sodium (Na), magnesium (Mg), aluminum (Al), silicon (Si), potassium (K), calcium (Ca), and/or the elements found in gases.

Gas molecules consist of atoms with covalent bonds between them. A **covalent bond** is a bond consisting of one or more pairs of electrons shared between two atoms. Gilbert Lewis (1875–1946) was the first to suggest that atoms could be held together by sharing electron pairs. Lewis proposed the use of **electron-dot symbols** (**Lewis symbols**) to describe the configuration of electrons in the outer shell (**valence shell**) of an atom. The maximum number of electrons in a valence shell for many elements is eight. When a molecule forms, each atom gains, loses, or shares electron pairs until its valence shell is full. This is called the **octet rule**, and a valence shell with eight electrons is called an **octet**. The first four elements of the periodic table – hydrogen, helium, lithium, and beryllium – are exceptions to the octet rule, since they cannot have more than two electrons in their valence shell. Elements in periods (rows) 3–7 of the periodic table violate the octet rule when

**Table 11.1.** Elements of Atmospheric Importance and Their Lewis Symbols

| Element | Lewis Symbol | No. $e^-$ | Valence | $m$ (g/mol) | Element | Lewis Symbol | No. $e^-$ | Valence | $m$ (g/mol) |
|---|---|---|---|---|---|---|---|---|---|
| Hydrogen | H• | 1 | 1 | 1.008 | Aluminum | •Al• (one more dot above) | 3 | 3 | 26.98 |
| Carbon | •C• (dots above and below) | 4 | 4 | 12.01 | Silicon | •Si• (dots above and below) | 4 | 4 | 28.09 |
| Nitrogen | •N• (dot above, pair below) | 5 | 3 | 14.01 | Sulfur | :S• (dot above, pair below) | 6 | 2 | 32.07 |
| Oxygen | :O• (dot above, pair below) | 6 | 2 | 16.00 | Chlorine | :Cl• (pairs above and below) | 7 | 1 | 35.45 |
| Fluorine | :F• (pairs above and below) | 7 | 1 | 19.00 | Potassium | K• | 1 | 1 | 39.10 |
| Sodium | Na• | 1 | 1 | 22.99 | Calcium | •Ca• | 2 | 2 | 40.08 |
| Magnesium | •Mg• | 2 | 2 | 24.30 | Bromine | :Br• (pairs above and below) | 7 | 1 | 79.90 |

No. $e^-$ is the number of valence-shell electrons, valence is the principal valence, and $m$ is the molecular weight.

they have more than eight electrons in their valence shell. Some other elements also violate the octet rule.

When an outer shell of an element has four electrons or fewer, the Lewis symbol is drawn by arranging the electrons around the element as single dots. The number of single dots is the number of unpaired electrons, or the **valence**, of the element. Each additional electron in an outer shell, beyond four, is paired with a preexisting electron, reducing the valence of the element. The **principal valence** of an element is the number of hydrogen atoms that can combine with one atom of the element. Table 11.1 shows the Lewis symbol, number of valence-shell electrons, and principal valence of several elements of atmospheric importance.

When atoms form covalent bonds during production of a molecule, the configuration of the molecule can be illustrated with a **Lewis structure**, which consists of a combination of Lewis symbols. Table B.3 of Appendix B shows the Lewis structures of many gases of atmospheric importance. The table is divided into inorganic and organic gases. **Inorganic gases** are those that contain any element, including hydrogen (H) or carbon (C), but not both H and C. **Organic gases** are those that contain both H and C, but may also contain other elements. Organic gases that contain only H and C are **hydrocarbons**. Hydrocarbons include alkanes, cycloalkanes, alkenes, alkynes, aromatics, and terpenes. When methane, which is fairly unreactive, is excluded from the list of hydrocarbons, the remaining hydrocarbons are **nonmethane hydrocarbons** (NMHCs). Oxygenated functional groups such as aldehydes, ketones, alcohols, acids, and nitrates, are added to hydrocarbons to produce **oxygenated hydrocarbons**. Nonmethane hydrocarbons and oxygenated hydrocarbons, together, are **reactive organic gases** (ROGs) or **volatile organic carbon** (VOC). Nonmethane hydrocarbons and **carbonyls** (aldehydes plus ketones), together, are **nonmethane organic carbon** (NMOC). **Total organic gas** (TOG) is the sum of ROGs and methane. Below the Lewis structures and characteristics of a few inorganic and organic gases are discussed.

### 11.1.1. Molecular Hydrogen ($H_2$)

Molecular hydrogen is emitted from volcanos and produced chemically in the atmosphere. It has a typical mixing ratio of about 0.6 ppmv in the troposphere and consists of two hydrogen atoms with a single covalent bond between them. Two Lewis structures for molecular hydrogen are

$$\mathrm{H:H} \qquad \mathrm{H{-}H}$$

The two dots and the line between atoms indicate a covalently bonding electron pair.

### 11.1.2. Molecular Oxygen ($O_2$)

$O_2$ is produced by green-plant photosynthesis and has a typical mixing ratio of 20.95% by volume. Its Lewis structures are

$$:\ddot{\mathrm{O}}::\ddot{\mathrm{O}}: \qquad \mathrm{O{=}O}$$

where the double bond indicates two shared electron pairs. In the second structure, all unshared pairs of electrons are ignored for convenience.

### 11.1.3. Molecular Nitrogen ($N_2$)

Sources of $N_2$ were discussed in Chapter 2. Its mixing ratio is typically 78.08% by volume. Molecular nitrogen has the Lewis structures

$$:\mathrm{N}:::\mathrm{N}: \qquad \mathrm{N{\equiv}N}$$

where the triple bond indicates three shared electron pairs.

### 11.1.4. Hydroxyl Radical (OH)

Molecular hydrogen and molecular oxygen have an even number of electrons and no unpaired electrons. Many other molecules have an odd number of electrons and one unpaired electron. Such molecules are **free radicals**. One free radical is the hydroxyl radical, which can be represented by the Lewis structures

$$:\dot{\underset{\cdot\cdot}{\mathrm{O}}}:\mathrm{H} \qquad \dot{\mathrm{O}}{-}\mathrm{H}$$

The single dot in the second structure indicates that the unpaired electron is associated with the oxygen atom. The hydroxyl radical is produced chemically in the atmosphere. It breaks down many other gases and is referred to as a **scavenger**. OH is discussed in more detail in Chapter 12.

### 11.1.5. Nitric Oxide (NO)

Nitric oxide is also a free radical. It is a colorless gas emitted from soils, plants, and combustion processes and produced by lightning and chemical reaction. Combustion sources include aircraft, automobiles, oil refineries, and biomass burning. The primary sink of NO is chemical reaction. A typical mixing ratio of NO in the background troposphere near sea level is 5 pptv. In the upper troposphere NO mixing ratios increase to 20–60 pptv. In urban regions, NO mixing ratios reach 0.1 ppmv in the early morning but decrease significantly by midmorning. The Lewis structures of NO can be drawn so that either N or O has seven electrons in its valence shell. Thus,

$$\cdot\ddot{N}::\ddot{O}: \longleftrightarrow {}^{-}:\ddot{N}::\dot{O}:{}^{+} \qquad \dot{N}=O \longleftrightarrow {}^{-}N=\ddot{O}{}^{\cdot\,+}$$

where each structure is a resonance structure. **A resonance structure** is one of two or more Lewis structures used to represent a molecule that cannot be represented correctly by one Lewis structure. Multiple Lewis structures arise because electrons are not local to one atom.

### 11.1.6. Nitrogen Dioxide (NO$_2$)

Nitrogen dioxide is a brown gas because it absorbs the shortest wavelengths of the sun's visible radiation. It absorbs almost all blue light and some green light, allowing the remaining green and all red light to scatter and transmit. The combination of red and some green light is brown. The major source of NO$_2$ is photochemical oxidation of NO. NO$_2$ is also produced by other reaction pathways and is emitted during combustion. NO$_2$ is more prevalent during midmorning than during midday or afternoon, since sunlight breaks NO$_2$ down past midmorning. Mixing ratios of NO$_2$ near sea level in the free troposphere range from 20 to 50 pptv. In the upper troposphere, mixing ratios increase to 30–70 pptv. In urban regions, they range from 0.1 to 0.25 ppmv. The resonance structures of NO$_2$ are

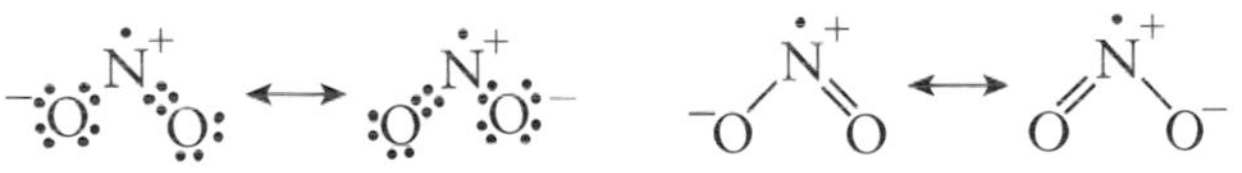

During nitrogen dioxide formation, a net negative charge is transferred to the oxygen atoms from the nitrogen atom, resulting in the charge distribution shown.

### 11.1.7. Ozone (O$_3$)

Ozone is a colorless gas that exhibits an odor, even in small concentrations. In urban areas, ozone affects human health in the short term by causing headache (>0.15 ppmv), chest pain (>0.25 ppmv), and sore throat and cough (>0.30 ppmv). Ozone decreases lung function for those who exercise steadily for over an hour while exposed to concentrations above 0.30 ppmv. Above 1 ppmv (high above ambient concentrations), ozone can temporarily narrow passages deep in the lung, increasing airway resistance and inhibiting breathing. Small decreases in lung function affect those with asthma, chronic bronchitis, and emphysema. Ozone may also accelerate the aging of lung tissue. Above 0.1 ppmv, ozone affects animals by increasing their susceptibility to bacterial infection. It also interferes with the growth of plants and trees and deteriorates organic materials, such as rubber, textile dyes and fibers, and some paints and coatings (USEPA 1978).

In the free troposphere, ozone mixing ratios range between 20 and 40 ppbv near sea level and between 30 and 70 ppbv at higher altitudes. In urban areas, ozone mixing ratios range from 0.01 ppmv at night to 0.35 ppmv during very smoggy afternoons, with typical values of 0.15 ppmv during moderately polluted afternoons. Ozone is chemically reactive but not a free radical. The resonance structures of ozone are

### 11.1.8. Carbon Monoxide (CO)

Carbon monoxide is a tasteless, colorless, and odorless gas that is toxic to humans and animals exposed to it for one hour at mixing ratios above about 700 ppmv. Exposure to 300 ppmv for one hour causes headaches. In urban regions away from freeways, CO mixing ratios are typically 2–10 ppmv. On freeways and in traffic tunnels, mixing ratios can rise to more than 100 ppmv. In the free troposphere, CO mixing ratios vary from 50 to 150 ppbv. A major source of CO is incomplete combustion by automobiles, trucks, and airplanes. Wood burning and grass burning are also important sources. Natural sources of CO are plants and biological activity in the oceans. Although CO is the most abundantly emitted gas in urban air, it does not play a large role in photochemical smog formation. Photochemical smog is characterized by the buildup of ozone and other oxidants. CO does not produce much ozone in urban air. The major sink of CO is chemical conversion to carbon dioxide ($CO_2$). CO can be represented by

$$^{-}\!:\!C\!:\!:\!:\!O\!:^{+} \qquad ^{-}C\equiv O^{+}$$

### 11.1.9. Carbon Dioxide ($CO_2$)

Carbon dioxide is an odorless and inert gas. Its major sources and sinks were described in Chapter 2. It is well mixed throughout the troposphere and stratosphere, with a current mixing ratio of about 360 ppmv. The Lewis structures of $CO_2$ are

$$:\!\ddot{O}\!:\!:\!C\!:\!:\!\ddot{O}\!: \qquad O=C=O$$

These structures indicate that $CO_2$ is a linear molecule but not a free radical. $CO_2$ is very stable: its lifetime against chemical destruction is over 100 years.

### 11.1.10. Sulfur Dioxide ($SO_2$)

Sulfur dioxide is a colorless gas that exhibits an odor and taste at high concentrations. It is emitted from coal-fired power plants, automobile tailpipes, and volcanos.

It is also produced chemically in the atmosphere from biologically emitted precursors, such as dimethylsulfide (DMS) and hydrogen sulfide ($H_2S$). In the background troposphere, $SO_2$ mixing ratios range from 20 pptv to 1 ppbv. In moderately polluted air, they range from 10 to 30 ppbv. $SO_2$ is removed from the atmosphere by gas-phase reaction, dissolution into cloud and rain drops, and deposition to the ground surface. Common Lewis structures of $SO_2$ are

### 11.1.11. Methane ($CH_4$)

Sources and sinks of methane (natural gas) were discussed in Chapter 2. It is one of the most unreactive hydrocarbons in the atmosphere. Because methane has a long lifetime against chemical loss (approximately 4 years), it is well diluted throughout the free troposphere, with a mixing ratio of about 1.7 ppmv. The Lewis structures of methane are

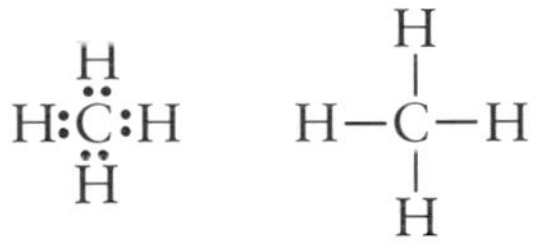

### 11.1.12. Peroxyacetyl Nitrate (PAN) ($CH_3C(O)OONO_2$)

PAN is an eye irritant, initially discovered in the laboratory as a product of smog-forming chemical reactions. Its peak mixing ratio of about 10–20 ppbv in polluted air occurs at the same time during the afternoon as does ozone's peak mixing ratio. PAN mixing ratio in the free troposphere ranges from 2 to 100 pptv. PAN can be represented as

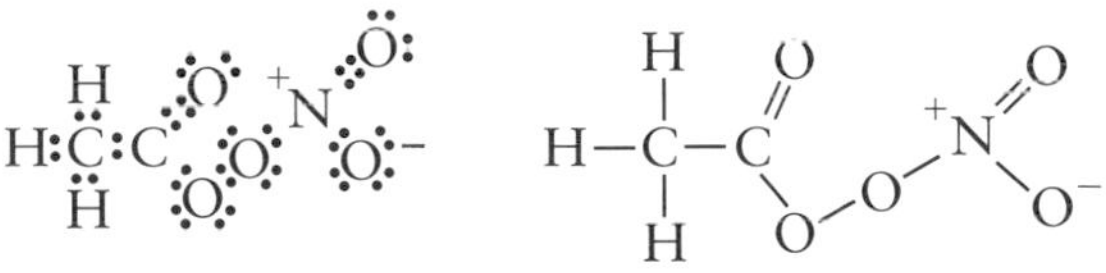

## 11.2. CHEMICAL REACTIONS AND PHOTOPROCESSES

A single chemical reaction with no intermediate products is an **elementary reaction.** Elementary homogeneous gas-phase chemical reactions in the atmosphere are conveniently divided into **photolysis reactions** (also called photoprocesses, photodissociation reactions, or photolytic reactions) and **chemical kinetic reactions.** Photolysis reactions are **unimolecular** (one-body) reactions initiated when a photon of radiation strikes a molecule and breaks it into two or more products. Elementary chemical kinetic reactions are **bimolecular** (two-body) or **termolecular** (three-body). Reactants and products of photolysis and kinetic reactions are neutral or charged atoms or molecules.

**Photolysis reactions** are unimolecular. An example of a photolysis reaction is

$$\dot{N}O_2 + h\nu \longrightarrow \dot{N}O + \dot{O}\cdot \qquad \lambda > 410\,\text{nm} \tag{11.1}$$

where $h\nu$ is a single photon of radiation, O [=O($^3P$)] is ground-state atomic oxygen, and paired electrons and bonds between atoms are not shown (only unpaired electrons are shown).

**Elementary bimolecular reactions** include thermal decomposition, isomerization, and standard collision reactions. Thermal decomposition and isomerization reactions occur when a reactant molecule collides with an air molecule. The kinetic energy of the collision elevates the reactant to a high enough vibrational energy state that it can decompose or isomerize. **Thermal decomposition** occurs when the excited reactant dissociates into two or more products. **Isomerization** occurs when the excited reactant changes chemical structure but not composition or molecular weight.

An example of a **bimolecular thermal decomposition reaction** is

$$N_2O_5 + M \longrightarrow \dot{N}O_2 + N\dot{O}_3 + M \tag{11.2}$$

where $N_2O_5$ is dinitrogen pentoxide, $NO_3$ is the nitrate radical, and M is the molecule that provides the collisional energy. In the atmosphere, M can be any molecule. Because oxygen and nitrogen, together, make up more than 99% of the gas molecules in the atmosphere, M is most likely oxygen or nitrogen. For rate calculation purposes, the concentration of M is usually set equal to the concentration of $N_2$, $O_2$, or $N_2 + O_2$, depending on how the rate coefficient was determined experimentally.

Since M in (11.2) does not change concentration, the reaction can be written in nonelementary form as

$$N_2O_5 \xrightarrow{M} \dot{N}O_2 + N\dot{O}_3 \tag{11.3}$$

Thermal decomposition reactions are temperature-dependent. At high temperatures, they proceed faster than at low temperatures. Isomerization reactions are similar to (11.2) and (11.3), except that an isomerization reaction has one product, which is another form of the reactant.

Elementary **bimolecular collision reactions** are the most common types of kinetic reaction and may occur between any two chemically active reactants that collide. A prototypical collision reaction is

$$CH_4 + \dot{O}H \longrightarrow \dot{C}H_3 + H_2O \tag{11.4}$$

where $CH_3$ is the methyl radical, and $H_2O$ is water vapor. In some cases, bimolecular reactions result in **collision complexes** that ultimately break into products. Such reactions have the form $A + B \rightleftharpoons AB^* \longrightarrow D + F$, where $AB^*$ is a molecule that has weak bonds and stays intact slightly longer than the characteristic time of the molecule's vibrations and rotations. Other reactions that may form collision complexes are pressure-dependent termolecular reactions.

**Termolecular reactions** often consist of pairs of elementary bimolecular reactions. Consider the termolecular **combination reaction**

$$\dot{N}O_2 + N\dot{O}_3 + M \longrightarrow N_2O_5 + M \tag{11.5}$$

The sequence of elementary bimolecular reactions resulting in (11.5) is $A+B \rightleftharpoons AB^*$ followed by $AB^* + M \longrightarrow AB + M$, where M is a **third body** whose purpose is to carry away energy released during the second reaction. In the absence of M, the energy release causes $AB^*$ to dissociate back to A and B. The purpose of M in (11.5) differs from that in (11.2), which was to provide collisional energy for the reaction. In both cases, M is probably $N_2$ or $O_2$, and its concentration may be taken as that of $N_2 + O_2$ for rate calculation purposes. (11.2) and (11.5) are pressure-dependent because the concentration of M is proportional to the air pressure. Since M in (11.5) does not change concentration, (11.5) can also be written as

$$\dot{N}O_2 + N\dot{O}_3 \xrightarrow{M} N_2O_5 \tag{11.6}$$

Elementary **termolecular collision reactions** are rare, since the probability that three trace gases collide and change form is not large. One possible reaction of this type is

$$\dot{N}O + \dot{N}O + O_2 \longrightarrow \dot{N}O_2 + \dot{N}O_2 \tag{11.7}$$

## 11.3. REACTION RATES

A **reaction rate** is the time rate of change of concentration of any reactant in a reaction. The rate of an elementary photolysis, collision, isomerization, thermal decomposition, or combination reaction equals a rate coefficient multiplied by the concentration of each reactant. The **rate coefficient** relates concentrations to the reaction rate. Elementary unimolecular reactions have **first-order rate coefficients,** meaning that such coefficients are multiplied by one reactant concentration. Elementary bimolecular and termolecular reactions have **second-** and **third-order** rate coefficients, respectively. If reactant concentrations are expressed in units of molecules of gas per cubic centimeter of air, the rate of reaction, regardless of whether it has a first-, second-, or third-order rate coefficient, is in units of molecules $cm^{-3}\ s^{-1}$. Rate expressions for reactions with first-, second-, and third-order rate coefficients are

$$\text{Rate} = k_F[A] \qquad \text{Rate} = k_S[A][B] \qquad \text{Rate} = k_T[A][B][C] \tag{11.8}$$

respectively, where brackets denote number concentration, and $k_F$, $k_S$, and $k_T$ are first-, second-, and third-order rate coefficients, in units of $s^{-1}$, $cm^3$ molecule$^{-1}$ $s^{-1}$, and $cm^6$ molecule$^{-2}$ $s^{-1}$, respectively. For a photolysis reaction, the rate expression is

$$\text{Rate} = J[\text{A}] \tag{11.9}$$

where $J$ is a first-order **photolysis rate coefficient** of species A ($s^{-1}$).

If the concentration of one reactant, such as [M], is invariant during a reaction, the concentration can be premultiplied by the rate coefficient. In such cases, second-order rate coefficients become **pseudo-first-order coefficients** and third-order coefficients become **pseudo-second-order coefficients**.

The time rate of change of concentration of a reactant equals the negative of its rate of reaction. For a photolysis reaction of the form $\text{A} + h\nu \longrightarrow \text{D} + \text{G}$, the loss rate of A is

$$\frac{d[\text{A}]}{dt} = -\text{Rate} = -J[\text{A}] \tag{11.10}$$

For a bimolecular thermal decomposition reaction of the form, $\text{A} \xrightarrow{\text{M}} \text{D} + \text{E}$, the loss rate of A is

$$\frac{d[\text{A}]}{dt} = -\text{Rate} = -k_\text{F}[\text{A}] \tag{11.11}$$

where $k_\text{F} = k_\text{S}[\text{M}]$ is a pseudo-first-order rate coefficient. For a bimolecular collision reaction of the form $\text{A} + \text{B} \longrightarrow \text{D} + \text{F}$ or a termolecular combination reaction of the form $\text{A} + \text{B} \xrightarrow{\text{M}} \text{E}$, the loss rates of A and B are

$$\frac{d[\text{A}]}{dt} = \frac{d[\text{B}]}{dt} = -\text{Rate} = -k_\text{S}[\text{A}][\text{B}] \tag{11.12}$$

where $k_\text{S}$ is a second-order rate coefficient for the first reaction, and $k_\text{S} = k_\text{T}[\text{M}]$ is a pseudo-second-order rate coefficient for the second reaction. For a bimolecular collision reaction of the form $\text{A} + \text{A} \longrightarrow \text{E} + \text{F}$, the loss rate of A is

$$\frac{d[\text{A}]}{dt} = -2\text{Rate} = -2k_\text{S}[\text{A}]^2 \tag{11.13}$$

where $k_\text{S}$ is a second-order rate coefficient. For a termolecular reaction, such as $\text{A} + \text{B} + \text{C} \longrightarrow \text{E} + \text{F}$, the loss rate of A, B, and C is

$$\frac{d[\text{A}]}{dt} = \frac{d[\text{B}]}{dt} = \frac{d[\text{C}]}{dt} = -\text{Rate} = -k_\text{T}[\text{A}][\text{B}][\text{C}] \tag{11.14}$$

For a termolecular reaction such as $2\text{A} + \text{B} \longrightarrow \text{E} + \text{F}$,

$$\frac{d[\text{A}]}{dt} = 2\frac{d[\text{B}]}{dt} = -2\text{Rate} = -2k_\text{T}[\text{A}]^2[\text{B}] \tag{11.15}$$

In general, reactions of the form aA + bB $\longrightarrow$ eE + fF have the rate expression

$$\text{Rate} = k_r[\text{A}]^{\text{a}}[\text{B}]^{\text{b}} \tag{11.16}$$

where $k_r$ denotes a rate coefficient of order a+b. The rates of loss of species A and B are

$$\frac{\text{d[A]}}{\text{d}t} = -\text{a Rate} = -\text{a}k_r[\text{A}]^{\text{a}}[\text{B}]^{\text{b}} \qquad \frac{\text{d[B]}}{\text{d}t} = -\text{b Rate} = -\text{b}k_r[\text{A}]^{\text{a}}[\text{B}]^{\text{b}} \tag{11.17}$$

respectively, and the rates of production of E and F are

$$\frac{\text{d[E]}}{\text{d}t} = \text{e Rate} = \text{e}k_r[\text{A}]^{\text{a}}[\text{B}]^{\text{b}} \qquad \frac{\text{d[F]}}{\text{d}t} = \text{f Rate} = \text{f}k_r[\text{A}]^{\text{a}}[\text{B}]^{\text{b}} \tag{11.18}$$

respectively. When M is treated as a reactant and product, such as is the reaction, A + B + M $\longrightarrow$ E + M, the rate of change of M is zero. Thus,

$$\frac{\text{d[M]}}{\text{d}t} = k_{\text{T}}[\text{A}][\text{B}][\text{M}] - k_{\text{T}}[\text{A}][\text{B}][\text{M}] = 0 \tag{11.19}$$

The rate of production of E and the rates of loss of A and B are nonzero in this reaction.

If M is nitrogen ($N_2$) or oxygen ($O_2$), the number concentration of M is obtained by multiplying the number concentration of dry-air molecules, $N_d = p_d/k_B T$ from (2.18), by the volume mixing ratio of nitrogen or oxygen, respectively. Thus,

$$[\text{M}] = N_{\text{N}_2} = \chi_{\text{N}_2} N_d \qquad [\text{M}] = N_{\text{O}_2} = \chi_{\text{O}_2} N_d \tag{11.20}$$

where all concentrations are in units of molecules per cubic centimeter. The volume mixing ratios of nitrogen and oxygen were given in Table 2.1 as $\chi = 0.7808$ and 0.2095, respectively. If M is $N_2$ plus $O_2$, then $[\text{M}] = N_d$.

**Example 11.1**

Find the number concentration of $N_2$ and $O_2$ when $T = 278$ K and $p_d = 920$ mb.

SOLUTION

From (2.18), $N_d = 2.40 \times 10^{19}$ molecules cm$^{-3}$. From (11.20), $N_{\text{N}_2} = 1.87 \times 10^{19}$ molecules cm$^{-3}$ and $N_{\text{O}_2} = 5.02 \times 10^{18}$ molecules cm$^{-3}$.

## 11.4. REACTION RATE COEFFICIENTS

### 11.4.1. Determining Rate Coefficients

The rate coefficient for a kinetic reaction is constant for a given temperature, pressure, and set of reactants. Rate coefficients are determined experimentally. A common method of determining coefficients for elementary bimolecular reactions at a given temperature is with **kinetic analysis**. With this method, a small amount of one substance is exposed to a large amount of another substance, and the rate of decay of the less abundant substance is observed. Consider the bimolecular reaction $A + B \longrightarrow D + F$. If a small quantity of A is exposed to a large quantity of B, the maximum loss of B is [A]. Since $[A] \ll [B]$, [B] can be held constant, and the rate of loss of A is approximately

$$\frac{d[A]_t}{dt} = -k_F[A]_t = -k_S[A]_t[B]_0 \tag{11.21}$$

where the subscript $t$ indicates that the concentration changes with time, the subscript 0 indicates that the concentration is fixed to its initial value, and $k_F = k_S[B]_0$ is a pseudo-first-order rate coefficient. Integrating (11.21) from time $t = 0$ to $h$ and from $[A]_0$ to $[A]_h$ and solving gives

$$k_S = -\frac{1}{[B]_0 h} \ln \frac{[A]_{t=h}}{[A]_0} \tag{11.22}$$

If the concentration of A at any time $t = h$ is measured, the rate coefficient of the reaction can be estimated from (11.22). For elementary uni- and termolecular reactions, similar analyses can be performed. For unimolecular reactions of the form $A \longrightarrow D + E$ and termolecular reactions of the form $A + B + C \longrightarrow E + F$, rate coefficients are estimated as

$$k_F = -\frac{1}{h} \ln \frac{[A]_{t=h}}{[A]_0} \qquad k_T = -\frac{1}{[B]_0[C]_0 h} \ln \frac{[A]_{t=h}}{[A]_0} \tag{11.23}$$

respectively. In the latter case, $[A]_0 \ll [B]_0$ and $[A]_0 \ll [C]_0$. More advanced methods of calculating reaction rate coefficients include fast flow systems, flash photolysis, molecular modulation, and pulse radiolysis techniques (e.g., Finlayson-Pitts and Pitts (1986)).

### 11.4.2. Temperature Dependence of Reactions

First-, second-, and third-order reaction rate coefficients vary with temperature. In many cases, the temperature dependence is estimated with the equation proposed by Arrhenius in 1889. The **Arrhenius equation** is found by integrating

$$\frac{d(\ln k_r)}{dT} = \frac{E_r}{R^* T^2} \tag{11.24}$$

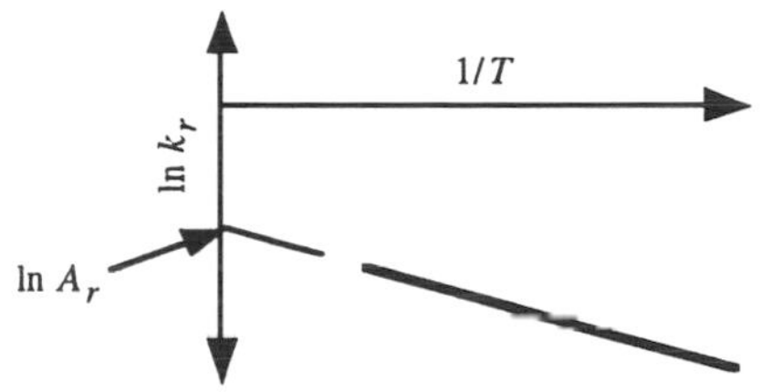


**Figure 11.1.** Plot of $\ln k_r$ versus $1/T$ (thick line). The slope of the line is $E_r/R^*$. The line can be extrapolated to $1/T \approx 0$ ($T \longrightarrow \infty$) to obtain $\ln A_r$.

where $k_r$ is the rate coefficient, $E_r$ is the activation energy of the reaction (J mole$^{-1}$), $T$ is temperature (K), and $R^*$ is the gas constant.

The **activation energy** is the smallest amount of energy required for reacting species to form an activated complex or transition state before forming products. Activation energies are determined by first integrating (11.24) as

$$\ln k_r = \ln A_r - \frac{E_r}{R^*T} \tag{11.25}$$

where $A_r$ is a constant of integration, called the **collisional prefactor** (frequency factory). The collisional prefactor is proportional to the frequency of those collisions with proper orientation for producing a reaction. It equals a collision frequency multiplied by a steric (efficiency) factor. The collision frequency depends on the relative size, charge, kinetic energies, and molecular weights of the reactant molecules. The **steric factor** gives the fraction of collisions that result in an effective reaction. For first- and second-order reactions, collisional prefactors and rate coefficients have units of s$^{-1}$ and cm$^3$ molecule$^{-1}$ s$^{-1}$, respectively.

The activation energy and collisional prefactor in (11.25) are found experimentally. If the rate coefficient ($k_r$) is measured at different temperatures, a graph of $\ln k_r$ versus $1/T$ can be plotted. $E_r$ and $A_r$ can be extracted from the plot, as shown in Fig. 11.1.

Solving (11.25) for $k_r$ gives

$$k_r = A_r \exp\left(-\frac{E_r}{R^*T}\right) = A_r \exp\left(\frac{C_r}{T}\right) \tag{11.26}$$

where $\exp(-E_r/R^*T)$ is the fraction of reactant molecules having the critical energy, $E_r$, required for the reaction to occur, and $C_r = -E_r/R^*$.

Although kept constant, the collisional prefactor is usually a weak function of temperature. When the activation energy is near zero, the collisional prefactor depends strongly on temperature, and the Arrhenius equation does not fully describe the temperature dependence. In such cases, a temperature factor $B_r$ is added to (11.26), so that

$$k_r = A_r \left(\frac{300}{T}\right)^{B_r} \exp\left(\frac{C_r}{T}\right) \tag{11.27}$$

$B_r$ is found by fitting the expression in (11.27) to data. Many combination reactions have the form of (11.27). Rate coefficients for such reactions are given in Table B.4 of Appendix B.

**Example 11.2**

At $T = 298$ K, the rate coefficients (cm$^3$ molecule$^{-1}$ s$^{-1}$ and cm$^6$ molecule$^{-2}$ s$^{-1}$) for a bi- and a termolecular reaction are, respectively,

$$\dot{N}O + O_3 \longrightarrow \dot{N}O_2 + O_2 \qquad k_1 = 1.80 \times 10^{-12} \exp(-1370/T) = 1.81 \times 10^{-14}$$

$$\dot{O}\cdot + O_2 + M \longrightarrow O_3 + M \qquad k_2 = 5.63 \times 10^{-34}(300/T)^{2.8} = 5.74 \times 10^{-34}$$

### 11.4.3. Pressure Dependence of Reactions

Thermal decomposition and combination reactions, both of which include an M ($O_2$, $N_2$, or $O_2 + N_2$), are **pressure-dependent** because [M] varies with pressure. Such reactions have the forms

$$A \xrightarrow{M} D + E \qquad A + B \xrightarrow{M} E \tag{11.28}$$

respectively. Overall rate coefficients for these reactions may be interpolated between a low- and a high-pressure limit rate coefficient. In some reactions, such as $O + O_2 + M \longrightarrow O_3 + M$, the rate coefficient is not interpolated.

The interpolation formula for the overall rate coefficient of a pressure-dependent reaction is

$$k_r = \frac{k_{\infty,T} k_{0,T}[M]}{k_{\infty,T} + k_{0,T}[M]} F_c^{\left[1+\left(\log_{10}\frac{k_{0,T}[M]}{k_{\infty,T}}\right)^2\right]^{-1}} \tag{11.29}$$

where $F_c$ is the **broadening factor,** $k_{0,T}$ is the low-pressure limit rate coefficient, and $k_{\infty,T}$ is the high-pressure limit rate coefficient. For the reactions in (11.28), $k_r$ has units of s$^{-1}$ and cm$^3$ molecules$^{-1}$ s$^{-1}$, respectively. Values of $k_{0,T}$, $k_{\infty,T}$, and $F_c$ are given in Table B.4 of Appendix B for applicable pressure-dependent reactions.

The **low-pressure limit rate coefficient** $k_{0,T}$ multiplied by [M], is a laboratory-determined coefficient of the overall reaction as [M] approaches zero. Thus,

$$k_{0,T}[M] = \lim_{[M] \longrightarrow 0} k_r \tag{11.30}$$

where [M] has units of molecules cm$^{-3}$, and $k_{0,T}$ has units of cm$^3$ molecule$^{-1}$ s$^{-1}$ or cm$^6$ molecule$^{-2}$ s$^{-1}$, respectively, for the reactions in (11.28).

The **high-pressure limit rate coefficient** $k_{\infty,T}$ is the rate of the overall reaction at infinite M concentration. Thus,

$$k_{\infty,T} = \lim_{[\mathrm{M}] \longrightarrow \infty} k_r \tag{11.31}$$

Since the high-pressure limit rate coefficients are independent of [M], the reactions corresponding to (11.28) at high pressure are

$$\mathrm{A} \longrightarrow \mathrm{D} + \mathrm{E} \qquad \mathrm{A} + \mathrm{B} \longrightarrow \mathrm{F} \tag{11.32}$$

respectively. The high-pressure limit rate coefficients have units of $s^{-1}$ and $cm^3$ $molecule^{-1}$ $s^{-1}$ for these two reactions, respectively.

**Example 11.3**

When $p_d = 140$ mb and $T = 216$ K (stratosphere), find the rate coefficient of

$$\dot{\mathrm{O}}\mathrm{H} + \dot{\mathrm{N}}\mathrm{O}_2 \xrightarrow{\mathrm{M}} \mathrm{HNO}_3$$

SOLUTION

From Table B.4 and (2.18), $[\mathrm{M}] = N_d = 4.69 \times 10^{18}$ molecules $cm^{-3}$. From Table B.4, $k_{0,T} = 2.60 \times 10^{-30}(300/T)^{2.9}$ $cm^6$ $molecule^{-2}$ $s^{-1}$, $k_{\infty,T} = 6.70 \times 10^{-11}(300/T)^{0.6}$ $cm^3$ $molecule^{-1}$ $s^{-1}$, and $F_c = 0.43$. Thus, at 216 K, $k_{0,T}[\mathrm{M}] = 2.47 \times 10^{-11}$, $k_{\infty,T} = 8.16 \times 10^{-11}$, and $k_r = 1.11 \times 10^{-11}$ $cm^3$ $molecule^{-1}$ $s^{-1}$.

Figure 11.2(a) and (b) show the pressure and temperature dependences of the rate of PAN decomposition by the reaction

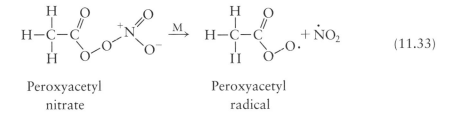

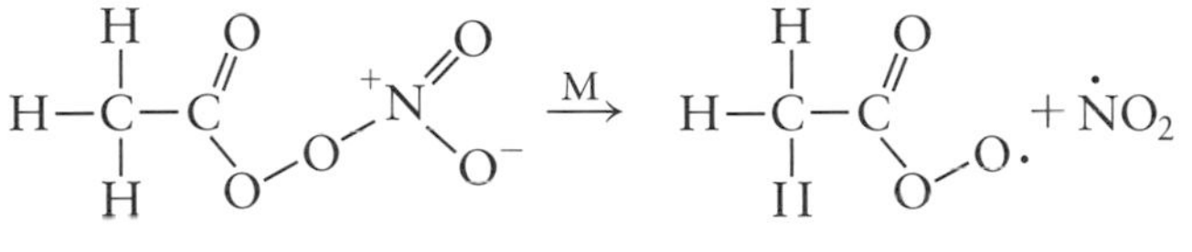

(11.33)

Peroxyacetyl nitrate — Peroxyacetyl radical

The graphs indicate that the pseudo-first-order rate coefficient for PAN decomposition varies by an order of magnitude when pressures range from 1 to 1000 mb. The rate coefficient varies by 15 orders of magnitude when temperatures range from 180 to 330 K. Temperature affects PAN thermal decomposition much more than does pressure.

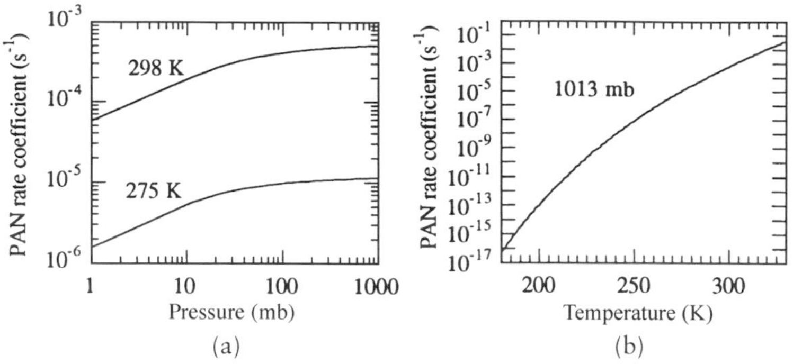


**Figure 11.2.** Rate coefficient of PAN decomposition (a) as a function of pressure for two temperatures and (b) as a function of temperature for one pressure.

### 11.4.4. Photolysis Reactions

Photolysis reactions instigate many atmospheric chemical pathways. From (10.113), the **photolysis coefficient** ($s^{-1}$) of gas $q$ producing product set $p$ was

$$J_{q,p} = \int_0^\infty 4\pi I_{p,\lambda} b_{a,g,q,\lambda,T} Y_{q,p,\lambda,T}\, d\lambda \tag{11.34}$$

where $4\pi I_{p,\lambda}$ is the actinic flux (photons $cm^{-2}\ \mu m^{-1}\ s^{-1}$) in the wavelength interval $d\lambda$ ($\mu$m), $b_{a,g,q,\lambda,T}$ is the average absorption cross section of the gas in the interval ($cm^2$ molecule$^{-1}$), and $Y_{q,p,\lambda,T}$ is the average quantum yield of the photoprocess in the interval (molecules photon$^{-1}$). Experimental absorption cross sections and quantum yields vary with wavelength and temperature. References for such data are given in Table B.4 of Appendix B.

Photolysis of a molecule may produce one or more sets of products. Photolysis of the nitrate radical, for example, produces two possible sets of products,

$$\dot{N}O_3 + h\nu \longrightarrow \begin{cases} \dot{N}O_2 + \dot{O}\cdot & 410\text{ nm} < \lambda < 670\text{ nm} \\ \dot{N}O + O_2 & 590\text{ nm} < \lambda < 630\text{ nm} \end{cases} \tag{11.35}$$

The probability of each set of products is embodied in the quantum yield, defined in Chapter 10. Whereas the absorption cross section of a gas is the same for each set of products, the quantum yield of a gas differs for each set. Figure 11.3(a) and (b) show predicted photolysis-coefficient profiles for several photoprocesses under specified conditions.

## 11.5. SETS OF REACTIONS

Atmospheric chemical problems require the determination of gas concentrations when many reactions occur at the same time. A difficulty arises because a species is usually produced and/or destroyed by several reactions. Consider the following

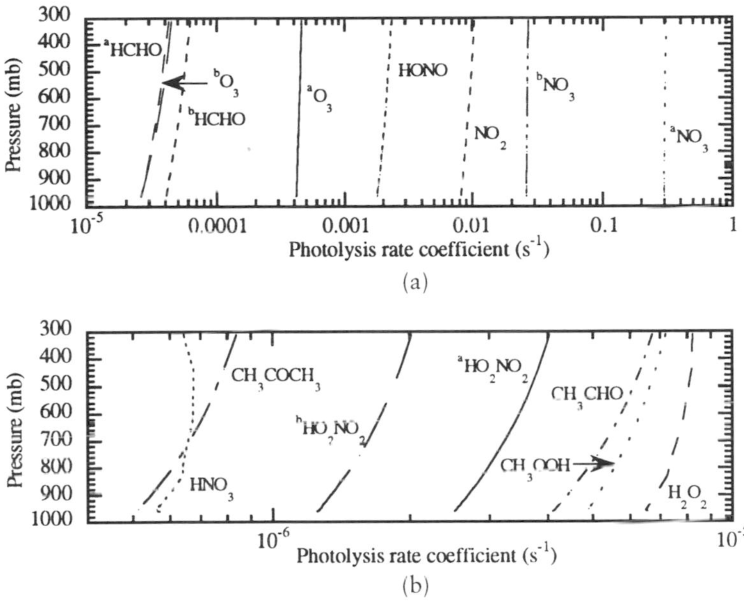


**Figure 11.3.** Predicted photolysis coefficient profiles at 11:30 a.m. on August 27, 1987, at Temecula, California (33.49°N, 117.22°W). Cross-section and quantum-yield data are referenced in Table B.4 of Appendex B. The photoprocesses are as follows;

| | | |
|---|---|---|
| $^{a}O_3 \longrightarrow O_2 + O$ | | $^{a}HCHO \longrightarrow 2H + CO$ |
| $^{a}NO_3 \longrightarrow NO_2 + O$ | | $^{b}HO_2NO_2 \longrightarrow HO_2 + NO_2$ |
| $^{b}O_3 \longrightarrow O_2 + O(^1D)$ | | $^{b}HCHO \longrightarrow H_2 + CO$ |
| $^{b}NO_3 \longrightarrow NO + O_2$ | | $^{a}HO_2NO_2 \longrightarrow OH + NO_3$ |
| $NO_2 \longrightarrow NO + O$ | | $HONO \longrightarrow OH + NO$ |
| $CH_3COCH_3 \longrightarrow CH_3 + COCH_3$ | | $H_2O_2 \longrightarrow 2OH$ |
| $HNO_3 \longrightarrow OH + NO_2$ | | $CH_3OOH \longrightarrow CH_3O + OH$ |

The rate coefficient for $HNO_3$ increases with height near 300 km because the absorption cross sections of $HNO_3$ decrease with decreasing temperature.

four reactions and corresponding rate expressions:

$$\dot{N}O + O_3 \longrightarrow \dot{N}O_2 + O_2 \qquad \text{Rate}_1 = k_1[NO][O_3] \tag{11.36}$$

$$\dot{O}\cdot + O_2 + M \longrightarrow O_3 + M \qquad \text{Rate}_2 = k_2[O][O_2][M] \tag{11.37}$$

$$\dot{N}O_2 + h\nu \longrightarrow \dot{N}O + \dot{O}\cdot \qquad \text{Rate}_3 = J[NO_2] \tag{11.38}$$

$$\dot{N}O_2 + O \longrightarrow \dot{N}O + O_2 \qquad \text{Rate}_4 = k_3[NO_2][O] \tag{11.39}$$

The time derivatives of NO, $NO_2$, O, and $O_3$ concentrations from the reactions are

$$\begin{aligned}\frac{d[NO]}{dt} &= P_c - L_c = \text{Rate}_3 + \text{Rate}_4 - \text{Rate}_1 \\ &= J[NO_2] + k_3[NO_2][O] - k_1[NO][O_3]\end{aligned} \tag{11.40}$$

$$\frac{d[NO_2]}{dt} = P_c - L_c = \text{Rate}_1 - \text{Rate}_3 - \text{Rate}_4 = k_1[NO][O_3] - J[NO_2] - k_3[NO_2][O] \tag{11.41}$$

$$\frac{d[O]}{dt} = P_c - L_c = \text{Rate}_3 - \text{Rate}_2 - \text{Rate}_4 = J[NO_2] - k_2[O][O_2][M] - k_3[NO_2][O] \tag{11.42}$$

$$\frac{d[O_3]}{dt} = P_c - L_c = \text{Rate}_2 - \text{Rate}_1 = k_2[O][O_2][M] - k_1[NO][O_3] \tag{11.43}$$

respectively. These equations are first-order, first-degree, homogeneous ordinary differential equations (ODEs), as shown in Table 6.1. In (11.37) and (11.39), $O_2$ is not affected significantly by the reactions, and no expression for the rate of change of $O_2$ is needed. M is neither created nor destroyed, and no expression for the rate of change of M is needed either. The $O_2$ and M concentrations are still included in the rate expression of (11.37). In (11.40)–(11.43), $P_c$ and $L_c$ are the **total rates of chemical production and loss**, respectively. In the case of NO,

$$P_c = J[NO_2] + k_3[NO_2][O] \qquad L_c = k_1[NO][O_3] \tag{11.44}$$

Thousands of reactions occur in the atmosphere simultaneously. Because computer speeds are limited, a model's chemical mechanism needs to be limited to the most important reactions. Whether a reaction is important can be determined from an analysis of its rate. The rate of a reaction depends on the reaction's rate coefficient and reactant concentrations. Low rate coefficients and reactant concentrations render a reaction unimportant. Rate coefficients vary with temperature, pressure, and/or solar radiation. Reactant concentrations vary with time of day, season, and location. If the rate coefficients and reactant concentrations are expected to be consistently small for a reaction during a simulation, the reaction can be ignored.

## 11.6. STIFF SYSTEMS

Gas and aqueous chemical reaction sets are **stiff**. Stiff systems of reactions are more difficult to solve numerically than nonstiff systems. A **stiff system** of reactions is one in which the **lifetimes** (or time scales) of species taking part in the reactions differ significantly from one another. Species lifetimes are expressed as *e*-folding or half-lifetimes. An ***e*-folding lifetime** is the time required for a species concentration to decrease to 1/*e* its original concentration. A **half-lifetime** is the time required for a species concentration to decrease to 1/2 its original concentration.

The **overall lifetime** of a species is determined by calculating the lifetime of the species against loss from individual reactions and applying

$$\tau_A = \frac{1}{\frac{1}{\tau_{A1}} + \frac{1}{\tau_{A2}} + \cdots \frac{1}{\tau_{An}}} \tag{11.45}$$

where $\tau_A$ is the total lifetime of species A, and $\tau_{A1}, \ldots, \tau_{An}$ are the lifetimes of A due to loss from reactions 1, ..., $n$, respectively. (11.45) applies to $e$-folding or half-lifetime calculations.

The $e$-folding lifetime of species A due to a unimolecular reaction of the form A $\longrightarrow$ products is calculated from the ODE, $d[A]/dt = -k_F[A]$. Integrating this equation from concentration $[A]_0$ at $t = 0$ to $[A]$ at $t = h$ gives $[A] = [A]_0 e^{-k_F h}$. The $e$-folding lifetime is the time at which $[A]/[A]_0 = 1/e = e^{-k_F h}$. This occurs when

$$\tau_{A1} = h = \frac{1}{k_F} \tag{11.46}$$

In the case of a bi- or termolecular reaction, the $e$-folding lifetime is calculated by assuming that each concentration, except for that of the species of interest, equals the initial concentration of the species. The integration is reduced to the that of a unimolecular reaction. For instance the rate of loss of A from the **bimolecular reaction** A + B $\longrightarrow$ products can be described by the linearized ODE, $d[A]/dt = -k_S[A][B]_0$, where $k_S$ is a second-order rate coefficient of the reaction, and $[B]_0$ is the initial concentration of species B. This equation gives A an $e$-folding lifetime of

$$\tau_{A2} = \frac{1}{k_S[B]_0} \tag{11.47}$$

Similarly, the loss rate of A from the **termolecular reaction** A + B + C $\longrightarrow$ product can be described by a linearized ODE that results in an $e$-folding lifetime for A of

$$\tau_{A3} = \frac{1}{k_T[B]_0[C]_0} \tag{11.48}$$

where $k_T$ is the rate coefficient of the reaction, and $[C]_0$ is the initial concentration of C.

The **half-lifetime** of a unimolecular reaction is determined in a manner similar to its $e$-folding lifetime. The half-lifetime is the time at which $[A]/[A]_0 = 1/2 = e^{-k_F h}$. This occurs when

$$\tau_{(1/2)A1} = h = \frac{0.693}{k_F} \tag{11.49}$$

Analogous equations for bi- and termolecular reactions are, respectively,

$$\tau_{(1/2)A2} = \frac{0.693}{k_S[B]_0} \qquad \tau_{(1/2)A3} = \frac{0.693}{k_T[B]_0[C]_0} \tag{11.50}$$

**Example 11.4**

The *e*-folding lifetimes of atmospheric gases vary significantly. The main loss of methane is to the reaction $CH_4 + OH \longrightarrow CH_3 + H_2O$, where $k_r = 6.2 \times 10^{-15}$ cm$^3$ molecule$^{-1}$ s$^{-1}$ at 298 K. When [OH] $= 5.0 \times 10^5$ molecule cm$^{-3}$, $\tau_{CH_4} = 1/k_r[OH] = 10.2$ years.

For $O(^1D) + M \longrightarrow O + M$, we have $k_r = 2.6 \times 10^{-11}$ cm$^3$ molecule$^{-1}$ s$^{-1}$ at 298 K. When $[M] = [N_2] = 1.9 \times 10^{19}$ molecules cm$^{-3}$, we have $\tau_{O(^1D)} = 1/k_r[M] = 2\times 10^9$ s.

Example 11.4 shows that the ratio of lifetimes between $CH_4$ and $O(^1D)$ is about 17 orders of magnitude. A system of equations that includes species with a wide range of lifetimes is said to be stiff. Stiffness depends on reaction rates and species concentrations. At low concentrations, some species lose their stiffness. For almost all atmospheric cases, the range in species lifetimes is large enough for chemical equations to be stiff.

## 11.7. SUMMARY

In this chapter, chemical species, structures, reactions, reaction rate coefficients, and reaction rates were discussed. Elementary reactions are unimolecular, bimolecular, or termolecular and give rise to first-order, second-order, and third-order rate coefficients, respectively. Photolysis reactions are unimolecular and give rise to first-order rate coefficients. Bimolecular reactions include thermal decomposition, isomerization, and basic collision reactions. Termolecular reactions include combination and collision reactions. A combination reaction consists of a pair of elementary bimolecular reactions. In the atmosphere, reactions occur simultaneously, and the lifetimes of species against chemical loss vary by orders of magnitude. Thus, ordinary differential equations arising from atmospheric chemistry problems are stiff.

## 11.8. PROBLEMS

**11.1.** When $T = 265$ K and $p_d = 223$ mb, calculate the second-order rate coefficient for

$$\dot{H} + O_2 \xrightarrow{M} H\dot{O}_2$$

**11.2.** When $T = 298$ K and $p_d = 1013$ mb, calculate the first-order rate coefficient for

$$N_2O_5 \xrightarrow{M} \dot{N}O_2 + N\dot{O}_3$$

Repeat for $T = 288$ K. Discuss temperature effects on the reaction rate coefficient.

**11.3.** Estimate the *e*-folding lifetimes of CO, NO, $O_3$, $SO_2$, $HNO_3$, ISOP (isoprene), and $HO_2$ against loss by OH if [OH] = $1.0 \times 10^6$ molecules $cm^{-3}$, $T = 288$ K, and $p_d = 1010$ mb. The rate coefficients are listed in Table B.4. of Appendix B. Order the species from shortest to longest lifetimes. Which species will most likely reach the stratosphere if only OH reaction is considered?

**11.4.** Write rate expressions for the reactions

$$\dot{Cl} + O_3 \longrightarrow Cl\dot{O} + O_2 \quad \text{and} \quad Cl\dot{O} + \dot{O}\cdot \longrightarrow \dot{Cl} + O_2$$

Write the time derivative of each species in the reactions, assuming the reactions are solved together. What is the expression for the steady-state concentration of ClO?

**11.5.** Given the following observed rate coefficients as a function of temperature, find the activation energy and collisional prefactor of the associated bimolecular reaction. Show your work. In the table, 1.303 (−14) means $1.303 \times 10^{-14}$.

| $T$ (K) | 278 | 288 | 298 | 308 | 318 |
|---|---|---|---|---|---|
| $k_r$($cm^3$ $molecule^{-1}$ $s^{-1}$) | 1.303 (−14) | 1.547 (−14) | 1.814 (−14) | 2.106 (−14) | 2.422 (−14) |

## 11.9. COMPUTER PROGRAMMING PRACTICE

**11.6.** Write a computer script to calculate the first-order rate coefficient for

$$N_2O_5 \xrightarrow{M} \dot{N}O_2 + N\dot{O}_3$$

as a function of temperature and pressure. Draw graphs of the rate coefficient versus temperature when $p_d = 1013$ mb and $p_d = 800$ mb, respectively, and versus pressure when $T = 298$ K and $T = 275$ K, respectively. Discuss the results.

**11.7.** Write a computer script to read in reactions and rate-coefficient data from a computer file. Use the script to calculate the rate coefficients for the first 15 reactions in Table B.4 of Appendix B when $T = 288$ K and $p_d = 980$ mb. Calculate reaction rates when $[O_3] = 2.45 \times 10^{12}$,[O] $= 1.0 \times 10^3$, $[O(^1D)] = 1.0 \times 10^{-3}$, [H] $= 1.0 \times 10^0$, [OH] $= 1.0 \times 10^6$, $[H_2] = 1.5 \times 10^{13}$, $[HO_2] = 1.0 \times 10^8$, and $[H_2O] = 2.0 \times 10^{17}$ molecules $cm^{-3}$.

# 12

# Urban, Free-Tropospheric, and Stratospheric Chemistry

THE importance of a chemical reaction varies with location and time. The free troposphere is affected primarily by inorganic, light organic, and some heavy organic reactions; urban regions are affected by inorganic, light organic, and heavy organic reactions; and the stratosphere is affected by inorganic, light organic, and halogen reactions. Reaction importance also varies between day and night and among seasons. In this chapter, chemical reaction pathways are described for the free troposphere, urban regions, and the stratosphere. Special attention is given to the marine sulfur cycle, ozone production in urban air, and ozone destruction cycles in the global and polar stratosphere. Heterogeneous reactions of gases on particle surfaces are also described.

## 12.1. FREE-TROPOSPHERIC PHOTOCHEMISTRY

Photochemistry in the troposphere outside of urban regions is governed primarily by inorganic and light organic chemical reactions. High-molecular-weight organic gases emitted anthropogenically, such as toluene and xylene, break down chemically over hours to a few days. Most of the free troposphere, especially at high altitudes and away from urban regions, is not affected by these gases. Air over the tropics and other vegetated regions is affected by reaction products of isoprene and other high-molecular-weight terpenes emitted from biogenic sources. Reaction pathways for these gases are discussed in Section 12.2. In the following subsections, inorganic and low-molecular-weight organic reaction pathways for the free troposphere are described.

### 12.1.1. Photostationary-State Relationship

In many regions of the troposphere, nitric oxide (NO), nitrogen dioxide ($NO_2$), and ozone ($O_3$) are strongly coupled by the reactions

$$\dot{N}O + O_3 \longrightarrow \dot{N}O_2 + O_2 \qquad (12.1)$$

$$\dot{N}O_2 + h\nu \longrightarrow \dot{N}O + \dot{O}\cdot \qquad \lambda < 420\,\text{nm} \qquad (12.2)$$

$$\dot{O}\cdot + O_2 + M \longrightarrow O_3 + M \qquad (12.3)$$

In the free troposphere, the mixing ratios of $O_3$ (20–60 ppbv) are much higher than those of NO (1–60 pptv) or $NO_2$ (5–70 pptv) (e.g., Singh et al. 1996), and (12.1) does not deplete ozone during day or night. At night in urban regions, NO mixing ratios may exceed those of $O_3$, and (12.1) can deplete local ozone.

Assuming (12.1) and (12.2) have rate coefficients $k_1$ and $J$, respectively, the time rate of change of the nitrogen dioxide concentration from these reactions is

$$\frac{d[NO_2]}{dt} = k_1[NO][O_3] - J[NO_2] \tag{12.4}$$

If the time rate of change of $NO_2$ is small compared to those of the other terms in (12.4), $NO_2$ is nearly in steady state and (12.4) simplifies to the **photostationary-state relationship,**

$$[O_3] = \frac{J[NO_2]}{k_1[NO]} \tag{12.5}$$

(12.5) does not state that ozone in the free troposphere is affected by [NO] and $[NO_2]$ only. Instead, it predicts a relationship among $[O_3]$, $[NO_2]$, and [NO]. If two of the three concentrations are known, the third can be found from (12.5).

**Example 12.1**

Find the photostationary-state mixing ratio of $O_3$ at midday when $p_d$ = 1013 mb, $T = 298$ K, $J \approx 0.01\ s^{-1}$, $\chi_{NO} = 5$ pptv, and $\chi_{NO_2} = 10$ pptv (typical free-tropospheric mixing ratios).

SOLUTION

At $T = 298$ K, $k_1 \approx 1.8 \times 10^{-14}\ cm^3\ molecule^{-1}\ s^{-1}$ from Table B.4. of Appendix B. Since the conversion from mixing ratio to number concentration is the same for each gas, $[NO_2]/[NO] = \chi_{NO_2}/\chi_{NO}$. From (12.5), $[O_3] = 1.1 \times 10^{12}$ molecules $cm^{-3}$. From (2.18), $N_d = 2.46 \times 10^{19}$ molecules $cm^{-3}$. Dividing $[O_3]$ by $N_d$ gives $\chi_{O_3} = 44.7$ ppbv, which is a typical free-tropospheric ozone mixing ratio.

Two important reactions, aside from (12.1)–(12.3), that affect ozone are

$$O_3 + h\nu \longrightarrow O_2 + \cdot\dot{O}(^1D) \qquad \lambda < 310\,nm \tag{12.6}$$

$$O_3 + h\nu \longrightarrow O_2 + \dot{O}\cdot \qquad \lambda > 310\,nm \tag{12.7}$$

where $O(^1D)$ is **excited atomic oxygen.** In the free troposphere, the *e*-folding lifetimes of ozone against destruction by these reactions are about 0.7 and 14 h, respectively. The lifetime of ozone against destruction by NO in (12.1) was 126 h under the conditions of Example 12.1. Thus, in the daytime free troposphere, photolysis destroys ozone faster than does reaction with NO. When the ozone concentration changes due to photolysis, the $[NO_2]/[NO]$ ratio in (12.5) changes to adapt to the new ozone concentration.

The photostationary-state relationship in (12.5) is useful for free-tropospheric analysis. Under urban conditions, it is useful only when organic-gas concentrations are low. Organic radicals provide sources of $NO_2$ not included in the rate expression for $NO_2$ given in (12.4). When organic-radical sources are large, as they are during the morning in urban air, the photostationary-state relationship does not hold.

When excited atomic oxygen forms, as in (12.6), it rapidly produces O by

$$\cdot\dot{O}(^1D) \xrightarrow{M} \dot{O}\cdot \qquad (12.8)$$

and O rapidly produces $O_3$ by (12.3). The reactions (12.3) and (12.6)–(12.8) cycle oxygen atoms quickly among $O(^1D)$, O, and $O_3$. Losses of O from the cycle, such as from conversion of NO to $NO_2$ by (12.1), are slower than transfers of O within the cycle.

### 12.1.2. Hydroxyl Radical

The hydroxyl radical (OH) is an important oxidizing agent in the atmosphere and decomposes (scavenges) many gases. A technique of estimating the OH concentration in the global troposphere is by determining the loss rate of methyl chloroform ($CH_3CCl_3$) from a balance of its global sources and atmospheric abundance (e.g., Singh 1995). The atmospheric abundance is obtained from measurements. The only source of $CH_3CCl_3$ is from known human-made emissions, and the major loss is reaction with OH. Smaller losses are transport to the oceans and stratosphere. Such a calculation reveals a global mean OH concentration of about $8 \times 10^5$ molecules $cm^{-3}$ (Singh 1995), corresponding to a mean $CH_3CCl_3$ *e*-folding lifetime of about 5 years. The OH concentration at any given location in the clean free troposphere usually ranges from $2 \times 10^5$ to $3 \times 10^6$ molecules $cm^{-3}$. When clean air is exposed to an urban plume, OH concentrations increase to $6 \times 10^6$ molecules $cm^{-3}$ or more due to the additional availability of oxides of nitrogen and reactive organic gases (Comes et al. 1997). In urban air, OH concentrations typically range from $1 \times 10^6$ to $1.0 \times 10^7$ molecules $cm^{-3}$.

The main free-tropospheric source of OH is

$$\cdot\dot{O}(^1D) + H_2O \longrightarrow 2\dot{O}H \qquad (12.9)$$

Sources of water vapor were discussed in Section 2.1.2.1. In the upper troposphere, $H_2O$ is scarce, limiting the ability of (12.9) to produce OH. Under such conditions, other reactions, which are normally minor, may be important sources of OH. These reactions are discussed in Section 12.1.10.

After its formation, OH reacts with inorganic and organic gases. The overall *e*-folding lifetime of OH against chemical destruction is about 0.1–1 s. The *e*-folding

**Table 12.1.** The *e*-Folding Lifetimes of Several Gases Against Destruction by OH in Clean Free-tropospheric Air When $N_{OH}$ is Low and High

| | *e*-folding Lifetime | | |
|---|---|---|---|
| **Reaction** | $N_{OH} = 5 \times 10^5$ molecules cm$^{-3}$ | $N_{OH} = 5 \times 10^6$ molecules cm$^{-3}$ | **Eq.** |
| $\dot{O}H + O_3 \longrightarrow H\dot{O}_2 + O_2$ | 346 days | 86.5 days | (12.10) |
| $\dot{O}H + H\dot{O}_2 \longrightarrow H_2O + O_2$ | 5.1 h | 1.3 h | (12.11) |
| $\dot{O}H + H_2O_2 \longrightarrow H\dot{O}_2 + H_2O$ | 13.6 days | 3.4 days | (12.12) |
| $\dot{O}H + \dot{N}O \xrightarrow{M} HONO$ | 2.4 days | 14 h | (12.13) |
| $\dot{O}H + \dot{N}O_2 \xrightarrow{M} HNO_3$ | 1.9 days | 11.4 h | (12.14) |
| $\dot{O}H + N\dot{O}_3 \longrightarrow H\dot{O}_2 + \dot{N}O_2$ | 1.2 days | 7.2 h | (12.15) |
| $\dot{O}H + HO_2NO_2 \longrightarrow H_2O + \dot{N}O_2 + O_2$ | 4.6 days | 1.2 days | (12.16) |
| $\dot{O}H + SO_2 \xrightarrow{M} HS\dot{O}_3$ | 26 days | 6.5 days | (12.17) |
| $\dot{O}H + CO \longrightarrow \dot{H} + CO_2$ | 111 days | 28 days | (12.18) |
| $\dot{O}H + CH_4 \longrightarrow H_2O + \dot{C}H_3$ | 10.2 yr | 2.6 yr | (12.19) |
| $\dot{O}H + C_2H_6 \longrightarrow H_2O + \dot{C}_2H_5$ | 93 days | 23 days | (12.20) |
| $\dot{O}H + C_3H_8 \longrightarrow H_2O + \dot{C}_3H_7$ | 21 days | 5.3 days | (12.21) |
| $\dot{O}H + CH_3OOH \longrightarrow H_2O + CH_3\dot{O}_2$ | 6.4 days | 1.6 days | (12.22) |

$T = 298$ K and $p_a = 1013$ mb. $HO_2$ = hydroperoxy radical, $H_2O_2$ = hydrogen peroxide, HONO = nitrous acid, $HNO_3$ = nitric acid, $NO_3$ = nitrate radical, $HO_2NO_2$ = peroxynitric acid, $CH_3$ = methyl radical, $HSO_3$ = bisulfite, $C_2H_6$ = ethane, $C_2H_5$ = ethoxy radical, $C_3H_8$ = propane, $C_3H_7$ = propoxy radical, and $CH_3COOH$ = methyl hydroperoxide.

lifetimes of gases reacting with OH vary, as shown in Table 12.1. The *e*-folding lifetimes of additional organic gases against destruction by OH are given in Table 12.5 (Section 12.2).

### 12.1.3. Hydroperoxy Radical

Like OH, $HO_2$ is a scavenger. Its concentrations are usually larger than those of OH. Whereas OH is present during daytime only, $HO_2$ is present during daytime and nighttime. Chemical sources of $HO_2$ include reactions of $O_3$, $H_2O_2$, and $NO_3$ with OH, given in (12.10), (12.12), and (12.15), respectively (Table 12.1), and

$$\dot{H} + O_2 \xrightarrow{M} H\dot{O}_2 \tag{12.23}$$

Sources of H include reactions of CO with OH, $O(^1D)$ with $H_2$, and OH with $H_2$.

The rate of $HO_2$ loss depends on NO mixing ratios. In the presence of high NO mixing ratios (>10 pptv), $HO_2$ most likely reacts with NO and $NO_2$. When NO mixing ratios are lower (3–10 pptv), $HO_2$ most likely reacts with ozone. When NO mixing ratios are very low (<3 pptv), $HO_2$ most likely reacts with itself (Finlayson-Pitts and Pitts 1986). The reactions corresponding to these three conditions are

| | | |
|---|---|---|
| $H\dot{O}_2 + \dot{N}O \longrightarrow \dot{O}H + \dot{N}O_2$<br>$H\dot{O}_2 + \dot{N}O_2 \xrightarrow{M} HO_2NO_2$ | >10 pptv NO | (12.24) |
| $H\dot{O}_2 + O_3 \longrightarrow \dot{O}H + 2O_2$ | 3–10 pptv NO | (12.25) |
| $H\dot{O}_2 + H\dot{O}_2 \longrightarrow H_2O_2 + O_2$ | <3 pptv NO | (12.26) |

### 12.1.4. Recycling of OH and $HO_2$

Some products of OH and $HO_2$ oxidation recycle back to OH and $HO_2$, respectively, by photolysis and thermal decomposition. **Nitrous acid** (HONO), produced during the day by the reaction of NO with OH given in (12.13) (Table 12.1), cycles back to OH by

$$HONO + h\nu \dashrightarrow \dot{O}H + \dot{N}O \qquad \lambda < 400\,\text{nm} \tag{12.27}$$

Because photolysis of HONO is rapid, HONO is destroyed soon after it forms. The destruction rate increases as sunlight becomes more intense during the day. Thus, HONO concentrations are high only during the early morning. After early morning, photolysis destroys HONO rapidly. During the night, HONO is not produced chemically, since OH, required for its production, is absent at night.

**Nitric acid** ($HNO_3$), produced by the reaction of $NO_2$ with OH given in (12.14) (Table 12.1), recycles back to OH by

$$HNO_3 + h\nu \longrightarrow \dot{O}H + \dot{N}O_2 \qquad \lambda < 335\,\text{nm} \tag{12.28}$$

The *e*-folding lifetime of nitric acid against destruction by photolysis is 15–80 days, depending on the day of the year and the latitude. Since this lifetime is fairly long, $HNO_3$ is a sink for OH over a short time. Nitric acid is soluble, and much of it dissolves in cloud or aerosol drops and reacts on aerosol surfaces. Over long periods, nitric acid photolysis regenerates OH and $NO_2$. $HNO_3$ mixing ratios range from 10 to 150 pptv in the free troposphere (e.g., Singh et al. 1996).

**Peroxynitric acid** ($HO_2NO_2$) is produced by the reaction of $NO_2$ with $HO_2$, given in (12.24). At high temperatures, it decomposes within seconds to reproduce $HO_2$ by

$$HO_2NO_2 \xrightarrow{M} H\dot{O}_2 + \dot{N}O_2 \tag{12.29}$$

$HO_2NO_2$ can photolyze by

$$HO_2NO_2 + h\nu \longrightarrow \begin{cases} H\dot{O}_2 + \dot{N}O_2 & \lambda < 330\text{ nm} \\ \dot{O}H \ \ + \dot{N}O_3 & \lambda < 330\text{ nm} \end{cases} \tag{12.30}$$

with *e*-folding lifetimes of 2 to 5 days. OH can scavenge $HO_2NO_2$ through (12.16) (Table 12.1). This reaction is slow in comparison with thermal decomposition.

**Hydrogen peroxide** ($H_2O_2$), produced by $HO_2$ self-reaction in (12.26), recycles to OH by

$$H_2O_2 + h\nu \longrightarrow 2\dot{O}H \qquad \lambda < 355\,\text{nm} \tag{12.31}$$

$H_2O_2$'s *e*-folding lifetime against photolysis by this reaction ranges from 1 to 2 days. Hydrogen peroxide is also scavenged by OH via (12.12) (Table 12.1). The *e*-folding lifetime of $H_2O_2$ against destruction by OH ranges from 3 to 14 days in the free troposphere.

### 12.1.5. Nighttime Nitrogen Chemistry

During nighttime in the free troposphere, concentrations of the **nitrate radical** ($NO_3$) and **dinitrogen pentoxide** ($N_2O_5$) build up. Both species are produced during day and night. During the day, sunlight breaks both down rapidly. $NO_3$ is an important nighttime scavenger of organic gases, as discussed in Section 12.2.

The nighttime cycle among $NO_2$, $NO_3$, $N_2O_5$, and $HNO_3$ is

$$\dot{N}O_2 + O_3 \longrightarrow N\dot{O}_3 + O_2 \tag{12.32}$$

$$\dot{N}O_2 + N\dot{O}_3 \overset{M}{\rightleftharpoons} N_2O_5 \tag{12.33}$$

$$N_2O_5 + H_2O \longrightarrow 2HNO_3 \tag{12.34}$$

(12.33) occurs in both directions. The forward reaction is a three-body, pressure-dependent reaction. The reverse reaction is a temperature-dependent thermal decomposition reaction. At high temperatures, such as during the day and in the lower atmosphere, the reverse reaction occurs within seconds. At low temperatures, such as during the night and at high altitudes, it takes hours to days or even months.

(12.34) is a **homogeneous reaction** (all species are in the gas phase) and has a low reaction rate. If $N_2O_5$ reacts on the surface of a particle coated with water or ice, the reaction produces $HNO_3$ adsorbed to the surface. This type of reaction is a **heterogeneous reaction**, discussed in Section 12.3.4, and is much faster than the corresponding homogeneous reaction so long as sufficient particle surfaces are present.

After sunrise, $NO_3$ photodissociates by

$$N\dot{O}_3 + h\nu \longrightarrow \begin{cases} \dot{N}O_2 + \dot{O}\cdot & 410\,\text{nm} < \lambda < 670\,\text{nm} \\ \dot{N}O + O_2 & 590\,\text{nm} < \lambda < 630\,\text{nm} \end{cases} \tag{12.35}$$

Both pathways give the nitrate radical a daytime *e*-folding lifetime of seconds. $N_2O_5$ photodissociates during the day by

$$N_2O_5 + h\nu \longrightarrow \dot{N}O_2 + N\dot{O}_3 \qquad \lambda < 385\,\text{nm} \tag{12.36}$$

The *e*-folding lifetime of $N_2O_5$ against loss by this reaction alone is a few hours.

### 12.1.6. Carbon Monoxide Oxidation

Ozone production in the free troposphere is enhanced by the presence of carbon monoxide (CO), methane ($CH_4$), and nonmethane organic gases. CO produces ozone by

$$CO + \dot{O}H \longrightarrow CO_2 + \dot{H} \qquad (12.37)$$

$$\dot{H} + O_2 \xrightarrow{M} H\dot{O}_2 \qquad (12.38)$$

$$\dot{N}O + H\dot{O}_2 \longrightarrow \dot{N}O_2 + \dot{O}H \qquad (12.39)$$

$$\dot{N}O_2 + h\nu \longrightarrow \dot{N}O + \dot{O}\cdot \qquad \lambda < 420 \text{ nm} \qquad (12.40)$$

$$\dot{O}\cdot + O_2 + M \longrightarrow O_3 + M \qquad (12.41)$$

Because the lifetime of CO against breakdown by (12.37) in free tropospheric air is 28–110 days, the rate of ozone production by this sequence is low. The sequence does not interfere with the photostationary-state relationship among $O_3$, NO, and $NO_2$.

### 12.1.7. Methane Oxidation

Methane, with a mixing ratio of 1.7 ppmv, is the most abundant organic gas in the earth's atmosphere. Table 12.1 indicates that its free-tropospheric *e*-folding lifetime is 2–10 years. This long lifetime has enabled it to mix uniformly up to an altitude of 15–20 km. From this height, upward, its mixing ratio gradually decreases. Methane's only important loss is hydroxyl radical oxidation. The methane-oxidation pathway produces ozone, but the incremental quantity of ozone produced is small compared to the photostationary quantity of ozone. Methane reaction with OH produces the **methyl radical** ($CH_3$) by

$$CH_4 + \dot{O}H \longrightarrow \dot{C}H_3 + H_2O \qquad (12.42)$$

The methyl radical reacts rapidly with $O_2$ to form the **methylperoxy radical** ($CH_3O_2$) and further products through the sequence

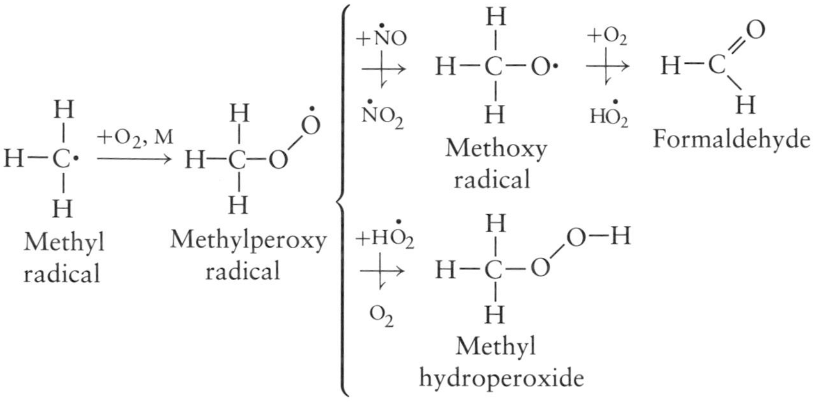


(12.43)

The first pathway in this sequence produces the **methoxy radical** ($CH_3O$), $NO_2$, and **formaldehyde** (HCHO). The *e*-folding lifetime of $CH_3O$ against destruction by $O_2$ is $10^{-4}$ s; thus, its conversion to formaldehyde is almost instantaneous. $NO_2$ produces $O_3$ by (12.2) and (12.3). Formaldehyde decomposition is discussed in Section 12.1.9. The second pathway in (12.43) produces **methyl hydroperoxide** ($CH_3OOH$), which stores OH and $HO_2$ radicals. Methyl hydroperoxide releases OH and HCHO during photolysis and releases $CH_3O_2$ during reaction with OH. The *e*-folding lifetime of $CH_3OOH$ against destruction by photolysis is 1.5–2.5 days, and that against destruction by OH is 1.6–6.4 days at 298 K (Table 12.1). The methyl hydroperoxide decomposition reactions are

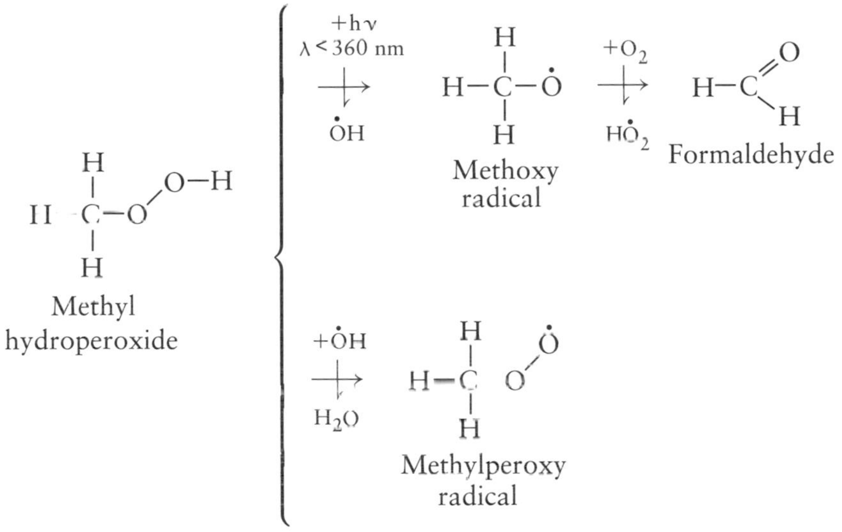


(12.44)

### 12.1.8. Ethane and Propane Oxidation

The primary nonmethane hydrocarbons in the free troposphere are ethane ($C_2H_6$), propane ($C_3H_8$), ethene ($C_2H_4$), and propene ($C_3H_6$). Free-tropospheric mixing ratios of these gases are 0–2.5 ppbv for ethane, 0–1.0 ppbv for propane, 0–1.0 ppbv for ethene, and 0–1.0 ppbv for propene (Singh et al. 1988; Bonsang et al. 1991). The primary oxidant of ethane, propane, and other alkanes is OH. Photolysis, reaction with $O_3$, reaction with $HO_2$, and reaction with $NO_3$ do not affect alkane concentrations significantly. In this subsection, ethane and propane oxidation by OH is discussed. Ethene and propene oxidation pathways are shown in Section 12.2.2.

Lifetimes of ethane and propane against decomposition by OH were given in Table 12.1 for free tropospheric conditions. The reaction of OH with propane is about 4 times faster than that with ethane.

The hydroxyl radical attacks **ethane** by hydrogen abstraction. The reaction pathway is

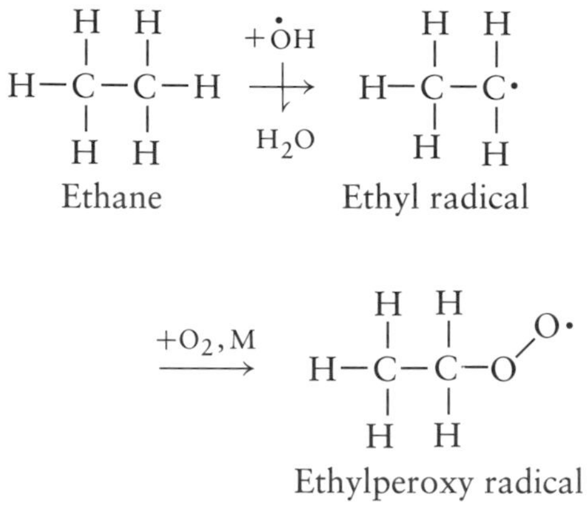


(12.45)

The **ethylperoxy radical** ($C_2H_5O_2$) takes one of two courses. The first leads to $NO_2$, $HO_2$, and **acetaldehyde** ($CH_3CHO$) formation. The second leads to **ethylperoxynitric acid** ($C_2H_5O_2NO_2$) formation. The reactions are

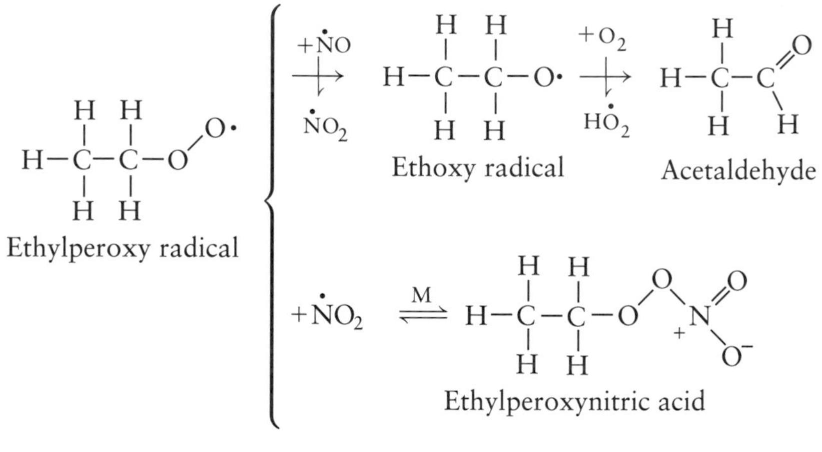


(12.46)

$NO_2$ forms ozone through (12.2) and (12.3). Acetaldehyde decomposition is discussed in Section 12.1.9. Ethylperoxynitric acid is a temporary sink for $NO_x$, since it decomposes thermally back to the ethylperoxy radical and $NO_2$. $C_2H_5O_2$ can also react with NO to form **ethyl nitrate** ($C_2H_5ONO_2$), but the *e*-folding lifetime of $C_2H_5O_2$ against this reaction is 48–192 days. The *e*-folding lifetime of $C_2H_5O_2$ against conversion to ethylperoxy nitric acid, $C_2H_5O_2NO_2$, is 1.7–6.7 days. The exthoxy radical ($C_2H_5O$) produced in (12.46) competitively reacts with NO and $NO_2$. Reaction with NO produces **ethyl nitrite** ($C_2H_5ONO$), and reaction with $NO_2$ produces ethyl nitrate.

An important pathway of **propane** oxidation by the hydroxyl radical is

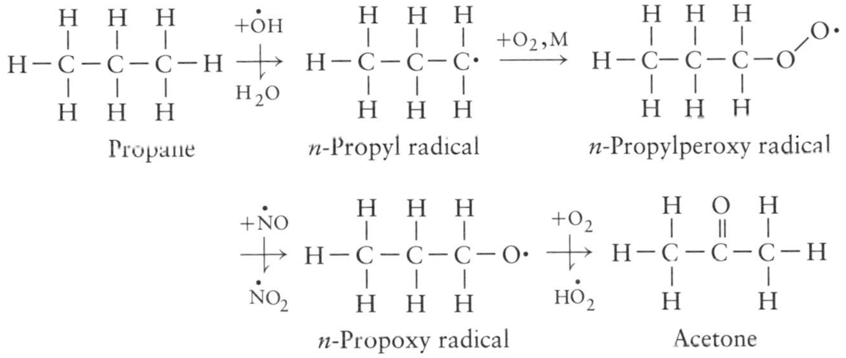


(12.47)

$NO_2$ from this reaction forms ozone. **Acetone** ($CH_3COCH_3$) is a long-lived species whose fate is discussed in Section 12.1.10. The **propylperoxy radical** ($C_3H_7O_2$) competitively reacts with NO to form **propyl nitrate** ($C_3H_7ONO_2$), but the reaction is slow. The **propoxy radical** ($C_3H_7O$) reacts with NO to form **propyl nitrite** ($C_3H_7ONO$) and with $NO_2$ to form propyl nitrate. The free-tropospheric *e*-folding lifetimes of $C_3H_7O$ against loss by the last two reactions are both 4–16 h.

### 12.1.9. Formaldehyde and Acetaldehyde Reactions

Formaldehyde is a carcinogen and an important ozone precursor. It is produced from (12.43) by methoxy-radical oxidation. It decomposes by photolysis, reaction with OH, reaction with $HO_2$, and reaction with $NO_3$. Acetaldehyde is a precursor to peroxyacetyl nitrate (PAN) and ozone. It is produced by ethoxy-radical oxidation and destroyed by photolysis, reaction with OH, reaction with $HO_2$, and reaction with $NO_3$.

#### 12.1.9.1. *Aldehyde Photolysis*

Formaldehyde and acetaldehyde photodissociates during the day by

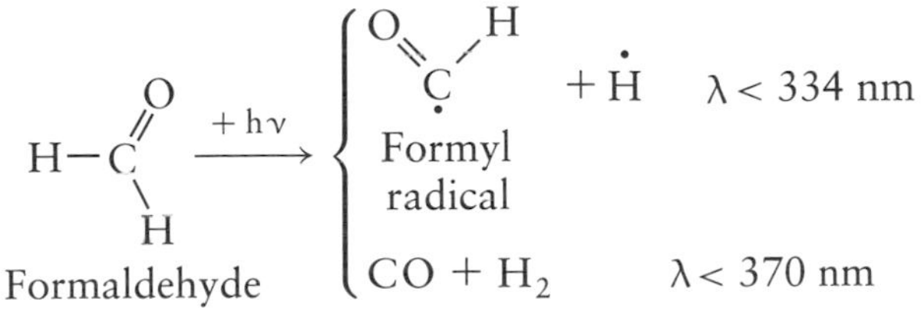


(12.48)

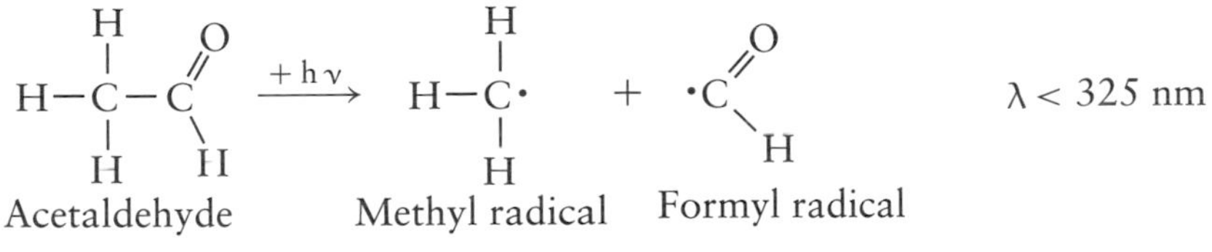


(12.49)

respectively, producing H, $CH_3$, and the **formyl radical** (HCO). H reacts with $O_2$ to form $HO_2$ by (12.23), which produces ozone by (12.39)–(12.41). $CH_3$ is oxidized in (12.43), producing $NO_2$, $HO_2$, HCHO, and $CH_3OOH$. HCO produces CO and $HO_2$ by

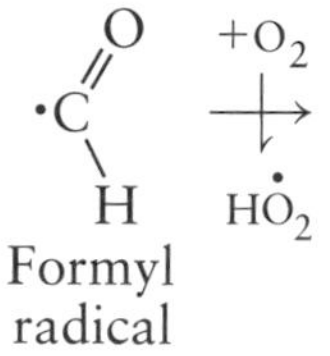


(12.50)

Both form ozone by CO oxidation as discussed in Section 12.1.6.

#### 12.1.9.2. *Aldehyde Reaction with the Hydroxyl Radical*

Formaldehyde and acetaldehyde react with the hydroxyl radical to produce ozone and peroxyacetyl nitrate, respectively. The HCHO–OH reaction is

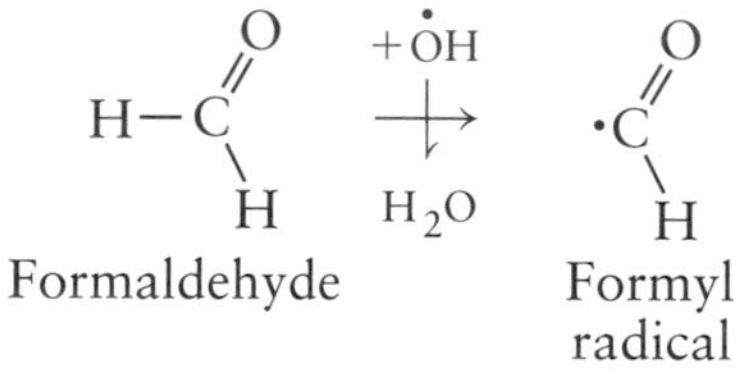


(12.51)

where HCO reacts further in (12.50). The $CH_3CHO$–OH reaction is

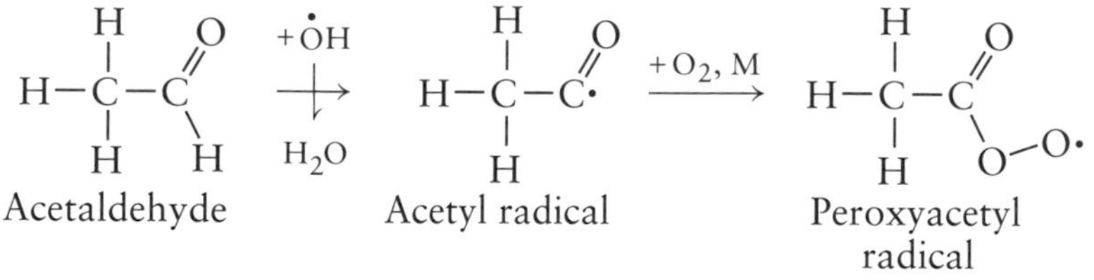


(12.52)

The **peroxyacetyl radical** [$CH_3C(O)OO$] reacts with NO and $NO_2$ by

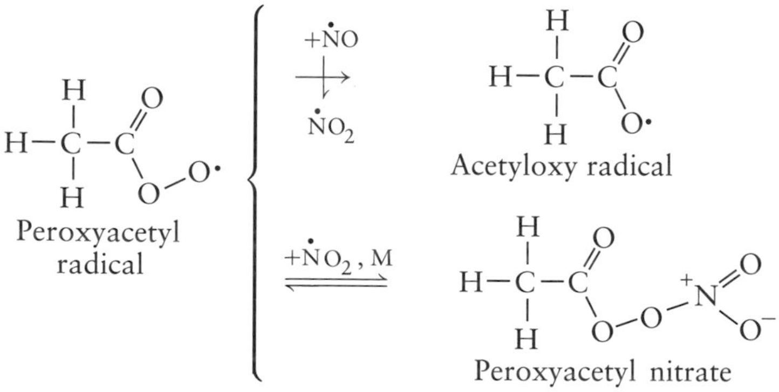


(12.53)

The first reaction produces $NO_2$, which forms ozone. The second produces **peroxyacetyl nitrate** (PAN). PAN decomposes at high temperatures with an *e*-folding lifetime of about 30 min at 298 K.

#### 12.1.9.3. *Aldehyde Reaction with Nitrate*

At night, acetaldehyde reacts with $NO_3$ by the sequence

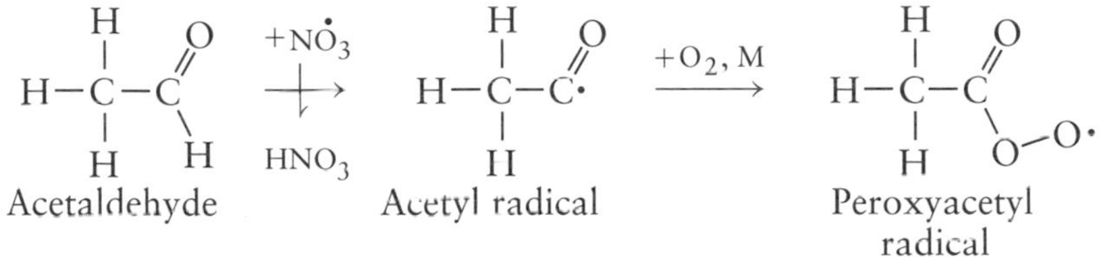


(12.54)

Since the peroxyacetyl radical forms PAN, as shown in (12.53), (12.54) is a nighttime source of PAN.

### 12.1.10. Acetone Reactions

Acetone is a long-lived ketone produced from the OH oxidation of propane [through (12.47)], *i*-butane, or *i*-pentane (e.g., Chatfield et al. 1987). The mixing ratio of acetone in the free troposphere is 200–700 pptv (Singh et al. 1995). Acetone decomposes by reaction with OH and photolysis. The OH reaction produces $CH_3COCH_2$ and $H_2O$. The *e*-folding lifetime of acetone against this reaction is 27 days when $T = 298$ K and $[OH] = 5 \times 10^5$ molecules $cm^{-3}$. The photolysis reaction is

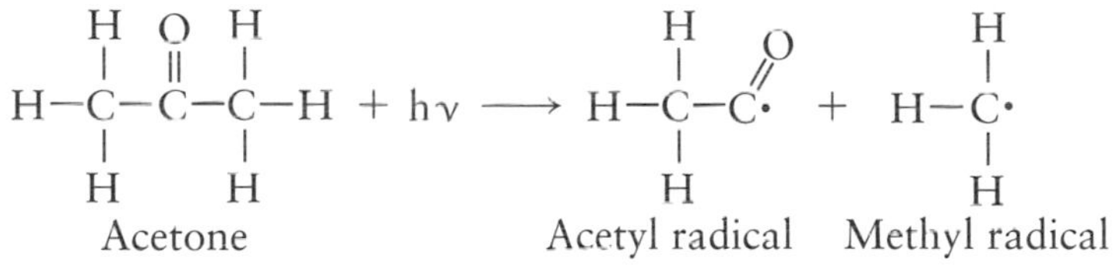


(12.55)

For the location and time given in Fig. 11.3(b), the *e*-folding lifetime of acetone against photolysis is 14–23 days, with the shorter lifetime corresponding to upper tropospheric conditions and the longer lifetime corresponding to lower tropospheric conditions. The acetyl radical from (12.55) forms PAN through (12.52)–(12.53). The methyl radical is oxidized in (12.43), producing $HO_2$, HCHO, and $CH_3OOH$, among other products. HCHO produces $HO_2$ through (12.48) and (12.50). $CH_3OOH$ decomposes, forming $HO_2$ and OH in the process, through (12.44). Thus, acetone photolysis yields primarily PAN, OH, $HO_2$, and intermediate products.

OH production from the $O(^1D)$–$H_2O$ reaction in the upper troposphere is less significant than in the lower troposphere because the concentration of $H_2O$ is low in the upper troposphere. In the upper troposphere, acetone photolysis and subsequent reaction may be an important source of OH (Singh et al. 1995). Other sources of OH in the upper troposphere are currently under investigation.

### 12.1.11. Sulfur Photochemistry

In the free troposphere, gases that contain sulfur (S) are important chemical reactants. Sulfur is emitted naturally in several forms. Anaerobic bacteria in marshes emit dimethyl sulfide (DMS) ($CH_3SCH_3$), dimethyl disulfide (DMDS) ($CH_3SSCH_3$), methanethiol ($CH_3SH$), and hydrogen sulfide ($H_2S$). Phytoplankton in the oceans emit DMS, DMDS, and other products. Volcanos emit carbonyl sulfide (OCS), carbon disulfide ($CS_2$), $H_2S$, and sulfur dioxide ($SO_2$). Most forms of sulfur ultimately oxidize to sulfur dioxide.

**Dimethyl sulfide** is produced by bacteria in certain soils and plants. It is also the most abundant sulfur-containing compound emitted from the oceans. DMS is produced from DMSP (dimethyl sulfonium propionate), which is emitted by many phytoplankton (Bates et al. 1994). When phytoplankton feed, DMSP is exuded and cleaved by enzymes to produce DMS and other products. Before DMS evaporates from the ocean surface, much of it chemically reacts or is consumed by microorganisms. Over the oceans, DMS mixing ratios vary between $<$10 pptv and 1 ppbv. The latter value occurs over eutrophic waters (Berresheim et al. 1995). Average near-surface DMS mixing ratios over oceans and land are about 100 and 20 pptv, respectively.

The major atmospheric chemical loss processes of DMS are through hydroxyl radical abstraction and addition. **Abstraction** is the process by which a radical, such as the hydroxyl radical, removes a hydrogen atom from a compound. **Addition** is the process by which a radical bonds to a compound. Above 285 K, hydroxyl radical abstraction is the dominant DMS–OH reaction pathway. Below 285 K, addition is more important (Hynes et al. 1986). Yin et al. (1990) and Tyndall and Ravishankara (1991) present oxidation pathways of DMS.

Hydroxyl radical abstraction of DMS results in the formation of the **methanethiolate radical** ($CH_3S$) and formaldehyde by the sequence

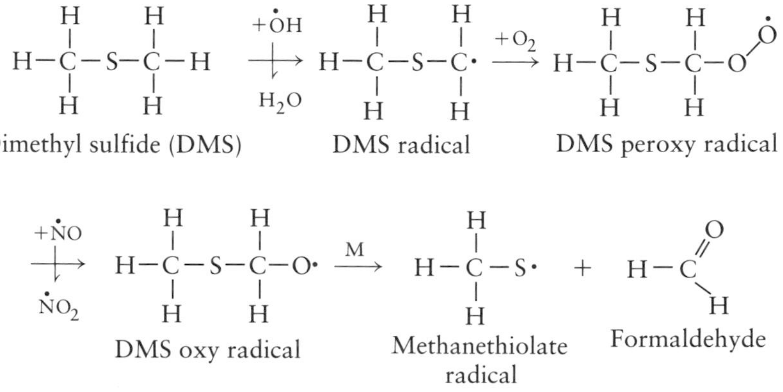


(12.56)

The methanethiolate radical reacts with $O_2$ by

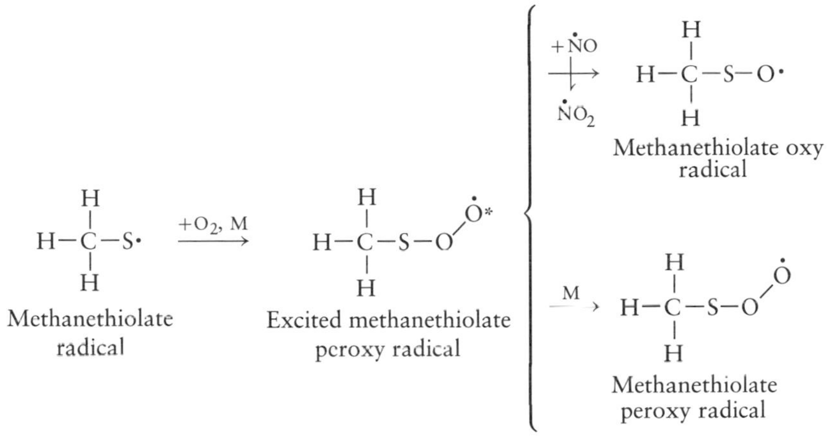


(12.57)

where the intermediate product is short-lived. The **methanethiolate oxy radical** ($CH_3SO$) decays to **sulfur monoxide** (SO) or forms the **methanethiolate peroxy radical** ($CH_3SO_2$) by

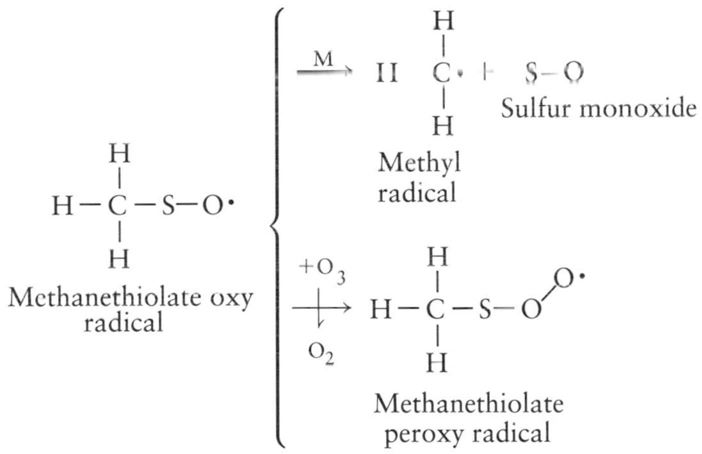


(12.58)

Sulfur monoxide forms sulfur dioxide by

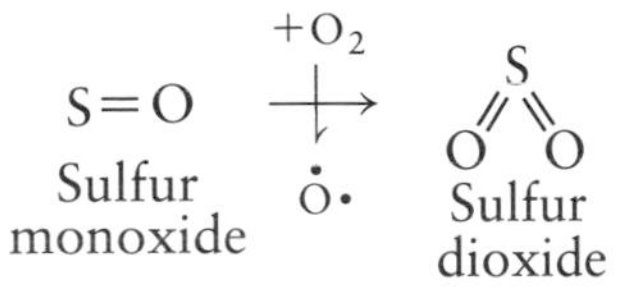


(12.59)

The *e*-folding lifetime of sulfur monoxide against destruction by $O_2$ is about 0.0005 s at 298 K and 1 atm pressure. $CH_3SO_2$ from (12.58) forms sulfur dioxide by

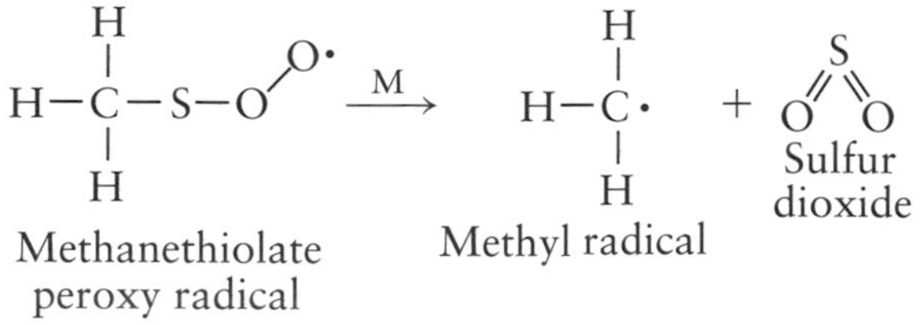


(12.60)

$CH_3SO_2$ also reacts with $O_3$, $NO_2$, and $HO_2$ to produce $CH_3SO_3$, which breaks down to $CH_3$ and $SO_3$ or abstracts a hydrogen atom from an organic gas to form $CH_3S(O)_2OH$ (**methanesulfonic acid**, MSA). In sum, OH abstraction of DMS results in the formation of sulfur dioxide, methanesulfonic acid, and other products.

The **DMS-addition pathway** initiates when OH bonds to the sulfur atom in DMS. The reaction sequence is

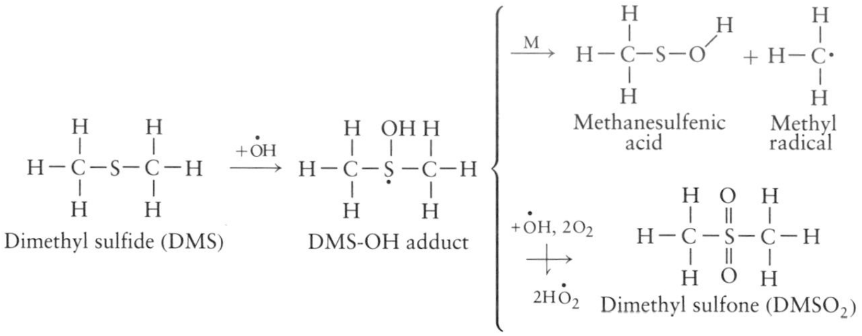


(12.61)

One product of this reaction, **dimethyl sulfone**, does not react further. Another product, **methanesulfenic acid** ($CH_3SOH$), produces $CH_3SO$ by

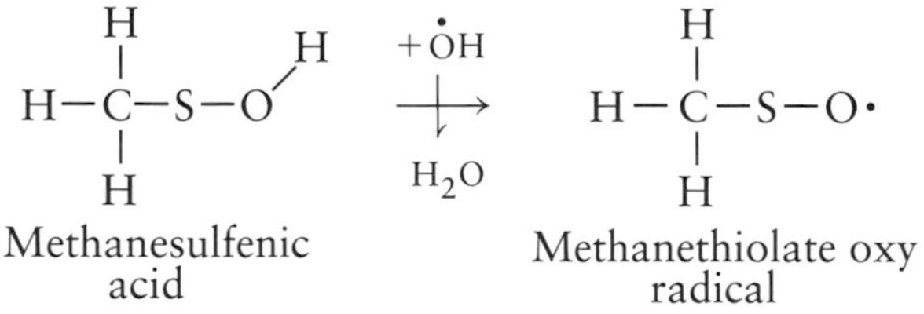


(12.62)

$CH_3SO$ reacts via (12.58)–(12.60) to produce sulfur dioxide.

**Dimethyl disulfide** (DMDS) oxidation proceeds almost exclusively by OH addition. The sequence is

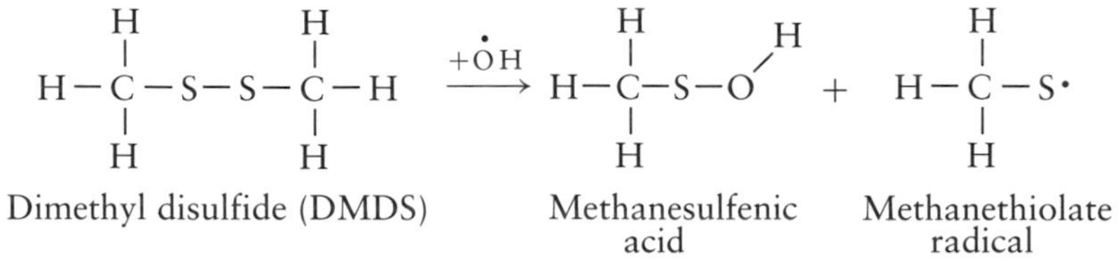


(12.63)

$CH_3SOH$ and $CH_3S$ react in (12.62) and (12.57), respectively, to form $SO_2$, MSA, and other products. DMDS also photolyzes by

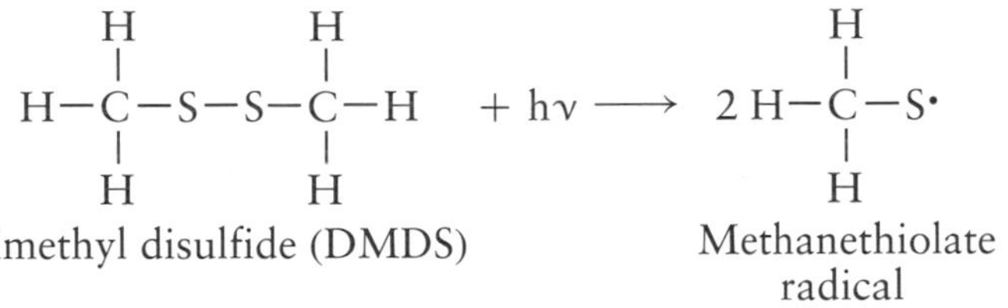

$$H-\underset{\underset{H}{|}}{\overset{\overset{H}{|}}{C}}-S-S-\underset{\underset{H}{|}}{\overset{\overset{H}{|}}{C}}-H \;+\; h\nu \;\longrightarrow\; 2\,H-\underset{\underset{H}{|}}{\overset{\overset{H}{|}}{C}}-S\cdot \tag{12.64}$$

Dimethyl disulfide (DMDS)        Methanethiolate radical

**Hydrogen sulfide** ($H_2S$), which has the odor of rotten eggs, is emitted from anaerobic soils, plants, paper manufacturing sources, and volcanos. It is also produced in the deep oceans but often does not escape from surface ocean water before it is oxidized. Over the remote ocean, $H_2S$ mixing ratios range from 5 to 15 pptv. Over shallow coastal waters, where more escapes the surface water, its mixing ratios reach 100–300 pptv. Over land, its mixing ratios are 5–150 pptv. Downwind of industrial sources, they increase to 1–100 ppbv (Berresheim et al. 1995). In the air, hydrogen sulfide reacts with OH to form the **hydrogen sulfide radical** (HS) by

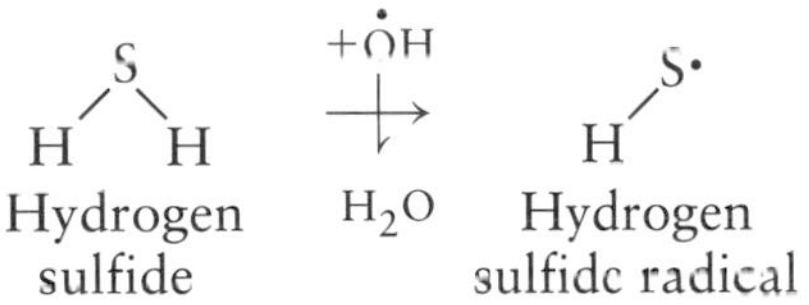

$$\underset{\underset{\text{Hydrogen sulfide}}{}}{\overset{S}{H\diagup\diagdown H}} \quad \overset{+\dot{O}H}{\underset{H_2O}{+\!\!\!\rightarrow}} \quad \underset{\underset{\text{Hydrogen sulfide radical}}{}}{\overset{S\cdot}{H\diagup}} \tag{12.65}$$

HS reacts with $O_3$, $NO_2$, and $O_2$. When $T = 298$ K, $p_d = 1013$ mb, $O_3 = 40$ ppbv, and $NO_2 = 50$ pptv, the *e*-folding lifetimes of HS against losses by $O_3$, $NO_2$, and $O_2$ are 0.27, 14, and 0.49 s, respectively. Thus, the HS–$O_2$ reaction is competitive with the HS–$O_3$ reaction under free-tropospheric conditions. The HS–$O_3$ and HS–$NO_2$ reactions produce HSO. HSO further reacts with $O_3$ and $NO_2$, producing $HSO_2$ in both cases. These reactions are slightly slower than the HS–$O_3$ and HS–$NO_2$ reactions, respectively. $HSO_2$ combines with $O_2$ to form $SO_2$ and $HO_2$. The HS–$O_2$ reaction is

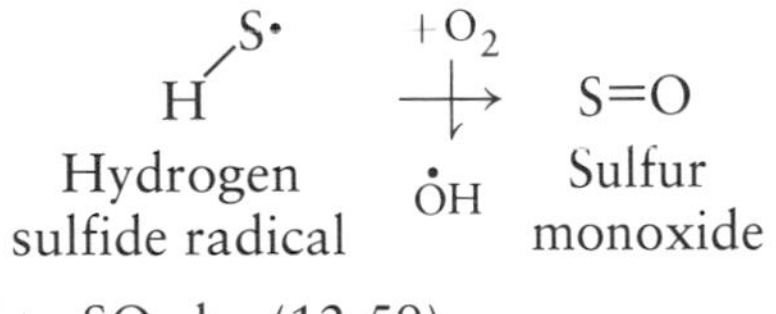

$$\underset{\underset{\text{Hydrogen sulfide radical}}{}}{\overset{S\cdot}{H\diagup}} \quad \overset{+O_2}{\underset{\dot{O}H}{+\!\!\!\rightarrow}} \quad \underset{\underset{\text{Sulfur monoxide}}{}}{S=O} \tag{12.66}$$

SO is rapidly oxidized to $SO_2$ by (12.59).

**Methanethiol** (methyl sulfide, $CH_3SH$) reacts with OH to produce the methanethilate radical by

$$H-\underset{\underset{H}{|}}{\overset{\overset{H}{|}}{C}}-S\overset{H}{\diagup} \quad \overset{+\dot{O}H}{\underset{H_2O}{+\!\!\!\rightarrow}} \quad H-\underset{\underset{H}{|}}{\overset{\overset{H}{|}}{C}}-S\cdot \tag{12.67}$$

Methanethiol                Methanethiolate radical

$CH_3S$ initiates $SO_2$ production in (12.57).

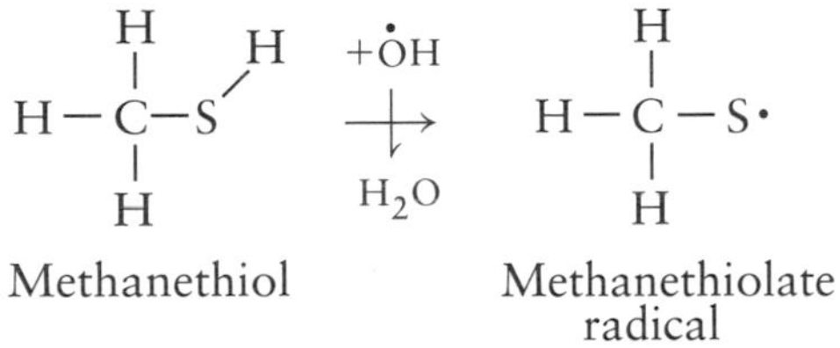

**Carbonyl sulfide** (OCS) and **carbon disulfide** ($CS_2$) are emitted from volcanos. Both react with OH, but the OCS–OH reaction is slow, and photolysis of OCS occurs only in the stratosphere. OCS mixing ratios have built up over time and, at 500 pptv, are the highest among background sulfur-containing compounds. Mixing ratios of OCS are relatively uniform between the surface and 15–20 km, at which point photolysis reduces them. The OH and photolysis reactions of OCS are, respectively,

$$\underset{\text{Carbonyl sulfide}}{O{=}C{=}S} + \dot{O}H \longrightarrow \underset{\text{Hydrogen sulfide radical}}{H{-}S\cdot} + CO_2 \qquad (12.68)$$

$$\underset{\text{Carbonyl sulfide}}{O{=}C{=}S} + h\nu \longrightarrow \underset{\text{Carbon monoxide}}{CO} + \underset{\text{Atomic sulfur}}{\dot{S}\cdot} \qquad \lambda < 260\ \text{nm} \qquad (12.69)$$

Atomic sulfur from (12.69) forms SO by

$$\underset{\text{Atomic sulfur}}{\dot{S}\cdot} \xrightarrow[\;-\dot{O}\cdot\;]{+O_2} \underset{\text{Sulfur monoxide}}{S{=}O} \qquad (12.70)$$

SO is further oxidized to $SO_2$ by (12.59).

Carbon disulfide reacts with OH and photolyzes. Because the $CS_2$–OH and $CS_2$ photolysis reactions are faster than the OCS–OH and OCS photolysis reactions, respectively, $CS_2$ mixing ratios of 2–200 pptv are lower than OCS mixing ratios in the background atmosphere. The $CS_2$–OH and $CS_2$ photolysis reactions are, respectively,

$$\underset{\text{Carbon disulfide}}{S{=}C{=}S} + \dot{O}H \longrightarrow \underset{\text{Hydrogen sulfide radical}}{H{-}S\cdot} + \underset{\text{Carbonyl sulfide}}{O{=}C{=}S} \qquad (12.71)$$

$$\underset{\text{Carbon disulfide}}{S{=}C{=}S} + h\nu \longrightarrow \underset{\text{Carbon monosulfide}}{{}^{-}C{\equiv}S^{+}} + \underset{\text{Atomic sulfur}}{\dot{S}\cdot} \qquad \lambda < 340\ \text{nm} \qquad (12.72)$$

**Carbon monosulfide** (CS) produces OCS by

$$\underset{\text{Carbon monosulfide}}{{}^{-}C{\equiv}S^{+}} + O_2 \longrightarrow \underset{\text{Carbonyl sulfide}}{O{=}C{=}S} + \dot{O}\cdot \qquad (12.73)$$

Once sulfur dioxide has been produced chemically or emitted into the atmosphere, it converts to the particle phase by one of two mechanisms. One is dissolution of $SO_2$ into water-containing particles and aqueous-phase oxidation to sulfate. This mechanism is discussed in Chapter 19. The other is conversion of $SO_2$ to sulfuric acid gas ($H_2SO_4$). Since $H_2SO_4$ has a low saturation vapor pressure, it easily condenses onto particles. The gas-phase conversion of sulfur dioxide to sulfuric acid requires three steps. The first is the conversion of $SO_2$ to **bisulfite** ($HSO_3$). Bisulfite quickly reacts with oxygen to form **sulfur trioxide** ($SO_3$). Sulfur trioxide then reacts with water to form **sulfuric acid.** The reaction sequence is

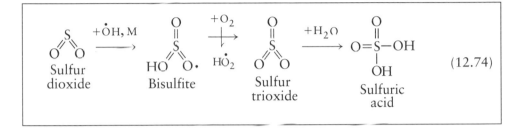

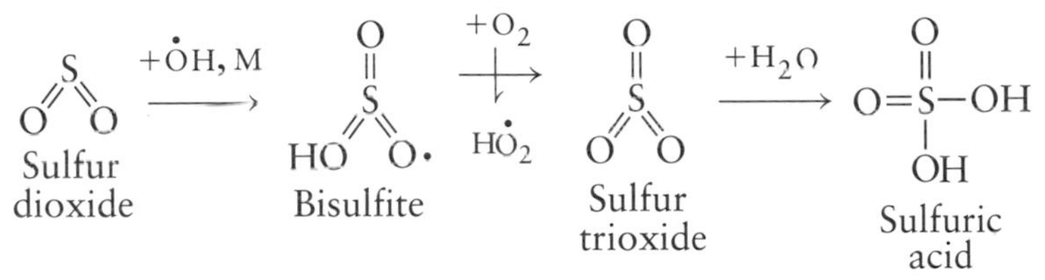

(12.74)

Gas-phase sulfuric acid condenses onto particles, attracting liquid water in the process. Once dissolved in liquid water, sulfuric acid dissociates into ions. Dissolution and dissociation are discussed in Chapter 18.

In sum, free-tropospheric chemistry is governed by inorganic and light organic reactions. Important inorganic constituents in the free troposphere are NO, $NO_2$, $O_3$, OH, $HO_2$, CO, and $SO_2$. Organic species of interest include $CH_4$, $C_2H_6$, $C_3H_8$, $C_2H_4$, $C_3H_6$, HCHO, $CH_3CHO$, and $CH_3COCH_3$. In the free troposphere, the relationship among NO, $NO_2$, and $O_3$ can be reasonably quantified with the photostationary-state relationship. Some gases in the free troposphere originate from natural sources while others originate from combustion and other emission sources in urban and rural regions. Urban photochemistry is discussed next.

## 12.2. URBAN PHOTOCHEMISTRY

Urban air pollution is divided into two major categories. When sunlight intensity and emissions from fossil-fuel combustion sources are high, the resulting air pollution is **photochemical smog.** When sulfur emissions from coal-fired power plants are high, sunlight is less intense, and the relative humidity is high, the air pollution is **London-type smog.** A classical site of photochemical smog is Los Angeles, California. Many cities exhibit photochemical smog, including Mexico City, Tokyo, Johannesburg, and Athens. Many cities in the midwest and eastern United States, southeastern Canada, and northern Europe, and elsewhere experience both London-type and photochemical smog.

Photochemical smog forms from reactions involving nitric oxide, nitrogen dioxide, reactive organic gases (ROGs), and sunlight. On a typical morning, NO and

ROGs are emitted from automobiles and other combustion sources. ROGs are transformed chemically to free radicals, denoted by ROG*. The radicals react with NO to form $NO_2$. Preexisting ozone also converts NO to $NO_2$. Sunlight breaks down nitrogen dioxide to nitric oxide and atomic oxygen. Finally, atomic oxygen reacts with molecular oxygen to form ozone. The basic reaction sequence is thus

$$\dot{N}O + ROG^* \longrightarrow \dot{N}O_2 + ROG^{**} \qquad (12.75)$$
$$\dot{N}O + O_3 \longrightarrow \dot{N}O_2 + O_2 \qquad (12.76)$$
$$\dot{N}O_2 + h\nu \longrightarrow \dot{N}O + \dot{O}\cdot \qquad \lambda < 420 \text{ nm} \qquad (12.77)$$
$$\dot{O}\cdot + O_2 + M \longrightarrow O_3 + M \qquad (12.78)$$

where ROG** is another radical formed from ROG*. The difference between ozone production in the free troposphere and in urban air is the contribution of ROG radicals that convert NO to $NO_2$ in urban air. Because ROG radicals compete with $O_3$ for NO in urban regions, the **photostationary-state approximation is usually not valid in urban air.** In the afternoons in urban air, the photostationary-state approximation is better than in the mornings, since ROG mixing ratios are lower in the afternoons.

Meteorological factors also affect the development of air pollution; Los Angeles is a textbook example. The Los Angeles basin is bordered on its southwestern side by the Pacific Ocean and on all other sides by mountain ranges. During the day, a **sea breeze** blows inland. The sea breeze is strongest in the afternoon, when the temperature difference between land and ocean is greatest. At night, a reverse **land breeze** (from land to sea) occurs, but it is often weak. Figure 12.1(a) shows the variation of sea- and land-breeze wind speeds at Hawthorne, near the coast in the Los Angeles basin, during a three-day period in 1987.

The sea breeze is instrumental in blowing primary pollutants, emitted mainly on the west side of the Los Angeles basin, towards the east side, where they arrive as secondary pollutants. A **primary pollutant** is a gas or particle emitted directly from a source, such as an automobile tailpipe. A **secondary pollutant** is a gas or particle that forms chemically or physically in the atmosphere. Along their trajectories through the basin, primary pollutants, such as NO, are converted to secondary pollutants, such as $O_3$. While NO mixing ratios peak on the west side of Los Angeles, as shown in Fig. 12.1(b), $O_3$ mixing ratios peak on the east side, as shown in Fig. 12.1(c). The west side of the basin is a **source region** and the east side is a **receptor region** of air pollution.

Other factors that exacerbate air pollution in the Los Angeles basin are its location relative to the Pacific high-pressure system and its exposure to sunlight. The Pacific high suppresses vertical motions of air, inhibiting clouds other than stratus from forming. The subsidence inversion caused by the high prevents pollution from rising easily over the mountains surrounding the basin. Because the basin is further south (about 34°N latitude) than most United States cities, Los Angeles receives

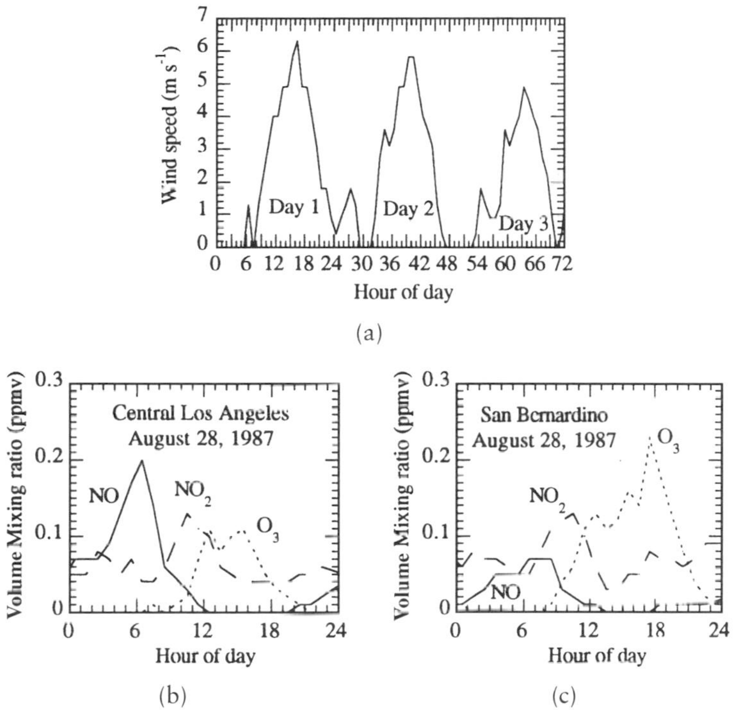


**Figure 12.1.** (a) Wind speeds at Hawthorne from August 26 to 28, 1987. The other panels show the evolution of the NO, $NO_2$, and $O_3$ mixing ratios at (b) central Los Angeles and (c) San Bernardino, on August 28. Central Los Angeles is closer to the coast than is San Bernardino. The sea breeze picks up during the day, sending primary pollutants, such as NO, from the western side of the Los Angeles basin (e.g., central Los Angeles) towards the eastern side (e.g., San Bernardino). As the pollution travels, ROG radicals convert NO to $NO_2$. Photolysis of $NO_2$ produces atomic oxygen, which forms ozone, a secondary pollutant.

more daily radiation than most cities, enhancing its rate of photochemical smog formation.

Important gases emitted in urban air include nitrogen oxides ($NO_x = NO + NO_2$), reactive organic gases, carbon monoxide, and sulfur oxides ($SO_x = SO_2 + SO_3$). Table 12.2 shows the quantity of emissions in the Los Angeles basin from stack and nonstack sources for several species and groups of species on a day in 1987. Table 12.3 shows the percentages of emission from each of several source categories.

The tables show that carbon monoxide, the most abundantly emitted gas in the basin, originated almost entirely (98 percent) from mobile sources, including cars, trucks, buses, off-road vehicles, trains, ships, and aircraft. Oxides of nitrogen were emitted 76 percent by mobile sources and 24 percent by stationary sources. Of all $NO_x$ emitted, about 95 percent was emitted as NO, and about 5 percent was

**Table 12.2.** Emissions for August 27, 1987, in a 400 × 150-km Region of the Los Angeles Basin

| | Emissions (tons $day^{-1}$) | | |
|---|---|---|---|
| **Substance** | **Nonstack** | **Stack** | **Total** |
| **Inorganics** | | | |
| Carbon monoxide (CO) | 9773 | 23 | 9796 |
| Nitric oxide (NO) | 700 | 54 | 754 |
| Nitrogen dioxide ($NO_2$) | 120 | 9 | 129 |
| Nitrous acid (HONO) | 6 | 0.5 | 6.5 |
| Sulfur dioxide ($SO_2$) | 96 | 13 | 109 |
| Sulfur trioxide ($SO_3$) | 4 | 0.5 | 4.5 |
| **Organics** | | | |
| Methane ($CH_4$) | 899 | 5 | 904 |
| Ethane ($C_2H_6$) | 72 | $\ll 1$ | 72 |
| Propane ($C_3H_8$) | 55 | $\ll 1$ | 55 |
| Paraffins (PAR) | 1268 | 4 | 1272 |
| Ethene ($C_2H_4$) | 171 | $\ll 1$ | 171 |
| Propene ($C_3H_6$) | 84 | $\ll 1$ | 84 |
| Olefins (OLE) | 58 | $\ll 1$ | 58 |
| Formaldehyde (HCHO) | 42 | $\ll 1$ | 42 |
| Other aldehydes (ALD2) | 66 | $\ll 1$ | 66 |
| Methanol ($CH_3OH$) | 5 | $\ll 1$ | 5 |
| Ethanol ($CH_3CH_2OH$) | 28 | $\ll 1$ | 28 |
| Acetone ($CH_3COCH_3$) | 22 | $\ll 1$ | 22 |
| Other ketones (KET) | 7 | $\ll 1$ | 7 |
| Toluene ($C_6H_5CH_3$) | 266 | $\ll 1$ | 266 |
| Xylene [$(CH_3)_2C_6H_4$] | 234 | $\ll 1$ | 234 |
| Isoprene ($C_5H_8$) | 47 | 0 | 47.2 |
| Total Organics | 3324 | 9 | 3333 |

Data from Allen and Wagner (1992). Mobile hot exhaust emissions of CO and organics were multiplied by 2.7 and 3.0, respectively, to account for underestimations in the inventory (Ingalls et al. 1989). Bond groups PAR, OLE, ALD2, and KET are described in Section 12.2.6.

emitted as $NO_2$. A small amount was probably emitted in the form of HONO. The thermal combustion reaction in automobiles that produces nitric oxide at high temperatures is

$$N{\equiv}N + O{=}O + \text{heat} \longrightarrow 2\,\dot{N}{=}O \qquad (12.79)$$

Table 12.3 shows that 38 percent of oxides of sulfur was emitted by stationary sources and 62 percent was emitted by mobile sources. Table 12.2 indicates that the mass of $SO_x$ emissions was one-eighth that of $NO_x$. Almost all $SO_x$ emissions were in the form of $SO_2$. Sulfur emissions are more important in the midwest and eastern United States than in Los Angeles, because many cities in the former regions have more coal-fired power plants.

Table 12.3 shows that mobile and stationary sources each accounted for 50 percent of ROGs emitted in the basin. Table 12.4 shows the most abundantly

Table 12.3. Percentages of Emission of Different Gases by Source Category

| Source | CO (%) | $NO_x$ (%) | $SO_x$ (%) | ROG (%) |
|---|---|---|---|---|
| **Stationary Sources** | | | | |
| Fuel combustion[a] | 1.56 | 22.1 | 17.28 | 1.22 |
| Waste burning[b] | 0.06 | 0.14 | 0.34 | 0.08 |
| Solvent use[c] | 0.00 | 0.02 | 0.00 | 33.72 |
| Petroleum processing, storage, transfer[d] | 0.13 | 0.75 | 14.40 | 7.75 |
| Industrial processes[e] | 0.14 | 0.96 | 5.66 | 2.99 |
| Misc. processes[f] | 0.11 | 0.06 | 0.06 | 4.19 |
| Total stationary sources | 2.00 | 24.03 | 37.74 | 49.95 |
| **Mobile Sources** | | | | |
| On-road vehicles[g] | 87.47 | 54.97 | 23.53 | 43.96 |
| Other mobile[h] | 10.53 | 21.00 | 38.73 | 6.09 |
| Total mobile sources | 98.00 | 75.97 | 67.26 | 50.05 |
| Total stationary and mobile | 100 | 100 | 100 | 100 |

Data from Chang et al. (1991).
[a]Agricultural, oil and gas production, petroleum refining, other manufacturing/industrial, electric utilities, other services and commerce, residential, other.
[b]Agricultural, range management, incineration, other.
[c]Dry cleaning, degreasing, architectural coating, other surface coating, asphalt paving, printing, consumer products, industrial solvent use, other.
[d]Oil and gas extraction, petroleum refining, petroleum marketing, other.
[e]Chemical, food, agricultural, mineral, metal, wood and paper processes.
[f]Pesticide application, farming operations, construction and demolition, entrained road dust – paved, entrained road dust – unpaved, unplanned fires, waste disposal, other.
[g]Light duty passenger, light- and medium-duty trucks, heavy-duty gas trucks, heavy-duty diesel trucks, motorcycles, heavy-duty diesel urban buses.
[h]Off-road vehicles, trains, ships, aircraft, mobile equipment, utility equipment.

emitted ROGs in Los Angeles. Once organics are emitted, they are broken down chemically into free radicals. **Six major processes break down hydrocarbons and other ROGs** – photolysis and reaction with OH, $HO_2$, O, $NO_3$, and $O_3$ (Finlayson-Pitts and Pitts 1986). OH and O are present during the daytime only, because they are short-lived and require photolysis for their production. $NO_3$ breaks down quickly during the day by photolysis; thus, it is important mostly at night. $O_3$ and $HO_2$ may be present during day and night.

OH is produced in urban air by some of the same reactions that produce it in the free troposphere. An early-morning source of OH in urban air is photolysis of HONO. HONO may be emitted from automobiles; thus, it is more abundant in urban air than in the free troposphere. Midmorning sources of OH in urban air are aldehyde (e.g., formaldehyde and acetaldehyde) photolysis and oxidation. Aldehydes are more abundant in urban than in free-tropospheric air. The major afternoon source of OH in urban air is reaction of $O(^1D)$ with $H_2O$. $O(^1D)$ is produced from $O_3$ photolysis. $O_3$ mixing ratios in urban air peak during the afternoon, and photolysis coefficients peak near noon. In sum, the three major reaction mechanisms that produce the hydroxyl radical in urban air are

**Table 12.4.** Organic Gases Emitted in the Greatest Quantity in the Los Angeles Basin

| | | | |
|---|---|---|---|
| 1. Methane | 10. Propylene | 19. Acetone | 28. Methylcyclohexane |
| 2. Toluene | 11. Chloroethylene | 20. *n*-Pentadecane | 29. Nonane |
| 3. Pentane | 12. Acetylene | 21. Cyclohexane | 30. Methyl alcohol |
| 4. Butane | 13. Hexane | 22. Methylethylketone | 31. 1-Hexane |
| 5. Ethane | 14. Propane | 23. Acetaldehyde | 32. Methylcyclopentane |
| 6. Ethylene | 15. Benzene | 24. Trimethylbenzene | 33. Methylpentane |
| 7. Octane | 16. Methyl chloroform | 25. Ethylbenzene | 34. Dimethylhexane |
| 8. Xylene | 17. Pentene | 26. Methylvinylketone | 35. Cyclopentene |
| 9. Heptane | 18. *n*-Butyl acetate | 27. Naphtha | |

Data from the south coast air quality management district (private communication) and from Pilinis and Seinfeld (1988).

*Early morning source*

$$\mathrm{HONO} + h\nu \longrightarrow \dot{\mathrm{O}}\mathrm{H} + \dot{\mathrm{N}}\mathrm{O} \qquad \lambda < 400\ \mathrm{nm} \tag{12.80}$$

*Midmorning source*

$$\mathrm{HCHO} + h\nu \longrightarrow \dot{\mathrm{H}} + \mathrm{H}\dot{\mathrm{C}}\mathrm{O} \qquad \lambda < 334\ \mathrm{nm} \tag{12.81}$$

$$\dot{\mathrm{H}} + \mathrm{O}_2 \xrightarrow{M} \mathrm{H}\dot{\mathrm{O}}_2 \tag{12.82}$$

$$\mathrm{H}\dot{\mathrm{C}}\mathrm{O} + \mathrm{O}_2 \longrightarrow \mathrm{H}\dot{\mathrm{O}}_2 + \mathrm{CO} \tag{12.83}$$

$$\dot{\mathrm{N}}\mathrm{O} + \mathrm{H}\dot{\mathrm{O}}_2 \longrightarrow \dot{\mathrm{N}}\mathrm{O}_2 + \dot{\mathrm{O}}\mathrm{H} \tag{12.84}$$

*Afternoon source*

$$\mathrm{O}_3 + h\nu \longrightarrow \mathrm{O}_2 + \cdot\dot{\mathrm{O}}(^1D) \qquad \lambda < 310\ \mathrm{nm} \tag{12.85}$$

$$\cdot\dot{\mathrm{O}}(^1D) + \mathrm{H}_2\mathrm{O} \longrightarrow 2\dot{\mathrm{O}}\mathrm{H} \tag{12.86}$$

ROGs emitted into urban air include alkanes, alkenes, alkynes, aldehydes, ketones, aromatics, and terpenes. Of these, photolysis breaks down aldehydes and ketones, OH reacts with all seven groups during the day, $HO_2$ reacts with aldehydes during the day and night, O reacts with alkenes and terpenes during the day, $NO_3$ reacts with alkanes, alkenes, aldehydes, aromatics, and terpenes during the night, and $O_3$ breaks down alkenes and terpenes during the day and night. Table 12.5 shows lifetimes of ROGs against breakdown by these six processes.

Each organic class listed above produces radicals that lead to ozone formation. Table 12.6 shows the most important ROGs in Los Angeles during the summer of 1987 in terms of a combination of concentration and ability to form ozone. The table shows that *m*- and *p*-xylene, both aromatic hydrocarbons, were most important in generating ozone. Other important ROGs included ethene, acetaldehyde, toluene, and formaldehyde. Aromatics, alkenes, and aldehydes are the most significant smog generators in urban air. Although alkanes are emitted in greater abundance than are other organics, alkanes are less reactive in producing ozone than are other organics.

**Table 12.5.** Estimated Lifetimes of Reactive Organic Gases Representing Alkanes, Alkenes, Alkynes, Aldehydes, Ketones, Aromatics, and Terpenes against Photolysis and Oxidation in Urban and Free-tropospheric Air

| | | Lifetime in Polluted Urban Air at Sea Level | | | | |
|---|---|---|---|---|---|---|
| ROG Species | Photolysis | [OH] $5 \times 10^6$ molec. $cm^{-3}$ | [$HO_2$] $2 \times 10^9$ molec. $cm^{-3}$ | [O] $8 \times 10^4$ molec. $cm^{-3}$ | [$NO_3$] $1 \times 10^{10}$ molec. $cm^{-3}$ | [$O_3$] $5 \times 10^{12}$ molec. $cm^{-3}$ |
| *n*-Butane | — | 22 h | 1000 y | 18 y | 29 d | 650 y |
| *trans*-2-Butene | — | 52 m | 4 y | 6.3 d | 4 m | 17 m |
| Acetylene | — | 3.0 d | — | 2.5 y | — | 200 d |
| Toluene | — | 9.0 h | — | 6 y | 33 d | 200 d |
| Isoprene | — | 34 m | — | 4 d | 5 m | 4.6 h |
| Formaldehyde | 7 h | 6.0 h | 1.8 h | 2.5 y | 2.0 d | 3200 y |
| Acetone | 23 d | 9.6 | — | — | — | — |
| | | **Lifetime in Free Tropospheric Air at Sea Level** | | | | |
| **ROG Species** | **Photolysis** | [OH] $5 \times 10^5$ molec. $cm^{-3}$ | [$HO_2$] $3 \times 10^8$ molec. $cm^{-3}$ | [O] $3 \times 10^3$ molec. $cm^{-3}$ | [$NO_3$] $5 \times 10^8$ molec. $cm^{-3}$ | [$O_3$] $1 \times 10^{12}$ molec. $cm^{-3}$ |
| *n*-Butane | — | 9.2 d | 6700 y | 480 y | 1.6 y | 3250 y |
| *trans*-2-Butene | — | 8.7 h | 27 y | 168 d | 1.3 h | 1.4 h |
| Acetylene | — | 30 d | — | 67 y | — | 2.7 y |
| Toluene | — | 3.8 d | — | 160 y | 1.8 y | 2.7 y |
| Isoprene | — | 5.7 h | — | 106 d | 1.7 h | 23 d |
| Formaldehyde | 7 h | 2.5 d | 11.7 h | 67 y | 40 d | 16,000 y |
| Acetone | 23 d | 96 d | — | — | — | — |

Estimated lifetimes for some species in urban air were recalculated from Finlayson-Pitts and Pitts (1986). Lifetimes of other species were obtained from rate-coefficient data. Photolysis rates were obtained from Figs. 11.3(a) and (b). Gas concentrations are typical, but not necessarily average values for each region. Units: m = minutes; h = hours, d = days, y = years, (—) = no data or insignificant loss.

**Table 12.6.** Ranking of the Most Abundant Species in Terms of Reactivity during the Summer Southern California Air Quality Study in 1987

| | | | |
|---|---|---|---|
| 1. *m*- and *p*-Xylene | 8. *o*-Xylene | 15. *m*-Ethyltoluene | 22. *p*-Ethyltoluene |
| 2. Ethene | 9. Butane | 16. Pentanal | 23. $C_4$ Olefin |
| 3. Acetaldehyde | 10. Methylcyclopentane | 17. Propane | 24. 3-Methylpentane |
| 4. Toluene | 11. 2-Methylpentane | 18. Propanal | 25. *o*-Ethyltoluene |
| 5. Formaldehyde | 12. Pentane | 19. *i*-Butane | |
| 6. *i*-Pentane | 13. 1,2,4-Trimethylbenzene | 20. $C_6$ Carbonyl | |
| 7. Propene | 14. Benzene | 21. Ethylbenzene | |

*Source:* (Lurmann et al. 1992). The ranking was determined by multiplying the weight fraction of each organic present in the atmosphere by a species-specific reactivity scaling factor developed by Carter (1991).

### 12.2.1. Alkanes

Table 12.6 shows that *i*-pentane and butane are the most effective alkanes with respect to the combination of concentration and reactivity, in producing ozone in Los Angeles air. As in the free troposphere, the main pathway of alkane decomposition in urban air is OH attack. Photolysis, reaction with $O_3$, and reaction with $HO_2$ do not affect alkane concentrations significantly, and reactions of alkanes with $NO_3$ are slow. Of all alkanes, methane is the least reactive and the least important with respect to urban air pollution. Methane is more important with respect to free-tropospheric and stratospheric chemistry. The oxidation pathways of methane was given in Section 12.1.7, and those of ethane and propane were given in Section 12.1.8.

### 12.2.2. Alkenes

Table 12.6 shows that alkenes, such as ethene and propene, are important ozone precursors in photochemical smog. Table 12.5 indicates that alkenes react most rapidly with OH, $O_3$, and $NO_3$. In the following subsections, these reaction pathways are discussed.

#### 12.2.2.1. *Alkene Reaction with the Hydroxyl Radical*

When ethene reacts with the hydroxyl radical, the radical substitutes into the double bond to produce an **ethanyl radical** in an OH addition process. The ethanyl radical further reacts to produce $NO_2$. The sequence is

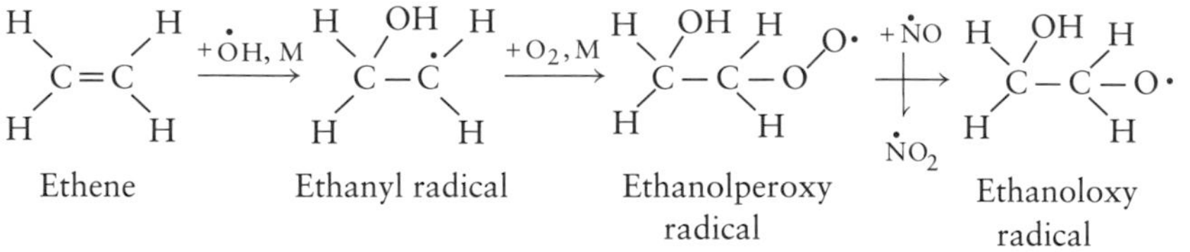


(12.87)

$NO_2$ produces ozone by (12.2) and (12.3). The **ethanoloxy radical** ($HOCH_2CH_2O$), a byproduct of ethene oxidation, produces formaldehyde and **glycol aldehyde** ($HOCH_2CHO$) by

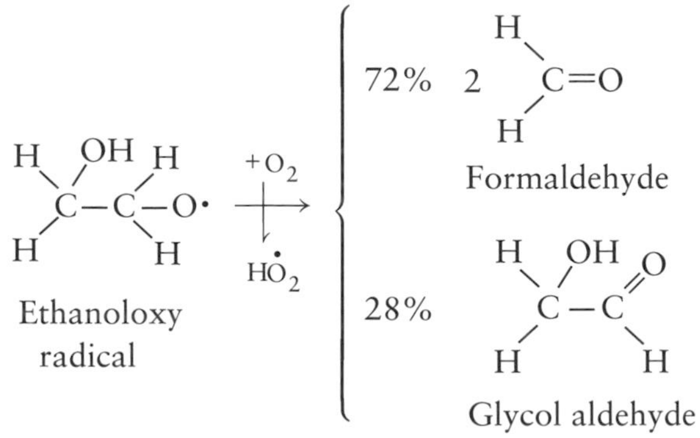


(12.88)

Formaldehyde decomposition produces ozone, as discussed in Section 12.1.9.1. Like other aldehydes, glycol aldehyde is decomposed by photolysis and reaction with OH.

#### 12.2.2.2. *Alkene Reaction with Ozone*

When ethene or propene reacts with ozone, the ozone substitutes into the double bond to form an unstable **ethene** or **propene molozonide**. The molozonide quickly decomposes to products that are also unstable. The reaction of ethene with ozone is

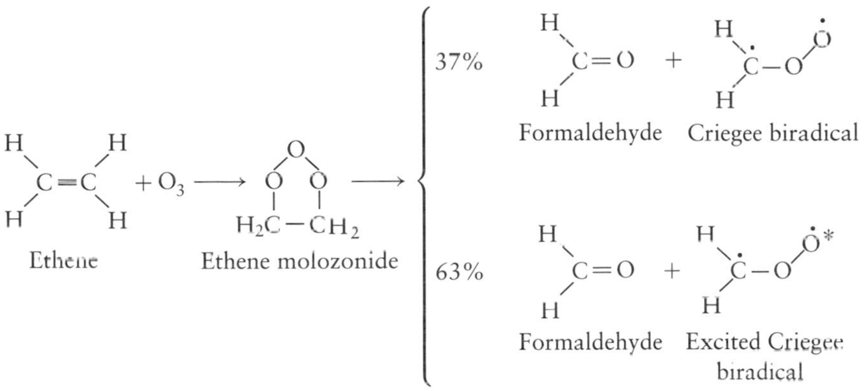


(12.89)

The **criegee biradical** ($H_2COO$) forms $NO_2$ by

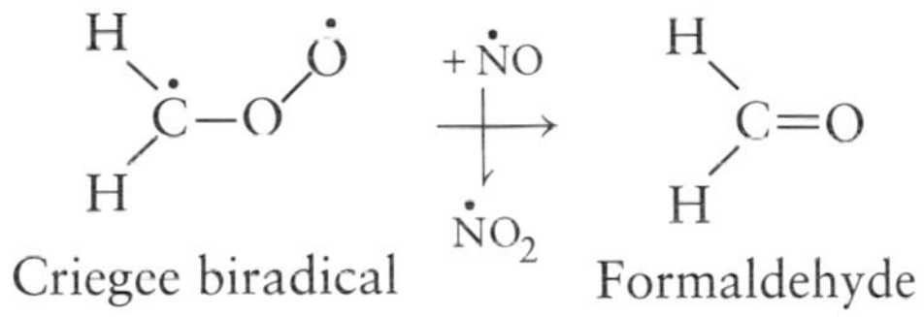


(12.90)

The **excited criegee biradical** ($H_2COO^*$) thermally decomposes by

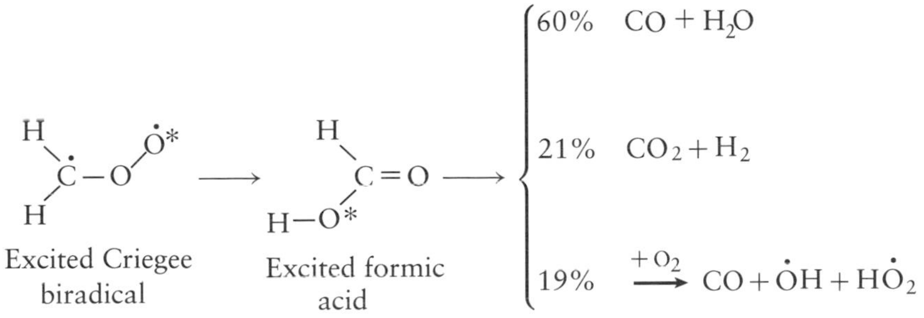


(12.91)

where the fractions, which vary with pressure and temperature, are valid at room temperature and surface pressure (Atkinson et al. 1997). In sum, ozone attack on ethene produces HCHO, $HO_2$, CO, and $NO_2$.

Ozone oxidizes **propene** by the sequence

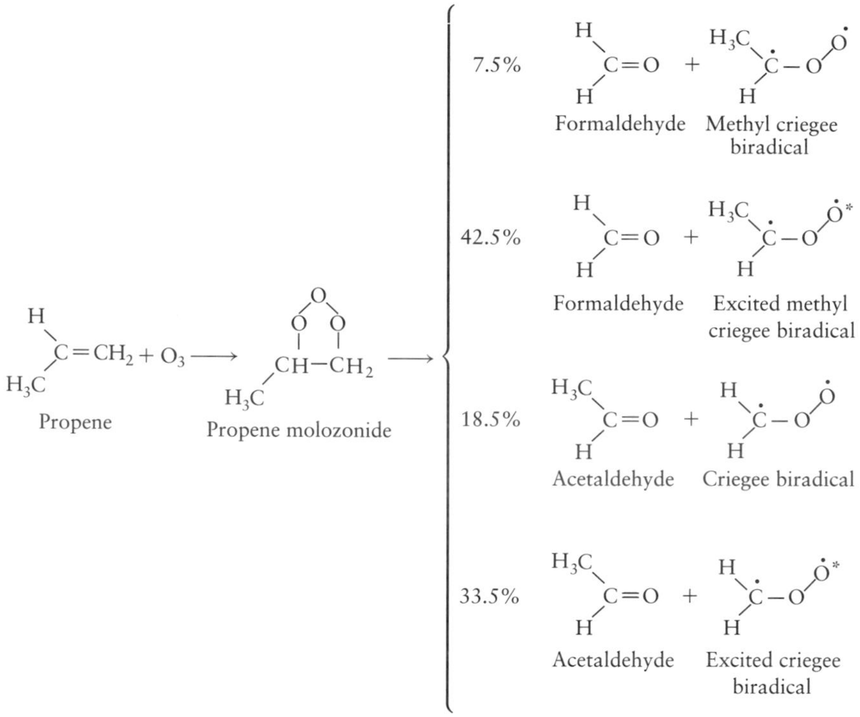


(12.92)

The **methyl criegee biradical** ($CH_3HCOO$) reacts with NO by

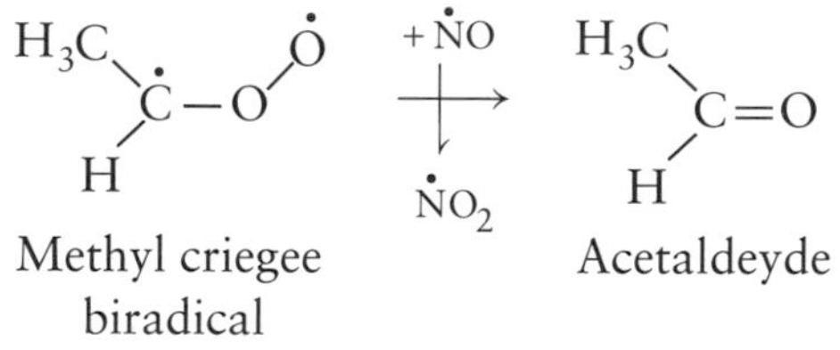


(12.93)

The **excited methyl criegee biradical** ($CH_3HCOO^*$) thermally decomposes via

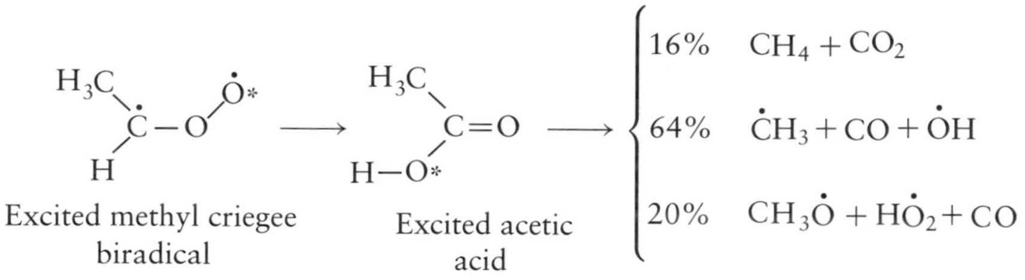


(12.94)

where the fractions are valid at room temperature and surface pressure (Atkinson et al. 1997). Thus, propene oxidation by ozone produces OH, $HO_2$ $NO_2$, HCHO, and $CH_3CHO$, which all react further to reform ozone.

### 12.2.2.3. *Alkene Reaction with Nitrate*

Table 12.5 shows that the reaction of $NO_3$ with an alkene is rapid. $NO_3$ is present primarily at night. Nighttime breakdown of alkenes leads to a morning buildup of organic peroxy radicals, which convert NO to $NO_2$. $NO_3$ oxidizes ethene and propene by

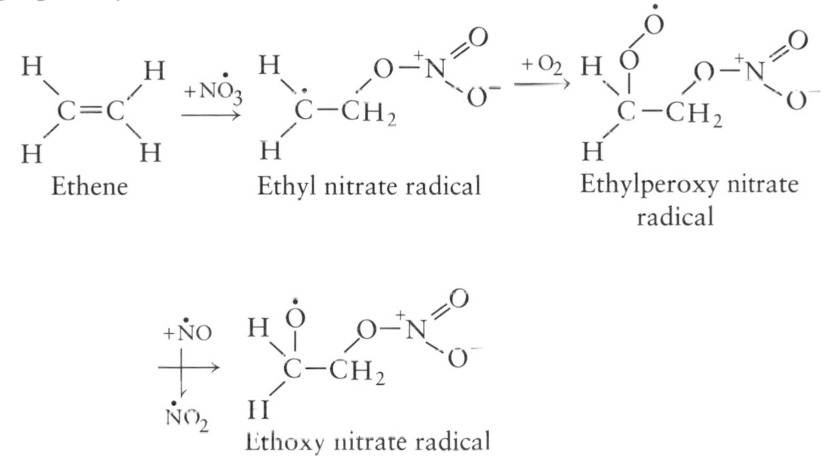


(12.95)

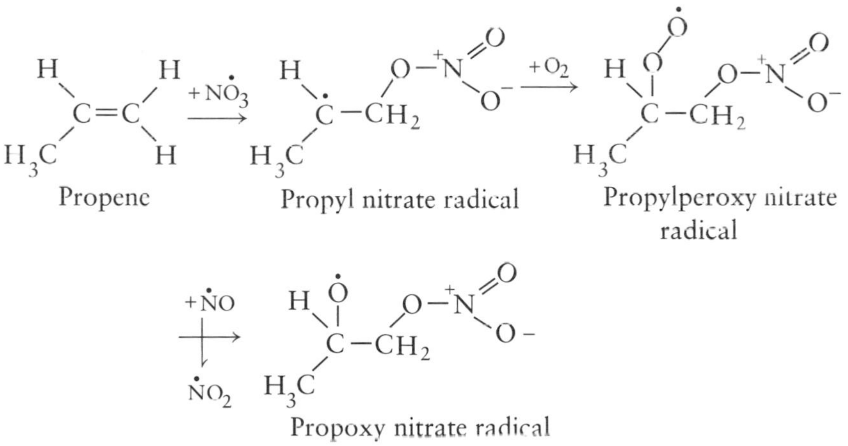


(12.96)

respectively. $NO_2$ produced from these reactions during the night leads to ozone formation the next morning.

### 12.2.3. Aromatics

Table 12.5 shows that **toluene** ($C_6H_5CH_3$), emitted during fuel combustion, is decomposed almost exclusively by OH abstraction or addition (e.g., Gery et al. 1989, 1985). The pathways are

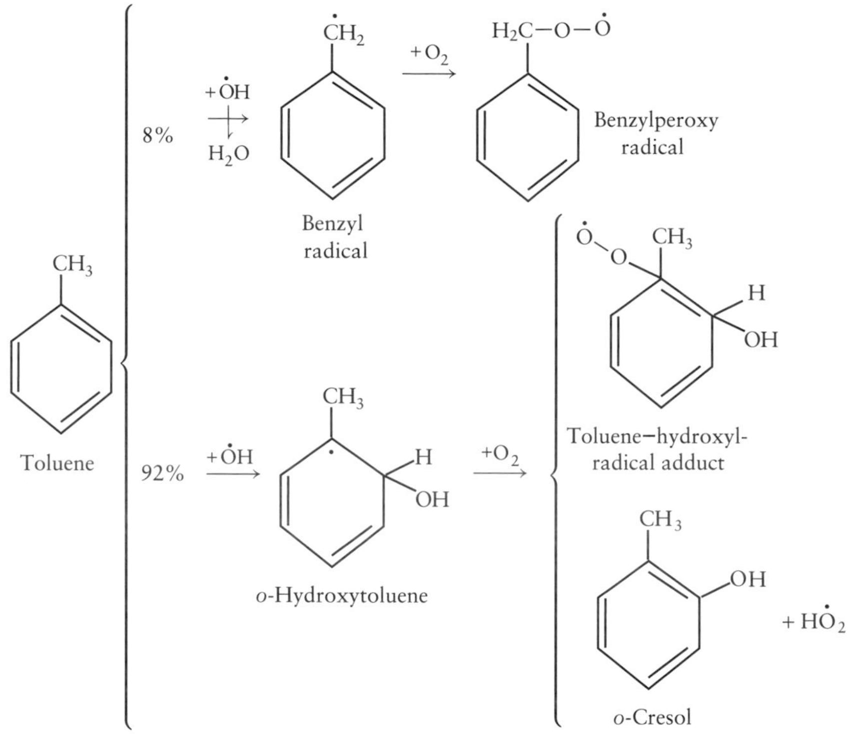


(12.97)

The benzylperoxy radical from the abstraction pathway reacts with NO by

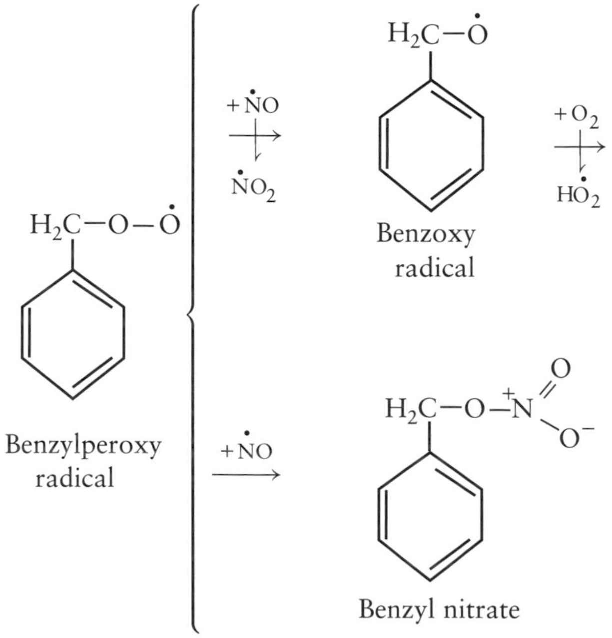


(12.98)

The toluene–hydroxyl-radical adduct from the toluene addition pathway forms $NO_2$ by

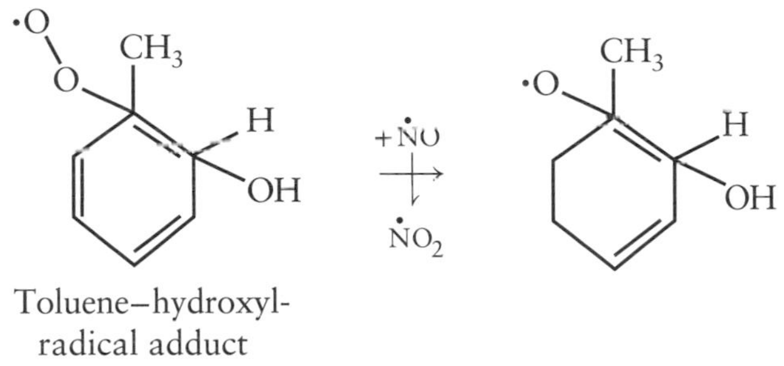


(12.99)

Cresol, from toluene addition, reacts with OH to form the **methylphenylperoxy radical** and **nitrocresol** by

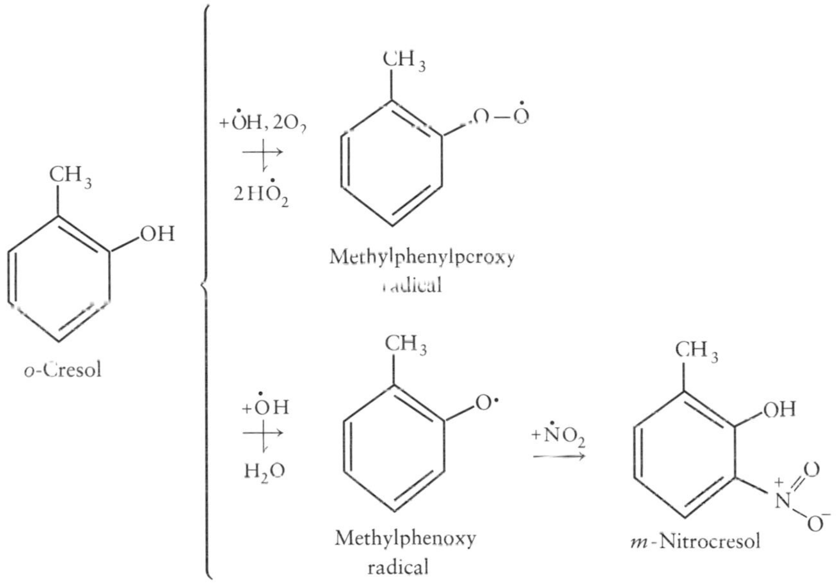


(12.100)

respectively. The methylphenylperoxy radical converts NO to $NO_2$. Nitrocresol readily condenses to the particulate phase, acting as a sink for $NO_x$.

### 12.2.4. Terpenes

The free troposphere and urban areas are affected by biogenic emissions of isoprene and other terpenes. **Biogenic emissions** are emissions produced from biological sources, such as plants, trees, algae, bacteria, and animals. **Isoprene** ($C_5H_8$) and other terpenes are emitted by plants and trees. Table 12.5 shows that OH, $O_3$, and $NO_3$ decompose isoprene. Although isoprene does not have a long chemical lifetime, some of its by-products do.

### 12.2.4.1. *Terpene Reaction with the Hydroxyl Radical*

The reaction pathways of isoprene with OH produce at least six peroxy radicals. The pathways are

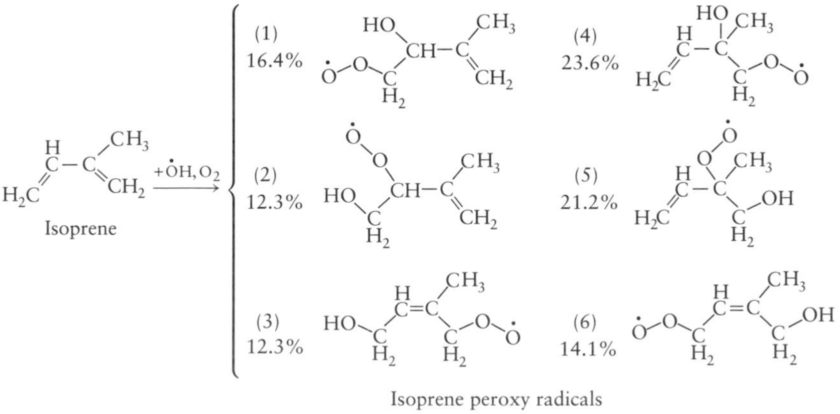


(12.101)

(Paulson and Seinfeld 1992). All six radicals convert NO to $NO_2$. The second and fifth radicals also create **methacrolein** and **methylvinylketone** by

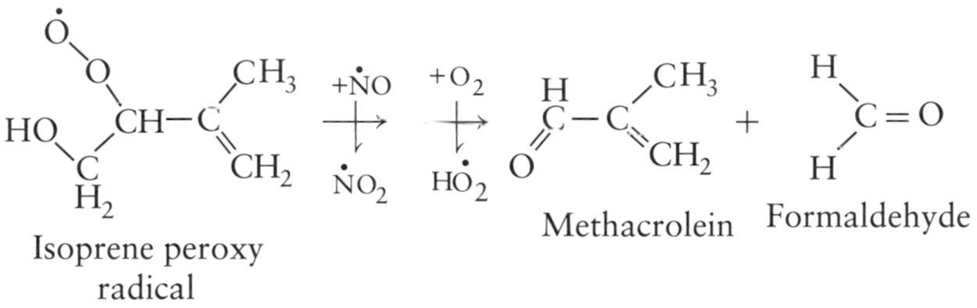


(12.102)

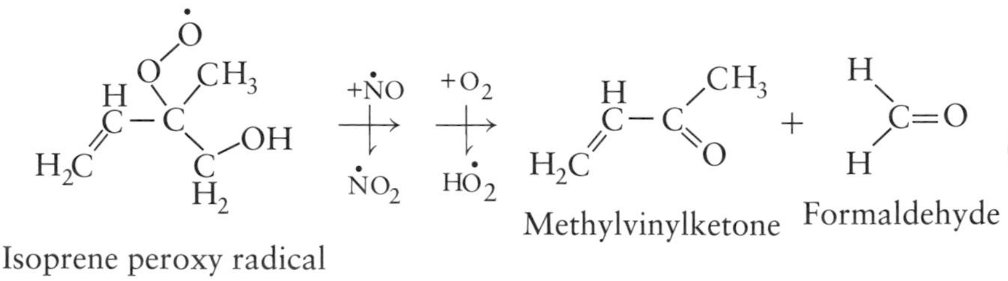


(12.103)

respectively. Methacrolein and methylvinylketone react with OH and $O_3$ to form additional products that convert NO to $NO_2$, as shown in Table B.4 of Appendix B.

### 12.2.4.2. *Terpene Reaction with Ozone*

The isoprene-$O_3$ reaction proceeds along one of several pathways, including

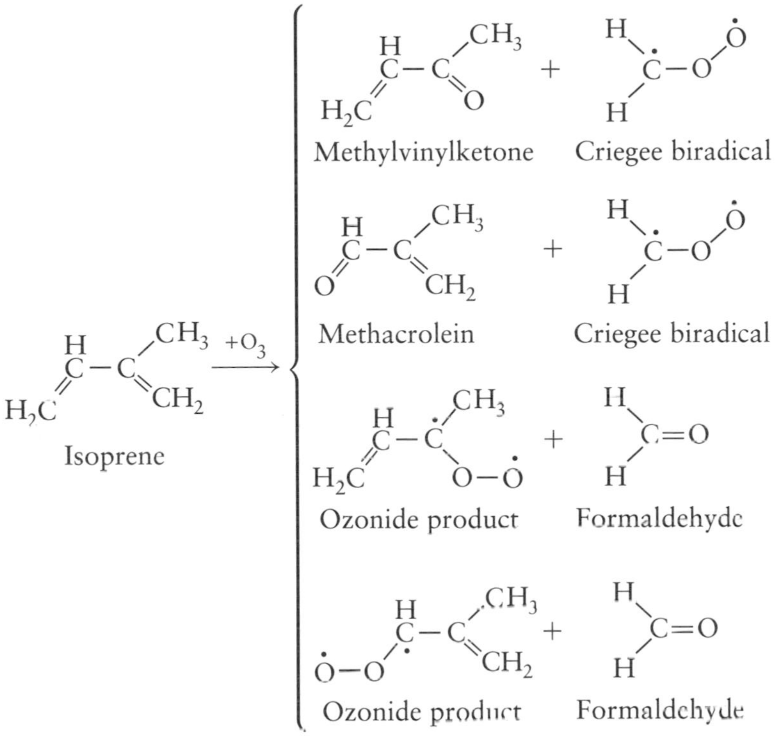


(12.104)

A criegee biradical reaction was given in (12.90). Formaldehyde reactions were shown in (12.48) and (12.51). Both reactions produce by-products that react further to form ozone.

### 12.2.5. Alcohols

Two alternative motor-vehicle fuels are methanol and ethanol. Methanol oxidation produces formaldehyde and ozone, and ethanol oxidation produces acetaldehyde, a precursor to PAN. The reaction of **methanol** with OH is

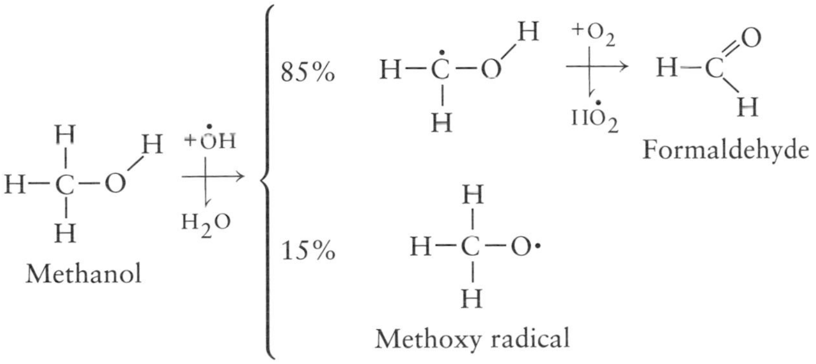


(12.105)

The faster of these reactions has an *e*-folding lifetime of 36 h when $[OH] = 1.0 \times 10^7$ molecules $cm^{-3}$; thus, the reaction is not rapid. The organic product of the first reaction is formaldehyde, and that of the second reaction is the methoxy radical, which quickly produces formaldehyde via (12.43). HCHO buildup from (12.105) is slow.

**Ethanol** oxidation by OH produces the branched reactions

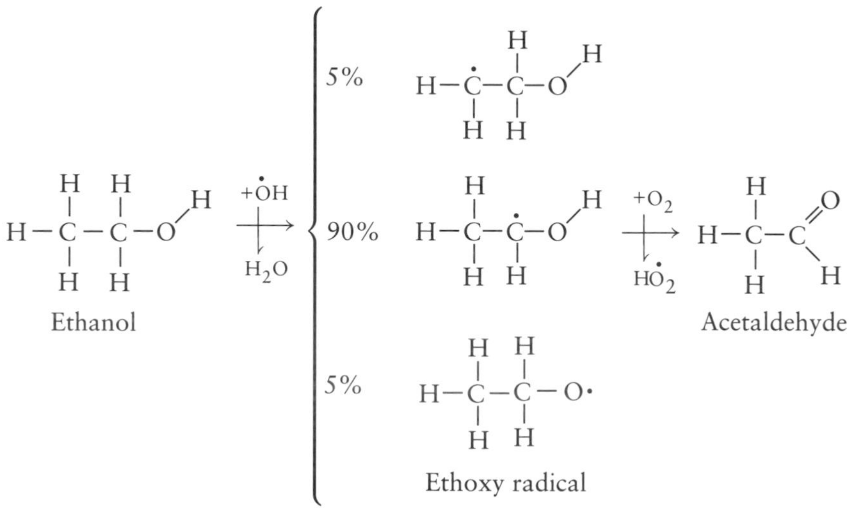


(12.106)

The fastest (middle) reaction has an *e*-folding lifetime of about 10 h when $[OH] = 1.0 \times 10^7$ molecules $cm^{-3}$. The organic oxidation product of the middle reaction is acetaldehyde, which produces PAN and ozone.

### 12.2.6. Condensed Mechanisms for Organic Chemistry

The number of chemical reactions involving organic gases in urban air is large and needs to be reduced for practical model simulations. Fewer reactions involving high-molecular-weight organics occur in the free troposphere, since such organics are chemically destroyed before reaching the free troposphere. High-molecular-weight free-tropospheric organics of importance include biogenically emitted isoprene and other terpenes. In the stratosphere, high-molecular-weight organics are nearly absent. Thus, organic chemical mechanisms need to be condensed, primarily for urban modeling studies. Three methods of reducing the number of organic reactions in a model are the carbon-bond lumping method (Whitten et al. 1980; Gery et al. 1989), the surrogate-species method (Atkinson et al. 1982, Lurmann et al. 1987), and the lumped-species method (Stockwell 1986).

With the **carbon-bond lumping method**, individual organic gases are segregated into one or more bond groups that have similar chemical reactivity. A butane molecule, which has four carbons connected by single bonds, is divided into four single carbon atoms, each represented by the **paraffin** (PAR) bond group. Cyclopentane, which has five single-bonded carbons, is broken into five PAR bond groups. All PAR bonds are assumed to have the same chemical reactivity, regardless of whether they originated from butane or cyclopentane. A terminal carbon atom pair with a double bond between the two atoms is represented by an **olefin** (OLE).

Nonterminal carbon-atom pairs with a double bond attached to one of the carbons and terminal two-carbon carbonyl groups [C–C(=O)H] are lumped as ALD2. Single-carbon ketone groups (C=O) are lumped as KET, 7-carbon aromatics are simulated as **toluene** (TOL), 8-carbon aromatics are simulated as ***m*-xylene** (XYL), and terpenes are grouped as **isoprene** (ISOP). **Methane** (CH4), **ethene** (ETH), **formaldehyde** (FORM), **methanol** (MEOH), **ethanol** (ETOH), **acetone** (AONE), and several other species are not lumped. Nonreactive carbon atoms in certain organic gases are labeled **unreactive** (UNR). In some cases, carbon atoms with a double or triple bond that have similar reactivity to carbon atoms with a single bond are labeled as PAR. Table 12.7 gives the carbon-bond representation of a few organic gases. Several of the reactions in Table B.4, Appendix B, include reactions among carbon-bond species.

With the **surrogate-species method**, all species of similar reactivity are grouped together. Propane and pentane are assumed to have the same reactivity as *n*-butane, and all three species are grouped as one surrogate species. With the **lumped-species method**, species of similar reactivity are lumped together, just as with the surrogate-species method. The difference is that with the surrogate-species method the reaction rate coefficient for each surrogate species is set equal to that of a particular gas. The reaction rate coefficient of a lumped species is determined before a model simulation by taking a mole-fraction-weighted average of the reaction rates of each species in the lumped group.

### 12.2.7. Summary of Urban Photochemistry

In sum, photochemical smog production is governed by emissions of oxides of nitrogen and reactive organic gases. Emitted gases, called primary pollutants, react in the presence of sunlight to produce secondary pollutants, such as ozone and peroxyacetyl nitrate. The radicals that break down emitted reactive organic gases are OH, $HO_2$, $O_3$, $NO_3$, and O. Photolysis also breaks down certain organics. Because reactive organic gas radicals compete with $O_3$ to produce $NO_2$ from NO, the photostationary-state relationship usually does not hold in urban air. Because gas-phase organic chemistry involves reactions among thousands of species, condensed reaction mechanisms have been developed to simplify the simulation of organic chemistry in numerical models.

## 12.3. STRATOSPHERIC PHOTOCHEMISTRY

Whereas ozone molecules in urban air are harmful to humans, animals, plants, trees, and structures, ozone in the stratosphere is beneficial in that it shields the earth from harmful ultraviolet radiation. Peak ozone mixing ratios in the stratosphere are usually greater than those in polluted urban air. Peak stratospheric mixing ratios are about 10 ppmv, while those in polluted urban air and free-tropospheric air are about 0.35 ppmv and 40 ppbv, respectively. The maximum number concentration of ozone molecules ($N_{O_3} = \chi_{O_3} N_d$, molecules cm$^{-3}$) in the stratosphere is close to the maximum number concentration in polluted urban air. Nevertheless,

**Table 12.7.** Carbon-Bond Representations of Several Organic Gases

| **Chemical Name**<br>**Carbon Bond Group**<br>**Chemical Structure** | | | |
|---|---|---|---|
| Ethane<br>0.4 PAR +1.6 UNR<br>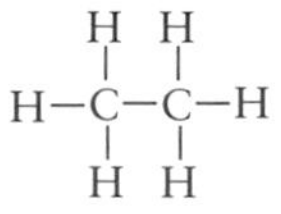 | *n*-Butane<br>4 PAR<br>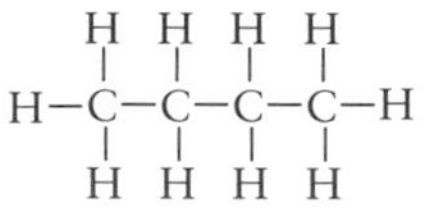 | 2,2,4-Trimethylpentane<br>8 PAR<br>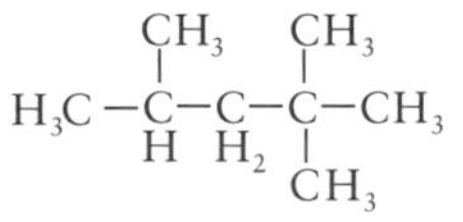 | Cyclopentane<br>5 PAR<br>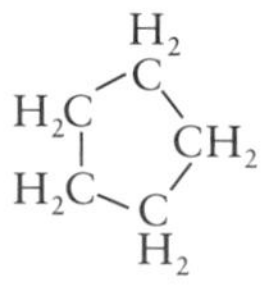 |
| Ethene<br>1 ETH<br>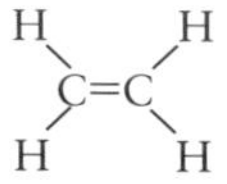 | *Trans* 2-butene<br>2 ALD2<br>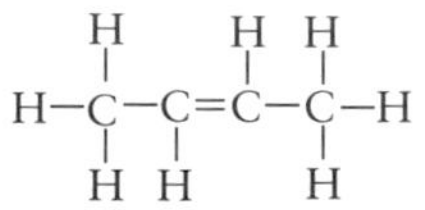 | Propene<br>1 PAR + 1 OLE<br>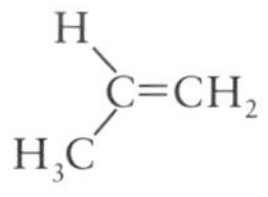 | Ethyne<br>1 PAR + 1 UNR<br>$H-C\equiv C-H$ |
| Formaldehyde<br>1 FORM<br>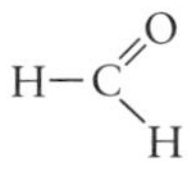 | Acetaldehyde<br>1 ALD2<br>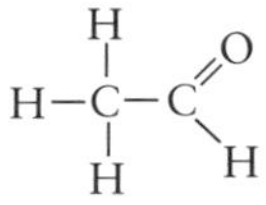 | Propionaldehyde<br>1 PAR + 1 ALD2<br>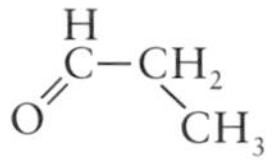 | Benzaldehyde<br>1 ALD2 + 5 UNR<br>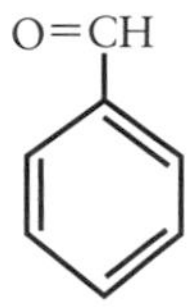 |
| Toluene<br>1 TOL<br>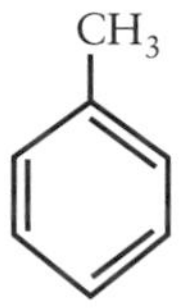 | Ethylbenzene<br>1 PAR + 1 TOL<br>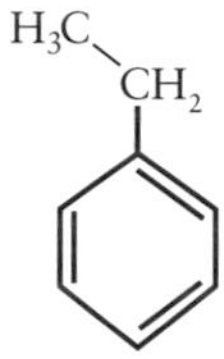 | *m*-Xylene<br>1 XYL<br>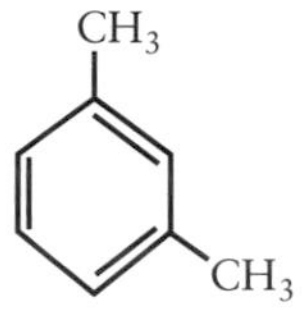 | 1,2,3-Trimethylbenzene<br>1 PAR + 1 XYL<br>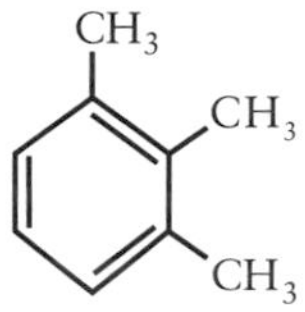 |
| Benzene<br>1 PAR + 5 UNR<br> | Methylethylketone<br>3 PAR + 1 KET<br>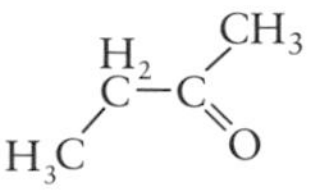 | Cyclopentene<br>1 PAR + 2 ALD2<br> | Cyclohexene<br>2 PAR + 2 ALD2<br>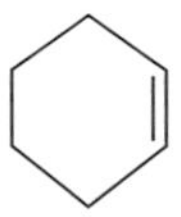 |

approximately 90 percent of all ozone molecules in the atmosphere reside in the stratosphere, because the stratosphere is approximately 32 to 40 km thick. Polluted urban regions are generally 0.5 to 1.5 km thick.

### 12.3.1. Background Stratospheric Photochemistry

In the absence of chlorine and other halogens, the photochemistry of the stratosphere has some similarities to that of the free troposphere. One difference is that photochemistry in the stratosphere is driven by the photolysis of molecular oxygen. Such photolysis is negligible in the troposphere. $O_2$ photolysis reactions are

$$O_2 + h\nu \longrightarrow \cdot\dot{O}(^1D) + \dot{O}\cdot \qquad \lambda < 175\ \text{nm} \qquad (12.107)$$
$$O_2 + h\nu \longrightarrow \dot{O}\cdot + \dot{O}\cdot \qquad 175 < \lambda < 245\ \text{nm} \qquad (12.108)$$

The first reaction is important only at the top of the stratosphere, since wavelengths shorter than 0.175 μm do not penetrate deeper into the stratosphere. Excited atomic oxygen from (12.107) is rapidly quenched to the ground state with

$$\cdot\dot{O}(^1D) \xrightarrow{M} \dot{O}\cdot \qquad (12.109)$$

As in the troposphere, ozone forms by

$$\dot{O}\cdot + O_2 + M \longrightarrow O_3 + M \qquad (12.110)$$

Thus, the primary difference between ozone formation in the stratosphere and in the free troposphere is that in the free troposphere atomic oxygen originates from nitrogen dioxide photolysis. In the stratosphere, it originates from molecular-oxygen photolysis. As in the free troposphere, ozone in the stratosphere photodissociates by

$$O_3 + h\nu \longrightarrow O_2 + \cdot\dot{O}(^1D) \qquad \lambda < 310\ \text{nm} \qquad (12.111)$$
$$O_3 + h\nu \longrightarrow O_2 + \dot{O}\cdot \qquad \lambda > 310\ \text{nm} \qquad (12.112)$$

In sum, ozone generation and destruction in the stratosphere are controlled by several Chapman-cycle reactions, discussed in Chapter 2.

Another difference between the troposphere and the stratosphere is the source of nitric oxide. In the troposphere, NO originates from biogenic and anthropogenic surface emissions and lightning. In the stratosphere, NO is produced by transport from the troposphere and by the first reaction in

$$N_2O + \cdot\dot{O}(^1D) \longrightarrow \begin{cases} 64\%\ 2\dot{N}O \\ 36\%\ N_2 + O_2 \end{cases} \qquad (12.113)$$

where $N_2O$ is nitrous oxide. **Nitrous oxide**, or laughing gas, is emitted into the stratosphere by transport from the troposphere or airplane emissions. In the troposphere, $N_2O$ is produced by lightning, emitted by bacteria in fertilizers, sewage, and the oceans, and emitted during combustion processes. Nitrous oxide is important, not only because it affects stratospheric ozone, but also because it is a greenhouse gas. Each nitrous oxide molecule is over 200 times more effective than a carbon dioxide molecule at absorbing infrared radiation, but the $N_2O$ mixing ratio in the atmosphere is only about 0.31 ppmv, which is more than 1000 times lower than that of $CO_2$.

(12.113) is responsible for less than 10 percent of $N_2O$ destruction. More than 90 percent of $N_2O$ destruction is due to photolysis by

$$N_2O + h\nu \longrightarrow N_2 + \cdot\dot{O}(^1D) \qquad \lambda < 240 \text{ nm} \tag{12.114}$$

At 25 km, the *e*-folding lifetime of $N_2O$ against photolysis is about 1.3 yr. $N_2O$ photolysis does not produce NO.

Nitric oxide in the stratosphere acts naturally and catalytically to reduce ozone concentrations. The reaction sequence is

$$\dot{N}O + O_3 \longrightarrow \dot{N}O_2 + O_2 \tag{12.115}$$
$$\dot{N}O_2 + \dot{O}\cdot \longrightarrow \dot{N}O + O_2 \tag{12.116}$$

$$\dot{O}\cdot + O_3 \longrightarrow 2O_2 \qquad \text{(net process)} \tag{12.117}$$

The net result of this sequence is that one molecule of ozone is destroyed, but neither NO nor $NO_2$ is lost. This type of sequences is a catalytic ozone destruction cycle. The cycle above is called the **$NO_x$ catalytic ozone destruction cycle**, where $NO_x = NO + NO_2$, and the catalyst is NO. The number of times the cycle is executed before $NO_x$ is removed from the cycle by reaction with another gas is the **chain length.** In the upper stratosphere, the chain length of this cycle is about $10^5$ (Lary 1997). Thus, $10^5$ molecules of $O_3$ are destroyed before one $NO_x$ molecule is removed from the cycle. In the lower stratosphere, the chain length decreases to near 10. The $NO_x$ catalytic cycle is largely natural. An increasing unnatural source of stratospheric $NO_x$ is stratospheric aircraft emissions.

Another catalytic ozone destruction cycle involves the hydroxyl radical. Reaction of $O(^1D)$ with $H_2O$, $CH_4$, and $H_2$ produce OH in the stratosphere by

$$\cdot\dot{O}(^1D) + \begin{cases} H_2O \longrightarrow 2\dot{O}H \\ CH_4 \longrightarrow \dot{C}H_3 + \dot{O}H \\ H_2 \longrightarrow \dot{H} + \dot{O}H \end{cases} \tag{12.118}$$

respectively. Once formed, OH can participate in an **$HO_x$ catalytic ozone destruction cycle**, where $HO_x = OH + HO_2$. $HO_x$ catalytic cycles are important in the lower stratosphere. The most effective $HO_x$ cycle, which has a chain length in the lower stratosphere of 1–40 (Lary 1997), is

$$\dot{O}H + O_3 \longrightarrow H\dot{O}_2 + O_2 \quad (12.119)$$
$$H\dot{O}_2 + O_3 \longrightarrow \dot{O}H + 2O_2 \quad (12.120)$$
$$2O_3 \longrightarrow 3O_2 \quad \text{(net process)} \quad (12.121)$$

$HO_x$ and $NO_x$ species can be removed temporarily from catalytic cycles by

$$H\dot{O}_2 + \dot{O}H \longrightarrow H_2O + O_2 \quad (12.122)$$
$$\dot{N}O_2 + \dot{O}H \xrightarrow{M} HNO_3 \quad (12.123)$$
$$H\dot{O}_2 + \dot{N}O_2 \xrightarrow{M} HO_2NO_2 \quad (12.124)$$

Nitric acid and peroxynitric acid, produced in the last two reactions, photodissociate, as shown in (12.28) and (12.30), respectively. Peroxynitric acid thermally decomposes by the reaction (12.29). In the stratosphere, thermal decomposition is slow, since temperatures are low.

Carbon monoxide and methane produce ozone in the stratosphere by the reactions (12.37)–(12.44). The contributions of CO and $CH_4$ to ozone production are small in comparison with the Chapman-cycle contribution to ozone production. The methane oxidation reaction

$$CH_4 + \dot{O}H \longrightarrow \dot{C}H_3 + H_2O \quad (12.125)$$

which leads to ozone production, is also an important source of water vapor in the stratosphere, since stratospheric water-vapor mixing ratios are low.

### 12.3.2. Chlorine and Bromine Photochemistry

Between 1979 and 1994, approximately 4–4.5 percent of the global stratospheric ozone layer was depleted between 60°N and 60°S latitude (WMO 1995). Ozone loss near the equator during this period was about 2 percent, and in the midlatitudes was near 5 percent. Between 1950 and 1980, no measurements from three ground-based stations in the Antarctic showed a springtime ozone depletion. Every spring since 1980, measurements of ozone have shown a depletion. Farman et al. (1985) reported depletions of more than 30 percent in the early 1980s. In October 1993, measurements over the South Pole indicated a 70 percent depletion of ozone. The areal extent of the ozone hole in 1993 was the size of North America. Molina and Rowland (1974) were the first to recognize that human-made chlorine compounds could destroy stratospheric ozone. Since then, scientists have strengthened the links among global ozone reduction, Antarctic ozone depletion, and the presence of chlorine and other halogenated compounds in the stratosphere.

In 1994, about 82 percent of chlorine entering the stratosphere originated from human emissions. Of the remaining emitted chlorine, about 15 percent was methyl

**Table 12.8.** Relative Emissions of Chlorine Compounds into the Stratosphere

| Chemical Formula | Trade Name | Chemical Name | Contribution Stratospheric Emissions (%) |
|---|---|---|---|
| | | **Anthropogenic Sources** | |
| $CF_2Cl_2$ | CFC-12 | Dichlorodifluoromethane | 28 |
| $CFCl_3$ | CFC-11 | Trichlorofluoromethane | 23 |
| $CCl_4$ | | Carbon tetrachloride | 12 |
| $CH_3CCl_3$ | | Methylchloroform | 10 |
| $CFCl_2CF_2Cl$ | CFC-113 | 1-Fluorodichloro, 2-difluorochloroethane | 6 |
| $CF_2ClH$ | HCFC-22 | Chlorodifluoromethane | 3 |
| | | **Natural Sources** | |
| $CH_3Cl$ | | Methyl chloride | 15 |
| HCl | | Hydrochloric acid | 3 |
| Total | | | 100 |

chloride ($CH_3Cl$), emitted from biogenic sources in the oceans, and 3 percent was hydrochloric acid (HCl), emitted from volcanos, evaporated from sea spray, and otherwise produced naturally. Table 12.8 lists anthropogenic and natural chlorine-containing compounds and their relative emissions into the stratosphere. Anthropogenically emitted compounds containing carbon and chlorine are **chlorocarbons**. Compounds that contain carbon, chlorine, and fluorine, such as $CFCl_3$ (CFC-11) and $CF_2Cl_2$ (CFC-12), are **chlorofluorocarbons** (CFCs). CFCs that have an H bonded to the carbon, such as $CF_2ClH$, are **hydrochlorofluorocarbons** (HCFCs).

In 1988, approximately $1.8 \times 10^6$ tons of chlorocarbons were emitted into the atmosphere. Because of an international agreement in 1987 called the **Montreal Protocol,** total emissions have decreased steadily since 1988. In 1993, chlorocarbon emissions were approximately $1.2 \times 10^6$ tons (WMO 1995).

Because most chlorocarbons have not caused serious human health concerns, they have been used widely. CFC-11 and -12 have been used as refrigerants and coolants, blowing agents in foam production, and aerosol propellants in spray cans. HCFC-22 has been used as a refrigerant and coolant. CFC-113, $CCl_4$, and $CH_3CCl_3$ have been used as solvents. $CCl_4$ has also been used as a grain fumigant, and $CH_3CCl_3$ has also been used as a degreasing agent.

CFCs are stable compounds in that their rates of chemical decomposition in the troposphere are slow. $CFCl_3$ and $CF_2Cl_2$ break down only after they diffuse above 15 to 20 km. At such altitudes, short ultraviolet wavelengths are available to photolyze these compounds by

$$\begin{array}{c}\mathrm{Cl}\\ |\\ \mathrm{F{-}C{-}Cl}\\ |\\ \mathrm{Cl}\end{array} + h\nu \longrightarrow \begin{array}{c}\mathrm{Cl}\\ |\\ \mathrm{F{-}C\cdot}\\ |\\ \mathrm{Cl}\end{array} + \dot{\mathrm{Cl}} \qquad \lambda < 250\ \mathrm{nm} \qquad (12.126)$$

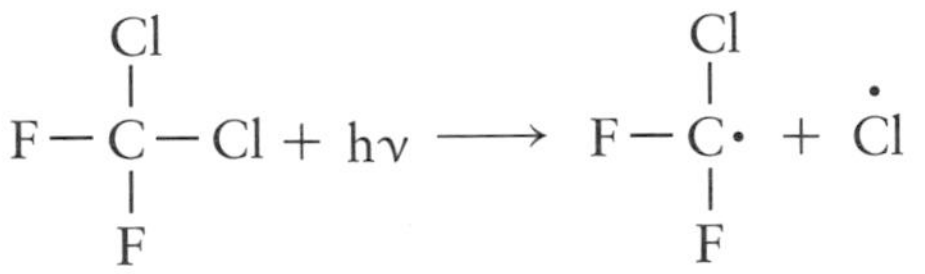 $\lambda < 230$ nm (12.127)

At 25 km, for example, the shortest *e*-folding lifetimes of $CFCl_3$ and $CF_2Cl_2$ against photolysis are on the order of 23 and 251 days, respectively (determined from Table B.4). Average lifetimes are on the order of 2 to 3 times these values. Why then are the overall lifetimes of $CFCl_3$ and $CF_2Cl_2$ between release at the surface and destruction in the middle stratosphere about 50 and 100 yr, respectively (Singh 1995)? The reason is that transport of CFCs from the troposphere to the middle stratosphere takes a long time. About 10 Mt of chlorine in the form of CFCs currently reside in the troposphere, and the transfer rate of CFC-chlorine from the troposphere to the middle stratophere is about 0.1 Mt per year. Thus, in this simplified scenario, the average time required for a CFC molecule to transfer from the troposphere to the middle stratosphere is about 100 yr.

Lifetimes of non-CFC chlorinated compounds are often shorter than those of CFCs. The lifetimes of HCFC-22, $CCl_4$, $CH_3CCl_3$, $CH_3Cl$, and HCl, between emissions and chemical destruction, are about 15, 70, 10, 1.5, and <0.1 yr, respectively. Non-CFCs have shorter lifetimes than do CFCs because some react faster with OH than do CFCs and others are more water soluble than are CFCs. Methyl chloride reacts with OH by

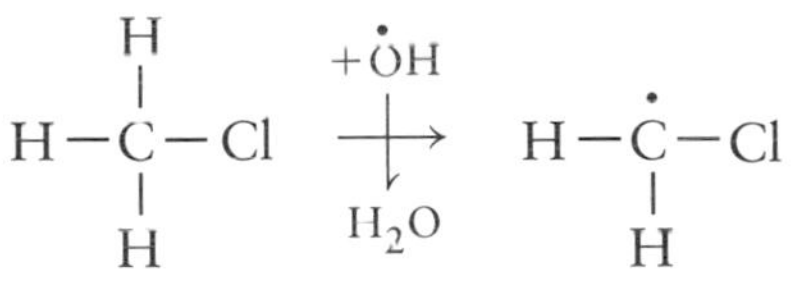 (12.128)

with an *e*-folding lifetime against OH abstraction of about 1.5 yr. The *e*-folding lifetime of HCl against OH abstraction in the troposphere is 15–30 days. HCl is also soluble in water and removed by clouds in the troposphere within minutes if conditions are right. In the stratosphere, non-CFCs photolyze at wavelengths similar to those of CFCs. Carbon tetrachloride and methyl chloride break down above 20 km by

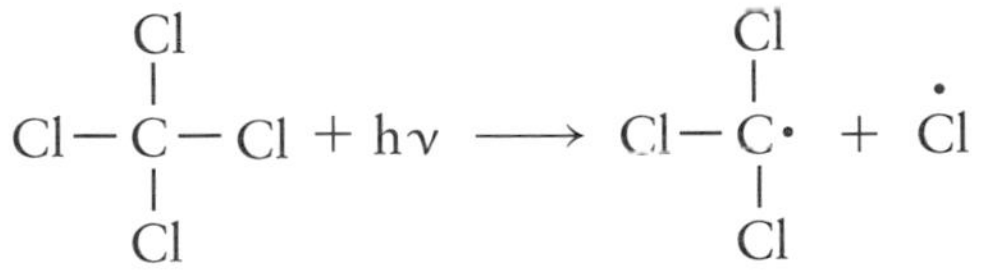 $\lambda < 250$ nm (12.129)

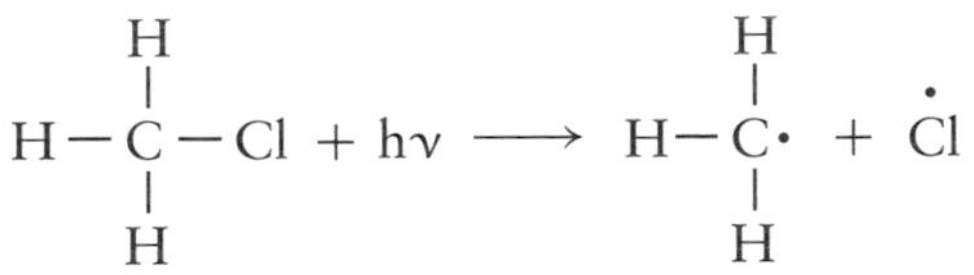 $\lambda < 220$ nm (12.130)

with *e*-folding lifetimes against photolysis at 25 km of 14 days and 2.4 yr, respectively.

Once released from their parent compounds in the stratosphere, active chlorine atoms from CFCs and non-CFCs react along one of several pathways. Chlorine reacts in a **catalytic ozone destruction cycle,**

$$\dot{Cl} + O_3 \longrightarrow Cl\dot{O} + O_2 \qquad (12.131)$$
$$Cl\dot{O} + \dot{O}\cdot \longrightarrow \dot{Cl} + O_2 \qquad (12.132)$$
$$\rule{6cm}{0.4pt}$$
$$\dot{O}\cdot + O_3 \longrightarrow 2O_2 \qquad \text{(net process)} \qquad (12.133)$$

in which chlorine atoms cycle to **chlorine monoxide** (ClO). Cl and ClO, together, are **active chlorine**. In midlatitudes, the chain length of this cycle increases from about 10 in the lower stratosphere to about 1000 in the middle and upper stratosphere (Lary 1997).

The primary removal mechanisms of active chlorine from the catalytic cycle are reactions that produce HCl, **chlorine nitrate** (ClONO$_2$), and, to a lesser extent, **hypochlorous acid** (HOCl). HCl and ClONO$_2$ are **chlorine reservoirs:** they withhold active chlorine from catalytic cycles. HOCl has a shorter lifetime than HCl or ClONO$_2$, and it is not considered a reservoir. Removal of Cl to HCl occurs by

$$\dot{Cl} + \begin{cases} CH_4 \longrightarrow HCl + \dot{C}H_3 \\ H\dot{O}_2 \longrightarrow HCl + O_2 \\ H_2 \longrightarrow HCl + \dot{H} \\ H_2O_2 \longrightarrow HCl + H\dot{O}_2 \end{cases} \qquad (12.134)$$

Removal of ClO to ClONO$_2$ and HOCl occurs by

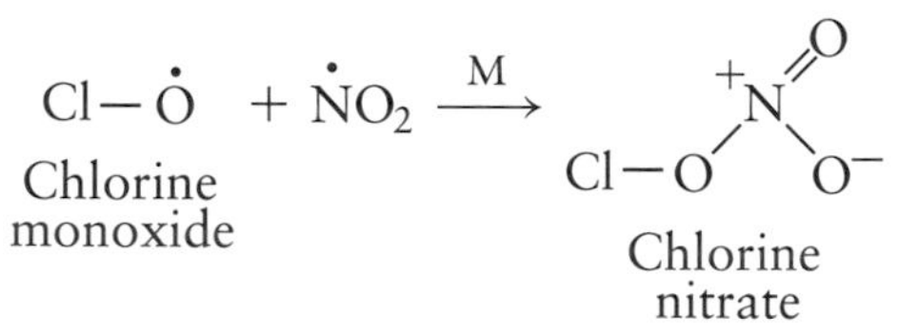

$$Cl-\dot{O} + \dot{N}O_2 \xrightarrow{M} \underset{\substack{\text{Chlorine} \\ \text{nitrate}}}{Cl-O-N^+(=O)O^-} \qquad (12.135)$$

Chlorine
monoxide

$$\underset{\substack{\text{Chlorine} \\ \text{monoxide}}}{Cl-\dot{O}} \xrightarrow[O_2]{+H\dot{O}_2} \underset{\substack{\text{Hypochlorous} \\ \text{acid}}}{Cl-O-H} \qquad (12.136)$$

respectively. At any time, about 1 percent of the chlorine in the stratosphere is in the form of active chlorine. Most of the rest is in the form of a chlorine reservoir.

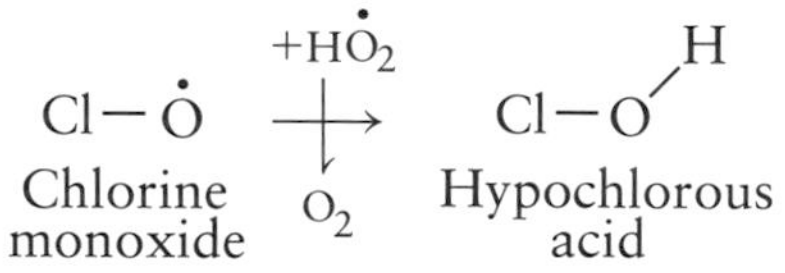

The HCl chlorine reservoir leaks slowly. HCl reproduces active chlorine by photolysis, reaction with OH, and reaction with O. The pertinent processes are

$$HCl + \begin{cases} h\nu \longrightarrow \dot{H} + \dot{C}l & \lambda < 220 \text{ nm} \\ \dot{O}H \longrightarrow \dot{C}l + H_2O \\ \dot{O}\cdot \longrightarrow \dot{C}l + \dot{O}H \end{cases} \qquad (12.137)$$

The *e*-folding lifetime of HCl against photolysis is about 1.5 yr at 25 km. HCl also diffuses back to the troposphere, where it can be absorbed by clouds.

The ClONO$_2$ and HOCl reservoirs leak primarily by the photolysis reactions

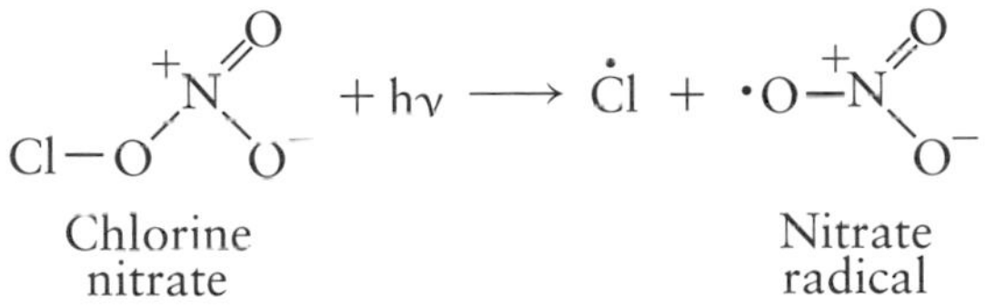

$$\underset{\substack{\text{Chlorine}\\\text{nitrate}}}{Cl-O\underset{O^-}{\overset{O}{\underset{|}{N^+}}}} + h\nu \longrightarrow \dot{C}l + \cdot O-\underset{O^-}{\overset{O}{\underset{|}{N^+}}} \qquad \lambda < 400 \text{ nm} \qquad (12.138)$$

$$\underset{\substack{\text{Nitrate}\\\text{radical}}}{}$$

$$Cl-O\overset{H}{\diagup} + h\nu \longrightarrow \dot{C}l + \dot{O}H \qquad \lambda < 375 \text{ nm} \qquad (12.139)$$

respectively. ClONO$_2$ Photodissociates with an *e*-folding lifetime of about 4.5 h at 25 km. HOCl photodissociates with an *e*-folding lifetime of only 11 min at 25 km. Thus both reactions are faster than the HCl photolysis reaction.

Like chlorine, bromine affects ozone in the stratosphere. The primary source of stratospheric bromine is **methyl bromide (CH$_3$Br)**, which is produced biogenically in the oceans and emitted as a soil fumigant. CF$_3$Br, emitted from fire extinguishers and as a fumigant, is another anthropogenically emitted bromine compound.

Methyl bromide photodissociates to produce Br above 20 km by

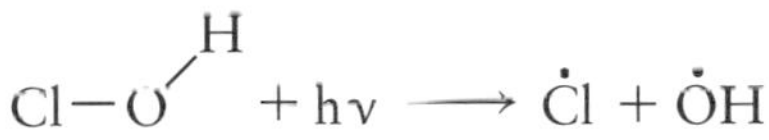

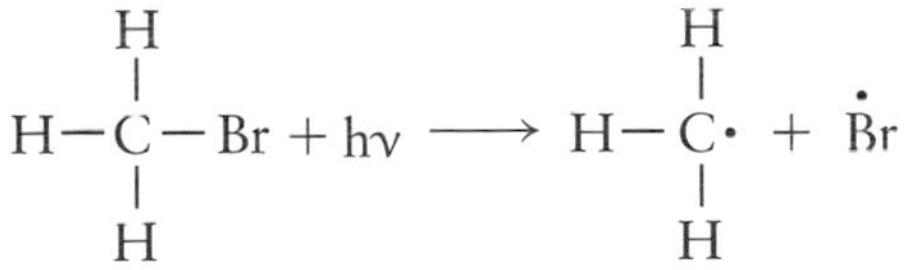

$$H-\underset{\substack{|\\H}}{\overset{\substack{H\\|}}{C}}-Br + h\nu \longrightarrow H-\underset{\substack{|\\H}}{\overset{\substack{H\\|}}{C}}\cdot + \dot{Br} \qquad \lambda < 260 \text{ nm} \qquad (12.140)$$

The *e*-folding lifetime of CH$_3$Br against loss by this reaction is about 10 days at 25 km.

In another **catalytic ozone destruction cycle**, bromine destroys ozone by the reactions

$$\dot{Br} + O_3 \longrightarrow Br\dot{O} + O_2 \qquad\qquad (12.141)$$
$$Br\dot{O} + \dot{O}\cdot \longrightarrow \dot{Br} + O_2 \qquad\qquad (12.142)$$

$$\overline{\dot{O}\cdot + O_3 \longrightarrow 2O_2 \qquad \text{(net process)} \qquad (12.143)}$$

where **bromine monoxide (BrO)** is the partner of Br in this cycle. The chain length of this cycle increases from about 100 at 20 km to about 10$^4$ at 40–50 km (Lary

1997). Atomic bromine is removed from the cycle by formation of **hydrobromic acid** (HBr) through

$$\dot{\mathrm{Br}} + \begin{cases} \mathrm{H\dot{O}_2} \longrightarrow \mathrm{HBr} + \mathrm{O_2} \\ \mathrm{H_2O_2} \longrightarrow \mathrm{HBr} + \mathrm{H\dot{O}_2} \end{cases} \qquad (12.144)$$

BrO reacts with nitrogen dioxide to form **bromine nitrate** ($BrONO_2$) by

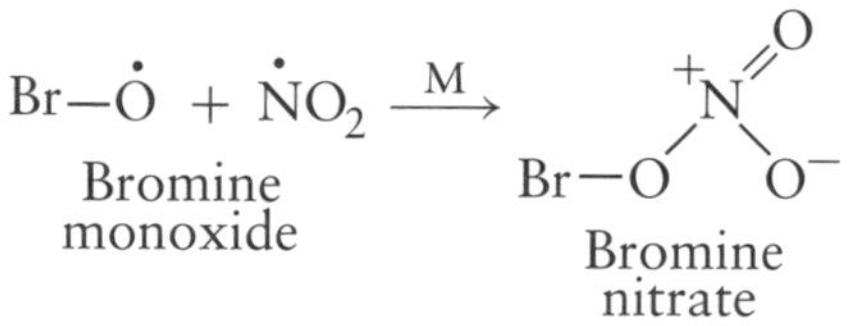

(12.145)

The HBr and $BrONO_2$ reservoirs leak through the respective reactions

$$\mathrm{HBr} + \dot{\mathrm{O}}\mathrm{H} \longrightarrow \dot{\mathrm{Br}} + \mathrm{H_2O} \qquad (12.146)$$

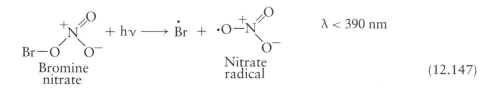

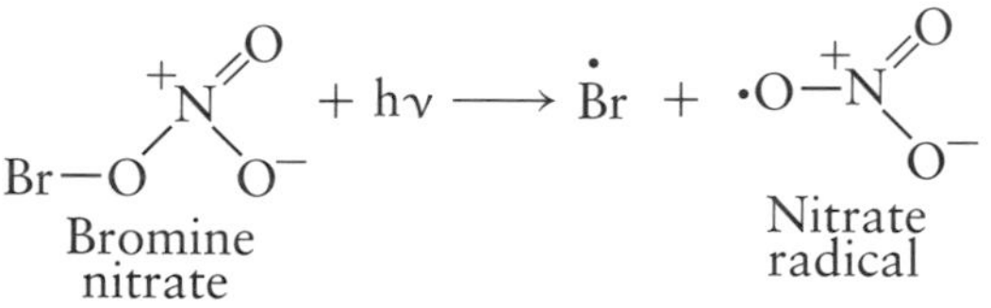

$\lambda < 390$ nm

(12.147)

The *e*-folding lifetime of $BrONO_2$ against the photolysis reaction is about 10 min at 25 km.

The presence of chlorine and bromine has decreased levels of ozone in the stratosphere. The reduction from 1980 to 1994 between 60°N and 60°S was from about 306 to 292 Dobson units (DU), or 4.5%. One **Dobson unit** equals 0.001 atm cm, which is approximately $2.7 \times 10^{16}$ molecules of ozone above a square centimeter of surface. If all molecules of ozone in a 1-$cm^2$ column, stretching from the surface to the top of the atmosphere, were counted, and the number were divided by $2.7 \times 10^{16}$, the result would be the column abundance of ozone in Dobson units.

The average global column abundance is now about 292 DU. The column abundance varies with latitude and season. Near the equator, it is 260–280 DU all year. Near 60°N and 60°S in March, it is around 430 and 300 DU, respectively. Near 60°N and 60°S in October, it is about 300 and 370 DU, respectively.

The following features of zonally averaged ozone column abundance occur: (1) an all-year **equatorial ozone minimum** due to upward motion of ozone-poor air from the lower stratosphere that displaces ozone-rich air aloft horizontally, (2) a northern hemisphere (**NH**) **spring maximum** over the North Pole due to downward motion of air over the pole, (3) a **SH spring minimum** over the South Pole due to chemical reactions of chlorine radicals with ozone, and (4) a **SH spring subpolar maximum** due to downward flow of air at the edge of the polar vortex (Tung and Yang 1988).

The ozone layer generally increases in thickness from the equator to poles, but it thins out near the South Pole during SH autumn, winter, and spring and near the North Pole during NH autumn. The highest column abundances occur over the North Pole in NH spring and at about 60°S latitude during SH spring. The spring peak in column abundance in the southern hemisphere does not extend to the Antarctic region itself, since the polar vortex, which is a circulation system that surrounds the Antarctic, prevents air from entering the Antarctic stratosphere during that time of year.

### 12.3.3. Ozone Regeneration Rates

Global stratospheric ozone depletion can be mitigated only when concentrations of halogenated compounds in the stratosphere decrease. Because many CFCs have long lifetimes, their natural removal takes many years, and no feasible engineering removal process is available.

As incremental quantities of chlorine are removed from the stratosphere, incremental quantities of ozone can regenerate by photochemistry and transport on a time scale of less than a year. Results from two global model simulations are shown here to support this hypothesis. In both cases, all ozone in the model atmosphere was removed on October 1, 1988. In the first case, ozone regeneration was simulated in the presence of chlorine, and in the second, chlorine was excluded from the simulation.

Figure 12.2 shows the changes in ozone column abundance during the simulations. In both cases, the globally averaged column abundance regenerated to about 150 DU after 10 days and 200 DU after 26 days. In the chlorine case, the column abundance regenerated to a maximum of 300.4 DU. After stabilizing, the column abundance varied between 281 and 300.4 DU. The observed globally averaged (90°S–90°N) column abundance of ozone in 1989 was approximately 300 DU (Bojkov and Fioletov 1995).

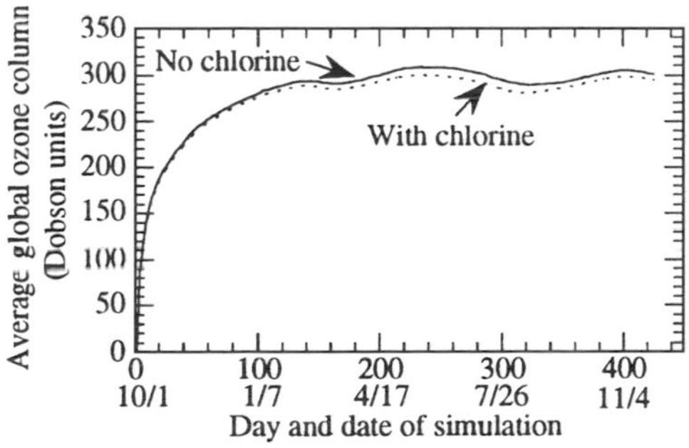


**Figure 12.2.** Change in ozone column abundance, averaged over the globe, during two global model simulations in which chlorine was present and absent, respectively. In both cases, ozone was initially removed from the model atmosphere.

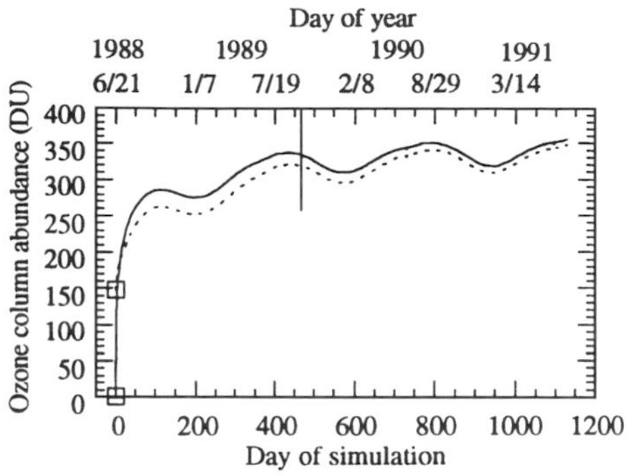


**Figure 12.3.** Results of a one-dimensional photochemical–radiative model for the change in column ozone at 34°N, over a three-year period for two cases. In the first case (solid line) all ozone in the column was removed initially. In the second case (dashed line), all ozone 23 km and below was removed initially. The boxes indicate the initial column abundance in each case. The line corresponds to September 26, 1989, the end time of Fig. 12.4(a) and (b).

In the no-chlorine case, the ozone layer regenerated to a maximum of 308.5 DU, but varied between 289.5 and 308.5 DU thereafter. The observed globally averaged ozone column abundance from 1964 to 1980 was about 306.4 DU (Bojkev and Fioletov 1995). A comparison of the chorine and the no-chlorine simulation indicates that chlorine caused a 2–3-percent decrease in globally averaged ozone column abundance through 1989. This difference is consistent with the 2-percent decrease in observed ozone between the pre-1980 period and 1989.

Figure 12.3 shows the rate of regeneration of the globally averaged (90°S–90°N) ozone column abundance when ozone was initialized with (1) 0 ppmv at all altitudes and (2) 0 ppmv below 23 km, in a one-dimensional simulation. The figure shows that when ozone was removed from the lower stratosphere only, regeneration took longer than when it was removed from the entire column. In the first case, ultraviolet radiation that dissociates oxygen was initially able to penetrate deep into the stratosphere and regenerate ozone there. In the second case, existing ozone at the top of the stratosphere inhibited deep penetration, slowing ozone reproduction in the lower stratosphere.

Figure 12.4(a) and (b) show the time-evolution of the vertical mixing ratio and number concentration, respectively, of ozone from June 21, 1988, through September 26, 1989, (464 days later) in the no-ozone case of Fig. 12.3. The graphs show that, within a few hours after the start of the simulation, ozone mixing ratios regenerated starting from the top of the stratosphere, but number concentrations regenerated starting from the bottom.

The results shown in this subsection indicate that, in the absence of chlorine and bromine, destroyed ozone re-forms rapidly in the stratosphere. Because chlorine

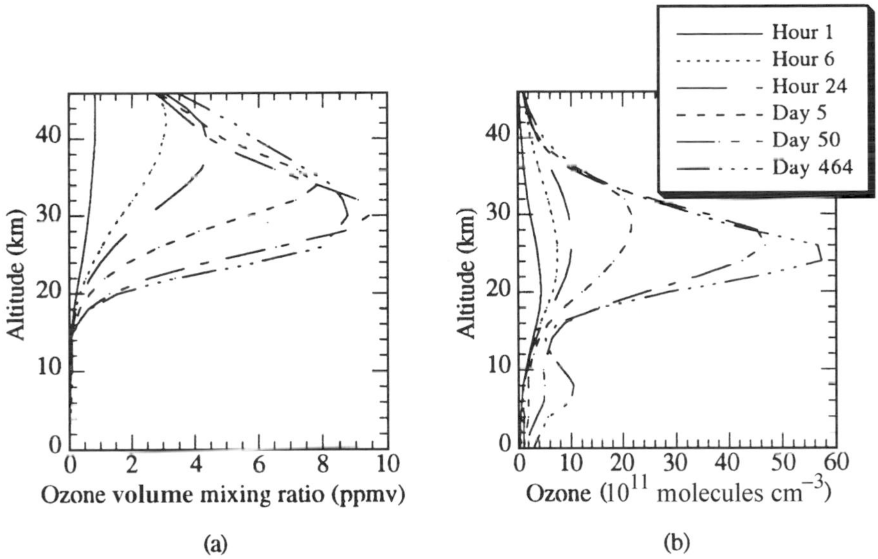


**Figure 12.4.** Time evolution of the vertical profile of ozone (a) volume mixing ratio and (b) number concentration at 34°N latitude, starting with zero ozone on June 21, 1988, as predicted by a one-dimensional model. Mixing ratios regenerated starting from the top of the stratosphere. Number concentrations regenerated starting from the bottom. The legend is the same for both parts.

and bromine are not easily removed from the stratosphere, current ozone reduction problems in the stratosphere can be solved only after existing chlorine and bromine are removed and emissions of these compounds are reduced.

### 12.3.4. Antarctic Ozone Depletion

Every September through November since 1980, the minimum ozone column abundance over the Antarctic has dropped steadily, as shown in Table 12.9. In 1994, the lowest measured column abundance was about 90 DU, which was 70 percent less than the average global value. The area over which ozone depletes (the **ozone hole**) increased yearly from 1981 to 1994, as shown in the table. The ozone hole in 1994 covered $23 \times 10^6$ km$^2$, which is almost the size of North America. Most Antarctic ozone depletion occurs between 14 and 18 km in altitude. Beginning in 1992, springtime ozone decreases were observed up to 24 km and down to 12–14 km. Because CFC emissions are declining due to the Montreal protocol, ozone reductions are expected to reverse themselves beginning in 1999, but the annual occurrence of the ozone hole is not expected to disappear until 2050 (Hofmann et al. 1997).

The ozone hole appears during the Southern Hemisphere spring over the Antarctic. A smaller **ozone dent** appears over the Arctic during the Northern Hemisphere spring (March–May). The hole and dent are caused by many factors. One factor linking global ozone reductions to polar ozone depletion is the presence of chlorine and bromine in the stratosphere.

**Table 12.9.** Minimum Ozone Column Abundances and Areal Extent of the Ozone Hole Over Antarctic Region from 1979 to 1994

| Year | Ozone Minima (DU) | Size ($10^6$ km$^2$) | Year | Ozone Minima (DU) | Size ($10^6$ km$^2$) |
|---|---|---|---|---|---|
| 1979 | 210 | 0 | 1987 | 121 | 19 |
| 1980 | 195 | 0.5 | 1988 | 179 | 8 |
| 1981 | 206 | 0 | 1989 | 124 | 18.5 |
| 1982 | 182 | 3 | 1990 | 126 | 17.5 |
| 1983 | 170 | 7 | 1991 | 110 | 18 |
| 1984 | 154 | 9 | 1992 | 121 | 21 |
| 1985 | 143 | 13 | 1993 | 86 | 22 |
| 1986 | 159 | 9.5 | 1994 | 90 | 23 |

Data from NASA Goddard Space Flight Center. For comparison, the area of the Antarctic is about $13 \times 10^6$ km$^2$ and the area of North America is about $24 \times 10^6$ km$^2$.

#### 12.3.4.1. *Polar Stratospheric Cloud Formation*

The ozone hole over the Antarctic appears in part because the Antarctic winter, from June to September, is very cold. Temperatures are low because much of the polar region is exposed to 24 h of darkness each day during winter, and a circulation system, the **polar vortex**, surrounds the Antarctic. The vortex traps cold air within the polar region and prevents the influx of warm air from outside the region.

Because temperatures are very low in the Antarctic stratosphere, optically thin clouds form. These clouds have few particles per unit volume of air in comparison with tropospheric clouds. Two major types of clouds form. When temperatures drop to below about 195 K, nitric acid and water vapor grow upon small sulfuric-acid–water aerosols (e.g., Toon et al. 1986). Initially it was thought that nitric acid and water molecules deposited to the ice phase in the ratio 1 : 3. Such ice crystals have the composition $HNO_3 \cdot 3H_2O$ and are called **nitric acid trihydrate** (NAT) crystals. More recently, it was found that these particles contain a variety of phases. Some contain **nitric acid dihydrate** (NAD) (Worsnop et al. 1993), and others contain supercooled liquid water, sulfuric acid, and nitric acid (e.g., Tabazadeh et al. 1994). Together, nitrate-containing cloud particles that form at temperatures below about 195 K in the winter polar stratosphere are caled **Type I polar stratospheric clouds** (PSCs).

When temperatures drop below the frost point, which is about 187 K for typical polar stratospheric conditions, a second type of cloud forms. These clouds contain pure water ice and are **Type II polar stratospheric clouds**. Usually, about 90 percent of PSCs are Type I and 10 percent are Type II (Turco et al. 1989). Typical diameters and number concentrations of a Type I PSC are 1 μm and $\leq$1 particle cm$^{-3}$, respectively, although diameters vary from 0.01 to 3 μm. Typical diameters and number concentrations of a Type II PSC are 20 μm and $\leq$0.1 particle cm$^{-3}$, respectively, although diameters vary from 1 to 100 μm.

#### 12.3.4.2. *PSC Surface Reactions*

Once PSC particles form, chemical reactions take place on their surfaces. Such reactions are **heterogeneous reactions** and occur after at least one gas has diffused to and adsorbed to a particle surface. **Adsorption** is a process by which a gas collides with and bonds to a surface. One type of bonding is **ion–dipole** bonding, which occurs when the surface is charged and the gas is polar. A second type is **dipole–dipole** bonding, which occurs when the surface and gas are polar. Once a gas molecule adsorbs to a surface, it can **desorb** (break away) from the surface. If it stays adsorbed to the surface, it can **diffuse** to another site on the surface. During the diffusion process, the adsorbed gas may collide and chemically react with another adsorbed molecule. A gas molecule suspended above a particle surface may also collide and react with the adsorbed molecule. In both cases, an adsorbed product is formed. This product can either desorb from the surface, diffuse on the surface, or participate in additional chemical reactions.

The primary heterogeneous reactions that occur on Type I and II PSC surfaces are

$$\mathrm{ClONO_2(g) + H_2O(a) \longrightarrow HOCl(g) + HNO_3(a)} \qquad (12.148)$$

$$\boxed{\mathrm{ClONO_2(g) + HCl(a) \longrightarrow Cl_2(g) + HNO_3(a)}} \qquad (12.149)$$

$$\mathrm{N_2O_5(g) + H_2O(a) \longrightarrow 2HNO_3(a)} \qquad (12.150)$$

$$\mathrm{N_2O_5(g) + HCl(a) \longrightarrow ClNO_2(g) + HNO_3(a)} \qquad (12.151)$$

$$\mathrm{HOCl(g) + HCl(a) \longrightarrow Cl_2(g) + H_2O(a)} \qquad (12.152)$$

In these reactions, (g) denotes a gas and (a) denotes an adsorbed species. Additional reactions exist for bromine. Laboratory studies show that HCl readily coats the surfaces of Type I and II PSCs. When $ClONO_2$, $N_2O_5$, or HOCl, impinge upon the surface of a Type I or II PSC, it can react with $H_2O$ or HCl already on the surface. The products of these reactions are adsorbed species, some of which stay adsorbed, but others of which desorb to the vapor phase. Nitric acid and water ice stay adsorbed, while $Cl_2$, HOCl, and $ClNO_2$ desorb after they are produced heterogeneously.

In sum, heterogeneous reactions convert relatively inactive forms of chlorine in the stratosphere, such as HCl and $ClONO_2$, to photochemically active forms, such as $Cl_2$, HOCl, and $ClNO_2$. This conversion process is **chlorine activation**. The most important heterogeneous reaction is (12.149), which generates $Cl_2$. (12.150) does not activate chlorine. Its only effect is to remove nitric acid from the gas phase. Other heterogeneous reactions also remove nitric acid. When nitric acid adsorbs to a Type II PSC, which is larger than a Type I PSC, the nitric acid can sediment out along with the PSC to lower regions of the stratosphere. This removal process is

**Table 12.10.** Estimated Reaction Probabilities for the Gases in the Reactions (12.147)–(12.151) on Type I and II PSC Surfaces

| Reaction | Type I PSCs | Type II PSCs |
|---|---|---|
| $ClONO_2(g) + H_2O(a)$ | 0.001 | 0.3 |
| $ClONO_2(g) + HCl(a)$ | 0.1 | 0.3 |
| $N_2O_5(g) + H_2O(a)$ | 0.0003 | 0.01 |
| $N_2O_5(g) + HCl(a)$ | 0.003 | 0.03 |
| $HOCl(g) + HCl(a)$ | 0.1 | 0.3 |

*Source:* DeMore et al. (1997) and references therein.

**denitrification.** Denitrification is important because it removes nitrogen that might otherwise either form additional Type I PSCs or return to the gas phase to tie up active chlorine as $ClONO_2$.

#### 12.3.4.3. *Reaction Probabilities*

Rate coefficients for (12.148)–(12.152) have the form

$$k_{s,q} = \frac{1}{4}\bar{v}_q \gamma_q a \tag{12.153}$$

where $\bar{v}_q$ is the **thermal velocity** of the impinging gas (cm $s^{-1}$), $\gamma_q$ is the **reaction probability** (also called the **uptake coefficient**) of the gas (unitless), and $a$ is the **surface-area concentration** (square centimeters of surface per cubic centimeter of air) of all particles on which reactions occur. (12.153) implicitly includes the concentration of the adsorbed reactant but not the gas reactant; thus, it is a pseudo-first-order rate coefficient ($s^{-1}$). When multiplied by the gas-phase reactant concentration, (12.153) gives a rate (molecules $cm^{-3}$ $s^{-1}$).

The thermal velocity of a gas is

$$\bar{v}_q = \left(\frac{8k_B T}{\pi \bar{M}_q}\right)^{1/2} \tag{12.154}$$

where $\bar{M}_q = m_q / A$ is the mass (g) of one gas molecule.

A **reaction probability** is the laboratory-measured fractional loss of a species from the gas phase due to reaction with the surface, and it takes account of diffusion of the gas to the surface as well as reaction with the surface. Estimated reaction probabilities for the gases in (12.148)–(12.152) on Type I and II PSCs are given in Table 12.10.

Reaction probabilities can be derived by considering adsorption, desorption, and reaction on a particle surface (e.g., Tabazadeh and Turco 1993; Adamson 1990). The rate of change of species concentration in the first molecular layer over a surface is

$$\frac{dn_{s,q}}{dt} = k_{a,q} p_q (n_m - n_{s,q}) - k_{d,q} n_{s,q} - k_{s,q} n_{s,q} \tag{12.155}$$

where $n_{s,q}$ is the surface-area concentration of adsorbed gas $q$ (molecules cm$^{-2}$), $k_{a,q}$ is the rate that gas molecules adsorb to the surface (molecules mb$^{-1}$ s$^{-1}$), $p_q$ is the partial pressure of the gas over the surface (mb), $n_m \approx 10^{15}$ sites cm$^{-2}$ is approximately the number of adsorption sites on the surface, $k_{d,q}$ is the rate that adsorbed molecules desorb from the surface (s$^{-1}$), and $k_{s,q}$ is the first-order rate coefficient for the loss of adsorbed species $q$ due to chemical reaction with another adsorbed species or with a gas (s$^{-1}$). (12.155) states that the change in concentration of molecules adsorbed to the surface of a particle equals an adsorption flux minus a desorption flux minus a reaction flux.

The **rate of molecular adsorption** is

$$k_{a,q} = \frac{\alpha_q \sigma_0}{2\pi \bar{M}_q k_B T} \exp\left(-\frac{E_{a,q} + \eta\theta_{f,q}}{R^* T}\right) \tag{12.156}$$

where $\alpha_q$ is the **mass accommodation coefficient** of the adsorbing gas, $\sigma_0 \approx 10^{-15}$ cm$^2$ is the molecular surface area of one site, $\bar{M}_q$ is the mass of one adsorbed molecule (g), $E_{a,q}$ is the activation energy of adsorption (J mole$^{-1}$), $\eta$ is a factor that allows $E_{a,q}$ to change with surface coverage, and $\theta_{f,q}$ is fractional surface coverage. The mass accommodation coefficient (sticking coefficient) is the fractional number of collisions of gas $q$ with a particle that results in the gas sticking to the particle's surface.

The **rate of molecular desorption** is

$$k_{d,q} = \nu_0 \exp\left(-\frac{\Delta G_{d,q} + E_{d,q}}{R^* T}\right) \tag{12.157}$$

where $\nu_0 \approx 10^{13}$ s$^{-1}$ is the frequency of atomic vibrations on the surface, $\Delta G_{d,q}$ is the change in Gibbs free energy of desorption (J mole$^{-1}$), and $E_{d,q}$ is the activation energy of desorption (J mole$^{-1}$). The **free energy of desorption** is related to the free energy of adsorption and entropy of adsorption by

$$-\Delta G_{d,q} = \Delta G_{a,q} = \Delta H_{a,q}(\theta_{f,q}) - T\,\Delta S_{a,q}(\theta_{f,q}) \tag{12.158}$$

where $\Delta H_{a,q}(\theta_{f,q})$ is the **enthalpy of adsorption** (J mole$^{-1}$), and $\Delta S_{a,q}(\theta_{f,q})$ is the **entropy of adsorption** (J mole$^{-1}$ K$^{-1}$), both of which depend on the fractional surface coverage. If the adsorbed species diffuses on the surface, the entropy of adsorption is

$$\Delta S_{a,q}(\theta_{f,q}) = -R^* \ln\theta_{f,q} + R^* \ln(\bar{M}_q T \sigma_0) - R^* \ln(T^{5/2} \bar{M}_q^{3/2}) + 96.65 \tag{12.159}$$

**Surface coverage** is the fraction of available surface sites filled with adsorbed molecules.The steady-state surface coverage of an adsorbed substance is obtained by setting (12.155) to zero. The result is

$$\theta_{f,q} = \frac{n_{s,q}}{n_m} = \frac{k_{a,q} p_q}{k_{d,q} + k_{s,q} + k_{a,q} p_q} \tag{12.160}$$

The adsorption minus desorption flux in (12.155) can be written in terms of a net diffusive flux from the gas phase and a reaction probability:

$$\tfrac{1}{4}\bar{v}_q \gamma_q N_q = k_{a,q} p_q (n_m - n_{s,q}) - k_{d,q} n_{s,q} \tag{12.161}$$

where $N_q = p_q / k_B T$ is the number concentration of gas molecules suspended over the particle surface (molecules cm$^{-3}$). Substituting (12.161) into (12.155) gives

$$\frac{dn_{s,q}}{dt} = \frac{1}{4}\bar{v}_q \gamma_q N_q - k_{s,q} n_{s,q} \tag{12.162}$$

Setting (12.162) equal to zero and solving for the steady-state reaction probability gives

$$\gamma_q = \frac{4k_{s,q} n_{s,q}}{\bar{v}_q N_q} = \frac{4k_{s,q} \theta_{f,q} n_m}{\bar{v}_q N_q} \tag{12.163}$$

Parameters required for $\theta_{f,q}$, such as $E_{a,q}$, $E_{d,q}$, $\eta$, and $k_{s,q}$, are difficult to obtain. Thus, laboratory-derived reaction probabilities are generally used and substituted into (12.153) to approximate $k_{s,q}$.

#### 12.3.4.4. *Springtime Polar Chemistry*

Chlorine activation occurs during the winter over the polar stratosphere. When the sun rises in early spring, Cl-containing gases created by PSC reactions photodissociate by

$$Cl_2 + h\nu \longrightarrow 2\dot{C}l \qquad \lambda < 450\,\text{nm} \tag{12.164}$$

$$HOCl + h\nu \longrightarrow \dot{C}l + \dot{O}H \qquad \lambda < 375\,\text{nm} \tag{12.165}$$

$$ClNO_2 + h\nu \longrightarrow \dot{C}l + \dot{N}O_2 \qquad \lambda < 370\,\text{nm} \tag{12.166}$$

Once Cl has been released, it attacks ozone. The catalytic cycle that destroys ozone in the springtime polar stratosphere differs from that shown in (12.131)–(12.133), which destroys ozone on a global scale. A polar stratosphere catalytic ozone destruction cycle is

$$2 \times (\dot{C}l + O_3 \longrightarrow Cl\dot{O} + O_2) \tag{12.167}$$
$$\dot{C}lO + \dot{C}lO \xrightarrow{M} Cl_2O_2 \tag{12.168}$$
$$Cl_2O_2 + h\nu \longrightarrow ClO\dot{O} + \dot{C}l \qquad \lambda < 360 \text{ nm} \tag{12.169}$$
$$ClO\dot{O} \xrightarrow{M} \dot{C}l + O_2 \tag{12.170}$$
$$2O_3 \longrightarrow 3O_2 \tag{12.171}$$

where $Cl_2O_2$ is **dichlorine dioxide** and ClOO is a **chlorine peroxy radical.** This mechanism, called the **dimer mechanism** (Molina and Molina 1986), is important in the springtime polar stratosphere because at that location and time the ClO required for (12.168) is concentrated enough for the reaction to proceed rapidly. A second cycle is

$$\dot{C}l + O_3 \longrightarrow Cl\dot{O} + O_2 \tag{12.172}$$
$$\dot{B}r + O_3 \longrightarrow Br\dot{O} + O_2 \tag{12.173}$$
$$Br\dot{O} + Cl\dot{O} \longrightarrow \dot{B}r + \dot{C}l + O_2 \tag{12.174}$$
$$2O_3 \longrightarrow 3O_2 \tag{12.175}$$

(McElroy et al. 1986), which is important in the polar lower stratosphere. In sum, chlorine activation and springtime photochemical reactions convert chlorine from reservoir forms, such as HCl and $ClONO_2$, to active forms, such as Cl and ClO, as shown in Fig. 12.5. The active forms of chlorine destroy ozone in catalytic cycles that regenerate active chlorine.

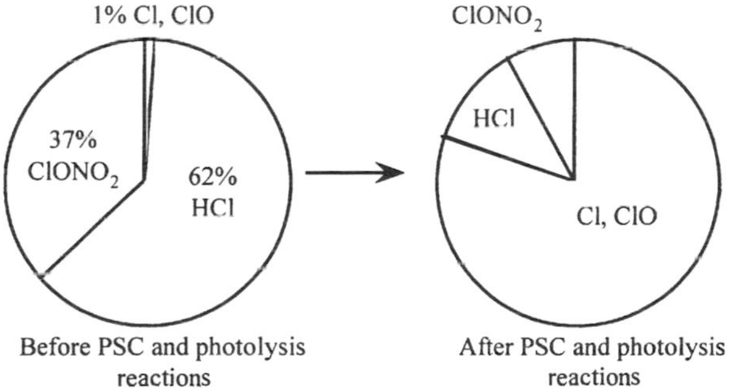


**Figure 12.5.** Pie chart showing conversion of chlorine reservoirs to active chlorine. During chlorine activation by PSCs, HCl, and $ClONO_2$ are converted to potentially active forms of chlorine, such as $Cl_2$, HOCl, and $ClNO_2$. These molecules are broken down by sunlight when the sun arises in springtime to form Cl. Cl reacts catalytically along with ClO to destroy ozone.

Every November, the Antarctic warms up sufficiently for the polar vortex to break down and PSCs to melt. Ozone from outside the polar region advects into the region. Ozone also regenerates chemically, and chlorine reservoirs of $ClONO_2$ and HCl reestablish themselves. Thus, the Antarctic ozone hole is an annual, regional phenomenon that is controlled primarily by the temperature of the polar stratosphere and the presence of chlorine and bromine. The radial extent of the hole has grown and minimum ozone values have decreased steadily over the last several years. The ozone dent over the Arctic is not so large as that over the Antarctic, because temperatures over the Arctic do not get so low as over the Antarctic. Thus, PSC formation is less widespread over the Arctic.

## 12.4. SUMMARY

In this chapter, chemistry of the free troposphere, urban areas, the global stratosphere, and the polar stratosphere were discussed. Important chemical equations and reactions pathways were given. The focus of many atmospheric pollution studies is on ozone formation or destruction. In urban regions, the problem is ozone formation. In the stratosphere, the problem is ozone destruction. The chemical pathways that form and destroy ozone in each region are different. In the free troposphere, ozone production is governed by NO reactions with $O_3$ and by $NO_2$ photolysis. In urban regions, ozone production is governed by NO reactions with ROGs and by $O_3$ and $NO_2$ photolysis. Ozone production in the stratosphere is governed by $O_2$ photolysis.

Table B.4 of Appendix B lists most of the reactions described in this chapter. In the next chapter, methods of solving chemical ordinary differential equations are described.

## 12.5. PROBLEMS

**12.1.** **(a)** Fill in Table 12.11, assuming that the photolysis rates of $NO_2$ at 8:30 a.m. and 1:30 p.m. were $J = 0.008$ s$^{-1}$ and $J = 0.01$ s$^{-1}$, respectively, and $T = 291$ K and $T = 298$ K at 8:30 a.m. and 1:30 p.m., respectively. Assume $p_d = 1005$ mb in both cases.

**(b)** Modeled ozone mixing ratios in Table 12.11 were calculated with a chemical model that included all the urban reactions, in Table B.4 of Appendix B, including organic reactions. Explain the difference between modeled and photostationary-state mixing ratios, if any, between the morning and the afternoon. Why do morning and afternoon values differ less at Costa Mesa than at other locations?

**12.2.** **(a)** Calculate the thermal velocity of chlorine nitrate when $T = 192$ K. At this temperature, what type of PSC might be present?

**(b)** Calculate the pseudo-first-order rate coefficient of chlorine nitrate on the surfaces of these cloud particles, assuming the surface is coated with

**Table 12.11.** Mixing-ratio Data from the Los Angeles Basin for Problem 12.1

| Station | $\chi_{NO}$ (ppmv) | $\chi_{NO_2}$ (ppmv) | $\frac{\chi_{NO_2}}{\chi_{NO}}$ | $\chi\chi_{O_{3,m}}$ (ppmv) modeled | $\chi_{O_{3,p}}$ (ppmv) photostat. | $\frac{\chi_{O_{3,m}} - \chi_{O_{3,p}}}{\chi_{O_{3,p}}} \times 100\%$ |
|---|---|---|---|---|---|---|
| | | | | **8:30 a.m.** | | |
| Anaheim | 0.1757 | 0.0760 | | 0.0058 | | |
| Costa Mesa | 0.0607 | 0.0446 | | 0.0119 | | |
| Hawthorne | 0.0654 | 0.0525 | | 0.0104 | | |
| La Habra | 0.0764 | 0.0501 | | 0.0092 | | |
| Lynwood | 0.1209 | 0.0625 | | 0.0073 | | |
| Pasadena | 0.1179 | 0.0740 | | 0.0089 | | |
| Reseda | 0.0704 | 0.0645 | | 0.0125 | | |
| Simi Valley | 0.0232 | 0.0355 | | 0.0204 | | |
| | | | | **1:30 p.m.** | | |
| Anaheim | 0.0088 | 0.0374 | | 0.0859 | | |
| Costa Mesa | 0.0047 | 0.0131 | | 0.0617 | | |
| Hawthorne | 0.0169 | 0.0394 | | 0.0471 | | |
| La Habra | 0.0108 | 0.0461 | | 0.0879 | | |
| Lynwood | 0.0096 | 0.0372 | | 0.0778 | | |
| Pasadena | 0.0321 | 0.0693 | | 0.0491 | | |
| Reseda | 0.0091 | 0.0495 | | 0.1216 | | |
| Simi Valley | 0.0031 | 0.0148 | | 0.0990 | | |

HCl. (Hint: Use a typical number concentration and diameter, given in the text, of the chosen PSC type).

**(c)** If $\chi_{ClONO_2} = 1.0$ ppbv initially, and only surface chemistry is considered, how much chlorine nitrate will remain after 15 days due to the reaction from Problem 12.2(b)? What is the resulting mixing ratio of $Cl_2$ if none existed initially?

**12.3.** Write the rate expressions for the reactions (12.167)–(12.170), and write the first-derivative expressions for each species (except for $O_2$ and M) in the reactions. Assume M is already included in the rate coefficients. What is the steady-state expression for the Cl concentration? If $T = 197$ K and $p_d = 25$ mb, calculate the rate coefficient of each kinetic reaction.

## 12.6. COMPUTER PROGRAMMING PRACTICE

**12.4.** **(a)** Input all the inorganic kinetic reactions and either the (i) chlorine, (ii) bromine, (iii) sulfur, or (iv) isoprene kinetic reactions from Table B.4 of Appendix B into the computer file developed for Problem 11.7. If the isoprene mechanism is chosen, many nonaromatic organic reactions should also be included to account for the chemistry of isoprene byproducts. Calculate rate coefficients for each reaction. For the chlorine and bromine cases, assume $T = 192$ K and $p_d = 25$ mb. For the isoprene and sulfur cases, assume $T = 298.15$ K and $p_d = 1013$ mb.

**(b)** Input the necessary photolysis reactions from Table B.4 of Appendix B to complete the mechanism chosen in part (a). If a stratospheric mechanism was chosen, input the stratospheric (25 km) peak photolysis rate coefficients from the table into the file. Otherwise, input the surface (0 km) peak rate coefficients. Write a script to scale the peak rate coefficients to any time of day. Assume the coefficients are zero before 6:00 a.m. and after 6:00 p.m. but vary as a sine function from 6:00 a.m. to 6:00 p.m, peaking at noon. Calculate rate coefficients for 9:00 a.m. and 5:00 p.m.

# 13

# Methods of Solving Chemical Ordinary Differential Equations

SEVERAL methods exist to solve chemical ordinary differential equations (ODEs). Those discussed in this chapter include analytical, Taylor series, forward Euler, backward Euler, simple exponential, quasi-steady-state, multistep implicit–explicit, backward differentiation (Gear), and family methods. Additional methods are hybrid predictor–corrector methods (e.g., Young and Boris 1977), parameterization methods (e.g., Jacob et al. 1989b), Runge–Kutta–Rosenbrock schemes (Hairer and Wanner 1991; Sandu et al. 1997), iterative backward Euler methods (e.g., Curtiss and Hirschfelder 1952; Shimazaki and Laird 1970; Rosenbaum 1976; Hertel et al. 1993), hybrid Newton–Raphson iterative schemes (Gong and Cho 1993), and Gauss–Seidel methods (Verwer 1994), among others. The choice of a method for solving atmospheric chemical problems depends on several factors, including stability, accuracy, mass conservation, positivity, and speed. These factors are discussed with respect to different solvers. Techniques of optimizing Gear's method over a large grid are also described.

## 13.1. CHARACTERISTICS OF CHEMICAL ODEs

Gas-phase chemical reactions are described by first-order, first-degree, homogeneous ordinary differential equations. Sets of gas-phase reactions are stiff in that the chemical *e*-folding lifetimes of individual species vary by many orders of magnitude. Because sets of chemical ODEs are stiff, some classical numerical methods are not useful for solving them. The fourth-order Runge–Kutta method (e.g., Press et al. 1992) and the Richardson extrapolation/Bulirsch–Stoer method (Stoer and Bulirsch 1980) are explicit and so are inefficient at solving stiff ODEs. An **explicit solution** is a solution obtained when derivatives are evaluated at the beginning of the current and previous time steps (times $t - h$, $t - 2h$, ...), but not at the end of the current step (time $t$). A **time step** (s) $h$ is the difference between the time of the current step ($t$) and the time of the previous step ($t - h$). When an explicit technique is used to solve a stiff set of equations, the time-step size is limited by the *e*-folding lifetime of the shortest-lived chemical. This lifetime may be $10^{-4}$ s or less. Time steps longer than the lifetime of the shortest-lived chemical may destabilize the solution scheme. When a time step is always small, integration of many reactions over days to months and over large model grids is impractical.

Solvers of stiff ODEs are **semiimplicit** in that their solutions at the end of a time step depend on derivatives evaluated at the end of the current time step, the beginning of the current time step, and the beginning of previous time steps (times $t$, $t - h$, $t - 2h$, ...). Semiimplicit solvers can take time steps much longer than the $e$-folding lifetime of the shortest-lived species and remain stable. Some accurate semiimplicit schemes used for solving stiff ODEs are Gear's method (Gear 1971) Runge–Kutta–Rosenbrock schemes (Kaps and Rentrop 1979; Hairer and Wanner 1991; Press et al. 1992), and semiimplicit Bulirsch–Stoer schemes (Bader and Deuflhard 1983; Press et al. 1992), among others.

### 13.1.1. Initial Value Problems

Problems requiring the use of chemical ordinary differential equations are **initial-value problems**. The initial concentration of each species is known at time $t = 0$. The solution to a set of chemical ODEs is a set of concentrations at a final time, $t_f > 0$. Solutions are found by integrating ODEs one time step at a time.

The concentration of a species $i$ at current time step $t$ is defined as $N_{i,t}$. The vector of concentrations of a set of species at the current time step is

$$\hat{N}_t = [N_{1,t}, N_{2,t}, \ldots, N_{i,t}, \ldots, N_{K,t}] \tag{13.1}$$

where $K$ is the number of species in the set. The concentration at one time step backward is $N_{i,t-h}$. When gases are considered, units of $N$ are molecules cm$^{-3}$.

At the beginning of the first time step of a chemical ODE problem, concentrations of all species from one time step backward are set to initial concentrations. Thus,

$$N_{i,t-h} = N_{i,0} \qquad \text{for } i = 1, \ldots, K \tag{13.2}$$

where $N_{i,0}$ is the initial number concentration of gas $i$. During the time step, the ODE scheme solves for all $N_{i,t}$. When values at time $t$ are found, they are set to values at time $t - h$, and the solution for the second time step is solved for.

With some techniques, solutions depend on concentrations from several time steps backward (times $t - h, t - 2h, \ldots$). If the general solution technique relies on values from two time steps backward, the first time step is solved for with a techniqe that requires only initial values. The second time step is solved for using initial values and solutions from the first time step.

### 13.1.2. Properties of ODE Solvers

For a chemical ODE solution scheme to be useful, it must be stable, accurate, mass-conserving, positive definite, and computationally fast. In Chapter 6, a numerical scheme was defined to be **stable** if the absolute-value difference between the numerical and the exact solution did not grow over time. This definition, and the definitions of unconditional stability, conditional stability, and unconditional instability given in Chapter 6, apply to ODE and PDE solution techniques alike.

Solutions must be **accurate** as well as stable. A method of testing a solver's accuracy is to compare a time-dependent solution from the solver with an exact solution. A **normalized gross error** (NGE) is then calculated as

$$\mathrm{NGE} = \frac{1}{N_{tim}} \sum_{j=1}^{N_{tim}} \left( \frac{1}{K_{s,t_j}} \sum_{i=1}^{K_{s,t_j}} \frac{|N_{i,t_j} - E_{i,t_j}|}{E_{i,t_j}} \right) \times 100\%, \tag{13.3}$$

where $N_{tim}$ is the number of time steps, $N_{i,t_j}$ and $E_{i,t_j}$ are predicted and exact concentrations, respectively, of species $i$ at time $t_j$, and $K_{s,t_j}$ is the number of exact concentrations above a minimum cutoff concentration at time $t_j$. Typical cutoff concentrations for gas chemistry range from $10^{-3}$ to $10^3$ molecules cm$^{-3}$. In (13.3), concentrations may be compared at evenly spaced time intervals, such as every one-half hour, instead of every time step. Exact solutions are obtained by solving the equations with an integrator of known high accuracy or with the given scheme using an extremely small time step. Good solvers of chemical ODEs produce NGEs < 1 percent.

A solver of chemical ODEs should be **mass-conserving**. A scheme is mass-conserving if the mass of each element (e.g., N, O, H, or C) summed over all species at the beginning of a simulation equals the mass of the element summed over all species at the end of the simulation, provided no external sources or sinks exist. Mass conservation cannot occur if individual reactions are not written in a mass-conserving manner. The reaction $NO + O_3 \longrightarrow NO_2 + O_2$ is written in a mass-conserving manner, but the reaction $NO + O_3 \longrightarrow O_2$ is not. If all reactions are mass-conserving, the ODE solution scheme has a chance of being mass-conserving. Explicit ODE solution schemes are usually mass-conserving, since the addition of mass to one species is accompanied by a loss of the same mass from another species. Since an inaccurate explicit scheme can still be mass-conserving, mass conservation does not imply accuracy. Implicit or semiimplicit schemes have the potential to be non-mass-conserving. If such a scheme is accurate, it is more likely to be mass-conserving than if it is inaccurate. **It is more important for a scheme to be accurate than exactly mass-conserving.**

Concentrations predicted by chemical ODE solvers must exceed or equal zero. This should not be a surprise, since in the atmosphere, gas concentrations exceed or equal zero. If a scheme always predicts nonnegative concentrations, it is **positive definite.** An accurate chemical ODE solver is usually positive definite, since correct solutions are always nonnegative. If a solver is unstable and/or inaccurate, concentrations may fall below zero. If at least one concentration falls below zero after a time step, the time step must be resolved with a shorter time step or new solution method. A good solver of chemical ODEs can predict the time step required to keep solutions stable and positive.

When used in atmospheric models, chemical ODE solvers should be **fast** as well as accurate. All chemical ODE solvers are exactly accurate at small enough time steps. A good solver can take long steps and maintain accuracy. For three-dimensional modeling, accuracy with long steps is generally not enough. Accurate solutions to chemical ODEs must be found over large model grids. In the following sections, methods of solving chemical ODEs are discussed in the light of their stability, accuracy, mass conservation, positive-definiteness, and/or speed.

## 13.2. ANALYTICAL SOLUTIONS TO ODEs

The most accurate solution to a set of chemical ODEs is an **analytical solution.** Analytical solutions to a single equation or a small set of equations are readily found. Suppose nitrogen dioxide is lost by the photolysis reaction,

$$\dot{N}O_2 + h\nu \longrightarrow \dot{N}O + \dot{O}\cdot \qquad (J) \tag{13.4}$$

The ODE describing this reaction is

$$\frac{d[NO_2]}{dt} = -J\,[NO_2] \tag{13.5}$$

Integrating (13.5) gives its analytical solution at time $t$ as

$$[NO_2]_t = [NO_2]_{t-h}e^{-Jh} \tag{13.6}$$

This equation states that the nitrogen dioxide concentration decays exponentially if only photolysis is considered.

**Example 13.1**

For an initial concentration of $[NO_2]_{t-h} = 10^{10}$ molecules cm$^{-3}$ and a photolysis coefficient of 0.02 s$^{-1}$, (13.6) reduces to $[NO_2]_t = 10^{10}e^{-0.02h}$, which is an exact solution for the conditions given.

While the solution to (13.5) was easily found, analytical solutions to a set of more than a few equations are usually impractical to obtain. Thus, sets of chemical ordinary differential equations are not solved analytically in atmospheric models.

## 13.3. TAYLOR SERIES SOLUTION TO ODEs

Because analytical solutions are difficult to derive for chemical ODEs, numerical solutions are needed. A useful method of solving sets of ODEs would appear to be an explicit Taylor series expansion of species concentrations. In an **explicit Taylor series expansion**, the concentration of species $i$ at time $t$ is approximated as

$$N_{i,t} = N_{i,t-h} + hN'_{i,t-h} + \frac{h^2}{2}N''_{i,t-h} + \frac{h^3}{6}N'''_{i,t-h} + \cdots \tag{13.7}$$

where $N'_{i,t-h}$, $N''_{i,t-h}$, and $N'''_{i,t-h}$ are the first, second, and third time derivatives, respectively, of the species concentration.

One difficulty with using the Taylor series expansion is that expressions for higher derivatives require combinations of many lower-derivative terms, increasing the computational burden significantly. Suppose three species – NO, $NO_2$, and $O_3$ – are considered. The explicit Taylor series expansion of the NO concentration is

$$[\mathrm{NO}]_t = [\mathrm{NO}]_{t-h} + h\frac{\mathrm{d}[\mathrm{NO}]_{t-h}}{\mathrm{d}t} + \frac{h^2}{2}\frac{\mathrm{d}^2[\mathrm{NO}]_{t-h}}{\mathrm{d}t^2} + \frac{h^3}{6}\frac{\mathrm{d}^3[\mathrm{NO}]_{t-h}}{\mathrm{d}t^3} + \cdots \quad (13.8)$$

Similar equations can be written for $O_3$ and $NO_2$. If one reaction,

$$\dot{\mathrm{N}}\mathrm{O} + \mathrm{O}_3 \longrightarrow \dot{\mathrm{N}}\mathrm{O}_2 + \mathrm{O}_2 \quad (13.9)$$

is considered, the first, second, and third time derivatives of NO concentration arising from this reaction are, respectively,

$$\frac{\mathrm{d}[\mathrm{NO}]}{\mathrm{d}t} = -k_b[\mathrm{NO}][\mathrm{O}_3] \quad (13.10)$$

$$\frac{\mathrm{d}^2[\mathrm{NO}]}{\mathrm{d}t^2} = -k_b\frac{\mathrm{d}[\mathrm{NO}]}{\mathrm{d}t}[\mathrm{O}_3] - k_b[\mathrm{NO}]\frac{\mathrm{d}[\mathrm{O}_3]}{\mathrm{d}t} \quad (13.11)$$

$$\frac{\mathrm{d}^3[\mathrm{NO}]}{\mathrm{d}t^3} = -k_b\frac{\mathrm{d}^2[\mathrm{NO}]}{\mathrm{d}t^2}[\mathrm{O}_3] - 2k_b\frac{\mathrm{d}[\mathrm{NO}]}{\mathrm{d}t}\frac{\mathrm{d}[\mathrm{O}_3]}{\mathrm{d}t} - k_b[\mathrm{NO}]\frac{\mathrm{d}^2[\mathrm{O}_3]}{\mathrm{d}t^2} \quad (13.12)$$

The first, second, and third derivatives of $O_3$ and $NO_2$ are related to those of NO by

$$\frac{\mathrm{d}[\mathrm{O}_3]}{\mathrm{d}t} = -\frac{\mathrm{d}[\mathrm{NO}_2]}{\mathrm{d}t} = \frac{\mathrm{d}[\mathrm{NO}]}{\mathrm{d}t} \qquad \frac{\mathrm{d}^2[\mathrm{O}_3]}{\mathrm{d}t^2} = -\frac{\mathrm{d}^2[\mathrm{NO}_2]}{\mathrm{d}t^2} = \frac{\mathrm{d}^2[\mathrm{NO}]}{\mathrm{d}t^2} \quad (13.13)$$
$$\frac{\mathrm{d}^3[\mathrm{O}_3]}{\mathrm{d}t^3} = -\frac{\mathrm{d}^3[\mathrm{NO}_2]}{\mathrm{d}t^3} = \frac{\mathrm{d}^3[\mathrm{NO}]}{\mathrm{d}t^3}$$

respectively. To solve for changes in concentration due to one reaction, many terms must be calculated. With the addition of more reactions, the number of terms increases.

A second problem with the explicit Taylor series method is that stiffness of the system of equations prevents a stable and accurate solution unless the time step used is small or the order of approximation is large. For these requirements to be met, an explicit Taylor series expansion requires many short time steps, many high-order terms, or both. In either case, the number of computations required to maintain stability in a model with $10^3$–$10^6$ grid cells is large.

## 13.4. FORWARD EULER SOLUTION TO ODEs

If concentrations are approximated with the first two terms of an explicit Taylor series expansion, the approximation is first order. The **order of approximation** was defined in Chapter 6 as the order of the lowest-order term in a Taylor series expansion that is neglected when the approximation is made. A first-order approximation of the explicit Taylor series expansion is the **forward Euler** solution, which is also explicit. The forward Euler solution for one species involved in a set of chemical ODEs is

$$N_{i,t} = N_{i,t-h} + hN'_{i,t-h} \tag{13.14}$$

where the first derivative of $N_i$ at time $t - h$ is a function of $N_i$ and all other concentrations at time $t - h$. In other words, $N'_{i,t-h} = f(\hat{N}_{t-h})$.

Suppose the concentrations of NO, $NO_2$, and $O_3$ are affected by the reactions,

$$\dot{N}O + O_3 \longrightarrow \dot{N}O_2 + O_2 \qquad (k_1) \tag{13.15}$$

$$\dot{N}O_2 + O_3 \longrightarrow N\dot{O}_3 + O_2 \qquad (k_2) \tag{13.16}$$

$$\dot{N}O_2 + h\nu \longrightarrow \dot{N}O + \dot{O}\cdot \qquad (J) \tag{13.17}$$

For these equations, the explicit first derivative of $NO_2$ is

$$\begin{aligned} N'_{NO_2,t-h} = {} & k_1[NO]_{t-h}[O_3]_{t-h} \\ & - k_2[NO_2]_{t-h}[O_3]_{t-h} - J[NO_2]_{t-h} \end{aligned} \tag{13.18}$$

Substituting (13.18) into (13.14) gives

$$\begin{aligned} [NO_2]_t = {} & [NO_2]_{t-h} + h(k_1[NO]_{t-h}[O_3]_{t-h} \\ & - k_2[NO_2]_{t-h}[O_3]_{t-h} - J[NO_2]_{t-h}) \end{aligned} \tag{13.19}$$

which is the forward Euler $NO_2$ concentration after one time step. This solution can be generalized by defining total production and loss rates, respectively, of $NO_2$ as

$$P_{c,NO_2,t-h} = k_1[NO]_{t-h}[O_3]_{t-h} \tag{13.20}$$

$$L_{c,NO_2,t-h} = k_2[NO_2]_{t-h}[O_3]_{t-h} + J[NO_2]_{t-h} \tag{13.21}$$

Substituting these terms into (13.19) gives another form of the forward Euler solution,

$$[NO_2]_t = [NO_2]_{t-h} + h(P_{c,NO_2,t-h} - L_{c,NO_2,t-h}) \tag{13.22}$$

This notation can be generalized for any species $i$ as

$$\boxed{N_{i,t} = N_{i,t-h} + h(P_{c,i,t-h} - L_{c,i,t-h})} \tag{13.23}$$

An advantage of the forward Euler solution, like the explicit Taylor series expansion, is that it is exactly mass-conserving. The forward Euler solution requires minimal time for computing production and loss terms. Because the order of approximation of the forward Euler solution is low, the time step required for maintaining stability is small, and many time steps are needed to complete a simulation. Because of this problem, the forward Euler approximation is not used on its own to solve atmospheric chemistry problems.

## 13.5. BACKWARD EULER SOLUTION TO ODEs

An approximation that avoids the instability of the forward Euler solution is the **linearized backward Euler** solution. A linearized solution is one in which concentrations of all species, except the species being solved for, are set to concentrations from previous time steps in derivative terms. The concentration of the species of interest is assumed to be unknown when used in a derivative. The linearized backward Euler solution for a species is

$$N_{i,t} = N_{i,t-h} + hN'_{i,t,t-h} \tag{13.24}$$

where the first derivative is evaluated for all species, except for the species of interest, at time $t-h$. Thus, $N'_{i,t,t-h} = f(N_{i,t}, \hat{N}_{t-h})$.

To illustrate, reactions (13.15)–(13.17) are reused. The **linearized first derivative** of $NO_2$ from these equations is

$$\begin{aligned} N'_{NO_2,t,t-h} = {} & k_1[NO]_{t-h}[O_3]_{t-h} \\ & - k_2[NO_2]_t[O_3]_{t-h} - J[NO_2]_t \end{aligned} \tag{13.25}$$

where $NO_2$ is evaluated at time $t$ in derivative terms. Substituting (13.25) into (13.24) gives

$$\begin{aligned} [NO_2]_t = {} & [NO_2]_{t-h} + h(k_1[NO]_{t-h}[O_3]_{t-h} \\ & - k_2[NO_2]_t[O_3]_{t-h} - J[NO_2]_t) \end{aligned} \tag{13.26}$$

The production and loss terms in this equation are

$$P_{c,NO_2,t-h} = k_1[NO]_{t-h}[O_3]_{t-h} \tag{13.27}$$

$$L_{c,NO_2,t,t-h} = k_2[NO_2]_t[O_3]_{t-h} + J[NO_2]_t \tag{13.28}$$

respectively. Dividing the loss term by $[NO_2]_t$ gives the **implicit loss coefficient,**

$$\Lambda_{c,NO_2,t-h} = \frac{L_{c,NO_2,t,t-h}}{[NO_2]_t} = k_2[O_3]_{t-h} + J \tag{13.29}$$

Substituting (13.27) and (13.29) back into (13.26) yields

$$[NO_2]_t = [NO_2]_{t-h} + h(P_{c,NO_2,t-h} - \Lambda_{c,NO_2,t-h}[NO_2]_t) \tag{13.30}$$

Solving this equation for $[NO_2]_t$ gives

$$[NO_2]_t = \frac{[NO_2]_{t-h} + hP_{c,NO_2,t-h}}{1 + h\Lambda_{c,NO_2,t-h}} \tag{13.31}$$

which is the linearized backward Euler concentration of $NO_2$ after one time step. The

**generalized, linearized backward Euler solution** for species $i$ is

$$N_{i,t} = \frac{N_{i,t-h} + hP_{c,i,t-h}}{1 + h\Lambda_{c,i,t-h}} \tag{13.32}$$

The advantage of the linearized backward Euler solution is that species concentrations cannot fall below zero. The disadvantage is that the method is not mass-conserving, because each reaction rate is linearized differently for different species. When ozone and nitric oxide are lost and nitrogen dioxide is produced, as in (13.15), the linearized rates are

$$\text{Rate} = k_1[\text{NO}]_{t-h}[\text{O}_3]_t \qquad (\text{O}_3 \text{ loss}) \tag{13.33}$$

$$\text{Rate} = k_1[\text{NO}]_t[\text{O}_3]_{t-h} \qquad (\text{NO loss}) \tag{13.34}$$

$$\text{Rate} = k_1[\text{NO}]_{t-h}[\text{O}_3]_{t-h} \qquad (\text{NO}_2 \text{ production}) \tag{13.35}$$

Although the rate of loss of NO should equal the rates of loss of $O_3$ and production of $NO_2$ to ensure mass conservation, (13.33)–(13.35) show that the rates differ in each case. Thus, mass is not conserved. Because mass is not conserved with the linearized backward Euler solution, errors accumulate over time.

## 13.6. SIMPLE EXPONENTIAL AND QUASI-STEADY-STATE SOLUTIONS TO ODEs

A method of solving ODEs, similar to the backward Euler, is the **simple exponential approximation.** This solution is obtained by integrating a linearized first derivative. The advantage of the simple exponential approximation is that resulting species concentrations cannot fall below zero. The disadvantage is that, like the backward Euler approximation, it is not mass-conserving and requires a small time step for accuracy to be maintained. Thus, the approximation is conditionally stable.

The simple exponential solution is illustrated with the linearized first derivative from the backward Euler example. Writing (13.30) in differential form gives

$$\frac{d[\text{NO}_2]_t}{dt} = P_{c,\text{NO}_2,t-h} - \Lambda_{c,\text{NO}_2,t-h}[\text{NO}_2]_t \tag{13.36}$$

Integrating this equation gives the **simple exponential solution,**

$$[\text{NO}_2]_t = [\text{NO}_2]_{t-h}e^{-h\Lambda_{c,\text{NO}_2,t-h}} + \frac{P_{c,\text{NO}_2,t-h}}{\Lambda_{c,\text{NO}_2,t-h}}\left(1 - e^{-h\Lambda_{c,\text{NO}_2,t-h}}\right) \tag{13.37}$$

which is written for any species $i$ as

$$N_{i,t} = N_{i,t-h}e^{-h\Lambda_{c,i,t-h}} + \frac{P_{c,i,t-h}}{\Lambda_{c,i,t-h}}\left(1 - e^{-h\Lambda_{c,i,t-h}}\right) \tag{13.38}$$

When the implicit loss coefficient is zero (no loss), (13.38) simplifies to $N_{i,t} = N_{i,t-h} + hP_{c,i,t-h}$. When the production term is zero, the solution simplifies to $N_{i,t} = N_{i,t-h}e^{-h\Lambda_{c,i,t-h}}$. When the implicit loss coefficient is large (short-lived species), the solution simplifies to $N_{i,t} = P_{c,i,t-h}/\Lambda_{c,i,t-h}$, which is the steady-state solution to the first-derivative equation.

The simple exponential solution is similar to the backward Euler solution, except that the former contains an analytical component and is slightly more accurate than the latter. Because the simple exponential approximation requires the computation of an exponent, it requires much more computer time, per time step, than does the backward Euler. The simple exponential approximation also requires the use of a small time step for the solution to be accurate.

A method related to the simple exponential is the **quasi-steady-state approximation** (QSSA) method (Hesstvedt et al. 1978). With this method, the parameter $h\Lambda_{c,i,t-h}$ is used to determine how a species concentration is solved during a time step. When $h\Lambda_{c,i,t-h} < 0.01$, the species' lifetime is long, and its final concentration is calculated with the forward Euler equation. When $0.01 \le h\Lambda_{c,i,t-h} \le 10$, the species' lifetime is moderate, and its concentration is calculated with the simple exponential method. When $h\Lambda_{c,i,t-h} > 10$, the species' lifetime is short, and its concentration is calculated with the steady-state equation. In sum,

$$N_{i,t} = \begin{cases} N_{i,t-h} + h(P_{c,i,t-h} - L_{c,i,t-h}) & h\Lambda_{c,i,t-h} < 0.01 \\ N_{i,t-h}e^{-h\Lambda_{c,i,t-h}} + \frac{P_{c,i,t-h}}{\Lambda_{c,i,t-h}}\left(1 - e^{-h\Lambda_{c,i,t-h}}\right) & 0.01 \le h\Lambda_{c,i,t-h} \le 10 \\ P_{c,i,t-h}/\Lambda_{c,i,t-h} & h\Lambda_{c,i,t-h} > 10 \end{cases} \quad (13.39)$$

Final concentrations for a time step are found after several iterations of (13.39). The iterative QSSA scheme can be accurate for many cases, but it is mass-conserving for only long-lived species.

In sum, the forward Euler scheme is mass-conserving, but stable for small time steps only; the backward Euler and simple exponential schemes are more stable, but not mass-conserving; and the iterative QSSA scheme is often stable, but generally not mass-conserving.

## 13.7. MULTISTEP IMPLICIT–EXPLICIT (MIE) SOLUTION TO ODEs

An iterative technique that takes advantage of the forward and backward Euler methods is the **multistep implicit–explicit** (MIE) method (Jacobson 1994). With this method, concentrations are estimated with an iterated backward Euler calculation, and the estimates are applied to reaction rates used in a forward Euler calculation of final concentrations. Upon iteration, the forward Euler solutions converge to backward Euler solutions. Since backward Euler solutions are always positive, the forward Euler solutions must converge to positive values as well. A technique related to the MIE method is merely iterating the backward Euler equation until convergence occurs for all species. Such a method requires significantly more iterations than does the MIE method.

The steps for determining final concentrations with the MIE method are described below. To illustrate, four species, NO, $NO_2$, $O_3$, and O, and two reactions,

$$\dot{N}O + O_3 \longrightarrow \dot{N}O_2 + O_2 \qquad (k_1) \tag{13.40}$$
$$O_3 + h\nu \longrightarrow O_2 + \dot{O}\cdot \qquad (J) \tag{13.41}$$

are considered. The change in $O_2$ concentration by these reactions is ignored.

The first step in the MIE solution is to set an initial backward Euler estimate and initial maximum estimate of each active species concentration (molecules cm$^{-3}$) equal to the initial concentration of the species. Thus,

$$N_{i,B,1} = N_{i,t-h} \qquad N_{i,\mathrm{MAX},1} = N_{i,t-h} \tag{13.42}$$

where $i$ is the species number, and the subscript $B, 1$ identifies the first estimate of the backward-Euler-method concentration. A maximum estimate is required to prevent estimated concentrations from blowing up to large values upon iteration and is updated each iteration.

Reaction rates are estimated by multiplying rate coefficients by backward-Euler-method concentrations. Examples of a two-body and a photolysis reaction rate are

$$R_{c,n,B,m} = k_n N_{i,B,m} N_{j,B,m} \qquad R_{c,n,B,m} = J_n N_{i,B,m} \tag{13.43}$$

respectively, where $R_c$ is the rate of reaction (molecules cm$^{-3}$ s$^{-1}$), $n$ is the reaction-rate number, $m$ is the iteration number, $k_n$ is the kinetic rate coefficient of the $n$th reaction, and $J_n$ is the photolysis rate coefficient of the $n$th reaction. In the two-reaction example, the backward-Euler-method rates are

$$R_{c,1,B,m} = k_1 [\mathrm{NO}]_{B,m} [\mathrm{O_3}]_{B,m} \qquad R_{c,2,B,m} = J\,[\mathrm{O_3}]_{B,m} \tag{13.44}$$

respectively. In (13.42), $m = 1$. In all subsequent equations, $m \geq 1$.

Estimated production rates, loss rates, and implicit loss coefficients are summed for each species from the reaction rates just calculated. The summed **backward-Euler-method chemical production rate** (molecules cm$^{-3}$ s$^{-1}$) of species $i$ is

$$P_{c,i,B,m} = \sum_{l=1}^{N_{prod,i}} R_{c,n_P(l,i),B,m} \tag{13.45}$$

where $N_{prod,i}$ is the number of reactions in which species $i$ is produced, and $R_{c,n_P(l,i),B,m}$ is the $l$th backward Euler production rate of species $i$. The array $n_P(l, i)$ gives the reaction number corresponding to the $l$th production term of species $i$. In the two-reaction example, the summed production rates are

$$P_{c,\mathrm{O},B,m} = J\,[\mathrm{O_3}]_{B,m} \qquad P_{c,\mathrm{NO_2},B,m} = k_1 [\mathrm{NO}]_{B,m} [\mathrm{O_3}]_{B,m} \tag{13.46}$$

where the two active species produced have only one production term each.

The **backward-Euler-method total chemical loss rate** (molecules cm$^{-3}$ s$^{-1}$) of a species is

$$L_{c,i,B,m} = \sum_{l=1}^{N_{loss,i}} R_{c,n_L(l,i),B,m} \tag{13.47}$$

where $N_{loss,i}$ is the number of reactions in which species $i$ is lost, and $R_{c,n_L(l,i),B,m}$ is the $l$th backward Euler loss rate of species $i$. The array $n_L(l,i)$ gives the reaction number corresponding to the $l$th loss term of species $i$. In the two-reaction example, the summed loss rates are

$$L_{c,\mathrm{NO},B,m} = k_1[\mathrm{NO}]_{B,m}[\mathrm{O_3}]_{B,m} \tag{13.48}$$
$$L_{c,\mathrm{O_3},B,m} = k_1[\mathrm{NO}]_{B,m}[\mathrm{O_3}]_{B,m} + J[\mathrm{O_3}]_{B,m}$$

The **backward-Euler-method implicit loss coefficient** (s$^{-1}$) of a species is

$$\Lambda_{c,i,B,m} = \frac{L_{c,i,B,m}}{N_{i,B,m}} \tag{13.49}$$

In the two-reaction example, the implicit loss coefficients are

$$\Lambda_{c,\mathrm{NO},B,m} = k_1[\mathrm{O_3}]_{B,m} \qquad \Lambda_{c,\mathrm{O_3},B,m} = k_1[\mathrm{NO}]_{B,m} + J \tag{13.50}$$

Since a reactant concentration can equal zero, computing implicit loss coefficient with (13.49) can result in a division by zero. If the coefficients are, instead, computed as in (13.50), division by zero is avoided.

Concentrations at iteration $m+1$ are estimated with the **backward Euler approximation,**

$$N_{i,B,m+1} = \frac{N_{i,t-h} + hP_{c,i,B,m}}{1 + h\Lambda_{c,i,B,m}} \tag{13.51}$$

Such estimates are used during the next iteration to calculate production and loss terms.

The **forward Euler approximation** is used to estimate final concentrations and to determine whether convergence has occurred. The forward Euler equation for one species is

$$N_{i,F,m+1} = N_{i,t-h} + h(P_{c,i,B,m} - L_{c,i,B,m}) \tag{13.52}$$

Convergence is now checked. **Convergence** is determined by first initializing a counter to zero before the first iteration of a time step. At the end of each iteration during the time step, the forward-Euler-method concentrations from (13.52) are checked to determine whether they all exceed or equal zero. If they do, the counter is incremented. If all concentrations $\geq 0$ for $N_P$ iterations in a row, convergence is said to have occurred. $N_P$ is a constant that depends on the number of reactions and their stiffness. Typical values are as low as 5 for large sets of equations and 30–50 for small sets for equations. If a single forward-Euler-method concentration $<0$ after a given iteration, the counter is reset to zero and iterations must continue until all concentrations $\geq 0$ for an additional $N_P$ iterations in a row.

For faster solutions, the criteria is modified so that, when $h\Lambda_{c,i,B,m} \geq L_T$ for a species during an iteration, the forward-Euler-method concentration of the species does not need to exceed zero for the counter to avoid being reset to zero. In such cases, the species is short-lived, and the backward Euler solution is more accurate than the forward Euler solution. $L_T$ is a constant between $10^2$–$10^6$. Values of $10^2$ speed solutions but increase errors. Values of $10^6$ slow solutions but decrease errors. In sum, the second convergence criterion requires that forward Euler concentrations of all species, except those that satisfy $h\Lambda_{c,i,B,m} \geq L_T$, exceed or equal zero for $N_P$ iterations in a row. When the modified criterion is met, the final concentrations are

$$N_{i,t} = \begin{cases} N_{i,B,m+1} & \text{(backward Euler)} \quad h\Lambda_{c,i,B,m} \geq L_T \\ N_{i,F,m+1} & \text{(forward Euler)} \quad h\Lambda_{c,i,B,m} < L_T \end{cases} \tag{13.53}$$

If the modified convergence criterion is not met, iterations continue. Before iterations continue, maximum estimated concentrations for each species are updated, and estimated concentrations are limited to the smaller of the current estimate and the maximum estimate from the last iteration. In other words,

$$N_{i,\mathrm{MAX},m+1} = \max(N_{i,B,m+1}, N_{i,t-h}) \tag{13.54}$$

$$N_{i,B,m+1} = \min(N_{i,B,m+1}, N_{i,\mathrm{MAX},m}) \tag{13.55}$$

for each species. Note that the value of $N_{i,\mathrm{MAX}}$ used in (13.55) is from the previous iteration while that calculated in (13.54) is for the current iteration. After these updates, the iteration returns to (13.43) until the second convergence criterion is satisfied.

Here, it is shown that **iterated backward Euler solutions converge to iterated forward Euler solutions and positive numbers**. At the end of any iteration, the backward-Euler-method concentration is

$$N_{i,B,m+1} = N_{i,t-h} + h(P_{c,i,B,m} - \Lambda_{c,i,B,m}N_{i,B,m+1}) \tag{13.56}$$

which is simply (13.51) rearranged. (13.56) must be positive, since (13.51) cannot be negative. At the end of the same iteration, the forward Euler method yields

$$N_{i,F,m+1} = N_{i,t-h} + h(P_{c,i,B,m} - \Lambda_{c,i,B,m}N_{i,B,m}) \tag{13.57}$$

where $\Lambda_{c,i,B,m}N_{i,B,m} = L_{c,i,B,m}$. For the forward and backward Euler solutions to converge to each other,

$$N_{i,B,m+1} = N_{i,B,m} \tag{13.58}$$

must be satisfied. This always occurs upon iteration of the backward Euler solution.

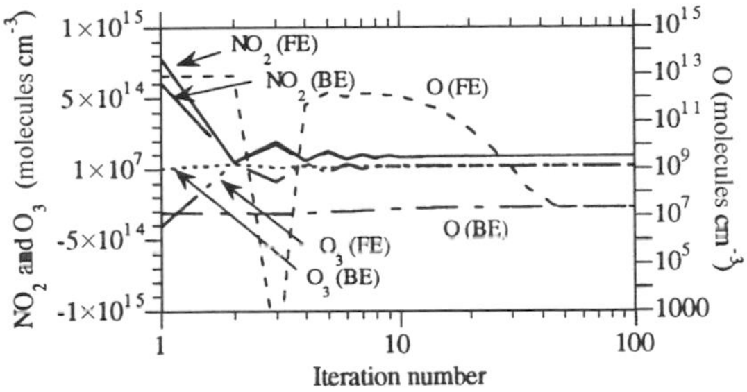


**Figure 13.1.** Concentrations of three species iterated until convergence with forward Euler (FE) and backward Euler (BE) calculations during a MIE simulation where $h = 10$ s.

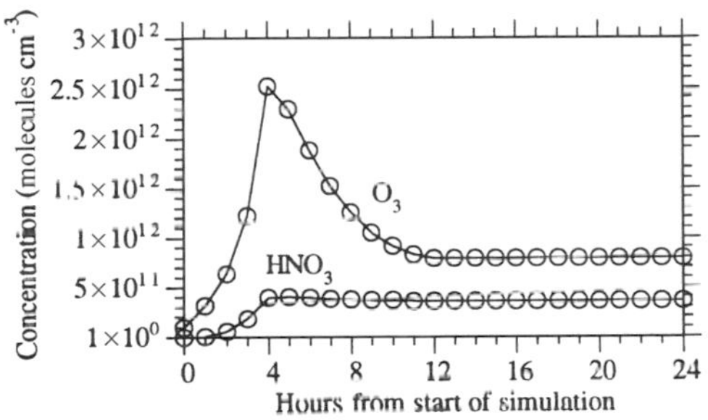


**Figure 13.2.** Comparison of MIE solution (circles) with an exact solution (lines) for two out of 92 species during a 24-h simulation period. The time step taken with MIE was 10 s. Photolysis rates changed every one-half hour during the first 12 h and were zero during the second 12 h. MIE and exact solutions are shown each hour.

Thus, when the backward Euler solution converges, the forward Euler solution must converge to the same value, and both converge to positive numbers.

Figure 13.1 shows convergence of forward Euler to backward Euler concentrations versus iteration number for four species. Figure 13.2 compares MIE simulation results with an exact solution for two species when a set of 92 ODEs was solved simultaneously. Figure 13.3 shows how a change in the time step affected the number of iterations and accuracy of the MIE solution in the 92-ODE case.

## 13.8. GEAR'S SOLUTION TO ODEs

A more advanced method of solving stiff chemical ODEs is **Gear's method** (Gear 1971). Gear's method is accurate and elegant. A drawback of the original method was its need to solve equations containing large matrices of partial derivatives. This drawback prevented the use of the original code in three-dimensional models. The application of **sparse-matrix** techniques improved the speed of Gear's code

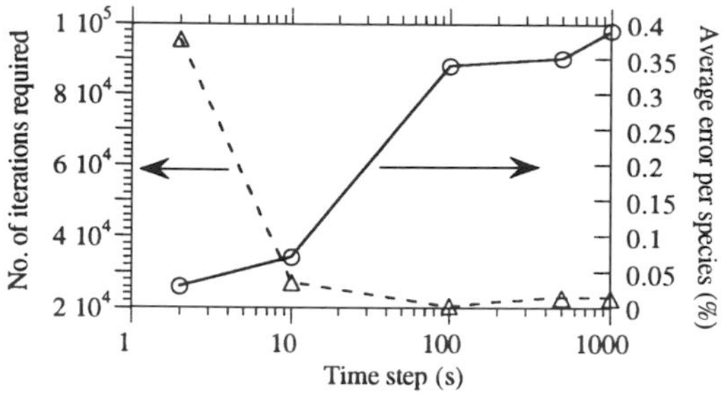


**Figure 13.3.** Effects of changing $h$ with the MIE method when a set of 92 ODEs were solved. Time steps of $h = 2$, 10, 100, 500, and 1000 s are compared. The simulation time interval was 10,000 s. Small time steps resulted in a low normalized gross error after 10,000 s but required more iterations to complete the interval.

**Table 13.1.** Values of $\alpha$ and $\beta$ for Use in (13.59)

| $s$ | $\alpha_{s,1}$ | $\alpha_{s,2}$ | $\alpha_{s,3}$ | $\alpha_{s,4}$ | $\alpha_{s,5}$ | $\alpha_{s,6}$ | $\beta_s$ |
|---|---|---|---|---|---|---|---|
| 1 | 1 | | | | | | 1 |
| 2 | 1 | 1/3 | | | | | 2/3 |
| 3 | 1 | 6/11 | 1/11 | | | | 6/11 |
| 4 | 1 | 35/50 | 10/50 | 1/50 | | | 24/50 |
| 5 | 1 | 225/274 | 85/274 | 15/274 | 1/274 | | 120/274 |
| 6 | 1 | 1624/1764 | 735/1764 | 175/1764 | 21/1764 | 1/1764 | 720/1764 |

tremendously (e.g., Sherman and Hindmarsh 1980; Hindmarsh 1983; Jacobson 1994, 1995, 1998a; Jacobson and Turco 1994), but the use of sparse-matrix techniques alone still does not permit the use of Gear's code in three-dimensional models. Only when sparse-matrix techniques are combined with computer optimization techniques can Gear's code be used to study complex problems in three dimensions (Jacobson 1994, 1995, 1998a; Jacobson and Turco 1994).

Gear's method solves the **backward differentiation formula** (BDF). The BDF is obtained by first discretizing the time derivative of an ODE with

$$\frac{dN_{i,t}}{dt} \approx \frac{N_{i,t} - \alpha_{s,1}N_{i,t-h} - \alpha_{s,2}N_{i,t-2h} \cdots - \alpha_{s,s}N_{i,t-sh}}{h\beta_s} = \frac{N_{i,t} - \sum_{j=1}^{s} \alpha_{s,j}N_{i,t-jh}}{h\beta_s} \quad (13.59)$$

where $s$ is the order of approximation of the method, and $\alpha$ and $\beta$ are scalar multipliers, given in Table 13.1, that depend on the order of the method. Rearranging

(13.59) for an individual species and set of species gives

$$0 = -N_{i,t} + \sum_{j=1}^{s} \alpha_{s,j} N_{i,t-jh} + h\beta_s \frac{dN_{i,t}}{dt} \tag{13.60}$$

$$0 = -\hat{N}_t + \sum_{j=1}^{s} \alpha_{s,j} \hat{N}_{t-jh} + h\beta_s \frac{d\hat{N}_t}{dt}$$

respectively. The solution to (13.60) is found by writing

$$\mathbf{P}_t(\hat{N}_{t,m+1} - \hat{N}_{t,m}) = -\hat{N}_{t,m} + \sum_{j=1}^{s} \alpha_{s,j} \hat{N}_{t-jh} + h\beta_s \hat{f}(\hat{N}_{t,m}) \tag{13.61}$$

where $\mathbf{P}_t$ is a **predictor matrix** at time $t$, $m$ is the **iteration number**, and $\hat{f}(\hat{N}_{t,m}) = d\hat{N}_{t,m}/dt$ is the first derivative of $N$ at iteration $m$. (13.61) can be rewritten as

$$\mathbf{P}_t \, \Delta\hat{N}_{t,m} = \hat{B}_{t,m} \tag{13.62}$$

where $\Delta\hat{N}_{t,m} = \hat{N}_{t,m+1} - \hat{N}_{t,m}$ and

$$\hat{B}_{t,m} = -\hat{N}_{t,m} + \sum_{j=1}^{s} \alpha_{s,j} \hat{N}_{t-jh} + h\beta_s \hat{f}(\hat{N}_{t,m}) \tag{13.63}$$

$\Delta\hat{N}_{t,m}$ is found from (13.62) in each iteration by matrix decomposition and backsubstitution. After each iteration, concentrations in (13.63) are updated with

$$\hat{N}_{t,m+1} = \hat{N}_{t,m} + \Delta\hat{N}_{t,m} \tag{13.64}$$

Upon iteration, $\Delta\hat{N}_{t,m}$ approaches zero. Convergence is checked with a local error test after each iteration within a time step and with a global error test after the completion of all iterations at the end of the step. In the **local error test**, a normalized root-mean-square error, determined as

$$NRMS_{t,m} = \sqrt{\frac{1}{K} \sum_{i=1}^{K} \left( \frac{\Delta N_{i,t,m}}{R_{tol} N_{i,t,1} + A_{tol,t}} \right)^2} \tag{13.65}$$

is compared with a parameter that varies with the order of approximation and iteration number. In (13.65), $\Delta N_{i,t,m}$ (molecules cm$^{-3}$) is the change in concentration of species $i$ at time step $t$ during iteration $m$, $K$ is the number of species, $R_{tol}$ is a constant **relative error tolerance**, $A_{tol,t}$ is an **absolute error tolerance** for time step $t$ (molecules cm$^{-3}$), and $N_{i,t,1}$ is a concentration at the start of a time step. The relative error tolerance controls errors relative to $N_{i,t,1}$, and the absolute error tolerance controls errors relative to fixed concentrations. For pure relative-error-tolerance control, $A_{tol,t} = 0$. For pure absolute-error-tolerance control, $R_{tol} = 0$.

Typically, $R_{tol} = 10^{-3}$. For gas chemistry, $A_{tol,t} \approx 10^3$–$10^7$ molecules cm$^{-3}$. Proper selection or prediction of $A_{tol,t}$ reduces the computer time in large models. Jacobson (1998a) gives a method of predicting $A_{tol,t}$.

If the local error test fails but $NRMS_{t,m}$ is decreasing relative to $NRMS_{t,m-1}$, iterations continue until the local error test succeeds. If $NRMS_{t,m}$ is increasing upon iteration, convergence is not occurring, and the matrix of partial derivatives is updated before iterations continue. If the convergence test fails again, the time step is reduced until the test succeeds. Once local convergence has occurred, a **global error test** is performed to check whether a cumulative normalized root-mean-square error, determined as

$$NRMS_t = \sqrt{\frac{1}{K}\sum_{i=1}^{K}\left(\frac{\sum_m \Delta N_{i,t,m}}{R_{tol}N_{i,t,1} + A_{tol,t}}\right)^2} \qquad (13.66)$$

exceeds a parameter value that depends on the order of approximation. In this equation, $\sum_m \Delta N_{i,t,m}$ is the net change in species concentration during a time step. If the global error check fails, a new time step is predicted at the same or one order lower approximation. If the error check continues to fail, the time step is reduced. If the global test succeeds, the time step was successful, and $\hat{N}_{i,m+1}$-values from the last iteration are set to final concentrations. Every few successful time steps, the time step and order of approximation are recalculated with a time-step estimation scheme.

The predictor matrix in (13.62) expands to

$$\mathbf{P}_t \approx I - h\beta_s J_t \qquad (13.67)$$

where $I$ is the identity matrix, $J_t$ is a Jacobian matrix of partial derivatives, and $\beta_s$, which depends on the order of the method, was defined earlier. The **Jacobian matrix** is

$$J_t = J(\hat{N}_{t,m}) = \left[\frac{\partial f_{i,t,m}}{\partial N_{k,t,m}}\right]_{i,k=1}^{K,K} \qquad (13.68)$$

where $K$ is the order of the matrix (number of species and ODEs). Matrices of partial derivatives for chemical ODEs are sparse, and sparse-matrix techniques are useful for treating them. Below, a matrix from a small equation set is shown, and a technique of reducing matrix computations is discussed.

The reactions used for the example matrix are

$$\dot{N}O + O_3 \longrightarrow \dot{N}O_2 + O_2 \qquad (k_1) \qquad (13.69)$$

$$\dot{O}\cdot + O_2 + M \longrightarrow O_3 + M \qquad (k_2) \qquad (13.70)$$

$$\dot{N}O_2 + h\nu \longrightarrow \dot{N}O + \dot{O}\cdot \qquad (J) \qquad (13.71)$$

The corresponding first derivatives of NO, $NO_2$, O, and $O_3$ are

$$\begin{aligned}
\frac{d[NO]}{dt} &= J[NO_2] - k_1[NO][O_3] \\
\frac{d[NO_2]}{dt} &= k_1[NO][O_3] - J[NO_2] \\
\frac{d[O]}{dt} &= J[NO_2] - k_2[O][O_2][M] \\
\frac{d[O_3]}{dt} &= k_2[O][O_2][M] - k_1[NO][O_3]
\end{aligned} \tag{13.72}$$

The corresponding partial derivatives are as follows:

*Partial derivatives of* NO

$$\frac{\partial^2[NO]}{\partial[NO]\partial t} = -k_1[O_3] \qquad \frac{\partial^2[NO]}{\partial[NO_2]\partial t} = J \qquad \frac{\partial^2[NO]}{\partial[O_3]\partial t} = -k_1[NO] \tag{13.73}$$

*Partial derivatives of* $NO_2$

$$\frac{\partial^2[NO_2]}{\partial[NO]\partial t} = k_1[O_3] \qquad \frac{\partial^2[NO_2]}{\partial[NO_2]\partial t} = -J \qquad \frac{\partial^2[NO_2]}{\partial[O_3]\partial t} = k_1[NO] \tag{13.74}$$

*Partial derivatives of* O

$$\frac{\partial^2[O]}{\partial[NO_2]\partial t} = J \qquad \frac{\partial^2[O]}{\partial[O]\partial t} = -k_2[O_2][M] \tag{13.75}$$

*Partial derivatives of* $O_3$

$$\begin{aligned}
&\frac{\partial^2[O_3]}{\partial[NO]\partial t} = -k_1[O_3] \qquad \frac{\partial^2[O_3]}{\partial[O]\partial t} = k_2[O_2][M] \\
&\frac{\partial^2[O_3]}{\partial[O_3]\partial t} = -k_1[NO]
\end{aligned} \tag{13.76}$$

All other partial derivatives are zero.

For the set of four active species, the predictor matrix ($\mathbf{P}_t$) has the form

$$\begin{array}{c} \\ NO \\ NO_2 \\ O \\ O_3 \end{array}
\begin{array}{cccc} NO & NO_2 & O & O_3 \end{array}$$

$$\begin{array}{c} NO \\ NO_2 \\ O \\ O_3 \end{array}
\begin{bmatrix}
1 - h\beta_s \frac{\partial^2[NO]}{\partial[NO]\,\partial t} & -h\beta_s \frac{\partial^2[NO]}{\partial[NO_2]\,\partial t} & -h\beta_s \frac{\partial^2[NO]}{\partial[O]\,\partial t} & -h\beta_s \frac{\partial^2[NO]}{\partial[O_3]\,\partial t} \\
-h\beta_s \frac{\partial^2[NO_2]}{\partial[NO]\,\partial t} & 1 - h\beta_s \frac{\partial^2[NO_2]}{\partial[NO_2]\,\partial t} & -h\beta_s \frac{\partial^2[NO_2]}{\partial[O]\,\partial t} & -h\beta_s \frac{\partial^2[NO_2]}{\partial[O_3]\,\partial t} \\
-h\beta_s \frac{\partial^2[O]}{\partial[NO]\,\partial t} & -h\beta_s \frac{\partial^2[O]}{\partial[NO_2]\,\partial t} & 1 - h\beta_s \frac{\partial^2[O]}{\partial[O]\,\partial t} & -h\beta_s \frac{\partial^2[O]}{\partial[O_3]\,\partial t} \\
-h\beta_s \frac{\partial^2[O_3]}{\partial[NO]\,\partial t} & -h\beta_s \frac{\partial^2[O_3]}{\partial[NO_2]\,\partial t} & -h\beta_s \frac{\partial^2[O_3]}{\partial[O]\,\partial t} & 1 - h\beta_s \frac{\partial^2[O_3]}{\partial[O_3]\,\partial t}
\end{bmatrix} \tag{13.77}$$

Substituting the partial derivatives from (13.73)–(13.76) into (13.77) gives

$$\begin{array}{c} \\ NO \\ NO_2 \\ O \\ O_3 \end{array}\begin{array}{c} \begin{array}{cccc} NO & NO_2 & O & O_3 \end{array} \\ \begin{bmatrix} 1-h\beta_s(-k_1[O_3]) & -h\beta_s(J) & 0 & -h\beta_s(-k_1[NO]) \\ -h\beta_s(k_1[O_3]) & 1-h\beta_s(-J) & 0 & -h\beta_s(k_1[NO]) \\ 0 & -h\beta_s(J) & 1-h\beta_s(-k_2[O_2][M]) & 0 \\ -h\beta_s(-k_1[O_3]) & 0 & -h\beta_s(k_2[O_2][M]) & 1-h\beta_s(-k_1[NO]) \end{bmatrix}\end{array} \quad (13.78)$$

In (13.78), 5 out of the 16 matrix positions contain zeros. Thus, the initial matrix fill-in is $\frac{11}{16}$. For larger chemical reaction sets, the order of the matrix is larger, but the initial percentage fill-in is much smaller. A typical set of 90 species and 200 chemical reactions has an initial fill-in of approximately 8 percent. The methodology of implementing a sparse-matrix technique can be applied to a matrix with any percentage of initial fill-in.

Substituting (13.78) into (13.62) gives

$$\begin{bmatrix} 1-h\beta_s(-k_1[O_3]) & -h\beta_s(J) & 0 & -h\beta_s(-k_1[NO]) \\ -h\beta_s(k_1[O_3]) & 1-h\beta_s(-J) & 0 & -h\beta_s(k_1[NO]) \\ 0 & -h\beta_s(J) & 1-h\beta_s(-k_2[O_2][M]) & 0 \\ -h\beta_s(k_1[O_3]) & 0 & -h\beta_s(k_2[O_2][M]) & 1-h\beta_s(-k_1[NO]) \end{bmatrix} \quad (13.79)$$

$$\times \begin{bmatrix} \Delta[NO]_{t,m} \\ \Delta[NO_2]_{t,m} \\ \Delta[O]_{t,m} \\ \Delta[O_3]_{t,m} \end{bmatrix} = \begin{bmatrix} B_{NO,t,m} \\ B_{NO_2,t,m} \\ B_{O,t,m} \\ B_{O_3,t,m} \end{bmatrix}$$

which is solved either by Gaussian elimination or by matrix decomposition and backsubstitution (e.g., Press et al. 1992). Because Gear's method reuses the same decomposed matrix for several backsubstitutions over the right side of (13.79), **decomposition and backsubstitution is favored over Gaussian elimination.**

During the matrix decomposition of (13.79), many multiplications by zero occur. One sparse-matrix technique is to **remove multiplications by zero** when such multiplications are known in advance. That is not enough. To reduce computer time significantly, the **matrix should be reordered** before multiplications by zero are eliminated. An efficient way of reordering is to place species with the most partial-derivative terms at the bottom and those with the fewest at the top of the matrix (Jacobson 1994). The use of this reordering scheme requires that partial pivoting be avoided during decomposition. Such a restriction is reasonable in that, if the decomposition should fail, the step size in the numerical solver can be reduced. Reordering (13.79) gives

$$\begin{bmatrix} 1-h\beta_s(-k_2[O_2][M]) & -h\beta_s(J) & 0 & 0 \\ 0 & 1-h\beta_s(-J) & -h\beta_s(k_1[O_3]) & -h\beta_s(k_1[NO]) \\ 0 & -h\beta_s(J) & 1-h\beta_s(-k_1[O_3]) & -h\beta_s(-k_1[NO]) \\ -h\beta_s(k_2[O_2][M]) & 0 & -h\beta_s(-k_1[O_3]) & 1-h\beta_s(-k_1[NO]) \end{bmatrix} \quad (13.80)$$

$$\times \begin{bmatrix} \Delta[O]_{t,m} \\ \Delta[NO_2]_{t,m} \\ \Delta[NO]_{t,m} \\ \Delta[O_3]_{t,m} \end{bmatrix} = \begin{bmatrix} B_{O,t,m} \\ B_{NO_2,t,m} \\ B_{NO,t,m} \\ B_{O_3,t,m} \end{bmatrix}$$

**Table 13.2.** Reduction in Array Space and Matrix Operations Before and After Reordering and Removing Multiplication by Zero for Three Chemistry Cases

| | Urban | | | Free Troposphere | | | Stratosphere | | |
|---|---|---|---|---|---|---|---|---|---|
| | | After Sparse Reductions | | | After Sparse Reductions | | | After Sparse Reductions | |
| Quantity | Initial | Day | Night | Initial | Day | Night | Initial | Day | Night |
| Order of matrix | 94 | 94 | 94 | 74 | 74 | 74 | 43 | 43 | 43 |
| Initial fill-in | 8836 | 735 | 712 | 5476 | 585 | 565 | 1849 | 292 | 267 |
| % of initial fill-in | 100 | 8.3 | 8.1 | 100 | 10.7 | 10.3 | 100 | 15.8 | 14.4 |
| Final fill-in | 8836 | 897 | 875 | 5476 | 728 | 710 | 1849 | 337 | 298 |
| % of final fill-in | 100 | 10.2 | 9.9 | 100 | 13.3 | 13.0 | 100 | 18.2 | 16.1 |
| Decomp. 1 | 272,459 | 1978 | 1894 | 132,349 | 1709 | 1627 | 25,585 | 709 | 581 |
| Decomp. 2 | 4371 | 458 | 440 | 2701 | 361 | 346 | 903 | 147 | 118 |
| Backsub. 1 | 4371 | 458 | 440 | 2701 | 361 | 346 | 903 | 147 | 118 |
| Backsub. 2 | 4371 | 345 | 341 | 2701 | 293 | 290 | 903 | 147 | 137 |

The initial and final fill-in are the initial and final numbers of matrix positions filled. The last four rows show the number of operations in each of four loops of matrix decomposition and backsubstitution. The first column in each case shows values before sparse-matrix reductions.

Reordering maximizes the number of multiplications by zero and minimizes matrix fill-in during decomposition and backsubstitution. Table 13.2 shows the reductions in computations from applying sparse matrix techniques to a large matrix.

Although sparse-matrix techniques speed Gear's solution method, another set of techniques is needed to further increase speed in large models. Such models require solutions to 50–250 chemical ODEs in each of $10^3$–$10^6$ grid cells. On the fastest supercomputers today, sparse-matrix techniques alone do not allow solutions of 200 ODEs in $10^5$ grid cells for extended simulation times. **Computer optimization techniques** are necessary to improve speeds beyond those obtained from sparse-matrix techniques.

One optimization technique useful on vector, scalar, and parallel computers is to divide the grid domain into blocks of 500 or so grid cells each. $10^5$ cells can be divided into 200 blocks of 500 cells each. Chemical ODEs can be solved in blocks of 500 cells at a time instead of serially, one at a time. When a grid is divided into blocks, every inner loop of the computer program should be the grid-cell loop, with length 500.

The following example (in Fortran) demonstrates how array references are minimized when the inner loop is the grid-cell loop. In this example, NBIMOLEC is the number (e.g., 150) of bimolecular reaction rates that are calculated, JSP1 is the species number of the first reactant, JSP2 is the species number of the second reactant, KTLOOP is the number (e.g., 500) of grid cells in a grid block, RRATE is the rate coefficient, CONC is a gas concentration, and TRATE is a reaction rate. In nested loop A, where the inner loop is the grid cell loop, only 150 references are made to arrays JPROD1 and JPROD2.

*Nested loop A*

```
    DO 100 NK          = 1, NBIMOLEC
       JSP1            = JPROD1(NK)
       JSP2            = JPROD2(NK)
       DO 100 K        = 1, KTLOOP
          TRATE(K,NK) = RRATE(K,NK) * CONC(K,JSP1) * CONC(K,JSP2)
100 CONTINUE
```

*Nested loop B*

```
    DO 100 K           = 1, KTLOOP
       DO 100 NK       = 1, NBIMOLEC
          JSP1         = JPROD1(NK)
          JSP2         = JPROD2(NK)
          TRATE(K,NK) = RRATE(K,NK) * CONC(K,JSP1) * CONC(K,JSP2)
100 CONTINUE
```

In nested loop B, where the grid-cell loop is on the outside, the number of references to JPROD1 and JPROD2 increases by a factor of 500. Thus, one pass through nested loop A is significantly faster than one pass through nested loop B, regardless of the machine. On vector machines, nested loop A also vectorizes better than does nested loop B.

When grid cells are grouped together, all cells are iterated together until the cell with the stiffest equations converges. This results in many unnecessary calculations, since computations are performed in some cells even after equations for the cell have converged. Excess iterations can be minimized by reordering cells each time interval, according to stiffness. If 20,000 cells are solved 500 at a time, the 500 stiffest cells can be placed together, the next 500 stiffest can be placed together, and so on. This way, cells that require the most iterations are grouped together.

To reorder cells according to stiffness each time interval, a **stiffness predictor** is needed. One such predictor is

$$S_p = \frac{1}{K}\sum_{i=1}^{K}\left(\frac{\mathrm{d}N_{i,t}/\mathrm{d}t}{N_{i,t} + A_{tol,t}}\right)^2 \tag{13.81}$$

(Jacobson 1995), where $S_p$ is proportional to stiffness, $K$ is the number of ODEs (and species), $A_{tol,t}$ is the absolute error tolerance at time $t$, $N_{i,t}$ is the species concentration, and $\mathrm{d}N_{i,t}/\mathrm{d}t$ is the first derivative of $N_{i,t}$. The larger the value of $S$, the stiffer the equations in the grid cell. Figure 13.4 shows how (13.81) predicted the stiffness in an example simulation.

In sum, Gear's method is accurate for solving atmospheric chemistry problems. When the method is combined with sparse-matrix and computer optimization techniques, the equations can be solved over large model grids.

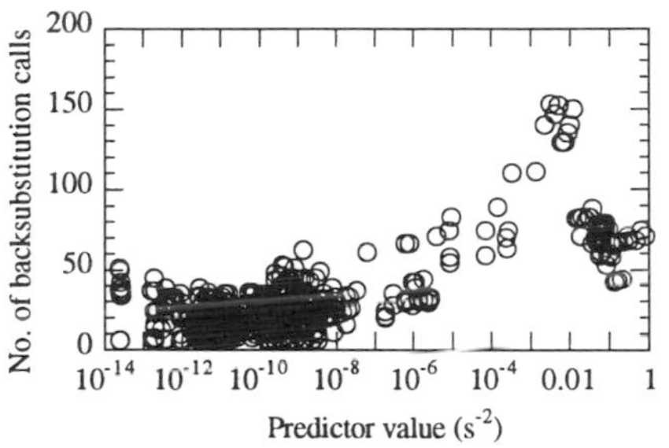


**Figure 13.4.** Resulting number of backsubstitution calls for different stiffness-predictor values from (13.81). The more backsubstitution calls, the greater the stiffness of the set of equations. The calculations were performed with the computer code SMVGEAR II (Jacobson 1995).

## 13.9. FAMILY SOLUTION TO ODEs

The last numerical method of solving chemical ODEs discussed is the **family method** (e.g., Crutzen 1971; Turco and Whitten 1974; Austin 1991; Elliott et al. 1993; Jacobson 1994). This method is less accurate than Gear's method for the same time step and must be tuned for each application. It is fast and useful for several applications.

The theory behind the family method is that some groups of gases, or **families**, exist in which gases in cycle quickly among each other. Example families are the **odd oxygen, odd hydrogen, odd nitrogen,** and **odd chlorine families.** Some of the major species in these families are

| | | |
|---|---|---|
| Odd oxygen: | $[O_T] = [O] + [O(^1D)] + [O_3] + [NO_2]$ | (13.82) |
| Odd hydrogen: | $[HO_T] = [OH] + [HO_2] + [H_2O_2]$ | (13.83) |
| Odd nitrogen: | $[NO_T] = [NO] + [NO_2] + [NO_3]$ | (13.84) |
| Odd chlorine: | $[Cl_T] = [Cl] + [ClO] + [ClO_2]$ | (13.85) |

Some species in the odd-oxygen family – atomic oxygen, excited atomic oxygen, and ozone – cycle among each other by

$$\begin{array}{ll} O_3 + h\nu \longrightarrow O_2 + \cdot\dot{O}(^1D) & O_2 + h\nu \longrightarrow \dot{O}\cdot + \dot{O}\cdot \\ \cdot\dot{O}(^1D) \xrightarrow{M} \dot{O}\cdot & \dot{O}\cdot + O_2 + M \longrightarrow O_3 + M \end{array} \qquad (13.86)$$

Cycling among these species is fast, whereas loss of O atoms out of the group is slower.

The family solution to ODEs for a time step requires four steps. Reaction rates are calculated with current species concentrations. Rates of production and loss of each species in a family are then summed across the family. The family concentration is advanced with a forward Euler approximation applied to the summed production and loss terms. Finally, species concentrations are repartitioned in the family for the next time step.

Suppose a system contains four species (A, B, C, and D) and two families,

$$[\mathrm{Fam}_1] = [\mathrm{A}] + [\mathrm{B}] + [\mathrm{C}] \qquad [\mathrm{Fam}_2] = [\mathrm{D}] \tag{13.87}$$

If the reactions

$$\mathrm{A} \rightarrow \mathrm{B} \quad (k_a) \qquad \mathrm{B} \rightarrow \mathrm{C} \quad (k_b) \qquad \mathrm{C} \rightarrow \mathrm{D} \quad (k_c) \tag{13.88}$$

occur in the system, the first derivatives of A, B, C, and D are

$$\frac{\mathrm{d}[\mathrm{A}]}{\mathrm{d}t} = -k_a[\mathrm{A}] \qquad \frac{\mathrm{d}[\mathrm{B}]}{\mathrm{d}t} = k_a[\mathrm{A}] - k_b[\mathrm{B}]$$
$$\frac{\mathrm{d}[\mathrm{C}]}{\mathrm{d}t} = k_b[\mathrm{B}] - k_c[\mathrm{C}] \qquad \frac{\mathrm{d}[\mathrm{D}]}{\mathrm{d}t} = k_c[\mathrm{C}] \tag{13.89}$$

Substituting (13.89) into the derivatives of (13.87) gives

$$\frac{\mathrm{d}[\mathrm{Fam}_1]}{\mathrm{d}t} = \frac{\mathrm{d}[\mathrm{A}]}{\mathrm{d}t} + \frac{\mathrm{d}[\mathrm{B}]}{\mathrm{d}t} + \frac{\mathrm{d}[\mathrm{C}]}{\mathrm{d}t} = -k_c[\mathrm{C}]$$
$$\frac{\mathrm{d}[\mathrm{Fam}_2]}{\mathrm{d}t} = \frac{\mathrm{d}[\mathrm{D}]}{\mathrm{d}t} = k_c[\mathrm{C}] \tag{13.90}$$

(13.90) can be discretized over a time step with

$$[\mathrm{Fam}]_t = [\mathrm{Fam}]_{t-h} + h\frac{\mathrm{d}[\mathrm{Fam}]_{t-h}}{\mathrm{d}t} \tag{13.91}$$

giving new family concentrations as

$$[\mathrm{Fam}_1]_t = [\mathrm{Fam}_1]_{t-h} - hk_c[\mathrm{C}]_{t-h} \qquad [\mathrm{Fam}_2]_t = [\mathrm{Fam}_2]_{t-h} + hk_c[\mathrm{C}]_{t-h} \tag{13.92}$$

respectively. After the family concentrations have been updated, the species concentrations must be extracted from them. One method of extracting species concentrations is to calculate and apply partitioning ratios. Such ratios are calculated by summing the concentrations of the species in the family as

$$[\mathrm{Fam}_1]_t = [\mathrm{A}]_t + [\mathrm{B}]_t + [\mathrm{C}]_t = [\mathrm{A}]_t\left(1 + \frac{[\mathrm{B}]_t}{[\mathrm{A}]_t} + \frac{[\mathrm{C}]_t}{[\mathrm{A}]_t}\right) \tag{13.93}$$

Individual species concentrations are found from the family concentrations with

$$[\mathrm{A}]_t = \frac{[\mathrm{Fam}_1]_t}{1 + \frac{[\mathrm{B}]_t}{[\mathrm{A}]_t} + \frac{[\mathrm{C}]_t}{[\mathrm{A}]_t}} \qquad [\mathrm{B}]_t = [\mathrm{A}]_t\frac{[\mathrm{B}]_t}{[\mathrm{A}]_t} \qquad [\mathrm{C}]_t = [\mathrm{A}]_t\frac{[\mathrm{C}]_t}{[\mathrm{A}]_t} \tag{13.94}$$

In these equations, the **partitioning ratios** $[\mathrm{B}]_t/[\mathrm{A}]_t$ and $[\mathrm{C}]_t/[\mathrm{A}]_t$ are currently unknown. In the case of species $D$, its final concentration is that of the family it resides in, since no other species is in the family. Thus, $[\mathrm{D}]_t = [\mathrm{Fam}_2]_t = hk_c[\mathrm{C}]_{t-h}$.

The simplest but least accurate way to estimate partitioning ratios is to assume that each species in the family is in steady state. With this method, the individual rates of reaction of species B and C are set to zero:

$$\frac{\mathrm{d}[\mathrm{B}]}{\mathrm{d}t} = k_a[\mathrm{A}] - k_b[\mathrm{B}] = 0 \qquad \frac{\mathrm{d}[\mathrm{C}]}{\mathrm{d}t} = k_b[\mathrm{B}] - k_c[\mathrm{C}] = 0 \tag{13.95}$$

The partitioning ratios of B and C are estimated as

$$\frac{[\mathrm{B}]_t}{[\mathrm{A}]_t} \approx \frac{[\mathrm{B}]}{[\mathrm{A}]} = \frac{k_a}{k_b} \qquad \frac{[\mathrm{C}]_t}{[\mathrm{A}]_t} \approx \frac{[\mathrm{C}]}{[\mathrm{A}]} = \frac{[\mathrm{C}]}{[\mathrm{B}]}\frac{[\mathrm{B}]}{[\mathrm{A}]} = \frac{k_b}{k_c}\frac{k_a}{k_b} = \frac{k_a}{k_c} \tag{13.96}$$

respectively. Substituting (13.96) into (13.94) gives the final species concentrations of A, B, and C, respectively, as

$$[\mathrm{A}]_t = \frac{[\mathrm{Fam}_1]_t}{1 + \frac{k_a}{k_b} + \frac{k_a}{k_c}} \qquad [\mathrm{B}]_t = \frac{[\mathrm{Fam}_1]_t \frac{k_a}{k_b}}{1 + \frac{k_a}{k_b} + \frac{k_a}{k_c}} \qquad [\mathrm{C}]_t = \frac{[\mathrm{Fam}_1]_t \frac{k_a}{k_c}}{1 + \frac{k_a}{k_b} + \frac{k_a}{k_c}} \tag{13.97}$$

A second way to estimate partitioning ratios is to linearize the ODEs of the individual species as

$$\frac{\mathrm{d}[\mathrm{A}]_t}{\mathrm{d}t} = -k_a[\mathrm{A}]_t \qquad \frac{\mathrm{d}[\mathrm{B}]_t}{\mathrm{d}t} = k_a[\mathrm{A}]_{t-h} - k_b[\mathrm{B}]_t \tag{13.98}$$
$$\frac{\mathrm{d}[\mathrm{C}]_t}{\mathrm{d}t} = k_b[\mathrm{B}]_{t-h} - k_c[\mathrm{C}]_t$$

and then to integrate each equation. The results are

$$[\mathrm{A}]_t \approx [\mathrm{A}]_{t-h}e^{-k_ah} \qquad [\mathrm{B}]_t \approx [\mathrm{B}]_{t-h}e^{-k_bh} + \frac{k_a[\mathrm{A}]_{t-h}}{k_b}(1 - e^{-k_bh}) \tag{13.99}$$
$$[\mathrm{C}]_t \approx [\mathrm{C}]_{t-h}e^{-k_ch} + \frac{k_b[\mathrm{B}]_{t-h}}{k_c}(1 - e^{-k_ch})$$

These estimates are substituted into the right sides of (13.94).

A third way to estimate partitioning ratios is to finite-difference the ODEs as

$$\frac{[\mathrm{A}]_t - [\mathrm{A}]_{t-h}}{h} = -k_a[\mathrm{A}]_t \qquad \frac{[\mathrm{B}]_t - [\mathrm{B}]_{t-h}}{h} = k_a[\mathrm{A}]_t - k_b[\mathrm{B}]_t \tag{13.100}$$
$$\frac{[\mathrm{C}]_t - [\mathrm{C}]_{t-h}}{h} = k_b[\mathrm{B}]_t - k_c[\mathrm{C}]_t$$

and then to rearrange the equations into a matrix equation as

$$\begin{bmatrix} 1 + hk_a & 0 & 0 \\ hk_a & 1 + hk_b & 0 \\ 0 & hk_b & 1 + hk_c \end{bmatrix} \begin{bmatrix} [\mathrm{A}]_t \\ [\mathrm{B}]_t \\ [\mathrm{C}]_t \end{bmatrix} = \begin{bmatrix} [\mathrm{A}]_{t-h} \\ [\mathrm{B}]_{t-h} \\ [\mathrm{C}]_{t-h} \end{bmatrix} \tag{13.101}$$

Solving the matrix equation for $[\mathrm{A}]_t$, $[\mathrm{B}]_t$, and $[\mathrm{C}]_t$ and substituting the results into the right sides of (13.94) gives final partitioned concentrations.

In the example above, the reactions were unimolecular. Most chemical reactions are bimolecular. When bimolecular reactions are used, the first derivatives need to be linearized for (13.101) to be solved. If the bimolecular reaction $\mathrm{A} + \mathrm{C} \rightarrow \mathrm{B} + \mathrm{D}$ ($k_{ac}$) is added to (13.88), the linearized finite-difference forms for A, B, and C are

$$\frac{[\mathrm{A}]_t - [\mathrm{A}]_{t-h}}{h} = -k_a[\mathrm{A}]_t - k_{ac}[\mathrm{A}]_t[\mathrm{C}]_{t-h} \tag{13.102}$$

$$\frac{[\mathrm{B}]_t - [\mathrm{B}]_{t-h}}{h} = k_a[\mathrm{A}]_t - k_b[\mathrm{B}]_t + 0.5k_{ac}([\mathrm{A}]_t[\mathrm{C}]_{t-h} + [\mathrm{A}]_{t-h}[\mathrm{C}]_t) \tag{13.103}$$

$$\frac{[\mathrm{C}]_t - [\mathrm{C}]_{t-h}}{h} = k_b[\mathrm{B}]_t - k_c[\mathrm{C}]_t - k_{ac}[\mathrm{A}]_{t-h}[\mathrm{C}]_t \tag{13.104}$$

respectively. The resulting matrix is (Jacobson 1994)

$$\begin{bmatrix} 1 + hk_a + hk_{ac}[\mathrm{C}]_{t-h} & 0 & 0 \\ -hk_a - 0.5(hk_{ac}[\mathrm{C}]_{t-h}) & 1 + hk_b & -0.5(hk_{ac}[\mathrm{A}]_{t-h}) \\ 0 & -hk_b & 1 + hk_c + hk_{ac}[\mathrm{A}]_{t-h} \end{bmatrix} \begin{bmatrix} [\mathrm{A}]_t \\ [\mathrm{B}]_t \\ [\mathrm{C}]_t \end{bmatrix} = \begin{bmatrix} [\mathrm{A}]_{t-h} \\ [\mathrm{B}]_{t-h} \\ [\mathrm{C}]_{t-h} \end{bmatrix} \tag{13.105}$$

Solving this matrix equation gives concentrations that can be used to estimate partitioning ratios.

The advantages of the family method are that it is fast, since it can use a long time step, and it may be accurate for moderate- to low-stiffness systems. The disadvantages are that the families need to be carefully set up and validated for each set of chemistry, and the accuracy of the method decreases with increasing stiffness. Although families are generally long-lived, the forward Euler solution for the family often becomes unstable. In such cases, the time step must be reduced and the family concentration must be recalculated. This reduces the speed advantage of the family method.

## 13.10. SUMMARY

In this chapter, methods of solving chemical first-order ordinary differential equations were described. These included analytical, explicit Taylor series, forward Euler, backward Euler, simple exponential, quasi-steady-state, multistep implicit-explicit, Gear, and family methods. A good solution scheme is stable, accurate, mass-conserving, positive definite, and fast. One code that fits these qualities is Gear's code combined with sparse matrix and vectorization techniques. Stiff ODE codes can be used to solve aqueous as well as gas-phase chemical equations. Such equations are discussed in Chapter 19.

## 13.11. PROBLEMS

**13.1.** Derive an analytical solution for the loss of molecular oxygen via the reaction $O_2 + h\nu \rightarrow O + O$. If oxygen is destroyed by this reaction but not re-created, how long will it take for its concentration to decrease to 10 percent of its initial value? Use a constant photolysis rate coefficient from Table 12.11.

**13.2.** Given reactions $A + h\nu \rightarrow B + C$ ($J_1$) and $B + h\nu \rightarrow C + C$ ($J_2$), where $J_1$ and $J_2$ are photolysis coefficients and where [A], [B], and [C] are concentrations of A, B, and C:

**(a)** Write out the chemical rate equations for A, B, and C (e.g., d[A]/dt...)

**(b)** Find the time-dependent analytical solutions for each species from the initial values $[A]_{t-h}$, $[B]_{t-h}$, and $[C]_{t-h}$. Assume $J_1$ and $J_2$ are constant in time.

**(c)** As time approaches infinity ($h \rightarrow \infty$), what is [C]? What is $[C]_{h\rightarrow\infty}$ when $J_1 = 0$?

## 13.12. COMPUTER PROGRAMMING PRACTICE

**13.3.** Write a program to solve chemical ODEs with the MIE algorithm. Test the program with the three reactions shown in Table 13.3, assuming that $T = 298$ K. Assume that the initial concentrations of $O_3$, $O(^3P)$, NO, and $NO_2$ are 0, 0, $10^{12}$, and $10^{10}$ molecules $cm^{-3}$, respectively. Do not solve for $O_2$. Use a time step of $h = 10$ s. The results should exactly match those shown in Fig. 13.5

**13.4.** Write a program to solve chemical ODEs with the simple exponential method. Test the program with the three reactions in Table 13.3, assuming $T = 298$ K. Assume that initial concentrations of $O_3$, $O(^3P)$, NO, and $NO_2$ are 0, 0, $10^{12}$, and $10^{10}$ molecules $cm^{-3}$, respectively. Find a time step size, if any, that gives the results shown in Fig. 13.5. Discuss what happens to the solution at increasingly larger time steps.

**13.5.** Repeat Problem 13.4 with a forward Euler approximation. Determine the time step size at which the solution becomes unstable (concentrations oscillate between positive and negative values or explode to large positive or negative numbers).

**13.6.** Repeat Problem 13.4 with a backward Euler approximation. Discuss differences between the results from Problems 13.5 and 13.6.

**Table 13.3.** Reactions and Rate Coefficients for Problem 13.3

| Reaction | Rate-Coefficient Expression |
|---|---|
| $O \xrightarrow{O_2, M} O_3$ | $k_1 = 1.4 \times 10^3 e^{1175/T}$ |
| $NO + O_3 \rightarrow NO_2 + O_2$ | $k_2 = 1.8 \times 10^{-12} e^{-1370/T}$ |
| $NO_2 + h\nu \rightarrow NO + O$ | $J = 1.7 \times 10^{-2}$ |

In the first reaction, only O is included in the rate-coefficient expression, since the rate coefficient has already been multiplied by $O_2$ and M. $T$ is in kelvin.

**Table 13.4.** Typical Volume-Mixing Ratios of Several Gases at Two Altitudes

| | Volume Mixing Ratio (Fraction) | | | Volume Mixing Ratio (Fraction) | |
|---|---|---|---|---|---|
| Gas | 25 km | 0 km | Gas | 25 km | 0 km |
| $N_2$ | 0.7808 | 0.7808 | $CH_3COOH$ | — | 3.0 (−9) |
| $O_2$ | 0.2095 | 0.2095 | $CH_3COCH_3$ | — | 7.5 (−10) |
| $O_3$ | 8.0 (−6) | 4.0 (−8) | $C_5H_8$ (ISOP) | — | 2.0 (−10) |
| $H_2$ | 6.0 (−7) | 6.0 (−7) | $CH_3Cl$ | 3.6 (−10) | 6.0 (−10) |
| $H_2O$ | 3.1 (−6) | 0.01 | $CH_3CCl_3$ | 2.0 (−11) | 1.4 (−10) |
| $H_2O_2$ | 2.1 (−10) | 3.0 (−9) | $CCl_4$ | 1.3 (−11) | 1.1 (−10) |
| NO | 1.1 (−9) | 5.0 (−12) | $CFCl_3$ | 5.0 (−11) | 2.7 (−10) |
| $NO_2$ | 1.9 (−9) | 4.0 (−11) | $CF_2Cl_2$ | 2.2 (−10) | 4.8 (−10) |
| $N_2O$ | 1.5 (−7) | 3.0 (−7) | $CF_2ClH$ | 7.0 (−11) | 1.2 (−10) |
| $HNO_3$ | 6.0 (−9) | 5.0 (−10) | $CFCl_2CF_2Cl$ | 3.0 (−11) | 7.0 (−11) |
| $HO_2NO_2$ | 3.4 (−10) | 1.0 (−11) | HCl | 1.3 (−9) | 9.0 (−11) |
| CO | 4.0 (−8) | 1.1 (−7) | $ClONO_2$ | 8.0 (−10) | — |
| $CO_2$ | 3.6 (−4) | 3.6 (−4) | HOCl | 3.3 (−11) | — |
| $CH_4$ | 1.0 (−6) | 1.7 (−6) | $CH_3Br$ | 1.0 (−11) | 1.2 (−11) |
| $C_2H_6$ | — | 6.0 (−10) | $BrONO_2$ | 2.0 (−12) | — |
| $C_3H_8$ | — | 4.8 (−11) | HBr | 3.0 (−12) | |
| $C_2H_4$ | — | 2.1 (−11) | $SO_2$ | 1.0 (−11) | 5.0 (−11) |
| $C_3H_6$ | — | 6.0 (−12) | $CH_3SCH_3$ | — | 1.0 (−10) |
| HCHO | — | 2.0 (−10) | $H_2S$ | — | 5.0 (−11) |
| $CH_3CHO$ | | 1.6 (10) | OCS | 1.0 (−10) | 5.0 (−10) |
| HCOOH | — | 1.8 (−9) | $CS_2$ | 2.0 (−11) | 1.0 (−10) |
| $CH_3OOH$ | — | 1.2 (−9) | | | |

6.0 (−6) means $6.0 \times 10^{-6}$. Multiply the volume mixing ratio of a gas (molecules of gas per molecule of dry air) by the number concentration of dry air (molecules $cm^{-3}$) to obtain the number concentration of the gas. The 0-km mixing ratios correspond to clean tropospheric conditions.

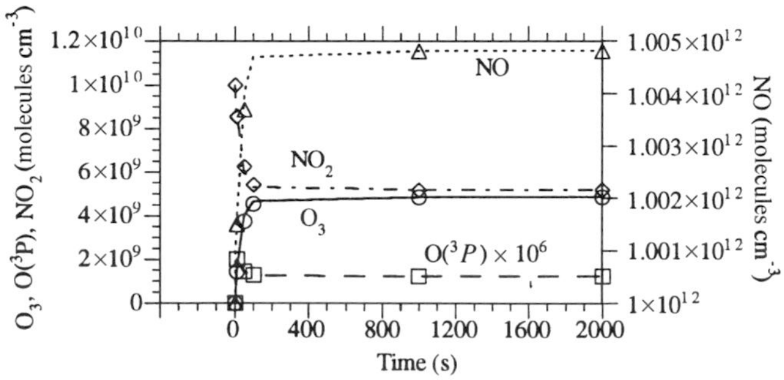


**Figure 13.5.** Exact and modeled solution to Problem 13.3. Exact concentrations are represented by continuous lines. Solutions from the MIE method are represented by symbols.

## 13.13. MODELING PROJECT

Using the program developed for Problem 13.3, read in the equations and rate coefficients from the computer file developed for Problems 12.4 and 11.7. Run a four-day simulation, starting at 5:00 a.m. on the first day. Calculate initial number concentrations for necessary species from volume mixing ratios shown in Table 13.4. If a stratospheric mechanism was chosen for Problem 12.4, assume $T = 192$ K and $p_d = 25$ mb, and use the mixing ratios at 25 km in Table 13.4. If a near-surface mechanism was chosen, assume $T = 298.15$ K and $p_d = 1013$ mb, and use the mixing ratios at 0 km, which correspond to clean lower-tropospheric conditions. Assume all initial mixing ratios not given in the table are zero. Once a baseline simulation has been run, vary temperature, pressure, and initial mixing ratios to test the sensitivity of model predictions to these parameters. Display time-series plots of species mixing ratios from the baseline and sensitivity test cases on the same graphs. Discuss results and their significance.

# 14

# Particle Components, Size Distributions, and Size Structures

PARTICLES in the atmosphere vary in size and composition. In this chapter, their importance and treatment in numerical models are discussed. The variation of particle concentration with size can be simulated with a continuous or discrete size distribution. Three continuous distributions discussed here are the lognormal, Marshall–Palmer, and modified gamma distributions. These distributions are discretized over a model size grid, which consists of size-bin increments in radius space. The variation of a size distribution over time is simulated with a size structure. Five structures discussed are the full-stationary, full-moving, quasi-stationary, hybrid, and moving-center structures. In later chapters, the treatment of time-dependent processes that affect particle size distributions and composition are examined. These include emissions, nucleation, coagulation, growth, evaporation, chemical equilibrium, aqueous chemistry, sedimentation, and dry deposition.

## 14.1. EFFECTS OF PARTICLES

Particles are agglomerations of molecules in the liquid and/or solid phases, suspended in air. **Aerosols, cloud drops**, and **raindrops** are considered particles. Aerosols differ from cloud drops and raindrops in that cloud drops and raindrops contain much more liquid water than do aerosols. Whereas almost all cloud drop and raindrop mass is liquid water, aerosols contain large fractions of other material as well. The sizes and number concentrations of aerosols, cloud drops, and raindrops also differ. Table 14.1 compares characteristics of gas molecules and different types of particles.

Aerosols directly and indirectly affect air quality, meteorology, and climate. Submicron aerosols (particles smaller than 1 μm in diameter) affect human health by penetrating to the deepest part of human lungs. Aerosols between 0.2 and 1.0 μm in diameter that contain sulfate, nitrate, and organic carbon scatter light efficiently. Aerosols smaller than 1.0 μm that contain elemental carbon absorb efficiently. Aerosol absorption and scattering are important because they affect radiative energy fluxes, air temperatures, and photolysis coefficients. Aerosols also serve as sites on which chemical reactions take place and as sinks in which some gas-phase species are removed from the atmosphere.

**Table 14.1.** Characteristics of Gases, Aerosols, Cloud Drops, and Raindrops

| | Typical Diameter μm | Number Concentration ($cm^{-3}$) | Mass Concentration ($\mu g\ m^{-3}$) |
|---|---|---|---|
| Gas molecules | 0.0005 | $2.45 \times 10^{19}$ | $1.2 \times 10^9$ |
| Small aerosols | <0.2 | $10^3$–$10^6$ | <1 |
| Medium aerosols | 0.2–1.0 | 1–$10^4$ | <250 |
| Large aerosols | 1.0–100 | <1–10 | <250 |
| Fog drops | 10–20 | 1–500 | $10^4$–$5 \times 10^5$ |
| Average cloud drops | 10–200 | <10–1000 | <$10^5$–$5 \times 10^6$ |
| Large cloud drops | 200 | <1–10 | <$10^5$–$5 \times 10^6$ |
| Drizzle | 400 | 0.1 | $10^5$–$5 \times 10^6$ |
| Small raindrops | 1000 | 0.01 | $10^5$–$5 \times 10^6$ |
| Medium raindrops | 2000 | 0.001 | $10^5$–$5 \times 10^6$ |
| Large raindrops | 8000 | <0.001 | $10^5$–$5 \times 10^6$ |

Data are for typical lower-tropospheric conditions.

### 14.1.1. Direct and Indirect Effects of Aerosols on Climate

The **direct** effects of aerosols on climate have been examined intensely. Some of the early studies revealed three conclusions. First, aerosols backscatter a fraction of incoming solar radiation, increasing the earth–atmosphere albedo and lowering surface air temperatures during the day. Second, some aerosols absorb solar radiation and reemit infrared radiation, decreasing the cooling effect of backscattering. Third, all aerosols absorb and reemit the earth's infrared radiation, decreasing the cooling effect caused by backscattering during the day and increasing surface air temperatures at night. (e.g., Rasool and Schneider 1971; Zdunkowski and McQuage 1972; Chylek and Coakley 1974; Weare et al. 1974; Zdunkowski et al. 1976; Penner et al. 1992; Charlson et al. 1991, 1992; Wigley and Raper 1992, Kiehl and Briegleb 1993; Charlock et al. 1993).

Scattering aerosols lower air temperatures near the surface during the day by decreasing solar fluxes to the ground. The resulting decreases in ground temperatures decrease mechanical and thermal turbulent energy fluxes to the boundary layer, reducing boundary layer temperatures. During the night, aerosols increase the net downward infrared flux of radiant energy to the ground, raising ground temperatures. The increase in ground temperatures increases heat transfer by mechanical turbulence to the boundary layer, increasing boundary-layer temperatures during the night.

The effect of aerosols on daytime cooling diminishes when aerosols contain material that absorbs solar radiation. Toon and Pollack (1976) quantified the globally averaged optical depth and single-scattering albedo of aerosols as $\approx$0.125 and $\approx$1.0, respectively. They assumed that aerosols contained primarily ammonium sulfate, sea salt, and basalt. The larger the single-scattering albedo, the greater the total scattering compared to absorption. A single-scattering albedo of nearly 1.0 indicates almost no absorption. Hansen et al. (1980) applied the optical depth and single-scattering albedo to a radiation code and found that the average surface air temperature decreased by about 1.3 K. When the single-scattering albedo was reduced below 0.83 by the addition of absorbing material to aerosols, a net warming of the surface occurred. Chylek et al. (1995) similarly showed that sulfate particles with more than 5 percent elemental carbon cause a much larger cooling than sulfate particles with less than 5 percent elemental carbon.

In urban regions, absorbing aerosols causes at least a short-term net warming of the air between day and night. Zdunkowski and McQuage (1972) used a one-dimensional aerosol–radiative model to find that aerosols warmed urban surface air over a one-day period, but cooled the air thereafter. A three-dimensional study by Jacobson (1997b) supported part of this hypothesis by showing that aerosols in the Los Angeles basin caused surface air temperatures to cool during the day and warm to a greater extent at night, yielding a net warming, averaged over day and night. The net warming was maintained for the length of the simulation period, which was two days.

The net direct effects of aerosols on surface air temperature vary with location because aerosol composition varies with location. Coastal urban aerosols are influenced by elemental carbon, organic carbon, metals from road dust, and sodium and chloride from sea spray. Gas-to-particle conversion in urban air produces secondary particles that contain organic carbon, sulfate, nitrate, and/or ammonium. Free tropospheric aerosols are dominated by the presence of sulfate.

Aerosols, which serve as cloud condensation nuclei, **indirectly** affect temperatures by altering the size distribution of cloud drops, thereby changing cloud albedo and lifetime (e.g., Twomey 1977, 1991; Charlson et al. 1987). When new particles are emitted or created in the atmosphere, water may condense on them, decreasing the average cloud drop size. The cross-sectional area, summed over a large number of small drops, is greater than that summed over a small number of large drops. Since reflectivity depends on cross-sectional area, a large total cross-sectional area increases cloud albedo, cooling air below the cloud during the day. The decrease in cloud drop size also slows the removal of drops by sedimentation or coalescence to raindrop sizes, increasing the cloud lifetime and time-integrated cloud albedo.

### 14.1.2. Direct Effects of Clouds on Climate

Clouds directly affect climate by reflecting and absorbing solar radiation and absorbing infrared radiation. Liou (1976) calculated that some thick clouds, such as nimbostratus and cumulonimbus, reflect 80–90 percent and absorb 10–20 percent of incident solar radiation, fair-weather cumulus clouds reflect about 68–85 percent and absorb 4–9 percent, and thin stratus clouds reflect about 45–72 percent and absorb 1–6 percent of incident radiation. Twomey (1976) predicted a maximum cloud absorption of incident solar radiation of about 20 percent. These values conflict with measurements (e.g., Reynolds et al. 1975) that show that clouds absorb as much as 30–52 percent of direct solar radiation.

Many theories have tried to explain the difference between predicted and observed absorption. Welch et al. (1980) suggested that solar radiation escaping from the sides of cumulus clouds may be mistaken for absorption. This theory did not hold, since absorptions of more than 20 percent have been observed in stratus clouds, in which side escape is small. Welch et al. (1980) also suggested that the inclusion in models of large drizzle drops and raindrops may account for observed absorption. Wiscombe et al. (1984) showed that the inclusion of such drops increased absorption by only 2–4 percent. Stephens and Tsay (1990) suggested that an unobserved water-vapor continuum may be present. Twomey (1977) suggested that aerosols within clouds may absorb some radiation. This may still be a possibility. The difference between modeled and observed solar absorption by clouds has been noticed for four decades and is called the **anomalous absorption paradox** (e.g., Wiscombe et al. 1984; Liou 1992; King 1993).

### 14.1.3. Effects of Aerosols and Clouds on Photolysis Coefficients

Aerosols and clouds affect actinic fluxes, which are used to calculate photolysis coefficients. Madronich (1987), van Weele and Duynkerke (1993), and de Arellano et al. (1994) showed that clouds increase the backscattered fraction of incident solar radiation, enhancing actinic fluxes above them. They also showed that clouds increase actinic fluxes within cloud layers and decrease actinic fluxes below cloud layers. In a study of the effects of aerosols on UV irradiance and photolysis coefficients, Wendisch et al. (1996) compared radiative predictions from a one-dimensional model with data to show that an enhanced aerosol layer within a temperature inversion reduced downward solar and UV irradiances and $NO_2$ photolysis rate coefficients in comparison with a background aerosol layer.

Jacobson (1997e) used a three dimensional model to study the effects of aerosols on photolysis coefficient profiles over polluted and clean air. Several findings of the study were the following: (1) aerosol absorption tends to decrease photolysis coefficients below the region of absorption, (2) aerosol scattering tends to enhance photolysis coefficients below and above the region of scattering, (3) the absorptive capabilities of aerosols decrease from the UV to the visible spectrum, (4) when aerosol UV absorption exceeds aerosol UV scattering, aerosols decrease photolysis coefficients for gases (e.g., $NO_2$) that dissociate in the UV spectrum, (5) when aerosol UV absorption is less than aerosol UV scattering, aerosols increase photolysis coefficients for UV-absorbing gases, (6) since aerosol visible scattering exceeds aerosol visible absorption, aerosols increase photolysis coefficients for gases (e.g., $NO_3$) that dissociate in the visible spectrum, (7) since aerosols above the boundary layer are stronger scatterers than absorbers in the visible and UV spectra, aerosols enhance visible and UV photolysis coefficients above the boundary layer, and (8) the greater the extinction due to aerosol absorption in the boundary layer, the less the photolysis coefficient enhancement due to scattering above the boundary layer.

## 14.2. AEROSOL, FOG, AND CLOUD COMPOSITION

Particles contain inorganic and organic substances in the solid, liquid, and/or dissolved phases. The most abundant substance in particles is typically liquid water. Because of its quantity and ability to dissolve other species, liquid water is considered a **solvent**. A species that dissolves in a solvent is a **solute**. Solute and solvent, together, make up a **solution**. Dissolved substances can dissociate or remain undissociated in solution. A **dissociated** species has a positive (cationic) or negative (anionic) electric charge. An **undissociated** species has a neutral charge. Suspended material, such as solids, may be mixed in a solution but are not part of the solution. Table B.5 of Appendix B lists typical cloud, fog, and aerosol components and some of their gas-phase precursors.

Inorganic substances within particles include sulfates, nitrates, ammonium, sodium, chloride, calcium, magnesium, potassium, iron, lead, zinc, manganese, copper, aluminum, silicon, bromide, and nickel. **Sulfate compounds** include aqueous sulfuric acid ($H_2SO_4$), bisulfate ($HSO_4^-$), and sulfate ($SO_4^{2-}$). When a particle is sufficiently dry, bisulfate or sulfate may chemically react with available cations to form solids, which precipitate from solution. Some sulfate-containing solids include ammonium sulfate ($[NH_4]_2SO_4$), ammonium bisulfate ($NH_4HSO_4$), sodium sulfate ($Na_2SO_4$), and sodium bisulfate ($NaHSO_4$).

Particulate **nitrate compounds** include aqueous nitric acid ($HNO_3$), nitrate ($NO_3^-$), solid ammonium nitrate ($NH_4NO_3$), and solid sodium nitrate ($NaNO_3$). **Ammonium compounds** include aqueous ammonia ($NH_3$), ammonium ($NH_4^+$), solid ammonium chloride ($NH_4Cl$), and other solids containing ammonium. **Sodium compounds** include sodium ($Na^+$), sodium chloride ($NaCl$), and other solids containing sodium. **Chloride compounds** include chloride ($Cl^-$) and solids containing chloride. Calcium, magnesium, potassium, iron, and other elements are also present as ions or solids.

An important inorganic particulate substance is **elemental carbon** (graphitic carbon). Elemental carbon is emitted during combustion of petroleum products, such as oil, coal, gasoline, diesel oil, and rubber. It strongly absorbs visible light and gives soot its blackish color. Soot also contains **organic carbon**. Organic carbon differs from elemental carbon in that the organic carbon contains at least carbon and hydrogen, and often oxygen and nitrogen as well. Elemental carbon contains only carbon atoms. The only source of elemental carbon in the atmosphere is emissions, but sources of particulate organic carbon include both gas-to-particle conversion and emissions. Emission sources of organic carbon are vegetation burning and fossil-fuel combustion. Organic carbon-containing particles emitted during vegetation burning often appear white or sandy. This contrasts with the blackish appearance of particles containing elemental carbon. Organic gases that readily condense include by-products of toluene, xylene, alkylbenzene, alkane, alkene, and biogenic hydrocarbon oxidation (Pandis et al. 1992).

## 14.3. DISCRETE SIZE DISTRIBUTIONS

The first steps in designing algorithms to simulate particle size and composition is to choose an initial particle size distribution. A particle **size distribution** gives the number concentration, surface-area concentration, volume concentration, or mass concentration of particles as a function of radius or diameter. Particle radii are usually expressed in micrometers (microns) for aerosols and cloud drops, and millimeters for raindrops.

A size distribution may be continuous or discrete (Friedlander 1977) A **continuous distribution** is one in which the variation of particle concentration with radius is represented by a continuous function. A **discrete distribution** is one in which concentrations are distributed over increments in radius space, called **size bins.** Once a discrete distribution has been chosen, a continuous distribution can be discretized over it.

A size bin has a lower and an upper edge radius (or diameter). All particles with radii between the lower and upper edge of a size bin are assumed to reside in the bin and to have the same diameter as each other. In reality, each particle in the atmosphere has a distinct diameter. The lower atmosphere contains on the order of $10^4$–$10^6$ particles per cubic centimeter of air, and particle diameters vary by over six orders of magnitude, from 0.001 μm (nucleated aerosols) to 5000 μm (raindrops). Each particle can contain from one to 1000 or more components.

**Example 14.1**

An idealized particle size distribution might consist of 10,000 particles of radius between 0.05 and 0.5 μm, 100 particles of radius between 0.5 and 5.0 μm, and 10 particles of radius between 5.0 and 50 μm. The particles can be naturally distributed into a model that uses three size bins.

The number of particle size bins and components in a model must be limited for two reasons. First, a typical three dimensional model contains $10^3$–$10^6$ grid cells, and the inclusion of too many particle sizes and/or components usually results in the need for more computer time than is available. Second, the minimum storage requirement for particle concentrations over a grid is one array with a dimension equal to the number of grid cells multiplied by the number of size bins and the number of components. This dimension can quickly surpass the central memory capabilities of many computers.

**Example 14.2**

With $10^5$ grid cells, 100 size bins, and 100 components per size bin, the central memory requirement, just to store one array for concentration, is one gigaword ($10^9$ words). This translates to 8 gigabytes when 1 word = 8 bytes.

Limiting the number of sizes (or size bins) in a model is often the easiest method of reducing the computational burden and memory limitations. It may cause problems because particles range in size over many orders of magnitude. To cover all particle sizes with a fixed number of bins, the bins need to be spread geometrically over the size range of interest. A discrete **geometric distribution**, called the **volume-ratio size distribution**, is discussed below. With this distribution, the volume of particles in a size bin equals the volume of particles in the next smallest size bin multiplied by a constant volume ratio, $V_{rat}$. Thus,

$$\upsilon_i = V_{rat}\upsilon_{i-1} \qquad \text{for} \quad i = 1, \ldots, N_B \tag{14.1}$$

where $\upsilon_i$ is the volume (cm$^3$) of a single particle in a size bin $i$, $\upsilon_{i-1}$ is the volume of a single particle in the next smaller size bin, and $N_B$ is the number of size bins (e.g., Toon et al. 1988). The volume in a size bin can also be expressed as

$$\boxed{\upsilon_i = \upsilon_1 V_{rat}^{i-1} \qquad \text{for} \quad i = 1, \ldots, N_B} \tag{14.2}$$

where $\upsilon_1$ is the volume of a single particle in the smallest size bin. Figure 14.1 illustrates how particle sizes are distributed in the volume ratio distribution.

For many model applications, it is convenient to consider particles as spherical. In reality, particles have a variety of geometries. Elemental carbon particles are porous and amorphous, whereas ice particles are crystalline. A highly resolved model can consider several particle types, each with a different size distribution, composition distribution, and shape distribution. Because many large three-dimensional models are limited by memory and available computer resources, particles are usually represented by one size distribution. When that is done, the particles are often assumed to be spherical, since that may be closer to an average particle shape in the atmosphere than any other shape. Spherical geometry is also easier to treat numerically than other geometries. When spherical particles are assumed, the average volume of a particle in any bin $i$ is related to its average diameter $d_i$ by $\upsilon_i = \pi d_i^3/6$.

In (14.1) and (14.2), $V_{rat}$ must be specified in advance. A disadvantage of the equations is that if $V_{rat}$ is specified incorrectly for a given $N_B$, the resulting volume of particles in the largest size bin, $\upsilon_{N_B}$, may be smaller or larger than desired. To

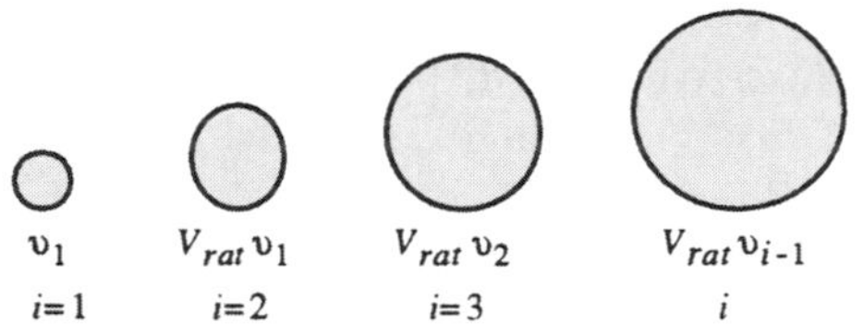


**Figure 14.1.** Variation in particle sizes with the volume-ratio size distribution. Below each particle is an expression for its volume.

avoid this problem, $V_{rat}$ can be precalculated from a specified $\upsilon_{N_B}$ with

$$V_{rat} = \left(\frac{\upsilon_{N_B}}{\upsilon_1}\right)^{1/(N_B-1)} = \left(\frac{d_{N_B}}{d_1}\right)^{3/(N_B-1)} \tag{14.3}$$

where $d_1$ is the diameter of particles in the smallest bin, and $d_{N_B}$ is the diameter of particles in the largest bin, assuming the particles are spherical. If, instead, $\upsilon_{N_B}$, $d_{N_B}$, and $V_{rat}$ are specified, but $N_B$ is not, then $N_B$ can be found with

$$N_B = 1 + \frac{\ln\left[(d_{N_B}/d_1)^3\right]}{\ln V_{rat}} \tag{14.4}$$

**Example 14.3**

(14.3) predicts that, for 30 size bins to cover a preset diameter range from 0.01 to 1000 μm, we have $V_{rat}$ =3.29.

**Example 14.4**

If $d_1 = 0.01$ μm, $d_{N_B} = 1000$ μm, and $V_{rat} = 4$, (14.4) implies that 26 size bins are needed. If $V_{rat} = 2$, then 51 size bins are needed to cover the same diameter range.

The volume width of a size bin in the volume-ratio distribution is derived by first assuming that the average volume of particles in a bin is

$$\upsilon_i = \tfrac{1}{2}(\upsilon_{i,hi} + \upsilon_{i,lo}) \tag{14.5}$$

where $\upsilon_{i,hi}$ and $\upsilon_{i,lo}$ are the volumes of the largest and smallest particles in the bin, respectively. The volume of the largest particles in the bin is limited by

$$\upsilon_{i,hi} = V_{rat}\upsilon_{i,lo} \tag{14.6}$$

Substituting (14.6) into (14.5) gives

$$\upsilon_{i,lo} = \frac{2\upsilon_i}{1 + V_{rat}} \tag{14.7}$$

The **volume width** of a size bin is

$$\Delta\upsilon_i = \upsilon_{i,hi} - \upsilon_{i,lo} = \frac{2\upsilon_{i+1}}{1+V_{rat}} - \frac{2\upsilon_i}{1+V_{rat}} = \frac{2\upsilon_i(V_{rat}-1)}{1+V_{rat}} \tag{14.8}$$

and the corresponding **diameter width** is

$$\Delta d_i = d_{i,hi} - d_{i,lo} = \left(\frac{6}{\pi}\right)^{1/3}\left(\upsilon_{i,hi}^{1/3} - \upsilon_{i,lo}^{1/3}\right) = d_i\, 2^{1/3}\frac{V_{rat}^{1/3} - 1}{(1+V_{rat})^{1/3}} \tag{14.9}$$

The average volume of one particle in a size bin is one parameter used in a discrete size distribution. Four other parameters are the number, volume, area, and mass concentrations of particles in the bin. The number concentration $n_i$ (particles $cm^{-3}$) is the number of particles in a size bin per unit volume of air, the volume concentration $v_i$ ($cm^3\ cm^{-3}$) is the volume of particles in a bin per unit volume of air, the surface-area concentration $a_i$ ($cm^2\ cm^{-3}$) is the surface area of particles in a bin per unit volume of air, and the mass concentration $m_i$ ($\mu g\ m^{-3}$) is the mass of particles in a bin per unit volume of air. **Number concentration** is related to volume and volume concentration in a bin by

$$n_i = \frac{v_i}{\upsilon_i} \tag{14.10}$$

The total number concentration of particles, summed over all size bins, is

$$N_D = \sum_{i=1}^{N_B} n_i \tag{14.11}$$

Individual particles usually consist of a mixture of a few to hundreds of components. For modeling purposes, the relationship between **total volume concentration** and **component volume concentrations** $v_{q,i}$ ($cm^3\ cm^{-3}$) is

$$v_i = \sum_{q=1}^{N_V} v_{q,i} \tag{14.12}$$

where $N_V$ is the number of components within each particle.

Assuming that particles are spherical, the surface-area concentration of particles in a size bin is

$$a_i = n_i 4\pi r_i^2 = n_i \pi d_i^2 \tag{14.13}$$

where $r_i = d_i/2$ is particle radius. The surface-area concentration is used for estimating particle surface areas available for gases to condense upon.

The **mass concentration** of particles in a size bin is

$$m_i = \sum_{q=1}^{N_V} m_{q,i} = c_m \sum_{q=1}^{N_V} \rho_q v_{q,i} = c_m \rho_{p,i} \sum_{q=1}^{N_V} v_{q,i} = c_m \rho_{p,i} v_i \tag{14.14}$$

where $m_{q,i} = c_m \rho_q v_{q,i}$ is the mass concentration (micrograms per cubic meter of air) of component $q$ in particles of size $i$, $c_m$ ($=10^{12}$) converts g $cm^{-3}$ to $\mu g\ m^{-3}$, $\rho_q$ is the mass density of $q$ (grams per cubic centimeter of component), and $\rho_{p,i}$ is the **volume-averaged mass density** of particles in bin $i$ (grams per cubic centimeter of particle). The volume-averaged density changes continuously and is found by equating the third and fourth terms in (14.14). The result is

$$\rho_{p,i} = \frac{\sum_{q=1}^{N_V} (v_{i,q}\rho_q)}{\sum_{q=1}^{N_V} v_{i,q}} \tag{14.15}$$

**Example 14.5**

If the mass concentrations of water and sulfate are $m_{q,i} = 3.0$ and $2.0\ \mu\text{g m}^{-3}$, respectively, in particles with an average diameter of $d_i = 0.5\ \mu\text{m}$, calculate the volume concentration of each component and the mass, volume, area, and number concentrations of whole particles in the size bin. Assume the densities of water and sulfate are $\rho_q = 1.0$ and $1.83\ \text{g cm}^{-3}$, respectively.

SOLUTION

From $m_{q,i} = c_m v_{q,i} \rho_q$, the volume concentrations of water and sulfate are $v_{q,i} = 3\times 10^{-12}$ and $1.09 \times 10^{-12}\ \text{cm}^3\ \text{cm}^{-3}$, respectively. From (14.14), $m_i = 5.0\ \mu\text{g m}^{-3}$. From (14.12), $v_i = 4.09 \times 10^{-12}\ \text{cm}^3\ \text{cm}^{-3}$. From $\upsilon_i = \pi d_i^3/6$, $\upsilon_i = 6.54 \times 10^{-14}\ \text{cm}^3$. From (14.10), $n_i = 62.5$ particles $\text{cm}^{-3}$, and from (14.13), $a_i = 4.8 \times 10^{-7}\ \text{cm}^2\ \text{cm}^{-3}$.

## 14.4. CONTINUOUS SIZE DISTRIBUTIONS

Once a discrete model size distribution has been laid out, the initial particle number, volume, and mass concentration must be distributed among model size bins. This can be accomplished by fitting measurements to a continuous size distribution, then discretizing the continuous distribution over the model bins. Three continuous distributions available for this procedure are the lognormal, Marshall–Palmer, and modified gamma distributions.

### 14.4.1. The Lognormal Distribution

A unimodal **lognormal distribution** is a bell-curve distribution on a log scale, as shown in Fig. 14.2(a). Figure 14.2(b) shows the same curve on a linear scale. When two, three, or four lognormal modes are superimposed on each other, the resulting lognormal distribution is bi-, tri-, or quadrimodal, respectively.

A lognormal distribution has a geometric mean diameter (geometric-mean mass, volume, surface area, or number diameter), a geometric standard deviation, and a total mass, volume, surface area, or number concentration associated with it. Fifty percent of the area under a lognormal distribution lies below the **geometric-mean diameter.** It is analogous to the mean diameter of a normal distribution. The **geometric standard deviation** is defined so that about 68 percent of the area under the lognormal distribution lies within one geometric standard deviation of the geometric-mean diameter. About 95 percent of the area falls within two standard

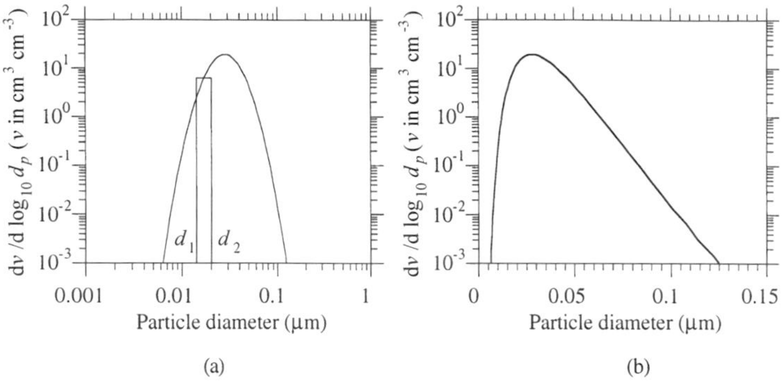


**Figure 14.2.** (a) A lognormal particle volume distribution. The volume concentration ($dv$) of material between any two diameters ($d_1$ and $d_2$) is estimated by multiplying the average value from the curve between the two diameters by ($\log_{10} d_2 - \log_{10} d_1$). (b) The lognormal curve shown in (a), drawn on a linear scale.

deviations of the geometric-mean diameter. The geometric standard deviation of a lognormal distribution is analogous to the standard deviation of a normal distribution.

#### 14.4.1.1. *Obtaining Parameters for a Lognormal Distribution*

Parameters for lognormal curves may be obtained from measurements. A low-pressure impactor (Hering et al. 1979) is an instrument that measures particle mass in seven size regimes (0.05–0.075, 0.075–0.12, 0.12–0.26, 0.26–0.5, 0.5–1.0, 1.0–2.0, and 2.0–4.0 μm in diameter). Component masses in each size regime can be determined by chemical extraction techniques.

Suppose a plot, such as that used for Fig. 14.2(a), shows that a set of impactor mass concentration data is lognormal. The **geometric-mean mass diameter** of the distribution can be found from

$$\ln \bar{D}_M = \frac{1}{\mathrm{M}_L} \sum_{j=1}^{7} (m_j \ln d_j) \qquad \mathrm{M}_L = \sum_{j=1}^{7} m_j \tag{14.16}$$

where $\mathrm{M}_L$ is the total mass concentration of particles (μg m$^{-3}$), summed over all impactor stages, and $m_j$ is the mass concentration of particles with average diameter $d_j$ in impactor stage $j$. The **geometric-mean volume diameter** of the distribution is

$$\ln \bar{D}_V = \frac{1}{\mathrm{V}_L} \sum_{j=1}^{7} (v_j \ln d_j) \qquad \mathrm{V}_L = \sum_{j=1}^{7} v_j \quad v_j = \frac{m_j}{c_m \rho_j} \tag{14.17}$$

where $V_L$ is the total volume concentration (cm$^3$ cm$^{-3}$) of particles, summed over all stages, $v_j$ is the volume concentration of particles in impactor stage $j$, and $\rho_j$ is the average density of particles in stage $j$. If the average density of particles is constant across all stages, $\bar{D}_V = \bar{D}_M$.

The geometric-mean surface-area diameter ($\bar{D}_A$) is always smaller than $\bar{D}_V$, and the geometric-mean number diameter ($\bar{D}_N$) is always smaller than $\bar{D}_A$. These latter parameters are given by

$$\ln \bar{D}_A = \frac{1}{A_L}\sum_{j=1}^{7}(a_j \ln d_j) \qquad A_L = \sum_{j=1}^{7} a_j \quad a_j = \frac{3m_j}{c_m \rho_j r_j} \tag{14.18}$$

$$\ln \bar{D}_N = \frac{1}{N_L}\sum_{j=1}^{7}(n_j \ln d_j) \qquad N_L = \sum_{j=1}^{7} n_j \quad n_j = \frac{m_j}{c_m \rho_j \upsilon_j} \tag{14.19}$$

where $A_L$ is the total area concentration (cm$^2$ cm$^{-3}$) of particles, summed over all stages, $a_j$ is the surface area concentration of particles in stage $j$, $N_L$ is the total number concentration (particles cm$^{-3}$) of particles, summed over all stages, and $n_j$ is the number concentration of particles in stage $j$.

From the geometric-mean diameters, the **geometric standard deviation** of the distribution is calculated from

$$\ln \sigma_g = \sqrt{\frac{1}{M_L}\sum_{j=1}^{7}\left(m_j \ln^2 \frac{d_j}{\bar{D}_M}\right)} = \sqrt{\frac{1}{V_L}\sum_{j=1}^{7}\left(v_j \ln^2 \frac{d_j}{\bar{D}_V}\right)}$$
$$= \sqrt{\frac{1}{A_L}\sum_{j=1}^{7}\left(a_j \ln^2 \frac{d_j}{\bar{D}_A}\right)} = \sqrt{\frac{1}{N_L}\sum_{j=1}^{7}\left(n_j \ln^2 \frac{d_j}{\bar{D}_N}\right)} \tag{14.20}$$

Thus, $\sigma_g$ is the same for a mass, volume, surface-area, or number distribution.

#### 14.4.1.2. *Creating a Model Lognormal Distribution*

Once parameters have been found for a lognormal mode, the mode can be discretized over a model size grid. The mass concentration of particles distributed from the continuous lognormal distribution to discrete model size bin $i$ is

$$m_i = \frac{M_L \Delta d_i}{d_i \sqrt{2\pi} \ln \sigma_g} \exp\left[-\frac{\ln^2(d_i/\bar{D}_M)}{2\ln^2 \sigma_g}\right] \tag{14.21}$$

where $\Delta d_i$ was defined in (14.9) for the volume-ratio size distribution. Once mass concentration is known, the volume concentration, number concentration and surface-area concentration for a size bin can be calculated from (14.14), (14.10), and (14.13), respectively. Alternatively, if $V_L$, $A_L$, or $N_L$ is known for a lognormal mode, the volume, surface-area, or number concentration in a size bin can be calculated from

$$v_i = \frac{V_L \Delta d_i}{d_i \sqrt{2\pi} \ln \sigma_g} \exp\left[-\frac{\ln^2(d_i/\bar{D}_V)}{2\ln^2 \sigma_g}\right] \tag{14.22}$$

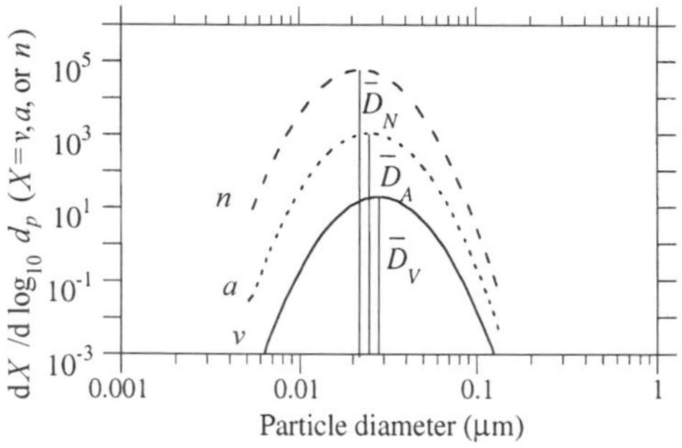


**Figure 14.3.** Number ($n$, particles cm$^{-3}$), area ($a$, cm$^2$ cm$^{-3}$), and volume ($v$, cm$^3$ cm$^{-3}$) concentrations and corresponding geometric-mean diameters for a lognormal distribution.

$$a_i = \frac{A_L \Delta d_i}{d_i \sqrt{2\pi} \ln \sigma_g} \exp\left[-\frac{\ln^2(d_i/\bar{D}_A)}{2\ln^2 \sigma_g}\right] \tag{14.23}$$

$$n_i = \frac{N_L \Delta d_i}{d_i \sqrt{2\pi} \ln \sigma_g} \exp\left[-\frac{\ln^2(d_i/\bar{D}_N)}{2\ln^2 \sigma_g}\right] \tag{14.24}$$

respectively. Figure 14.3 shows a plot of number, surface area, and volume concentrations for a lognormal distribution calculated from (14.22)–(14.24).

The total volume and number concentrations in a lognormal mode are related by

$$V_L = \int_0^\infty v_d \, \mathrm{d}d = \frac{\pi}{6} \int_0^\infty n_d d^3 \, \mathrm{d}d = \frac{\pi}{6} \bar{D}_N^3 \exp\left(\frac{9}{2} \ln^2 \sigma_g\right) N_L \tag{14.25}$$

Thus, if $N_L$ is known, $V_L$ and $v_i$ can be found from (14.25) and (14.22), respectively.

#### 14.4.1.3. *Multiple Lognormal Modes and Particle Components*

Particle distributions in the atmosphere are described by up to four lognormal modes. Such modes may include a nucleation mode, two subaccumulation modes, and a coarse-particle mode. The **nucleation mode** (geometric-mean diameter <0.1 μm) contains nucleated or small emitted particles. Sulfuric acid and water homogeneously nucleate together under the right conditions. Hot vapors from combustion nucleate quickly to form small particles. Automobiles also emit particles in the nucleation mode that contain elemental and organic carbon. Small nucleated or emitted particles increase in size by coagulation and growth.

Growth and coagulation move particles into the **accumulation mode**, where diameters are 0.1–2 μm. Some of these particles are lost by rainout and washout, but they are too light to sediment out significantly. Particles in the nucleation and accumulation modes, together, are **fine particles**. The accumulation mode often consists of two sub modes with geometric-mean diameters near 0.2 μm and 0.5–0.7 μm, respectively (Hering and Friedlander 1982; John et al. 1989). Hering and

Friedlander observed that the mass-median diameter of particles containing sulfate in an urban area was about 0.20 μm on dry days ( $f_r$ = 17–68 percent) and 0.54 μm on moist days ( $f_r$ = 26–100 percent).

The **coarse-particle mode** consists of particles larger than 2-μm diameter. These particles originate from wind-blown dust, sea spray, volcanos, plants, and other sources and are generally heavy enough to sediment out (Whitby 1978). The emission sources and deposition sinks of fine particles differ from those of coarse particles. Fine particles usually do not grow by condensation much larger than 1 μm, indicating that coarse mode particles originate primarily from emissions.

The mass of individual components within particles can be described with lognormal modes different from those of the whole particle. Mylonas et al. (1991) found that the distribution of organonitrates in ambient aerosols is typically bimodal. Venkataraman and Friedlander (1994) found that ambient elemental carbon had two distinct modes, one between 0.05 and 0.12 μm and the other between 0.5 and 1.0 μm in diameter. Sulfate in coarse-mode particles has been observed in Los Angeles to have a larger geometric-mean mass diameter than nitrate (Noll et al. 1990). Sodium and chloride appear mostly in large sea-spray particles. Kritz and Rancher (1980) measured the size distribution of particles containing NaCl over the ocean, and found the geometric-mean volume diameter of these particles to be about 6.9 μm.

To initialize a model size distribution with several particle components, each with multiple lognormal modes, the following steps may be taken. First, lognormal paramcters, such as the total mass concentration $M_{L,q,k}$, geometric-mean mass diameter $\bar{D}_{M,q,k}$, and geometric standard deviation $\sigma_{g,q,k}$, are obtained for each mode $k$ and component $q$. Second, the mass concentration of each component ($m_{q,i,k}$) is distributed into each size bin $i$ from each lognormal mode with (14.21). When a highly resolved model size distribution is used (e.g., as $V_{rat}$ approaches unity), (14.21) conserves mass. Thus,

$$\sum_{i=1}^{N_B} m_{q,i,k} \approx M_{L,q,k} \tag{14.26}$$

Third, the mass concentration of component $q$ in bin $i$ is summed over all lognormal modes. When four lognormal modes are used, the summation is

$$m_{q,i} = m_{q,i,1} + m_{q,i,2} + m_{q,i,3} + m_{q,i,4} \tag{14.27}$$

Fourth, the volume concentration of component $q$ in size bin $i$ is calculated from the summed mass concentration as

$$v_{q,i} = \frac{m_{q,i}}{c_m \rho_q} \tag{14.28}$$

The number concentration in bin $i$ is determined from

$$n_i = \frac{1}{\upsilon_i} \sum_{i=1}^{N_V} v_{q,i} \tag{14.29}$$

which was obtained by substituting (14.12) into (14.10).

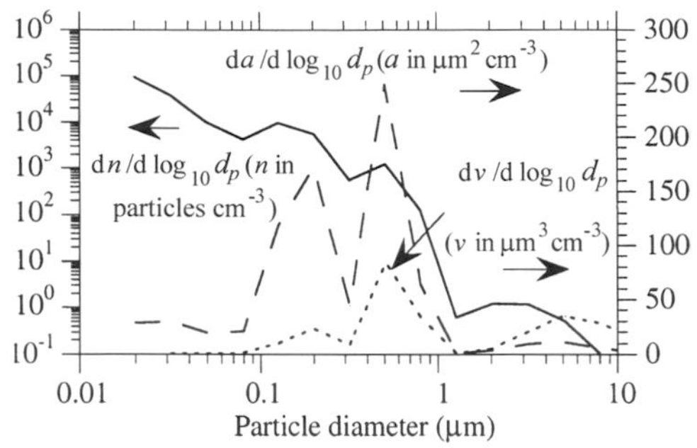


**Figure 14.4.** Number-, area-, and volume-concentration size distribution of particles at Claremont, California, on the morning of August 27, 1987. Sixteen model size bins and four lognormal modes were used to simulate the distribution (Jacobson 1997a).

**Table 14.2.** Lognormal Parameters for Typical Continental Aerosols

| Parameter | Nuclei Mode | Accumulation Mode | Coarse Particle Mode |
|---|---|---|---|
| $\sigma_g$ | 1.7 | 2.03 | 2.15 |
| $N_L$ (particles cm$^{-3}$) | $7.7 \times 10^4$ | $1.3 \times 10^4$ | 4.2 |
| $\bar{D}_N$ (μm) | 0.013 | 0.069 | 0.97 |
| $A_L$ (μm$^2$ cm$^{-3}$) | 74 | 535 | 41 |
| $\bar{D}_S$ (μm) | 0.023 | 0.19 | 3.1 |
| $V_L$ (μm$^3$ cm$^{-3}$) | 0.33 | 22 | 29 |
| $\bar{D}_V$ (μm) | 0.031 | 0.31 | 5.7 |

*Source*: Whitby (1978).

Figure 14.4 shows a quadrimodal distribution, fitted from data at Claremont, California, for the morning of August 27, 1987. The particles contained organic carbon, elemental carbon, ammonium, nitrate, sodium, chloride, liquid water, and crustal material (e.g., Mg, Al, Si, K, Ca, and Fe). All four modes are most noticeable in the number concentration distribution. The nucleation mode is less noticeable in the area concentration distribution and invisible in the volume concentration distribution.

Table 14.2 shows trimodal lognormal parameters for a size distribution that describes typical continental particles. The table shows that the nucleation mode has the highest particle number concentration, the accumulation mode has the highest surface-area concentration, and the coarse mode has the highest volume concentration of particles.

### 14.4.2. Marshall–Palmer Distribution

A second continuous size distribution, used to parameterize raindrop sizes, is the **Marshall–Palmer distribution** (Marshall and Palmer 1948). In discretized form, the distribution gives the drop number concentration (particles cm$^{-3}$) between diameters $d_i$ (μm) and $d_i + \Delta d_i$ as

$$n_i = \Delta d_i \mathrm{n}_0 e^{-\lambda_r d_i} \tag{14.30}$$

where $\Delta d_i \mathrm{n}_0$ is the value of $n_i$ at $d_i = 0$, and $\lambda_r$ is an empirical parameter that depends on the rainfall rate. Marshall and Palmer found $\mathrm{n}_0 = 8.0 \times 10^{-6}$ particles cm$^{-3}$ μm$^{-1}$ and $\lambda_r = 4.1 \times 10^{-3} R^{-0.21}$ μm$^{-1}$, where $R$ is the rainfall rate in millimeters per hour. Rainfall rates are typically 1–25 mm h$^{-1}$. If (14.30) is applied to a volume-ratio size distribution, $\Delta d_i$ is found from (14.9). The total number concentration (particles cm$^{-3}$) and liquid water content (g m$^{-3}$) in a Marshall–Palmer distribution are $n_T = \mathrm{n}_0/\lambda_r$ and $w_L = 10^{-6}\rho_w \pi \mathrm{n}_0/\lambda_r^4$, respectively, where $\rho_w$ is the liquid-water density (g cm$^{-3}$) (Pruppacher and Klett 1997).

**Example 14.6**

Estimate the number concentration of raindrops, in the diameter range $d_i = 1.0$ mm to $d_i + \Delta d_i = 2.0$ mm of a Marshall-Palmer distribution, when the rainfall rate is $R = 5$ mm h$^{-1}$. What are the total number concentration and liquid water content in the distribution?

SOLUTION

From (14.30), $n_i = 4.3 \times 10^{-4}$ particles cm$^{-3}$ in the given size range. The total number concentration and liquid water content are thus $n_T = 0.0027$ particles cm$^{-3}$ and $w_L = 0.34$ g m$^{-3}$, respectively.

### 14.4.3. Modified Gamma Distribution

A third type of continuous size distribution is the **modified gamma distribution** (Deirmendjian 1969), which is used to approximate cloud drop and raindrop number concentration as a function of size. In discretized form, the distribution gives the drop number concentration (particles cm$^{-3}$) in size bin $i$ as

$$n_i = \Delta d_i A_g r_i^{\alpha_g} \exp\left[-\frac{\alpha_g}{\gamma_g}\left(\frac{r_i}{r_{c,g}}\right)^{\gamma_g}\right] \tag{14.31}$$

where $\Delta d_i$ is the diameter width of the bin (μm), centered around mean radius $r_i$ (μm), $r_{c,g}$ is a critical radius (μm), around which the entire size distribution is centered, and $A_g$, $\alpha_g$, and $\gamma_g$ are parameterized coefficients obtained from observation. Table 14.3 shows parameters for several cloud types.

**Table 14.3.** Modified Gamma Parameters, Liquid-Water Content, and Total Drop Number Concentration for Several Cloud Types

| Cloud Type | $A_g$ | $\alpha_g$ | $\gamma_g$ | $r_{c,g}$ (μm) | Liquid-water Content (g m$^{-3}$) | Number Conc. (particles cm$^{-3}$) |
|---|---|---|---|---|---|---|
| Stratocumulus base | 0.2823 | 5.0 | 1.19 | 5.33 | 0.141 | 100 |
| Stratocumulus top | 0.19779 | 2.0 | 2.46 | 10.19 | 0.796 | 100 |
| Stratus base | 0.97923 | 5.0 | 1.05 | 4.70 | 0.114 | 100 |
| Stratus top | 0.38180 | 3.0 | 1.3 | 6.75 | 0.379 | 100 |
| Nimbostratus base | 0.080606 | 5.0 | 1.24 | 6.41 | 0.235 | 100 |
| Nimbostratus top | 1.0969 | 1.0 | 2.41 | 9.67 | 1.034 | 100 |
| Cumulus congestus | 0.5481 | 4.0 | 1.0 | 6.0 | 0.297 | 100 |
| Light rain | $4.97\times10^{-8}$ | 2.0 | 0.5 | 70.0 | 1.17 | 0.01 |

When (14.31) is summed over all sizes, the result equals the value in the last column. *Source*: Welch et al. (1980).

**Example 14.7**

Use the modified gamma size distribution equation to approximate the number concentration of drops between 14 and 16 μm in diameter at the base of a stratus cloud.

SOLUTION

The mean radius of this seze bin is $r_{c,g} \approx 7.5$ μm, and the diameter width of the bin is $\Delta d_i = 2$ μm. From Table 14.3 and (14.31), the resulting number concentration of stratus base drops in the size bin is $n_i = 19.46$ particles cm$^{-3}$.

## 14.5. EVOLUTION OF SIZE DISTRIBUTIONS OVER TIME

A **size structure** is defined here as a size distribution that evolves over time. Five size structures discussed below are the full-stationary, full-moving, quasistationary, hybrid, and moving-center structures. In all structures, each size bin contains $n_i$ particles cm$^{-3}$, each particle in a size bin has the same average volume ($\upsilon_i$) and composition as each other particle in the bin, and $v_i = n_i\upsilon_i$.

### 14.5.1. Full-Stationary Structure

In the **full-stationary size structure** (e.g., Turco et al. 1979; Gelbard and Seinfeld 1980), the average volume of particles in a size bin ($\upsilon_i$) stays constant, but $n_i$ and $v_i$ change throughout a model simulation. During growth, $\upsilon_i$ does not increase; instead, $n_i$ and $v_i$ decrease in one size bin and increase in a larger bin. Thus, particles grow numerically, not by increasing in volume, but by moving to larger size bins.

An advantage of the full-stationary structure is that it covers a wide size distribution in diameter space with a relatively few size bins. During nucleation, new particles are placed into an existing size bin, usually the smallest. During emissions, new particles are placed in or partitioned between existing size bins. During coagulation, particles from one size bin collide with and stick to particles from another bin, forming a larger particle, a fraction of which is partitioned by number between two bins. During transport, particles in one size bin and grid cell advect and replace particles with the same $\upsilon_i$ in adjacent grid cells. The full-stationary structure is convenient for nucleation, emissions, coagulation, and transport because $\upsilon_i$ is always constant.

A disadvantage of the full-stationary size structure is that when growth occurs, information about the original composition of the growing particle is lost. When liquid water condenses onto aerosols to form a fog, the original aerosol material from many small size bins merges into a few large bins, and the composition of the material in each large bin is averaged. Upon evaporation of the fog, aerosol material does not "remember" where it originated from, and evaporation results in a distribution of core aerosol material different from the initial distribution. Figure 14.5 demonstrates this problem.

A second disadvantage of the full-stationary structure is that particle growth leads to numerical diffusion in diameter space just as Eulerian advection leads to numerical diffusion in horizontal space. When particles in size bin A grow, the number of particles in a larger bin (e.g., bin B) increases. The new particles in bin B are assumed to have the same average diameter, $d_B$, as all other particles in the

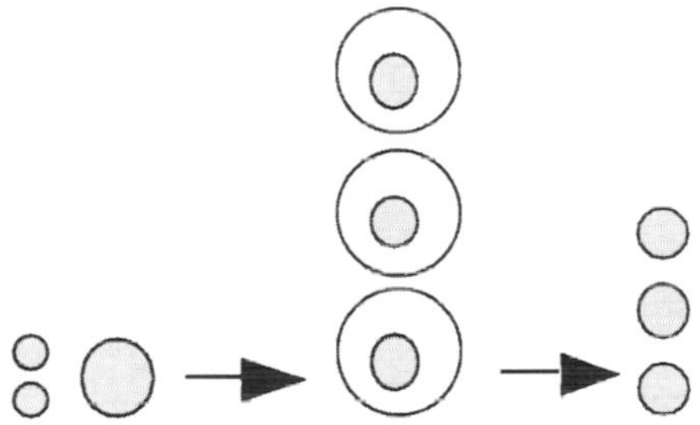

**Figure 14.5.** Demonstration of a problem with the full-stationary size structure. Suppose two particles exist in one size bin, one particle exists in a second size bin, and all three particles are composed of the same material. Suppose also that liquid water nucleates and condenses on all particles, causing them to grow to the same total size. In a full-stationary size structure, the three particles are moved to one size bin with volume $\upsilon_i$. Since the composition of all particles in a bin must be the same, the core volume of each particle must be averaged with those of all other particles in the bin. If the water now evaporates, the resulting distribution differs from the original distribution although it should not.

bin. In reality, the new particles in bin B may be smaller than $d_B$. Nevertheless, during the next time step, new particles in bin B can grow to the next larger bin, C, where the average diameter is $d_C$. In reality, they should have grown from a diameter smaller than $d_B$ to a diameter smaller than $d_C$. The artificial spreading of the distribution into bin C is numerical diffusion.

### 14.5.2. Full-Moving Structure

The extreme alternative to the full-stationary size structure is a **full-moving size structure** (e.g., Mordy 1959; Gelbard 1990; Jacobson 1994, 1997a). This structure may be initialized with a volume ratio or another size distribution. During growth, the number concentration ($n_i$) of particles in a size bin does not change. Instead, the volume ($\upsilon_i$) changes. Once $\upsilon_i$ changes, the average volume of particles in one size bin no longer equals $V_{rat}$ multiplied by the average volume of particles in a previous size bin and particle volumes no longer necessarily increase with increasing size-bin number. The full-moving structure is analogous to Lagrangian horizontal advection, just as the full-stationary size structure is analogous to Eulerian horizontal advection.

One advantage of the full-moving size structure is that core particle material is preserved during growth. The problem shown in Fig. 14.5 for a full-stationary structure is not a problem for the full-moving structure, as shown in Fig. 14.6.

The second advantage of the full-moving structure is that it eliminates numerical diffusion during growth. When one particle grows from $d_i = 0.1$ to 0.15 μm in diameter in a full-moving structure, the particle physically attains this diameter and is not partitioned by volume between adjacent size bins. Particles in a full-moving structure grow to their exact sizes, eliminating numerical diffusion. When two particles coagulate, they form a larger particle whose total volume is partitioned between the total volume of particles in two adjacent size bins (Jacobson 1994). Thus, coagulation is treated in the same way as in the full-stationary structure. For coagulation in the full-moving structure, size bins are reordered from smallest to largest each time step, since growth changes particle volumes. Figure 14.7 illustrates

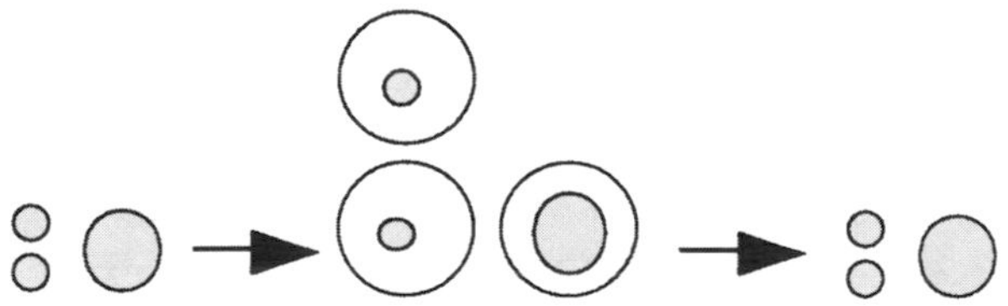

**Figure 14.6.** Preservation of an aerosol distribution upon growth and evaporation in a full-moving structure. When water condenses onto particles of different sizes, all particles may grow to the same total volume, but the core within each particle is not averaged with those of other particles of the same size. Upon evaporation, the original core distribution is preserved.

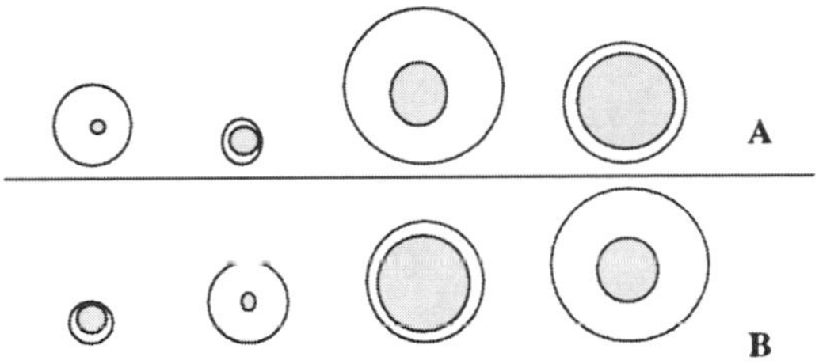


**Figure 14.7.** Reordering of particle size bins for coagulation in a full-moving structure. After growth, particles in each bin have new volumes, as illustrated by row A (the inner circle in each particle is core material, and the outer shell is volatile material). During coagulation, particle bins are reordered from smallest to largest total volume (row B), and coagulation proceeds in the same way as with a full-stationary structure.

bin reordering during coagulation.

The full-moving structure has disadvantages. Because it contains a finite number of size bins and all bins can grow to any volume, nucleation, coagulation, and transport cause problems. Suppose all size bins grow to large fog-sized drops. In such a case, a size bin no longer exists for nucleated particles. Similarly, if all particles grow to the same size, no larger size bin remains for particles to coagulate into. Transport also causes problems, since if particles in grid cell *X* and size bin *i* grow to volume $\upsilon_{X,i}$, and particles in cell *Y* and bin *i* grow to volume $\upsilon_{Y,i}$, then transport of particles from cell *X* to *Y* results in new particles in cell *Y* with volumes averaged between $\upsilon_{X,i}$ and $\upsilon_{Y,i}$. Averaging diminishes benefits of nondiffusive growth. Because of the problems described, the full-moving structure is not used for three-dimensional modeling.

### 14.5.3. Quasistationary Structure

The **quasistationary size structure** (Jacobson 1997a) is similar to the full-stationary structure in most respects. With a quasistationary structure, particle volumes grow to their exact sizes during one time step, but the adjusted volumes are fitted back onto a stationary grid at each step in a volume- and number-conserving manner. The simplest way to partition an adjusted volume is to split it between two adjacent stationary size bins. Suppose that particles in bin *i* grow by condensation to actual volume $\upsilon'_i$. This volume can be partitioned between two adjacent fixed size bins, *j* and *k*, where $\upsilon_j \le \upsilon'_i < \upsilon_k$. The partitioning is performed in a way that conserves number concentration ($n_i = n_j + n_k$) and volume concentration ($n_i\upsilon'_i = n_j\upsilon_j + n_k\upsilon_k$), where $n_j$ and $n_k$ are the number concentration of grown particles added to size bins *j* and *k*, respectively. The solution to this set of two equations and two unknowns is

$$n_j = n_i \frac{\upsilon_k - \upsilon_i'}{\upsilon_k - \upsilon_j} \qquad n_k = n_i \frac{\upsilon_i' - \upsilon_j}{\upsilon_k - \upsilon_j} \tag{14.32}$$

Like the full-stationary structure, the quasistationary structure allows practical treatment of nucleation, emission, and transport in three dimensions. Like the full-moving structure, it allows particles to grow to their exact sizes during a time step. Because it partitions number and volume concentration after growth with (14.32), the quasistationary structure is **numerically diffusive** and should be used cautiously.

### 14.5.4. Hybrid Structure

The **hybrid structure** (Jacobson 1994; Jacobson and Turco 1995) is a useful tool for box Lagrangian modeling. With this structure, volatile material (e.g., water vapor) and involatile core material (e.g., elemental carbon, soil, or sodium) are treated separately. Core material is treated over a full-stationary size structure, and volatile material is treated over a full-moving structure. The average core volume of particles in a size bin remains constant, but the total volume varies. When two particles coagulate, they become one particle whose core volume (not total volume) is partitioned between two adjacent size bins. During condensation of volatile material, particles grow to their exact sizes, eliminating numerical diffusion. Because particles do not move to a different size bin during growth, their core volumes are preserved upon evaporation.

The hybrid structure has a problem with respect to transport over an Eulerian grid. Suppose particles in a small core bin in one grid cell grow to cloud-sized drops while those in the same bin in an adjacent cell do not. Transport from one cell to the next results in unrealistic averaging of the volatile material. Thus, the hybrid structure, which is useful for box or Lagrangian modeling, is less useful for Eulerian modeling.

### 14.5.5. Moving-Center Structure

The **moving-center structure** (Jacobson 1997a) maintains advantages of the full-stationary structure, nearly eliminates numerical diffusion during growth, and is useful for three-dimensional modeling. With this structure, size-bin edges are fixed but size-bin centers change. Thus, $\upsilon_{i,hi}$, $\upsilon_{i,lo}$, and $\Delta\upsilon_i$ are fixed, and $\upsilon_i$ varies between $\upsilon_{i,hi}$ and $\upsilon_{i,lo}$. Because size-bin edges are fixed, nucleation, emissions, coagulation, and transport are treated in the same way as in the full-stationary structure. A difference is that, with the moving-center structure, particles in a size bin are allowed to grow to their exact sizes. If the average volume of a particle in a bin grows larger than the high-edge volume of the bin, all particles in the bin are moved to a single size bin bounding the average volume. Because all particles are moved to the same size bin and not fractionated among two or more size bins, numerical diffusion is reduced. Some diffusion can occur, since the volume of particles moved to the larger (or smaller) bin is averaged with the volume of

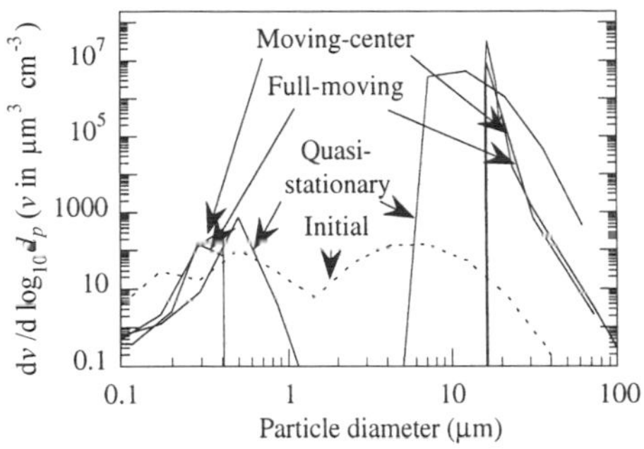


**Figure 14.8.** Comparison of moving-center, full-moving, and quasistationary size structure after growth of water onto aerosols to form cloud-sized drops. Growth occurred when the relative humidity was increased to 100.002% and replenished every 60-s time step for a 10-min period. Results from the moving-center and full-moving distributions were nondiffusive and nearly identical. The quasistationary structure was very diffusive.

particles already in the bin. Figure 14.8 shows that numerical diffusion from the moving-center structure was small during a case of cloud-drop growth.

When nucleation and emissions occur in the moving-center structure, new particles enter the size bin containing their volume, and new and current particles in the bin are averaged by volume. When transport occurs, particles in a bin move and replace particles in adjacent grid cells with the same high- and low-edge diameters. When two particles coagulate, they form one particle that is partitioned by volume between the current average volume of two adjacent size bins.

A full-moving structure eliminates numerical diffusion during growth. As shown in Fig. 14.8, the moving-center structure is also nondiffusive, but the quasistationary structure is diffusive. Because the moving-center structure treats nucleation, emissions, coagulation, and transport realistically, it is more useful than the full-moving structure in three dimensions.

## 14.6. SUMMARY

In this chapter, particle composition and size distributions were introduced. Particles can contain hundreds of components, many of which form from gas-to-particle conversion and chemical reaction. Size distributions are used to describe number, surface area, volume, or mass concentrations of particles versus diameter. A model size structure simulates the change of a size distribution over time. Size structures discussed include the full-stationary, full-moving, quasistationary, hybrid, and moving-center structures. The full-moving structure eliminates numerical diffusion during growth, but treats nucleation, emissions, and transport unrealistically and is not used for Eulerian simulations. The full-stationary and quasi-stationary structures treat these processes well, although both lead to numerical diffusion during growth. The hybrid structure eliminates numerical diffusion during growth but treats transport unrealistically. The moving-center structure treats nucleation, coagulation, and transport like the full-stationary structure does but minimizes numerical diffusion during growth. All structures have some drawbacks.

## 14.7. PROBLEMS

**14.1.** Calculate the number concentration of raindrops from the Marshall–Palmer distribution in the diameter ranges (a) 200–300 μm and (b) 800–900 μm, when $R = 25$ mm $h^{-1}$. Why are the number concentrations different in the two size ranges?

**14.2.** Compare the number concentrations of drops between 18 and 22 μm in diameter at the base and at the top of a nimbostratus cloud using a modified gamma distribution. Why do you think the concentrations differ in the two cases?

## 14.8. COMPUTER PROGRAMMING PRACTICE

**14.3.** Write a program to find $\upsilon_i$, $\upsilon_{i,lo}$, $\upsilon_{i,hi}$, $d\upsilon_i$, and $dd_i$ for a volume-ratio size distribution when $d_1 = 0.005$ μm, $d_{N_B} = 500$ μm, and $V_{rat} = 1.5$. Print a table of results.

**14.4.** **(a)** Calculate $\bar{D}_M$ and $\sigma_g$ for the size distribution resulting from the data in the accompanying table. The data give a typical distribution of emitted elemental carbon particles over a freeway.

**(b)** Using $\bar{D}_M$ and $\sigma_g$ from (a), replot the size distribution on a curve of $d_i$ versus $dm/(M_L\, d\log_{10} d_p)$, where $d\log_{10} d_p = d\log_{10} d_i = \log_{10} d_{i+1} - \log_{10} d_i$, $M_L$ is the total mass concentration (μg $m^{-3}$), and $dm$ is the mass concentration in each size bin (subscript $i$ refers to the model distribution size bin and subscript $j$ refers to the data distribution size bin). To obtain the curve, use a volume-ratio size distribution with $V_{rat} = 1.5$, and assume the particles are spherical. Choose appropriate values for $d_1$ and $N_B$.

**(c)** If elemental carbon is emitted primarily in small particles, why is it observed in larger particles away from emission sources?

| Stage $j$ | Stage Diameter $d_j$ (μm) | Mass between $d_j$ and $d_{j-1}$ (μg $m^{-3}$) | Stage $j$ | Stage Diameter $d_j$ (μm) | Mass between $d_j$ and $d_{j-1}$ (μg $m^{-3}$) |
|---|---|---|---|---|---|
| 1 | 4 | | 5 | 0.26 | 0 |
| 2 | 2 | 0 | 6 | 0.12 | 2.61 |
| 3 | 1 | 0 | 7 | 0.075 | 47.1 |
| 4 | 0.5 | 0 | 8 | 0.05 | 10.5 |

# 15

# Aerosol Emissions and Nucleation

Two processes that increase the number concentration of particles in the atmosphere, emissions and nucleation, are discussed in this chapter. Horizontal advection, vertical convection/eddy diffusion, and sedimentation also move particles into or remove particles from a region. Particle emissions originate from natural and human sources. Globally, emission rates from natural sources exceed those from human sources. In many urban areas, the reverse is true. Nucleation may occur on the surface of an existing particle (heterogeneous nucleation) or in the absence of preexisting surfaces (homogeneous nucleation). Nucleation may also involve the agglomeration of one, two, or more types of molecules.

## 15.1. EMISSIONS

New particles emitted into the atmosphere are **primary particles**. Particles produced by a gas-to-particle conversion process, such as nucleation, are **secondary particles**. Primary particles are emitted naturally or anthropogenically.

An important natural source of new particles is volcanic eruptions. **Volcanos** eject particles containing silicate, sulfate, chloride, fluoride, water, and/or other material into the air. Volcanos also emit secondary-particle precursor gases, such as sulfur dioxide ($SO_2$) and carbonyl sulfide (OCS). If OCS is injected into the stratosphere, its photodissociation products react to form sulfur dioxide, which nucleates with water to form sulfuric-acid–water aerosols. Over time, a layer of sulfuric-acid–water particles, called the **Junge layer**, has formed in the stratosphere by this process (Junge 1961). The average diameter of these particles is about 0.14 μm. Sulfate is the dominant particle constituent, not only in the stratosphere, but also the upper troposphere. More than 97 percent of particles in the lower stratosphere and 91–94 percent of particles in the upper troposphere contain oxygen and sulfur in detectable quantities (Sheridan et al. 1994).

Other natural emission sources of new particles include wood fires, bacterial decomposition, weathering, and wind injection. **Wood fires** emit particles containing organic carbon, elemental carbon, chloride, and other substances. They also emit reactive organic gases, some of which convert to the particle phase. **Bacteria** break down dead organic matter, such as plant litter or animal tissue, into new particles. **Weathering** occurs when wind, water, chemical reactions, or heat disintegrates soil or rocks containing silicon, aluminum, iron, or other substances. **Wind** lifts dust,

**Table 15.1.** Concentrations of Major Constituents in Sea Water

| Constituent | Sea-Water Conc. ($mg\ L^{-1}$) | Constituent | Sea-Water Conc. ($mg\ L^{-1}$) |
|---|---|---|---|
| Water | $1.00 \times 10^6$ | Sulfur | $9.05 \times 10^2$ |
| Sodium | $1.08 \times 10^4$ | Calcium | $4.12 \times 10^2$ |
| Chloride | $1.94 \times 10^4$ | Potassium | $3.99 \times 10^2$ |
| Magnesium | $1.29 \times 10^3$ | Carbon | $2.80 \times 10^1$ |

Source: Lide (1993).

sand particles, pollen, sea spray, and other particles from the ground or sea surface into the air.

Sea spray and spume drop emissions are also important particle sources. Sea **spray** forms when winds and waves force air bubbles to burst at the sea surface (Woodcock 1953). **Spume drops** form when wind tears off wave crests to form large drops. Sea-spray emission parameterizations have been developed by Monahan et al. (1986), Andreas (1992), and Wu (1993), among others. Sea spray initially contains all the substances of sea water. About 3.5 percent of sea-spray weight is sea salt, 85 percent of which is sodium chloride (Graedel and Weschler 1981). Table 15.1 shows the concentrations of the major constituents in sea water. The table shows that the chloride-to-sodium mass ratio in seawater is about 1.8 (also see Duce and Hoffman 1976; Hitchcock et al. 1980).

When sea water is emitted as burst bubble drops or spume drops, the chloride-to-sodium ratio in the drops decreases to <1.0–1.7 because hydrochloric acid (HCl) is removed by dehydration or acidification (Ericksson 1960; Duce 1969; Martens et al. 1973; Hitchcock et al. 1980; Kritz and Rancher 1980). **Dehydration** occurs when water from a drop evaporates. As water evaporates, HCl becomes undersaturated at the drop surface and evaporates as well. Sodium does not become undersaturated at the drop surface because its saturation vapor pressure is much lower than that of hydrochloric acid. **Acidification** occurs when acids, such as sulfuric and nitric acid, enter a drop. The anions from these acids (e.g., $HSO_4^-$, $SO_4^{2-}$, $NO_3^-$) decrease the solubility of chloride ($Cl^-$), forcing chloride to evaporate as HCl. Hitchcock et al. (1980) observed that some marine aerosols that react with nitric and sulfuric acid lose up to 100 percent of their chloride.

A stratosphere source of new particles is **meteoric debris**. Most meteorites disintegrate before they reach 80-km altitude. Those that reach the stratosphere contain iron, titanium, and aluminum, among other elements. The net contribution of meteorites to particles in the stratosphere is small (Sheridan et al. 1994; Pruppacher and Klett 1997).

Human sources of particles include coal, oil, gasoline, and wood combustion, chemical processing, and mechanical resuspension. Resuspension occurs when automobiles, tractors, and other moving vehicles kick up dust. In urban areas, elemental and organic carbon emissions from combustion are large sources of particles. Mechanical resuspension of silicon, aluminum, and iron is also important. Table 15.2 gives estimated particulate emissions over a $400 \times 150$-km region of the

**Table 15.2.** Emission Rates of Aerosol Constituents in the Los Angeles Basin as a Function of Particle Diameter

| | Emission Rate (Kg day $^{-1}$) | | | | | |
|---|---|---|---|---|---|---|
| Substance | <1 μm | 1–2.5 μm | 2.5–10 μm | >10 μm | All sizes | Total % |
| Other (O, H, etc.) | 147,884 | 86,431 | 380,221 | 712,049 | 1,326,585 | 53.167 |
| Silicon | 37,312 | 37,086 | 183,641 | 166,527 | 424,566 | 17.015 |
| Organic carbon | 28,462 | 9,363 | 69,967 | 58,111 | 165,903 | 6.649 |
| Aluminum | 14,550 | 14,644 | 67,991 | 58,216 | 155,401 | 6.228 |
| Iron | 7,090 | 7,189 | 37,210 | 38,947 | 90,436 | 3.624 |
| Calcium | 5,587 | 5,511 | 32,619 | 34,028 | 77,745 | 3.116 |
| Sulfates | 45,922 | 894 | 3,998 | 3,122 | 53,936 | 2.162 |
| Potassium | 7,364 | 3,586 | 16,266 | 18,989 | 46,205 | 1.852 |
| Elemental carbon | 28,467 | 1,095 | 7,247 | 7,429 | 44,238 | 1.773 |
| Unknown | 9,919 | 6,745 | 11,110 | 11,903 | 39,677 | 1.590 |
| Chloride | 11,318 | 814 | 4,535 | 4,796 | 21,463 | 0.860 |
| Titanium | 1,048 | 877 | 4,241 | 4,716 | 10,882 | 0.436 |
| Sulfur | 618 | 573 | 3,216 | 2,129 | 6,536 | 0.262 |
| Carbonate ion | 306 | 162 | 2,514 | 1,879 | 4,861 | 0.195 |
| Sodium | 569 | 233 | 2,080 | 1,916 | 4,798 | 0.192 |
| Manganese | 899 | 521 | 1,511 | 1,824 | 4,755 | 0.191 |
| Phosphorous | 130 | 286 | 1,660 | 1,148 | 3,224 | 0.129 |
| Nitrates | 1,237 | 147 | 935 | 782 | 3,101 | 0.124 |
| Zinc | 226 | 154 | 729 | 674 | 1,783 | 0.071 |
| Lead | 173 | 156 | 758 | 653 | 1,740 | 0.070 |
| Barium | 79 | 88 | 544 | 856 | 1,567 | 0.063 |
| Ammonium | 841 | 51 | 120 | 136 | 1,148 | 0.046 |
| Strontium | 25 | 42 | 308 | 364 | 739 | 0.030 |
| Vanadium | 94 | 66 | 274 | 280 | 714 | 0.029 |
| Copper | 132 | 60 | 203 | 208 | 603 | 0.024 |
| Cobalt | 127 | 52 | 158 | 212 | 549 | 0.022 |
| Nickel | 130 | 48 | 132 | 158 | 468 | 0.019 |
| Chromium | 87 | 26 | 158 | 176 | 447 | 0.018 |
| Rubidium | 11 | 12 | 91 | 100 | 214 | 0.009 |
| Zirconium | 6 | 10 | 80 | 110 | 206 | 0.008 |
| Lanthanum | 26 | 7 | 52 | 65 | 150 | 0.006 |
| Bromine | 68 | 5 | 30 | 24 | 127 | 0.005 |
| Arsenic | 26 | 3 | 10 | 19 | 58 | 0.002 |
| Cadmium | 9 | 2 | 23 | 16 | 50 | 0.002 |
| Antimony | 5 | 3 | 15 | 23 | 46 | 0.002 |
| Yttrium | 2 | 3 | 17 | 22 | 44 | 0.002 |
| Tin | 6 | 7 | 15 | 14 | 42 | 0.002 |
| Indium | 5 | 1 | 11 | 12 | 29 | 0.001 |
| Mercury | 2 | 1 | 11 | 13 | 27 | 0.001 |
| Molybdenum | 3 | 1 | 7 | 9 | 20 | <0.001 |
| Silver | 5 | 2 | 6 | 6 | 19 | <0.001 |
| Palladium | 3 | 1 | 5 | 10 | 19 | <0.001 |
| Selenium | 3 | 0 | 1 | 2 | 6 | <0.001 |
| Gallium | 0 | 0 | 5 | 0 | 5 | <0.001 |
| Totals | 350,776 | 176,958 | 834,725 | 1,132,673 | 2,495,132 | 100.00 |
| % of total | 14.06 | 7.09 | 33.45 | 45.40 | 100.00 | |

Emissions of several species may be significantly overestimated. The component "Other" comprises oxygen, hydrogen, and other components not included in the aerosol analysis (e.g., the oxygen atoms in oxides of silicon, aluminum, and iron). Data were provided by Allen and Wagner (1992).

Los Angeles Basin for a day in August 1987. This inventory includes particulate emissions in four size regimes: <1-, 1–2.5-, 2.5–10-, and >10-μm diameter.

Of the emissions in Table 15.2, about 91 percent originated from area sources, 7 percent originated from mobile source, and 3 percent originated from point sources. Some of the largest sources of $PM_{10}$ (particles smaller then 10 μm in diameter) were dust from paved (32 percent) and unpaved (29.7 percent) roads, and construction/demolition emissions (23.5 percent).

Table 15.2 shows that about 21 tons per day of particulate chloride were emitted in Los Angeles in August 1987. The important sources of land-based chloride are forest burning, gasoline combustion, agricultural burning, fireplace burning, chemical manufacturing, and soil dust (CARB 1988; Saxena et al. 1993). Gaseous hydrochloric acid is also emitted during coal combustion and waste incineration. Almost all (98 percent) of anthropogenic HCl(g) emissions are from coal combustion (Saxena et al. 1993).

The table shows that only 3 tons of particulate nitrate were emitted per day in 1987 in Los Angeles. The major atmospheric source of particulate nitrate is not emissions but gas-to-particle conversion of nitric acid gas. The table also shows that about 28 tons per day of elemental carbon <1 μm in diameter were emitted. Much of this originated from automobile exhaust. About 85 percent of elemental carbon and other particulate material originating from automobile combustion (e.g., organic carbon) reside in particles less than 0.12 μm in diameter (Venkataraman et al. 1994).

Globally, natural sources contribute more to particle loading than do anthropogenic sources. Natural sources emit 250–1610 Tg/yr of primary particles and enough gases to produce 345–2080 Tg/yr of secondary-particle material. Anthropogenic sources emit 6–224 Tg/yr of primary particles and enough gases to produce 140–396 Tg/yr of secondary-particle material (Jaenicke 1988; Pruppacher and Klett 1997; Wolf and Hidy 1997). The relative contribution of anthropogenic sources to global particle loadings is increasing. In urban regions, anthropogenic sources exceed natural sources of emitted particles.

## 15.2. NUCLEATION

Nucleation is a process by which gas molecules aggregate to form clusters. If the radius of a cluster reaches a critical size, the cluster becomes stable and can grow further by condensation. Nucleation is either homogeneous or heterogeneous and either homomolecular, binary, or ternary. **Homogeneous nucleation** occurs when one or more gases nucleate without the aid of a surface. Heterogeneous nucleation occurs when one or more gases nucleates on a preexisting surface. **Homomolecular nucleation** occurs when a single gas nucleates, **binary nucleation** occurs when two gases, such as sulfuric acid and water, nucleate, and **ternary nucleation** occurs when three gases, such as sulfuric acid, water, and ammonia, nucleate. Nucleation of more than three gases can occur as well. Ternary and higher-order nucleation is not discussed here.

### 15.2.1. Homogeneous Nucleation Rates from Classical Theory

Equations based on **classical nucleation theory** are used to predict homogeneous and heterogeneous nucleation rates. Although other theories exist and classical theory has been criticized for inaccurately simulating nucleation in the atmosphere, the other theories are often too computationally intensive to be used in a large model. With respect to homogeneous nucleation, classical theory predicts a production rate of new particles. The rate is derived from

$$\Delta G = 4\pi r_p^2 \sigma_p - \frac{4}{3}\pi r_p^3 \rho_p \frac{R^* T}{m_q} \ln S_q \tag{15.1}$$

which states that the **change in Gibbs free energy** ($\Delta G$)(J) during a nucleation of a cluster of molecules equals a surface-tension term minus a saturation-ratio term. In the equation, $r_p$ is the radius of the nucleating cluster (cm), $\rho_p$ is the mass density of the condensed cluster (g cm$^{-3}$), $R^*$ is the gas constant (8.314 J mole$^{-1}$ K$^{-1}$), $m_q$ is the molecular weight of the condensing gas (g mole$^{-1}$), $\sigma_p$ is the surface tension of the condensed cluster (J cm$^{-2}$), and $S_q$ is the saturation ratio of the gas. The **saturation ratio** of a gas is its partial pressure divided by its saturation vapor pressure ($S_q = p_q / p_{q,s}$).

**Surface tension** is a force per unit distance that tends to reduce the surface area of a body. The force arises because molecules in a body are equally attracted in all directions by other molecules, except at the surface, where a net inward attraction exists. Work needs to be done to increase the surface area against the surface-tension force arising from the inward attraction. The greater the tension, the greater the work, or **surface energy**, required to bring molecules to the surface from within the body.

Surface tension acts parallel to the boundary surface and is expressed in units of energy per unit area, which simplifies to force per unit distance. The surface tension of liquid water at an air–water interface has been studied as a function of temperature by Dorsch and Hacker (1951) and Gittens (1969). In the former case, measurements were taken at temperatures above and below 0°C. In the latter case, measurements were taken above 0°C. Both studies found that the surface tension of liquid water decreases with increasing temperature. Pruppacher and Klett (1997) fit a polynomial expression to Dorsch and Hacker's data for supercooled liquid water, valid for the temperature range −40 to 0°C. The polynomial fit, together with a fit to the data of Gittens for temperatures above 0°C, gives the surface tension (dyn cm$^{-1}$) as

$$\sigma_{w/a} = \sigma_p = \begin{cases} \sum_{n=0}^{6} a_n T_c^n & -40 \le T_c < 0 \\ 76.1 - 0.155 T_c & 0 \le T_c < 40 \end{cases} \tag{15.2}$$

where $\sigma_{w/a}$ is the surface tension of water against air, $T_c$ is the temperature in degrees Celsius, $a_0 = 75.93, a_1 = 0.115, a_2 = 0.06818, a_3 = 6.511 \times 10^{-3}, a_4 =$

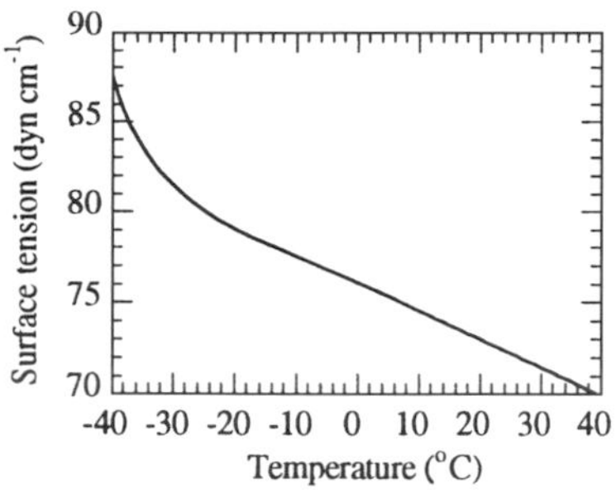


**Figure 15.1.** Surface tension of liquid water against air, obtained from (15.2).

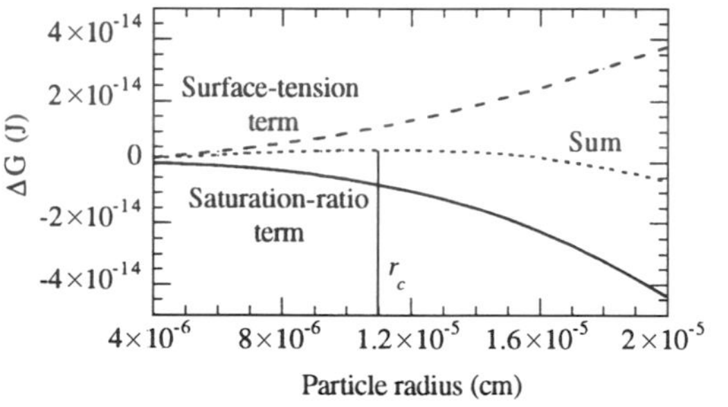


**Figure 15.2.** The net change in free energy as a function of radius is the sum of the surface tension and the supersaturation terms, as described in the text. The critical radius is the radius at which the net change in free energy with radius is maximized. Conditions for this curve are $T$ = 288 K, $\sigma_p$ = 7.5 x $10^{-6}$ J cm$^{-2}$, $S$ = 1.01, $\rho_p$ = 1.0 g cm$^{-3}$, $m_q$ = 18 g mole$^{-1}$ (water).

$2.933 \times 10^{-4}$, $a_5 = 6.283 \times 10^{-6}$, and $a_6 = 5.285 \times 10^{-8}$. Figure 15.1 shows a plot of (15.2). The surface tension should decrease with decreasing radius of curvature, but researchers have found that the surface tension of a liquid water cluster with 13 molecules is anywhere from 0 to 40 percent less than that of a flat liquid water surface (Pruppacher and Klett 1997).

Figure 15.2 depicts the terms on the right side of (15.1) as a saturation-ratio term and a surface-tension term, respectively. When molecules are added to a cluster, they condense, releasing latent heat (saturation-ratio term). The addition of molecules increases the energy required to expand the surface area of the cluster against the force of surface tension. The surface-tension term in Fig. 15.2 represents the energy required to expand the surface, and the saturation term represents the energy available from latent-heat release. When the sum of these two terms is increasing, the addition of a molecule to the cluster consumes more energy than it adds, and the cluster becomes more unstable. When the sum of these terms is decreasing, the

**Table 15.3.** Critical Radii and Number of Water Molecules in a Critical Cluster

| Saturation Ratio $S$ | Critical Radius (μm) | Number of Molecules |
|---|---|---|
| 1 | $\infty$ | $\infty$ |
| 1.01 | 0.11 | $2.03 \times 10^8$ |
| 1.10 | 0.011 | $2.32 \times 10^5$ |
| 1.5 | 0.0028 | $3.01 \times 10^3$ |
| 2 | 0.0016 | $6.03 \times 10^2$ |
| 5 | 0.0007 | 48 |
| 10 | 0.00048 | 16 |

$T = 288$ K, $\sigma_p = 7.5 \times 10^{-6}$ J cm$^{-2}$, $\rho_p = 1.0$ g cm$^{-3}$, and $m_q = 18$ g mole$^{-1}$.

addition of a molecule adds more energy than it consumes, and the cluster becomes more stable. When the sum reaches a maximum, the addition of a molecule causes no net energy change. The radius at which the maximum occurs is the **critical radius.** Nucleation occurs if a cluster contains enough molecules for the cluster radius to exceed the critical radius. At larger radii, condensation readily occurs on the nucleated cluster. At smaller radii, the cluster is unstable and may evaporate.

The critical radius can be found by setting $\mathrm{d}\,\Delta G/\mathrm{d}r_p = 0$ in (15.1). The result is

$$\frac{\mathrm{d}\,\Delta G}{\mathrm{d}r_p} = 8\pi r_p \sigma_p - 4\pi r_p^2 \rho_p \frac{R^* T}{m_q} \ln S_q = 0 \tag{15.3}$$

When $\mathrm{d}\,\Delta G/\mathrm{d}r_p$ is positive, the cluster is unstable; when it is zero or negative, the cluster is stable. Solving (15.3) for the radius gives the **critical radius** as

$$r_c = \frac{2\sigma_p m_q}{\rho_p R^* T \ln S_q} \tag{15.4}$$

Dividing the volume of a critical cluster $(4\pi r_c^3/3)$ by the volume of one gas molecule $[m_q/(\rho_p A)]$, where $A$ is Avogadro's number (molecules mole$^{-1}$), gives the number of molecules in a nucleated cluster as

$$n_c = \frac{32\pi \sigma_p^3 m_q^2 A}{3\rho_p^2 (R^* T \ln S_q)^3} \tag{15.5}$$

Table 15.3 shows critical radii and number of water molecules in a nucleated cluster for several saturation ratios.

An expression for the **homogeneous homomolecular nucleation rate** of particles (particles cm$^{-3}$ s$^{-1}$) from classical theory is

$$J_{hom} = 4\pi r_c^2 \beta_x Z_n N_x \exp\left(\frac{-\Delta G_{hom}^*}{k_B T}\right) \tag{15.6}$$

where

$$\Delta G^*_{hom} = 4\pi r_c^2 \sigma_p / 3 \tag{15.7}$$

is found by combining (15.1) with (15.3), $N_x$ is the number concentration (molecules cm$^{-3}$) of nucleating gas $x$, and $N_x \exp(-\Delta G^*_{hom}/k_B T)$ is an equilibrium number concentration (particles cm$^{-3}$) of nucleated clusters of critical radius $r_c$. The equilibrium cluster concentration is an ideal cluster number per unit volume of air that exists when the number of molecules added to a cluster equals that lost from the cluster. At theoretical equilibrium, the number of nucleated clusters of radius $r_c$ does not change.

In homogeneous nucleation theory, the number of nucleated clusters is not always in equilibrium. Instead, some clusters form and others evaporate over time. (15.6) accounts for the overall rate of change of cluster concentration. In that equation,

$$\beta_x = N_x \left( \frac{k_B T}{2\pi \bar{M}_x} \right)^{1/2} \tag{15.8}$$

is the number of gas molecules striking a unit surface area per second (molecules cm$^{-2}$ s$^{-1}$), where $\bar{M}_x = m_x / A$ is the mass (g) of one molecule of $x$. Units of $k_B$ are $1.3807 \times 10^{-23}$ J K$^{-1}$. In (15.8), the value of $k_B$ is $1.3807 \times 10^{-16}$ g cm$^{-2}$ s$^{-2}$ K$^{-1}$ molecule$^{-1}$.

(15.6) includes the **Zeldovich nonequilibrium factor** (Zeldovich 1942),

$$Z_n = \frac{\bar{M}_x}{2\pi r_c^2 \rho_x} \sqrt{\frac{\sigma_p}{k_B T}} \tag{15.9}$$

which takes account of the remaining differences between equilibrium and nonequilibrium cluster concentrations. Its value is less than unity; thus, time-dependent nucleation results in a cluster concentration below the equilibrium level (Pruppacher and Klett 1997).

Calculations with (15.6) indicate that water vapor, alone, does not homogeneously nucleate in the atmosphere, since its saturation ratio rarely exceeds 1.02. Because the saturation ratio of water is small, its critical radius of cluster formation is large, and (15.7) predicts a large Gibbs-free-energy change. For large values of $\Delta G$, (15.6) predicts a zero nucleation rate. Water vapor readily nucleates heterogeneously, since the free energy required to nucleate heterogeneously is much less than that required to nucleate homogeneously. The free energy required for heterogeneous nucleation is discussed shortly.

In (15.6), small changes in free energy affect the nucleation rate exponentially. Similar changes in the prefactor affect the rate linearly. A small error in the predicted saturation ratio affects the critical radius through (15.4), which affects the Gibbs free energy through (15.7). The error in free energy is amplified in the exponential factor of (15.6), causing large errors in nucleation-rate predictions. Because the saturation ratio continuously changes and is difficult to calculate precisely, nucleation rates from classical theory may be several orders of magnitude off.

Classical theory predicts the **homogeneous binary nucleation rate** as

$$J_{hom} = 4\pi r_c^2 \beta_y N_x \exp\left(\frac{-\Delta G^*_{hom}}{k_B T}\right) \tag{15.10}$$

(particles $cm^{-3}$ $s^{-1}$) where two gases, $x$ and $y$, are assumed to nucleate together. Gas $x$ is assumed to have a much larger concentration than gas $y$ ($N_x \gg N_y$). In the atmosphere, sulfuric acid commonly nucleates homogeneously with water vapor, which has a much higher number concentration than does sulfuric acid. For sulfuric-acid–water binary nucleation, gas $x$ is water vapor and gas $y$ is sulfuric acid.

In equilibrium, the number concentration of nucleated clusters of critical radius, $r_c$, is about $(N_x N_y)\exp(-\Delta G^*_{hom}/k_B T)$. Since $N_x \gg N_y$, this expression reduces to $N_x \exp(-\Delta G^*_{hom}/k_B T)$, which is used in (15.10). Because the flux of molecules to nucleating clusters is limited by the flux of the gas with the lowest concentration, $\beta$ is written in terms of gas $y$ in (15.10). The equation for $\beta_y$ (molecules $cm^{-2}$ $s^{-1}$) is

$$\beta_y = N_y \left(\frac{k_B T}{2\pi \bar{M}_y}\right)^{1/2} \tag{15.11}$$

where $\bar{M}_y$ is the mass of one molecule of gas $y$.

For basic homogeneous binary nucleation calculations, (15.10) does not include a Zeldovich nonequilibrium factor. The saturation ratio, surface tension, and particle density used to calculate $\Delta G^*_{hom}$ must be estimated as weighted averages between the two nucleating gases.

### 15.2.2. Heterogeneous Nucleation Rates from Classical Theory

During **heterogeneous nucleation**, a critical embryo forms on the surface of an existing particle, as shown in Fig. 15.3. The ability of the embryo to survive breakup depends on the **contact angle** $\theta_c$, which is the angle at which an embryo contacts a substrate at its surface. If the contact angle is 0°, the surface is completely covered, or **wetted**, with the embryo. If the contact angle is 180°, the surface is nonwettable, and no embryo forms. If water readily wets a surface, the surface is **hydrophilic**. Otherwise, it is **hydrophobic**. Contact angles of water vapor vary from surface to surface. The contact angle of water over sand is 43–52°, over soil is 65.2–68.9°, and over silver iodide is 9–17° (Pruppacher and Klett 1997). The contact angle of water over a substrate is calculated from Young's relation as

$$\theta_c = \cos^{-1}\left(\frac{\sigma_{S,a} - \sigma_{S,w}}{\sigma_{w,a}}\right) \tag{15.12}$$

where $\sigma_{S,a}$, $\sigma_{S,w}$, and $\sigma_{w,a}$ are the surface tensions of the substrate against air, the substrate against water, and water against air, respectively.

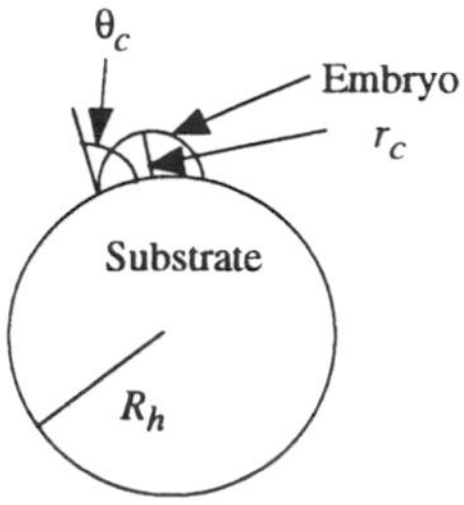


**Figure 15.3.** Formation of a critical embryo on the surface of an existing particle. $\theta_c$ is the contact angle, $R_h$ is the radius of the particle, and $r_c$ is the critical radius.

The free energy required to heterogeneously nucleate is approximated with

$$\Delta G^*_{het} = \Delta G^*_{hom} \, f_h(x_h, m_h) \tag{15.13}$$

where $f_h$ is a correction factor that depends on $x_h = R_h/r_c$, the ratio of the radius of the host particle to the critical radius, and on $m_h = \cos\theta_c$, the cosine of the contact angle. Fletcher (1958) parameterized the correction factor as

$$f_h(x_h, m_h) = 1 + \left(\frac{1 - m_h x_h}{g_h}\right)^3 + x_h^3\left[2 - 3\left(\frac{x_h - m_h}{g_h}\right) + \left(\frac{x_h - m_h}{g_h}\right)^3\right] + 3m_h x_h^2\left(\frac{x_h - m_h}{g_h} - 1\right) \tag{15.14}$$

where $g_h = \sqrt{1 + x_h^2 - 2m_h x_h}$. Figure 15.4 shows a plot of $f_h$ versus $x_h$ and $\theta_c$. When $\theta_c = 0°$, $f_h$ and $\Delta G^*_{het}$ are minimized. When $\theta_c = 180°$, $f_h$ and $\Delta G^*_{het}$ are maximized. The factors $f_h$ and $\Delta G^*_{het}$ can have small values even when $x_h < 1$, indicating that clusters may form when the host particle is smaller than the nucleating cluster.

A classical-theory expression for the **heterogeneous nucleation rate** (number of embryos cm$^{-2}$ s$^{-1}$) of new particles on a surface is

$$J_{het} = 4\pi r_c^2 \beta_y \beta_x \tau \exp\left(\frac{-\Delta G^*_{het}}{k_B T}\right) \tag{15.15}$$

where $\tau = \tau_0 \exp(E/R^* T)$ is the characteristic time (s) that a gas molecule spends on the surface before bouncing off. In the latter expression, $\tau_0$ is the characteristic time of adsorption (s), which depends on the nature of the surface and the condensing gas(es), and $E$ is the heat of adsorption (J mole$^{-1}$), which depends on the

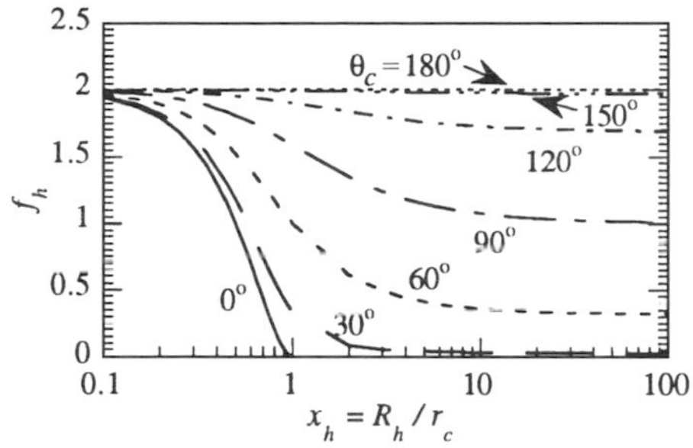


**Figure 15.4.** Values of Fletcher's correction factor as a function of $x_h$ for seven different contact angles $\theta_c$.

surface type and properties of the condensed gas. For water nucleating on solid carbon particles, $\tau_0 \approx 2.4 \times 10^{-16}$ s and $E \approx 45{,}188$ J mole$^{-1}$ (Hamill et al. 1982).

$\beta_x$ and $\beta_y$ are the numbers of molecules of gases $x$ and $y$, respectively, striking the substrate surface per second (molecules cm$^{-2}$ s$^{-1}$) and can be found from (15.8) and (15.11). If gas $x$ differs from gas $y$, the nucleation is heterogeneous binary. If the two are the same, it is heterogeneous homomolecular. Common components of binary nucleation are sulfuric acid and water vapor, since sulfuric acid has a low saturation vapor pressure, and water readily equilibrates with sulfuric acid when sulfuric acid condenses.

The product $\beta_x \tau \leq n_m$ is the number of molecules of gas $x$ on a unit area of the particle surface (molecules cm$^{-2}$), where $n_m \approx 10^{15}$ sites cm$^{-2}$ is the maximum number of sites per unit area available on a surface, as discussed in Section 12.3.4.3. Gas $x$ is chosen because, in the case of binary nucleation, it is assumed to be in the larger concentration. $J_{het}$ gives the rate per unit surface area at which embryos form when molecules of gas $y$ transfer to a surface at rate $\beta_y$ and combine with molecules of $x$ adsorbed to the surface.

### 15.2.3. Parameterized Homogeneous Nucleation Rate

The most important homogeneous nucleation process in the free atmosphere is binary nucleation of sulfuric acid and water. Homogeneously nucleated sulfuric-acid–water particles are typically 3–20 nm in diameter. In the remote atmosphere (e.g., over the ocean), homogenous nucleation generally produces $>10^4$ particles cm$^{-3}$ in this size range. Because classical theory is unreliable for predicting nucleation rates in the atmosphere, nucleation parameterizations have been developed to replace classical theory. In one parameterization, the binary homogeneous nucleation rate of sulfuric-acid–water particles was estimated for the remote marine boundary layer as

$$J_{hom} = 10^{7.0 - (64.24 + 4.7 f_r) + (6.13 + 1.95 f_r) \log_{10} N_{H_2SO_4}} \tag{15.16}$$

(e.g., Russell et al. 1994), where $J_{hom}$ is the rate of production of new sulfuric-acid–water particles (particles cm$^{-3}$ s$^{-1}$), $f_r$ is the relative humidity (expressed as a fraction here), and $N_{H_2SO_4}$ is the number concentration (molecules $cm^{-3}$) of gas-phase sulfuric acid molecules. This particular parameterization does not have a temperature dependence. Thus, it may induce errors when temperatures vary from the standard sea-level temperature.

**Example 15.1.**
Estimate the homogeneous nucleation rate of sulfuric-acid–water particles over the remote ocean when $f_r = 90$ percent and when $H_2SO_4(g)$ concentrations are (a) 0.005 and (b) 0.05 μg $m^{-3}$.

SOLUTION
In the first case, the number concentration of gas-phase sulfuric acid molecules is

$$N_{H_2SO_4} = \left(\frac{0.005\,\mu g}{m^3}\right)\left(\frac{m^3}{10^6\,cm^3}\right)\left(\frac{mole}{98 \times 10^6\,\mu g}\right) \times \left(\frac{6.02 \times 10^{23}\,molecules}{mole}\right) = 3.1 \times 10^7 \frac{molecules}{cm^3}$$

Substituting this result into (15.16) gives $J_{hom} = 2.6 \times 10^{-5}$ particles $cm^{-3}$ $s^{-1}$. In the second case, $J_{hom} = 3.1 \times 10^{-5}$ particles $cm^{-3}$ $s^{-1}$. A factor-of-10 increase in $N_{H_2SO_4}$ between the two cases increased the nucleation rate by 10 orders of magnitude. In reality, $H_2SO_4(g)$ condenses on particles as they nucleate, reducing the amount of gas available for further nucleation.

## 15.3. SUMMARY

In this chapter, aerosol emissions and nucleation were discussed. Emissions originate from human and natural sources. In urban regions, human emissions dominate. Globally, natural sources dominate. Some natural emission sources include volcanos and sea spray emissions. Homogeneous nucleation is a second means of new particle production. Such nucleation can be homomolecular or binary. Homogeneous and heterogeneous nucleation rates can be estimated with classical nucleation theory, but this theory may lead to large errors in nucleation rates when applied to the atmosphere. Empirical parameterizations of nucleation rates may be better in some cases. Once nucleated, small particles collide and coalesce during coagulation, discussed next.

## 15.4. PROBLEMS

**15.1.** Given identical vapor concentrations of water at both locations, would you expect more particles to nucleate heterogeneously over the ocean or over land? Why? Once particles have nucleated, would you expect the particles to grow larger over the ocean or over land? Why?

**15.2.** If the gas concentrations of sulfuric acid and nitric acid are the same, why should sulfuric acid and water homogeneously nucleate more readily than nitric acid and water?

**15.3.** **(a)** Calculate the critical radius of a nucleating water drop when the saturation ratio is (i) 4.0 and (ii) 1.10. Assume $\rho_p = 1.0\ \mathrm{g\ cm^{-3}}$ and $T = 298.15$ K in both cases.

**(b)** Using the information from (a), calculate the classical-theory homogeneous nucleation rate of liquid water drops in both cases.

**(c)** If the air temperature in part (a)(i) decreased from 298.15 to 283.15 K, what would the new nucleation rate be? (Be sure to calculate the change in saturation ratio.)

**15.4.** For the two cases in Problem 15.3 (a), calculate heterogeneous nucleation rates when $\theta_c = 90°$, $R_h = 0.5\ \mu\mathrm{m}$, $\tau_0 = 2.4 \times 10^{-16}$ s, $E = 45{,}188\ \mathrm{J\ mole^{-1}}$, and $T = 298.15$ K.

## 15.5. COMPUTER PROGRAMMING PRACTICE

**15.5** Write a computer script to calculate $r_c$, $J_{hom}$, and $J_{het}$ for water. Perform calculations at $T = 298$ K for $f_r = 100$–400 percent and at $f_r = 400$ percent for $T = 233$–315 K. Plot the results. Use (2.56) to calculate $p_{v,s}$. Use information from Problems 15.3 and 15.4 where needed.

# 16

# Coagulation

COAGULATION occurs when two particles collide and stick together, reducing the number concentration but conserving the volume concentration of particles in the atmosphere. More small particles than large particles are lost to larger sizes by coagulation, because the atmosphere contains many more small particles than large particles, as shown in Tables 14.1 and 14.2. Simulating coagulation in a model is important; if coagulation is neglected, erroneously large aerosol number concentrations will be predicted. Even if the total aerosol mass concentration in a model is correct, the mass concentration will be spread among too many particles. Coagulation is also important for determining the number concentration of particles that exceed the critical radius required for cloud-drop activation and determining the evolution of nucleated and newly emitted particles. In fogs and clouds, where the number concentration of drops with large cross sections is significant, coagulation (coalescence) also affects the size distribution.

To practically simulate coagulation in an Eulerian model, a numerical scheme must be fast and accurate. One scheme is described here. A solution for homogeneous particles over a single size distribution is first derived. The solution is then expanded to include any number of size distributions and particles of any composition.

## 16.1. FULLY IMPLICIT COAGULATION

The derivation of the semiimplicit solution begins with the fully implicit, integro-differential coagulation equation (Muller 1928)

$$\frac{\partial n_\upsilon}{\partial t} = \frac{1}{2}\int_0^\upsilon \beta_{\upsilon-\bar{\upsilon},\bar{\upsilon}} n_{\upsilon-\bar{\upsilon}} n_{\bar{\upsilon}}\, d\bar{\upsilon} - n_\upsilon \int_0^\infty \beta_{\upsilon,\bar{\upsilon}} n_{\bar{\upsilon}}\, d\bar{\upsilon} \tag{16.1}$$

where $\upsilon - \bar{\upsilon}$ and $\bar{\upsilon}$ are the volumes of two coagulating particles, $\upsilon$ is the volume of the new, coagulated particle, $n$ is the time-dependent number concentration (particles cm$^{-3}$) of particles of volume $\upsilon$, $\upsilon - \bar{\upsilon}$, or $\bar{\upsilon}$, and $\beta$ is the coagulation kernel (rate coefficient) of the two colliding particles (cm$^3$ particle$^{-1}$ s$^{-1}$). (16.1) states that the change in number concentration of particles of volume $\upsilon$ equals the rate at which particles of volume $\upsilon - \bar{\upsilon}$ coagulate with particles of volume $\bar{\upsilon}$ minus the rate at which particles of volume $\upsilon$ are lost due to coagulation with particles

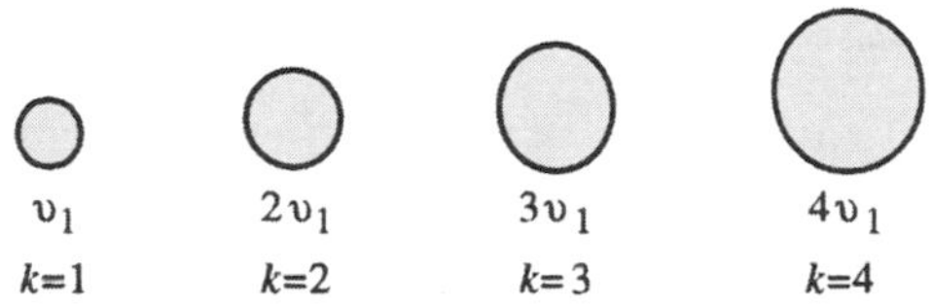


**Figure 16.1.** A monomer size distribution. The volume of particles in any size bin $k$ equals the volume of particles in the smallest bin multiplied by $k$.

of all sizes. The first integral in (16.1) is multiplied by one-half to eliminate double counting of production terms.

The coagulation equation can be written in terms of a **monomer size distribution**. In a monomer distribution, the volume of each particle in size bin $k$ equals the volume of a particle in the smallest size bin multiplied by $k$, as shown in Figure 16.1. In the monomer distribution, every possible particle volume that is a multiple of the smallest volume is accounted for. Thus, coagulation always results in particles moving to an exact bin instead of being partitioned between two bins. For monomer size bins, the coagulation equation is rewritten as

$$\frac{\partial n_k}{\partial t} = \frac{1}{2}\sum_{j=1}^{k-1} \beta_{k-j,j} n_{k-j} n_j - n_k \sum_{j=1}^{\infty} \beta_{k,j} n_j \tag{16.2}$$

where $j$ and $k$ are size-bin indices. Particles in bin $k$ are produced when particles in bin $k - j$ coagulate with particles in bin $j$. Bin $k$ corresponds to particles with volume $\upsilon_1 k$, and bin $j$ corresponds to particles with volume $\upsilon_1 j$.

(16.2) can be written in fully implicit finite-difference form as

$$\frac{n_{k,t} - n_{k,t-h}}{h} = \frac{1}{2}\sum_{j=1}^{k-1} P_{k,j} - \sum_{j=1}^{\infty} L_{k,j} \tag{16.3}$$

where $h$ is the time step, subscripts $t$ and $t - h$ are the final and initial times, respectively, and

$$P_{k,j} = \beta_{k-j,j} n_{k-j,t} n_{j,t} \qquad L_{k,j} = \beta_{k,j} n_{k,t} n_{j,t} \tag{16.4}$$

are production and loss rates, respectively. Note that $P_{k,j} = L_{k-j,j}$. Substituting (16.4) into (16.3) and rearranging gives

$$n_{k,t} = n_{k,t-h} + \frac{1}{2} h \sum_{j=1}^{k-1} \beta_{k-j,j} n_{k-j,t} n_{j,t} - h \sum_{j=1}^{\infty} \beta_{k,j} n_{k,t} n_{j,t} \tag{16.5}$$

which represents a coagulation equation for one monomer size bin $k$.

## 16.2. SEMIIMPLICIT COAGULATION

(16.5) can be solved with an iterative ordinary differential equation solver or other technique (e.g., Suck and Brock 1979; Gelbard and Seinfeld 1980; Tsang and Brock 1982; Friedlander 1983; Strom et al. 1992; Fassi-Fihri et al. 1997). The scheme discussed here (Jacobson et al. 1994; Toon et al. 1988; Turco et al. 1979) assumes that $n_{j,t-h}$ can be substituted for $n_{j,t}$ in the production and loss terms of (16.4). The terms are now written in semiimplicit form as

$$P_{k,j} = \beta_{k-j,j} n_{k-j,t} n_{j,t-h} \qquad L_{k,j} = \beta_{k,j} n_{k,t} n_{j,t-h} \tag{16.6}$$

respectively. The use of these terms allows a noniterative solution to coagulation. Combining (16.3) and (16.6), gives the **semiimplicit coagulation solution** for the number concentration of monomer particles after a time step as

$$n_{k,t} = n_{k,t-h} + \frac{1}{2} h \sum_{j=1}^{k-1} \beta_{k-j,j} n_{k-j,t} n_{j,t-h} - h \sum_{j=1}^{\infty} \beta_{k,j} n_{k,t} n_{j,t-h} \tag{16.7}$$

which can be rewritten as

$$n_{k,t} = \frac{n_{k,t-h} + \frac{1}{2} h \sum_{j=1}^{k-1} \beta_{k-j,j} n_{k-j,t} n_{j,t-h}}{1 + h \sum_{j=1}^{\infty} \beta_{k,j} n_{j,t-h}} \tag{16.8}$$

In this equation, $k$ varies from 1 to any desired number of monomer size bins. Although (16.8) correctly accounts for the reduction in particle number when two particles coagulate (reducing the number by one-half), it does not conserve volume. (16.5) correctly accounts for number and volume, but is fully implicit. In order **to conserve volume** (which coagulation physically does) though giving up some accuracy in number, (16.8) can be rederived in terms of volume concentrations of monomer particles and rewritten as

$$v_{k,t} = \frac{v_{k,t-h} + h \sum_{j=1}^{k-1} \beta_{k-j,j} v_{k-j,t} n_{j,t-h}}{1 + h \sum_{j=1}^{\infty} \beta_{k,j} n_{j,t-h}} \tag{16.9}$$

where $v_{k,t} = \upsilon_k n_{k,t}$ from (14.10). (16.9) satisfies the volume-conservation requirement, $\upsilon_{k-j} P_{k,j} = \upsilon_{k-j} L_{k-j,j}$, for each $k$ and $j$.

(16.9) is used to solve coagulation over a monomer size distribution. It is usually desirable to solve coagulation over an arbitrary size distribution, such as the volume-ratio distribution. As discussed in Chapter 14, a volume ratio size distribution is defined so that the volume of a particle in one size bin equals the volume of a particle in the next smallest size bin multiplied by a constant, $V_{rat}$.

Regardless of the size distribution, the semiimplicit solution can be altered to treat coagulation. The solution for particles of uniform composition is found by defining the volume of an intermediate particle that results when a particle in size bin $i$ collides and sticks to a particle in size bin $j$ as

$$V_{i,j} = \upsilon_i + \upsilon_j \tag{16.10}$$

The intermediate particle has volume between those of two arbitrary model size bins, $k$ and $k+1$, and needs to be partitioned between the two bins. This is done by defining the volume fraction of $V_{i,j}$ that is partitioned to each model bin $k$ as

$$f_{i,j,k} = \begin{cases} \left(\frac{\upsilon_{k+1}-V_{i,j}}{\upsilon_{k+1}-\upsilon_k}\right)\frac{\upsilon_k}{V_{i,j}} & \upsilon_k \le V_{i,j} < \upsilon_{k+1} \quad k < N_B \\ 1 - f_{i,j,k-1} & \upsilon_{k-1} < V_{i,j} < \upsilon_k \quad k > 1 \\ 1 & V_{i,j} \ge \upsilon_k \quad k = N_B \\ 0 & \text{all other cases} \end{cases} \tag{16.11}$$

(Jacobson et al. 1994). The volume fractions in (16.11) are independent of the size distribution. They work with a monomer distribution (where all values of $f$ are 1 or 0), a volume-ratio distribution, and an arbitrary distribution. When fractions are used, the semiimplicit, volume-conserving coagulation solution for total particle volume concentration becomes

$$v_{k,t} = \frac{v_{k,t-h} + h\sum_{j=1}^{k}\left(\sum_{i=1}^{k-1} f_{i,j,k}\beta_{i,j}v_{i,t}n_{j,t-h}\right)}{1 + h\sum_{j=1}^{N_B}(1 - f_{k,j,k})\beta_{k,j}n_{j,t-h}} \tag{16.12}$$

Values of $f_{i,j,k}$ in (16.12) are frequently zero. During computer simulations, all multiplications by such zeros are eliminated. The series of equations must be solved in the order $k = 1, \ldots, N_B$. No production occurs into the first bin, $k = 1$, since for it $k - 1 = 0$ in the numerator of (16.12). Thus, all $v_{i,t}$ terms are known when $v_{k,t}$ is calculated.

(16.12) can be rewritten in terms of number concentration ($n$) and particle volume ($\upsilon$) by substituting the relationship $v = n\upsilon$ throughout. The resulting volume-conserving expression for particle number concentration after a time step is

$$n_{k,t} = \frac{n_{k,t-h} + (h/\upsilon_k)\sum_{j=1}^{k}\left(\sum_{i=1}^{k-1} f_{i,j,k}\upsilon_i\,\beta_{i,j}n_{i,t}n_{j,t-h}\right)}{1 + h\sum_{j=1}^{N_B}(1 - f_{k,j,k})\beta_{k,j,t-h}\,n_{j,t-h}} \tag{16.13}$$

(16.12) gives the coagulation solution for the volume concentration of total particles. If particles have multiple components, the change in volume concentration of each component $q$ is

$$v_{q,k,t} = \frac{v_{q,k,t-h} + h\sum_{j=1}^{k}\left(\sum_{i=1}^{k-1} f_{i,j,k}\beta_{i,j}v_{q,i,t}n_{j,t-h}\right)}{1 + h\sum_{j=1}^{N_B}(1 - f_{k,j,k})\beta_{k,j}n_{j,t-h}} \tag{16.14}$$

The advantage of a semiimplicit over a fully implicit equation is that the semiimplicit equation yields an immediate volume-conserving, stable solution for coagulation with any time step. A fully implicit equation requires the use of an iterative ordinary-differential-equation solver. The disadvantage of a semiimplicit equation is that particle number is not exactly conserved. By increasing the resolution of the size distribution, the error in number concentration approaches zero while the solution remains noniterative and volume-conserving.

## 16.3. COAGULATION OVER MULTIPLE PARTICLE DISTRIBUTIONS

The semiimplicit equations can be used to solve coagulation, not only over a single size distribution with multiple components, but also over multiple size distributions, each with multiple components. Figure 16.2 shows an example of three particle size distributions, each with a different composition. For simplicity, suppose all distributions are volume-ratio size distributions. Also suppose that each distribution has the same number of size bins, and all bins with the same index number across the distributions contain particles of the same volume, but different composition. Particles in one distribution, called the **internally mixed** (I) distribution, must contain all components present in all other distributions, but may contain additional components. The simpler distributions are **externally mixed** (E) distributions. Figure 16.2 shows an example of two E and one I distribution, and Table 16.1 shows how coagulation among distributions affects number and volume concentration.

An assumption made here is that, when an E distribution **heterocoagulates** (an $E_1$ particle coagulates with an $E_2$ or an I particle), the resulting particle enters a multicomponent mixture (I). In reality, when two particles of different composition

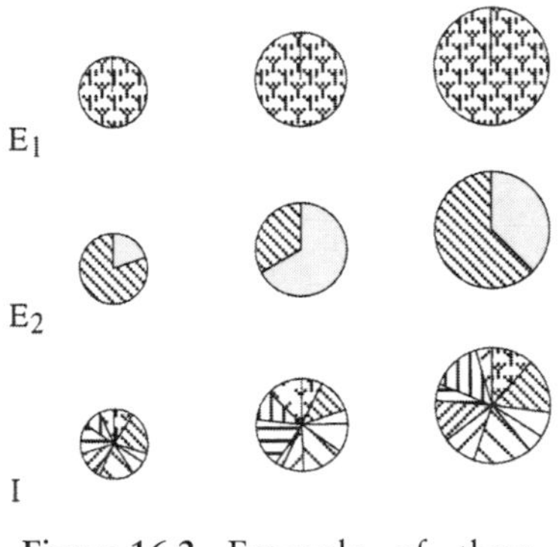


**Figure 16.2.** Example of three size distributions, each with the same number of size bins, but with particles of different composition. Distributions $E_1$ and $E_2$ are externally mixed, and I is internally mixed. I particles contain all components in the E particles, but may contain additional components.

**Table 16.1.** Changes in Particle Number and Volume during Coagulation among Different Size Distributions

| | Number | | | Volume | | |
|---|---|---|---|---|---|---|
| Coagulating Pairs | $E_1$ | $E_2$ | I | $E_1$ | $E_2$ | I |
| $E_1$ with $E_1$ | (−) | — | — | (0) | — | — |
| $E_1$ with $E_2$ | (−) | (−) | (+) | (−) | (−) | (+) |
| $E_1$ with I | (−) | — | (0) | (−) | — | (+) |
| I with I | — | — | (−) | — | — | (0) |

$E_1$ and $E_2$ are externally mixed distributions, and I is an internally mixed distribution. (+), increase upon coagulation; (−), decrease upon coagulation; (0), no change upon coagulation; —, no effect.

collide, they form a third particle of unique composition. Allowing for every combination of particle composition is computationally impractical beyond a small number of particles and spatial grid cells. Suppose a model contains 11 particle distributions, each with 1–72 components. Coagulation among the distributions results in 2047 distribution (the sum of combinations of 11 taken $1, 2, \ldots, 11$ at a time). If each resulting distribution has an average of 20 components and 30 size bins, and if the model has 10,000 grid cells, then 12.3 billion pieces of information need to be tracked at each time step. This memory requirement is beyond the capacity of most computers. By assuming that, when an E particle heterocoagulates, it produces an I particle, the number of resulting distributions in the above case is reduced from 2047 to 12, and the required memory is reduced significantly.

The semiimplicit coagulation solution for E particles differs from that for I particles. For each $N = 1, \ldots, N_X$ E-type distribution, the volume concentration of component $q$ in size bin $k$ after a time step of coagulation is

$$v_{q,Nk,t} = \frac{v_{q,Nk,t-h} + h\sum_{j=1}^{k}\left(\sum_{i=1}^{k-1} f_{i,j,k}\beta_{Ni,Nj}v_{q,Ni,t}n_{Nj,t-h}\right)}{1 + h\sum_{j=1}^{N_B}\left[(1-f_{k,j,k})\beta_{Nk,Nj}n_{Nj,t-h} + \sum_{\substack{M=1\\M\neq N}}^{N_T}\beta_{Nk,Mj}n_{Mj,t-h}\right]} \tag{16.15}$$

where $N_T$ is the number of E plus I distributions. For each E distribution, (16.15) is solved for each size bin $k = 1, \ldots, N_B$ and for each volume component $q = 1, N_V$. All implicit (time $t$) values on the right side of (16.15) are known when used. The final particle number concentration in bin $k$ of each E-type distribution is

$$n_{Nk,t} = \frac{n_{Nk,t-h} + \frac{h}{\upsilon_k}\sum_{j=1}^{k}\left(\sum_{i=1}^{k-1} f_{i,j,k}\beta_{Ni,Nj}\upsilon_i n_{Ni,t}n_{Nj,t-h}\right)}{1 + h\sum_{j=1}^{N_B}\left[(1-f_{k,j,k})\beta_{Nk,Nj}n_{Nj,t-h} + \sum_{\substack{M=1\\M\neq N}}^{N_T}\beta_{Nk,Mj}n_{Mj,t-h}\right]} \tag{16.16}$$

where $\upsilon_i$ and $\upsilon_k$ are the volumes of one particle in bin $i$ and $k$, respectively.

For the one I distribution, denoted by subscript $N$, the volume concentration of each component $q$ in size bin $k$ after a time step of coagulation is

$$v_{q,Nk,t} = \frac{v_{q,Nk,t-h} + h\sum_{M=1}^{N_T}\left[\sum_{j=1}^{k} n_{Mj,t-h}\left(\sum_{i=1}^{k-1} f_{i,j,k}\beta_{Ni,Mj}v_{q,Ni,t} + \sum_{\substack{I=1\\ I\neq M\\ M\neq N}}^{N_X}\sum_{i=1}^{k} f_{i,j,k}\beta_{Ii,Mj}v_{q,Ii,t}\right)\right]}{1 + h\sum_{\substack{M=1\\ k\neq N_B}}^{N_T}\left(\sum_{j=1}^{N_B}(1-f_{k,j,k})\beta_{Nk,Mj}n_{Mj,t-h}\right)} \tag{16.17}$$

The corresponding final number concentration in each bin is

$$n_{Nk,t} = \frac{n_{Nk,t-h} + \frac{h}{\upsilon_k}\sum_{M=1}^{N_T}\left[\sum_{j=1}^{k} n_{Mj,t-h}\left(\sum_{i=1}^{k-1} f_{i,j,k}\beta_{Ni,Mj}\upsilon_i n_{Ni,t} + \sum_{\substack{I=1\\ I\neq M\\ M\neq N}}^{N_X}\sum_{i=1}^{k} f_{i,j,k}\beta_{Ii,Mj}\upsilon_i n_{Ii,t}\right)\right]}{1 + h\sum_{\substack{M=1\\ k\neq N_B}}^{N_T}\left(\sum_{j=1}^{N_B}(1-f_{k,j,k})\beta_{Nk,Mj}n_{Mj,t-h}\right)} \tag{16.18}$$

(16.17) and (16.18) must be solved for each size bin $k = 1, \ldots, N_B$ and each volume component $q = 1, \ldots, N_V$, after (16.15) and (16.16) have been solved.

(16.15)–(16.18) are a generalized, volume-conserving semiimplicit coagulation solution for multicomponent, multitype aerosols. To conserve volume at the boundaries, (16.17) and (16.18) assume that particles cannot coagulate out of the largest size bin of I particles. They can coagulate out of the largest bin of E particles to become I particles. Upon heterocoagulation of E + I particles, all coagulating E particles become I particles.

## 16.4. COAGULATION KERNEL

The overall coagulation kernel is the sum of the kernels due to Brownian diffusion, convective Brownian diffusion enhancement, gravitational collection, turbulent inertial motion, and turbulent shear. These kernels are discussed below.

### 16.4.1. Particle Flow Regimes

Coagulation kernels depend on several parameters, including the Knudsen number for air, mean free path of an air molecule, thermal velocity in air, and particle Reynolds number. The **Knudsen number for air** is

$$\mathrm{Kn}_{a,i} = \frac{\lambda_a}{r_i} \tag{16.19}$$

where $\lambda_a$ is the mean free path of an air molecules, and $r_i$ is the radius of a particle of size $i$. The **mean free path of an air molecule** is the average distance an air molecule must travel before it encounters another air molecule by random motion and exchanges momentum with it. It is analogous to the mixing length of an eddy, defined in Chapter 8, and may be calculated as

$$\lambda_a = \frac{2\eta_a}{\rho_a \bar{v}_a} = \frac{2\nu_a}{\bar{v}_a} \tag{16.20}$$

where $\eta_a$ is the dynamic viscosity of air (g cm$^{-1}$ s$^{-1}$), $\rho_a$ is the density of air (g cm$^{-3}$), $\bar{v}_a$ is the thermal velocity of an air molecule (cm s$^{-1}$), and $\nu_a = \eta_a/\rho_a$ is the kinematic viscosity of air, defined in (4.59). The greater the air density, the closer molecules are to each other, and the shorter the mean free path of a molecule. The dynamic viscosity of air was given empirically as a function of temperature in (4.55). The average **thermal velocity of an air molecule**, derived from (2.2), is

$$\bar{v}_a = \left(\frac{8k_B T}{\pi \bar{M}}\right)^{1/2} \tag{16.21}$$

where $\bar{M} = m_d/A$ is the average mass of one air molecule. The thermal velocity depends on the temperature. The higher the temperature of the air, the greater the kinetic energy and velocity of an average air molecule.

**Example 16.1.**

Find $\lambda_a$ and $\text{Kn}_{a,i}$ when $T$ = 288 K, $p_a$ = 1013 mb, the air is dry, and $r_i$ = 0.1 μm.

SOLUTION

From (16.21), $\bar{v}_a = 4.59 \times 10^4$ cm s$^{-1}$. From (4.55), $\eta_a = 1.79 \times 10^{-4}$ g cm$^{-1}$ s-1. Since the air is dry, $R_m = R'$ and $\rho_a$ = 0.00123 g cm$^{-3}$ from (2.30). Substituting these values into (16.20) gives $\lambda_a = 6.34 \times 10^{-6}$ cm. From (16.19), $\text{Kn}_{a,i}$ = 0.63.

When a particle is large relative to the mean free path of an air molecule ($\text{Kn}_{a,i} \ll 1$), a particle is likely to be intercepted by many molecules, and its resistance to motion is due to air viscosity. This Knudsen-number regime is called the **continuum regime**, because particles see air as a continuum. When the mean free path of an air molecule is large relative to the particle size ($\text{Kn}_{a,i} \gg 10$), a particle is likely to be intercepted by relatively few air molecules, and its resistance to motion is due primarily to the inertia of air molecules hitting it. This Knudsen-number regime is called the **free molecular regime**, because particle motion is governed by free-molecular kinetics. Between the continuum and free-molecular regimes is the **transition regime**.

The **particle Reynolds number** gives the ratio of the inertial force exerted by a particle to the viscous force exerted by air. If the inertial force is due to the particles'

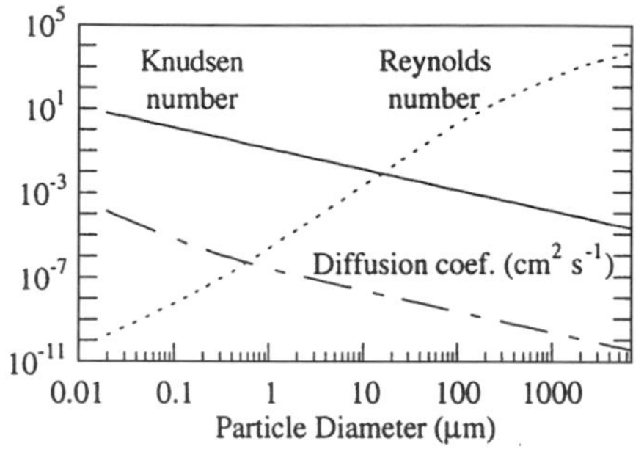


**Figure 16.3.** Knudsen number for air, Reynolds number, and diffusion coefficient of falling particles as a function of diameter when $T = 292$ K, $p_a = 999$ mb, and $\rho_p = 1.0$ g cm$^{-3}$.

sedimentation velocity, the Reynolds number (dimensionless) is

$$\mathrm{Re}_i = 2r_i V_{f,i}/\nu_a \tag{16.22}$$

where $V_{f,i}$ is the **terminal fall velocity of the particle** (cm s$^{-1}$), calculated in Chapter 20. Figure 16.3 shows Reynolds and Knudsen numbers for falling particles under specified conditions.

When particle radius is less than the mean free path of an air molecule, the particle motion through the air is called **slip flow,** because particles slip through the air with little viscous resistance. Slip flow occurs mostly in the free-molecular regime. Pure slip flow for falling particles occurs at low Reynolds numbers (e.g., $\mathrm{Re}_i < 10^{-6}$), since small, falling particles have a low terminal fall velocity, as shown in Fig. 16.3.

When the size of a particle is larger than that of an air molecule, but the inertial force imparted by the particle is smaller than the viscous force exerted by air, the flow of the particle through air is **Stokes flow**, named after George Stokes (1819–1903). Stokes flow occurs within the continuum regime, since in Stokes flow the particle radius is larger than the mean free path of air. Stokes flow for falling particles occurs at moderate Reynolds numbers (e.g., $10^{-2} < \mathrm{Re}_i < 1$). At higher Reynolds numbers, the inertial force imparted by a particle becomes larger than the viscous force exerted by air. At smaller Reynolds number, Stokes flow approaches slip flow.

### 16.4.2. Individual Kernels

**Brownian motion** is the random movement of particles suspended in a fluid. **Brownian coagulation** is the process by which particles diffuse, collide, and coalesce in the atmosphere due to random motion. When two particles collide due to Brownian motion, they may or may not stick together, depending on the efficiency of coalescence. The efficiency of coalescence depends on particle shape, composition, ambient relative humidity, and other factors. Because the kinetic energy of small particles is relatively small in comparison to that of large particles at a given temperature, the likelihood that bounce-off occurs when small-particles collide is low, and the coalescence efficiency of small-particle coagulation is approximately unity (Pruppacher and Klett 1997). Since Brownian diffusion affects primarily small particles, the coalescence efficiency for Brownian diffusion is approximately unity.

The coagulation kernel due to diffusion and collision resulting from Brownian motion depends on the flow regime. In the **continuum regime**, the **Brownian kernel** (cm$^3$ particle$^{-1}$ s$^{-1}$) for particles of size $i$ coagulating with particles of size $j$ is

$$\beta^{B}_{i,j} = 4\pi(r_i + r_j)(D_{p,i} + D_{p,j}) \tag{16.23}$$

where $D_{p,i}$ and $D_{p,j}$ are the diffusion coefficients of particles $i$ and $j$, respectively. A **particle diffusion coefficient** is calculated as

$$D_{p,i} = \frac{k_B T}{6\pi r_i \eta_a} G_i \tag{16.24}$$

where $k_B$ is Boltzmann's constant, $T$ is the Kelvin temperature, and $G_i = 1 + \mathrm{Kn}_{a,i}\alpha_{a,i}$ is the **Cunningham slip-flow correction** (Cunningham 1910). In the absence of the Cunningham correction factor, (16.24) is called the **Stokes–Einstein equation**; it is valid for Stokes flow only. At Reynolds numbers above or below the Stokes flow regime, the Stokes–Einstein equation induces an error (Fuchs 1964). When Reynolds numbers are below the Stokes flow regime, they are in the slip-flow regime, where particle resistance to motion is relatively small, since molecules move freely through air. The Cunningham correction allows for the reduced resistance in the slip-flow regime.

Knudsen and Weber (1911) approximated $\alpha_{a,i} = A' + B'\exp(-C'/\mathrm{Kn}_{a,i})$. With this equation, the Cunningham slip-flow correction becomes

$$G_i = 1 + \mathrm{Kn}_{a,i}[\mathrm{A}' + \mathrm{B}'\exp(-C'/\mathrm{Kn}_{a,i})] \tag{16.25}$$

Knusden and Weber estimated A′, B′, and C′ based on experiments with glass beads falling in air. In a more relevant experiment, Millikan (1923) estimated A′, B′, and C′ as 0.864, 0.29, and 1.25, respectively, based on measurements of oil drops falling in air. Kasten (1968) gives a more recent set of values for A′, B′, and C′ as 1.249, 0.42, and 0.87, respectively. (16.25) predicts that, at low Knudsen numbers (large particle radii), $G_i$ approaches zero, and the particle diffusion coefficient approaches that from the Stokes–Einstein equation. At high Knudsen numbers (small particle radii), $G_i$ increases, increasing the particle diffusion coefficient. Figure 16.3 shows the corrected particle diffusion coefficient as a function of particle diameter. The figure shows that, at very low particle diameters (and Reynolds numbers), diffusion coefficients are slightly enhanced by the reduced resistance to particle motion.

In the **free-molecular regime**, the Brownian kernel (cm$^3$ particle$^1$ s$^{-1}$) for particles of size $i$ coagulating with particles of size $j$ is based on the kinetic theory of gases and given by

$$\beta_{i,j}^{B} = \pi(r_i + r_j)^2\left(\bar{v}_{p,i}^2 + \bar{v}_{p,j}^2\right)^{1/2} \tag{16.26}$$

where

$$\bar{v}_{p,i} = \left(\frac{8k_B T}{\pi \bar{M}_{p,i}}\right)^{1/2} \tag{16.27}$$

is the **thermal velocity** and $\bar{M}_{p,i}$ is the mass of one particle of size $i$. (16.27) indicates that the kinetic energy of light (small) particles exceeds that of heavy (large) particles at a given temperature.

For particles in the **transition regime**, the Brownian kernel can be calculated with the interpolation formula of Fuchs (1964),

$$\beta_{i,j}^{B} = \frac{4\pi(r_i + r_j)(D_{p,i} + D_{p,j})}{\frac{r_i + r_j}{r_i + r_j + \left(\delta_i^2 + \delta_j^2\right)^{1/2}} + \frac{4(D_{p,i} + D_{p,j})}{\left(\bar{v}_{p,i}^2 + \bar{v}_{p,j}^2\right)^{1/2}(r_i + r_j)}} \tag{16.28}$$

where $\delta_i$ is the mean distance from the center of a sphere reached by particles leaving the sphere's surface and traveling a distance of particle mean free path $\lambda_{p,i}$. These two parameters are, respectively,

$$\delta_i = \frac{(2r_i + \lambda_{p,i})^3 - \left(4r_i^2 + \lambda_{p,i}^2\right)^{3/2}}{6r_i\lambda_{p,i}} - 2r_i \qquad \lambda_{p,i} = \frac{2D_{p,i}}{\pi \bar{v}_{p,i}} \tag{16.29}$$

(16.28) simplifies to (16.23) for small Knudsen numbers and to (16.26) for large Knudsen numbers.

When large particles fall in the atmosphere, eddies created in their wake enhance diffusion of other particles to their surfaces. The coagulation kernel due to this process, called **convective Brownian diffusion enhancement**, is parameterized as

$$\beta_{i,j}^{DE} = \begin{cases} \beta_{i,j}^{B}\ 0.45\mathrm{Re}_i^{1/3}\ \mathrm{Sc}_{p,j}^{1/3} & \mathrm{Re}_i \le 1 \\ \beta_{i,j}^{B}\ 0.45\mathrm{Re}_i^{1/2}\ \mathrm{Sc}_{p,j}^{1/3} & \mathrm{Re}_i > 1 \end{cases} \tag{16.30}$$

(Pruppacher and Klett 1997), where $\mathrm{Sc}_{p,j}$ is the **particle Schmidt number**. The Schmidt number gives the ratio of viscous to diffusive forces and is calculated as

$$\mathrm{Sc}_{p,i} = \nu_a / D_{p,i} \tag{16.31}$$

(16.30) states that the larger the particle and the higher its fall velocity, the greater the effect of diffusion enhancement.

A third coagulation kernel is that for gravitational collection (differential fall velocities). When two particles of different sizes fall, one may catch up and collide with the other. The kinetic energy of the larger particle is higher, increasing the chance that collision will result in a bounce-off rather than a coalescence. Thus, the **coagulation kernel for gravitational collection** includes a coalescence (collection) efficiency term and is estimated as

$$\beta_{i,j}^{GC} = E_{i,j}\pi(r_i + r_j)^2 |V_{f,i} - V_{f,j}| \tag{16.32}$$

where $E_{i,j}$ is the **coalescence** (collection) **efficiency**. A parameterization for the coalescence efficiency is

$$E_{i,j} = \frac{60 E_{V,i,j} + E_{A,i,j}\mathrm{Re}_j}{60 + \mathrm{Re}_j} \tag{16.33}$$

$$E_{V,i,j} = \begin{cases} \left(1 + \frac{0.75 \ln(2\mathrm{K}_{i,j})}{\mathrm{K}_{i,j} - 1.214}\right)^{-2} & \mathrm{K}_{i,j} > 1.214 \\ 0 & \mathrm{K}_{i,j} < 1.214 \end{cases} \qquad E_{A,i,j} = \frac{\mathrm{K}_{i,j}^2}{(\mathrm{K}_{i,j} + 0.5)^2} \tag{16.34}$$

(Ludlum 1980) where $\mathrm{K}_{i,j} = V_{f,i}|V_{f,j} - V_{f,i}|/r_j g$ is the dimensionless **Stokes number**. (16.33) simplifies to $E_{V,i,j}$ when $\mathrm{Re}_j \ll 1$ and to $E_{A,i,j}$ when $\mathrm{Re}_j \gg 1$. Gravitational collection is the most important mechanisum by which cloud drops form raindrops in warm clouds (clouds with temperature above 0°C).

Two more kernels are those for turbulent inertial motion and turbulent shear. When **turbulent inertial motion** causes particles of different sizes to accelerate differentially, the resulting coagulation kernel may be written as

$$\beta_{i,j}^{TI} = \frac{\pi \varepsilon_k^{3/4}}{g \nu_a^{1/4}} (r_i + r_j)^2 |V_{f,i} - V_{f,j}| \tag{16.35}$$

(Saffman and Turner 1956). In this equation, $\varepsilon_k$ is the rate of dissipation of turbulent kinetic energy per gram of medium (cm$^2$ s$^{-3}$). Values for $\varepsilon_k$ vary between 3 and 2000 cm$^2$ s$^{-3}$, with lower values corresponding to clear or partly cloudy air and larger values corresponding to strong cumulus convection. A typical value for clear air is 5 cm$^2$ s$^{-3}$ (Pruppacher and Klett 1997).

When wind shear in turbulent air causes particles moving with the air to collide, the coagulation kernel is parameterized with the kernel for **turbulent shear**,

$$\beta_{i,j}^{TS} = \left(\frac{8\pi\varepsilon_k}{15\nu_a}\right)^{1/2} (r_i + r_j)^3 \tag{16.36}$$

(Saffman and Turner 1956). Two small particles are more likely to collide because of turbulent shear than because of turbulent inertial motion. A large particle is more likely to collide with a small particle because of turbulent inertial motion than because of turbulent shear. Brownian motions dominate both processes so long as at least one particle is small.

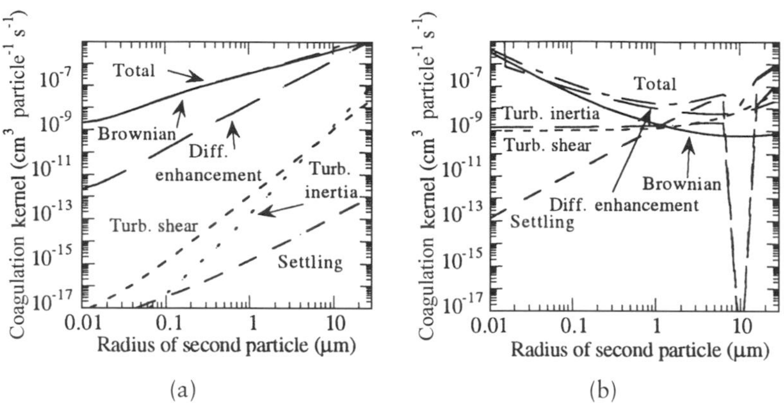


**Figure 16.4.** (a) Coagulation kernels for five processes when a particle 0.01 μm in radius coagulates with particles of different sizes at 298 K. (b) Coagulation kernels for five processes when a particle 10 μm in radius coagulates with particles of different sizes at 298 K. The dip at 10 μm results because the difference in fall velocities is zero at that point. The real width of the dip is narrower, but the resolution of the size bins used for the graph was coarse.

Figure 16.4(a) and (b) show how each of five coagulation kernel varies when particles of 0.01 and 10 μm, respectively, coagulate with particles of all sizes. The curves indicate that, for small particles, Brownian coagulation is always dominant. For larger particles, other kernels become important. Inclusion of the remaining kernels in simulating cloud microphysical processes is necessary. When two particles have the same size, their fall velocities are identical, and their coagulation kernels for turbulent inertial motions and gravitational collection are zero, as seen in Fig 16.4(b).

## 16.5. COMPARISON WITH ANALYTICAL SOLUTIONS

Here, the semiimplicit coagulation solution is compared with analytical and numerical solutions. Smoluchowski's (1918) analytical solution assumes that particles are initially **monodisperse** (all have the same size) and the coagulation kernel is constant. After initialization, coagulation occurs over a monomer size distribution. Atmospheric particles are really **polydisperse** (particle number concentrations vary with size). Nevertheless, Smoluchowski's analytical solution is useful for checking the accuracy of numerical coagulation schemes. Smoluchowski's solution is

$$n_{k,t} = \frac{n_{T,t-h}(0.5h\beta n_{T,t-h})^{k-1}}{(1 + 0.5h\beta n_{T,t-h})^{k+1}} \tag{16.37}$$

where $n_{k,t}$ is the final concentration of particles in monomer size bin $k$ after time $h$, $n_{T,t-h}$ is the initial number concentration of monodisperse particle, the exponents $k-1$ and $k+1$ refer to the size-bin number, the subscripts $t$ and $t-h$ refer to the

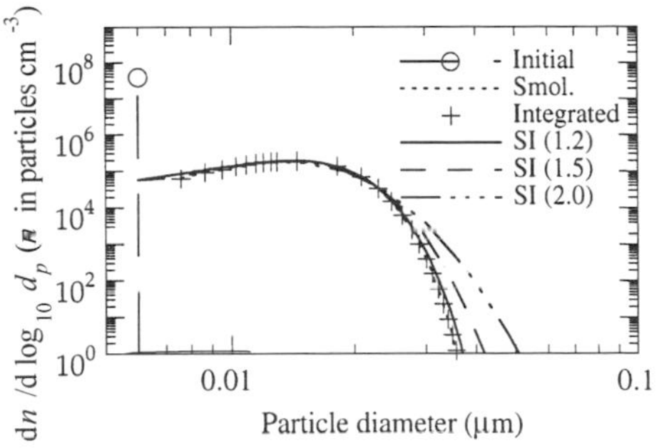


**Figure 16.5.** Comparison of Smoluchowski's analytical solution with an integrated numerical solution and with three semiimplicit solutions (SI), each with a different value of $V_{rat}$ (given in parentheses). The initial number concentration of monodisperse particles was $10^7$ particles cm$^{-3}$ and $T = 298$ K. The simulation time was 12 h. Remaining conditions are described in the text.

final and initial times, respectively, and

$$\beta = \frac{8k_B T}{3\eta_a} \tag{16.38}$$

is the constant coagulation kernel. (16.37) is solved over one time step $h$ of any size. Figure 16.5 compares Smoluchowski's solution with an integrated and three semiimplicit solutions of (16.37) over a full-stationary size structure that assumes a volume-ratio size distribution. The semiimplicit solutions improved with higher resolution (smaller $V_{rat}$). The integrated solution matched the analytical solution exactly, but the computer times required for the semiimplicit solutions were much less than that required for the integrated solution.

Another analytical solution is that for a self-preserving size distribution. Such a distribution has initial number concentration in each size bin $i$ of

$$n_{i,t-h} = \frac{n_{T,t-h}\,\Delta\upsilon_i}{\upsilon_p}\exp\left(\frac{\upsilon_i}{\upsilon_p}\right) \tag{16.39}$$

where $n_{T,t-h}$ is the total number concentration of particles among all size bins, $\upsilon_i$ is the volume of a single particle in size bin $i$, $\upsilon_p$ is the volume at which the peak number concentration occurs in the distribution, and $\Delta\upsilon_i$ is the volume width of a size bin. For a volume-ratio size distribution, $\Delta\upsilon_i$ is calculated with (14.8).

After time $h$, the solution for coagulation over the self-preserving distribution is

$$n_{i,t} = \frac{n_{T,t-h}\Delta\upsilon_i/\upsilon_p}{(1+0.5h\beta n_{T,t-h})^2}\exp\left(-\frac{\upsilon_i/\upsilon_p}{1+0.5h\beta n_{T,t-h}}\right) \tag{16.40}$$

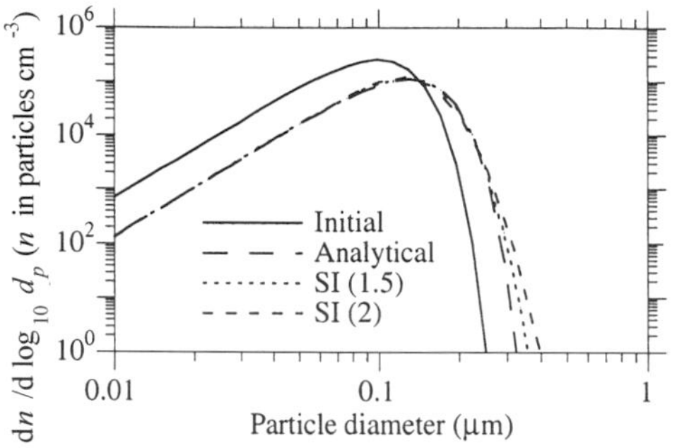


**Figure 16.6.** Comparison of self-preserving analytical solution with two semiimplicit solutions (SI) each with a different value of $V_{rat}$ (given in parentheses). The conditions were $n_{T,t-h} = 10^6$ cm$^{-3}$, $T = 298$ K, and $\upsilon_p = 0.1$ μm. The simulation time was 12 h.

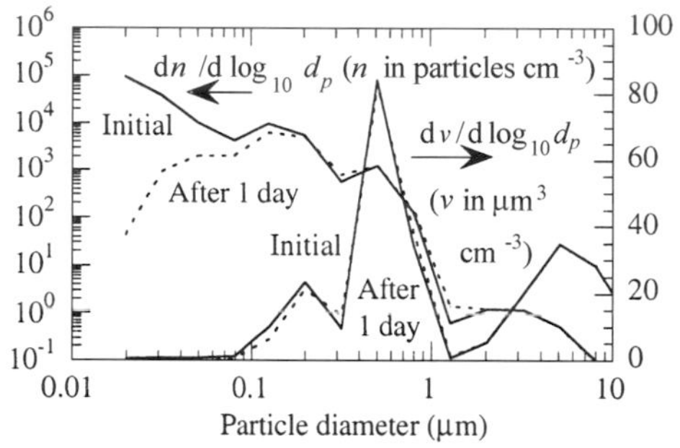


**Figure 16.7.** Estimated change in size-distributed aerosol number and volume at Claremont over a 24-h period when coagulation alone was considered.

where the coagulation kernel is given in (16.38). Figure 16.6 shows a comparison of the self-preserving analytical solution with semiimplicit solutions.

## 16.6. APPLICATION OF COAGULATION EQUATIONS

Coagulation is important in urban regions because it **affects the number concentration of particles less than 0.2 $\mu$m in size** over the course of a day. To test this hypothesis, the semiimplicit coagulation scheme was applied to a realistic initial size distribution over a 24-h period, without interruption by other processes. The predicted changes in particle number and volume concentrations are shown in Fig. 16.7. The results indicate that coagulation, when considered alone, affected the number concentration of small particles. Since 85 percent of elemental carbon and other material originating from automobile combustion resides in particles less than 0.12 μm in diameter (Venkataraman et al. 1994), simulating coagulation near automobile sources is important, especially for predicting the aerosol size distribution near these sources. Since homogeneously nucleated particles are much smaller than 0.1 μm in diameter, simulating coagulation near nucleation sources is important for predicting size distributions near these sources.

## 16.7. SUMMARY

Coagulation can be simulated exactly or approximately. Exact solutions include analytical and integrated numerical solutions. One approximate solution is a semi implicit solution. Analytical solutions are useful only under specific conditions, such as when the coagulation kernel is constant. Integrated numerical solutions are more robust than analytical solutions although they require significant computer time. A noniterative solution, such as the semiimplicit solution, is fast and reliably accurate when the size-bin resolution is high. The technique also allows simulation of coagulation among several particle size distributions and among particles with multiple components. The coagulation kernel depends on Brownian diffusion, enhancement of Brownian motion due to eddies, differences in fall velocities, turbulent shear, and turbulent inertial motion. Kernels were given for each of these processes. Coagulation primarily affects the size distribution of particles smaller than 0.2 μm in diameter. Effects of coagulation coupled to growth are discussed in Chapters 17 and 18.

## 16.8. PROBLEMS

**16.1.** Calculate the coagulation kernel due to Brownian motion when the diameter of one particle is 0.1 μm and the diameter of the other is 1.0 μm. Compare the result with the kernel obtained when both particles are 1.0 μm in diameter. Assume $T = 298$ K, $p_a = 990$ mb, the air is dry, and the particle density is $\rho_p = 1.2$ g cm$^{-3}$.

## 16.9. COMPUTER PROGRAMMING PRACTICE

**16.2.** Using the constant coagulation kernel from (16.38), plot Smoluchowski's analytical solution on a $dn/d\log_{10} d_p$ graph when $T = 298$ K and $h = 12$ h ($dn = n_i$ and $d\log_{10} d_p = \log_{10} d_{i+1} - \log_{10} d_i$ for each $i$). Assume that the initial total particle number concentration is $10^6$ particles cm$^{-3}$ and all initial particles have diameter 0.006 μm.

**16.3.** Write a computer script to calculate the coagulation kernel due to Brownian motion as a function of temperature and particle size. Use a volume-ratio size distribution, with $d_1 = 0.001$ μm, $d_{N_B} = 10$ μm, and $V_{rat} = 1.5$. Assume $p_a =$ 1013 mb, the air is dry, and the particle mass density is $\rho_p = 1.5$ g cm$^{-3}$. Calculate the coagulation kernels when (a) $T = 298$ K, the diameter of one particle ($d_a$) is 0.001 μm, and the diameter of the other particle ($d_b$) varies from 0.001 to 10 μm; (b) $T = 298$ K, $d_a = 10$ μm, and $d_b$ varies from 0.001 to 10 μm;

(c) the diameters of both particles are 0.001 μm and $T$ varies from 190 to 320 K in increments of 5 K; and (d) $d_a = 0.001$ μm, $d_b = 10$ μm, and $T$ varies from 190 to 320 K in increments of 5 K. Plot and interpret the results.

**16.4.** Write a computer program to simulate semiimplicit coagulation with (16.14). Use the program with the coagulation kernel and conditions described in Problem 16.2 to simulate Smoluchowski's solution. Compare the results with those from Problem 16.2.

# 17

# Condensation, Evaporation, Deposition, and Sublimation

A PARTICLE grows in size by the mass transfer of gas to its surface followed by conversion of the gas at the surface by condensation, deposition, chemical reaction, or dissolution. In a model, complicated surface interactions between the gas phase and particle surfaces are often simplified because surface geometries and compositions are difficult to resolve and they vary among particles. In this chapter, condensation and deposition equations are discussed. Equations for water-vapor condensation onto a single, homogeneous drop are first derived. The growth equations are extended to other gases, to a population of drops, and to particles with multiple components. Numerical solutions to growth equations are given. An application of the growth equations and an extension of the equations to ice deposition are also described. Dissolution and surface reaction are discussed in Chapter 18.

## 17.1. FLUXES TO AND FROM A SINGLE DROP

**Condensation** occurs when a gas, such as water vapor, diffuses and changes state to a liquid on a homogeneously or heterogeneously nucleated particle surface. **Deposition** occurs when a gas diffuses and changes state to a solid on a nucleated surface. These growth processes increase the volume of particles. **Evaporation** and **sublimation** are the reverse of condensation and deposition, respectively. In this section, growth equations for the condensation of water vapor onto a single homogeneous liquid water drop are derived.

From **Fick's law of diffusion**, the rate of change of mass of a single, spherical, homogeneous liquid water drop can be described as

$$\frac{\mathrm{d}\mathsf{m}}{\mathrm{d}t} = 4\pi R^2 D_v \frac{\mathrm{d}\rho_v}{\mathrm{d}R} \tag{17.1}$$

where $\mathsf{m}$ is the mass of the drop (g), $R$ is the radial distance from the center of the drop (cm), $D_v$ is the molecular diffusion coefficient of water vapor in air (cm$^2$ s$^{-1}$), $\rho_v$ is the density of water vapor (g cm$^{-3}$), and $\mathrm{d}\rho_v/\mathrm{d}R$ is the radial gradient of vapor density. Integrating (17.1) from the drop surface, where $\rho_v = \rho_{v,r}$ and $R = r$, to infinity, where $\rho_v = \rho_v$ and $R = \infty$, gives

$$\frac{\mathrm{d}\mathsf{m}}{\mathrm{d}t} = 4\pi r D_v(\rho_v - \rho_{v,r}) \tag{17.2}$$

If $\rho_v > \rho_{v,r}$, vapor diffuses to and condenses on the drop surface. If $\rho_v < \rho_{v,r}$, condensate evaporates from and diffuses away from the surface.

Because latent heat is released when water vapor condenses, the temperature at the drop surface increases, creating a gradient between the surface and surrounding air. The gradient is reduced by conduction of energy through the air. The equation for the heating rate at the drop surface due to conduction is

$$\frac{dQ_r^*}{dt} = -4\pi R^2 \kappa_d \frac{dT}{dR} \tag{17.3}$$

where $dQ_r^*/dt$ is the **conductive heating rate at the drop surface** (J s$^{-1}$), $\kappa_d$ is the thermal conductivity of air (J cm$^{-1}$ s$^{-1}$ K$^{-1}$), and the temperature gradient is negative, since temperature decreases with increasing distance from the drop during condensation. Integrating (17.3) from the drop surface, where $T = T_r$, to infinity where $T = T$, gives

$$\frac{dQ_r^*}{dt} = 4\pi r \kappa_d (T_r - T) \tag{17.4}$$

(17.2) and (17.4) were first written in a similar form by James Clerk Maxwell in 1877 (Maxwell 1890). The two equations are related to temperature at the drop surface by

$$m c_W \frac{dT_r}{dt} = L_e \frac{dm}{dt} - \frac{dQ_r^*}{dt} \tag{17.5}$$

where $c_W$ is the specific heat of liquid water (J g$^{-1}$ K$^{-1}$), and $L_e$ is the latent heat of evaporation of water (J g$^{-1}$). Combining (17.4) with (17.5) under steady-state conditions ($dT_r/dt = 0$) gives

$$L_e \frac{dm}{dt} = 4\pi r \kappa_d (T_r - T) \tag{17.6}$$

The remainder of this derivation originates from Mason (1971) (see also Rogers and Yau 1989). Combining the equation of state at saturation, $p_{v,s} = \rho_{v,s} R_v T$, with the Clausius–Clapeyron equation, $dp_{v,s}/dT = \rho_{v,s} L_e / T$ from (2.51), gives

$$\frac{d\rho_{v,s}}{\rho_{v,s}} = \frac{L_e}{R_v} \frac{dT}{T^2} - \frac{dT}{T} \tag{17.7}$$

Integrating this equation from infinity, where $T = T$ and $\rho_{v,s} = \rho_{v,s}(T)$, to the drop surface, where $T = T_s$ and $\rho_{v,s} = \rho_{v,s}(T_r)$, yields

$$\ln \frac{\rho_{v,s}(T_r)}{\rho_{v,s}(T)} = \frac{L_e}{R_v} \frac{(T_r - T)}{T T_r} - \ln \frac{T_r}{T} \tag{17.8}$$

Since $T \approx T_r$, (17.8) simplifies to

$$\frac{\rho_{v,s}(T_r) - \rho_{v,s}(T)}{\rho_{v,s}(T)} = \frac{L_e}{R_v} \frac{(T_r - T)}{T^2} - \frac{T_r - T}{T} \tag{17.9}$$

Substituting $T_r - T$ from (17.6) into (17.9) gives

$$\frac{\rho_{v,s}(T_r) - \rho_{v,s}(T)}{\rho_{v,s}(T)} = \frac{L_e}{4\pi r \kappa_d T}\left(\frac{L_e}{R_v T} - 1\right)\frac{\mathrm{dm}}{dt} \tag{17.10}$$

Dividing (17.2) by $\rho_{v,s}(T)$, rearranging, adding the result to (17.10), and assuming $\rho_{v,r} \approx \rho_{v,s}(T_r)$ results in

$$\frac{\rho_v - \rho_{v,s}(T)}{\rho_{v,s}(T)} = \left[\frac{L_e}{4\pi r \kappa_d T}\left(\frac{L_e}{R_v T} - 1\right) + \frac{1}{4\pi r D_v \rho_{v,s}(T)}\right]\frac{\mathrm{dm}}{dt} \tag{17.11}$$

Substituting $\rho_{v,s} = \rho_{v,s}(T)$, $\rho_{v,s} = p_{v,s}/R_v T$, and $\rho_v = p_v/R_v T$ into (17.11) and solving for $\mathrm{dm}/dt$ gives the **mass-flux form** of the growth equation for a single, homogeneous liquid water drop as

$$\frac{\mathrm{dm}}{dt} = \frac{4\pi r D_v (p_v - p_{v,s})}{\frac{D_v L_e p_{v,s}}{\kappa_d T}\left(\frac{L_e}{R_v T} - 1\right) + R_v T} \tag{17.12}$$

This equation can be modified for any gas by replacing $L_e, D_v, p_v, p_{v,s}$, and $R_v$ with $L_{e,q}, D_q, p_q, p_{q,s}$, and $R^*/m_q$, respectively. It can also be applied to particles in a size bin by adding the subscript $i$ to m and $r$. The molecular diffusion coefficient ($D_q$), thermal conductivity ($\kappa_d$), and saturation vapor pressure ($p_{q,s}$) also depend on size and surface characteristics. $D_q$ and $\kappa_d$ are affected by the collision geometry between vapor molecules and a particle surface and by the probability that molecules stick to a surface upon collision. The saturation vapor pressure is affected by surface curvature, the presence of solute on a surface, and radiative heating/cooling. For now, $D_q, \kappa_d, p_{q,s}$ are modified by adding the subscript $i$ to denote size dependence, and a prime to denote a modified value. With the above substitutions in (17.12), the mass-flux form of the growth equation for one homogeneous drop of size $i$, made of substance $q$, becomes

$$\frac{\mathrm{dm}_i}{dt} = \frac{4\pi r_i D'_{q,i}(p_q - p'_{q,s,i})}{\frac{D'_{q,i} L_{e,q} p'_{q,s,i}}{\kappa'_{d,i} T}\left(\frac{L_{e,q} m_q}{R^* T} - 1\right) + \frac{R^* T}{m_q}} \tag{17.13}$$

Sometimes, the assumption $L_{e,q} m_q / R^* T - 1 \approx L_{e,q} m_q / R^* T$ is made. If $T = 273$ K, $L_{e,q} m_q / R^* T \approx 19.8$ for water vapor, and the assumption induces an error of 1/19.8. Because (17.13) is a better approximation, it is used hereafter.

The change in mass of a spherical drop is related to its change in radius by

$$\frac{\mathrm{dm}_i}{dt} = 4\pi r_i^2 \rho_{p,i} \frac{dr_i}{dt} \tag{17.14}$$

where $\rho_{p,i}$ is the drop density (g cm$^{-3}$). Combining (17.14) with (17.13) gives the

**radius flux form** of the growth equation for a single, homogeneous, spherical drop as

$$r_i \frac{dr_i}{dt} = \frac{D'_{q,i}\left(p_q - p'_{q,s,i}\right)}{\frac{D'_{q,i} L_{e,q} \rho_{p,i} p'_{q,s,i}}{\kappa'_{d,i} T}\left(\frac{L_{e,q} m_q}{R^* T} - 1\right) + \frac{R^* T \rho_{p,i}}{m_q}} \tag{17.15}$$

The time rate of change in mass of a spherical drop is related to its change in volume by $\mathrm{d}m_i/\mathrm{d}t = \rho_{p,i}\,\mathrm{d}\upsilon_i/\mathrm{d}t$, and radius is related to volume by $r_i = (3\upsilon_i/4\pi)^{1/3}$. Combining these expressions with (17.13) gives the **volume flux form** of the growth equation for a single, homogeneous, spherical drop as

$$\frac{d\upsilon_i}{dt} = \frac{(48\pi^2 \upsilon_i)^{1/3} D'_{q,i}\left(p_q - p'_{q,s,i}\right)}{\frac{D'_{q,i} L_{e,q} \rho_{p,i} p'_{q,s,i}}{\kappa'_{d,i} T}\left(\frac{L_{e,q} m_q}{R^* T} - 1\right) + \frac{R^* T \rho_{p,i}}{m_q}} \tag{17.16}$$

## 17.2. CORRECTIONS TO GROWTH PARAMETERS

Above, an effective molecular diffusion coefficient, thermal conductivity of dry air, and saturation vapor pressure were used but not defined. These parameters are discussed here.

### 17.2.1. Corrections to the Molecular Diffusion Coefficient

**Molecular diffusion** is the movement of molecules due to their kinetic energy and redirection due to their collision with other molecules. Molecules in a body with a temperature greater than 0 K have an average kinetic energy and thermal velocity given in (12.154). When a molecule moves, it collides with other molecules, which redirect it along an arbitrary path. The pace at which a molecule spreads by diffusion is proportional to its thermal velocity and the distance between collisions. Thermal velocity is proportional to the square root of temperature, and the distance between collisions is inversely proportional to the density of air. An expression for the **molecular diffusion coefficient** of a trace gas in air (cm$^2$ s$^{-1}$) with these characteristics is

$$D_q = \frac{3}{8 A d_q^2 \rho_a} \sqrt{\frac{R^* T m_a}{2\pi}\left(\frac{m_q + m_a}{m_q}\right)} \tag{17.17}$$

(e.g., Chapman and Cowling 1970; Davis 1983), where $d_q$ is the diameter of gas molecule $q$. A typical value is $d_q = 4.5 \times 10^{-8}$ cm (4.5 Å).

**Example 17.1**

Calculate the molecular diffusion coefficient of carbon monoxide in air when the temperature is $T = 288$ K and air pressure is $p_a = 1013$ mb. Assume the air is dry.

SOLUTION

Since the air is dry, $R_m = R'$. From the equation of state, (2.30), $\rho_a = 0.00123$ g cm$^{-3}$. The molecular weight of carbon monoxide (CO) is $m_q = 28$ g mole$^{-1}$. Substituting these and remaining values into (17.17) gives the molecular diffusion coefficient of CO at standard temperature and pressure as $D_q = 0.12$ cm$^2$ s$^{-1}$.

When used in growth equations, the molecular diffusion coefficient must be corrected for collision geometry, sticking probability, and ventilation. The **corrected molecular diffusion coefficient** (cm$^2$ s$^{-1}$) which takes account of these effects is

$$D'_{q,i} = D_q \omega_{q,i} F_{q,i} \tag{17.18}$$

where $\omega_{q,i}$ is the correction for collision geometry and sticking probability, and $F_{q,i}$ is the correction for ventilation.

The **correction for collision geometry** is a function of the ratio of the distance between molecular collisions of a diffusing gas and the size of the particle it is diffusing to. The **correction for sticking probability** takes account of the ability of a gas to stick to a particle surface once the gas has diffused to the surface. An expression for the two corrections, together is

$$\omega_{q,i} = \left\{ 1 + \left[ \frac{1.33 + 0.71\,\mathrm{Kn}_{q,i}^{-1}}{1 + \mathrm{Kn}_{q,i}^{-1}} + \frac{4(1 - \alpha_{q,i})}{3\alpha_{q,i}} \right] \mathrm{Kn}_{q,i} \right\}^{-1} \tag{17.19}$$

(Fuchs and Sutugin 1971; Pruppacher and Klett 1997), where

$$\mathrm{Kn}_{q,i} = \frac{\lambda_q}{r_i} \tag{17.20}$$

is the **Knudsen number of the condensing gas** with respect to particles of size $i$. This Knudsen number gives the ratio of the mean free path of the condensing gas to the size of the particle of interest. The mean free path of a gas is the average distance a trace gas molecule travels before colliding and exchanging momentum with an air molecule.

If the Knudsen number is large, the particle of interest is small relative to the distance between molecular collisions, and the particle is intercepted relatively infrequently by gas molecules. If the Knudsen number is small, the particle is large relative to the distance between molecular collisions, and the particle is frequently intercepted by gas molecules. In sum, (17.19) states that, in the absence of sticking probability,

$$\omega_{q,i} \rightarrow \begin{cases} 0 & \text{as } \mathrm{Kn}_{q,i} \rightarrow \infty \quad \text{(small particles)} \\ 1 & \text{as } \mathrm{Kn}_{q,i} \rightarrow 0 \quad \text{(large particles)} \end{cases} \tag{17.21}$$

In a dilute mixture, where the number concentration of a trace-gas is much less than that air of all, the **mean free path of a trace gas** is

$$\lambda_q = \frac{m_a}{\pi A d_q^2 \rho_a} \sqrt{\frac{m_a}{m_a + m_q}} \tag{17.22}$$

(Jeans 1954; Davis 1983). In (17.22), the mean free path is inversely proportional to the air density. The greater the air density, the shorter the distance between collisions of a gas molecule with an air molecule. Combining (12.154), (17.17), and (17.22) gives

$$\lambda_q = \frac{32 D_q}{3\pi \bar{v}_q}\left(\frac{m_a}{m_a + m_q}\right) \approx \frac{3 D_q}{\bar{v}_q} \tag{17.23}$$

where $\bar{v}_q$ is the **thermal velocity of a trace-gas molecule** (cm $s^{-1}$) and the last approximation follows from assuming $m_q \ll m_a$ (e.g., Fuchs and Sutugin 1970).

In (17.19), $\alpha_{q,i}$ is the **mass accommodation coefficient (sticking coefficient)** of the gas, which is the fractional number of collisions of gas $q$ with particles of size $i$ that result in the gas sticking to the surface of the particle. Mass accommodation coefficients differ from **reaction probabilities**, discussed in Chapter 12, in that reaction probabilities take account of molecular diffusion to and reaction with a material on a particle surface. Mass accommodation coefficients take account of only adsorption of a gas onto the surface.

Accommodation coefficients for soluble species range from 0.01 to 1.0 (Mozurkewich et al. 1987; Jayne et al. 1990; Van Doren et al. 1990; Chameides and Stelson 1992). Van Doren et al. (1990) found that the mass accommodation coefficient for nitric acid ($HNO_3$) on liquid water varied from 0.07 at 268 K to 0.193 at 293 K, the coefficient for hydrochloric acid (HCl) on liquid water varied from 0.064 at 294 K to 0.177 at 274 K, and the coefficient for dinitrogen pentoxide ($N_2O_5$) on liquid water varied from 0.04 at 282 K to 0.061 at 271 K.

To correct for the increased rate of vapor and energy transfer to the upstream surface of a large particle, a **ventilation factor for vapor** is included in the corrected molecular diffusion coefficient equation. The ventilation factor is the vapor mass flux to (or from) a drop when the drop is moving, dividing by the flux when it is at rest. Increased vapor transfer occurs when a large drops falls, creating eddies that sweep additional vapor towards the particle. The ventilation factor does not affect small particles. An expression for the ventilation factor of water vapor is

$$F_{q,i} = \begin{cases} 1 + 0.108 x_{q,i}^2 & x_{q,i} \le 1.4 \\ 0.78 + 0.308 x_{q,i} & x_{q,i} > 1.4 \end{cases} \qquad x_{q,i} = \mathrm{Re}_i^{1/2}\,\mathrm{Sc}_q^{1/3} \tag{17.24}$$

(Pruppacher and Klett 1997). It depends on the particle Reynolds number, defined in (16.22), and the **gas Schmidt number,**

$$\mathrm{Sc}_q = \frac{\nu_a}{D_q} \tag{17.25}$$

which is the ratio of the kinematic viscosity of air to the molecular diffusion coefficient of a gas. Since the particle Reynolds number is proportional to a particle's radius and its fall velocity, the larger a particle is and the faster it falls, the greater the ventilation factor.

### 17.2.2. Corrections to the Thermal Conductivity of Air

Conductive energy transfer to and from a particle surface is affected by some of the same correction factors that affect vapor transfer. These factors are embodied in the **corrected thermal conductivity** term

$$\kappa'_{d,i} = \kappa_d \omega_{h,i} F_{h,i} \tag{17.26}$$

where $\kappa_d$ is the uncorrected thermal conductivity of dry air (J cm$^{-1}$ s$^{-1}$ K$^{-1}$), from (2.3), $\omega_{h,i}$ is the correction factor for collision geometry and sticking probability, and $F_{h,i}$ is the thermal ventilation factor. The correction factor is

$$\omega_{h,i} = \left\{1 + \left[\frac{1.33 + 0.71\,\mathrm{Kn}_{h,i}^{-1}}{1 + \mathrm{Kn}_{h,i}^{-1}} + \frac{4(1-\alpha_h)}{3\alpha_h}\right]\mathrm{Kn}_{h,i}\right\}^{-1} \tag{17.27}$$

where

$$\mathrm{Kn}_{h,i} = \frac{\lambda_h}{r_i} \tag{17.28}$$

is the **Knudsen number for energy** with respect to particles of size $i$, which depends on the thermal mean free path $\lambda_h$ (cm). This term is the average distance an air molecule travels before exchanging energy by conduction with another molecule. It is quantified as

$$\lambda_h = \frac{3D_h}{\bar{v}_a} \tag{17.29}$$

where $D_h$ is the **molecular thermal diffusivity**, defined in (8.10), and $\bar{v}_a$ is the **thermal velocity of air**, defined in (16.21). The molecular thermal diffusivity depends on the thermal conductivity of air, the specific heat of air, and the air density.

In (17.27), $\alpha_h$ is the **thermal accommodation coefficient**, interpreted as the fraction of molecules bouncing off the surface of a drop that have acquired the temperature of the drop (Pruppacher and Klett 1997). An equation for $\alpha_h$ is

$$\alpha_h = \frac{T_m - T}{T_s - T} \tag{17.30}$$

where $T_m$ is the temperature of vapor molecules leaving the surface of a drop, $T_s$ is the temperature of the surface, and $T$ is the temperature of the ambient vapor. For water-vapor growth, $\alpha_h$ has a typical value of 0.96.

The **thermal ventilation factor** has a similar meaning to that for vapor. When a large particle falls through the air, it creates eddies that sweep additional energy to the surface of the particle. An empirical equation for the ventilation factor is

$$F_{h,i} = \begin{cases} 1 + 0.108x_{h,i}^2 & x_{h,i} \le 1.4 \\ 0.78 + 0.308x_{h,i} & x_{h,i} > 1.4 \end{cases} \qquad x_{h,i} = \mathrm{Re}_i^{1/2}\,\mathrm{Pr}^{1/3} \tag{17.31}$$

(Pruppacher and Klett 1997), where the unitless **Prandtl number**,

$$\mathrm{Pr} = \eta_a c_{p,m} / \kappa_d \tag{17.32}$$

is proportional to the ratio of the dynamic viscosity of air to its thermal conductivity.

### 17.2.3. Corrections to the Saturation Vapor Pressure

Expressions for the saturation vapor pressure of water over dilute, flat liquid water and ice surfaces were given in (2.56) and (2.58), respectively. Curvature, the presence of solute, and radiative cooling affect the saturation vapor pressure over a surface.

#### 17.2.3.1. *Curvature Effect*

The saturation vapor pressure increases over a curved surface relative to over a flat surface due to the **curvature (Kelvin) effect**. The surface of a small spherical particle is more curved than that of a large spherical particle, and the corresponding saturation vapor pressure over a small particle is greater than that over a large particle. The saturation vapor pressure over a curved surface is greater than that over a flat surface because molecules desorb more readily from a curved surface than from a flat surface. Surface tension also plays a role in the curvature effect. The greater the surface tension of a particle, the more likely a molecule will desorb from the surface, and the greater the saturation vapor pressure.

The saturation vapor pressure over a curved, dilute surface relative to that over a flat, dilute surface is

$$\frac{p'_{q,s,i}}{p_{q,s}} = \exp\left(\frac{2\sigma_p m_p}{r_i R^* T \rho_{p,i}}\right) \approx 1 + \frac{2\sigma_p m_p}{r_i R^* T \rho_{p,i}} \tag{17.33}$$

where $p_{q,s}$ is the saturation vapor pressure of gas $q$ over a flat, dilute surface, $p'_{q,s,i}$ is the saturation vapor pressure over a curved dilute surface, $\sigma_p$ is the average particle surface tension, $\rho_{p,i}$ is the average particle density, and $m_p$ is the average particle molecular weight. For liquid water, $p_{q,s} = p_{v,s}$, and for ice, $p_{q,s} = p_{v,I}$. Expressions for $p_{v,s}$ and $p_{v,I}$ were given in (2.56) and (2.58), respectively. An expression for the surface tension of liquid water was given in (15.2). The second expression in (17.33) was obtained by noting that the exponent in the first term is small, and $e^x \approx 1 + x$ for small $x$.

#### 17.2.3.2. *Solute Effect*

A second factor that affects the saturation vapor pressure, particularly of liquid water, is the presence of solute dissolved in water. Some solutes, such as sodium chloride, ammonium sulfate, sulfuric acid, and nitric acid, dissociate in solution. Solute molecules replace liquid water on the surface of a drop, and if the saturation vapor pressure of water over the solute is less than that over liquid water, the presence of solute reduces the overall saturation vapor pressure of the drop. The saturation vapor pressure of water over solute is less than that over liquid water because liquid-water molecules bond to solute molecules by **hydration**. When liquid water hydrates with a solute, water vapor condenses to replace the hydrated water, reducing the saturation vapor pressure over the solute surface. The reduction is roughly proportional to the number of moles of solute on the surface and is called the **solute effect**. The solute effect affects primarily small particles, since small particles have a higher concentration of solute than do large particles.

The saturation vapor pressure over a flat liquid surface containing solute relative to that over a flat liquid surface without solute is approximated by **Raoult's law**,

$$\frac{p'_{q,s,i}}{p_{q,s}} = \frac{n_w}{n_w + n_s} \tag{17.34}$$

where $n_s$ is the number of moles of solute in the drop, and $n_w$ is the number of moles of liquid water in the drop. For dilute solutions, $n_w \gg n_s$, and (17.34) simplifies to

$$\frac{p'_{q,s,i}}{p_{q,s}} \approx 1 - \frac{n_s}{n_w} \tag{17.35}$$

The number of moles of solute in solution is approximately $n_s = i_v M_s / m_s$, where $M_s$ is the mass of solute in the particle (g), $m_s$ is the molecular weight of the solute (g mole$^{-1}$), and $i_v$ is the degree of dissociation of the solute into ions, called the **van't Hoff factor**. The factor $i_v$ gives the actual number of moles of ions that dissociate from one mole of solute. For sodium chloride (NaCl), which dissociates in solution to $Na^+$ and $Cl^-$, $i_v = 2$. For ammonium sulfate [$(NH_4)_2SO_4$], which dissociates to $2NH_4^+$ and $SO_4^{2-}$, $i_v = 3$. For a nondissociating solute, $i_v = 1$. The more ions present on the surface, the greater the reduction in saturation vapor pressure.

The number of moles of liquid water in a drop, used in (17.35), is approximately

$$n_w = \frac{M_w}{m_v} \approx \frac{4\pi r_i^3 \rho_w}{3m_v} \tag{17.36}$$

where $M_w$ is the mass of liquid water in the drop (g), $m_v$ is the molecular weight of water (g mole$^{-1}$), and $\rho_w$ is the density of liquid water (g cm$^{-3}$). This equation assumes that $M_w \gg M_s$. Substituting $n_s = i_v M_s / m_s$ and (17.36) into (17.35) gives the solute effect over a dilute liquid water solution as

$$\frac{p'_{q,s,i}}{p_{q,s}} \approx 1 - \frac{3m_v i_v M_s}{4\pi r_i^3 \rho_w m_s} \tag{17.37}$$

For substances other than water, saturation vapor pressures are also affected by the presence of solute, as discussed in Chapter 18.

#### 17.2.3.3. *Köhler Equation*

Taking the product of the right sides of (17.33) and (17.37) and eliminating the last term of the result, which is small, gives the **Köhler equation** (Köhler 1936),

$$S'_{q,i} = \frac{p'_{q,s,i}}{p_{q,s}} \approx 1 + \frac{2\sigma_p m_p}{r_i R^* T \rho_{p,i}} - \frac{3m_v i_v M_s}{4\pi r_i^3 \rho_w m_s} \tag{17.38}$$

where $S'_{q,i}$ is the **saturation ratio at equilibrium.** The Köhler equation relates the saturation vapor pressure of water over a curved surface containing solute to that over a flat surface without solute.

The saturation ratio at equilibrium differs from the **ambient saturation ratio,** defined as

$$S_q = \frac{p_q}{p_{q,s}} \tag{17.39}$$

where $p_q$ is the partial pressure of the condensing gas. The ambient saturation ratio gives the partial pressure of a gas relative to its saturation vapor pressure over a flat, dilute surface. When $S_q > S'_{q,i}$, the air is supersaturated with vapor over particles of size $i$, and excess vapor condenses onto size-$i$ particles. When $S_q < S'_{q,i}$, the air is undersaturated with vapor over particles of size $i$, and condensate on these particles evaporates. When $S_q = S'_{q,i}$, the air is exactly saturated with vapor over particles of size $i$, and neither condensation nor evaporation occurs. In such a case, $p_q = p'_{q,s,i}$.

Figure 17.1 shows how the curvature and solute effects affect the equilibrium saturation ratio. In the figure, the two effects are added, and the resulting curve is the equilibrium saturation curve, plotted as a function of particle radius. The radius at which the equilibrium saturation ratio is maximum is the **critical radius for growth** ($r^*$). The equilibrium saturation ratio at this radius is the **critical saturation ratio** ($S^*$). The curve shows that drops smaller than the critical radius cannot grow

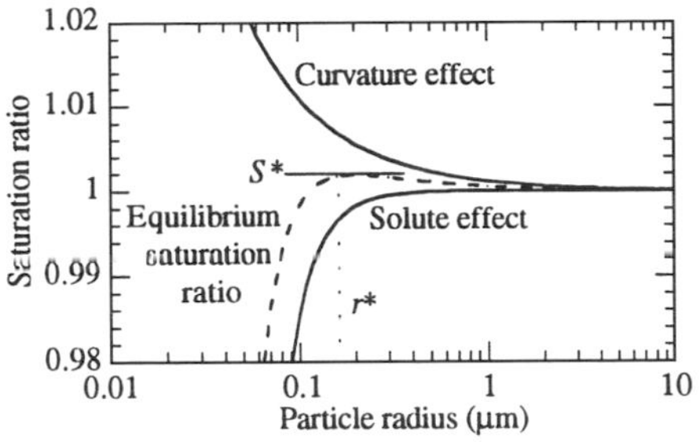


**Figure 17.1.** Example of how curvature and solute effects affect the equilibrium saturation ratio $S'_{q,i}$. The sum of the curvature and solute curves (dashed line) is $S'_{q,i}$. The curvature effect increases the saturation vapor pressure over small drops, increasing $S'_{q,i}$. The solute effect decreases the saturation vapor pressure over small drops, decreasing $S'_{q,i}$. The maximum $S'_{q,i}$ is the critical saturation ratio $S^*$, which occurs at the critical radius $r^*$.

to cloud-sized drops (near 5 μm in radius) unless the ambient saturation ratio exceeds the critical saturation ratio. Drops larger than the critical radius can grow to cloud-sized drops if the ambient saturation ratio exceeds the equilibrium saturation ratio over such drops.

In sum, growth of a particle to cloud-sized drops can occur if $r_i > r^*$ and $S_q > S'_{q,i}$ or if $r_i < r^*$ and $S_q > S^*$. In such cases a particle **activates** into a **cloud condensation nucleus** (CCN). Once activated, the particle continues to grow if the above conditions are maintained. When $r_i < r^*$ and $S^* > S_q > S'_{q,i}$, particles are not activated and can grow no larger than $r^*$. When $S_q < S'_{q,i}$, particles are not activated.

The critical radius and critical saturation ratio are found by rewriting (17.38) as

$$S'_{q,i} = 1 + \frac{a}{r_i} - \frac{b}{r_i^3} \qquad a = \frac{2\sigma_p m_p}{R^* T \rho_{p,i}} \qquad b = \frac{3 m_v i_v M_s}{4\pi \rho_w m_s} \tag{17.40}$$

Taking the derivative of $S'_{q,i}$, setting the derivative equal to zero, solving for the radius, and substituting the result back into (17.40) gives the critical radius for growth and critical saturation ratio as

$$r^* = \sqrt{\frac{3b}{a}} \qquad S^* = 1 + \sqrt{\frac{4a^3}{27b}} \tag{17.41}$$

respectively. This derivation assumes that the solute mass is the same in all particles. In reality, the solute mass varies in each particle.

A **supersaturation** is a saturation ratio minus one. Table 17.1 gives critical radii and critical supersaturations (expressed as percentages) when particles contain

**Table 17.1.** Critical Radii for Growth and Critical Supersaturations for Water Drops Containing Sodium Chloride and Ammonium Sulfate When Solute Mass Varies at 275 K

| Solute Mass | Sodium Chloride | | Ammonium Sulfate | |
|---|---|---|---|---|
| (g) | $r^*$ (μm) | $S^* - 1$ (%) | $r^*$ (μm) | $S^* - 1$ (%) |
| 0 | 0 | $\infty$ | 0 | $\infty$ |
| $10^{-18}$ | 0.019 | 4.1 | 0.016 | 5.1 |
| $10^{-16}$ | 0.19 | 0.41 | 0.16 | 0.51 |
| $10^{-14}$ | 1.9 | 0.041 | 1.6 | 0.051 |
| $10^{-12}$ | 19 | 0.0041 | 16 | 0.0051 |

either sodium chloride or ammonium sulfate. In the atmosphere, ambient supersaturations rarely exceed 1% (saturation ratio of 1.01) and are usually much smaller. Since the number concentration of cloud drops (1000 $cm^{-3}$) is much less than that of background particles ($10^4$–$10^6$ $cm^{-3}$), only a small percentage of background particles activates into cloud drops. When ambient supersaturations are less than the critical supersaturation, only particles larger than the critical radius for growth activate into cloud condensation nuclei. The number concentration of these particles is often on the order of the number concentration of cloud drops.

Table 17.1 shows that, in the absence of solute, all particles have potential to activate to CCN so long as the ambient saturation ratio exceeds the equilibrium saturation ratio. In such cases, when particles containing solute are larger than the critical radius, large particles are more likely to grow than are small particles. Large particles are favored because the difference between the ambient saturation ratio and the equilibrium saturation ratio is larger for large particles than for small particles (the difference between the partial pressure and saturation vapor pressure of water vapor is greater for large particles). The larger the difference between the partial pressure and saturation vapor pressure, the greater the growth rate of a particle.

#### 17.2.3.4. *Radiative Cooling Effect*

When large particles emit or absorb electromagnetic energy, their surface temperatures and saturation vapor pressures change. Large water particles usually cool faster by releasing infrared radiation than they heat up by absorbing solar radiation. Small liquid water particles do not absorb much solar radiation, simply because their solar absorption efficiencies are negligible, as shown in Fig. 10.18, and the ability of a particle to absorb depends on its absorption efficiency. When a large water-containing particle cools, its saturation vapor pressures decreases, since saturation vapor pressure is proportional to temperature. The decrease in saturation vapor pressure over large water-containing drops is the **radiative cooling effect.**

The saturation vapor pressure over a flat, dilute surface that experiences radiative cooling relative to that over a flat, dilute surface that does not is

$$\frac{p'_{q,s,i}}{p_{q,s}} \approx 1 + \frac{L_{e,q} m_q H_{r,i}}{4\pi r_i R^* T^2 \kappa'_{d,i}} \tag{17.42}$$

where

$$H_{r,i} = (\pi r_i^2) 4\pi \int_0^\infty Q_a(m_\lambda, \alpha_{i,\lambda})(I_\lambda - B_\lambda)\, d\lambda \tag{17.43}$$

is the **radiative cooling/heating rate** (W) of the particle (e.g., Toon et al. 1989b). To calculate the radiative cooling/heating rate, the single-particle absorption efficiency $Q_a$, incoming spectral radiance $I_\lambda$, and Planckian spectral radiance emitted by the particle surface ($B_\lambda$) need to be known. Absorption efficiencies, which depend on the complex index of refraction and size parameter ($m_\lambda$ and $\alpha_{i,\lambda}$), were discussed in Section 10.4.3. A method of determining radiance was given in Section 10.8.3. The Planck function was given in (10.4). Radiative heating rates can be positive or negative.

### 17.2.3.5. *Overall Effects*

For condensation of water vapor onto a liquid-water surface, the overall equilibrium saturation ratio is obtained by taking the product of the right sides of (17.38) and (17.42) and eliminating small terms. The result is

$$S'_{q,i} = \frac{p'_{q,s,i}}{p_{q,s}} \approx 1 + \frac{2\sigma_p m_p}{r_i R^* T \rho_{p,i}} - \frac{3 m_v i_v M_s}{4\pi r_i^3 \rho_w m_s} + \frac{L_{e,q} m_q H_{r,i}}{4\pi r_i R^* T^2 \kappa'_{d,i}} \tag{17.44}$$

The solute-effect term in (17.44) is valid only for liquid-water growth. When gases other than water, such as nitric acid, hydrochloric acid, and ammonia, dissolve into liquid-water-containing particles, interactions among ions also affect saturation vapor pressures. The effects of such interactions may be estimated by solving equilibrium equations, as described in Chapter 18. For nonwater species, the radiative cooling term is ignored, since most particle constituents have mass loadings too small to affect the saturation vapor pressure noticeably. Also, whereas all particle constituents absorb and emit infrared radiation, few absorb solar radiation, except at large particle size. Removing the solute and radiative terms from (17.44) gives the overall equilibrium saturation for nonwater species as

$$S'_{q,i} = \frac{p'_{q,s,i}}{p_{q,s}} \approx 1 + \frac{2\sigma_p m_p}{r_i R^* T \rho_{p,i}} \tag{17.45}$$

## 17.3. FLUXES TO A PARTICLE WITH MULTIPLE COMPONENTS

The equations in Section 17.1 gave the change in mass, volume, and radius of a single, homogeneous particle. Particles often contain many components. Treating growth and evaporation for a particle with multiple components is more difficult than treating them for a particle with one component. With one component, the volume of a spherical drop changes proportionately to the one-third power of the volume of the condensed material already in the drop, as shown in (17.16). With multiple components, the volume changes proportionately to the one-third power of the volume of the sum of components in the drop. Growth equations for individual components in a drop are derived below.

The total volume and mass of one particle, in which one component grows, can be defined as

$$\upsilon_{i,t} = \upsilon_{q,i,t} + \upsilon_{i,t-h} - \upsilon_{q,i,t-h} \tag{17.46}$$

$$\mathrm{m}_{i,t} = \rho_{p,i,t}\upsilon_{i,t} = \rho_{p,q}\upsilon_{q,i,t} + \rho_{p,i,t-h}\upsilon_{i,t-h} - \rho_{p,q}\upsilon_{q,i,t-h} \tag{17.47}$$

respectively, where $\upsilon_{q,i,t}$ is the volume of growing component $q$ in size bin $i$ at current time $t$, $\upsilon_{i,t-h}$ is the initial total volume of the particle before growth, $\upsilon_{q,i,t-h}$ is the initial volume of component $q$ in the particle before growth (at time $t-h$), $\rho_{p,i,t-h}$ is the mass density (g cm$^{-3}$) of the total particle before growth, and $\rho_{p,q}$ is the mass density (g cm$^{-3}$) of component $q$, which is assumed constant. In reality, the mass density of a component in a particle may vary. The time derivative of (17.47) is

$$\frac{\mathrm{dm}_{i,t}}{\mathrm{d}t} = \rho_{p,i,t}\frac{\mathrm{d}\upsilon_{i,t}}{\mathrm{d}t} = \rho_{p,q}\frac{\mathrm{d}\upsilon_{q,i,t}}{\mathrm{d}t} \tag{17.48}$$

where $\mathrm{d}\upsilon_{i,t-h}/\mathrm{d}t = \mathrm{d}\upsilon_{q,i,t-h}/\mathrm{d}t = 0$. Combining (17.48) and (17.46) with (17.16) gives the volume rate of change of one component in a multicomponent particle as

$$\frac{\mathrm{d}\upsilon_{q,i,t}}{\mathrm{d}t} = \frac{\left[48\pi^2(\upsilon_{q,i,t} + \upsilon_{i,t-h} - \upsilon_{q,i,t-h})\right]^{1/3} D'_{q,i}(p_q - p'_{q,s,i})}{\dfrac{D'_{q,i}L_{e,q}\rho_{p,q}p'_{q,s,i}}{\kappa'_{d,i}T}\left(\dfrac{L_{e,q}m_q}{R^*T} - 1\right) + \dfrac{R^*T\rho_{p,q}}{m_q}} \tag{17.49}$$

## 17.4. FLUXES TO A POPULATION OF PARTICLES

(17.49) gave the flux of one gas to **one** particle that contains many components. The fluxes of individual gases to **many** particles, each with different size and composition, need to be taken into account. In a model, particles are often segregated into size bins, and each particle in a bin is assumed to have the same volume and composition as each other particle in the bin. The volume (cm$^3$) of one particle is related to its volume concentration (cm$^3$ cm$^{-3}$) and number concentration (particles cm$^{-3}$) by $\upsilon_i = v_i/n_i$. The volume of an individual component in a particle is related to its volume concentration and the number concentration of particles by $\upsilon_{q,i,t} = v_{q,i,t}/n_{i,t-h}$. Substituting expressions of this type into (17.49) gives the **volume-concentration form** of the growth equation for a component in particles of size $i$ as (Jacobson and Turco 1995)

$$\frac{dv_{q,i,t}}{dt} = \frac{n_{i,t-h}^{2/3}[48\pi^2(v_{q,i,t} + v_{i,t-h} - v_{q,i,t-h})]^{1/3} D'_{q,i}(p_q - p'_{q,s,i})}{\frac{D'_{q,i} L_{e,q} \rho_{p,q} p'_{q,s,i}}{\kappa'_{d,i} T}\left(\frac{L_{e,q} m_q}{R^* T} - 1\right) + \frac{R^* T \rho_{p,q}}{m_q}} \tag{17.50}$$

(17.50) can be simplified by defining

$$p_q = C_q R^* T \qquad p'_{q,s,i} = C'_{q,s,i} R^* T \tag{17.51}$$

where $C_q$ is the gas-phase mole concentration (moles cm$^{-3}$) of component $q$, and $C'_{q,s,i}$ is an effective surface vapor concentration (moles cm$^{-3}$). Combining (17.50) with (17.51) and adding more time subscripts gives

$$\frac{dv_{q,i,t}}{dt} \tag{17.52}$$
$$= n_{i,t-h}^{2/3}[48\pi^2(v_{q,i,t} + v_{i,t-h} - v_{q,i,t-h})]^{1/3} D^{eff}_{q,i,t-h} \frac{m_q}{\rho_{p,q}}(C_{q,t} - C'_{q,s,i,t-h})$$

where

$$D^{eff}_{q,i,t-h} = \frac{D'_{q,i}}{\frac{m_q D'_{q,i} L_{e,q} C'_{q,s,i,t-h}}{\kappa'_{d,i} T}\left(\frac{L_{e,q} m_q}{R^* T} - 1\right) + 1} \tag{17.53}$$

is an **effective molecular diffusion coefficient** (cm$^2$ s$^{-1}$). For water vapor, $C'_{q,s,i,t-h}$ is large, and the left side of the denominator of (17.53) is approximately unity. For other gases, $C'_{q,s,i,t-h}$ is usually orders of magnitude smaller than for water vapor, making the left side of the denominator $\ll 1$. Thus, for gases other than water vapor, the left side of the denominator can be ignored, and (17.53) simplifies to

$$D^{eff}_{q,i,t-h} \approx D'_{q,i} = D_q \omega_{q,i} F_{q,i} = \frac{D_q F_{q,i}}{1 + \left[\frac{1.33 + 0.71\,\mathrm{Kn}_{q,i}^{-1}}{1 + \mathrm{Kn}_{q,i}^{-1}} + \frac{4(1 - \alpha_{q,i})}{3\alpha_{q,i}}\right]\mathrm{Kn}_{q,i}} \tag{17.54}$$

(17.52) is the growth equation for a single gas to a population of spherical particles of size $i$. The equation yields the change in volume concentration of a particle with many components when one component in the particle grows or evaporates. The equation is written for each particle size bin and for each volatile component $q$. Since gases condense on particles in several size bins, a **gas conservation equation** must be written to accompany (17.52). Such an equation is

$$\frac{dC_{q,t}}{dt} = -\frac{\rho_{p,q}}{m_q}\sum_{i=1}^{N_B}\frac{dv_{q,i,t}}{dt} \tag{17.55}$$

## 17.5. SOLUTIONS TO GROWTH EQUATIONS

Many methods have been developed to solve condensation and evaporation equations. These include finite-element methods (Varoglu and Finn 1980; Tsang and

Brock 1986; Tsang and Huang 1990), discrete-size-bin methods (e.g., Gelbard and Seinfeld 1980; Toon et al. 1988; Rao and McMurry 1989), the cubic spline method (e.g., Middleton and Brock 1976), modified upwind difference methods (e.g., Smolarkiewicz 1983; Tsang and Korgaonkar 1987), and moment methods (e.g., Friedlander 1983; Whitby 1985; Lee 1985; Brock et al. 1986). Some of these methods conserve mass between the gas and particle phases, and others do not. Some methods reduce or eliminate numerical diffusion during growth, and others are more diffusive. Some methods are computationally more efficient than others.

Below, two recent solutions to condensation/evaporation equations are given. Both solutions conserve mass between the gas and aerosol phases and reduce or eliminate numerical diffusion during growth when used in conjunction with a hybrid, full-moving, or moving-center size structure. Although the first method gives exact solutions, it is iterative and requires more computational time than the second. Both schemes have been used in three-dimensional models. The second method, which is noniterative but unconditionally stable, may be more desirable when computer time is limited.

### 17.5.1. Integrated Numerical Solution

Together, (17.52) and (17.55) constitute $N_B+1$ nonlinear, first-order ordinary differential equations (ODEs). A method of solving these equations is to integrate them with a sparse-matrix ODE solver, such as that discussed in Section 13.8. An advantage of such a solver is that the matrix of partial derivatives arising from (17.52) and (17.55) is sparse. When $N_B=3$, a matrix of partial derivatives arising from this set of equations is β∂

$$\begin{array}{c} \\ v_{q,1,t} \\ v_{q,2,t} \\ v_{q,3,t} \\ C_{q,t} \end{array}\begin{array}{c} \begin{array}{cccc} v_{q,1,t} & v_{q,2,t} & v_{q,3,t} & C_{q,t} \end{array} \\ \left[\begin{array}{cccc} 1-h\beta_s\dfrac{\partial^2 v_{q,1,t}}{\partial v_{q,1,t}\,\partial t} & 0 & 0 & -h\beta_s\dfrac{\partial^2 v_{q,1,t}}{\partial C_{q,t}\,\partial t} \\ 0 & 1-h\beta_s\dfrac{\partial^2 v_{q,2,t}}{\partial v_{q,2,t}\,\partial t} & 0 & -h\beta_s\dfrac{\partial^2 v_{q,2,t}}{\partial C_{q,t}\,\partial t} \\ 0 & 0 & 1-h\beta_s\dfrac{\partial^2 v_{q,3,t}}{\partial v_{q,3,t}\,\partial t} & -h\beta_s\dfrac{\partial^2 v_{q,3,t}}{\partial C_{q,t}\,\partial t} \\ -h\beta_s\dfrac{\partial^2 C_{q,t}}{\partial v_{q,1,t}\,\partial t} & -h\beta_s\dfrac{\partial^2 C_{q,t}}{\partial v_{q,2,t}\,\partial t} & -h\beta_s\dfrac{\partial^2 C_{q,t}}{\partial v_{q,3,t}\,\partial t} & 1-h\beta_s\dfrac{\partial^2 C_{q,t}}{\partial C_{q,t}\,\partial t} \end{array}\right] \end{array} \tag{17.56}$$

The partial derivatives in this matrix are

$$\frac{\partial^2 v_{q,i,t}}{\partial v_{q,i,t}\,\partial t} = \frac{1}{3}\left(\frac{n_{i,t-h}}{v_{q,i,t}}\right)^{2/3}\left(48\pi^2\right)^{1/3} \times D^{eff}_{q,i,t-h}\frac{m_q}{\rho_{p,q}}\left(C_{q,t}-C'_{q,s,i,t-h}\right) \qquad i=1,\ldots,N_B \tag{17.57}$$

**Table 17.2.** Reduction in Array Space and Number of Matrix Operations Before and After Sparse-Matrix Techniques Are Applied for Growth/Evaporation ODES

| Quantity | Value | |
|---|---|---|
| | Initial | After Sparse Reductions |
| Order of matrix ($N_B + 1$) | 17 | 17 |
| No. of initial matrix positions filled | 289 | 49 |
| Percent of initial positions filled | 100 | 17 |
| No. of final matrix positions filled | 289 | 49 |
| Percent of final positions filled | 100 | 17 |
| No. of operations in decomp. 1 | 1496 | 16 |
| No. of operations in decomp. 2 | 136 | 16 |
| No. of operations in backsub. 1 | 136 | 16 |
| No. of operations in backsub. 2 | 136 | 16 |

The last four rows show the number of operations in each of four loops of decomposition and backsubstitution.

$$\frac{\partial^2 v_{q,i,t}}{\partial C_{q,t}\,\partial t} = n_{i,t-h}^{2/3}\left[48\pi^2(v_{q,i,t} + v_{i,t-h} - v_{q,i,t-h})\right]^{1/3} \times D^{\mathrm{eff}}_{q,i,t-h}\frac{m_q}{\rho_{p,q}} \qquad i = 1, \ldots, N_B \tag{17.58}$$

$$\frac{\partial^2 C_{q,t}}{\partial v_{q,i,t}\,\partial t} = -\frac{\rho_{p,q}}{m_q}\frac{\partial^2 v_{q,i,t}}{\partial v_{q,i,t}\,\partial t} \qquad i = 1, \ldots, N_B \tag{17.59}$$

$$\frac{\partial^2 C_{q,t}}{\partial C_{q,t}\,\partial t} = -\frac{\rho_{p,q}}{m_q}\sum_{i=1}^{N_B}\frac{\partial^2 v_{q,i,t}}{\partial C_{q,t}\,\partial t} \tag{17.60}$$

(17.56) does not require reordering, since it is already ordered from the fewest partial-derivative entries at the top of the matrix to the most at the bottom. Table 17.2 confirms that the current ordering of the matrix is optimal. It shows statistics from a case with 16 size bins ($N_B + 1 = 17$). The table shows that no matrix fill-in was required during matrix decomposition and backsubstitution.

An ODE integrator gives an accurate and relatively rapid solution to the growth equations if sparse-matrix techniques are used. An integrator still requires iteration; thus, in a model with $10^4$–$10^6$ grid cells, faster solutions may be desired.

### 17.5.2. Analytical Predictor of Condensation

Another method of solving growth equations is the **analytical predictor of condensation** (APC) scheme (Jacobson 1997c). This scheme requires no iteration, conserves mass exactly, and is unconditionally stable. The solution is obtained by holding the volume on the right side of (17.52) constant for a time step. Thus,

$$\frac{dv_{q,i,t}}{dt} = n_{i,t-h}^{2/3}(48\pi^2 v_{i,t-h})^{1/3} D_{q,i,t-h}^{eff} \frac{m_q}{\rho_{p,q}}\left(C_{q,t} - C'_{q,s,i,t-h}\right) \tag{17.61}$$

This equation is modified by defining a mass-transfer coefficient ($s^{-1}$), an effective surface vapor mole concentration, and a volume concentration as

$$k_{q,i,t-h} = n_{i,t-h}^{2/3}(48\pi^2 v_{i,t-h})^{1/3} D_{q,i,t-h}^{eff} = n_{i,t-h} 4\pi r_{i,t-h} D_{q,i,t-h}^{eff} \tag{17.62}$$

$$C'_{q,s,i,t-h} = S'_{q,i,t-h} C_{q,s,i,t-h} \tag{17.63}$$

$$v_{q,i,t} = m_q c_{q,i,t}/\rho_{p,q} \tag{17.64}$$

respectively. In (17.63), $C_{q,s,i,t-h} = p_{q,s,t-h}/R^*T$ is an uncorrected surface vapor mole concentration (moles $cm^{-3}$), and $S'_{q,i,t-h}$ is the equilibrium saturation ratio of the condensing gas. In (17.64), $c_{q,i,t}$ is the particle-phase mole concentration of component $q$ (moles per cubic centimeter of air). Equilibrium saturation ratios were defined for liquid water in (17.44) and other gases in (17.45). Substituting (17.62)–(17.64) into (17.61) and (17.55) gives

$$\frac{dc_{q,i,t}}{dt} = k_{q,i,t-h}\left(C_{q,t} - S'_{q,i,t-h} C_{q,s,i,t-h}\right) \qquad i = 1, \ldots, N_B \tag{17.65}$$

$$\frac{dC_{q,t}}{dt} = -\sum_{i=1}^{N_B}\left[k_{q,i,t-h}\left(C_{q,t} - S'_{q,i,t-h} C_{q,s,i,t}\right)\right] \tag{17.66}$$

respectively. (17.65) and (17.66) represent $N_B + 1$ ordinary differential equations.

The noniterative solution to the growth equations is obtained by integrating (17.65) for a final aerosol concentration. The result is

$$c_{q,i,t} = c_{q,i,t-h} + h k_{q,i,t-h}\left(C_{q,t} - S'_{q,i,t-h} C_{q,s,i,t-h}\right) \tag{17.67}$$

where the final gas mole concentration, $C_{q,t}$, is currently unknown. Final aerosol and gas concentrations are constrained by the mass-balance equation,

$$C_{q,t} + \sum_{i=1}^{N_B}(c_{q,i,t}) = C_{q,t-h} + \sum_{i=1}^{N_B}(c_{q,i,t-h}) = C_{tot} \tag{17.68}$$

Substituting (17.67) into (17.68) and solving for the gas concentration gives

$$C_{q,t} = \frac{C_{q,t-h} + h\sum_{i=1}^{N_B}\left(k_{q,i,t-h} S'_{q,i,t-h} C_{q,s,i,t-h}\right)}{1 + h\sum_{i=1}^{N_B} k_{q,i,t-h}} \tag{17.69}$$

The concentration from this equation cannot fall below zero, but can increase above the total mass of the species in the system. In such cases, gas concentration is limited by $C_{q,t} = \min(C_{q,t}, C_{tot})$. This value is substituted into (17.67), which must be limited by $c_{q,i,t} = \max(c_{q,i,t}, 0)$ to prevent evaporation beyond total mass

**Table 17.3.** Demostration of Unconditional Stability of the APC Scheme

| Time Step Size (s) | $C_{q,t}$ ($\mu$g m$^{-3}$) | $c_{q,1,t}$ ($\mu$g m$^{-3}$) | $c_{q,2,t}$ ($\mu$g m$^{-3}$) | $c_{3,q,t}$ ($\mu$g m$^{-3}$) |
|---|---|---|---|---|
| 0.1 | 1.0 | 2.18 | 5.36 | 1.46 |
| 10 | 1.0 | 2.18 | 5.36 | 1.46 |
| 60 | 1.0 | 2.18 | 5.36 | 1.46 |
| 600 | 1.0 | 2.18 | 5.36 | 1.46 |
| 7200 | 1.0 | 2.18 | 5.36 | 1.46 |

Assume a gas transfers to and from three particle size bins, where $k_{q,i,t-h} = 0.00833$, 0.001667, and 0.00667 s$^{-1}$ for the respective bins, $S'_{q,i,t-h} = 1.0$ for all three bins, $c_{q,i,t-h} = 2$, 5, and 0 $\mu$g m$^{-3}$ for the respective bins, $C_{q,t-h} = 3$ $\mu$g m$^{-3}$, and $C_{q,s,i,t-h} = 1$ $\mu$g m$^{-3}$. The table shows concentrations after 4 h predicted by the APC scheme when each of five time step sizes was used. Regardless of the time step size, the APC solution was stable and mass-conserving.

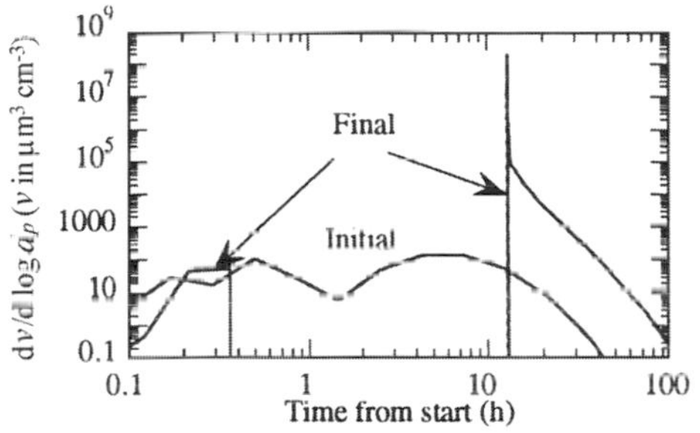


**Figure 17.2.** Comparison of APC solution, when $h = 10$ s, with an exact solution of condensational growth over 16 size bins. The two solutions lie almost exactly on top of each other. At the start of the simulation, the relative humidity was increased from 90 to 100.002 percent and replenished every 10 s back to 100.002 percent for a 5-min period.

evaporation in each bin. To ensure mass conservation when the latter limit is used, the final gas concentration is

$$C_{q,t} = C_{tot} - \sum_{i=1}^{N_B} c_{q,i,t} \tag{17.70}$$

The APC scheme is unconditionally stable, since all final concentrations are bounded between 0 and $C_{tot}$, regardless of the time step. Table 17.3 demonstrates the unconditional stability and mass conservation of the scheme.

Figure 17.2 compares an APC scheme prediction with an exact solution for a case of cloud drop growth from aerosols. The figure shows that the APC scheme matched the exact solution almost exactly for this set of conditions.

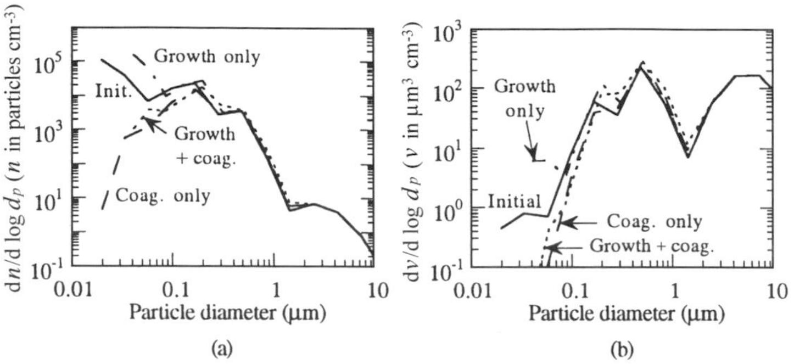


**Figure 17.3.** Change in aerosol (a) number concentration and (b) volume concentration after 8 h when coagulation alone, growth alone, and growth coupled to coagulation were considered. Only $H_2SO_4$(g) condensed. Initial $H_2SO_4$(g), 50 μg m$^{-3}$. Initial particle number concentration, 45,070 particles cm$^{-3}$. $T = 298$ K.

## 17.6. EFFECTS OF CONDENSATION ON COAGULATION

Condensation increases particle volume, and coagulation reduces particle number. Because both processes occur simultaneously, it is difficult to estimate the effect of each experimentally. A model can separate the relative effects. Results from one such simulation are shown in Fig. 17.3(a) and (b). Simulations of coagulation alone, coagulation coupled with growth, and growth alone were performed for an urban size distribution consisting of four lognormal modes (Jacobson 1997a). Two modes had geometric-mean diameters near 0.2 and 0.5–0.7 μm, respectively (Hering and Friedlander 1982; John et al. 1989). Simulations were performed over an 8-h period. The only growth process considered was sulfuric acid condensation.

The figures show that coagulation alone decreased the number concentration of particles smaller than 0.2 μm in diameter and slightly increased the number concentration of particles near 0.3 μm in diameter. Growth alone caused the initial size distribution to shift to the right. The shifting of particles to larger diameters caused the volume concentration of those particles to increase as well, as shown in Fig. 17.3(b). When coagulation was combined with growth, the number concentration of particles smaller than 0.2 μm decreased, and the number concentration of particles between 0.2 and 0.5 μm increased. In sum, **growth plus coagulation pushed particles to slightly larger sizes than did growth alone or coagulation alone.**

Figure 17.4 shows the growth-only and growth plus coagulation solutions from Fig. 17.3(a) when the moving-center and full-moving size structures, described in Chapter 14, were used. Since the full-moving structure is nondiffusive during growth, the figure implies that the moving-center size structure was also rather nondiffusive.

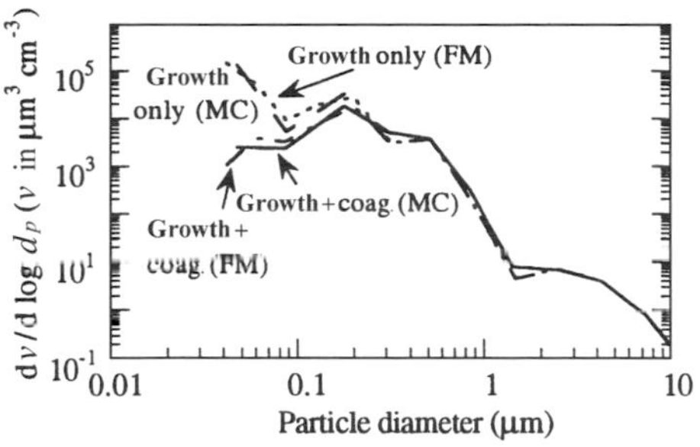


**Figure 17.4.** Comparison of results when using the moving-center (MC) versus full-moving (FM) size structures for simulations shown in Fig. 17.3(a).

## 17.7. ICE CRYSTAL GROWTH

The growth equation for an ice crystal is similar to that for a liquid water drop. The rate of mass growth of a single ice crystal, obtained by modifying (17.13), is

$$\frac{\mathrm{d}m_i}{\mathrm{d}t} = \frac{4\pi\chi_i D'_{v,i}\left(p_q - p'_{v,I,i}\right)}{\frac{D'_{v,i} L_s p'_{v,I,i}}{\kappa'_{d,i} T}\left(\frac{L_s}{R_v T} - 1\right) + R_v T} \tag{17.71}$$

where $\chi_i$ is the crystal **electrical capacitance** (cm), $p'_{v,I,i}$ is the effective saturation vapor pressure of water over ice, and $L_s$ is the latent heat of sublimation (J g$^{-1}$).

The **electrical capacitance**, analogous to the radius in the liquid drop growth equation, is a function of crystal shape only. The crystal shape changes with temperature, relative humidity, and crystal size. The electrical capacitance is indirectly a function of these variables. Parameterizations of electrical capacitance as a function of crystal shape include

$$\chi_i = \begin{cases} \mathrm{a}_{c,i}/2 & \text{sphere} \\ \mathrm{a}_{c,i}e_{c,i}/\ln[(1+e_{c,i})a_{c,i}/\mathrm{b}_{c,i}] & \text{prolate spheroid} \\ \mathrm{a}_{c,i}e_{c,i}/\sin^{-1} e_{c,i} & \text{oblate spheroid} \\ \mathrm{a}_{c,i}/\ln(4\mathrm{a}^2_{c,i}/\mathrm{b}^2_{c,i}) & \text{needle} \\ \mathrm{a}_{c,i}e_{c,i}/\ln[(1+e_{c,i})/(1-e_{c,i})] & \text{column} \\ \mathrm{a}_{c,i}e_{c,i}/(2\sin^{-1} e_{c,i}) & \text{hexagonal plate} \\ \mathrm{a}_{c,i}/\pi & \text{thin plate} \end{cases} \tag{17.72}$$

(e.g., Harrington et al. 1995), where $\mathrm{a}_{c,i}$ and $\mathrm{b}_{c,i}$ are the lengths of the major and minor semiaxes (cm), respectively, and $e_{c,i} = \sqrt{1 - \mathrm{b}^2_{c,i}/\mathrm{a}^2_{c,i}}$. The **major** and **minor semiaxes** are the maximum and minimum dimensions, respectively, across a crystal. A prolate spheroid is elongated at the poles and thin at the equator. An oblate spheroid is flattened at the poles and bulging at the equator. The electrical capacitance of a sphere is its radius.

The effective saturation vapor pressure in (17.71) is $p'_{v,I,i} = S'_{v,i} p_{v,I}$, where $p_{v,I}$ is the saturation vapor pressure over a flat ice surface, given in (2.58), and $S'_{v,i}$ is the equilibrium saturation ratio over ice, analogous to (17.44). (17.71) requires an effective molecular diffusion coefficient and thermal conductivity, given by (17.18) and (17.26), respectively. The ventilation factor used in these equations must be modified for ice crystals. An empirical expression for the ventilation factors of a falling oblate spheroid crystal is

$$F_{q,i}, F_{h,i} = \begin{cases} 1 + 0.14x^2 & x < 1.0 \\ 0.86 + 0.28x & x \geq 1.0 \end{cases} \tag{17.73}$$

where $x = x_{q,i}$ for vapor and $x = x_{h,i}$ for energy (Pruppacher and Klett 1997). Both $x_{q,i}$ and $x_{h,i}$ were given in (17.24) and (17.31), respectively. (17.71) can be solved by integration or with the APC scheme discussed earlier.

## 17.8. SUMMARY

In this chapter, the growth equation for a single, homogeneous, spherical liquid drop was derived. The equation assumed that, during condensation, gas diffuses to and energy conducts away from a particle surface. The reverse occurs during evaporation. The equation was expanded to a population of particles and to particles with multiple components. Two numerical solutions to growth equations – an exact numerical solution and a semianalytical solution – were discussed. The growth equations were used to test the importance of condensation relative to coagulation. The conclusion was that growth plus coagulation moved particles to larger sizes than did growth alone or coagulation alone. An ice-crystal depositional growth equation was also given. In the next chapter, the discussion of growth is extended to particles in which trace gases dissolve.

## 17.9. PROBLEMS

**17.1.** **(a)** Solve (17.65) and (17.66) simultaneously (analytically) for $c_{q,i,t}$ and $C_{q,t}$. Assume one size bin is present and all variables are constant, except $c_{q,i,t}$ and $C_{q,t}$. (Hint: $C_{q,t} + c_{q,i,t} = C_{q,t-h} + c_{q,i,t-h}$, since $i = 1$.)

**(b)** Discretize the time derivative in the equations from part (a) in first-order, backward-difference form, and write out the right side of both equations implicitly. Solve the resulting equations for $c_{q,i,t}$ and $C_{q,t}$.

**17.2.** Can condensation of a species occur on particles of some sizes while evaporation occurs on particles of other sizes? Is so, under what conditions would this occur?

**17.3.** In Fig. 17.1, particles above what radius can grow to cloud-sized drops? Under what condition can all particles grow larger than the critical radius? Is this condition likely to occur in the atmosphere? Why or why not?

**17.4.** For what temperature range is (17.71) valid? Can (17.13) be used to simulate the growth of supercooled liquid water?

## 17.10. COMPUTER PROGRAMMING PRACTICE

**17.5.** Write a computer script to calculate the equilibrium saturation ratio versus diameter and temperature for water vapor over a pure liquid water drop from (17.44). Ignore the solute and radiative heating effects. Calculate the Kelvin term for $d_i = 0.001$ to 10 μm, when $T = 290$ K, and plot the results. Calculate the term for $d_i = 0.01$ μm when $T$ varies from 233 K to 315 K. Discuss the results.

**17.6.** **(a)** Write a computer script to calculate $\rho_a$, $\eta_a$, $\nu_a$, $D_q$, $\omega_{q,i}$, $\bar{v}_q$, $\lambda_q$, $\text{Kn}_{q,i}$, $N_{\text{Re},i}$, $N_{\text{Sc},q}$, $F_{q,i}$, and $D'_{q,i}$ assuming that water vapor condenses onto a particle of radius $r_i = 0.2$ μm. Assume also that $T = 298$ K and $p_a = 1013$ mb. For the fall velocity required in the Reynolds-number calculation, assume $V_{f,i} = 2r_i^2 \rho_p g/(9\eta_a)$. Use additional data from the text, where necessary. Assume the air is dry.

**(b)** Using information from Problem 17.2(a), calculate $\kappa_d$, $\omega_{h,i}$, $\bar{v}_a$, $\lambda_h$, $\text{Kn}_{h,i}$, $N_{\text{Pr}}$, $F_{h,i}$, and $\kappa'_{d,i}$ with the computer script.

**(c)** Using information from Problems 17.2(a) and (b), calculate $L_e$, $p_{q,s}$, $S'_{q,i}$, $p'_{q,s,i}$, $C'_{q,s,i}$, and $D^{eff}_{q,i}$ with the computer script. Assume that only curvature affects $S'_{q,i}$.

**(d)** Repeat Problem 17.2(a), (b), and (c) for $T = 273$ K. Comment on the results.

**(e)** Repeat Problems 17.2(a), (b), and (c) for $r_i = 1.0$ μm. Comment on the results.

**17.7.** Write a computer script to solve (17.65) and (17.66) with the APC scheme. Assume three size bins are present ($N_B = 3$). Also assume $k_{q,1,t-h} = 0.000833$ s$^{-1}$, $k_{2,q,t-h} = 0.001667$ s$^{-1}$, $k_{q,3,t-h} = 0.00667$ s$^{-1}$, $c_{q,i,t-h} = 0$ for all $i$, $S'_{q,i,t-h} = 1$ for all $i$, $C_{q,t-h} = 10$, $C_{q,s,i,t-h} = 2$, and concentration units are μg m$^{-3}$. Plot the mass concentration of particles in each size bin versus time for a 50-min simulation period, using $h = 1$ min. Compare the result with an explicit solution to (17.65) and (17.66), when $h = 0.01$ s. Is this simulation representative of growth under atmospheric conditions? Why or why not?

# 18

# Chemical Equilibrium and Dissolution Processes

THE change in size and composition of an aerosol depends on several processes, including nucleation, emissions, coagulation, condensational growth, depositional growth, dissolutional growth, heterogeneous reaction, molecular dissociation, solid precipitation, irreversible chemistry, sedimentation, dry deposition, and advection. Many of these processes are tightly coupled. Condensation and deposition at the gas–particle interface were discussed in Chapter 17, and irreversible reactions are examined in Chapter 19. In this chapter, dissolutional growth, molecular dissociation, solid precipitation, and heterogeneous reaction are described. These four processes can be simulated with reversible (equilibrium) chemical reactions. Dissolution can also be simulated as a nonequilibrium process. In the sections that follow, terminology, equilibrium equations, activity-coefficient equations, the water equation, and methods of solving equilibrium and nonequilibrium equations are given.

## 18.1. DEFINITIONS

An important process that takes place at gas–particle interfaces is dissolution. **Dissolution** occurs when a gas, suspended over a particle surface, adsorbs to and dissolves in liquid on the surface. The liquid in which the gas dissolves is a **solvent**. A solvent makes up the bulk of a solution, and in particles, liquid water is most often the solvent. In some cases, such as when sulfuric acid combines with water to form particles, the concentration of sulfuric acid exceeds that of liquid water, and sulfuric acid is the solvent. In this chapter, liquid water is assumed to be the solvent in all cases.

A gas, liquid, or solid that dissolves in solution is a **solute**. Together, solute and solvent make up a **solution**, which is a homogeneous mixture of substances that can be separated into individual components upon a change of state (e.g., freezing). A solution may contain many solutes. Suspended material (e.g., solids) may also be mixed throughout a solution. Such material is not considered part of the solution.

The ability of a gas to dissolve in water depends on the solubility of the gas. The **solubility** is the maximum amount of a gas that can dissolve in a given amount of solvent at a given temperature. Solutions usually contain solute other than the dissolved gas. The solubility of a gas depends on the quantity of other solutes, since such solutes affect the thermodynamic activity of the dissolved gas in solution. (Thermodynamic activity is discussed shortly.) If water is saturated with a dissolved gas, and if the solubility of the gas changes due to a change in composition of the solution, the dissolved gas can **evaporate** from the solution to the gas phase. In solution, dissolved molecules may dissociate. **Dissociation** is the process by which a dissolved molecule breaks into simpler components, namely ions. Dissociation products of a dissolved molecule may combine with other components and **precipitate** out of solution as solids.

Another process that occurs at gas–particle interfaces is **surface (heterogeneous) reaction**. Heterogeneous reaction, first discussed in Section 12.3.4, occurs when a gas collides with a particle surface. A reaction may occur if the gas molecule collides with a molecule adsorbed to the surface or if the molecule adsorbs to the surface, diffuses on the surface, then reacts with another adsorbed molecule. Some surface reactions result in solid formation. Suppose nitric acid gas ($HNO_3$) collides with and adsorbs to a solid particle surface, and an ammonia gas ($NH_3$) molecule collides with and heterogeneously reacts with the adsorbed nitric acid. This reaction may result in the formation of solid ammonium nitrate ($NH_4NO_3$), which is adsorbed to the surface.

Gas dissolution, molecular dissociation, solid precipitation, and heterogeneous reaction can be described by **reversible chemical reactions** (thermodynamic equilibrium reactions). The rates of these reactions in the forward and reverse directions are usually fast. **Irreversible chemical reactions** act only in the forward direction and are described by first-order ordinary differential equations similar to the gas-phase reactions discussed in Chapter 12. Gas dissolution can be treated as a nonequilibrium or equilibrium process, both of which are discussed in this chapter.

## 18.2. EQUILIBRIUM EQUATIONS AND RELATIONS

In this section, equilibrium equations are discussed, and temperature-dependent equilibrium coefficients are derived.

### 18.2.1. Equilibrium Reactions

An **equilibrium reaction** describes a reversible chemical process among solids, liquids, ions, and or gases. The change in the number of moles $\mathrm{d}n_i$ of each reactant $i = \mathrm{A, B, D, E}$, etc. during such a process is described by

$$\mathrm{d}n_\mathrm{D}\mathrm{D} + \mathrm{d}n_\mathrm{E}\mathrm{E} + \cdots \rightleftharpoons \mathrm{d}n_\mathrm{A}\mathrm{A} + \mathrm{d}n_\mathrm{B}\mathrm{B} + \cdots \tag{18.1}$$

Each equilibrium reaction must conserve mass. Thus,

$$\sum_i k_i(\mathrm{d}n_i)m_i = 0 \tag{18.2}$$

where $m_i$ is the molecular weight of species $i$, $k_i = +1$ for products, and $k_i = -1$ for reactants. Dividing each $\mathrm{d}n_i$ by the smallest $\mathrm{d}n_i$ among all species in (18.1) gives a set of dimensionless stoichiometric coefficients $\nu_i$ that can be substituted into (18.1) to yield

$$\nu_D D + \nu_E E + \cdots \rightleftharpoons \nu_A A + \nu_B B + \cdots \quad (18.3)$$

which is the general form of an equilibrium reaction. In this subsection, several types of equilibrium reactions are discussed.

Reversible dissolution reactions occur at gas-solution interfaces, such as at air-ocean, air-cloud drop, and air-aerosol interfaces. Such reactions have the form

$$AB(g) \rightleftharpoons AB(aq) \quad (18.4)$$

where (g) indicates a gas, (aq) indicates a species dissolved in solution, and the stoichiometric coefficients are unity in this case. In (18.4), the gas and dissolved (solution) phases of species AB are assumed to be in equilibrium with each other at the gas-solution interface; thus, the number of molecules of AB transferring from the gas to the solution is assumed to equal the number of molecules transferring in the reverse direction. Some **dissolution reactions** are

$$\begin{aligned} HCl(g) &\rightleftharpoons HCl(aq) \quad (18.5) \\ HNO_3(g) &\rightleftharpoons HNO_3(aq) \\ CO_2(g) &\rightleftharpoons CO_2(aq) \\ NH_3(g) &\rightleftharpoons NH_3(aq) \\ H_2SO_4(g) &\rightleftharpoons H_2SO_4(aq) \end{aligned}$$

In equilibrium, almost all sulfuric acid is partitioned to the aqueous phase; thus, the last relation in (18.5) is rarely used. Instead, sulfuric acid transfer to the aqueous phase is treated as a diffusion-limited condensational growth process (Chapter 17).

Once in solution, the dissolved species in (18.5) usually dissociate reversibly into ions. Substances that undergo partial or complete dissociation in solution are **electrolytes**. The degree of dissociation of an electrolyte depends on the acidity of solution, the strength of the electrolyte, and the concentrations of other ions in solution.

The **acidity** of a solution is a measure of the concentration of **hydrogen ions** (protons or $H^+$ ions) in solution. Acidity is measured in terms of

$$pH = -\log_{10}[H^+] \quad (18.6)$$

where $[H^+]$ is the **molarity** of $H^+$ (moles of $H^+$ per liter of solution). The more acidic the solution, the higher the molarity of protons, and the lower the pH. Protons in solution are donated by acids that dissolve. Such acids include $H_2CO_3(aq)$, $HCl(aq)$, $HNO_3(aq)$, and $H_2SO_4(aq)$. The abilities of acids to dissociate into protons and anions vary. $HCl(aq)$, $HNO_3(aq)$, and $H_2SO_4(aq)$ dissociate readily, whereas $H_2CO_3(aq)$ does not. The former species are **strong acids**, and the latter species is a **weak acid**. Since all acids are electrolytes, a strong acid is a **strong**

**electrolyte** (i.e., it dissociates significantly) and a weak acid is a **weak electrolyte.** Hydrochloric acid is a strong acid and strong electrolyte in water because it almost always dissociates completely above a pH of −6 by the reversible reaction

$$HCl(aq) \rightleftharpoons H^+ + Cl^- \tag{18.7}$$

**Sulfuric acid** is also a strong acid and electrolyte and dissociates to another strong acid, bisulfate ($HSO_4^-$), above a pH of −3. **Busulfate** dissociates significantly to **sulfate** ($SO_4^{2-}$) above a pH of +2. Sulfuric acid and bisulfate dissociation reactions are

$$\begin{aligned} H_2SO_4(aq) &\rightleftharpoons H^+ + HSO_4^- \\ HSO_4^- &\rightleftharpoons H^+ + SO_4^{2-} \end{aligned} \tag{18.8}$$

respectively. Another strong acid, **nitric acid,** dissociates significantly to nitrate above a pH of −1 by

$$HNO_3(aq) \rightleftharpoons H^+ + NO_3^- \tag{18.9}$$

**Carbon dioxide** is a weak acid and electrolyte; it converts to **carbonic acid,** which dissociates significantly to bicarbonate above a pH of only +6. Dissociation of **bicarbonate** occurs above a pH of +10. Carbon dioxide and bicarbonate dissociation reactions are

$$\begin{aligned} CO_2(aq) + H_2O(aq) \rightleftharpoons H_2CO_3(aq) &\rightleftharpoons H^+ + HCO_3^- \\ HCO_3^- &\rightleftharpoons H^+ + CO_3^{2-} \end{aligned} \tag{18.10}$$

respectively.

Whereas acids provide hydrogen ions, **bases** provide **hydroxide ions** ($OH^-$). Such ions react with hydrogen ions to form neutral water via

$$H_2O(aq) \rightleftharpoons H^+ + OH^- \tag{18.11}$$

An important basic substance in aerosols is ammonia. **Ammonia** reacts with water in solution to form **ammonium** and the hydroxide ion by

$$NH_3(aq) + H_2O(aq) \rightleftharpoons NH_4^+ + OH^- \tag{18.12}$$

Since some strong electrolytes, such as HCl(aq) and $HNO_3$(aq), dissociate completely in atmospheric particles, the undissociated forms of these species are sometimes ignored in equilibrium models. Instead, gas–ion equilibrium equations replace the combination of gas–liquid and liquid–ion equations. For example,

$$HCl(g) \rightleftharpoons H^+ + Cl^- \tag{18.13}$$

$$HNO_3(g) \rightleftharpoons H^+ + NO_3^- \tag{18.14}$$

replace the equation pairs, (18.5)/(18.7) and (18.5)/(18.9), respectively.

Under the right conditions, ions may precipitate from solution to form **solid electrolytes.** If the liquid-water content in a solution increases sufficiently, existing solid electrolytes may dissociate into ions. Examples of **solid precipitation/dissociation reactions** for ammonium-containing electrolytes include

$$
\begin{aligned}
NH_4Cl(s) &\rightleftharpoons NH_4^+ + Cl^- \\
NH_4NO_3(s) &\rightleftharpoons NH_4^+ + NO_3^- \\
(NH_4)_2SO_4(s) &\rightleftharpoons 2NH_4^+ + SO_4^{2-}
\end{aligned}
\tag{18.15}
$$

Examples of such reactions for sodium-containing electrolytes are

$$
\begin{aligned}
NaCl(s) &\rightleftharpoons Na^+ + Cl^- \\
NaNO_3(s) &\rightleftharpoons Na^+ + NO_3^- \\
Na_2SO_4(s) &\rightleftharpoons 2Na^+ + SO_4^{2-}
\end{aligned}
\tag{18.16}
$$

If the relative humidity is low, a gas may react heterogeneously with another adsorbed gas on a particle surface to form a solid. Such reactions can be simulated as **gas–solid equilibrium reactions**, such as

$$
\begin{aligned}
NH_4Cl(s) &\rightleftharpoons NH_3(g) + HCl(g) \\
NH_4NO_3(s) &\rightleftharpoons NH_3(g) + HNO_3(g)
\end{aligned}
\tag{18.17}
$$

In sum, equilibrium relationships describe aqueous–ion, ion–ion, ion–solid, gas–solid, or gas–ion reversible reactions. Relationships can be written for other interactions as well. Table B.7 of Appendix B lists several equilibrium reactions of atmospheric interest.

### 18.2.2. Equilibrium Relations and Coefficients

Thermodynamic activities in a reversible reaction, such as (18.3), are related to an equilibrium coefficient by

$$
\prod_i \{a_i\}^{k_i \nu_i} = \frac{\{A\}^{\nu_A}\{B\}^{\nu_B}\ldots}{\{D\}^{\nu_D}\{E\}^{\nu_E}\ldots} = K_{eq}(T) \tag{18.18}
$$

where $K_{eq}(T)$ is a temperature-dependent **equilibrium coefficient**, $\{a_i\}$ is the **thermodynamic activity** of species $i$, $\{A\}$, etc., are individual thermodynamic activities, $k_i = +1$ for products, and $k_i = -1$ for reactants.

An activity is a relative quantity of a substance and is determined differently for each phase state. The activity of a gas over a particle surface is its saturation vapor pressure (atm). Thus,

$$
\{A(g)\} = p_{A,s} \tag{18.19}
$$

The activity of an ion in solution or of an undissociated electrolyte is its **molality** $\mathbf{m}_A$ (moles of solute per kilogarm of solvent) multiplied by its **activity coefficient** $\gamma$

(unitless). Thus,

$$\{A^+\} = \mathbf{m}_{A^+}\gamma_{A^+} \qquad \{A(aq)\} = \mathbf{m}_A\gamma_A \tag{18.20}$$

respectively. An **activity coefficient** represents the deviation from ideal behavior of a solution. It is a dimensionless parameter by which the molality of a species in solution is multiplied to give the species' thermodynamic activity. In an ideal, infinitely dilute solution, the activity coefficient of a species is unity. In a nonideal, concentrated solution, activity coefficients may exceed or be less than unity. **Debye and Hückel** showed that, in sufficiently dilute solutions, where ions are far apart, the deviation of molality from thermodynamic activity is caused by Coulombic (electric) forces of attraction and repulsion. At high concentrations, ions are close together, and ion–ion interactions affect activity coefficients more than do Coulombic forces.

The activity of liquid water in a particle is the ambient relative humidity. Thus,

$$\{H_2O(aq)\} = a_w = \frac{p_v}{p_{v,s}} = f_r \tag{18.21}$$

where $a_w$ denotes the activity of water, $p_v$ is the partial pressure of water vapor, $p_{v,s}$ is the saturation vapor pressure of water over a bulk liquid surface, and $f_r$ is the relative humidity, expressed as a fraction. For pure liquid water in equilibrium with water vapor, $p_v = p_{v,s}$, and $a_w$ simplifies to unity, which is the water activity in a dilute solution. When liquid water contains solute, some of the solute replaces liquid water on the surface of the drop. Water molecules bond to the solute (discussed in Section 18.8), requiring more water to condense from the vapor phase to maintain saturation over the solution, lowering the pressure exerted by water vapor over the drop surface, decreasing the relative humidity over the surface, and decreasing the water activity.

Solids are not in solution, and their concentrations do not affect the molalities or activity coefficients of electrolytes in solution. The activity of a pure solid is

$$\{A(s)\} = 1 \tag{18.22}$$

which is the mole fraction of a pure solid.

(18.18) is derived by minimizing the change in Gibbs free energy of a system. The Gibbs free energy (J) is defined as

$$G^* = H^* - TS^* = U^* + p_a V - TS^* \tag{18.23}$$

where $H^* = U^* + p_a V$ is the **enthalpy** (J), $T$ is the absolute temperature, $S^*$ is the **entropy** ($J\,K^{-1}$), $U^*$ is the internal energy (J), $p_a$ is the pressure, and $V$ is the volume of a system. The change in Gibbs free energy is a measure of the maximum amount of useful work that may be obtained from a change in enthalpy or entropy in the system. It is calculated as

$$dG^* = d(H^* - TS^*) = dU^* + p_a\,dV + V\,dp_a - T\,dS^* - S^*\,dT \tag{18.24}$$

where $dS^* = dQ^*/T$. Thus, the change in entropy of a system equals the incremental energy added to or removed from the system divided by the absolute temperature.

The internal energy was introduced in Chapter 2. For a system in which no changes in chemical composition occur, it is written as

$$dU^* = dQ^* - p_a\, dV = T\, dS^* - p_a\, dV \tag{18.25}$$

For a system in which reversible chemical reactions of the form shown in (18.1) occur, it must be modified to

$$dU^* = T\, dS^* - p_a\, dV + \sum_i k_i (dn_i)\mu_i \tag{18.26}$$

where $\mu_i$ (J mole$^{-1}$) is the chemical potential of species $i$. **Chemical potential** is a measure of the intensity of a substance and is a function of temperature and pressure. Differences in chemical potential between two substances can lead to chemical reactions between them. Substituting (18.26) into (18.24) gives

$$dG^* = V\, dp_a - S^*\, dT + \sum_i k_i (dn_i)\mu_i \tag{18.27}$$

When temperature and pressure are held constant, (18.27) simplifies to

$$dG^* = \sum_i k_i (dn_i)\mu_i \tag{18.28}$$

From this equation, it can be seen that the chemical potential of a substance is its change in Gibbs free energy per unit change in its number of moles, or its partial molar free energy, when temperature and pressure are held constant. The chemical potential can also be defined in terms of a standard chemical potential and perturbation term. Thus,

$$\mu_i = \left(\frac{\partial G_i^*}{\partial n_i}\right)_{T,p_a} = \mu_i^{\circ}(T) + R^* T \ln\{a_i\} \tag{18.29}$$

where $dG_i^*$ is the change in Gibbs free energy of an individual species at constant temperature and pressure. In the second expression, $\mu_i^{\circ}$ (J mole$^{-1}$), which varies with temperature, is the standard chemical potential of a substance, and $R^*T \ln\{a_i\}$ accounts for the deviation in chemical potential from the standard state due to the activity of the substance.

Thermodynamic equilibrium occurs when $dG^* = 0$. Setting (18.28) equal to zero and dividing through by the smallest value of $dn_i$ among all species in the reaction under consideration gives

$$\sum_i k_i \nu_i \mu_i = 0 \tag{18.30}$$

where $\nu_i$ is the dimensionless stoichiometric coefficient. Substituting (18.29) into (18.30), assuming $T = T_0 = 298.15$ K for now, defining $\Delta_f G_i^{\circ} = \mu_i^{\circ}(T_0)$ as the **standard molal Gibbs free energy of formation** (J mole$^{-1}$) of substance $i$ at $T = T_0$, and noting that $\sum_i k_i \nu_i \ln\{a_i\} = \ln \prod_i \{a_i\}^{k_i \nu_i}$ gives

$$\sum_i k_i \nu_i \mu_i^{\circ}(T_0) + R^* T_0 \sum_i k_i \nu_i \ln\{a_i\} = \sum_i k_i \nu_i \Delta_f G_i^{\circ} + R^* T_0 \ln \prod_i \{a_i\}^{k_i \nu_i} = 0 \tag{18.31}$$

Rearranging (18.31) yields

$$\prod_i \{a_i\}^{k_i \nu_i} = \exp\left(-\frac{1}{R^* T_0} \sum_i k_i \nu_i \Delta_f G_i^\circ\right) \tag{18.32}$$

The right side of (18.32) is the **equilibrium coefficient** at $T = T_0$. Thus,

$$K_{eq}(T_0) = \exp\left(-\frac{1}{R^* T_0} \sum_i k_i \nu_i \Delta_f G_i^\circ\right) \tag{18.33}$$

which is independent of species concentration. Substituting (18.33) into (18.32) when $T = T_0$ gives the equilibrium coefficient-equation shown in (18.18). Values of $\Delta_f G_i^\circ$ are given in Table B.6 of Appendix B for several substances.

## 18.3. TEMPERATURE DEPENDENCE OF THE EQUILIBRIUM COEFFICIENT

The temperature dependence of the equilibrium coefficient can be estimated by solving the **van't Hoff equation,**

$$\frac{d \ln K_{eq}(T)}{dT} = \frac{1}{R^* T^2} \sum_i k_i \nu_i \Delta_f H_i \tag{18.34}$$

where $\Delta_f H_i$ is the **molal enthalpy of formation** (J mole$^{-1}$) of a substance. The Van't Hoff equation is similar in form to the Arrhenius equation given in Chapter 11. $\Delta_f H_i$ can be approximated as a function of temperature with

$$\Delta_f H_i \approx \Delta_f H_i^\circ + c_{p,i}^\circ (T - T_0) \tag{18.35}$$

where $\Delta_f H_i^\circ$ is the **standard molal enthalpy of formation** (J mole$^{-1}$) (at $T_0$ = 298.15 K) and $c_{p,i}^\circ$ is the **standard molal heat capacity** (standard specific heat) **at constant pressure** (J mole$^{-1}$ K$^{-1}$). Although $c_{p,i}^\circ$ varies slightly with temperature, (18.35) assumes that it does not. Combining (18.34) and (18.35) and writing the result in integral form gives

$$\int_{T_0}^{T} d \ln K_{eq}(T) = \int_{T_0}^{T} \frac{1}{R^* T^2} \sum_i k_i \nu_i \left[\Delta_f H_i^\circ + c_{p,i}^\circ (T - T_0)\right] dT \tag{18.36}$$

Integrating yields the temperature-dependent expression for the equilibrium coefficient,

$$K_{eq}(T) = K_{eq}(T_0) \exp\left\{-\sum_i k_i \nu_i \left[\frac{\Delta_f H_i^\circ}{R^* T_0}\left(\frac{T_0}{T} - 1\right) + \frac{c_{p,i}^\circ}{R^*}\left(1 - \frac{T_0}{T} + \ln \frac{T_0}{T}\right)\right]\right\} \tag{18.37}$$

where $K_{eq}(T_0)$ is the equilibrium coefficient at $T = T_0$ found in (18.33). Values of $\Delta_f H_i^\circ$ and $c_{p,i}^\circ$ are measured experimentally and given in Table B.6 of

Appendix B for several substances. Table B.7 of Appendix B gives temperature-dependent equilibrium-coefficient expressions for several reactions of the form shown in (18.37) derived from the data in Table B.6.

**Example 18.1**

Determine the equilibrium coefficient expression for $Na_2SO_4(s) \rightleftharpoons Na^+ + SO_4{}^{2-}$. Applying data from Table B.6 gives

$$\sum_i k_i \nu_i \Delta_f G_i^\circ = 1820 \frac{\mathrm{J}}{\mathrm{mole}} \qquad \sum_i k_i \nu_i \Delta_f H_i^\circ = -2430 \frac{\mathrm{J}}{\mathrm{mole}}$$

$$\sum_i k_i \nu_i c_{p,i}^\circ = -328.4 \frac{\mathrm{J}}{\mathrm{mole\ K}}$$

Substituting these expressions into (18.37) and (18.33) yields

$$K_{eq}(T) = 0.4799 \exp\left\{0.9802\left(\frac{T_0}{T} - 1\right) - 39.497\left[1 + \ln\left(\frac{T_0}{T}\right) - \frac{T_0}{T}\right]\right\} \frac{\mathrm{mole}^2}{\mathrm{kg}^2}$$

## 18.4. FORMS OF EQUILIBRIUM-COEFFICIENT EQUATIONS

Equation (18.18) expresses an equilibrium reaction in terms of thermodynamic activities and an equilibrium coefficient. In this section, activity expressions are substituted for activities to give equilibrium-coefficient equations for several types of reactions.

For a dissolution reaction, such as $HNO_3(g) \rightleftharpoons HNO_3(aq)$, the equilibrium coefficient relates the pressure exerted by a gas at the gas–liquid interface to the molality of the dissolved gas in solution. The equilibrium coefficient expression for the $HNO_3$ reaction is

$$\frac{\{HNO_3(aq)\}}{\{HNO_3(g)\}} = \frac{\mathbf{m}_{HNO_3(aq)} \gamma_{HNO_3(aq)}}{p_{HNO_3(g),s}} = K_{eq}(T) \frac{\mathrm{moles}}{\mathrm{kg\ atm}} \tag{18.38}$$

where $p_{HNO_3(g),s}$ is the saturation vapor pressure of nitric acid (atm), $\mathbf{m}_{HNO_3(aq)}$ is the molality of nitric acid in solution (moles $kg^{-1}$), and $\gamma_{HNO_3(aq)}$ is the activity coefficient of dissolved, undissociated nitric acid (unitless). In Chapter 2, a saturation vapor pressure was defined as the partial pressure of a gas when the gas is in equilibrium with a particle surface at a given temperature. Since $HNO_3(g)$ is in equilibrium with the particle surface in this case, its saturation vapor pressure, not its partial pressure, is used in (18.38).

Over a dilute solution, the pressure exerted by a gas is proportional to the molality of the dissolved gas. This is **Henry's law**. For a dilute solution, $\gamma_{HNO_3(aq)} = 1$, and (18.38) obeys Henry's law. In a dilute or concentrated solution, $K_{eq}(T)$ in (18.38) is called a **Henry's constant**. Henry's constants, like other equilibrium coefficients, are temperature-dependent. Henry's constants for gases dissolving in liquid water are given in Table B.7. Species with large Henry's constants include $NH_3$, $HNO_3$,

HCl, $H_2SO_4$, $H_2O_2$, and $SO_2$. For these and many other gases $\sum_i k_i \nu_i \Delta_f H_i^\circ < 0$. In such cases, (18.37) predicts that solubilities and Henry's constants increase with decreasing temperature.

In solution, some substances dissociate. The dissociation reaction for nitric acid has the form $HNO_3(aq) \rightleftharpoons H^+ + NO_3^-$. The equilibrium coefficient expression for this reaction is

$$\frac{\{H^+\}\{NO_3^-\}}{\{HNO_3(aq)\}} = \frac{m_{H^+}\gamma_{H^+} m_{NO_3^-}\gamma_{NO_3^-}}{m_{HNO_3(aq)}\gamma_{HNO_3(aq)}} \tag{18.39}$$

$$= \frac{m_{H^+} m_{NO_3^-}\gamma^2_{H^+,NO_3^-}}{m_{HNO_3(aq)}\gamma_{HNO_3(aq)}} = K_{eq}(T)\frac{\text{moles}}{\text{kg}}$$

The activity coefficients in (18.39) are mixed activity coefficients because they are determined by considering a mixture of all dissociated and undissociated electrolytes in solution. The parameters $\gamma_{H^+}$ and $\gamma_{HNO_3^-}$ are single-ion mixed activity coefficients, and $\gamma_{H^+,NO_3^-}$ is a **mean** (geometric-mean) **mixed activity coefficient.** When $HNO_3(aq)$, $H^+$, and $NO_3^-$ are alone in solution, $\gamma_{H^+}$ and $\gamma_{NO_3^-}$ are single-ion binary activity coefficients, and $\gamma_{H^+,NO_3^-}$ is a **mean** (geometric-mean) **binary activity coefficient.** Activity coefficients for single ions are difficult to measure, because single ions cannot be isolated from a solution. Single-ion activity coefficients are easier to estimate mathematically. Mean binary activity coefficients are measured in the laboratory. Mean mixed activity coefficients can be estimated from mean binary activity coefficient data through a mixing rule.

A geometric mean activity coefficient, $\gamma_\pm$, is related to single-ion activity coefficients by

$$\gamma_\pm = \left(\gamma_+^{\nu_+}\gamma_-^{\nu_-}\right)^{1/(\nu_+ + \nu_-)} \tag{18.40}$$

where $\gamma_+$ and $\gamma_-$ are the activity coefficients of a cation and anion, respectively, and $\nu_+$ and $\nu_-$ are the stoichiometric coefficients of the cation and anion, respectively. In the reaction for (18.39), $\nu_+ = 1$ and $\nu_- = 1$. When $\nu_+ = 1$ and $\nu_- = 1$, the electrolyte is **univalent.** When $\nu_+ > 1$ or $\nu_- > 1$, the electrolyte is **multivalent.** When $\nu_+ = \nu_-$, the electrolyte is **symmetric**; otherwise, it is **nonsymmetric.** In all cases, a dissociation reaction must satisfy the charge balance requirement $z_+\nu_+ + z_-\nu_- = 0$, where $z_+$ is the positive charge on the cation and $z_-$ is the negative charge on the anion.

**Example 18.2**

$HNO_3(aq)$ dissociates by the reaction $HNO_3(aq) \rightleftharpoons H^+ + NO_3^-$. Since $\nu_+ = 1$ and $\nu_- = 1$, $HNO_3(aq)$ is a univalent, symmetric electrolyte. Since $z_+ = +1$ and $z_- = -1$, nitric acid dissociation satisfies $z_+\nu_+ + z_-\nu_- = 0$.

$Na_2SO_4(s)$ dissociates by the reaction, $Na_2SO_4(s) \rightleftharpoons 2Na^+ + SO_4^{2-}$. Since $\nu_+ = 2$ and $\nu_- = 1$, $Na_2SO_4(s)$ is a multivalent, nonsymmetric electrolyte. Since $z_+ = +1$ and $z_- = -2$, sodium sulfate dissociation satisfies $z_+\nu_+ + z_-\nu_- = 0$.

Rearranging (18.40) gives

$$\gamma_{\pm}^{\nu_+ + \nu_-} = \gamma_+^{\nu_+}\gamma_-^{\nu_-} \tag{18.41}$$

which is the form of the mean activity coefficient used in (18.39). Below, equilibrium coefficient expressions for several reactions are given.

$$\begin{aligned}
\mathrm{Na_2SO_4(s)} &\rightleftharpoons \mathrm{2Na^+ + SO_4{}^{2-}} \\
\frac{\{\mathrm{Na^+}\}^2\{\mathrm{SO_4{}^{2-}}\}}{\{\mathrm{Na_2SO_4(s)}\}} &= \frac{\mathbf{m}^2_{\mathrm{Na^+}}\gamma^2_{\mathrm{Na^+}}\mathbf{m}_{\mathrm{SO_4{}^{2-}}}\gamma_{\mathrm{SO_4{}^{2-}}}}{1.0} \\
&= \mathbf{m}^2_{\mathrm{Na^+}}\mathbf{m}_{\mathrm{SO_4{}^{2-}}}\gamma^3_{\mathrm{2Na^+,SO_4{}^{2-}}} = K_{eq}(T)\,\frac{\mathrm{moles}^3}{\mathrm{kg}^3}
\end{aligned} \tag{18.42}$$

$$\begin{aligned}
\mathrm{HSO_4{}^-} &\rightleftharpoons \mathrm{H^+ + SO_4{}^{2-}} \\
\frac{\{\mathrm{H^+}\}^2\{\mathrm{SO_4{}^{2-}}\}}{\{\mathrm{H^+}\}\{\mathrm{HSO_4{}^-}\}} &= \frac{\mathbf{m}^2_{\mathrm{H^+}}\gamma^2_{\mathrm{H^+}}\mathbf{m}_{\mathrm{SO_4{}^{2-}}}\gamma_{\mathrm{SO_4{}^{2-}}}}{\mathbf{m}_{\mathrm{H^+}}\gamma_{\mathrm{H^+}}\mathbf{m}_{\mathrm{HSO_4{}^-}}\gamma_{\mathrm{HSO_4{}^-}}} \\
&= \frac{\mathbf{m}_{\mathrm{H^+}}\mathbf{m}_{\mathrm{SO_4{}^{2-}}}\gamma^3_{\mathrm{2H^+,SO_4{}^{2-}}}}{\mathbf{m}_{\mathrm{HSO_4{}^-}}\gamma^2_{\mathrm{H^+,HSO_4{}^-}}} = K_{eq}(T)\,\frac{\mathrm{moles}}{\mathrm{kg}}
\end{aligned} \tag{18.43}$$

$$\begin{aligned}
\mathrm{NH_3(g) + HNO_3(g)} &\rightleftharpoons \mathrm{NH_4{}^+ + NO_3{}^-} \\
\frac{\{\mathrm{NH_4{}^+}\}\{\mathrm{NO_3{}^-}\}}{\{\mathrm{NH_3(g)}\}\{\mathrm{HNO_3(g)}\}} &= \frac{\mathbf{m}_{\mathrm{NH_4{}^+}}\gamma_{\mathrm{NH_4{}^+}}\mathbf{m}_{\mathrm{NO_3{}^-}}\gamma_{\mathrm{NO_3{}^-}}}{p_{\mathrm{NH_3(g)},s}\,p_{\mathrm{HNO_3(g)},s}} \\
&= \frac{\mathbf{m}_{\mathrm{NH_4{}^+}}\mathbf{m}_{\mathrm{NO_3{}^-}}\gamma^2_{\mathrm{NH_4{}^+,NO_3{}^-}}}{p_{\mathrm{NH_3(g)},s}\,p_{\mathrm{HNO_3(g)},s}} = K_{eq}(T)\,\frac{\mathrm{moles}^2}{\mathrm{kg}^2\,\mathrm{atm}^2}
\end{aligned} \tag{18.44}$$

$$\begin{aligned}
\mathrm{NH_3(aq) + H_2O(aq)} &\rightleftharpoons \mathrm{NH_4{}^+ + OH^-} \\
\frac{\{\mathrm{NH_4{}^+}\}\{\mathrm{OH^-}\}}{\{\mathrm{NH_3(aq)}\}\{\mathrm{H_2O(aq)}\}} &= \frac{\mathbf{m}_{\mathrm{NH_4{}^+}}\gamma_{\mathrm{NH_4{}^+}}\mathbf{m}_{\mathrm{OH^-}}\gamma_{\mathrm{OH^-}}}{\mathbf{m}_{\mathrm{NH_3(aq)}}\gamma_{\mathrm{NH_3(aq)}} \times f_r} \\
&= \frac{\mathbf{m}_{\mathrm{NH_4{}^+}}\mathbf{m}_{\mathrm{OH^-}}\gamma^2_{\mathrm{NH_4{}^+,OH^-}}}{\mathbf{m}_{\mathrm{NH_3(aq)}}\gamma_{\mathrm{NH_3(aq)}} \times f_r} = K_{eq}(T)\,\frac{\mathrm{moles}}{\mathrm{kg}}
\end{aligned} \tag{18.45}$$

## 18.5. MEAN BINARY ACTIVITY COEFFICIENTS

The mean binary activity coefficient of an electrolyte, which is a function of molality and temperature, can be determined from measurements or theory. Measurements of binary activity coefficients for several species at 298.15 K have been made. Theoretical parameterizations have also been developed to predict binary activity coefficients.

A classic parameterization is **Pitzer's method** (Pitzer and Mayorga 1973; Pitzer 1991), which estimates the mean binary activity coefficient of an electrolyte

**Table 18.1.** Pitzer Parameters for Three Electrolytes

| **Electrolyte** | $\beta_{12}^{(1)}$ | $\beta_{12}^{(2)}$ | $C_{12}^{\gamma}$ |
|---|---|---|---|
| HCl | 0.1775 | 0.2945 | 0.0012 |
| $HNO_3$ | 0.1119 | 0.3206 | 0.0015 |
| $NH_4NO_3$ | −0.0154 | 0.112 | −0.000045 |

*Source:* Pilinis and Seinfeld (1987).

at 298.15 K with

$$\ln \gamma_{12b}^{0} = Z_1 Z_2 f^{\gamma} + \mathbf{m}_{12}\frac{2\nu_1\nu_2}{\nu_1+\nu_2}B_{12}^{\gamma} + \mathbf{m}_{12}^2\frac{2(\nu_1\nu_2)^{3/2}}{\nu_1+\nu_2}C_{12}^{\gamma} \tag{18.46}$$

where $\gamma_{12b}^{\circ}$ is the mean binary activity coefficient of electrolyte 1–2 (cation 1 plus anion 2) at reference temperature 298.15 K, $Z_1$ and $Z_2$ are the absolute value of the charges of cation 1 and anion 2, respectively; $\mathbf{m}_{12}$ is the molality of the electrolyte dissolved in solution, and $\nu_1$ and $\nu_2$ are the stoichiometric coefficients of the dissociated ions. The remaining quantities in (18.46) are

$$f^{\gamma} = -0.392\left[\frac{\mathbf{I}^{1/2}}{1+1.2\mathbf{I}^{1/2}} + \frac{2}{1.2}\ln(1+1.2\mathbf{I}^{1/2})\right] \tag{18.47}$$

$$B_{12}^{\gamma} = 2\beta_{12}^{(1)} + \frac{2\beta_{12}^{(2)}}{4\mathbf{I}}\left[1 - e^{-2\mathbf{I}^{1/2}}(1 + 2\mathbf{I}^{1/2} - 2\mathbf{I})\right] \tag{18.48}$$

where $\mathbf{I}$ is the solution's ionic strength (moles kg$^{-1}$). **Ionic strength** is a measure of interionic effects resulting from attraction and repulsion among ions. It is calculated from

$$\mathbf{I} = \frac{1}{2}\left(\sum_{i=1}^{N_C}\mathbf{m}_{2i-1}Z_{2i-1}^2 + \sum_{i=1}^{N_A}\mathbf{m}_{2i}Z_{2i}^2\right) \tag{18.49}$$

where $N_C$ and $N_A$ are the numbers of cations and anions, respectively, in solution. Odd-numbered subscripts refer to cations, and even-numbered subscripts refer to anions. In the case of one electrolyte, such as HCl(aq) alone in solution, $N_C = 1$ and $N_A = 1$. The quantities, $\beta_{12}^{(1)}$, $\beta_{12}^{(2)}$, and $C_{12}^{(\gamma)}$ are empirical parameters derived from measurements. **Pitzer parameters** for three electrolytes are shown in Table 18.1.

Pitzer's method predicts mean binary activity coefficients at 298.15 K from physical principals. Its limitation is that the coefficients are typically accurate only at molalities up to about 6. Figure 18.1 shows a comparison of mean binary activity coefficients predicted by Pitzer's method to those measured by Hamer and Wu (1972). The measured data are accurate to higher molalities.

Important information about chloride and nitrate in atmospheric aerosols can be derived from Fig. 18.1. When sea-spray particles containing chloride encounter nitric acid gas, the nitric acid enters the particles as nitrate, displacing the chloride

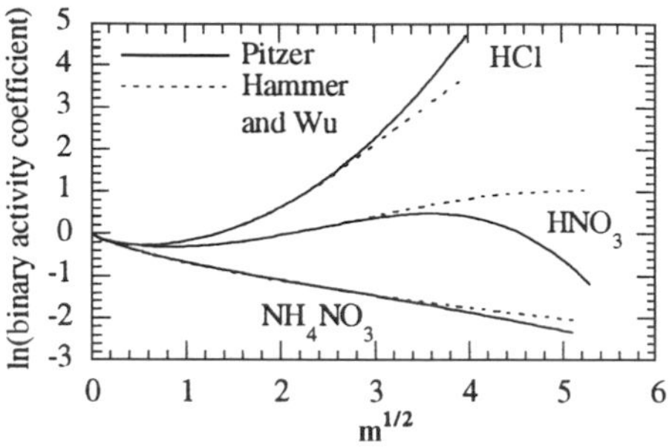


**Figure 18.1.** Comparison of mean binary activity coefficients measured by Hamer and Wu (1972) with those computed from Pitzer's method.

to the gas phase. This process can remove nearly all chloride from sea spray particles over the ocean, as discussed in Section 15.1. Upon initial observation of the effective Henry's constants of HCl(g) and $HNO_3$(g), it is not obvious that nitrate should force chloride out of solution. The effective Henry's-constant expressions for the two species at 298.15 K, obtained from Table B.7, are

$$\frac{\mathbf{m}_{H^+}\mathbf{m}_{Cl^-}\gamma^2_{H^+,Cl^-}}{p_{HCl(g),s}} = 1.97 \times 10^6 \qquad \frac{\mathbf{m}_{H^+}\mathbf{m}_{NO_3^-}\gamma^2_{H^+,NO_3^-}}{p_{HNO_3(g),s}} = 2.51 \times 10^6 \qquad (18.50)$$

The equilibrium coefficients are comparable for the two reactions, but the binary-activity-coefficient data in Fig. 18.1 together with (18.50) indicate that, at the same molality, $\mathbf{m}_{H^+}\mathbf{m}_{Cl^-}/p_{HCl(g)}$ is much smaller than $\mathbf{m}_{H^+}\mathbf{m}_{NO_3^-}/p_{HNO_3(g)}$. The large binary activity coefficient of HCl decreases HCl's solubility, causing it to evaporate. Thus, if $HNO_3$(aq) and HCl(aq) are initially present in particles at the same time, nitrate is more likely to stay in solution than is chloride.

Whether mean binary activity coefficients at 298.15 K are determined from measurements or theory, they can be parameterized with a polynomial fit of the form

$$\ln \gamma^0_{12b} = B_0 + B_1\mathbf{m}_{12}^{1/2} + B_2\mathbf{m}_{12} + B_3\mathbf{m}_{12}^{3/2} + \cdots \qquad (18.51)$$

where $B_0, B_1, \ldots$ are fitting coefficients. Polynomial coefficients for several electrolytes are given in Table B.9 of Appendix B. Polynomial fits are used to simplify and speed up the use of binary activity coefficient calculations in computer programs.

## 18.6. TEMPERATURE DEPENDENCE OF MEAN BINARY ACTIVITY COEFFICIENTS

The temperature dependence of solute mean binary activity coefficients can be derived from thermodynamic principles. An expression for the binary activity coefficient of electrolyte 1–2 at temperature $T$ is (Harned and Owen 1958)

$$\ln \gamma_{12b}(T) = \ln \gamma_{12b}^0 + \frac{T_L}{(\nu_1+\nu_2)R^* T_0}\left(\phi_L + \mathbf{m}\frac{\partial \phi_L}{\partial \mathbf{m}}\right) + \frac{T_C}{(\nu_1+\nu_2)R^*}\left(\phi_{c_p} + \mathbf{m}\frac{\partial \phi_{c_p}}{\partial \mathbf{m}} - \phi_{c_p}^0\right) \tag{18.52}$$

where $T_0$ is the reference temperature (298.15 K), $R^*$ is the gas constant (J $\text{mole}^{-1}$ $\text{K}^{-1}$), $\phi_L$ is the relative apparent molal enthalpy (J $\text{mole}^{-1}$) of the electrolyte at molality $\mathbf{m}$ (with subscript 12 omitted), $\phi_{c_p}$ is the apparent molal heat capacity (J $\text{mole}^{-1}$ $\text{K}^{-1}$) at molality $\mathbf{m}$, $\phi_{c_p}^0$ is the apparent molal heat capacity at infinite dilution, and

$$T_L = \frac{T_0}{T} - 1 \qquad T_C = 1 + \ln\left(\frac{T_0}{T}\right) - \frac{T_0}{T} \tag{18.53}$$

The relative apparent molal enthalpy equals the negative of the heat of dilution. Table B.9 lists several sources of experimental data for heat of dilution and apparent molal heat capacity. With these data, polynomials of the form

$$\phi_L = U_1\mathbf{m}^{1/2} + U_2\mathbf{m} + U_3\mathbf{m}^{3/2} + \cdots \tag{18.54}$$

$$\phi_{c_p} = \phi_{c_p}^0 + V_1\mathbf{m}^{1/2} + V_2\mathbf{m} + V_3\mathbf{m}^{3/2} + \cdots$$

can be constructed. The apparent relative molal enthalpy is defined as $\phi_L = \phi_H - \phi_H^0$, where $\phi_H$ is the apparent molal enthalpy, and $\phi_H^0$ is the apparent molal enthalpy at infinite dilution, which occurs when $\mathbf{m} = 0$. At $\mathbf{m} = 0$, $\phi_H = \phi_H^0$ and $\phi_L = 0$.

(18.51)–(18.54) can be combined to give temperature-dependent mean-binary-activity-coefficient polynomials of the form

$$\ln \gamma_{12b}(T) = F_0 + F_1\mathbf{m}^{1/2} + F_2\mathbf{m} + F_3\mathbf{m}^{3/2} + \cdots \tag{18.55}$$

where

$$F_0 = B_0 \qquad F_j = B_j + G_j T_L + H_j T_C \quad \text{for } j = 1, \ldots \tag{18.56}$$

$$G_j = \frac{0.5(j+2)U_j}{(\nu_1+\nu_2)R^* T_0} \qquad H_j = \frac{0.5(j+2)V_j}{(\nu_1+\nu_2)R^*} \tag{18.57}$$

With sufficient data, many temperature- and molality-dependent mean binary activity coefficients can be written in terms of (18.55)–(18.57). Table B.9 lists $B$'s, $G$'s and $H$'s for 10 electrolytes and maximum molalities associated with the fits. Since polynomial fits wander when the molality increases much beyond the limit of the given data, activity coefficients from beyond the last valid molality should be used cautiously when modeling.

Determining temperature-dependent binary activity coefficients of bisulfate and sulfate is more difficult. Figure 18.2 shows calculated activity coefficients as a

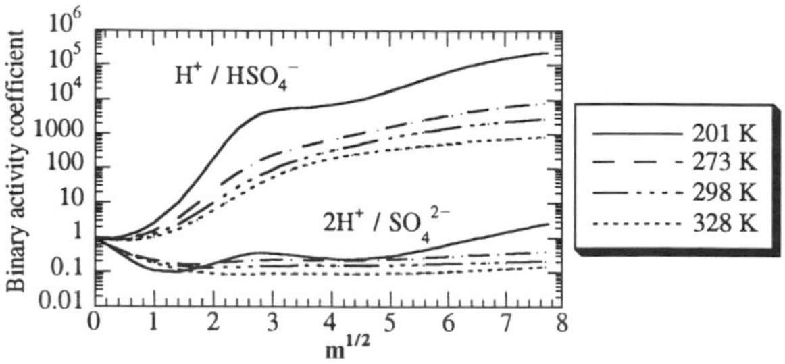


**Figure 18.2.** Binary activity coefficients of sulfate and bisulfate, alone in solution (Jacobson et al. 1996b), found by combining equations from Clegg and Brimblecombe (1995) with (72) and (73) of Stelson et al. (1984) in a Newton–Raphson iteration. The results are valid for 0–40 m total $H_2SO_4$.

function of temperature and molality and discusses how they were determined. Such coefficients can be written in polynomial or tabular form.

Activity coefficients are tabulated or written as simple polynomials to speed the solution of equilibrium calculations over a large model domain. During a model time interval, the temperature is usually known. Thus, (18.56)–(18.57) can be calculated once for each species at the beginning of each time interval. During the iteration of equilibrium equations, only (18.55) needs to be solved to find binary activity coefficients.

## 18.7. MEAN MIXED ACTIVITY COEFFICIENTS

In atmospheric particles, electrolytes, such as dissolved sulfuric acid, nitric acid, hydrochloric acid, ammonia, and sodium chloride, often coexist in solution. In such cases, mean mixed activity coefficients are approximated with an empirical mixing rule that is a function of mean binary activity coefficients and takes into account interactions among ions. One mixing rule is **Bromley's method** (Bromley 1973), which gives the mean activity coefficient of electrolyte 1–2 in a mixture as

$$\log_{10}\gamma_{12m}(T) = -A_\gamma \frac{Z_1 Z_2 \mathbf{I}_m^{1/2}}{1+\mathbf{I}_m^{1/2}} + \frac{Z_1 Z_2}{Z_1+Z_2}\left(\frac{W_1}{Z_1} + \frac{W_2}{Z_2}\right) \tag{18.58}$$

where $A_\gamma$ is the **Debye-Hückel parameter** (0.392 at 298 K), $Z_1$ and $Z_2$ are the absolute-value charges of cation 1 and anion 2, respectively, $\mathbf{I}_m$ is the total ionic strength of the mixture, and

$$W_1 = Y_{21}\left(\log_{10}\gamma_{12b}(T) + A_\gamma \frac{Z_1 Z_2 \mathbf{I}_m^{1/2}}{1+\mathbf{I}_m^{1/2}}\right) + Y_{41}\left(\log_{10}\gamma_{14b}(T) + A_\gamma \frac{Z_1 Z_4 \mathbf{I}_m^{1/2}}{1+\mathbf{I}_m^{1/2}}\right) + \cdots \tag{18.59}$$

$$W_2 = X_{12}\left(\log_{10}\gamma_{12b}(T) + A_\gamma \frac{Z_1 Z_2 \mathbf{I}_m^{1/2}}{1+\mathbf{I}_m^{1/2}}\right) + X_{32}\left(\log_{10}\gamma_{32b}(T) + A_\gamma \frac{Z_3 Z_2 \mathbf{I}_m^{1/2}}{1+\mathbf{I}_m^{1/2}}\right) + \cdots \tag{18.60}$$

are functions of all electrolytes in solution. In the last equations,

$$Y_{21} = \left(\frac{Z_1+Z_2}{2}\right)^2 \frac{\mathbf{m}_{2,m}}{\mathbf{I}_m} \qquad X_{12} = \left(\frac{Z_1+Z_2}{2}\right)^2 \frac{\mathbf{m}_{1,m}}{\mathbf{I}_m} \tag{18.61}$$

Similar expressions can be written for $X_{32}$, $X_{52}$, ..., $Y_{41}$, $Y_{61}$, .... In these equations, $\gamma_{12b}(T)$, $\gamma_{14b}(T)$, $\gamma_{32b}(T)$, ... are temperature-dependent mean binary activity coefficients, odd-numbered subscripts refer to cations, and even-numbered subscripts refer to anions. Thus, $\mathbf{m}_{1,m}$ and $\mathbf{m}_{2,m}$ are molalities of a cation and anion, respectively, in the mixture.

When a mean binary activity coefficient is used in (18.59) and (18.60), it is determined at a binary electrolyte molality $\mathbf{m}_{12,b}$ that results in the current ionic strength of the mixture, $\mathbf{I}_m$. The molality of binary electrolyte 1–2 is found from

$$\mathbf{I}_m = \tfrac{1}{2}\left(\mathbf{m}_{1,b}Z_1^2 + \mathbf{m}_{2,b}Z_2^2\right) = \tfrac{1}{2}\left(\nu_+\mathbf{m}_{12,b}Z_1^2 + \nu_-\mathbf{m}_{12,b}Z_2^2\right) \tag{18.62}$$

where $\mathbf{m}_{1,b} = \nu_+\mathbf{m}_{12,b}$ and $\mathbf{m}_{2,b} = \nu_-\mathbf{m}_{12,b}$ are the molalities of a cation and an anion, alone in solution, that results in the total ionic strength of the mixture. Solving (18.62) for $\mathbf{m}_{12,b}$ gives

$$\mathbf{m}_{12,b} = \frac{2\mathbf{I}_m}{\nu_+ Z_1^2 + \nu_- Z_2^2} \tag{18.63}$$

The molality of the binary electrolyte in (18.63) is used to determine $\gamma_{12b}(T)$ from the polynomial fit given in (18.55).

## 18.8. THE WATER EQUATION

Interaction between the solvent and a solute in solution is **solvation**. Solvation occurs when a solvent bonds to cations, anions, or nonelectrolytes (such as sucrose) in solution. When the solvent is liquid water, the bonding is called **hydration**. During hydration of a cation, the lone pair of electrons on the oxygen atom of a water molecule bonds to the cation end of the ion dipole. During hydration of an anion, the water molecule attaches to the anion end of the ion dipole via hydrogen bonding. Several water molecules can hydrate to each ion.

When liquid-water molecules bond to ions in solution, water vapor condenses to maintain saturation over the solution surface, increasing the liquid-water content. The liquid-water content is a unique function of electrolyte molality and sub-100-percent relative humidity. As the relative humidity increases up to 100 percent, hydration increases the aerosol liquid-water content. The liquid-water content also increases with increasing solute molality in solution. Above 100-percent relative humidity, particles grow rapidly by condensation. When particles are large and

**Table 18.2.** Test of ZSR Prediction Accuracy for a Mixture of Sucrose (Species *a*) and Mannitol (Species *b*) at Two Water Activities

| Case | $\mathbf{m}_{x,a}$ | $\mathbf{m}_{y,a}$ | $\mathbf{m}_{x,m}$ | $\mathbf{m}_{y,m}$ | $\frac{\mathbf{m}_{x,m}}{\mathbf{m}_{x,a}}+\frac{\mathbf{m}_{y,m}}{\mathbf{m}_{y,a}}$ |
|---|---|---|---|---|---|
| 1 | 0.7751 | 0.8197 | 0.6227 | 0.1604 | 0.9990 |
| 2 | 0.9393 | 1.0046 | 0.1900 | 0.8014 | 1.0000 |

*Source:* Stokes and Robinson (1966).

dilute, the volume of water added to them by hydration is small compared to the volume of water already present. Thus, hydration does not affect water content much when the relative humidity exceeds 100 percent.

A convenient method of estimating aerosol liquid-water content as a function of electrolyte molality and sub-100-percent relative humidity is the Zdanovskii–Stokes–Robinson (**ZSR**) equation (Stokes and Robinson 1966). The equation can be applied to electrolytes or nonelectrolytes. The simplest form of the equation is

$$\frac{\mathbf{m}_{x,m}}{\mathbf{m}_{x,a}}+\frac{\mathbf{m}_{y,m}}{\mathbf{m}_{y,a}}=1 \tag{18.64}$$

where $\mathbf{m}_{x,a}$ and $\mathbf{m}_{y,a}$ are the molalities of solutes $x$ and $y$ each alone in solution at a given water activity, and $\mathbf{m}_{x,m}$ and $\mathbf{m}_{y,m}$ are the molalities of $x$ and $y$ mixed together in solution at the same water activity. The **water activity** for atmospheric particles was defined in (18.21) to be the relative humidity. Table 18.2 gives molalities of sucrose and mannitol, alone and mixed in water, for two cases. The resulting molalities satisfy (18.64).

(18.64) can be generalized for a mixture of any number of components with

$$\sum_k \frac{\mathbf{m}_{k,m}}{\mathbf{m}_{k,a}}=1 \tag{18.65}$$

where the summation is over all solutes, $\mathbf{m}_{k,m}$ is the molality of solute $k$ in a solution containing all solutes at the ambient relative humidity, and $\mathbf{m}_{k,a}$ is the molality of solute $k$ alone in solution at the ambient relative humidity.

Experimental data for water activity as a function of binary electrolyte molality are available (e.g., Robinson and Stokes 1955; Pitzer and Mayorga 1973; Cohen et al. 1987a,b; Tang and Munkelwitz 1994; Tang 1997). Such data can be fitted to polynomial expressions of the form.

$$\mathbf{m}_{k,a}=Y_{0,k}+Y_{1,k}a_w+Y_{2,k}a_w^2+Y_{3,k}a_w^3+\cdots \tag{18.66}$$

where $a_w$ is the water activity (relative humidity expressed as a fraction), and the $Y$'s are polynomial coefficients, listed in Table B.10 of Appendix B for several electrolytes. Figures 18.3(a) and (b) show curves of molality versus water activity for some of these electrolytes.

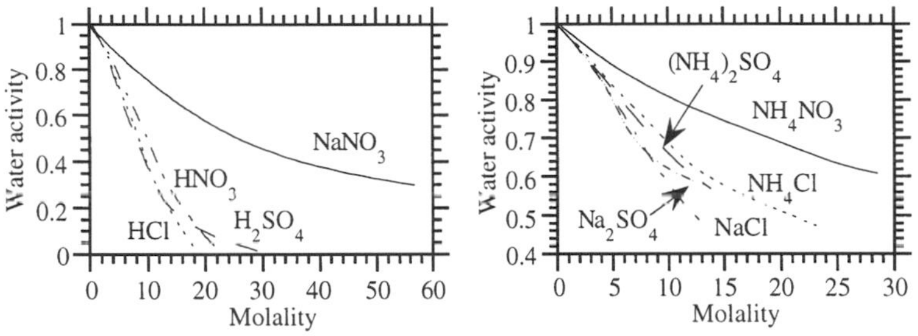


(a) (b)

**Figure 18.3.** Water activities of several electrolytes at 298.15 K as a function of molality. The curves were obtained from (18.66) and the coefficients in Table B.10 of Appendix B.

In comparison to the temperature dependence of the solute binary activity coefficients, the temperature dependence of the water binary activity coefficient is small for lower tropospheric conditions. This temperature dependence can be rewritten from Harned and Owen (1958) as

$$\ln a_w(T) = \ln a_w^0 - \frac{m_v \mathbf{m}_{k,a}^2}{1000 R^*}\left(\frac{T_L}{T_0}\frac{\partial \phi_L}{\partial \mathbf{m}_{k,a}} + T_C \frac{\partial \phi_{c_P}}{\partial \mathbf{m}_{k,a}}\right) \tag{18.67}$$

If the water activity at the reference temperature is extracted from (18.66) and expressed as

$$\ln a_w^0 = A_0 + A_1 \mathbf{m}_{k,a}^{1/2} + A_2 \mathbf{m}_{k,a} + A_3 \mathbf{m}_{k,a}^{3/2} + \cdots \tag{18.68}$$

then (18.67), (18.68), and (18.54) can be combined to form the polynomial

$$\ln a_w(T) = A_0 + A_1 \mathbf{m}_{k,a}^{1/2} + A_2 \mathbf{m}_{k,a} + E_3 \mathbf{m}_{k,a}^{3/2} + E_4 \mathbf{m}_{k,a}^2 + \cdots \tag{18.69}$$

where

$$E_l = A_l - \frac{0.5(l-2)m_v}{1000 R^*}\left(\frac{T_L}{T_0} U_{l-2} + T_C V_{l-2}\right) \qquad \text{for} \quad l = 3, \ldots \tag{18.70}$$

The $U$'s and $V$'s are polynomial coefficients from (18.54). Since the $A$-terms are not affected by temperature, temperature affects the water-activity polynomial in (18.69) starting with the fourth term, $E_3$. In (18.55) temperature affected the solute activity polynomial starting with the second term.

**Example 18.3**

At high molalities (above 10) and ambient temperatures (273–310 K), temperature affects water activity only slightly. At $\mathbf{m} = 16$, HCl has a water activity of 0.09 when $T = 273$ K and 0.11 when $T = 310$ K. At lower molalities, temperature has even less of an effect.

**Table 18.3.** Three Sets of Hypothetical Electrolyte Concentrations That May Result When 6 μmole m$^{-3}$ of $H^+$, 6 μmole m$^{-3}$ $Na^+$, 7 μmole m$^{-3}$ of $Cl^-$, and 5 μmole m$^{-3}$ of $NO_3^-$ are Combined

| Case | $c_{HCL,m}$ | $c_{HNO_3,m}$ | $c_{NaCl,m}$ | $c_{NaNO_3,m}$ |
|---|---|---|---|---|
| 1 | 6 | 0 | 1 | 5 |
| 2 | 4 | 2 | 3 | 3 |
| 3 | 1 | 5 | 6 | 0 |

Case 1 correspond to the solution from (18.72). Units are micromoles per cubic meter.

In a model of mixed aerosols, the water equation is rewritten from (18.65) as

$$c_w = \frac{1000}{m_v} \sum_{i=1}^{N_C} \left( \sum_{j=1}^{N_A} \frac{c_{i,j,m}}{\mathbf{m}_{i,j,a}} \right) \tag{18.71}$$

where $c_w$ is the **particle liquid-water content** in units of mole concentration (moles of $H_2O(aq)$ per cubic centimeter of air), $m_v$ is the molecular weight of water (g mole$^{-1}$), 1000 converts grams to kilograms, $c_{i,j,m} = \mathbf{m}_{i,j,a} c_w m_v / 1000$ is the number of moles of electrolyte pair $i$, $j$ per cubic centimeter of air in a solution containing all solutes at the ambient relative humidity, and $\mathbf{m}_{i,j,a}$ is the molality of the electrolyte pair alone in solution at the ambient relative humidity, found from (18.66). The subscript $i$, $j$ in (18.71) replaces the subscript $k$ in (18.66).

In a model, the mole concentrations of electrolyte pairs are not usually known, but those of individual ions are. Thus, ions must be combined into hypothetical electrolyte pairs for (18.71) to be solved. The easiest way to recombine ions to find hypothetical electrolyte concentrations in a mixture is to execute the equations

$$c_{i,j,m} = \min\left(\frac{c_{i,m}}{\nu_i}, \frac{c_{j,m}}{\nu_j}\right) \qquad c_{i,m} = c_{i,m} - \nu_i c_{i,j,m} \qquad c_{j,m} = c_{j,m} - \nu_j c_{i,j,m} \tag{18.72}$$

in succession, for each undissociated electrolyte pair $i$–$j$, where $c_{i,j,m}$ is the mole concentration of undissociated electrolyte $i$–$j$ (e.g., $HNO_3$, HCl), and $c_{i,m}$ and $c_{j,m}$ are cation and anion mole concentrations, respectively. Applying (18.72) to Example 18.4 gives case 1 of Table 18.3.

**Example 18.4**

In a mixture, the combination of ion concentrations to form hypothetical electrolyte concentrations for use in the water equation can be arbitrary. Suppose a solution contains 6 μmoles m$^{-3}$ of $H^+$, 6 μmoles m$^{-3}$ $Na^+$, 7 μmoles m$^{-3}$ of

$Cl^-$, and 5 μmoles $m^{-3}$ of $NO_3^-$. This set of concentrations is charge-conserving (i.e., the sum, over all ions, of mole concentration multiplied by ion charge equals zero), which is a requirement for any aqueous system. The four ions can combine in many ways to form electrolytes: $HNO_3$, HCl, $NaNO_3$, and NaCl. Recombination is limited by the mass-balance constraints

$$c_{H^+,m} = c_{HNO_3,m} + c_{HCl,m} \qquad c_{Na^+,m} = c_{NaNO_3,m} + c_{NaCl,m}$$

$$c_{Cl^-,m} = c_{HCl,m} + c_{NaCl,m} \qquad c_{NO_3^-,m} = c_{HNO_3,m} + c_{NaNO_3,m}$$

Table 18.3 shows three sets of electrolyte concentrations that satisfy these constraints.

## 18.9. EXAMPLE EQUILIBRIUM PROBLEM

Equilibrium equations, activity-coefficient equations, and the water equation are solved together in an atmospheric model to estimate particle composition and liquid-water content. In this section, a set of equations required to solve a basic equilibrium problem is given. In this problem, HCl(g) is assumed to equilibrate with dissolved $Cl^-$ at a particle surface, and dissolved sulfuric acid is assumed to exist as $HSO_4^-$ or $SO_4^{2-}$. The reactions are

$$HCl(g) \rightleftharpoons H^+ + Cl^- \qquad HSO_4^- \rightleftharpoons H^+ + SO_4^{2-} \tag{18.73}$$

The **equilibrium-coefficient expressions** for these reactions are

$$\frac{\mathbf{m}_{H^+,eq}\mathbf{m}_{Cl^-,eq}\gamma^2_{H^+,Cl^-,eq}}{p_{HCl,s,eq}} = K_{eq}(T) \tag{18.74}$$

$$\frac{\mathbf{m}_{H^+,eq}\mathbf{m}_{SO_4^{2-},eq}\gamma^3_{2H^+,SO_4^{2-},eq}}{\mathbf{m}_{HSO_4^-,eq}\gamma^2_{H^+,HSO_4^-,eq}} = K_{eq}(T)$$

respectively, where the subscript *eq* indicates the value is at equilibrium, and the subscript *s* indicates a saturation vapor pressure. The activity coefficients are non-linear functions of $H^+$, $Cl^-$, $HSO_4^-$ and $SO_4^{2-}$ mole concentrations.

When the reactions in (18.73) are solved, masses of individual atoms (e.g., chlorine, sulfur) must be conserved. The **mass balance constraints** are

$$C_{HCl(g),eq} + c_{Cl^-,eq} = C_{HCl(g),t-h} + c_{Cl^-,t-h} \tag{18.75}$$

$$c_{HSO_4^-,eq} + c_{SO_4^{2-},eq} = c_{HSO_4^-,t-h} + c_{SO_4^{2-},t-h} \tag{18.76}$$

where $C$ is the mole concentration of a gas (moles $cm^{-3}$), $c$ is the mole concentration of a particle component, and $t - h$ indicates an initial value. Saturation vapor pressures and molalities are related to mole concentrations by

$$p_{HCl,s,eq} = C_{HCl(g),s,eq}R^*T \qquad \mathbf{m}_{Cl^-,eq} = \frac{1000c_{Cl^-,eq}}{c_{w,eq}m_v} \tag{18.77}$$

respectively. The **charge balance equation** for this problem requires

$$c_{\mathrm{Cl}^-,eq} + c_{\mathrm{HSO_4}^-,eq} + 2c_{\mathrm{SO_4}^{2-},eq} = c_{\mathrm{H}^+,eq} \tag{18.78}$$

and the ZSR water equation is

$$c_{w,eq} = \frac{1000}{m_w}\left(\frac{c_{\mathrm{H}^+,\mathrm{Cl}^-,m}}{\mathbf{m}_{\mathrm{H}^+,\mathrm{Cl}^-,a}} + \frac{c_{\mathrm{H}^+,\mathrm{HSO_4}^-,m}}{\mathbf{m}_{\mathrm{H}^+,\mathrm{HSO_4}^-,a}} + \frac{c_{2\mathrm{H}^+,\mathrm{SO_4}^{2-},m}}{\mathbf{m}_{2\mathrm{H}^+,\mathrm{SO_4}^{2-},a}}\right) \tag{18.79}$$

The molality of species alone in solution ($\mathbf{m}_a$) are determined from an empirical functions of the relative humidity, such that given in (18.66). The quantities $c_{\mathrm{H}^+,\mathrm{Cl}^-,m}$, etc., are hypothetical mole concentrations of electrolyte pairs, constrained by

$$c_{\mathrm{H}^+,eq} = c_{\mathrm{H}^+,\mathrm{Cl}^-,m} + c_{\mathrm{H}^+,\mathrm{HSO_4}^-,m} + 2c_{2\mathrm{H}^+,\mathrm{SO_4}^{2-},m} \tag{18.80}$$

$$c_{\mathrm{Cl}^-,eq} = c_{\mathrm{H}^+,\mathrm{Cl}^-,m} \qquad c_{\mathrm{HSO_4}^-,eq} = c_{\mathrm{H}^+,\mathrm{HSO_4}^-,m}$$

$$c_{\mathrm{SO_4}^{2-},eq} = c_{2\mathrm{H}^+,\mathrm{SO_4}^{2-},m}$$

This problem requires the solution of two equilibrium equations, two mass balance equations, a charge balance equation, a water equation, and activity-coefficient equations. These equations or a variation of them can be solved with an iterative Newton–Raphson method (e.g., Press et al. 1992), an iterative bisectional Newton method (e.g., Pilinis and Seinfeld 1987; Kim et al. 1993a,b), an iterative method that minimizes the Gibbs free energy (Bassett and Seinfeld 1983, 1984; Wexler and Seinfeld 1990, 1991), an approximation method (e.g., Saxena et al. 1986), or a mole-fraction-based thermodynamic model that parameterizes ion pair and triplet interactions (Clegg et al. 1997). Most of these methods are mass- and charge-conserving.

## 18.10. METHOD OF SOLVING EQUILIBRIUM EQUATIONS

Another method used to solve equilibrium problems is the **mass flux iteration** (MFI) method (Jacobson 1994, Jacobson et al. 1996b, Villars 1959). This method can converge thousands of equilibrium equations simultaneously, cannot produce negative concentrations, and is mass- and charge-conserving at all times. The only constraints are that the equilibrium equations must be mass- and charge-conserving, and the system must start in charge balance. For example, the equation $HNO_3(aq) = H^+ + NO_3^-$ conserves mass and charge. The charge balance constraint allows initial charges to be distributed among all dissociated ions, but the initial sum, over all species, of charge multiplied by molality must equal zero. The simplest way to initialize charge is to set all ion molalities to zero. Initial mass in the system can be distributed arbitrarily, subject to the charge balance constraint. If the total nitrate in the system is known to be, say, 20 μg $m^{-3}$, the nitrate can initially be distributed in any proportion among $HNO_3(aq)$, $NO_3^-$, $NH_4NO_3(s)$, etc.

The MFI method requires the solution of one equilibrium equation at a time by iteration. A system of equations is solved by iterating all equations many times. Suppose a system consists of a single aerosol size bin and 15 equations representing the equilibrium chemistry within that bin. At the start, the first equation is iterated. When the first equation converges, the updated and other initial concentrations are used as inputs into the second equation. This continues until the last equation has converged. At that point, the first equation is no longer converged, since concentrations used in it have changed. Thus, the iteration sequence must be repeated over all equations several times until the concentrations no longer change upon more iteration.

Equilibrium among multiple particle size bins and the gas phase is solved for in a similar manner. Suppose a system consists of several size bins, equations per bin, and gases that equilibrate with dissolved molecules in each bin. Each gas' saturation vapor pressure over a particle surface is assumed to equal the gas' partial pressure, which is a single value. In reality, the saturation vapor pressure differs over every particle surface. In order to take variations in saturation vapor pressure over particle surfaces into account, nonequilibrium gas-aerosol transfer equations, discussed in Section 18.13, must solved.

Gas-particle equilibrium over multiple size bins is solved by iterating each equilibrium equation, including gas solution equations, starting with the first size bin. Updated gas concentrations from the first bin affect the equilibrium distribution in subsequent bins. After the last size bin has been iterated, the sequence is repeated in reverse order (to speed convergence), from the last to first size bin. The marches back and forth among size bins continue until gas and aerosol concentrations do not change upon more iteration.

To demonstrate a solution of one equilibrium equation, an example where two gases equilibrate with two ions is shown. The sample equation has the form of (18.3), with the gases on the left side of the equation. The first step is to calculate $Q_d$ and $Q_n$, the smallest ratio of mole concentration to the stoichiometric coefficient among species appearing in the denominator and numerator, respectively, of (18.18). Thus,

$$Q_d = \min\left(\frac{C_{D,1}}{\nu_D}, \frac{C_{E,1}}{\nu_E}\right) \qquad Q_n = \max\left(\frac{c_{A,1}}{\nu_A}, \frac{c_{B,1}}{\nu_B}\right) \tag{18.81}$$

where $C$ and $c$ are gas- and particle-phase mole concentrations (moles cm$^{-3}$), respectively, and subscript 1 refers to an initial value. If an equilibrium equation contains a solid, the solid's concentration is included in (18.81).

After two parameters, $z_1 = 0.5(Q_d + Q_n)$ and $\Delta x_1 = Q_d - z_1$, are initialized, the iteration begins by adding the mass flux factor $\Delta x$, which may be positive or negative, to each mole concentration in the numerator and subtracting it from each mole concentration in the denominator of the equilibrium equation. Thus,

$$\begin{aligned} c_{A,l+1} &= c_{A,l} + \nu_A \Delta x_l & c_{B,l+1} &= c_{B,l} + \nu_B \Delta x_l \\ C_{D,l+1} &= C_{D,l} - \nu_D \Delta x_l & C_{E,l+1} &= C_{E,l} - \nu_E \Delta x_l \end{aligned} \tag{18.82}$$

Starting with (18.82), iteration numbers are referred to by subscripts $l$ and $l+1$. If the equation contains a solid, the change of the solid's concentration is calculated with (18.82) (solid, aqueous, and ionic mole concentrations are identified with a $c$). (18.82) shows that, during each iteration, mass and charge are transferred either from reactants to products or vice versa. This transfer continues until $\Delta x = 0$. Thus, the scheme conserves mass and charge each iteration.

Third, a ratio comparing activities to the equilibrium coefficient is calculated as

$$F = \frac{\mathbf{m}_{A,l+1}^{\nu_A}\mathbf{m}_{B,l+1}^{\nu_B}\gamma_{AB,l+1}^{\nu_A+\nu_B}}{p_{D,l+1}^{\nu_D}p_{E,l+1}^{\nu_E}}\frac{1}{K_{eq}(T)} \tag{18.83}$$

For this calculation, mole concentrations are converted to molalities (for dissolved gas molecules) or atmospheres (for gases). In the case of solids, the activities are unity and do not appear in (18.83). Mean mixed activity coefficients (e.g., $\gamma_{AB,l+1}$) are updated before each iteration of an equilibrium equation and converge over time. The liquid-water content $c_w$, which also converges, is updated with (18.71) before each iteration of an equilibrium equation or more sporadically.

The fourth step in the procedure is to recalculate $z$ for the next iteration as $z_{l+1} = 0.5\,z_l$. The final step is to check convergence with the criterion

$$F = \begin{cases} > 1 & \rightarrow \quad \Delta x_{l+1} = -z_{l+1} \\ < 1 & \rightarrow \quad \Delta x_{l+1} = +z_{l+1} \\ = 1 & \rightarrow \quad \text{convergence} \end{cases} \tag{18.84}$$

At each nonconvergence, $\Delta x$ is updated, the iteration number is advanced, and the iteration sequence restarts with (18.82). All molalities eventually converge to positive numbers.

## 18.11. SOLID FORMATION AND DELIQUESCENCE RELATIVE HUMIDITY

Solids can form within a particle by precipitation or on its surface by chemical reaction. **Precipitation** is the formation of an insoluble compound from a solution and can be simulated as a reversible equilibrium process. The equilibrium coefficient of a precipitation reaction is a **solubility product**. A solid precipitates from solution when the product of its reactant ion concentrations and mean activity coefficient exceeds its solubility product. For the reaction $NH_4NO_3(s) \rightleftharpoons NH_4^+ + NO_3^-$, precipitation of ammonium nitrate occurs when

$$\mathbf{m}_{NH_4^+}\mathbf{m}_{NO_3^-}\gamma_{NH_4^+,NO_3^-}^2 > K_{eq}(T) \tag{18.85}$$

Solid formation on a particle surface by heterogeneous reaction of gases can be simulated as an equilibrium process with $NH_4NO_3(s) \rightleftharpoons NH_3(g) + HNO_3(g)$. The solid may form when one gas adsorbs to a surface and the other gas collides and reacts with the adsorbed gas or when both gases adsorb to the surface, diffuse on the surface, and collide. A solid is assumed to form on the surface from the gas

**Table 18.4.** DRHs and CRHs of Several Electrolytes at 298 K

| Electrolyte | DRH (%) | CRH (%) | Electrolyte | DRH (%) | CRH (%) |
|---|---|---|---|---|---|
| NaCl | 75.28[a] | 47[c] | $(NH_4)_2SO_4$ | 79.97[a] | 37–40[b] |
| $Na_2SO_4$ | 84.2[b] | 57–59[b] | $NH_4HSO_4$ | 40.0[b] | 0.05–22[b] |
| $NaHSO_4$ | 52.0[d] | <0.05[d] | $NH_4NO_3$ | 61.83[a] | 25–32[d] |
| $NaNO_3$ | 74.5[d] | 0.05–30[b] | $(NH_4)_3H(SO_4)_2$ | 69[b] | 35–44[b] |
| $NH_4Cl$ | 77.1[a] | 47[e] | KCl | 84.26[a] | 62[c] |

[a]Robinson and Stokes (1955). [b]Tang and Munkelwitz (1994). [c]Tang (1997). [d]Tang (1996). [e]Cohen et al. (1987a).

phase when

$$p_{NH_3(g),s}\, p_{HNO_3(g),s} > K_{eq}(T) \tag{18.86}$$

For the examples above, solid formation can be simulated with the MFI solution method. When a solid forms during iteration of a gas–solid or ion–solid equilibrium reaction, $F$ from (18.83) converges to 1.0. The solid, ion, and/or gas concentrations are then updated with (18.82). When a solid does not form, $F$ cannot converge, and (18.82) predicts no net changes in concentrations.

The process by which an initially dry particle lowers its saturation vapor pressure and takes up liquid water is **deliquescence**. If a particle consists of an initially solid electrolyte at a given relative humidity, and the relative humidity increases, the electrolyte does not take on liquid water by hydration until the **deliquescence relative humidity** (DRH), is reached. At the DRH, water rapidly hydrates with the electrolyte, dissolving the solid, and increasing the liquid-water content of the particle. Above the DRH, the solid phase no longer exists, and the particle takes up additional liquid water to maintain equilibrium.

If a particle consists of an initially aqueous electrolyte, and the relative humidity decreases below the DRH, water evaporates, but dissolved ions in solution do not necessarily **precipitate** (crystallize) immediately. Instead, the solution is supersaturated and remains so until solid nucleation occurs. The relative humidity at which nucleation occurs and an initially aqueous electrolyte becomes crystalline is the **crystallization relative humidity** (CRH). The CRH is always less than or equal to the DRH. Table 18.4 shows the DRHs and CRHs of several electrolytes at 298 K.

Some electrolytes, such as $NH_3$, $HNO_3$, HCl, and $H_2SO_4$ do not have a solid phase at room temperature. These substances, therefore, do not have a DRH or a CRH.

For an initially dry mixture of two or more electrolytes, the relative humidity above which the mixture takes on liquid water is the mutual deliquescence relative humidity (MDRH). Above the MDRH, mixed particles may consist of one or more solids in equilibrium with the solution phase. The MDRH of a mixture is always lower than the DRH of any electrolyte within the mixture. (Wexler and Seinfeld 1991, Tang and Munkelwitz 1993).

## 18.12. EQUILIBRIUM-SOLVER RESULTS

Figures 18.4 and 18.5 show results from simulations of equilibrium composition determined by EQUISOLV, a computer program that uses the MFI method. In Fig. 18.4, equilibrium aerosol composition as a function of NaCl concentration is illustrated. The figure shows that nearly all ammonia gas dissolved and formed ammonium at low NaCl concentration. As NaCl concentration increased, liquid-water and nitrate concentrations increased. This was expected, since an increase in NaCl concentration increases liquid-water content due to hydration, enabling more dissolution and dissociation of nitric acid. Sulfate existed in the solution phase at all times.

Figure 18.5 shows a simulation of the change in aerosol composition as a function of relative humidity. At low relative humidities, aerosols contained solid ammonium nitrate and solid ammonium sulfate. When the relative humidity increased to the MDRH of just less than 62 percent (in this case), which is near the DRH

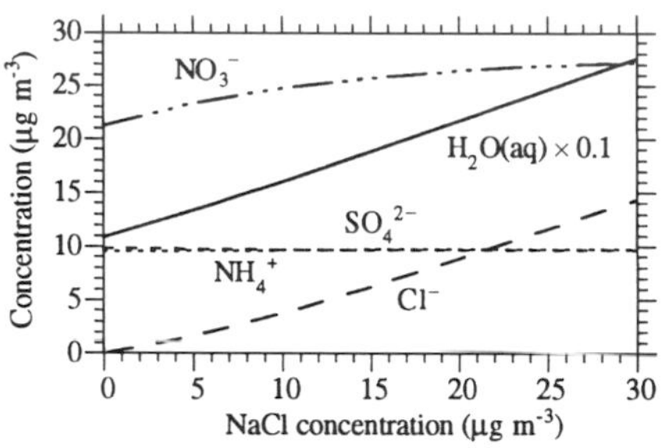


**Figure 18.4.** Aerosol composition versus NaCl concentration when the relative humidity was 90 percent. Other initial conditions were $H_2SO_4(aq)$, 10 μg m$^{-3}$; HCl(g), 0 μg m$^{-3}$; $NH_3(g)$, 10 μg m$^{-3}$; $HNO_3(g)$, 30 μg m$^{-3}$; and $T = 298$ K. NaCl dissolves and dissociates completely.

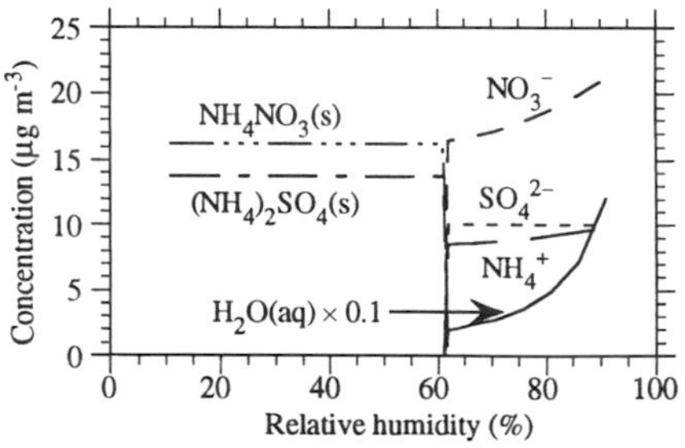


**Figure 18.5.** Aerosol composition versus relative humidity. Initial conditions were $H_2SO_4(aq)$, 10 μg m$^{-3}$; HCl(g), 0 μg m$^{-3}$; $NH_3(g)$, 10 μg m$^{-3}$; $HNO_3(g)$, 30 μg m$^{-3}$; and $T = 298$ K.

of ammonium nitrate, liquid water condensed onto the particle, dissolving both solids. Although its DRH is about 80 percent, ammonium sulfate dissolved at the MDRH. This is possible because the MDRH of a mixture is less than the DRH of any electrolyte in the mixture. As the relative humidity increased past the MDRH, the liquid-water content of the solution increased, increasing the rate of dissolution of nitric acid and ammonia from the gas phase.

## 18.13. NONEQUILIBRIUM BETWEEN GAS AND PARTICLE PHASES

Transfer between the gas and particle phases is time-dependent and not instantaneous. Thus, a gas' saturation vapor pressure over a particle surface does not equal the gas' partial pressure. Under many conditions, gas-particle equilibrium equations should not be used to solve for the partial pressure of a gas. Instead, transfer between the gas and particle phases should be simulated with time-dependent growth equations. In Chapter 17, solutions to time-dependent (nonequilibrium) condensational and depositional growth equations were given. In the following subsection, dissolutional growth solutions are described.

### 18.13.1. Growth Due to Dissolution

When a gas transfers to a particle surface and dissolves in liquid water on the surface, the process is **dissolutional growth**. Growth equations that describe dissolution are similar to (17.65) and (17.66), which describe condensation. A difference is that, for a dissolving gas, the saturation vapor pressure depends on the quantity of gas dissolved, the composition of solution, and the temperature. For a condensing gas, the saturation vapor pressure depends primarily on the temperature. A saturation vapor pressure was defined earlier as the partial pressure of a gas when the gas is in equilibrium with a particle surface at a given temperature. During dissolution, the saturation vapor pressure of gas A can be determined from the equilibrium relationship $A(g) \rightleftharpoons A(aq)$. In terms of an arbitrary gas $q$ the Henry's constant equation for this reaction is

$$p_{q,s,i} = \frac{\mathbf{m}_{q,i}}{H_q} \tag{18.87}$$

where $p_{q,s,i}$ is the saturation vapor pressure (atm) of gas $q$ over a particle in size bin $i$, $\mathbf{m}_{q,i}$ is the molality of dissolved gas $q$ in particles of size $i$ (moles kg$^{-1}$), and $H_q = K_{eq,q}$ is the Henry's constant (moles kg$^{-1}$ atm$^{-1}$). Activity coefficients of undissociated, dissolved gases are often assumed to be unity. (18.87) states that, as more gas dissolves, its saturation vapor pressure over a particle surface increases.

Saturation vapor pressure is related to saturation vapor mole concentration, $C_{q,s,i}$ (moles cm$^{-3}$), by

$$p_{q,s,i} = C_{q,s,i}\, 1000 R^* T \tag{18.88}$$

where $R^*$ is the gas constant in L atm mole$^{-1}$ K$^{-1}$, and $T$ is the temperature (K).

Molality (moles $kg^{-1}$) is related to mole concentration in solution by

$$\mathbf{m}_{q,i} = \frac{1000 c_{q,i}}{m_v c_{w,i}} \tag{18.89}$$

where $c_{q,i}$ is the mole concentration of dissolved gas $q$ in size bin $i$ (moles $cm^{-3}$), $c_{w,i}$ is the mole concentration of liquid water in bin $i$, and $m_v$ is the molecular weight of water (g $mole^{-1}$). Substituting (18.87) and (18.89) into (18.88) gives the saturation vapor concentration of a soluble species over a dilute solution as

$$C_{q,s,i} = \frac{p_{q,s,i}}{1000 R^* T} = \frac{\mathbf{m}_{q,i}}{1000 R^* T H_q} = \frac{c_{q,i}}{m_v c_{w,i} R^* T H_q} = \frac{c_{q,i}}{H'_{q,i}} \tag{18.90}$$

where $H'_{q,i} = m_v c_{w,i} R^* T H_q$.

When a gas dissolves in solution, it may dissociate, precipitate, or otherwise react reversibly. These effects are accounted for by solving equilibrium equations together with growth equations. Substituting (18.90) into the growth equation from (17.65) and adding $(\mathrm{d}c_{q,i,t}/\mathrm{d}t)_{eq}$, an equilibrium term that represents the rapid change in particle concentration due to reversible reactions (dissociation, precipitation, etc.), gives

$$\left(\frac{\mathrm{d}c_{q,i,t}}{\mathrm{d}t}\right)_{ge,eq} = k_{q,i,t-h}\left(C_{q,t} - S'_{q,i,t-h}\frac{c_{q,i,t}}{H'_{q,i,t-h}}\right) + \left(\frac{\mathrm{d}c_{q,i,t}}{\mathrm{d}t}\right)_{eq} \tag{18.91}$$

The subscript $t$ identifies a current concentration, and the subscript $t-h$ identifies an initial value. The mass transfer rate $k_{q,i,t-h}$ was given in (17.62). The equation for mass conservation between the gas phase and all size bins of the particle phase, corresponding to (18.91), is

$$\frac{\mathrm{d}C_{q,t}}{\mathrm{d}t} = -\sum_{i=1}^{N_B}\left[k_{q,i,t-h}\left(C_{q,t} - S'_{q,i,t-h}\frac{c_{q,i,t}}{H'_{q,i,t-h}}\right)\right] \tag{18.92}$$

The saturation-vapor-pressure terms in (18.91) and (18.92) should be modified when dissolution of strong acids and bases is simulated. Strong acids or bases usually dissociate immediately. Hydrochloric acid (HCl), nitric acid ($HNO_3$), and ammonia ($NH_3$) dissociate to $H^+ + Cl^-$, $H^+ + NO_3^-$, and $NH_4^+ + OH^-$, respectively. Because dissociation is rapid, the gas-aqueous equilibrium equations and equilibrium-coefficient relationships for these species can be rewritten from (18.87) as

$$HCl(g) \rightleftharpoons H^+ + Cl^- \qquad \frac{\mathbf{m}_{H^+,i}\mathbf{m}_{Cl^-,i}\gamma^2_{i,H^+/Cl^-}}{p_{HCl(g),s,i}} = K_{HCl} \tag{18.93}$$

$$HNO_3(g) \rightleftharpoons H^+ + NO_3^- \qquad \frac{\mathbf{m}_{H^+,i}\mathbf{m}_{NO_3^-,i}\gamma^2_{i,H^+/NO_3^-}}{p_{HNO_3(g),s,i}} = K_{HNO_3} \tag{18.94}$$

$$NH_3(g) + H^+ \rightleftharpoons NH_4^+ \qquad \frac{\mathbf{m}_{NH_4^+,i}\gamma_{i,NH_4^+}}{p_{NH_3(g),s,i}\mathbf{m}_{H^+,i}\gamma_{i,H^+}} = K_{NH_3} \tag{18.95}$$

respectively, where the $K$'s are equilibrium coefficients for the given reactions, listed in Table B.7. The last equation was obtained by combining

$$NH_3(aq) + H_2O(aq) \rightleftharpoons NH_4^+ + OH^- \qquad H^+ + OH^- \rightleftharpoons H_2O(aq) \quad (18.96)$$

The $\gamma$'s in (18.93) and (18.94) are mean activity coefficients of an electrolyte pair in a mixture containing many pairs. The single-ion activity-coefficient ratio in (18.95) is calculated in terms of mixed binary activity coefficients with

$$\frac{\gamma_{i,NH_4^+}}{\gamma_{i,H^+}} = \frac{\gamma_{i,NH_4^+}\gamma_{i,NO_3^-}}{\gamma_{i,H^+}\gamma_{i,NO_3^-}} = \frac{\gamma^2_{i,NH_4^+NO_3^-}}{\gamma^2_{i,H^+/NO_3^-}} = \frac{\gamma_{i,NH_4^+}\gamma_{i,Cl^-}}{\gamma_{i,H^+}\gamma_{i,Cl^-}} = \frac{\gamma^2_{i,NH_4^+/Cl^-}}{\gamma^2_{i,H^+/Cl^-}} \quad (18.97)$$

Combining (18.93)–(18.95) with (18.88) and (18.89) gives

$$C_{Cl^-,s,i} = \frac{p_{Cl^-,s,i}}{1000R^*T} = \frac{\mathbf{m}_{H^+,i}\mathbf{m}_{Cl^-,i}\gamma^2_{i,H^+/Cl^-}}{1000R^*TK_{HCl}}$$
$$= \frac{c_{Cl^-,i}\mathbf{m}_{H^+,i}\gamma^2_{i,H^+/Cl^-}}{m_v c_{w,i} R^*TK_{HCl}} = \frac{c_{Cl^-,i}}{H'_{Cl^-,i}} \quad (18.98)$$

$$C_{NO_3^-,s,i} = \frac{p_{NO_3^-,s,i}}{1000R^*T} = \frac{\mathbf{m}_{H^+,i}\mathbf{m}_{NO_3^-,i}\gamma^2_{i,H^+/NO_3^-}}{1000R^*TK_{HNO_3}}$$
$$= \frac{c_{NO_3^-,i}\mathbf{m}_{H^+,i}\gamma^2_{i,H^+/NO_3^-}}{m_v c_{w,i} R^*TK_{HNO_3}} = \frac{c_{NO_3^-,i}}{H'_{NO_3^-,i}} \quad (18.99)$$

$$C_{NH_4^+,s,i} = \frac{p_{NH_4^+,s,i}}{1000R^*T} = \frac{\mathbf{m}_{NH_4^+,i}\gamma_{i,NH_4^+}}{\mathbf{m}_{H^+,i}\gamma_{i,H^+}1000R^*TK_{NH_3}}$$
$$= \frac{c_{NH_4^+,i}\gamma_{i,NH_4^+}}{\mathbf{m}_{H^+,i}\gamma_{i,H^+}m_v c_{w,i} R^*TK_{NH_3}} = \frac{c_{NH_4^+,i}}{H'_{NH_4^+,i}} \quad (18.100)$$

where

$$H'_{Cl^-,i} = \frac{m_v c_{w,i} R^*TK_{HCl}}{\mathbf{m}_{H^+,i}\gamma^2_{i,H^+/Cl^-}} \quad (18.101)$$

$$H'_{NO_3^-,i} = \frac{m_v c_{w,i} R^*TK_{HNO_3}}{\mathbf{m}_{H^+,i}\gamma^2_{i,H^+/NO_3^-}} \quad (18.102)$$

$$H'_{NH_4^+,i} = \frac{\mathbf{m}_{H^+,i}\gamma_{i,H^+}m_v c_{w,i} R^*TK_{NH_3}}{\gamma_{i,NH_4^+}} \quad (18.103)$$

Substituting (18.98)–(18.100) into (18.91) gives

$$\frac{dc_{Cl^-,i}}{dt} = k_{HCl,i}\left(C_{HCl} - S'_{Cl^-,i}\frac{c_{Cl^-,i}}{H'_{Cl^-,i}}\right) + \left(\frac{dc_{Cl^-,i}}{dt}\right)_{eq} \quad (18.104)$$

$$\frac{dc_{NO_3^-,i}}{dt} = k_{HNO_3,i}\left(C_{HNO_3} - S'_{NO_3^-,i}\frac{c_{NO_3^-,i}}{H'_{NO_3^-,i}}\right) + \left(\frac{dc_{NO_3^-,i}}{dt}\right)_{eq} \quad (18.105)$$

$$\frac{dc_{NH_4^+,i}}{dt} = k_{NH_3,i}\left(C_{NH_3} - S'_{NH_4^+,i}\frac{c_{NH_4^+,i}}{H'_{NH_4^+,i}}\right) + \left(\frac{dc_{NH_4^+,i}}{dt}\right)_{eq} \quad (18.106)$$

The equations (18.104)–(18.106) are solved most readily by operator-splitting the growth from the equilibrium terms. Some values used in the equations, including $\mathbf{m}_{H^+}$, $\gamma_{i,\pm}$, and $(dc_{q,i}/dt)_{eq}$, are determined from an equilibrium calculation. The remaining concentrations are determined from a solver of nonequilibrium growth equations, discussed in Section 18.14. Operator-splitting is necessary because the simultaneous solution to growth coupled to equilibrium is too computationally intensive for practical application over a large model grid. Since growth equations are coupled tightly to equilibrium equations, the time interval between calculations of dissolutional growth and equilibrium equations must be small.

### 18.13.2. Growth Due to Surface Adsorption and Reaction

When a particle surface is solid or contains only a few monolayers of liquid water, gases may diffuse to and adsorb to the surface. The adsorbed species may desorb back to the gas phase, diffuse on the surface, or react chemically on the surface. Some gases likely to adsorb and react are nitric acid, hydrochloric acid, and ammonia. The time rate of change of adsorbed ammonia concentration in the first molecular layer over a surface is

$$\frac{dn_{NH_3,s}}{dt} = k_{a,NH_3} p_{NH_3} (n_m - n_{NH_3,s}) - k_{d,NH_3} n_{NH_3,s} + \sum_j P_{c,NH_3,j} - \sum_k L_{c,NH_3,k} \qquad (18.107)$$

where $n_{NH_3,s}$ is the surface concentration of adsorbed ammonia (molecules per square centimeter of surface), $k_{a,NH_3}$ is the rate that gas molecules adsorb to the surface (molecules $mb^{-1}$ $s^{-1}$), $p_{NH_3}$ is the partial pressure of ammonia (mb), $n_m \approx 10^{15}$ sites per square centimeter is the maximum number of adsorption sites on the surface, $k_{d,NH_3}$ is the rate that adsorbed molecules desorb from the surface ($s^{-1}$), $P_{c,j,NH_3}$ is the $i$th chemical production term for adsorbed ammonia (molecules $cm^{-2}$ $s^{-1}$), and $L_{c,k,NH_3}$ is the $j$th chemical loss term for adsorbed ammonia (molecules $cm^{-2}$ $s^{-1}$). Adsorption and desorption rates were defined in Section 12.3.4.3. (18.107) states that the time rate of change of adsorbed molecule concentration equals an adsorption flux minus a desorption flux plus a net chemical reaction flux.

Once adsorbed to a surface, ammonia can react chemically with other adsorbed species or directly with gases. Two reactions that can occur on a surface are

$$NH_3(a) + HNO_3(a) \rightleftharpoons NH_4NO_3(s) \qquad (18.108)$$

$$NH_3(a) + HCl(a) \rightleftharpoons NH_4Cl(s)$$

These reactions proceed in the forward and reverse directions.

The difficulty with solving (18.107) is that many variables in the equation are hard to obtain because they depend on surface type, surface composition, temperature, and relative humidity. Experimental data as a function of each parameter are sparse. Thus, surface reaction processes are usually simplified. One simplification is to assume that gases are in equilibrium with solids on a particle surface. In such a case, (18.108) reduces to

$$NH_3(g) + HNO_3(g) \rightleftharpoons NH_4NO_3(s) \tag{18.109}$$
$$NH_3(g) + HCl(g) \rightleftharpoons NH_4Cl(s)$$

If these equations are solved, nonequilibrium gas-particle transfer equations are ignored.

Another method of simulating surface reactions is to assume that gases transfer to particle surfaces and equilibrate with each other to form solids on the surfaces. This process can be carried out by solving (18.104)–(18.106), where the equilibrium term takes account of ion–solid or liquid–solid equilibrium reactions. When liquid water covers the particle surface, solids form by gas dissolution followed by precipitation. When the liquid-water content approaches zero, the saturation-vapor-pressure terms in (18.104)–(18.106) are replaced by constant saturation vapor pressures determined from gas–solid equilibrium-coefficient relations, such as those corresponding to the reactions in (18.109). In such cases, dissolutional growth equations become condensational growth equations and can be solved with the APC scheme discussed in Chapter 17.

## 18.14. SOLUTION TO GROWTH EQUATIONS FOR A SOLUBLE SPECIES

When growth and equilibrium equations are operator-split, the two processes are solved separately but sequentially. An equilibrium solution method was given in Section 18.10. (18.91), without the equilibrium term, and (18.92) represent $N_B + 1$ growth/evaporation expressions. Like condensation equations, dissolution equations may be solved with an iterative ODE solver. They may also be solved with the **analytical predictor of dissolution** (APD) scheme (Jacobson 1997c), described below. This scheme requires no iteration, conserves mass, and is unconditionally stable.

The noniterative solution is obtained by assuming that the final concentration of component $q$ in size bin $i$ is calculated by integrating (18.91). The result is

$$c_{q,i,t} = \frac{H'_{q,i,t-h} C_{q,t}}{S'_{q,i,t-h}} + \left( c_{q,i,t-h} - \frac{H'_{q,i,t-h} C_{q,t}}{S'_{q,i,t-h}} \right) \exp\left( -\frac{h S'_{q,i,t-h} k_{q,i,t-h}}{H'_{q,i,t-h}} \right) \tag{18.110}$$

where $C_{q,t}$ is currently unknown. As with the APC scheme discussed in Chapter 17, the final aerosol and gas concentrations are constrained by

$$C_{q,t} + \sum_{i=1}^{N_B} (c_{q,i,t}) = C_{q,t-h} + \sum_{i=1}^{N_B} (c_{q,i,t-h}) = C_{tot} \tag{18.111}$$

**Table 18.5.** Stability Test of the APD Scheme

| Time Step Size (s) | $C_{q,t}$ ($\mu$g m$^{-3}$) | $c_{q,1,t}$ ($\mu$g m$^{-3}$) | $c_{q,2,t}$ ($\mu$g m$^{-3}$) | $c_{q,3,t}$ ($\mu$g m$^{-3}$) |
|---|---|---|---|---|
| 0.1 | 0.769 | 3.08 | 3.08 | 3.08 |
| 10 | 0.769 | 3.08 | 3.08 | 3.08 |
| 60 | 0.769 | 3.08 | 3.08 | 3.08 |
| 600 | 0.769 | 3.08 | 3.08 | 3.08 |
| 7200 | 0.769 | 3.08 | 3.08 | 3.08 |

Assume a gas dissolves in three particle size bins, where $k_{q,i,t-h} =$ 0.00333, 0.00833, and 0.0117 s$^{-1}$ for the respective bins. Also $H'_{q,i} = 4.0$, $S'_{q,i,t-h} = 1.0$, and $c_{q,i,t-h} = 0\ \mu\text{g m}^{-3}$ for the bins, and $C_{q,t-h} = 10\ \mu\text{g m}^{-3}$. The table shows concentrations after 2 h predicted by the APD scheme when five time steps sizes were used. In all cases, the solution was stable and mass-conserving.

Substituting (18.110) into (18.111) and solving for $C_{q,t}$ gives

$$C_{q,t} = \frac{C_{q,t-h} + \sum_{i=1}^{N_B}\left\{c_{q,i,t-h}\left[1 - \exp\left(-\frac{hS'_{q,i,t-h}k_{q,i,t-h}}{H'_{q,i,t-h}}\right)\right]\right\}}{1 + \sum_{i=1}^{N_B}\left\{\frac{H'_{q,i,t-h}}{S'_{i,q,t-h}}\left[1 - \exp\left(-\frac{hS'_{q,i,t-h}k_{q,i,t-h}}{H'_{q,i,t-h}}\right)\right]\right\}} \tag{18.112}$$

which is the final gas concentration. This concentration is substituted into (18.110) to give the final particle component concentration for a time step. The APD scheme is unconditionally stable, since all final concentrations are bounded between 0 and $C_{tot}$, regardless of the time step. Table 18.5 illustrates the unconditional stability and mass conservation of the APD scheme.

## 18.15. SIMULATIONS UNDER ATMOSPHERIC CONDITIONS

Simulation results for condensation and dissolution coupled to equilibrium are shown here for two cases representing typical conditions in coastal urban air. In both cases, the initial particle size distribution and composition contained four modes – one nucleation, two subaccumulation, and one coarse-particle mode. Initially, the particles were assumed to contain only sulfuric acid, sodium chloride, elemental carbon, and nonreacting organic carbon. Sixteen size bins were used. The diameter of the smallest bin was 0.02 μm, and the volume ratio of adjacent size bins was 5.0. Figure 18.6 shows the initial distribution of water and ionic components in the particles when the relative humidity and temperature were 90 percent and 298 K, respectively.

For the first case, dissolutional growth was coupled to equilibrium at $f_r = 90$ percent. Dissolution growth was solved for with two schemes, the APD scheme and SMVGEAR II, the sparse-matrix, vectorized Gear-type ordinary-differential-equation solver discussed in Chapter 12. In both cases, growth solutions were coupled to equilibrium solutions from EQUISOLV. Figure 18.7 compares results from APD–EQUISOLV with those from SMVGEAR II–EQUISOLV when the time interval between growth and equilibrium was 1 s in both cases. The figure shows

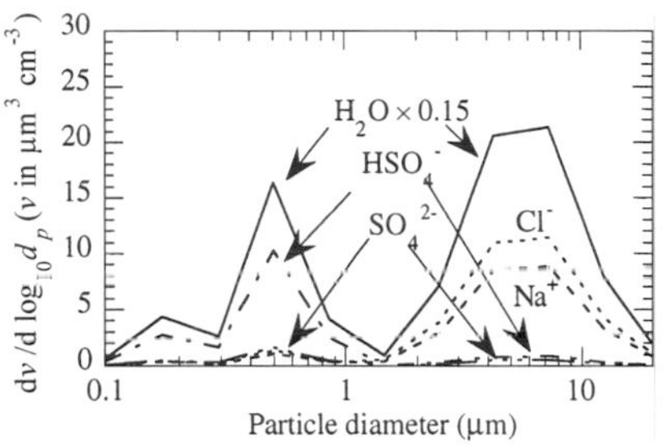


**Figure 18.6.** Initial model size distribution obtained by solving equilibrium equations when $f_r$ = 90 percent and $T$ = 298 K, as described in the text.

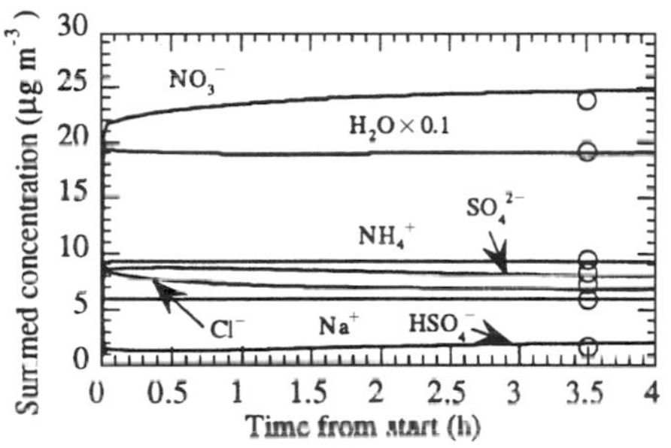


**Figure 18.7.** Time-series comparison of APD–EQUISOLV with SMVGEAR II–EQUISOLV results when a 1-s interval was used between growth and equilibrium. The solutions lie on top of each other. The circles are equilibrium solutions, found from EQUISOLV alone. Initial conditions were $f_r$ = 90 percent; $T$ = 298 K; $HNO_3$(g), 30 μg m$^{-3}$; $NH_3$(g), 10 μg m$^{-3}$; and HCl(g), 0 μg m$^{-3}$.

that the results from the two schemes matched exactly, and growth-equilibrium solutions nearly matched a pure equilibrium solution after 2 h, indicating that the gas and aerosol phases were almost in equilibrium when the relative humidity was high.

For the second case, cloud drop formation with gas dissolution was simulated. Growth processes accounted for included dissolution of $HNO_3$(g), $NH_3$(g), and HCl(g) and condensation of $H_2SO_4$(g) and $H_2O$(g). Conditions for the simulation are described in the caption for Fig. 18.8. The figure shows a time-series comparison of the APD–APC–EQUISOLV solution with the SMVGEAR II–EQUISOLV solution when the time interval between growth and equilibrium was 10 s. The solutions from the two schemes matched for almost all species. Figure 18.9 shows the size distribution of liquid water initially, after growth, and after evaporation for the cases shown in Fig. 18.8.

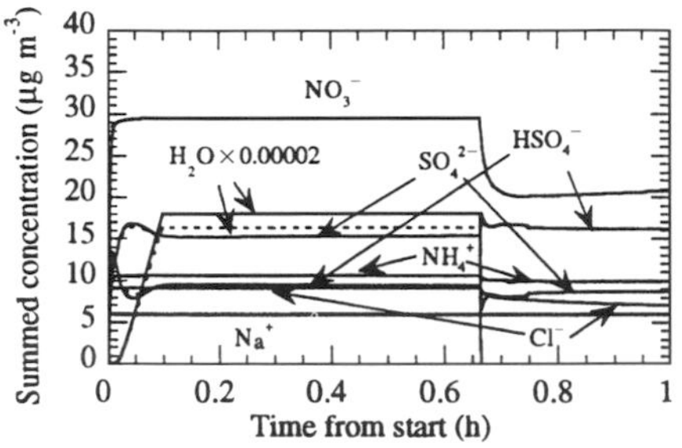


**Figure 18.8.** Time-series comparisons of APD–APC–EQUISOLV results with SMVGEAR II–EQUISOLV results when a 10-s time interval was used between growth and equilibrium for both schemes. The solutions lie nearly on top of each other for all species except liquid water. Initial conditions were $T = 298$ K; $HNO_3$(g), 30 μg $m^{-3}$; $NH_3$(g), 10 μg $m^{-3}$; HCl(g), 0 μg $m^{-3}$; and $H_2SO_4$(g), 15 μg $m^{-3}$. At time zero, the relative humidity was increased from 90 to 100.001 percent. It was reset to 100.001 percent after every 10 s of growth calculation for the first 5 min. After 40 min, the relative humidity was reduced to 90 percent.

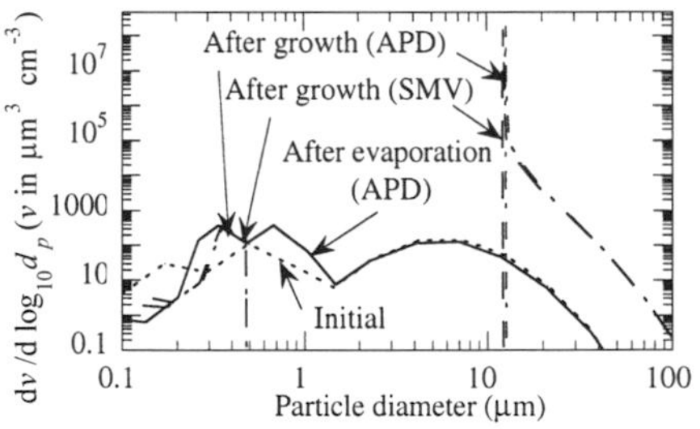


**Figure 18.9.** Model size distribution initially, after growth, and after evaporation obtained from the simulation shown in Fig. 18.8. Growth solutions for the APD–APC–EQUISOLV (APD) and SMVGEAR–EQUISOLV (SMV) schemes are shown.

## 18.16. SUMMARY

In this chapter, chemical equilibrium, activity coefficient, and water content equations were discussed and applied. A method of iterating equilibrium, activity coefficient, and water content equations simultaneously was also given, and temperature-dependent expressions for mean binary activity coefficients were shown. Mean mixed activity coefficients were calculated from mean binary activity coefficients with a mixing rule. A technique of solving nonequilibrium dissolutional growth equations and coupling growth to equilibrium was also discussed. When particles contain sufficient liquid water, irreversible aqueous chemical reaction should be accounted for, as discussed next.

## 18.17. PROBLEMS

**18.1.** What is the purpose of calculating mean mixed activity coefficients? Why can't binary activity coefficients be used for all simulations?

**18.2.** How are the thermodynamic activities of gases, liquids, ions, and solids related to their mole concentrations? Write out expressions for each case.

**18.3.** List the experimental data required to solve an equilibrium problem that includes gas, solution, and solid phases. Ignore coupling to growth equations.

**18.4.** What is the mathematical relationship between $Q_d$ and $Q_n$ in Section 18.10 when $\Delta x = 0$?

**18.5.** **(a)** Solve (18.91) and (18.92) together analytically, excluding the equilibrium term, for $c_{q,i,t}$ and $C_{q,t}$. Assume only one size bin is present and all variables, except $c_{q,i,t}$ and $C_{q,t}$, are constant. (Hint: $C_{q,t} + c_{q,i,t} = C_{q,t-h} + c_{q,i,t-h}$, since $i = 1$.)

**(b)** Discretize the time derivatives in the equations from part (a) in first-order, backward-difference form, and write out the right sides of both equations implicitly. Find the resulting solutions for $c_{q,i,t}$ and $C_{q,t}$.

## 18.18. COMPUTER PROGRAMMING PRACTICE

**18.6.** Write a computer script to solve the following problem. Assume uniformly-sized atmospheric particles contain dissolved nitric acid ($H^+$, $NO_3^-$), dissolved hydrochloric acid ($H^+$, $Cl^-$), and liquid water. Assume $f_r = 82$ percent, $T = 298$ K, and the mass concentrations of nitrate ($NO_3^-$) and chloride ($Cl^-$) at equilibrium in each particle are both 4 micrograms per cubic meter of air.

**(a)** Convert the units of chloride and nitrate to moles per cubic centimeter (mole concentration), and determine the mole concentration of $H^+$ by charge balance, assuming $H^+$ is the only cation and $Cl^-$ and $NO_3^-$ are the only anions in solution. Use the mole balance to determine mole concentrations of $HNO_3$ and HCl dissolved in solution.

**(b)** Calculate the liquid-water content of air (micrograms of $H_2O$ per cubic meter of air) from the water equation using the water activity data in Table B.10.

**(c)** Using the liquid-water content, convert all particle component concentrations to units of molality (moles of solute per kilogarm of water), and calculate the pH of the mixed solution, assuming molarity equals molality for convenience.

**(d)** Calculate the binary activity coefficients, $\gamma_{H^+,NO_3^-}$ and $\gamma_{H^+,Cl^-}$, using Pitzer's method. For each calculation, assume $H^+$–$NO_3^-$ and $H^+$–$Cl^-$ pairs are alone in solution. Assume $\mathbf{m}_{NO_3^-}$ and $\mathbf{m}_{Cl^-}$ are the same as in part (c). Assume $\mathbf{m}_{H^+} = \mathbf{m}_{NO_3^-}$ and $\mathbf{m}_{H^+} = \mathbf{m}_{Cl^-}$ in each of the two cases, since the pairs are alone in solution.

**(e)** Assume $H^+$–$NO_3^-$ and $H^+$–$Cl^-$ pairs are alone in solution. Calculate the gas-phase partial pressures (in atmospheres) of HCl and $HNO_3$ from (18.50).

**18.7.** Assume for some anion $B^-$ that HB(g) dissolves in particles of three sizes, and its effective Henry's constant for each size bin is $H'_{i,HB} = 4$. In solution, HB(aq) dissociates into $H^+$ and $B^-$ with a dissociation constant $K_{eq} = 8$, written in generic units. Assume that the activity coefficients are unity, and ignore the Kelvin effect.

**(a)** Write the Henry's law and dissociation equilibrium-coefficient relationships for each size bin.

**(b)** Write the charge-balance constraint for each size bin and the mass-balance equation for B between the gas and aerosol phases. Assume $C_{tot} = 10\ \mu g\ m^{-3}$ is the total amount of B in the system.

**(c)** Solve for the equilibrium gas mole concentration of HB(g) and aerosol mole concentrations of $H^+$, $B^-$, and HB(aq) in each size bin.

**(d)** Describe how you would solve this problem if the gas phase were not in equilibrium with the particle phase (i.e., write out and describe the equations you would solve). You do not have to solve the equations.

# 19

# Aqueous Chemistry

AQUEOUS reactions differ from chemical equilibrium reactions in that the former are not reversible and the latter are. Aqueous reactions are similar to gas-phase reactions in that both are described by first-order, ordinary differential equations. Aqueous reactions are important when cloud drops or raindrops are present. In such case, the liquid-water content of air exceeds $10^5$ μg $m^{-3}$ and aqueous reactions affect drop composition over a period of minutes to tens of minutes. When only aerosols are present, the liquid-water content of air is usually less than 500 μg $m^{-3}$, and aqueous reactions affect aerosol composition over periods of hours to days. Since many aqueous reactants originate as gases, aqueous chemistry should be simulated in conjunction with gas chemistry and gas-to-particle conversion. In this chapter, the importance of aqueous chemistry, a description of aqueous reactions, and a method of solving aqueous chemistry together with growth processes is discussed.

## 19.1. SIGNIFICANCE OF AQUEOUS CHEMICAL REACTIONS

Gas uptake and reaction in particles involves several steps. In the first step, soluble gases diffuse to and dissolve in liquid water on a particle surface. Some of the dissolved molecules may then dissociate reversibly to ions. Undissociated molecules and dissociated ions can diffuse through the particle. As they diffuse, irreversible chemical reactions take place. Some of the dissolved molecules and reaction products may diffuse back to the particle surface and evaporate back to the gas phase.

Gas uptake and irreversible (aqueous) reactions in liquid-containing particles affect particle composition. Such reactions affect cloud drop composition more than they affect aerosol composition. In both cases, when a species reacts in the aqueous phase, more of the species is drawn from the gas phase to replace the reacted molecules. Modeling studies have shown that clouds reduce gas-phase ozone by this process (e.g., Walcek et al. 1997; Liang and Jacob 1997).

Some of the most important aqueous reactions are those that convert dissolved sulfur dioxide [$SO_2$(aq)] and its dissociation products, bisulfite and sulfite, to dissolved sulfuric acid [$H_2SO_4$(aq)] and its dissociation products, bisulfate and sulfate. In solution, dissolved sulfur dioxide and its dissociation products constitute the **S(IV) family**, and dissolved sulfuric acid and its products constitute the **S(VI) family**. The species in these families and their Lewis structures are given in Table 19.1.

**Table 19.1.** Names, Formulae, and Lewis Structures of S(IV) and S(VI) Species

| S(IV) Family | | S(VI) Family | |
|---|---|---|---|
| **Chemical Name and Formula** | **Chemical Structure** | **Chemical Name and Formula** | **Chemical Structure** |
| Sulfur dioxide(aq) $SO_2(aq)$ | 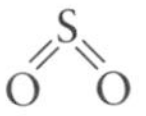 | | |
| Sulfurous acid(aq) $H_2SO_3(aq)$ | 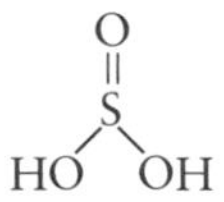 | Sulfuric acid(aq) $H_2SO_4(aq)$ | 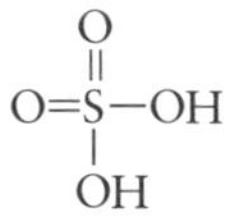 |
| Bisulfite ion $HSO_3^-$ | 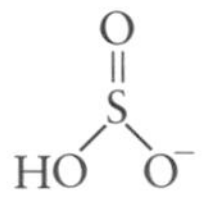 | Bisulfate ion $HSO_4^-$ | 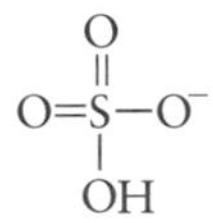 |
| Sulfite ion $SO_3^{2-}$ | 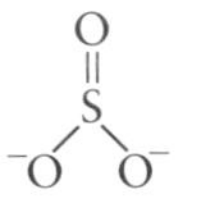 | Sulfate ion $SO_4^{2-}$ | 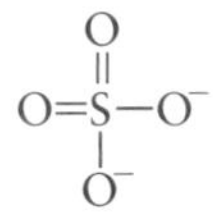 |

$SO_2(aq)$ is produced when $SO_2(g)$ dissolves in liquid water. $SO_2(g)$ is less soluble than $HCl(g)$ or $HNO_3(g)$, but more soluble than many other gases. Chemical and emission sources of $SO_2(g)$ werc discussed in Chapter 12. Once dissolved, $SO_2(aq)$ dissociates to $HSO_3^-$ and $SO_3^{2-}$. These ions are oxidized irreversibly to $HSO_4^-$ and $SO_4^{2-}$ primarily by hydrogen peroxide ($H_2O_2$), the hydroxyl radical (OH), molecular oxygen ($O_2$) (catalyzed by $Fe^{3+}$, $Mn^{2+}$), and ozone ($O_3$) (Hoffmann and Calvert 1985; Jacob et al. 1989a; Pandis and Seinfeld, 1989).

When $SO_2(aq)$ dissociates and reacts, $SO_2(g)$ dissolves to replenish it. The faster S(IV) oxidation occurs, the faster $SO_2(g)$ is removed from the gas phase, and the faster S(VI) forms. Because the rate of change of mass of a particle due to dissolution depends on the volume of liquid water present, dissolution is a **volume-limited** process.

Dissolution and reaction of sulfur dioxide is one mechanism by which S(VI) accumulates in particles. A second mechanism is condensation of sulfuric acid gas onto particles. Upon condensation, $H_2SO_4(aq)$ dissociates to $HSO_4^-$ and $SO_4^{2-}$. Condensation accounts for much of the increase of S(VI) in particles when the relative humidity is below 100 percent. Below 70-percent relative humidity, nearly all S(VI) production in particles is due to condensation, as opposed to dissolution/reaction. Because the rate of change of mass of a particle due to condensation is proportional to the radius of the particle, as shown in (17.13), condensation is a **radius-limited** process.

Figures 19.1 and 19.2 demonstrate that dissolution and aqueous reaction can convert S(IV) to S(VI) molecules within minutes when cloud drops are present. The first figure compares changes in $H_2O$, $HSO_4^-$, and $SO_4^{2-}$ concentrations over

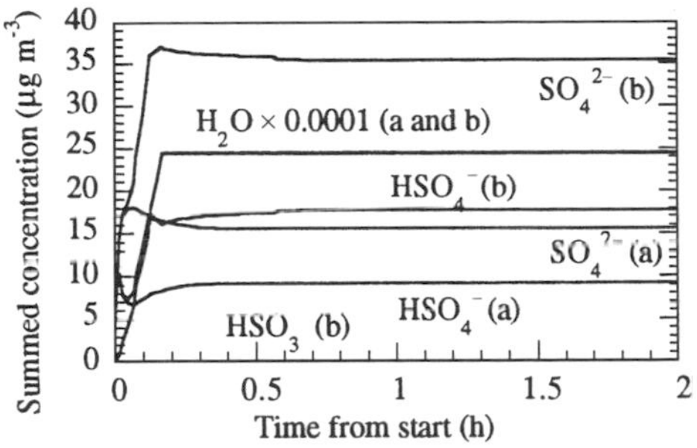


**Figure 19.1.** Time-series comparison of gas concentrations when cloud drops grew and sulfur dioxide (a) did not and (b) did dissolve and react within the drops. $H_2SO_4$ condensed in both cases, but $SO_2$ dissolved and reacted in the second case only. Initial conditions were $SO_2$(g), 52 μg m$^{-3}$; $H_2O_2$(g), 10 μg m$^{-3}$; $HNO_3$(g), 30 μg m$^{-3}$; $NH_3$(g), 10 μg m$^{-3}$; HCl(g), 0 μg m$^{-3}$; $H_2SO_4$(g), 1.5 μg m$^{-3}$; $H_2SO_4$(aq), 10 μg m$^{-3}$; NaCl(aq), 15 μg m$^{-3}$; and $T = 298$ K. The cloud was formed by increasing the relative humidity to 100.01 percent at the start, solving the growth equations with a 1-s time step, and replenishing the relative humidity to 100.01 percent every time step for the first 10 min of simulation. After 10 min, the relative humidity was allowed to relax to its calculated value.

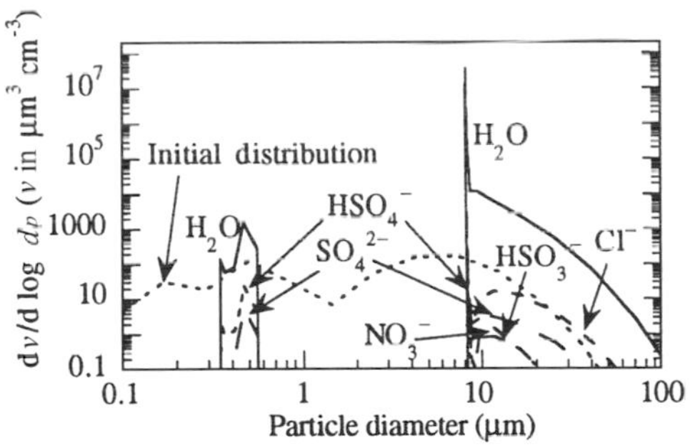


**Figure 19.2.** Initial distribution and final particle composition for case (b) in Fig. 19.1.

time in cloud drops when $SO_2$ dissolution and aqueous reaction were and were not, respectively, included in the calculations. Within 20 min, cloud drops dissolved and converted more than 30 out of an initial 52 μg m$^{-3}$ of $SO_2$(g). Figure 19.2 gives the resulting size distributions of individual species after the two-hour simulation.

The conversion rate of S(IV) to S(VI) molecules depends on the liquid-water content, S(IV) oxidant concentration, and pH. Aerosol liquid-water contents can range from <100 to 250 μg $m^{-3}$ at 90-percent relative humidity, as shown in Fig. 18.3. Fog liquid-water contents are generally less than 0.2 g $m^{-3}$. Cloud liquid-water contents, which range from 0.2 to 2 g $m^{-3}$, exceed those of fogs (Pruppacher and Klett 1997). Thus, a cloud has $10^3$–$10^4$ times more liquid water than does an aerosol plume. The Henry's constants of $H_2O_2$, $HO_2$, OH, and $O_3$ are moderate to low, as shown in Table B.7. Because of its high liquid-water content, a cloud can absorb all four gases. Because of its low liquid-water content, an aerosol plume can absorb only $H_2O_2$, which has a moderate Henry's constant and moderate concentration, in significant quantities. Thus, S(IV) conversion in an aerosol plume is often limited by the concentration of $H_2O_2$(g). In a cloud, the greater abundance of $H_2O_2$, $HO_2$, OH, and $O_3$ in solution permits S(IV) to convert to S(VI) faster than in an aerosol plume.

Over a range of liquid-water contents (200 μg $m^{-3}$ to 2 g $m^{-3}$), $H_2O_2$ and $O_3$ are the major aqueous-phase sinks of S(IV) at pH $\leq 6$ and pH $> 6$, respectively. When the liquid-water content exceeds 0.2 g $m^{-3}$ and pH $\approx 5$, or when $H_2O_2$ is depleted, OH also reduces S(IV) (Liang and Jacobson 1998). S(IV) oxidation is slow in aerosols, but it can be enhanced at low temperatures (273 K) if the relative humidity is high.

Conversion of S(IV) to S(VI) in clouds contributes to **acid rain**. Acid rain is prevalent in the northeastern United States, eastern Canada, Scandinavia, and Eastern Europe. It arises when coal-fired power plants emit $SO_2$(g), which is advected from west to east in midlatitudes. As the plume travels, some $SO_2$(g) dissolves, dissociates, and oxidizes in cloud drops, ultimately forming S(VI). $SO_2$(g) also oxidizes to $H_2SO_4$(g), which condenses onto drops. The addition of S(VI) to cloud drops decreases pH. The natural pH of cloud and rainwater is 5.6. The slight acidity is due to dissolution of background carbon dioxide gas. Addition of S(VI) to cloud drops and raindrops increases their acidity by decreasing the pH to 2.0–5.5.

## 19.2. COMMON REACTIONS

The source of S(IV) in a liquid drop is diffusion of $SO_2$(g) to the drop surface and dissolution of $SO_2$(g) into the drop. At the air–drop interface, the relevant reversible reaction is

$$SO_2(g) \rightleftharpoons SO_2(aq) \qquad (19.1)$$

In solution, sulfur dioxide reaches equilibrium with other S(IV) species by

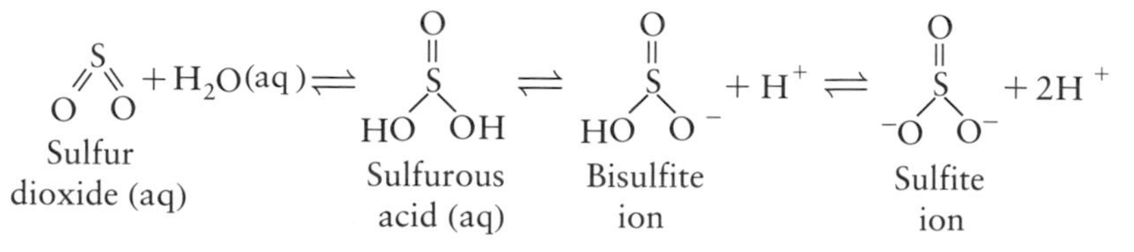

(19.2)

At a pH of 2–7, most S(IV) exists as $HSO_3^-$. Thus, aqueous reactions involving S(IV) are often written in terms of loss or production of $HSO_3^-$. At pH > 2, most S(VI) exists as $SO_4^{2-}$. Thus, reactions involving S(VI) are often written in terms of $SO_4^{2-}$ production.

$H_2O_2(aq)$ is the primary S(IV) oxidant when pH ≤ 6. Its reaction with bisulfite is

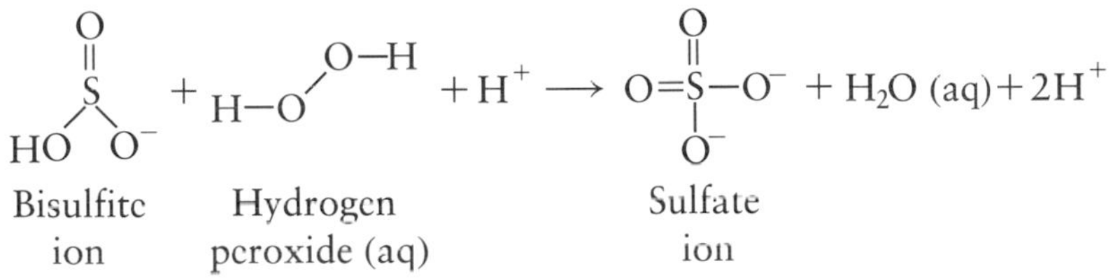

$$\text{Bisulfite ion} + \text{Hydrogen peroxide (aq)} + H^+ \longrightarrow \text{Sulfate ion} + H_2O \text{ (aq)} + 2H^+ \quad (19.3)$$

which consumes S(IV) within tens of minutes if the $H_2O_2(aq)$ concentration exceeds that of S(IV). If the S(IV) concentration exceeds that of $H_2O_2(aq)$, then $H_2O_2(aq)$ is depleted within minutes (Jacob 1986). The reaction rate of (19.3) increases with decreasing pH. The rate coefficient multiplied by [$H^+$] is 9 times greater at pH = 0 than at pH = 4, as seen in Table B.8. A cloud drop typically has a pH of 2.5 to 5.6. An aerosol often has a pH of −1 to +2. The lower pH in aerosols speeds the S(IV)–$H_2O_2$–$H^+$ reaction in aerosols relative to in clouds. But, because aerosols contain less $H_2O_2$ than do clouds, aerosols are still less efficient at converting S(IV) than are clouds.

Sources of $H_2O_2(aq)$ in clouds ⌐ ⌐ its dissolution by

$$H_2O_2(g) \rightleftharpoons H_2O_2(aq) \quad (19.4)$$

and aqueous production by

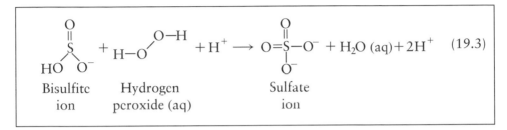

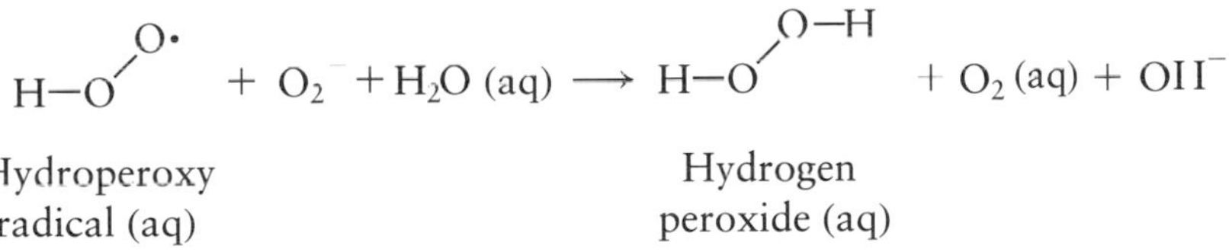

$$\text{Hydroperoxy radical (aq)} + O_2^- + H_2O \text{ (aq)} \longrightarrow \text{Hydrogen peroxide (aq)} + O_2 \text{ (aq)} + OH^- \quad (19.5)$$

where $HO_2(aq)$ is the aqueous hydroperoxy radical, $O_2^-$ is the peroxy ion, $O_2(aq)$ is aqueous molecular oxygen, and $OH^-$ is the hydroxide ion. $HO_2(aq)$ is produced by dissolution of $HO_2(g)$ followed by aqueous reaction. In solution, it maintains equilibrium with the peroxy ion by $HO_2(aq) \rightleftharpoons H^+ + O_2^-$.

Aside from its reaction with S(IV), loss processes of $H_2O_2(aq)$ are photolysis,

$$H_2O_2(aq) + h\nu \longrightarrow 2\dot{O}H(aq) \quad (19.6)$$

and reaction with the hydroxyl radical,

$$H_2O_2(aq) + \dot{O}H(aq) \longrightarrow H_2O(aq) + H\dot{O}_2(aq) \quad (19.7)$$

At pH > 6, which occurs only in clouds drops that contain basic substances, such as ammonium or sodium, the most important oxidant of S(IV) is ozone. The relevant S(IV)–$O_3$ reaction in this pH region is

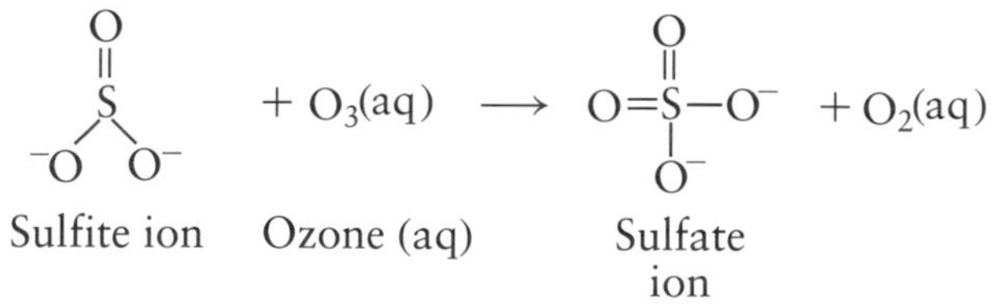


(19.8)

When the liquid-water content exceeds 0.2 g $m^{-3}$ and pH ≈ 5, or when $H_2O_2$ is depleted, OH can also oxidize S(IV) by the reaction

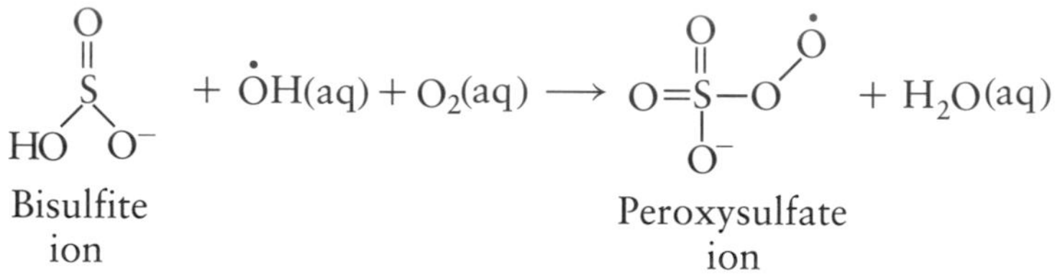


(19.9)

The solubility of OH(aq) is low; thus, its aqueous sources generally exceed its gas-phase sources (Jacob 1986). Aqueous reactions producing and destroying OH(aq) are given in Table B.8. Sinks of $SO_5^-$, include its reactions with $O_2^-$, $HCOO^-$, and other $SO_5^-$ molecules, as shown in Table B.8. $SO_5^-$ catalyzes production of peroxymonosulfate by

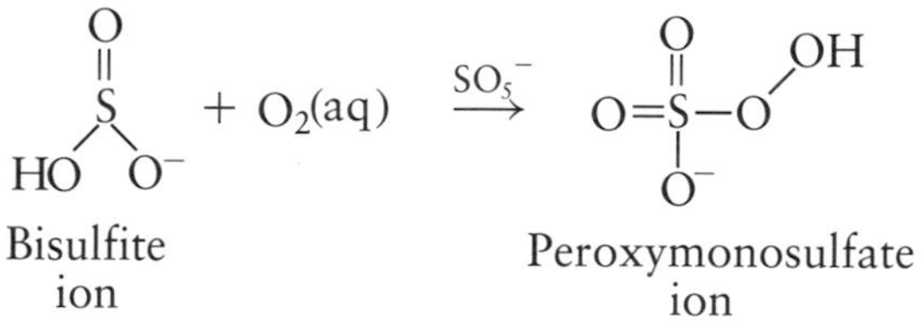


(19.10)

Peroxymonosulfate oxidizes S(IV) via

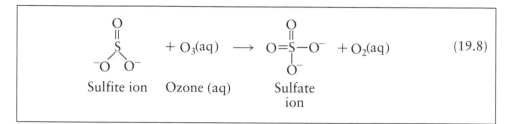

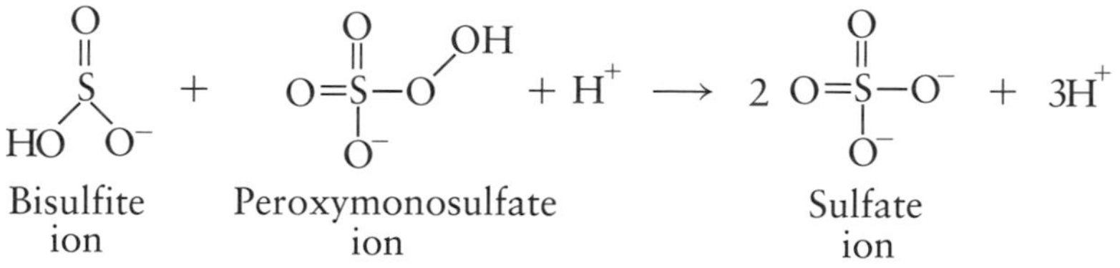


(19.11)

Another important oxidant of S(IV) is $O_2$(aq), catalyzed by **transition metals,** such as Fe(III) and Mn(II) (Jacob et al. 1989a). About 85–90 percent of soluble iron appears as Fe(II) in the form of $Fe^{2+}$, and about 10–15 percent appears as Fe(III). Although Fe(II) reacts with oxidants, such as OH(aq), $HO_2$(aq), and $H_2O_2$(aq), it does not readily catalyze S(IV) oxidation reactions. The solubility of Fe(III) depends on pH. At high pH, Fe(III) is in the form of precipitated $Fe(OH)_3$(s), which does not catalyze S(IV) oxidation, or in the form of soluble sulfate and hydroxy complexes, which do. At low pH, Fe(III) is primarily in the form of a catalyst, $Fe^{3+}$. The primary catalytic reaction involving Fe(III) is

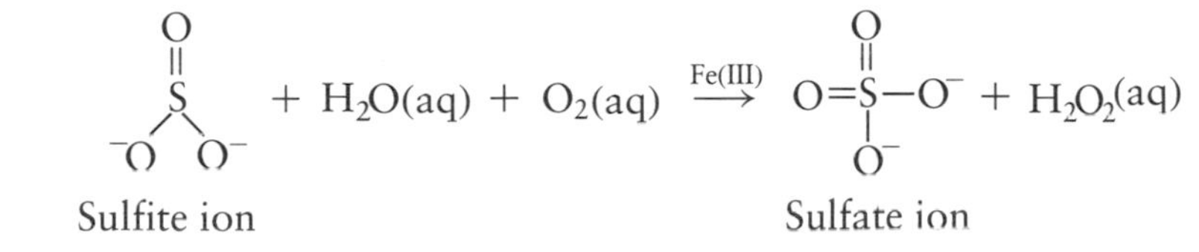

$$
\underset{\text{Sulfite ion}}{\overset{\displaystyle\underset{\overset{\displaystyle S}{\overset{|}{\phantom{}}}}{\overset{\|}{O}}}{\underset{^-O\quad O^-}{}}} + H_2O(aq) + O_2(aq) \xrightarrow{\text{Fe(III)}} \underset{\text{Sulfate ion}}{O=\overset{\displaystyle\overset{O}{\|}}{\underset{\displaystyle\underset{O^-}{|}}{S}}-O^-} + H_2O_2(aq) \qquad (19.12)
$$

Mn(III), which is usually in the form of $Mn(OH)^{2+}$ at high pH and in the form of $Mn^{3+}$ at low pH, does not catalyze S(IV) oxidation significantly. Mn(II) is usually in the form of $Mn^{2+}$ and catalyzes S(IV) by

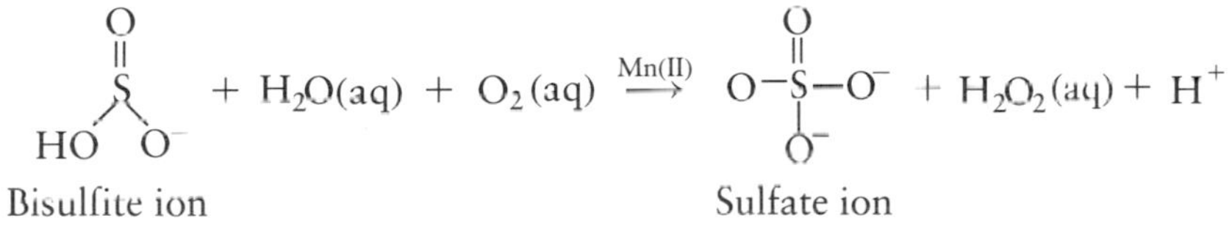

$$
\underset{\text{Bisulfite ion}}{\overset{\displaystyle\overset{O}{\|}}{\underset{HO\quad O^-}{S}}} + H_2O(aq) + O_2(aq) \xrightarrow{\text{Mn(II)}} \underset{\text{Sulfate ion}}{O-\overset{\displaystyle\overset{O}{\|}}{\underset{\displaystyle\underset{O^-}{|}}{S}}-O^-} + H_2O_2(aq) + H^+ \qquad (19.13)
$$

Carbonyl compounds and carboxylic acids, present in cloud- and rainwater (e.g., Munger et al. 1989) also react with S(IV) in solution. Important carbonyl compounds in solution include formaldehyde (HCHO) and acetaldehyde ($CH_3CHO$). Important aqueous carboxylic acids include formic acid (HCOOH) and acetic acid ($CH_3COOH$). The sources of these aqueous species are gas-phase emissions and/or oxidation followed by dissolution. Combustion and gas-phase reaction produce gas-phase formaldehyde and acetaldehyde. Combustion, vegetation, soil decomposition, biomass burning, and chemical reaction produce gas-phase formic and acetic acids.

Gas phase formaldehyde readily dissolves in cloud- and rainwater. Dissolved formaldehyde reacts reversibly to form **methylene glycol** [$H_2C(OH)_2$] by

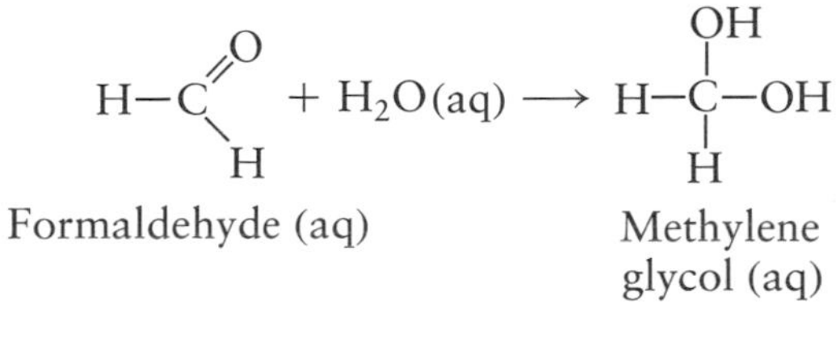

$$
\underset{\text{Formaldehyde (aq)}}{H-C\overset{\displaystyle O}{\underset{\displaystyle H}{\big\Vert}}} + H_2O(aq) \longrightarrow \underset{\substack{\text{Methylene}\\\text{glycol (aq)}}}{H-\overset{\displaystyle\overset{OH}{|}}{\underset{\displaystyle\underset{H}{|}}{C}}-OH} \qquad (19.14)
$$

At high pH, HCHO(aq) can also react with the sulfite ion to form **hydroxymethanesulfonate** (HMSA), an S(IV)–HCHO(aq) adduct, by the reaction,

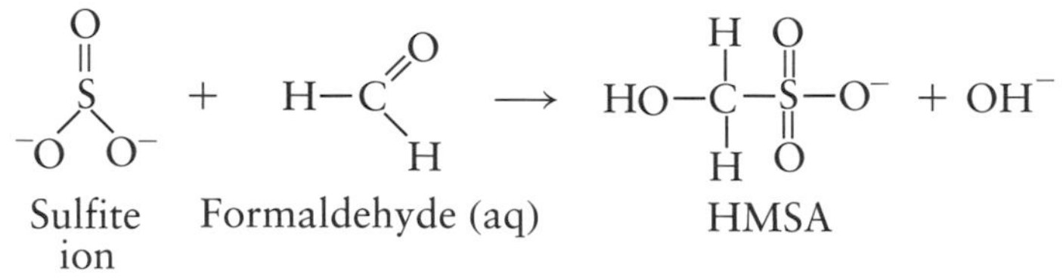

HMSA is lost by

$$
HO-\overset{\underset{\displaystyle H}{|}}{\underset{\displaystyle H}{\overset{\displaystyle |}{C}}}-\overset{\underset{\displaystyle O}{\|}}{\overset{\displaystyle O}{\|}}{S}-O^- + \overset{\centerdot}{O}H(aq)+O_2(aq) \longrightarrow H-C\overset{\displaystyle O}{\underset{\displaystyle H}{\diagup}} + O{=}\overset{\underset{\displaystyle O^-}{|}}{S}-O\overset{\centerdot}{O} \quad (19.16)
$$

HMSA             Formaldehyde    Peroxysulfate
                          (aq)            ion
                                      $+ H_2O$ (aq)

Methylene glycol from (19.14) is lost by

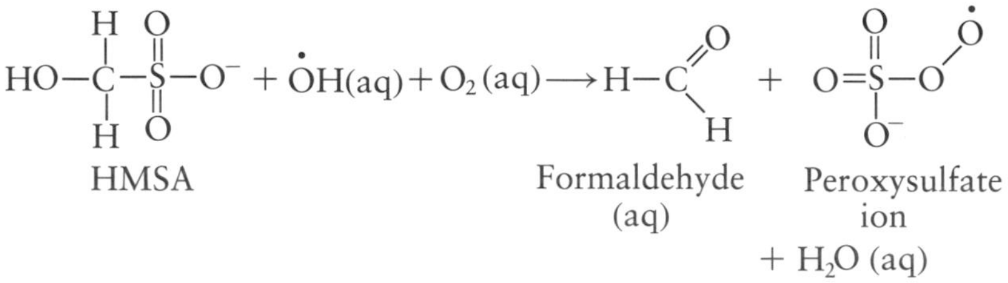

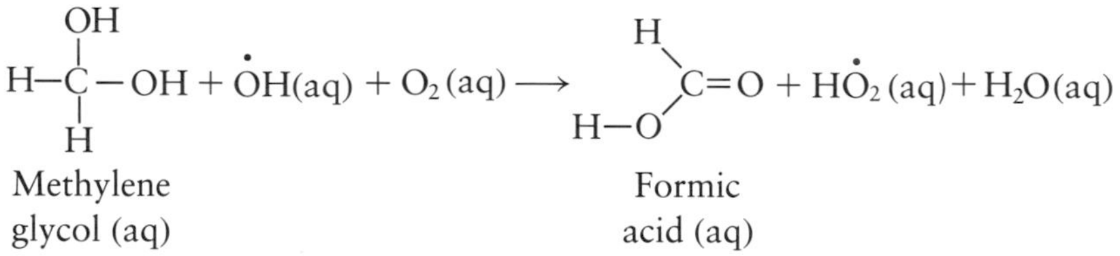

$$
H-\overset{\underset{\displaystyle H}{|}}{\overset{\displaystyle OH}{|}}{C}-OH + \overset{\centerdot}{O}H(aq) + O_2(aq) \longrightarrow \overset{\displaystyle H}{\underset{\displaystyle H-O}{\diagdown}}C{=}O + H\overset{\centerdot}{O}_2(aq)+H_2O(aq) \quad (19.17)
$$

Methylene                            Formic
glycol (aq)                           acid (aq)

The production of formic acid from this reaction is more significant than its production from the gas phase. Since formic acid is a weak acid, it can give up a proton in water to form **formate** by

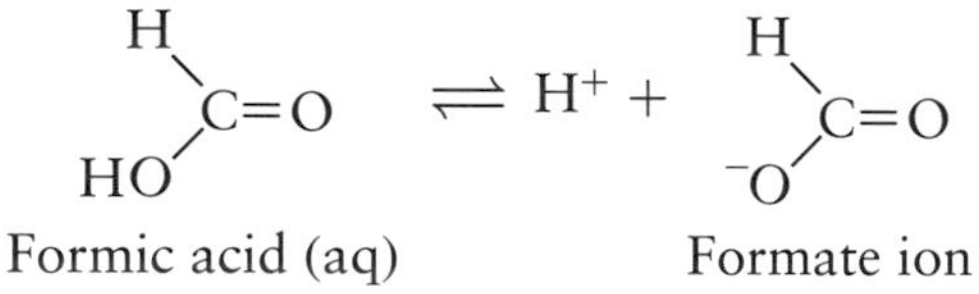

$$
\overset{\displaystyle H}{\underset{\displaystyle HO}{\diagdown}}C{=}O \rightleftharpoons H^+ + \overset{\displaystyle H}{\underset{\displaystyle {}^-O}{\diagdown}}C{=}O \quad (19.18)
$$

Formic acid (aq)                      Formate ion

Formic acid and formate are oxidized in solution by the reactions

$$
\overset{\displaystyle H}{\underset{\displaystyle HO}{\diagdown}}C{=}O + \overset{\centerdot}{O}H(aq) + O_2(aq) \longrightarrow CO_2(aq) + H\overset{\centerdot}{O}_2(aq) + H_2O(aq) \quad (19.19)
$$

Formic acid
(aq)

$$
\overset{\displaystyle H}{\underset{\displaystyle {}^-O}{\diagdown}}C{=}O + \overset{\centerdot}{O}H(aq) + O_2(aq) \longrightarrow CO_2(aq) + H\overset{\centerdot}{O}_2(aq) + OH^- \quad (19.20)
$$

Formate ion
(aq)

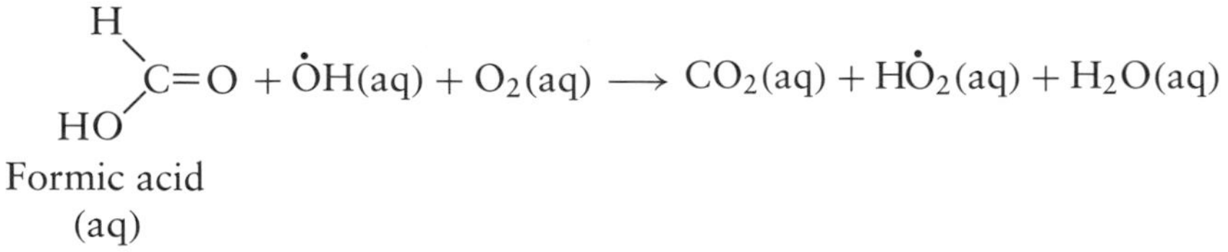

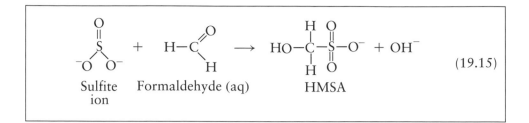

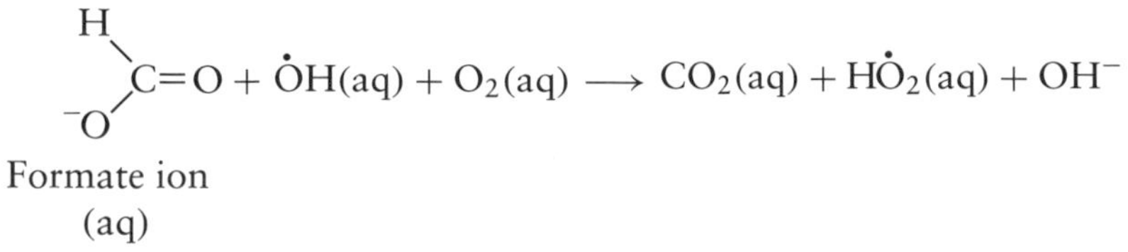

respectively. Because these reactions, coupled with (19.18), are relatively fast, they may consume formic acid before it evaporates.

Other soluble organics include the methylperoxy radical ($CH_3O_2$), methyl hydroperoxide ($CH_3OOH$), acetic acid ($CH_3COOH$), and peroxyacetic acid ($CH_3C(O)$-$OOH$). For these species, relevant aqueous reactions appear in Table B.8, and equilibrium reactions appear in Table B.7.

Other soluble inorganics include nitric acid ($HNO_3$), ammonia ($NH_3$), and hydrochloric acid (HCl). Under most conditions, $HNO_3(aq)$, $NH_3(aq)$, $NO_3^-$, and $NH_4^+$ concentrations are unaffected by aqueous reactions (Pandis and Seinfeld 1989). The chemistry of these species is governed by reversible equilibrium reactions, discussed in Chapter 18.

Chloride ($Cl^-$), the dissociation product of HCl(aq), reacts irreversibly in solution. It maintains equilibrium with the aqueous chlorine atom (Cl), the dichloride ion ($Cl_2^-$), and the chlorine hydroxide radical ($ClOH^-$) through the reactions

$$Cl_2^- \rightleftharpoons \dot{C}l(aq) + Cl^- \tag{19.21}$$

$$ClOH^- \rightleftharpoons Cl^- + \dot{O}H(aq) \tag{19.22}$$

(Jayson et al. 1973). Cl(aq) from (19.21) further reacts with hydrogen peroxide by

$$H_2O_2(aq) + \dot{C}l(aq) \longrightarrow Cl^- + H\dot{O}_2(aq) + H^+ \tag{19.23}$$

and $Cl_2$ reacts with S(IV) by

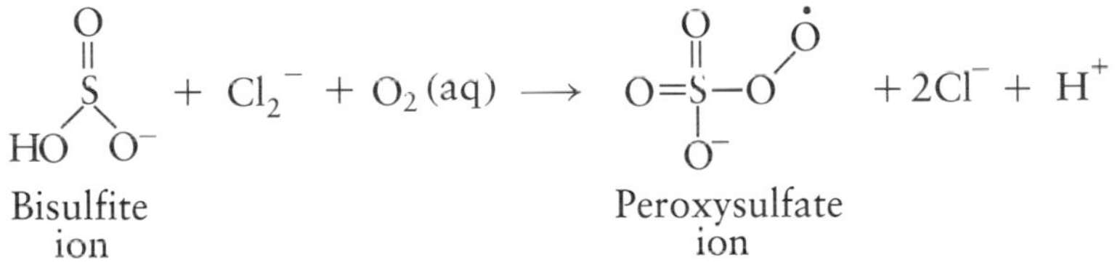


(19.24)

(19.24) is less important than other reactions that consume S(IV).

Table B.8 lists additional aqueous-phase chemical reactions in cloud drops and raindrops.

## 19.3. DIFFUSION WITHIN A DROP

In a cloud drop or raindrop, the concentration of a dissolved species varies as a function of drop radius. A typical diffusion coefficient for a solute diffusing through a liquid-water drop at 25°C is $D_{aq,q} = 2 \times 10^{-5}$ cm$^2$ s$^{-1}$ (e.g., Wilke and Chang 1955; Himmelblau 1964; Jacob 1986; Schwartz 1986). The corresponding characteristic time scale (s) **for aqueous-phase diffusion** is (Crank 1975)

$$\tau_{ad,q} = \frac{r_i^2}{\pi^2 D_{aq,q}} \tag{19.25}$$

**Example 19.1**

From (19.25), the characteristic time for diffusion within a 30-μm-diameter cloud drop is about 0.05 s. The characteristic time for diffusion within a 10-μm-diameter drop is about 0.001 s. Thus, for many purposes, diffusion can be assumed instantaneous.

Although the time scales for diffusion are small, chemical reactions often occur on even smaller time scales. Jacob (1986) found that $O_3(aq)$, $NO_3(aq)$, $OH(aq)$, $Cl(aq)$, $SO_4^-$, $CO_3^-$, and $Cl_2^-$ had chemical lifetimes shorter than diffusion transport times. These species react near the surface of a drop before they can diffuse to the center. When diffusion and chemical reaction are considered, the time rate of change of concentration of species $q$ in size bin $i$ as a function of radius $r$ is

$$\left(\frac{dc_{q,i,r}}{dt}\right)_{ad,aq} = D_{aq,q}\frac{1}{r^2}\frac{\partial}{\partial r}\left(r^2\frac{\partial c_{q,i,r}}{\partial r}\right) + P_{c,q,i,r} - L_{c,q,i,r} \qquad (19.26)$$

where $P_{c,q,i,r}$ and $L_{c,q,i,r}$ are the net aqueous chemical production and loss rates (moles cm$^3$ s$^{-1}$), respectively, of $q$ in size bin $i$ a distance $r$ from the center of the drop, and the subscript $ad, aq$ indicates that aqueous diffusion is coupled with aqueous chemistry. At the drop center, $\partial c_{q,i,r}/\partial r = 0$. At the center and the surface, $c_{q,i,r}$ has an initial value.

## 19.4. SOLVING GROWTH AND AQUEOUS CHEMICAL ODEs

Aqueous chemical reactions are similar to gas reactions in that both are described by first-order ODEs that can be solved with the methods discussed in Chapter 13. Aqueous-phase chemical ODEs are more difficult to solve than gas-phase chemical ODEs for two reasons. First, the lifetimes of aqueous species against chemical loss are usually shorter than those of gases. Thus, aqueous reactions are stiffer and more computationally demanding to solve than are gas reactions. Second, when dissolved molecules are destroyed by chemical reactions, they are rapidly replenished by dissolution from the gas phase to maintain saturation over the particle surface. When $H_2O_2(aq)$ reacts in solution, $H_2O_2(g)$ dissolves to replace lost $H_2O_2(aq)$ to satisfy (19.4). Thus, in a model that resolves aerosol, cloud, and raindrops over multiple size bins, aqueous ODEs must be solved within each bin, and growth ODEs must be solved between the gas phase and all bins. Gas chemistry is effectively solved in one size bin.

To make matters more complex, many dissolving molecules dissociate reversibly in solution. Thus, dissolution, equilibrium, and aqueous reaction processes must be solved nearly simultaneously. The rate of change of particle-phase concentration due to these processes is

$$\left(\frac{dc_{q,i,t}}{dt}\right)_{ge,eq,aq} = \left(\frac{dc_{q,i,t}}{dt}\right)_{ge} + \left(\frac{dc_{q,i,t}}{dt}\right)_{eq} + \left(\frac{dc_{q,i,t}}{dt}\right)_{aq} \qquad (19.27)$$

where $c_{q,i,t}$ is the mole concentration of species $q$ in bin $i$ at time $t$, and subscripts $ge$, $eq$, and $aq$ identify growth/evaporation, equilibrium, and aqueous chemistry, respectively. The aqueous-chemistry term in (19.27) is the same as that in (19.26), except that diffusion is assumed to be instantaneous and particles are assumed to be well mixed in (19.27). The mass conservation equation corresponding to (19.27) is

$$\frac{\mathrm{d}C_{q,t}}{\mathrm{d}t} = -\sum_{i=1}^{N_B}\left(\frac{\mathrm{d}c_{q,i,t}}{\mathrm{d}t}\right)_{ge} \tag{19.28}$$

where $C_{q,t}$ is the gas mole concentration of species $q$, and $N_B$ is the number of size bins. Together, (19.27) and (19.28) make up a large set of stiff ODEs.

An exact solution to (19.27) and (19.28) is difficult to obtain. The equilibrium term is not an ordinary differential equation. It is a reversible equation solved with an iterative method, such as the MFI scheme discussed in Chapter 18. A reversible equation of the form $\mathrm{D} + \mathrm{E} \rightleftharpoons \mathrm{A} + \mathrm{B}$, can be written as a combination of two ODEs, $\mathrm{D} + \mathrm{E} \rightarrow \mathrm{A} + \mathrm{B}$ and $\mathrm{A} + \mathrm{B} \rightarrow \mathrm{D} + \mathrm{E}$. These equations are very stiff and time-consuming to solve. Instead, $\mathrm{D} + \mathrm{E} \rightleftharpoons \mathrm{A} + \mathrm{B}$ is usually simulated as a reversible equation that is operator-split from the other terms in (19.27).

The growth and aqueous-chemistry terms in (19.27) have been solved together in several box and one-dimensional modeling studies (e.g., Schwartz 1984; Chameides 1984; Jacob 1986; Pandis and Seinfeld 1989; Bott and Carmichael 1993; Sander et al. 1995). In some cases, the equations have been solved with an integrator of stiff ODEs. In others, they have been solved with a forward Euler method taking a small time step. Here, a method of solving the equations in a three-dimensional model is discussed.

With this method, species involved in dissociation and hydration reactions are grouped in **families**. Since such reactions are rapid and reversible, family species cycle rapidly among each other. Changes in family concentrations are slower than changes in species concentrations. Here, the S(IV), S(VI), $HO_{2,T}$, $CO_{2,T}$, $HCHO_T$, $HCOOH_T$, and $CH_3COOH_T$ families are considered. Family concentrations (moles per cubic centimeter of air) in size bin $i$ are

$$c_{\mathrm{S(IV)},i} = c_{\mathrm{SO_2(aq)},i} + c_{\mathrm{HSO_3^-},i} + c_{\mathrm{SO_3^{2-}},i} \tag{19.29}$$

$$c_{\mathrm{S(VI)},i} = c_{\mathrm{H_2SO_4(aq)},i} + c_{\mathrm{HSO_4^-},i} + c_{\mathrm{SO_4^{2-}},i} \tag{19.30}$$

$$c_{\mathrm{HO}_{2,T},i} = c_{\mathrm{HO_2(aq)},i} + c_{\mathrm{O_2^-},i} \tag{19.31}$$

$$c_{\mathrm{CO}_{2,T},i} = c_{\mathrm{CO_2(aq)},i} + c_{\mathrm{HCO_3^-},i} + c_{\mathrm{CO_3^{2-}},i} \tag{19.32}$$

$$c_{\mathrm{HCHO}_T,i} = c_{\mathrm{HCHO(aq)},i} + c_{\mathrm{H_2C(OH)_2},i} \tag{19.33}$$

$$c_{\mathrm{HCOOH}_T,i} = c_{\mathrm{HCOOH(aq)},i} + c_{\mathrm{HCOO^-},i} \tag{19.34}$$

$$c_{\mathrm{CH_3COOH}_T,i} = c_{\mathrm{CH_3COOH(aq)},i} + c_{\mathrm{CH_3COO^-},i} \tag{19.35}$$

respectively. Some species, such as $H_2O_2$(aq) and $O_3$(aq), are self-contained in their own families. $H_2O_2$(aq) dissociates significantly only at a pH higher than observed in cloud drops, and $O_3$(aq) does not dissociate in solution. Growth–aqueous-chemical ODEs can be obtained for families by combining (18.91) with chemical production and loss terms. For the S(IV) family, the resulting equation for a size bin is

$$\frac{dc_{S(IV),i}}{dt} = k_{S(IV),i}\left(C_{SO_2(g)} - S'_{S(IV),i}\frac{c_{SO_2(aq),i}}{H'_{SO_2(aq),i}}\right) + P_{c,S(IV),i} - L_{c,S(IV),i} \quad (19.36)$$

where $k_{S(IV),i}$ is the coefficient for mass transfer of $SO_2$ to particles of size $i$ ($s^{-1}$) from (17.62), $C_{SO_2(g)}$ is the concentration of $SO_2(g)$ (moles $cm^{-3}$), $S'_{S(IV),i}$ is the saturation ratio from (17.45), $P_{c,S(IV),i}$ is the aqueous chemical production rate of S(IV) in size bin $i$ (moles $cm^{-3}$ $s^{-1}$), $L_{c,S(IV),i}$ is the aqueous chemical loss rate of S(IV) in bin $i$ (moles $cm^{-3}$ $s^{-1}$), and

$$H'_{SO_2(aq),i} = m_v c_w, \; R^* T H_{SO_2} \quad (19.37)$$

is the effective Henry's constant of $SO_2$. The symbols in (19.37) were defined in Section 18.13.1. The concentrations $c_{SO_2(aq),i}$ and $c_{S(IV),i}$ are related by

$$c_{SO_2(aq),i} = c_{S(IV),i}\frac{\mathbf{m}^2_{H^+,i}}{\mathbf{m}^2_{H^+,i} + \mathbf{m}_{H^+,i}K_{1,S(IV)} + K_{1,S(IV)}K_{2,S(IV)}} \quad (19.38)$$

where $\mathbf{m}$ is molality (mole $kg^{-1}$), $K_{1,S(IV)}$ is the dissociation constant for $SO_2 + H_2O \rightleftharpoons H^+ + HSO_3^-$, and $K_{2,S(IV)}$ is the dissociation constant for $HSO_3^- \rightleftharpoons H^+ + SO_3^{2-}$. (19.38) was derived by combining (19.29), (18.89), and

$$\frac{\mathbf{m}_{H^+,i}\mathbf{m}_{HSO_3^-,i}\gamma^2_{i,H^+/HSO_3^-}}{\mathbf{m}_{SO_2(aq),i}} = K_{1,S(IV)} \qquad \frac{\mathbf{m}_{H^+,i}\mathbf{m}_{SO_3^{2-},i}\gamma^2_{i,H^+/HSO_3^{2-}}}{\mathbf{m}_{HSO_3^-,i}\gamma^2_{i,H^+/HSO_3^-}} = K_{2,S(IV)} \quad (19.39)$$

The activity coefficients in (19.39) are assumed to be unity, an assumption valid for dilute solutions (e.g., cloud drops). Substituting (19.38) into (19.36) and adding time subscripts gives

$$\frac{dc_{S(IV),i,t}}{dt} = k_{S(IV)i,t-h}\left(C_{SO_2(g),t} - S'_{S(IV),i,t-h}\frac{c_{S(IV),i,t}}{H'_{S(IV),i,t-h}}\right) + P_{c,S(IV),i,t} - L_{c,S(IV),i,t} \quad (19.40)$$

where

$$H'_{S(IV),i,t-h} = m_v c_{w,i,t-h} R^* T H_{SO_2}\left(1 + \frac{K_{1,S(IV)}}{\mathbf{m}_{H^+,i,t-h}} + \frac{K_{1,S(IV)}K_{2,S(IV)}}{\mathbf{m}^2_{H^+,i,t-h}}\right)$$

The chemical production and loss rates of a species $q$ are determined from

$$P_{c,q,i,t} = \sum_{l=1}^{N_{prod,q}} R_{c,n_P(l,q),t} \qquad L_{c,q,i,t} = \sum_{l=1}^{N_{loss,q}} R_{c,n_L(l,q),t} \quad (19.41)$$

where $N_{prod,q}$ ($N_{loss,q}$) is the number of reactions in which species $q$ is produced (lost), $R_{c,n_P(l,q),t}$ ($R_{c,n_L(l,q),t}$) is the $l$th production (loss) rate of $q$, and $n_P(l, q)$ [$n_L(l, q)$] gives the reaction number corresponding to the $l$th production [loss] term of species $q$. The terms in (19.41) are similar to those in (13.45) and (13.47), respectively. S(IV) has no important aqueous chemical production sources but several loss sources. The chemical loss rate of S(IV) is determined, in part, as

$$L_{c,S(IV),i,t} = k_a c_{S(IV),i,t} c_{H_2O_2,i,t} c_{H^+,i,t} + k_b c_{S(IV),i,t} c_{HO_{2,T},i,t} + \cdots \qquad (19.42)$$

where the $k$'s are third- or second-order rate coefficients (cm$^6$ mole$^{-2}$ s$^{-1}$ or cm$^3$ mole$^{-1}$ s$-1$, respectively). The third-order coefficient is

$$k_a = k_{a,1}\alpha_{1,S(IV)} \qquad (19.43)$$

where $k_{a,1}$ is the rate coefficients for the reaction $HSO_3^- + H_2O_2(aq) + H^+$ given in Table B.8 of Appendix B (converted from M$^{-2}$ s$^{-1}$ to cm$^6$ mole$^{-2}$ s$^{-1}$), and

$$\alpha_{1,S(IV)} = \frac{\mathbf{m}_{H^+,i,t-h} K_{1,S(IV)}}{\mathbf{m}^2_{H^+,i,t-h} + \mathbf{m}_{H^+,i,t-h} K_{1,S(IV)} + K_{1,S(IV)} K_{2,S(IV)}} \qquad (19.44)$$

The second-order coefficient is

$$k_b = [k_{b,1}\alpha_{1,S(IV)} + k_{b,2}\alpha_{2,S(IV)}]\alpha_{0,HO_{2,T}} + [k_{b,3}\alpha_{1,S(IV)} + k_{b,4}\alpha_{2,S(IV)}]\alpha_{1,HO_{2,T}} \qquad (19.45)$$

where $k_{b,1}$, $k_{b,2}$, $k_{b,3}$, and $k_{b,4}$ are the rate coefficients for the reactions $HSO_3^-$ + $HO_2(aq)$, $SO_3^{2-} + HO_2^-(aq)$, $HSO_3^- + O_2^-$, and $SO_3^{2-} + O_2$, respectively, and

$$\alpha_{2,S(IV)} = \frac{K_{1,S(IV)} K_{2,S(IV)}}{\mathbf{m}^2_{H^+,i,t-h} + \mathbf{m}_{H^+,i,t-h} K_{1,S(IV)} + K_{1,S(IV)} K_{2,S(IV)}} \qquad (19.46)$$

$$\alpha_{0,HO_{2,T}} = \frac{\mathbf{m}_{H^+,i,t-h}}{\mathbf{m}_{H^+,i,t-h} + K_{1,HO_{2,T}}} \qquad \alpha_{1,HO_{2,T}} = \frac{K_{1,HO_{2,T}}}{\mathbf{m}_{H^+,i,t-h} + K_{1,HO_{2,T}}} \qquad (19.47)$$

$K_{1,HO_{2,T}}$ in (19.47) is the equilibrium coefficient for $HO_2(aq) \rightleftharpoons H^+ + O_2^-$.

The gas conservation equation corresponding to (19.40) is

$$\boxed{\frac{dC_{SO_2(g),t}}{dt} = -\sum_{i=1}^{N_B}\left\{K_{S(IV),i,t-h}\left[C_{SO_2(g),t} - S'_{S(IV),i,t-h}\frac{c_{S(IV),i,t}}{H'_{S(IV),i,t-h}}\right] + P_{c,S(IV),i,t} - L_{c,S(IV),i,t}\right\}} \qquad (19.48)$$

Together, (19.40) and (19.48) make up a set of ODEs whose size depends on the number of dissolving gases, particle size bins, and species taking part in chemical reactions. This set of equations can be solved effectively in a three-dimensional model with a sparse-matrix solver. Table 19.2 shows that such a solver reduced the

**Table 19.2.** Reduction in Array Space and Number of Matrix Operations Before and After the Use of Sparse-Matrix Techniques for ODEs Arising from 10 Gases Transferring to 16 Size Bins, and 11 Aqueous-Chemistry Reactions Occurring among 11 Species within Each Bin

| | Value | |
|---|---|---|
| **Quantity** | **Initial** | **After Sparse Reductions** |
| Order of matrix | 186 | 186 (100%) |
| No. of initial matrix positions filled | 34,596 | 1226 (3.5%) |
| No. of final matrix positions filled | 34,596 | 2164 (6.3%) |
| No. of operations in decomp. 1 | 2,127,685 | 6333 (0.3%) |
| No. of operations in decomp. 2 | 17,205 | 1005 (5.8%) |
| No. of operations in backsub. 1 | 17,205 | 1005 (5.8%) |
| No. of operations in backsub. 2 | 17,205 | 973 (5.6%) |

The last four rows show the number of operations in each of four loops of matrix decomposition and backsubstitution. The percentages are values in the second column relative to those in the first. The solver used was SMVGEAR II (Jacobson 1995, 1998a).

number of multiplications during one call to a matrix decomposition subroutine by over two million (99.7 percent) for a given set of species, reactions, and size bins.

## 19.5. SUMMARY

In this chapter, aqueous chemical reactions and their effects on aerosol, cloud drop, and raindrop composition were discussed. Aqueous reactions are important because they affect the rate of chemical conversion of S(IV) to S(VI) species in solution. S(IV) species include sulfur dioxide, bisulfite, and sulfite. S(VI) species include sulfuric acid, bisulfate, and sulfate. The rate of S(IV) conversion to S(VI) depends on liquid-water content, S(IV) oxidant concentrations, and pH. Conversion is faster in cloud drops than in aerosols because clouds have higher liquid-water contents than do aerosols. Numerical methods of solving aqueous-chemistry ODEs are the same as those for solving gas-chemistry ODEs. Aqueous ODEs are more difficult to solve because they are stiffer and more tightly coupled to other processes than are gas-phase ODEs.

## 19.6. PROBLEMS

**19.1.** What are the two primary sources of S(VI) in particles? How might changes in relative humidity affect the rate of growth of particles due to each source?

**19.2.** Why do aerosols convert S(IV) species to S(VI) species at a lower rate than do clouds?

**19.3.** Draw curves of relative concentration versus radius when the time scale for diffusion in a drop is (a) less than, (b) equal to, (c) greater than that for chemical reaction loss of an aqueous species. Assume the relative concentration at the drop center is unity.

**19.4.** Why are aqueous-phase reactions generally stiffer than gas-phase reactions?

**19.5.** Calculate the *e*-folding lifetimes of $HSO_3^-$ against loss by reaction with $O_3(aq)$, $H_2O_2(aq)$, and $OH(aq)$, respectively. Assume $\chi_{O_3(g)}$, $\chi_{H_2O_2(g)}$, and $\chi_{OH(g)}$ are 0.1, $8 \times 10^{-4}$, and $4.0 \times 10^{-8}$ ppmv, respectively. Estimate aqueous molalities of these species using Henry's constant from Table B.7. Assume that $T = 280$ K, $p_d = 950$ mb, the solution is dilute, and pH = 4.5.

## 19.7. COMPUTER PROGRAMMING PRACTICE

**19.6.** Write a computer script to read reactions and rate coefficients. Use the script to calculate rate coefficients for all reactions in Table B.8 when $T = 298$ K and $T = 273$ K.

# 20

# Sedimentation and Dry Deposition

SEDIMENTATION occurs when particles in the atmosphere sink due to the force of gravity. Gases also sediment, but their weights are so small that their sedimentation velocities are negligible. A typical gas molecule has a diameter of 0.5–1 nm (nanometer). Such diameters result in sedimentation (fall) velocities of 10–30 cm per year. Because these velocities are low, and because the slightest turbulence moves gas molecules much further in seconds than sedimentation moves them in years, gas sedimentation velocities are not important over short time periods. In fact, they are important only over time scales of thousands to hundreds of thousands of years. Small-particle sedimentation is also of limited significance in the atmosphere over short time periods. Sedimentation of large particles is of greater importance, especially when cloud drops and raindrops are considered.

**Dry deposition** removes gases and particles at air–surface interfaces. Dry deposition occurs when gases or particles contact a surface and stick to or react with the surface. Sedimentation is one mechanism by which particles contact surfaces. Others, which affect gases and particles, are molecular diffusion, turbulent diffusion, and advection.

## 20.1. SEDIMENTATION

The rate of sedimentation of a particle in air is determined by equating the downward force of gravity to the opposing force of drag created as the particle descends through the air, as illustrated in Fig. 20.1.

**Stokes flow** occurs when a particle falling through the air is larger than the mean free path of a gas molecule but small enough so that the inertial force it imparts is less than the viscous force exerted by air. During Stokes flow, the drag force of air on a sphere is governed by **Stokes' law**,

$$F_D = 6\pi r_i \eta_a V_{f,i} \tag{20.1}$$

where $r_i$ is the particle radius (cm), $\eta_a$ is the dynamic viscosity of air (g cm$^{-1}$ s$^{-1}$), and $V_{f,i}$ is the **sedimentation (fall) velocity** (cm s$^{-1}$). An expression for the dynamic viscosity of air in units of kg m$^{-1}$ s$^{-1}$ was given in (4.55).

For particles smaller than the mean free path of air, the motion through air is called **slip flow**. For such flow, Stokes' law can be modified with the Cunningham slip-flow correction and Knudsen–Weber term, defined in (16.25) as $G_i = 1 +$

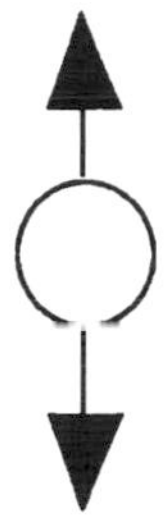


**Figure 20.1.** Vertical forces acting on a particle.

$\mathrm{Kn}_{a,i}[A' + B'\exp(-C'\mathrm{Kn}_{a,i}^{-1})]$. With this correction, the drag force becomes

$$F_D = \frac{6\pi r_i \eta_a V_{f,i}}{G_i} = \frac{6\pi r_i \eta_a V_{f,i}}{1 + \mathrm{Kn}_{a,i}\left[\mathrm{A}' + \mathrm{B}'\exp\left(-C'\mathrm{Kn}_{a,i}^{-1}\right)\right]} \tag{20.2}$$

The smaller the particle, the larger the Knudsen number for air, the larger $G_i$, and the smaller the force of drag on a falling sphere. For Stokes flow, $G_i$ approaches unity.

The downward force of gravity, net of buoyancy, acting on a spherical particle is

$$F_G = \frac{4}{3}\pi r_i^3(\rho_p - \rho_a)g \tag{20.3}$$

where $\rho_p$ is the density of the particle (g cm$^{-3}$), $\rho_a$ is the density of air (g cm$^{-3}$), and $g$ is the gravitational acceleration (cm s$^{-2}$).

Equating the drag and gravitational forces for a falling spherical particle and solving for the fall velocity gives

$$V_{f,i}^{est} = \frac{2r_i^2(\rho_p - \rho_a)g}{9\eta_a} G_i \tag{20.4}$$

In this equation, the superscript *est* indicates that the fall velocity is an estimated value, since this equation will be used shortly in an iterative procedure.

Without the Cunningham factor, (20.4) is the **Stokes terminal velocity**. The Stokes terminal velocity is accurate for Stokes flow only. The Cunningham correction factor permits the use of (20.4) for slip flow and Stokes flow. Both types of flow occur at low particle Reynolds number. The Reynolds number, first defined in (16.22), is written here as

$$\mathrm{Re}_i^{est} = 2r_i V_{f,i}^{est}/\nu_a \tag{20.5}$$

When $\mathrm{Re}_i^{est} < 1$, Stokes or slip flow occurs. Larger Reynolds numbers correspond to larger particle radii. At such radii, the Stokes terminal velocity is not accurate, and the drag force needs to be corrected again. Beard (1976) studied the fall velocities of large liquid water drops in air and found that the correction factor for such particles can be parameterized as a function of the physical properties of

**Table 20.1.** Coefficients for Polynomial Fits Used in (20.6)

| | |
|---|---|
| $B_0 = -3.18657$ | $E_0 = -5.00015$ |
| $B_1 = 0.992696$ | $E_1 = 5.23778$ |
| $B_2 = -0.00153193$ | $E_2 = -2.04914$ |
| $B_3 = -0.000987059$ | $E_3 = 0.475294$ |
| $B_4 = -0.000578878$ | $E_4 = -0.0542819$ |
| $B_5 = 0.0000855176$ | $E_5 = 0.00238449$ |
| $B_6 = -0.00000327815$ | |

*Source:* Beard (1976).

the drop and its surroundings. Such properties include temperature, pressure, air density, air viscosity, particle density, surface tension, and gravity.

Beard's parameterization is iterative. The first step is to estimate the fall velocity and Reynolds number with (20.4) and (20.5), respectively. The second step is to recalculate the Reynolds number for one of three flow regimes: **slip flow** around a rigid sphere at low Reynolds numbers, **continuum flow** past a rigid sphere at moderate Reynolds numbers, and **continuum flow** around a nonspherical, equilibrium-shaped drop at large Reynolds numbers. In sum,

$$\mathrm{Re}_i^{final} = \begin{cases} 2r_i V_{f,i}^{est}/\nu_a & \mathrm{Re}_i^{est} < 0.01 \\ G_i \exp(B_0 + B_1 X + B_2 X^2 + \cdots) & 0.01 < \mathrm{Re}_i^{est} < 300 \\ N_P^{1/6} G_i \exp(E_0 + E_1 Y + E_2 Y^2 + \cdots) & \mathrm{Re}_i^{est} > 300 \end{cases} \quad (20.6)$$

where

$$X = \ln\left[\frac{32 r_i^3(\rho_p - \rho_a)\rho_a g}{3\eta_a^2}\right] \qquad Y = \ln\left[\tfrac{4}{3}\mathrm{N}_{Bo}\mathrm{N}_P^{1/6}\right] \quad (20.7)$$

are functions of the physical properties of particles and the atmosphere. These equations contain the dimensionless **physical-property number** and **Bond number,**

$$N_P = \frac{\sigma_{w/a}^3 \rho_a^2}{\eta_a^4(\rho_p - \rho_a)g} \qquad N_{Bo} = \frac{4r_i^2(\rho_p - \rho_a)g}{\sigma_{w/a}} \quad (20.8)$$

respectively, where $\sigma_{w/a}$ is the surface tension of water against air, defined in (15.2). The Bond number measures the relative strength of the drag and surface tension forces for a drop at terminal velocity (Pruppacher and Klett 1997). The $B$ and $E$ coefficients in (20.6) are given in Table 20.1. After the final Reynolds number is found, a **final fall velocity** is calculated as

$$V_{f,i} = V_{f,i}^{final} = Re_i^{final}\nu_a/2r_i \quad (20.9)$$

Low Reynolds numbers correspond to small particle diameters (<20 μm), moderate Reynolds numbers correspond to diameters ranging from those of large cloud drops (20 μm) to those of small raindrops (1 mm), and large Reynolds numbers

Table 20.2. Time for Particles to Fall 1 km in the Atmosphere by Sedimentation

| Particle Diameter (μm) | Time to Fall 1 km | Particle Diameter (μm) | Time to Fall 1 km |
|---|---|---|---|
| 0.02 | 228 yr | 5.0 | 14.5 days |
| 0.1 | 36 yr | 10.0 | 3.6 days |
| 0.5 | 3.2 yr | 20.0 | 23 h |
| 1.0 | 328 days | 100.0 | 1.1 h |
| 2.0 | 89 days | 1000.0 | 4 min |
| 4.0 | 23 days | 5000.0 | 1.8 min |

Data obtained from the full velocities in Fig. 20.3.

correspond to raindrop diameters (1–5 mm). In the low-Reynolds-number regime, (20.9) simplifies to the Stokes terminal velocity with the Cunningham correction factor. Liquid water drops smaller than 1 mm in diameter are essentially spherical, and neither surface tension nor internal viscosity affects their shape or sedimentation velocity significantly. Large raindrops are affected by turbulence and vary in shape from oblate to prolate spheroids.

Beard's parameterization is valid for liquid water drops falling in air. Fall-velocity parameterizations for columnar crystals, platelike crystals, oblate spheroids, hexagonal crystals, hailstones, and graupel are given in Pruppacher and Klett (1997). (20.9) is reasonable for nonspherical particles so long as sedimentation is less important than other model processes during a simulation period. The time for a particle to fall a distance $\Delta z$ due to sedimentation is $\tau_{f,i} = \Delta z / V_{f,i}$. Most aerosols of atmospheric importance are 0.1–1-μm in diameter. Table 20.2 indicates that these particle require 36 yr-328 days, respectively, to drop 1 km. If a model simulates 0.1–1-μm nonspherical particles for a period of months to a few years, the use of (20.9) should not induce much error, since other processes affect particles over this time period much more than does the assumption of sphericity in the sedimentation velocity equation.

Emitted volcanic particles range in diameter from <0.1 to >100 μm. Particles 100 μm in diameter take 1.1 h to fall 1 km. The only volcanic particles that survive more than a few months before falling out are those smaller than 4.0 μm. Such particles require no less than 23 d to fall 1 km. These particles are often spherical and contain liquid water, dissolved sulfuric acid, and dissolved hydrochloric acid. Large volcanic particles are often nonspherical but fall from the atmosphere quickly.

Fall velocities are useful for calculating not only downward fluxes of particles but also coagulation kernels (Section 16.4) and dry deposition velocities (discussed next).

## 20.2. DRY DEPOSITION

**Dry deposition** occurs when gases or particles are removed at an air–surface interface (e.g., Sehmel 1980). Gases or particles can deposit onto trees, buildings, grass, the ocean surface, car windows, or any other surface. A gas is removed from the air

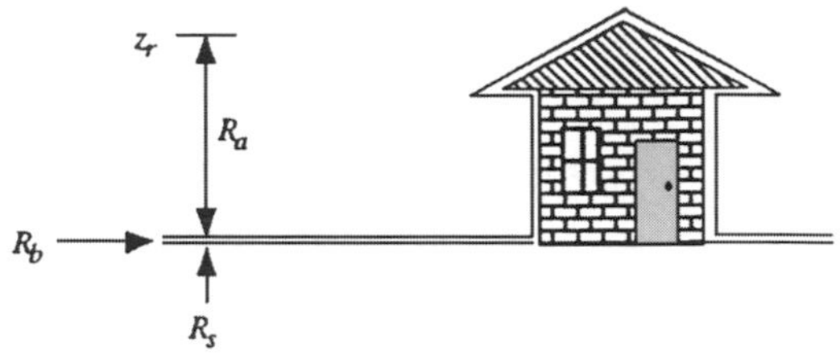


**Figure 20.2.** Locations where gas dry-deposition resistances apply. The figure is not to scale.

when it impacts and sticks to or reacts with a surface. Dry-deposition velocities are usually calculated as the inverse sum of a series of resistances (Wesely and Hicks 1977; Slinn et al. 1978; McRae et al. 1982; Walcek et al. 1986; Russell et al. 1993). The **dry-deposition velocity of a gas** (m $s^{-1}$) is

$$V_{d,gas} = (R_a + R_b + R_s)^{-1} \tag{20.10}$$

where $R_a$ is the aerodynamic resistance between a reference height (about 10 m above the surface) and the laminar sublayer just above the surface, $R_b$ is the resistance to molecular diffusion through a 0.1 to 0.01-cm-thick laminar sublayer, and $R_s$ is the resistance to chemical, biological, and physical interactions between the surface and the gas once the gas has collided with the surface. Resistances have units of time per unit distance (e.g., s $m^{-1}$). Figure 20.2 depicts the region near the surface where the three resistances listed apply.

The aerodynamic resistance is a property of turbulent transfer between a reference height and the laminar sublayer; it depends on atmospheric stability. The resistance to molecular diffusion depends on the gas molecular diffusion coefficient, air viscosity, air conductivity, air temperature, and air density. The resistance to surface interactions depends on physical and chemical properties of the surface and the depositing gas.

Dry-deposition velocities of particles differ from those of gases in two ways. First, particles are heavier than gases; thus, particle sedimentation velocities must be included in their dry-deposition equation. Second, because of their weight, particles tend to stay on surfaces once they deposit; thus, the surface resistance term may be ignored. The **dry-deposition velocity of a particle** (m $s^{-1}$) of size $i$ is

$$V_{d,part,i} = (R_a + R_b + R_a R_b V_{f,i})^{-1} + V_{f,i} \tag{20.11}$$

where the fall velocity is in meters per second.

For gas and particle dry deposition, the **aerodynamic resistance** (s $m^{-1}$) can be approximated from similarity theory as

$$R_a = \frac{\int_{z_{0,q}}^{z_r} \phi_h \frac{dz}{z}}{k u_*} \tag{20.12}$$

where $k$ is the unitless von Kármán constant (0.40), $u_*$ is the friction velocity (m $s^{-1}$), $z$ is the height above the surface (m), $z_r$ is a reference height (10 m), $z_{0,q}$ is the surface roughness length of gas or particle $q$ (m), and $\phi_h$ is the dimensionless potential temperature gradient. A dimensionless potential temperature gradient is used instead of a dimensionless wind shear because the transport of trace gases resembles the transport of energy more than it does that of momentum. The friction velocity, defined in Chapter 8, is a measure of the vertical turbulent flux of horizontal momentum. A large value of $u_*$ corresponds to increased turbulence, decreased resistance of gases and particles to vertical turbulent motions, and increased deposition velocities. The surface roughness length of species $q$ ($z_{0,q}$) can be found from (8.9) by substituting the diffusion coefficient of the gas or particle of interest for that of water vapor.

Integrals of $\phi_h$ were given in (8.26) as a function of the Monin–Obukhov length $L$, which relies on $u_*$ and the turbulent potential temperature scale $\theta_*$. A noniterative method of determining $u_*$, $L$, and $\theta_*$ was discussed in Section 8.6.4. During the day, winds are stronger, the atmosphere is more unstable, $u_*$ is larger, the integral of $\phi_h$ is smaller, and $R_a$ is smaller than during the night. Because resistance to turbulent transfer is smaller during the day than night, dry-deposition velocities are generally larger during the day than the night.

The reciprocal of an aerodynamic resistance is a **transfer velocity**, which gives the rate at which a gradient in concentration of a species is transferred between $z_{0,q}$ and $z_r$. (8.31) gave the vertical turbulent flux of moisture (kg $m^{-2}$ $s^{-1}$) from Monin–Obukhov similarity theory as $E_f = -\rho_a q_* u_*$. Substituting (8.33) and (20.12) into (8.31), and assuming species $q$ is water vapor, we obtain another expression for the **turbulent moisture flux**, in terms of aerodynamic resistance, as

$$E_f = -\frac{\rho_a k u_* [\bar{q}_v(z_r) - \bar{q}_v(z_{0,v})]}{\int_{z_{0,v}}^{z_r} \phi_h \frac{dz}{z}} = -\frac{\rho_a}{R_a}[\bar{q}_v(z_r) - \bar{q}_v(z_{0,v})] \tag{20.13}$$

where $1/R_a$ (m $s^{-1}$) is the transfer velocity. The transfer velocity is larger under unstable than under stable conditions.

The **resistance to molecular diffusion** in the laminar sublayer (s $m^{-1}$) is

$$R_b = \ln\left(\frac{z_{0,m}}{z_{0,q}}\right)\frac{(\mathrm{Sc}/\mathrm{Pr})^{2/3}}{k u_*} \tag{20.14}$$

where $z_{0,m}$ is the surface roughness length for momentum, defined in Chapter 8. For vegetated surfaces, $z_{0,m}/z_{0,q}$ is approximately 100. For snow, ice, water, and bare soil, $z_{0,m}/z_{0,q}$ ranges from <1 to 3. For rough surfaces, $z_{0,m}/z_{0,q}$ can be as large as $10^5$ (Garratt and Hicks 1973; Brutsaert 1991; Ganzeveld and Lelieveld 1995). The Schmidt number for particles is $\mathrm{Sc}_{p,i} = \nu_a/D_{p,i}$ and for gases is $\mathrm{Sc}_q = \nu_a/D_q$. The Prandtl number is $\mathrm{Pr} = \eta_a c_{p,m}/\kappa_d$. These three terms were defined in (16.31), (17.25), and (17.32), respectively.

The **resistance of gases to surface interactions** is difficult to quantify because many chemical and biological processes occur on surfaces and in plant canopies. The surface-resistance parameterization discussed here was developed by Wesely

(1989), with modifications by Walmsley and Wesely (1996). In this model, the total surface resistance (s m$^{-1}$) was approximated as a function of individual resistances by

$$R_s = \left( \frac{1}{R_{stom} + R_{meso}} + \frac{1}{R_{cut}} + \frac{1}{R_{conv} + R_{exp}} + \frac{1}{R_{canp} + R_{soil}} \right) \quad (20.15)$$

where $R_{stom}$ is the leaf stomata resistance, $R_{meso}$ is the leaf mesophyll resistance, $R_{cut}$ is the leaf cuticle resistance, $R_{conv}$ is the resistance due to buoyant convection in canopies, $R_{exp}$ is the resistance due to leaves, twigs, bark, or other exposed surfaces in the lower canopy, $R_{canp}$ is the resistance due to canopy height and density, and $R_{soil}$ is the resistance due to soil and leaf litter at the ground surface.

The stomata, mesophyll, and cuticle resistances account for removal of gases by deposition into openings on leaf surfaces, into liquid water within leaves, and onto outer leaf surfaces, respectively. A plant **cuticle** is a waxy outer covering of a plant surface, formed from **cutin**. The cuticle is relatively impermeable to water vapor, preventing significant water loss from the leaf. Water vapor, carbon dioxide, oxygen, and other gases can pass through the cuticle into or out of plant leaves through **stomata**. Stomata are openings, scattered throughout the cuticle, that extend into the leaf's **epidermis**, or underlying skin. Stomata are important for allowing carbon dioxide to enter the middle leaf, called the **mesophyll**. The mesophyll contains empty spaces and chloroplasts. **Chloroplasts** are cells in the leaf in which photosynthesis takes place. During photosynthesis, carbon dioxide and water are converted to molecular oxygen and organic material. The oxygen escapes the mesophyll to the atmosphere through the stomata. Liquid water, brought through **xylem** from the plant's roots to the mesophyll, evaporates from the surfaces of chloroplasts into the open space of the mesophyll. This vapor diffuses, often through the stomata, to the open atmosphere. Gases from the atmosphere can also diffuse through the stomata to the mesophyll, where they may deposit onto chloroplast surfaces.

The **leaf stomata resistance** (s m$^{-1}$) in (20.15) refers to the loss of a gas due to deposition within stomata. The stomata resistance depends on the specific plant species, but a parameterization that represents an average resistance for many species is (Baldocchi et al. 1987; Wesely 1989)

$$R_{stom,q} = R_{min} \left[ 1 + \left( \frac{200}{S_f + 0.1} \right)^2 \right] \frac{400}{T_{a,c}(40 - T_{a,c})} \frac{D_v}{D_q} \quad (20.16)$$

where $R_{min}$ is the minimum bulk canopy stomata resistance for water vapor, $S_f$ is the surface solar irradiance (W m$^{-2}$), $T_{a,c}$ is the surface air temperature (°C), and $D_v/D_q$ is the ratio of the molecular diffusion coefficient of water vapor to that of gas $q$. Minimal resistances for water vapor are given in Table B.11 for different land-use types and seasons. Diffusion-coefficient ratios for different gases are given in Table B.11. In (20.16), $T_{a,c}$ ranges from 0 to 40°C. Outside this range, $R_{stom,q}$ is assumed to be infinite. During the night, stomata close, and the resistance to

**Table 20.3.** Cuticle Resistances $R_{cut}$ (s $m^{-1}$) for $SO_2$, $O_3$, and Any Gas $q$

| | $R_{cut,i,SO_2}$ | | | |
|---|---|---|---|---|
| $i$ | Nonurban | Urban | $R_{cut,i,O_3}$ | $R_{cut,i,q}$ |
| dew | 100 | 50 | $\left(\frac{1}{3000}+\frac{1}{3R_{cut,0}}\right)^{-1}$ | $\left(\frac{1}{3R_{cut,dry,q}}+10^{-7}H_q^*+\frac{f_{0,q}}{R_{cut,dew,O_3}}\right)^{-1}$ |
| rain | $\left(\frac{1}{5000}+\frac{1}{3R_{cut,0}}\right)^{-1}$ | 50 | $\left(\frac{1}{1000}+\frac{1}{3R_{cut,0}}\right)^{-1}$ | $\left(\frac{1}{3R_{cut,dry,q}}+10^{-7}H_q^*+\frac{f_{0,q}}{R_{cut,rn,O_3}}\right)^{-1}$ |

"Urban" refers to areas with little vegetation, such as urban areas. Source: Wesely (1989), Walmsley and Wesely (1996).

deposition increases to infinity. (20.16) simulates this effect, predicting that $R_{stom,q}$ approaches $10^6$ when $S_f = 0$.

Leaf mesophyll resistance depends on the ability of a gas to dissolve and react in liquid water on the surface of plant cells within the mesophyll of a leaf. The ability of a gas to dissolve depends on its effective Henry's constant $H_q^*$ and on a reactivity factor for gases, $f_{0,q}$. The **effective Henry's constant** is the ratio of a dissolved and dissociated substance's molarity to its gas-phase partial pressure. The reactivity factor accounts for the ability of the dissolved gas to oxidize biological substances once in solution. A value of $f_{0,q} = 1$ indicates the dissolved gas is highly reactive, and $f_{0,q} = 0$ indicates it is nonreactive. The **leaf mesophyll resistance** is estimated as

$$R_{meso,q} = \left(\frac{H_q^*}{3000} + 100\, f_{0,q}\right)^{-1} \qquad (20.17)$$

where values of $H_q^*$ and $f_{0,q}$ are given in Table B.12 for several gases.

The **leaf cuticle resistance** accounts for the deposition of gases to leaf cuticles when vegetation is healthy, and to outside surfaces in the canopy when vegetation is not. This resistance depends on whether the cuticle is moist or dry. The leaf cuticle resistance over a sufficiently dry surface (s $m^{-1}$) is approximated as

$$R_{cut} = R_{cut,dry,q} = R_{cut,0}\left(10^{-5}H_q^* + f_{0,q}\right)^{-1} \qquad (20.18)$$

where $R_{cut,0}$ is given in Table B.11 for several seasonal and land-use categories. When dew or rain is present, the leaf cuticle resistances for sulfur dioxide, ozone, and other gases are modified from (20.18) to expressions shown in Table 20.3.

The **resistance of gases to buoyant convection** (s $m^{-1}$) is estimated for all gases as

$$R_{conv} = 100\left(1 + \frac{1000}{S_f + 10}\right)\frac{1}{1 + 1000 s_t} \qquad (20.19)$$

where $s_t$ is the slope of the local terrain, in radians. The **resistance due to deposition** on leaves, twigs, bark, and other exposed surface (s $m^{-1}$) is estimated for any gas $q$ as

$$R_{exp,q} = \left(\frac{10^{-5}H_q^*}{R_{exp,SO_2}} + \frac{f_{0,q}}{R_{exp,O_3}}\right)^{-1} \qquad (20.20)$$

where $R_{exp,SO_2}$ and $R_{exp,O_3}$ are given in Table B.11.

The **in-canopy resistance** (s m$^{-1}$) is a function of canopy height and one-sided leaf area index. The **one-sided leaf area index** (LAI) is calculated by integrating the foliage area density from the surface to a canopy height $h_c$. The **foliage area density** is the area of plant surface per unit volume of air. Thus, the LAI measures a canopy area density. Table B.11 contains estimated values of in-canopy resistance from Wesely (1989). Another parameterization of this resistance is

$$R_{canp} = \frac{b_c h_c \text{LAI}}{u_*} \tag{20.21}$$

where $b_c = 14$ m$^{-1}$, obtained from data (Erisman et al. 1994).

The **resistance due to soil and leaf litter at the ground surface** (s m$^{-1}$) is estimated as

$$R_{soil,q} = \left( \frac{10^{-5} H_q^*}{R_{soil,\mathrm{SO_2}}} + \frac{f_{0,q}}{R_{soil,\mathrm{O_3}}} \right)^{-1} \tag{20.22}$$

where $R_{soil,\mathrm{SO_2}}$ and $R_{soil,\mathrm{O_3}}$ are also given in Table B.11.

## 20.3. DRY-DEPOSITION AND SEDIMENTATION CALCULATIONS

Figure 20.3 shows components of the particle dry-deposition velocity versus diameter. For small particles, the nonsedimentation component of the dry-deposition velocity exceeds the sedimentation component, because particle diffusion affects small particles more than does gravity. For large particles, sedimentation dominates, because the effect of gravity exceeds that of particle diffusion. The total dry-deposition velocity is smallest for particles of about 1 μm in diameter; neither diffusion nor gravity dominates for particles near that size. The sedimentation velocity levels off at large diameters due to the physical-property correction term in the sedimentation equation. At small particle diameters, the slope of the sedimentation velocity also decreases due to the correction for particle resistance to motion.

**Example 20.1**

The sedimentation velocities shown in Fig. 20.3 can be used to estimate the time required for particles to fall a given distance in the atmosphere. The figure shows that a 0.1-μm-diameter particle with a sedimentation velocity of 0.0001 cm s$^{-1}$ takes 31.7 yr to fall 1 km in the atmosphere. A 1-mm-diameter raindrop with a sedimentation velocity of 500 cm s$^{-1}$ takes only 3.3 min to fall 1 km. Dry deposition affects particles near surfaces only.

Figure 20.4 shows the calculated particle diffusion coefficient, Knudsen number for air, Reynolds number, and particle total dry-deposition velocity as a function

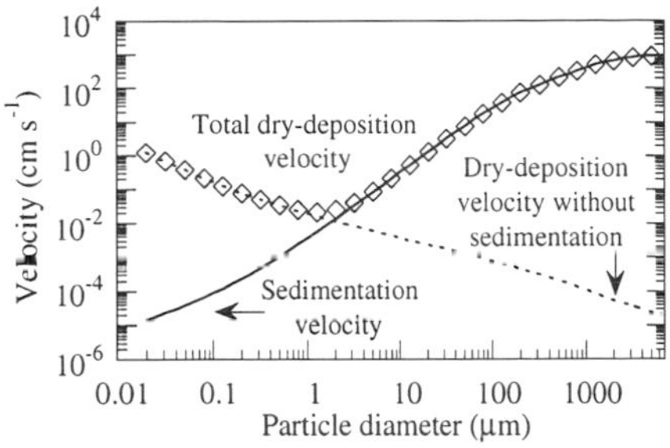


**Figure 20.3.** Particle total dry-deposition, velocity and its components versus particle diameter. Conditions were $z_r = 10$ m, $z_{0,m} = 0.0001$ m, $T(z_r) = 290.1$ K, $T(z_{0,m}) = 290$ K, $p_a(z_r) = 999$ mb, $|\mathbf{v}_h(z_r)| = 10$ m s$^{-1}$, and $\rho_p = 1.0$ g cm$^{-3}$. The near-ground temperature was taken at $z_{0,m}$ instead of $z_{0,h}$. Surface resistance was ignored.

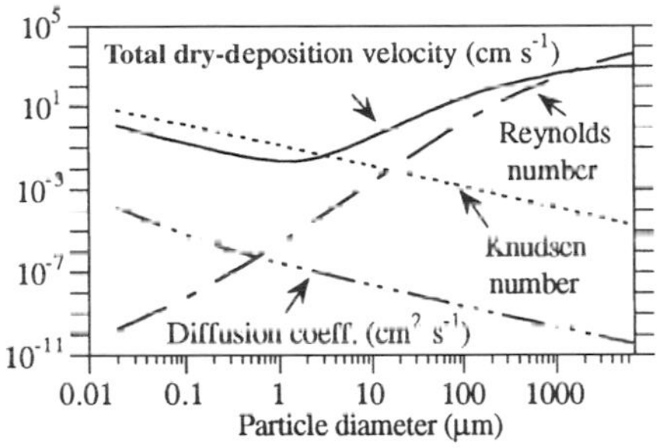


**Figure 20.4.** Particle total dry-deposition velocity from (20.11), Reynolds number, Knudsen number for air, and particle diffusion coefficient versus particle diameter. Conditions were $z_r = 10$ m, $z_{0,m} = 0.0001$ m, $T(z_r) = 292$ K, $T(z_{0,m}) = 290$ K, $p_a(z_r) = 999$ mb, $|\mathbf{v}_h(z_r)| = 10$ m s$^{-1}$, and $\rho_p = 1.0$ g cm$^{-3}$. The near-ground temperature was taken at $z_{0,m}$ instead of $z_{0,h}$.

of size, for the conditions described in the figure. The figure shows that the diffusion coefficient decreased with increasing particle size, as expected, since the diffusion coefficient is inversely proportional to the radius of the particle, as shown in (16.24). The Knudsen number also decreased with increasing size, since it is inversely proportional to the radius, as shown in (16.19). The Reynolds number increased with size because it is proportional to the radius and fall velocity, both of which increased with size. The dry-deposition velocity decreased then increased with size, as in Fig. 20.3.

Figure 20.5(a) and (b) show gas deposition velocities versus surface resistance, molecular weight, wind speed, and surface roughness length for momentum in

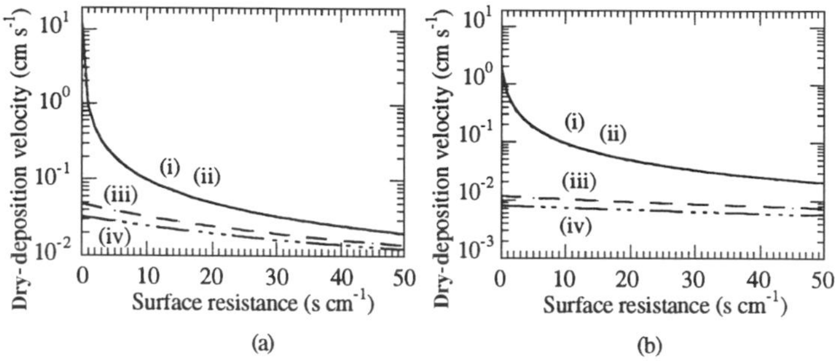


**Figure 20.5.** Gas dry-deposition velocities versus surface resistance when (a) $z_{0,m} = 3$ m and (b) $z_{0,m} = 0.01$ m, and when the molecular weight and wind speed were (i) 10 g mole$^{-1}$, 10 m s$^{-1}$; (ii) 130 g mole$^{-1}$, 10 m s$^{-1}$; (iii) 10 g mole$^{-1}$, 0 m s$^{-1}$; and (iv) 130 g mole$^{-1}$, 0 m s$^{-1}$. Also, $z_r = 10$ m, $T(z_r) = 288$ K, $T(z_{0,m}) = 290$ K, and $p_a(z_r) = 999$ mb.

an unstable atmosphere. The curves show that surface resistance affected dry-deposition velocities most when the surface roughness length was large and the wind speed was high. Molecular weight had relatively little effect on dry-deposition velocities, except when the wind speed was zero. Surface roughness length affected dry-deposition velocities primarily when the surface resistance was low. Wind speed affected dry-deposition velocities at both roughness lengths tested.

## 20.4. SUMMARY

Particle sedimentation velocities and gas and particle dry-deposition velocities were discussed in this chapter. Sedimentation is affected by gravity and air viscosity. At very small and very large particle diameters, sedimentation velocities are modified by particle resistance to motion and physical-property effects, respectively. Sedimentation velocities decrease with decreasing particle size, but dry-deposition velocities increase with decreasing size due to enhanced molecular diffusion at small sizes. Particle dry-deposition velocities are influenced by meteorological variables and diffusion at small diameters and by sedimentation at large diameters. Gas dry-deposition velocities are influenced by meteorological variables and surface interactions. Such interactions can be parameterized as a function of leaf-stomata, leaf-mesophyll, leaf-cuticle, buoyant-convective, exposed-surface, canopy, and soil resistances.

## 20.5. PROBLEMS

**20.1.** Briefly describe the three major resistances affecting gas dry-deposition velocities. What condition(s) give(s) maximum values for each resistance?

**20.2.** How do increases in surface heating and instability, respectively, affect individual components of surface resistance?

## 20.6. COMPUTER PROGRAMMING PRACTICE

**20.3.** Terrain significantly affects particle dry-deposition velocities. Write a computer script to calculate dry-deposition velocities of spherical particles 0.01 μm in diameter when $z_{0,m} = 0.0001$ m (ocean) and when $z_{0,m} = 5$ m (urban area). At $z_{0,m} = 0.0001$ m, assume $u_* = 0.48$ m s$^{-1}$ and $L = +180$ m. When $z_{0,m} = 5$ m, assume $u_* = 0.8$ m s$^{-1}$ and $L = +3000$ m. Also, $p_d =$ 1000 mb and $T = 292$ K at $z_r = 10$ m, and $\rho_p = 1$ g cm$^{-3}$.

**20.4.** Submicron atmospheric particles require significant time to fall out of the atmosphere. Write a computer script to calculate how long a 0.1-μm- and a 10-μm-diameter liquid water particle take to sediment from an altitude of 2 km to 1 km. Assume $T = 280$ K and $p_d = 850$ mb are constant. Assume drops are spherical and $\rho_p = 1$ g cm$^{-3}$.

**20.5.** Write a computer script to calculate the time required for molecules of ozone and of carbon monoxide to dry-deposit to the surface from a reference height of $z_r = 10$ m, assuming $R_s = 500$ s m$^{-1}$ for $O_3$ and $R_s = 3000$ s m$^{-1}$ for CO. Assume $p_d = 1000$ mb and $T = 290$ K at the ground surface. Also assume that, at $z_r = 10$ m, $p_d = 999$ mb, $T = 288$ K, $u = 7$ m s$^{-1}$, and $v = 7$ m s$^{-1}$. Let $z_{0,m} = 1$ m. Use the temperature, pressure, etc. at 10 m to calculated other atmospheric variables.

**20.6.** Write a computer script to compare the dry-deposition velocity of a 10-μm-diameter particle with that of a 0.1 μm diameter particle to determine which is influenced more by surface resistance. Assume $V_f = 0.0031$ m s$^{-1}$ for the 10-μm particle and $V_f = 8.6 \times 10^{-7}$ m s$^{-1}$ for the 0.1-μm particle. Assume the other conditions are the same as in Problem 20.5.

**20.7.** Meteorological conditions significantly affect gas dry-deposition velocities when the resistance to surface interactions ($R_s$) is small. Write a computer script to calculate the difference in the dry deposition velocity of water vapor when $R_s = 100.0$ s m$^{-1}$ in the following two cases: (a) when $u = 15$ m s$^{-1}$, $v = 15$ m s$^{-1}$, $p_d = 999$ mb, and $T = 292$ K at $z_r = 10$ m, and (b) when $u = 2$ m s$^{-1}$, $v = 2$ m s$^{-1}$, $p_d = 999$ mb, and $T = 288$ K at $z_r = 10$ m. In both cases, assume $p_d = 1000$ mb and $T = 290$ K at the surface and $z_{0,m} = 0.1$ m.

**20.8.** Write a computer script to calculate surface resistance with equations in Section 20.2. (a) Calculate the total surface resistance for $SO_2$ during summer over a coniferous forest when $S_f = 600$ W m$^{-2}$, $T_{a,c} = 15$°C, and $s_t = 0$. (b) Assume $S_f$ varies as a sine function between 6 a.m. and 6 p.m., peaking with $S_f = 800$ W m$^{-2}$ at noon and equaling zero at night. Plot surface resistance versus time for a 24-h period, assuming the same remaining conditions in part (a). (c) Repeat part (b) for $O_3$. Comment on the diurnal variations of surface resistances for both gases.

# 21

# Model Design, Application, and Testing

In previous chapters, physical processes in the atmosphere and numerical methods to simulate those processes were described. Here, steps in model design, application, and testing are discussed. The most important of these steps are to define the goals of a modeling study, select appropriate algorithms, obtain sufficient input and emissions data, compare model predictions with data, and analyze results. To illustrate these steps, the design of an existing model is briefly discussed, and statistical and graphical comparisons of predictions from the model to data are shown. This chapter integrates numerical methods discussed in previous chapters with model development and analysis procedures.

## 21.1. STEPS IN MODEL FORMULATION

Model design, application, and testing require several steps. These include (1) defining and understanding the problem of interest, (2) determining the spatial and temporal scale of the problem, (3) determining the dimension of the model, (4) selecting the physical, chemical, and/or dynamical processes to simulate, (5) selecting variables, (6) selecting a computer architecture, (7) codifying and implementing algorithms, (8) optimizing the model on a computer architecture, (9) selecting time steps and intervals (10) setting initial conditions, (11) setting boundary conditions, (12) obtaining input data, (13) obtaining ambient data for comparison, (14) interpolating input data and model predictions, (15) developing statistical and graphical techniques, (16) comparing results with data, (17) running sensitivity tests and analyzing the results, and (18) improving algorithms. Each of these steps is discussed below.

### 21.1.1. Defining the Purpose of the Model

The first step in model development is to define the scientific, regulatory, or computational problem of interest. Example scientific topics include determining the effects of (1) aerosols on global climate, (2) aerosols on stratospheric ozone reduction, (3) aerosols on radiative absorption by clouds, (4) moisture and temperature fields on severe weather patterns, (5) soil moisture on wind patterns, (6) oceans on carbon dioxide loadings, (7) carbon dioxide buildup on global circulation patterns and climate, (8) clouds on climate, (9) clouds on tropospheric ozone concentrations, (10) urban pollution on regional and global gas loadings, (11) urban pollution buildup on local meteorology, and (12) aircraft emissions on ozone concentrations.

Some topics of interest to regulators and scientists include estimating the effects of (1) pollutant source controls on human exposure, (2) chlorofluorocarbon emissions regulations on stratospheric ozone, and (3) carbon dioxide emissions regulations on global warming. Computational topics of interest include (1) determining the most efficient computer architectures, in terms of speed and memory, to run a model on and (2) developing numerical and computational techniques to speed algorithms on existing architectures.

### 21.1.2. Determining Scales of Interest

The second step in model development is to determine the spatial and temporal scale of interest. Spatial scales include the molecular, micro-, meso-, synoptic, and global scales, as shown in Table 1.1. Urban air pollution studies are simulated over a micro- to mesoscale domain, regional acid deposition is simulated over a meso- to synoptic scale domain, and global climate change studies are simulated over a global scale domain. Meteorological events occur over all scales. With respect to time, urban air pollution events are simulated over periods of hours to days, regional acid deposition events are simulated over periods of days to weeks, and climate change events are simulated over periods of months to hundreds of years and beyond.

### 21.1.3. Determining the Dimension of the Model

The third step in model development is to determine whether a zero-, one-, two-, or three-dimensional model is required. Three-dimensional models are ideal, but because such models require enormous computer-time and central-memory resources, zero-, one-, and two-dimensional models are often used instead.

A **zero-dimensional** (0-D) **model** is a **box model** in which chemical and/or physical transformations occur. Gases and particles may enter or leave the box from any side. Since all material in the box is assumed to mix instantaneously, the concentration of each gas and particle is uniform throughout the box. A standard box model is fixed in space. A **parcel model** is a box model that moves through space along the direction of the wind. Emission enter the box at different locations and times. Because a parcel model moves in a Lagrangian sense, it is also called a **Lagrangian trajectory model**. Box models have been used to simulate photochemical reactions that occur in smog chambers, fog production in a controlled environment, and chemical and physical interactions between aerosols and gases. Parcel models have been used to trace changes in an air parcel as it travels from ocean to land and through the polar vortex.

A **one-dimensional** (1-D) **model** is a set of adjacent box models, usually piled vertically. The advantages of a vertical 1-D model over a pure box model are that radiative transfer and vertical transport can be treated in a 1-D model. In a 1-D model, gases and particles in each layer are permitted to attenuate solar radiation as it travels through the column. A unidirectional advection–diffusion equation can also be used to transport species in the column. The disadvantage of a 1-D model compared to a 2-D or 3-D model is that vertical velocities in a 1-D model are crudely estimated, especially because they do not allow for horizontal variation in the wind. A 1-D model either ignores gas, particle, and potential temperature fluxes through horizontal boundaries or roughly parameterizes them. One-dimensional models have been used to simulate cloud formation, vertical profiles of gas concentrations, and vertical profiles of radiative fields. Figures 12.3 and 12.4 show results from a 1-D photochemical-radiative model.

A **two-dimensional** (2-D) **model** is a set of 1-D models connected side by side. 2-D models can lie in the $x$–$y$, $x$–$z$, or $y$–$z$ planes. Advantages of a 2-D over a 1-D model are that transport can be treated more realistically and a larger spatial region can be simulated in a 2-D model. 2-D models have been used to simulate transport, chemistry, and dynamics on a global scale (e.g., Garcia et al. 1992). A global 2-D model may stretch from the South to the North Pole and vertically. South-north and vertical winds in such a model are predicted or estimated from observations at each latitude and vertical layer. Zonally averaged winds (west–east winds, averaged over all longitudes for a given latitude and vertical layer) are needed in a 2-D global model. Such winds are found prognostically or diagnostically. Prognostic velocities are obtained by writing an equation of motion for the average west–east velocity at each latitude and altitude. Diagnostic velocities are obtained by writing prognostic equations for south–north and vertical velocities, then extracting the average zonal velocity from the continuity equation for air.

A **three-dimensional** (3-D) **model** is a set of horizontal 2-D models layered on top of one another. The advantage of a 3-D over a 2-D model is that dynamics and transport can be treated more realistically in a 3-D model. The disadvantage is that a 3-D model requires significantly more computer time and memory than does a 2-D model. In many cases, computer time is not a limitation. Studies of urban air pollution are readily carried out in 3-D, since simulation periods are generally only a few days. Computer-time limitations are most apparent for studies of global-scale problems that last months to years. Because 3-D models represent dynamical and transport processes better than do 0-, 1-, and 2-D models, 3-D models should be used when computer time requirements are not a hindrance.

### 21.1.4. Selecting Processes

An important step in model development is to select the physical, chemical, and dynamical equations for the model and the best available tools to solve the equations. An ideal model includes every conceivable process, each simulated with the most accurate solver. Because computer speed and memory are limited, either the number of processes simulated or the accuracy of individual solutions must be limited.

Six major groups of processes simulated in atmospheric models are (1) meteorological, (2) transport, (3) cloud, (4) radiative, (5) gas, and (6) aerosol processes. When a model is developed, it is necessary to decide whether one or more of the groups can be excluded from the simulation or replaced by observations.

Many air pollution models interpolate meteorological fields from observations, ignore the affects of clouds, and/or ignore the effects of aerosols. The advantage of using interpolated meteorological fields is that, if sufficient data are available, an interpolated field is more accurate than a prognostic field. The disadvantages are that observed meteorological data are usually available at only a few locations, most of which are near the surface. Thus, data for elevated model layers are scarce or nonexistent. Also, a model that uses a preexisting data base to predict meteorology is not prognostic, since it can simulate only past events.

Many models ignore the effects of clouds and/or aerosols. In some urban locations, clouds do not form during the summer; thus, cloud formation can be neglected. The effects of aerosols in unpolluted air are often neglected, when insoluble gases such as $O_3$, NO, and $NO_2$ are simulated. When aerosols grow to cloud-sized drops, their effects on ozone are more significant and should be simulated. In the stratosphere, ice crystals and volcanic aerosols catalyze reactions that release chlorine, which reduce ozone concentrations. Thus, simulating particles in the stratosphere is important for determining ozone concentrations when volcanic aerosols or polar stratospheric clouds are present.

### 21.1.5. Selecting Variables

Once processes in a model are selected, variables must be chosen. If the model treats meteorology prognostically, some of the variables required include air temperature, air pressure, air density, horizontal velocity, vertical velocity, geopotential, water-vapor mass mixing ratio, liquid-water mass mixing ratio, and/or ice mass mixing ratio. If the model includes trace gases, prognostic variables include the concentration of each gas. If the model includes aerosols, prognostic variables include the particle number concentration and component volume concentration in each size bin. If radiative calculations are performed, prognostic variables may include heating rates. Many intermediate variables are also stored, but not permanently, in a model. Photolysis coefficients, extinction coefficients, particle growth rates, gas dry-deposition velocities, entrainment rates, and pressure-gradient forces are variables that are stored temporarily.

When model variables, especially gas and aerosol variables, are selected, computer central-memory limits must be considered. At a minimum, an atmospheric model contains arrays with

$$N_g = (N_M + N_G + N_B N_V + N_R) N_c \tag{21.1}$$

floating-point values, where $N_M$ is the number of meteorological variables taken into account, $N_G$ is the number of gas-phase species, $N_B$ is the number of particles size bins, $N_V$ is the number of volume components in a size bin, $N_R$ is the number of radiative components, and $N_c$ is the total number of grid cells in the model. (21.1) gives the minimum central-memory requirement of a model. Additional central memory will be needed for work arrays.

**Example 21.1**

If $N_M = 10$, $N_G = 100$, $N_B = 20$, $N_V = 30$, $N_R = 2$, and $N_c = 50{,}000$, (21.1) implies that $N_g = 3.56 \times 10^7$. Thus, the model requires over 35 million words of central memory (almost 284 million bytes, since 1 word = 8 bytes) just to store values of variable. If only meteorological variables are included, $N_g = 5 \times 10^5$.

### 21.1.6. Selecting a Computer Architecture

The next step in model development is to select a computer architecture for building the model on. Example architectures are single-processor scalar, single-processor vector, multiprocessor shared-memory scalar, multiprocessor shared-memory vector, and multiprocessor distributed-memory scalar architectures.

A **scalar processor** is a processor that operates on variables in a loop, one at a time. In the Fortran loop

```
      DO 100 I    = 1, 150
        DVAR(I)   = AVAR(I)*BVAR(I)                    (21.2)
100   CONTINUE
```

150 multiplications are performed in sequence on a scalar processor. A **single-processor scalar machine** has one processor performing scalar operations.

A **vector processor** is a processor that operates on several variables within a loop at the same time. In the loop (21.2), a vector processor may break down variables AVAR and BVAR each into two registers of 64 and one register of 22 values. Elements from the first register of each AVAR and BVAR are loaded into a functional unit that multiplies values of AVAR and BVAR. At least seven values of AVAR and BVAR can be loaded into the functional unit at the same time. Thus, several operations are carried out simultaneously on a vector processor. Vector processors operate only on the inner loop of a nested loop. A **single-processor vector machine** has one processor performing vector operations.

The fastest vector-processing machine is far faster than the fastest scalar-processing machine if the code on the vector-processing machine is optimized. For a code to run fast on a vector machine, every inner loop must be optimized for the vector processor. Such an optimized code is often efficient on a scalar processor as well.

A multiprocessor machine, also called a parallel-processor machine, is one in which several processors operate on different parts of a program simultaneously. It is analogous to a network of many computers hooked up to each other and operating on different parts of a program at the same time. On a multiprocessor machine, each of the 150 multiplications in loop (21.2) can be distributed to a separate processor. A **shared-memory multiprocessor machine** is a multiprocessor machine in which memory is common to all processors. A **distributed-memory multiprocessor machine** is a multiprocessor machine in which memory is allocated to each processor and not shared by all processors.

### 21.1.7. Coding the Model

The next step is to codify the model on the chosen architecture. Although some coding may preexist, new coding is often required. A 3-D chemical-transport model may consist of a chemistry solver, a transport algorithm, an emissions algorithm, and a deposition algorithm. All but the emissions algorithm may preexist. The modeler must develop an algorithm to treat emissions and link the other processes to emissions.

Several steps can be taken to improve the appearance and usefulness of a code to others. First, the code should be sufficiently commented and referenced. Comments are important, not only for the model developer, who may need to reedit the model after several years, but also for others who may use and/or modify the model. Comments include definitions, units of variables, and descriptions of equations. References are important, so that users of the model can check the origin of equations or numerical algorithms, either to understand them or to determine whether they should be replaced.

Indenting and selecting appropriate variable names is also important. Indenting nested loops (e.g., DO...CONTINUE or IF...THEN in Fortran) makes the code more readable. Selecting descriptive variable names enables a user to follow the code easily. The variable names TEMPK, PRESS, UWIND, and VWIND are easier to follow than T, P, U, and V, respectively. Variables should be no fewer than three characters in length because one- or two-character names are difficult to search for. A search for all occurrences of T may generate a list of thousands of T's.

Major processes in a code should be controlled with on-off switches. Control of a process can be determined by a statement such as

```
IF (ISWITCH.EQ.1) CALL SUBROUTA                    (21.3)
```

where ISWITCH is a variable equal to either 1 or 0.

A code should be easy to modify. If a chemical integrator reads equations from a data set, and the equations can be changed without the integrator being changed, the code is easy to modify. If the integrator itself must be modified with each change of a chemical equation, the code is difficult to modify.

### 21.1.8. Optimizing the Model

Optimizing the speed and central-memory requirements of a model is another important step in its development. One optimization method is to improve the speed of individual loops on a given computer. Vectorizing a loop on a vector machine improves the speed of the loop. Another optimization method is to improve the speed of an algorithm without changing its accuracy. Implementing sparse-matrix techniques into an ordinary differential-equation solver reduces the number of calculations without affecting the accuracy of solution.

Reducing central-memory requirements is important when the computer used has central-memory limits. One method of reducing memory requirements is to prevent work arrays from having a dimension equal to the size of the grid. Arrays for concentration and primary meteorological variables must have one dimension equal to the size of the grid, but other arrays do not. A way to limit the size of work arrays is to divide the total model grid into **blocks of grid cells** where operations are carried out one block at a time. In such cases, one dimension of each work array is reduced to the number of grid cells in a block. A second method of reducing memory requirements is to write variable values to data files instead of storing them as arrays. The tradeoff is that additional computer time is required to read and write these files, and hard-disk memory is required to store the files.

### 21.1.9. Setting Time Steps and Intervals

Selecting model time steps and intervals is the next step in model development. It depends on desired accuracy, computer-time requirements, and stability requirements. For a 5-km × 5-km horizontal grid, a typical time step for explicit dynamical calculations is 5 s. For a global grid of $5^\circ \times 5^\circ$, it is 300 s. For chemistry, the time step is variable with some solution methods and fixed with others. The Gear integration method, discussed in Chapter 13, uses a predicted time step that varies from $<10^{-4}$ to 900 s, depending on the stiffness of the system of equations. The MIE method, as described in the same chapter, uses a fixed time step of 1, 10, 30, or 60 s. Generally, the longer the fixed time step, the less accurate the solution and the less computer time required.

Other processes, such as cloud, aerosol, radiative, and transport processes, use fixed or variable time steps. Aerosol chemical processes require small, variable time steps to maintain accuracy. Some physical processes, such as coagulation, are slow enough to allow long time steps, such as 300–900 s or more.

Time intervals between operator-split processes must be selected. The time interval may be chosen as the longest time step of any single process in a model. If chemistry requires a variable time step but transport requires a 300-s time step, a natural time interval is 300 s. Different groups of processes use different intervals. While the interval between gas chemistry and transport may be 300 s, that between gas chemistry and aerosol physics may be 900 s.

### 21.1.10. Setting Initial Conditions

The choice of initial conditions in a model depends on available data. If the model simulation is for a period during which many observed concentrations and meteorological variables were measured, initial conditions may be set by interpolating such measurements spatially at the time corresponding to the beginning of the simulation. When data are not available, initial conditions should be set carefully, and sensitivity tests should be run to test the effects of different initial conditions. In the absence of data to initialize, predictions often become erroneous.

One way to set meteorological conditions when initial data are absent is to run a preliminary model simulation that ends at the time of the start of the simulation of interest. The preliminary simulation also requires initial values that may not be correct. At the end of the presimulation, fields for temperature, pressure, velocity, and other parameters are available for every grid cell. Whether such values are accurate is open to question. Initializing a model with known data is the best way to reduce uncertainty.

### 21.1.11. Setting Boundary Conditions

In most atmospheric models, surface boundary conditions are needed. At the model top and surface, vertical velocities are usually set to zero, and variables are not transported through the boundaries by mean vertical velocities. At the surface, heat and moisture fluxes from the ground and gas/aerosol emission fluxes enter the model. Dry deposition and sedimentation remove gases and aerosols from the bottom layer.

In global models, the west–east grid generally wraps around on itself; thus, lateral boundary conditions are not needed. At the poles in a spherical-coordinate global model, south–north velocities may be set to zero so that all material travels around polar singularities. In regional models, lateral horizontal boundary conditions are needed. Lateral boundary conditions for meteorological variables are described in Chapter 7. Boundary conditions for gases and particles often require concentrations from outside the boundary.

In many models, gas mixing ratios outside a lateral boundary are held constant, and such mixing ratios are allowed to advect into the model domain. In reality, gas mixing ratios outside a boundary vary during the day and night due to chemistry. Thus, solving time-dependent chemical equations for gas concentrations in a virtual row or column outside a lateral boundary improves estimates of gas fluxes into the model domain.

Aerosol concentrations outside a lateral boundary are difficult to estimate and are usually held constant. In reality, such values are affected by time-dependent physical and chemical processes, such as coagulation, growth, and aqueous chemistry. Time-dependent equations for the change in aerosol concentration can be written for a virtual row or column outside a lateral boundary.

### 21.1.12. Input Data

Input data, such as topographical, land-use, soil moisture, emissions, chemical-rate, absorption cross-section, and activity-coefficient data, are needed to run a model. **Topographical data** are necessary for calculating surface geopotential and graphing model predictions. **Land-use data** may be used to estimate average grid-cell values of the surface roughness length for momentum, soil specific heat, soil density, soil porosity, surface albedo, and leaf area index. If each grid cell in a model is broken down into fractional land-use types, an average surface roughness length for momentum in the cell can be calculated as

$$z_{0,m,c} = f_{l,1} z_{0,m,1} + f_{l,2} z_{0,m,2} + \cdots + f_{l,i} z_{0,m,i} \cdots \tag{21.4}$$

where $z_{0,m,i}$ is the surface roughness length for momentum for land-use type $i$, and $f_{l,i}$ is the fraction of the grid cell consisting of land-use type $i$. Similar equations can be written for other parameters. The fractions in (21.4) satisfy $f_{l,1} + f_{l,2} + \cdots + f_{l,i} + \cdots = 1$. Surface roughness lengths are used to estimate eddy diffusion coefficients and dry deposition velocities.

**Soil moisture (soil liquid water) data** are needed for calculating surface temperatures in a soil model. Soil moisture varies with relative humidity, air temperature, soil temperature, soil porosity, and soil specific heat. Ideal sources of soil moisture data for model initialization are measurements from remote sensing instruments, which are generally housed on an aircraft or satellite. Such instruments include the long-wavelength microwave-radiometer (LWMR), the scanning-multichannel microwave-radiometer (SMMR), and the synthetic-aperture radar (SAR). These instruments provide parameters, such as emissivities in the case of the LWMR or soil backscattering coefficients in the case of the SAR, from which soil moisture can be derived. Soil moisture data can also be derived from evaporation rates extracted from surface temperature, albedo, and vegetation index data measured remotely (e.g., van den Hurk et al. 1997) or from brightness temperature data measured remotely (Lakshmi et al. 1997). Although remote-sensing methods for obtaining soil moisture data have been developed, such data are currently available only for limited times and areas.

**Emissions data** are required to simulate gas or particle pollution buildup. Emissions data must be accurate to properly simulate urban air pollution, since emissions make up the bulk of the source gases and particles entering an urban atmosphere. Urban emissions inventories are often prepared by state and local agencies. Summed emissions rates from two such inventories were given in Tables 12.2 and 15.2. Global emissions rates are more uncertain because taking into account every gas or aerosol source in each grid cell of a global model requires a significant effort.

**Chemical-rate coefficient data** are essential for simulating gas, aqueous, or reversible chemical reactions. Such data include temperature- and/or pressure-dependent uni-, bi-, and termolecular rate coefficients for gas- or aqueous-phase chemical reactions. Examples of these data are given in Tables B.4 and B.8. Rate-coefficient data also include equilibrium-coefficient data, such as those given in Table B.7.

**Absorption cross-section data** are important for estimating photolysis coefficients of gases and aqueous species and for calculating extinction coefficients due to gas absorption. Sources of such data are given in Table B.4.

**Activity-coefficient data** are necessary for properly simulating reversible chemistry in concentrated aerosols. Examples of temperature-dependent activity-coefficient data are shown in Table B.9.

### 21.1.13. Ambient Data

Confidence in a model's performance is gained only if model predictions stand up against rigorous comparisons with data. Data used for comparing results against are ideally an extension of the data used for initializing the model. In other words, if time-dependent observational data are available, data for the time corresponding to the beginning of the model simulation should be used to initialize the model, and data for all subsequent times should be used to compare model results against.

Data should be compared with predictions for as many parameters and locations as possible. If meteorological and gas variables are predicted by a model, predictions should be compared with observed temperatures, pressures, velocities, and concentrations. Comparisons should be made at as many horizontal and vertical locations as possible.

Data for times after the start of a simulation should not be used to force or nudge a model towards a better solution. Nudging model results by data assimilation prevents a model from being prognostic. In other words, if a known result is used to push the model to that result, the model cannot be used to predict future events. Even when past events are simulated, the use of nudging or data assimilation impairs the evaluation of the accuracy of the model.

### 21.1.14. Interpolating Data and Model Results

Input and output data are often interpolated in a model. Ambient data are interpolated to each grid cell in order to initialize the model. Emissions or land-use data are sometimes interpolated between coordinate systems. Model output for each grid cell are interpolated to locations where ambient measurements are taken. A basic interpolation method is discussed for each of these cases.

#### 21.1.14.1. *Interpolating from Scattered Points to a Fixed Point*

Scattered ambient data are often interpolated horizontally and vertically to each grid cell to initialize a model. Goodin et al. (1979) provide a review of several interpolation methods. One method for horizontally interpolating scattered data to a fixed point is **inverse-square interpolation.** With this method, each datum value is weighed by the inverse square of the distance of the datum location to the point of interest. Only data within a **domain of influence** are considered. The domain of influence is a circular area of a given **radius of influence.** Figure 21.1(a) shows an example point O and its radius of influence, $r_I$. The figure also shows five locations within the domain of influence – A, B, C, D, and E – where data values are known.

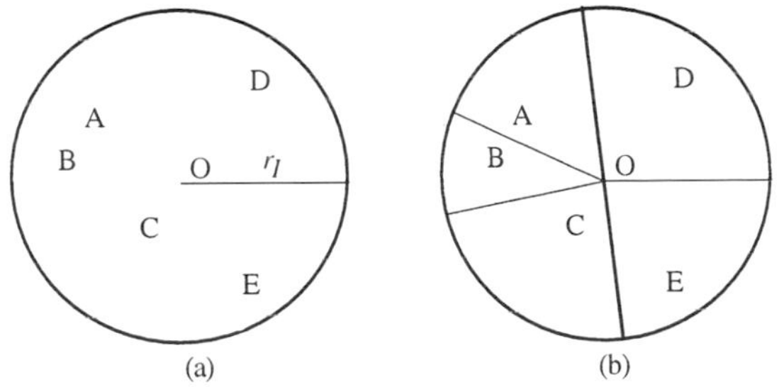


**Figure 21.1.** (a) Domain of influence around point O. The letters A, B, C, D, and E represent locations where data are available for interpolation to point O. (b) Division of the domain of influence into sectors.

The value at point O is interpolated from data at the other points with

$$V_O \approx \frac{V_A d_{AO}^{-2} + V_B d_{BO}^{-2} + V_C d_{CO}^{-2} + V_D d_{DO}^{-2} + V_E d_{EO}^{-2}}{d_{AO}^{-2} + d_{BO}^{-2} + d_{CO}^{-2} + d_{DO}^{-2} + d_{EO}^{-2}} \tag{21.5}$$

where $V_A, V_B, \ldots$ are known data values at points A, B, ..., $V_O$ is the value interpolated to point O, and $d_{AO}, d_{BO}$, etc. are the distances from points A to O, B to O, etc..

Advantages of inverse-square interpolation are that it weighs nearby data points more than distant data points and is simple to implement. A disadvantage is that when two points, such as A and B in Fig. 21.1(a), are close together, they are each weighed the same as any other point having the same distance from O. A method to reduce this problem is to multiply each datum value by an angular distance (in radians) found by drawing a line from point O through the midpoint between each pair of data locations, as shown in Fig. 21.1(b). The angles $(\theta_A, \theta_B, \ldots)$ are constrained by $\theta_A + \theta_B + \theta_C + \theta_D + \theta_E = 2\pi$ Including the angle-dependence in (21.5) gives

$$V_O \approx \frac{\theta_A V_A d_{AO}^{-2} + \theta_B V_B d_{BO}^{-2} + \theta_C V_C d_{CO}^{-2} + \theta_D V_D d_{DO}^{-2} + \theta_E V_E d_{EO}^{-2}}{\theta_A d_{AO}^{-2} + \theta_B d_{BO}^{-2} + \theta_C d_{CO}^{-2} + \theta_D d_{DO}^{-2} + \theta_E d_{EO}^{-2}} \tag{21.6}$$

With this equation, a datum value at point D now carries approximately the same weight as the data values at points A and B combined.

#### **21.1.14.2.** *Interpolating from One Coordinate System to Another*

When data are given in one coordinate system and a model uses a different coordinate system, the data must be interpolated. Emissions and soil moisture data for urban modeling are often developed on a Universal Transverse Mercator (UTM) grid projection, which has rectangular grid cells. Models are often run on a spherical (geographic) grid. When a UTM grid is laid on top of a geographic grid, UTM cell boundaries cross geographic cell boundaries at random locations; thus, interpolation is needed between the two grids.

The first interpolation step is to determine the area of each UTM grid cell that lies within a geographic grid cell. Since no formulae are available to calculate such areas, the areas must be obtained by physical integration. Each UTM cell (usually $5 \times 5$ km in area) can be divided into 10,000 or more smaller cells (minicells), each 50 m$\times$50 m in area. Each minicell contains the same data values as the larger UTM cell that it lies in. The latitude and longitude of each corner of each minicell can be found from UTM-to-geographic conversion formulae (U.S. Department of the Army 1958). Once the geographic boundaries of each minicell have been found, the number of minicells from each large UTM cell that fall within a geographic cell is counted, and the sum is denoted as $N_{U,G}$. The cumulative and average interpolated values of a variable in each geographical cell are then determined with

$$V_{G,c} = \sum_{U=1}^{M_G} \left( \frac{N_{U,G} A_M V_U}{A_U} \right) \qquad V_{G,a} = \frac{1}{A_G} \sum_{U=1}^{M_G} (N_{U,G} A_M V_U) \tag{21.7}$$

respectively, where $M_G$ is the number of large UTM cells overlapping part of a geographic cell, $A_M$ is the area of a UTM minicell, $A_U$ is the area of a large UTM cell, $A_G$ is the area of a geographic cell, and $V_U$ is the datum value in the large UTM cell.

Cumulative values are used to interpolate emissions data. When emissions rates in a geographic cell are desired, they must be summed over all UTM minicells that lie within a geographic cell. The first term in (21.7) gives such a sum. Average values are used to interpolate most other variables. When the average soil moisture in a geographic cell is desired, it is averaged over all UTM minicells that lie within a geographic cell. The second term in (21.7) gives such an average. The interpolation method described above can be applied to any two coordinate systems. The precision of the method improves when smaller and smaller minicells are used.

#### 21.1.14.3. *Interpolating from Fixed Points to a Random Point*

A model generally produces output at regularly spaced locations, such as at the horizontal center or edge of a grid cell. Observational data, used to compare model results against, are located at scattered locations throughout a model grid. Interpolating model predictions from fixed points to locations where data are measured is necessary. If grid cells are roughly rectangular, model values at data locations can be estimated with **bilinear interpolation** (e.g., Press et al. 1992).

Figure 21.2 shows the locations of known and unknown variables when bilinear interpolation is used. In this figure, values at points B, C, D, and E are known and produced from model output. A model value must be interpolated from these points to point O in order to compare the model value with a datum value at point O.

With bilinear interpolation, the model value interpolated to point O is

$$V_O \approx \frac{A_B V_B + A_C V_C + A_D V_D + A_E V_E}{A_B + A_C + A_D + A_E} \tag{21.8}$$

where $V_B$, $V_C$, ... are model values at points B, C, ..., and $A_B$, $A_C$, ... are the

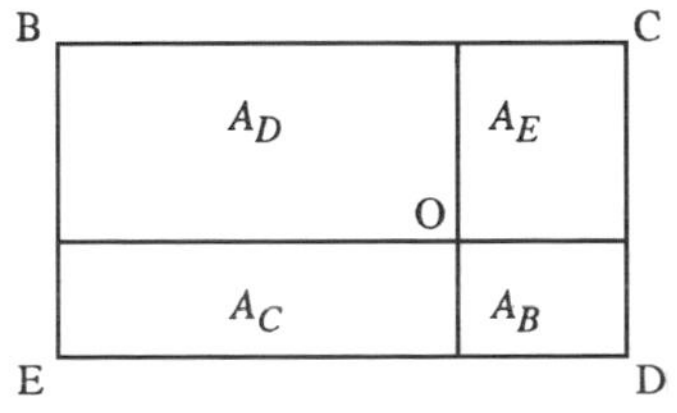


**Figure 21.2.** Location of point O in a rectangle with points B, C, D, and E at the corners.

rectangular areas shown in Fig. 21.2. Equation (21.8) states that, as location O approaches location B, area $A_B$ increases in size, increasing the weight of B on the value at O.

### 21.1.15. Statistics and Graphics

Numerical models produce output that is difficult to analyze number by number. Statistics and graphics are useful to evaluate model performance, find bugs, and study model results.

Common statistical measures are the normalized gross error, normalized bias, paired peak estimation accuracy, and temporally paired peak estimation accuracy (Tesche 1988). The **overall normalized gross error** (NGE) compares the absolute value of a prediction with an observation, summed and averaged over all observations and times. Thus,

$$\text{NGE} = \frac{1}{N_{tim}N_{obs}} \sum_{j=1}^{N_{tim}} \left( \sum_{i=1}^{N_{obs}} \frac{\left| P_{x_i,t_j} - O_{x_i,t_j} \right|}{O_{x_i,t_j}} \right) \tag{21.9}$$

where $N_{tim}$ is the number of times that observations were taken, $N_{obs}$ is the number of observations taken at each time, $P$ is a predicted value, $O$ is an observed value, $x_i$ is the location of site $i$, and $t_j$ is the $j$th time period. In this equation, all observations and predictions are paired in space and time; thus, they are compared at the same time and location. Because low measured values are often uncertain, (21.9) is applied when the observation is larger than a threshold, called a **cutoff level**. With respect to ozone mixing ratios in urban air, cutoff levels are typically between 0.02 and 0.05 ppmv.

Normalized gross errors can be measured at a single location for all times or at a single time for all locations. The **location-specific NGE** is

$$\text{NGE}_x = \frac{1}{N_{tim}} \sum_{j=1}^{N_{tim}} \frac{\left| P_{x,t_j} - O_{x,t_j} \right|}{O_{x,t_j}} \tag{21.10}$$

where $x$ is the location of interest. The **time-specific NGE** is

$$\text{NGE}_t = \frac{1}{N_{obs}} \sum_{i=1}^{N_{obs}} \frac{\left| P_{x_i,t} - O_{x_i,t} \right|}{O_{x_i,t}} \tag{21.11}$$

where $t$ is the time of interest.

The last two parameters are useful for isolating locations or times of poor predictions. If the location-specific NGEs at 29 out of 30 observations sites are 25 percent but the NGE is 1000 percent at the 30th site, the overall NGE is 57.5 percent. Thus, the overall NGE does not give complete information about the model performance. The location-specific NGE, in this case, shows that the model performance at 29 sites was much better than at the 30th site. With this information, the modeler can focus efforts on determining why the model performance at the 30th site was poor.

Another statistical measure is **normalized bias** (NB),

$$\mathrm{NB} = \frac{1}{N_{tim}N_{obs}}\sum_{j=1}^{N_{tim}}\left(\sum_{i=1}^{N_{obs}}\frac{P_{x_i,t_j} - O_{x_i,t_j}}{O_{x_i,t_j}}\right) \tag{21.12}$$

The NB indicates whether a parameter is over- or underpredicted, on average, in comparison with the data. The NB does not indicate whether the model performance is accurate.

A measure of the magnitude of the spread around the mean value of a distribution is the variance. The **biased variance**, or square of the standard deviation of a distribution, is defined as

$$\sigma_u^2 = \frac{1}{N}\sum_{i=1}^{N}(V_i - \bar{V})^2 \tag{21.13}$$

where $N$ is the total number of data values, $\bar{V}$ is the mean of all data values, and $V_i$ is the $i$th data value. The biased variance of the time-specific normalized gross error is

$$\sigma^2_{u,\mathrm{NGE}_t} = \frac{1}{N_{obs}}\sum_{i=1}^{N_{obs}}\left(\frac{\left|P_{x_i,t} - O_{x_i,t}\right|}{O_{x_i,t}} - \mathrm{NGE}_t\right)^2 \tag{21.14}$$

This value gives the spread of gross errors around the mean gross error taken at a specific time. The **unbiased variance** is the same as the biased variance, except the summation in the unbiased variance is divided by $N_{obs} - 1$ instead of $N_{obs}$.

The **paired peak accuracy** (PPA) identifies how well a model predicts a peak observed parameter value at the time and location of the peak. It is given by

$$\mathrm{PPA} = \frac{P_{\hat{x},\hat{t}} - O_{\hat{x},\hat{t}}}{O_{\hat{x},\hat{t}}} \tag{21.15}$$

where the circumflex ( ^ ) indicates that a value is taken at the time and location of the peak observed value. The **temporally paired peak accuracy** (TPPA) identifies how well the model predicts the peak observed value at the same time of the peak, but at any other location. It is

$$\mathrm{TPPA} = \frac{P_{x,\hat{t}} - O_{x,\hat{t}}}{O_{x,\hat{t}}} \tag{21.16}$$

The TPPA is less useful than the PPA in that, even if the TPPA is zero, the location of the predicted peak may be far from that of the observed peak.

Another method of judging the accuracy of a model is with **time-series plots**, which are graphical comparisons of predictions with observations for one parameter at one location over the time period of the simulation. Parameters frequently compared are gas concentration, particle concentration, temperature, relative humidity, wind speed, wind direction, and solar radiation, among others. Example plots are shown in Section 21.2. Like location-specific NGEs, time-series plots are often better indicators of model performance than are overall NGEs.

A third method of judging the accuracy of a model is by spatial comparisons of model predictions with data at a given time. Two-dimensional contour plots of predictions can be laid on top of two-dimensional maps of observations. Such comparisons allow a modeler to judge whether predictions at a given time are similar to or different from observations. Although the interpretation of these plots is subjective, they are useful for estimating accuracy and whether serious flaws exist in the model.

The statistical and graphical techniques discussed above are used to judge model performance when data are available for comparison. Graphical displays are also useful for discovering bugs. A large and unrealistic perturbation in temperature in a three-dimensional plot may suggest that a bug has infected the model. Brilliant graphics, though, should not be used to argue the validity or performance of a model. Without evaluation against data, model results have little reliability. When graphical results are presented without attempt to validate the model, the results should be viewed with caution.

### 21.1.16. Simulations

After a model has been developed and data have been obtained, simulations can be run. When a simulation is first started, it usually does not run to completion, because bugs still exist in the program. Debugging can take hours to weeks, depending on the number and severity of bugs and on the debugging experience of the programmer. Nevertheless, bugs are usually ironed out, and a baseline simulation can be performed.

During a baseline simulation, predictions and statistical comparisons to data should be gathered, stored, and/or printed out. The primary purpose of model development is to study a scientific or regulatory issue, and the baseline simulation should be designed with this study in mind.

### 21.1.17. Sensitivity Tests

After the baseline simulation is run, sensitivity tests should be carried out to gauge the effect of different assumptions on model performance. The results of such tests should be compared with data and results from the baseline simulation.

For regional modeling, common sensitivity tests include testing changes in boundary conditions, initial conditions, and emissions. One test is to set all inflow gas and aerosol concentrations at horizontal boundaries equal to zero and compare the results with the baseline case and with data. Another test is to set all initial gas and particle concentrations equal to zero. A third is to adjust the emissions inventory to estimate the effects of possible under- or overpredictions of emissions on model results. On a global scale, similar sensitivity tests for emissions and initial conditions can be run.

### 21.1.18. Improving the Model

A modeler may find that simulation results deteriorate over time because of poor numerical treatment or physical representations in the model. In such cases, better numerical algorithms or sets of equations may be needed. Modelers are continually improving and updating their models.

## 21.2. EXAMPLE MODEL SIMULATIONS

To demonstrate the steps involved in model design, application, and testing, a set of urban air pollution simulations is discussed. The purposes of the simulations were to test the effects of aerosols on surface air temperatures and to test the accuracy of a model against data. The model used, GATORM (Gas, Aerosol, Transport, Radiation, and Meteorological Model), includes the processes shown in Fig. 1.1. Here, results from an application of the model by Jacobson (1997a,b) to an air pollution episode in the Los Angeles basin for August 27–28, 1987, are shown. In the following subsections, model grids, model variables, ambient data, emission data, initial conditions, and boundary conditions are briefly discussed before results are shown and analyzed.

### 21.2.1. Model Grid

For the simulations, the grid contained 55 west–east by 38 south–north grid cells and had a southwest corner at 33.06°N latitude and 119.1°W longitude. Grid spacing was 0.05 degrees west–east (about 4.6 km) by 0.045 degrees south–north (about 5.0 km). Horizontal spherical coordinates were used. In the vertical, 20 sigma-pressure coordinate layers were used for meteorological calculations, and 14 were used for all other calculations. The bottom eight layers (below 850 mb – about 1.5 km) were the same for all processes. The model top was set to 250 mb (about 10.3 km).

### 21.2.2. Model Variables and Time Steps

Variables solved for in the model included horizontal and vertical velocity, air pressure, air temperature, relative humidity, 106 gas concentrations, 16 aerosol size bin number concentrations, and 73 aerosol component volume concentrations per size bin. The time step for meteorology was 6 s, the time step for transport was 300 s, the time step for chemistry varied from $<10^{-4}$ to 900 s, the time step for radiation was 900 s, and the time steps for aerosol processes varied from $<10^{-4}$ to 900 s. The aerosol species included 18 solids, 24 liquids, and 30 ions, and one category of *residual* material. The 16 size bins used ranged from 0.014 to 74 μm in diameter.

### 21.2.3. Ambient and Emissions Data

August 27–28, 1987, was simulated because the number of available ambient measurements was large and detailed emissions inventories were prepared for that period. Ambient measurements were available for near-surface mixing ratios of ozone, carbon monoxide, nitrogen dioxide, nitric oxide, sulfur dioxide, reactive organic gases, methane, ammonia, and nitric acid. Surface data were available for temperature, dew point, relative humidity, sea-level pressure, wind speed, wind direction, and solar radiation. Aerosol measurements of elemental carbon, organic carbon, sodium, chloride, ammonium, nitrate, sulfate, and total aerosol mass in the sub-2.5-μm and sub-10-μm size regimes were also available. The aerosol and gas emissions inventories extended over a region 325 km east–west by 180 km north–south, with a resolution of 5 km in each direction.

### 21.2.4. Initial Conditions

Initial vertical and horizontal profiles of temperature, dew point, and pressure were interpolated from 12 sounding sites in and outside the basin for the early morning of August 27, 1987. Sea surface temperatures were interpolated each hour from buoy data. Winds were initialized with zero velocities to ensure mass conservation and to avoid startup waves near mountain regions. The differential heating and cooling over spatially varying topography created pressure gradients that forced winds to generate. The Coriolis force, pressure-gradient force, and turbulent fluxes affected the equations of motion over time. Diabatic heating and heat advection influenced changes in potential temperature, which affected pressure gradients. Initial gas and aerosol concentration were interpolated from observations available at 04:30 on August 27th.

### 21.2.5. Boundary Conditions

Outside the horizontal boundaries, initial gas and aerosol concentrations were interpolated from data. Photochemical calculations were performed on gas concentrations outside the boundary to simulate their time variation during the model run. Aerosol concentrations outside the boundary were fixed at low initial values.

### 21.2.6. Results from Baseline Simulation

A baseline simulation was run from 04:30 PST, August 27 to 0:30 PST, August 29, 1987. Table 21.1 shows statistical results from the simulation. The statistics indicate that the NGEs for sulfate, sodium, light absorption, surface solar radiation, temperature, relative humidity, sulfur dioxide gas, formaldehyde, and ozone were the lowest among the parameters compared. The NGEs for nitrate and ammonia gas were largest.

Figure 21.3 shows time-series plots of predictions against observations for several parameters and locations. In some plots, three curves are shown. The third curve is a prediction from a simulation in which aerosol processes were turned off in the model. The other two curves are predictions from the baseline simulation and observations. The ammonia figures, for example, show that the inclusion of aerosols in the model was necessary for properly predicting ammonia gas mixing ratios.

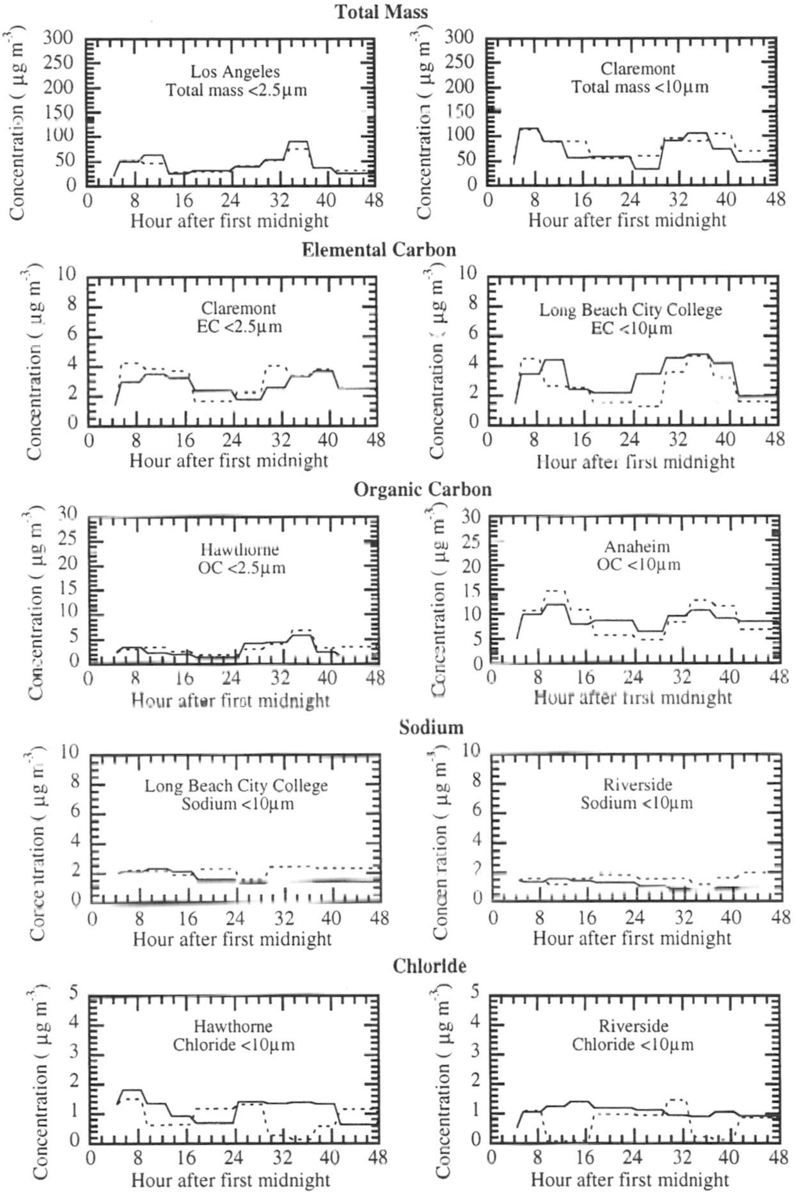


**Figure 21.3.** Time-series comparisons of baseline (gas and aerosol processes included) model predictions (solid lines) with data (short-dashed lines) from 04:30 PST August 27 to 0:30 PST August 29, 1987. In the case of formaldehyde, circles are observed values. Most observational data were given as an average over a 4-h interval. In such cases, model predictions were averaged over the same interval. In plots with three curves, the third curve (long-dashed lines) is the predicted-value curve with gas, but not aerosol processes, turned on.

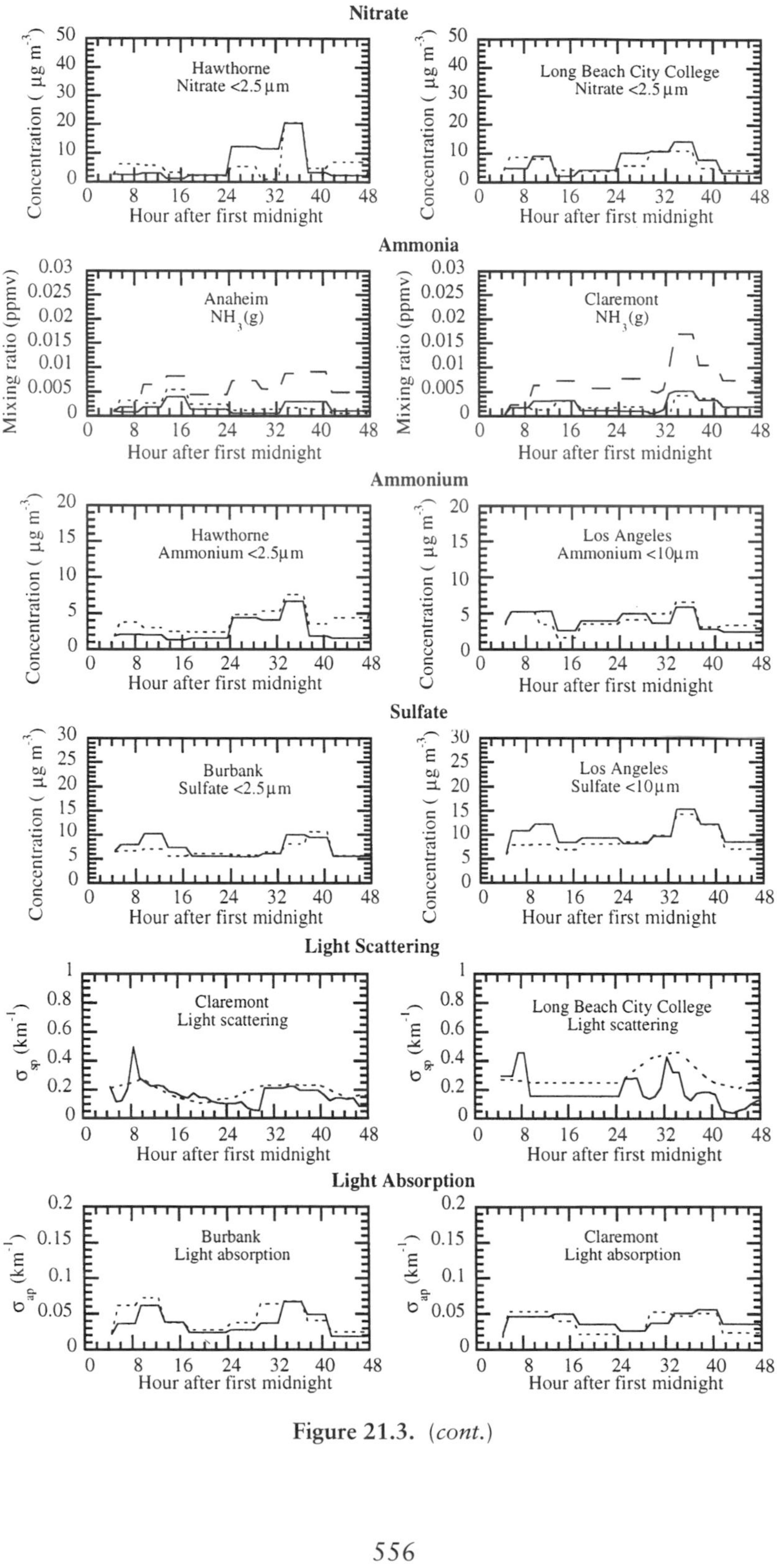


**Figure 21.3.** *(cont.)*

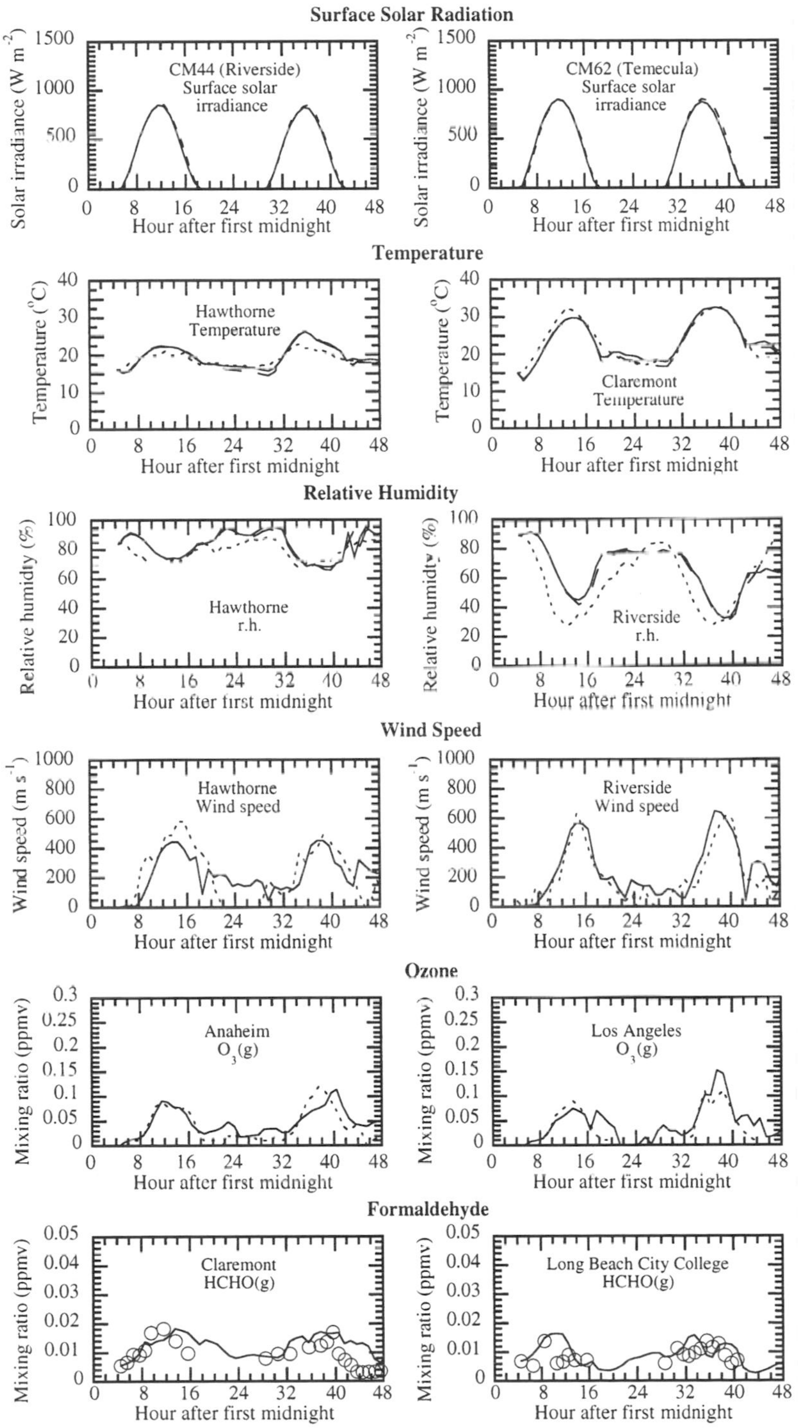


**Figure 21.3.** *(cont.)*

**Table 21.1.** Normalized Gross Errors (NGEs) and Normalized Biases (NBs) after 44 h for the Baseline Simulation

| Parameter | No. of Comparisons | NGE (%) | NB (%) | Parameter | No. of Comparisons | NGE (%) | NB (%) |
|---|---|---|---|---|---|---|---|
| $Mass_{10}$ | 360 | 50.1 | 9.3 | $Na_{10}$ | 360 | 36.0 | −30.2 |
| $Mass_{2.5}$ | 356 | 43.9 | −8.1 | $Cl_{10}$ | 190 | 46.8 | 16.0 |
| $EC_{10}$ | 356 | 50.6 | 16.2 | $O_3$ (g) | 571 | 27.8 | −6.6 |
| $EC_{2.5}$ | 356 | 57.5 | 29.9 | $SO_2$ (g) | 339 | 35.4 | −24.2 |
| $OC_{10}$ | 352 | 45.4 | 0.33 | $NH_3$ (g) | 269 | 69.3 | −25.6 |
| $OC_{2.5}$ | 352 | 49.0 | −44.1 | $HNO_3$ (g) | 109 | 54.6 | 22.3 |
| $NH4_{10}$ | 325 | 45.7 | −30.2 | HCHO (g) | 61 | 45.8 | 32.9 |
| $NH4_{2.5}$ | 321 | 55.2 | −0.15 | Temperature | 628 | 0.63 | 0.09 |
| $SO4_{10}$ | 304 | 26.3 | −52.3 | Rel. hum. | 358 | 21.6 | 4.2 |
| $SO4_{2.5}$ | 360 | 28.4 | −8.3 | Solar rad. | 50 | 7.9 | −3.0 |
| $NO3_{10}$ | 360 | 69.8 | 3.7 | $\sigma_{sp}$ | 125 | 43.0 | −13.8 |
| $NO3_{2.5}$ | 360 | 67.8 | 18.4 | $\sigma_{ap}$ | 255 | 39.8 | 16.3 |

The subscripts 10 and 2.5 indicate the mass of the species that resides in particles <10 μm and <2.5 μm in diameter, respectively. "Mass" is total particle mass, "EC" is elemental carbon, "OC" is organic carbon, "Solar rad." is the surface solar irradiance (W $m^{-2}$), $\sigma_{sp}$ is the extinction coefficient due to particle scattering, and $\sigma_{ap}$ is the extinction coefficient due to particle absorption. Cutoff mixing ratios were 50 ppbv for $O_3$(g), 5 ppbv for $SO_2$(g), 5 ppbv for HCHO(g), 1 ppbv for $NH_3$(g), and 3 ppbv for $HNO_3$(g). Other cutoff levels were 0.5 μg $m^{-3}$ for sub-10-μm chloride, 2.0 μg $m^{-3}$ for sub-2.5-μm ammonium, 0.02 $km^{-1}$ for $\sigma_{ap}$, 10 W $m^{-2}$ for solar irradiance, and 0 for all other parameters.

Other statistics from the 44-h simulation indicate that aerosols increased nighttime surface air temperatures by about 0.77 K, decreased daytime temperatures by about 0.08 K, and increased overall temperatures (day plus night) by 0.43 K. Nighttime temperatures increased due to aerosol absorption of infrared radiation from the earth's surface. Daytime temperatures decreased due to reduced solar radiation from aerosol scattering that exceeded warming due to aerosol absorption.

### 21.2.7. Results From Sensitivity Tests

Two sensitivity tests are discussed. In the first, aerosol concentrations entering the model domain from all lateral boundaries were set to zero. In the second, lateral and initial concentrations of both gases and aerosols were set to zero.

The first sensitivity test was run to test the reliability of the boundary conditions chosen for the baseline simulation. The test demostrated that setting particulate concentrations outside the lateral boundaries to zero did not change results significantly for most species. The average change was less than ±1 percent. Errors for some species, such as sodium and chloride, increased in the sensitivity test, indicating that boundary conditions affected results primarily for species with high concentrations over the ocean.

The second sensitivity test was run to test whether the choice of initial conditions for the baseline simulation made a difference. When initial concentrations of gas and aerosol species were set to zero for this test, the results degraded for most parameters. Ozone gross errors increased by over 50 percent, fine sulfate prediction errors doubled, and errors for several other particulate components increased. The removal of lateral boundary and initial values turned overpredictions into underpredictions for some species, but caused degradation in prediction results for most. Thus, initializing concentrations with realistic values was important for maintaining accuracy.

## 21.3. SUMMARY

In this chapter, model design, application, and testing were discussed. Important steps in developing and analyzing a model are to define the purpose of the model, select appropriate algorithms for simulation, obtain sufficient input data, and compare model results with data. Other steps include determining an appropriate computer architecture, optimizing the computer code, setting initial and boundary conditions, running sensitivity tests, improving numerics, and improving physical parameterizations. Simulations of a pollution episode indicated that a model can reasonably simulate atmospheric processes so long as data, such as initial and emissions data, are available.

## 21.4. PROBLEMS

**21.1.** Interpolate data values 3, 7, 11, 4, and 6, located at geographic points (−119.2° W, 32.5°N), (−119.4°W, 32.6°N), (−118.9°W, 32.3°N), (−119.0°W, 32.4°N), and (−119.1°W, 32.3°N), respectively, to the location (−119.3°W, 32.4°N) with the inverse-square method. What are the main advantage and the main disadvantage of this method?

**21.2.** Interpolate data values 4, 8, 12, and 3, located at geographic points (−117.3°W, 34.5°N), (−117.3°W, 34.6°N), (−116.9°W, 34.6°N), and (−116.9°W, 34.5°N), respectively, to the location (−117.2°W, 34.53°N) with bilinear interpolation. For purposes of calculating areas, assume rectangular shapes. Interpolate the data with the inverse-square method. Discuss differences in results, if any.

## 21.5. COMPUTER PROGRAMMING PRACTICE

**21.1.** Write a computer script to calculate the time-specific normalized gross error, time-specific normalized bias, variance, standard deviation, peak prediction accuracy, and temporally paired peak predication accuracy of the data in Table 21.2.

**Table 21.2.** Predicted and Observed $O_3$ Mixing Ratios at 17 Stations at 10:30 a.m

| Station | A | B | C | D | E | F | G | H | I | J | K | L | M | N | O | P | Q |
|---|---|---|---|---|---|---|---|---|---|---|---|---|---|---|---|---|---|
| Prediction | 7.2 | 8.4 | 9.6 | 8.2 | 6.5 | 4.3 | 3.2 | 3.8 | 4.8 | 5.9 | 6.1 | 7.0 | 7.7 | 8.2 | 9.2 | 8.7 | 7.3 |
| Observation | 6.1 | 7.2 | 8.3 | 8.5 | 7.4 | 5.8 | 4.9 | 4.8 | 3.5 | 5.2 | 7.2 | 8.1 | 5.5 | 5.3 | 6.3 | 7.1 | 5.1 |

Mixing ratios are in units of parts per hundred million by volume.

APPENDIX A

# Conversions, Constants, and Symbols

## A.1. CONVERSIONS AND CONSTANTS

Système Internationale (SI) and centimeter–gram–second (CGS) units are used. Table A.1 lists SI units and their conversion to CGS units. The conversions after the table include conversions to English units as well.

**Table A.1.** Systéme Internationale (SI) and Centimeter–Gram–Second (CGS) Units

| Quantity | SI Base or Derived Unit | CGS Unit Equivalent |
|---|---|---|
| Length | meter (m) | $10^2$ cm (centimeters) |
| Mass | kilogram (kg) | $10^3$ g (grams) |
| Temperature | kelvin (K) | K |
| Time | second (s) | s |
| Force | newton (N) | $10^5$ dyn (dynes) |
| Pressure | pascal (Pa) | 10 dyn cm$^{-2}$ |
| Energy | joule (J) | $10^7$ erg (erg) |
| Power | watt (W) | $10^7$ erg s$^{-1}$ |

### A.1.1. Pressure Conversions

| | | | |
|---|---|---|---|
| 1 bar | = $10^3$ mb | = 0.986923 atm | = $10^5$ N m$^{-2}$ |
| | = $10^5$ J m$^{-3}$ | = $10^5$ Pa | = $10^5$ kg m$^{-1}$ s$^{-2}$ |
| | = $10^6$ dyn cm$^{-2}$ | = $10^6$ g cm$^{-1}$ s$^{-2}$ | = 750.06 torr |
| | = 750.06 mm Hg | | |
| 1 atm | = 1.01325 bar | = 760 torr | = 760 mm Hg |

### A.1.2. Energy Conversions

| | | | |
|---|---|---|---|
| 1 J | = 1 N m | = $10^7$ erg | = 1 W s |
| | = $10^4$ mb cm$^3$ | = $10^7$ dyn cm | = 0.239 cal |
| | = 1 kg m$^2$ s$^{-2}$ | = $10^7$ g cm$^2$ s$^{-2}$ | = $10^{-5}$ bar m$^3$ |
| | = $6.25 \times 10^{18}$ eV | = 1 C V | |

### A.1.3. Velocity Conversions

| | | | |
|---|---|---|---|
| 1 m s$^{-1}$ | = 100 cm s$^{-1}$ | = 3.6 km h$^{-1}$ | = 1.9459 knots |
| | = 2.2378 mi h$^{-1}$ | | |

### A.1.4. Constants

| | | |
|---|---|---|
| $A$ | = Avogadro's number | $= 6.02213 \times 10^{23}$ molecules mole$^{-1}$ |
| $c$ | = speed of light | $= 2.99792 \times 10^{8}$ m s$^{-1}$ |
| $c_{p,d}$ | = specific heat of dry air at constant pressure | = 1004.67 J kg$^{-1}$ K$^{-1}$ |
| | = 1.00467 J g$^{-1}$ K$^{-1}$ | = 1004.67 m$^{2}$ s$^{-2}$ K$^{-1}$ |
| | = 0.240 cal g$^{-1}$ K$^{-1}$ | |
| $c_{v,d}$ | = specific heat of dry air at constant volume | = 717.63 J kg$^{-1}$ K$^{-1}$ |
| | = 0.718 J g$^{-1}$ K$^{-1}$ | = 717.63 m$^{2}$ s$^{-2}$ K$^{-1}$ |
| | = 0.171 cal g$^{-1}$ K$^{-1}$ | |
| $c_{p,V}$ | = specific heat of water vapor at constant pressure (298.15 K) | = 1865.1 J kg$^{-1}$ K$^{-1}$ |
| $c_{v,V}$ | = specific heat of water vapor at constant volume (298.15 K) | = 1403.2 J kg$^{-1}$ K$^{-1}$ |
| $c_W$ | = specific heat of liquid water | = 4185.5 J kg$^{-1}$ K$^{-1}$ |
| | = 4.1855 J g$^{-1}$ K$^{-1}$ | = 4185.5 m$^{2}$ s$^{-2}$ K$^{-1}$ |
| | = 1.00 cal g$^{-1}$ K$^{-1}$ | |
| $\bar{F}_s$ | = mean solar constant | = 1365 W m$^{-2}$ |
| $g$ | = effective gravity at surface of earth | = 9.81 m s$^{-2}$ |
| | = 981 cm s$^{-2}$ | |
| $G$ | = universal gravitational constant | $= 6.6720 \times 10^{-11}$ m$^{3}$ kg$^{-1}$ s$^{-2}$ |
| | $= 6.6720 \times 10^{-8}$ cm$^{3}$ g$^{-1}$ s$^{-2}$ | |
| h | = Planck's constant | $= 6.6260755 \times 10^{-34}$ J s |
| $k$ | = von Kármán constant | = 0.40 (0.35 – 0.41) |
| $k_B$ | = Boltzmann's constant ($R^*/A$) | $= 1.3807 \times 10^{-23}$ JK$^{-1}$ |
| | $= 1.3807 \times 10^{-23}$ kg m$^{2}$ s$^{-2}$ K$^{-1}$ molecule$^{-1}$ | $= 1.3625 \times 10^{-22}$ cm$^{3}$ atm K$^{-1}$ |
| | $= 1.3807 \times 10^{-16}$ g cm$^{2}$ s$^{-2}$ K$^{-1}$ molecule$^{-1}$ | $= 3.299 \times 10^{-24}$ cal K$^{-1}$ |
| | $= 1.3807 \times 10^{-19}$ cm$^{3}$ mb K$^{-1}$ molecule$^{-1}$ | $= 1.3625 \times 10^{-25}$ L atm K$^{-1}$ molecule$^{-1}$ |
| | $= 1.3625 \times 10^{-22}$ cm$^{3}$ atm K$^{-1}$ molecule$^{-1}$ | $= 1.3807 \times 10^{-25}$ m$^{3}$ mb K$^{-1}$ molecule$^{-1}$ |
| $L_p$ | = luminosity of the sun's photosphere | $= 3.9 \times 10^{26}$ W |
| $m_d$ | = molecular weight of dry air | = 28.966 g mole$^{-1}$ |
| $m_v$ | = molecular weight of water | = 18.02 g mole$^{-1}$ |
| $\bar{M}$ | = mass of an air molecule ($m_d/A$) | $= 4.8096 \times 10^{-26}$ kg |
| | $= 4.8096 \times 10^{-23}$ g | |
| $M_e$ | = mass of the earth | $= 5.98 \times 10^{24}$ kg |
| | $= 5.98 \times 10^{27}$ g | |
| $R^*$ | = universal gas constant | = 8.3145 J mole$^{-1}$ K$^{-1}$ |
| | = 8.3145 kg m$^{2}$ s$^{-2}$ mole$^{-1}$ K$^{-1}$ | = 0.083145 m$^{3}$ mb mole$^{-1}$ K$^{-1}$ |
| | $= 8.3145 \times 10^{7}$ g cm$^{2}$ s$^{-2}$ mole$^{-1}$ K$^{-1}$ | = 0.08206 L atm mole$^{-1}$ K$^{-1}$ |
| | $= 8.3145 \times 10^{4}$ cm$^{3}$ mb mole$^{-1}$ K$^{-1}$ | $= 8.3145 \times 10^{7}$ erg mole$^{-1}$ K$^{-1}$ |

| | | |
|---|---|---|
| | $= 82.06\ \text{cm}^3\ \text{atm mole}^{-1}\ \text{K}^{-1}$ | $= 1.987\ \text{cal mole}^{-1}\ \text{K}^{-1}$ |
| $R'$ | = gas constant for dry air ($R^*/m_d$) | $= 287.04\ \text{J kg}^{-1}\ \text{K}^{-1}$ |
| | $= 0.28704\ \text{J g}^{-1}\ \text{K}^{-1}$ | $= 2.8704\ \text{m}^3\ \text{mb kg}^{-1}\ \text{K}^{-1}$ |
| | $= 2870.4\ \text{cm}^3\ \text{mb g}^{-1}\ \text{K}^{-1}$ | $= 287.04\ \text{m}^2\ \text{s}^{-2}\ \text{K}^{-1}$ |
| | $= 2.8704 \times 10^6\ \text{cm}^2\ \text{s}^{-2}\ \text{K}^{-1}$ | |
| $R_e$ | = radius of the earth | $= 6.371 \times 10^6\ \text{m}$ |
| $R_p$ | = radius of the sun | $= 6.96 \times 10^8\ \text{m}$ |
| $\bar{R}_{es}$ | = mean earth–sun distance | $= 1.5 \times 10^{11}\ \text{m}$ |
| $R_v$ | = gas constant for water vapor ($R^*/m_v$) | $= 461.40\ \text{J kg}^{-1}\ \text{K}^{-1}$ |
| | $= 0.46140\ \text{J g}^{-1}\ \text{K}^{-1}$ | $= 4.6140\ \text{m}^3\ \text{mb kg}^{-1}\ \text{K}^{-1}$ |
| | $= 4614.0\ \text{cm}^3\ \text{mb g}^{-1}\ \text{K}^{-1}$ | $= 461.40\ \text{m}^2\ \text{s}^{-2}\ \text{K}^{-1}$ |
| | $= 4.6140 \times 10^6\ \text{cm}^2\ \text{s}^{-2}\ \text{K}^{-1}$ | |
| $T_p$ | = temperature of sun's photosphere | $= 5796\ \text{K}$ |
| $\gamma$ | $= c_{p,d}/c_{v,d}$ | $= 1.4$ |
| $\kappa$ | $= R'/c_{p,d}$ | $= 0.286$ |
| $\sigma_B$ | = Stefan–Boltzmann constant | $= 5.67051 \times 10^{-8}\ \text{W m}^{-2}\ \text{K}^{-4}$ |
| $\Omega$ | = angular velocity of the earth | $= 7.292 \times 10^{-5}\ \text{s}^{-1}$ |

## A.2. LIST OF SYMBOLS

### a, A

| | |
|---|---|
| $\mathbf{a}_{c,i}$ | = length of major semiaxis of ice crystal (cm) |
| $\mathbf{a}_i$, $\mathbf{a}_l$, $\mathbf{a}_c$, $\mathbf{a}_r$ | = inertial, local, earth's centripetal, Coriolis acceleration vectors ($\text{m s}^{-2}$) |
| $a$, $a_i$ | = surface-area concentration of particles, of particles in size bin $i$ ($\text{cm}^2\ \text{cm}^{-3}$) |
| $a_q$, $a_w$ | = activity of species $q$ (variable units), of water (fraction) |
| $\mathbf{a}_R$ | = local centripetal acceleration vector ($\text{m s}^{-2}$) |
| $a_\lambda$ | = absorptivity of a substance at wavelength $\lambda$ (fraction) |
| $A$ | = Avogadro's number (molecules $\text{mole}^{-1}$) |
| $A_{e,0}$ | = average earth–atmosphere albedo (—) |
| $A_g$ | = prefactor in modified gamma size distribution |
| $A_L$ | = particle area concentration in lognormal mode ($\text{cm}^2\ \text{cm}^{-3}$) |
| $A_r$ | = collisional prefactor in chemical rate coefficient (various units) |
| $A_s$ | = surface area on a sphere intercepted by radiation ($\text{cm}^2$) |
| $A_\gamma$ | = Debye–Hückel parameter |

### b, B

| | |
|---|---|
| $b_{a,g,q,\lambda}$, $b_{a,a,q,\lambda}$ | = absorption cross section of a gas, of an aerosol at wavelength $\lambda$ ($\text{cm}^2$) |
| $\text{b}_{c,i}$ | = length of minor semiaxis of ice crystal of size $i$ (cm) |
| $b_{sg,\lambda}$, $b_{sp,i,\lambda}$ | = scattering cross section of a gas, of a particle at wavelength $\lambda$ ($\text{cm}^2$) |
| $B$ | = buoyancy factor (—) |
| $B_r$ | = exponent in reaction-rate-coefficient expression (—) |
| $B_{\lambda,T}$ | = Planck function ($\text{W m}^{-2}\ \mu\text{m}^{-1}\ \text{sr}^{-1}$) |

## c, C

| | |
|---|---|
| $c$ | = speed of light (cm $s^{-1}$) |
| $\mathbf{c}_g$ | = group velocity vector for atmospheric wave (m $s^{-1}$) |
| $c_{g,x}$, $c_{g,y}$, $c_{g,z}$ | = group speeds of atmospheric wave in $x$-, $y$-, and $z$-directions (m $s^{-1}$) |
| $c_{p,d}$, $c_{p,m}$, $c_{p,q}$, $c_{p,V}$ | = specific heat of dry air, moist air, gas $q$, water vapor at constant pressure (J $kg^{-1}$ $K^{-1}$) |
| $c^{\circ}_{p,i}$ | = standard molal heat capacity at constant pressure (J $mole^{-1}$ $K^{-1}$) |
| $c_G$, $c_I$, $c_S$, $c_W$ | = specific heat of soil–water–air mixture, ice, pure soil, liquid water (J $kg^{-1}$ $K^{-1}$) |
| $c_q$, $c_w$, $c_{i,j,m}$ | = mole concentration of species $q$, liquid water, electrolyte $i$, $j$, (moles $cm^{-3}$) |
| $c_s$ | = speed of sound (m $s^{-1}$) |
| $c_{v,d}$, $c_{v,m}$, $c_{v,V}$ | = specific heat of dry air, moist air, water vapor at constant volume (J $kg^{-1}$ $K^{-1}$) |
| $c_\alpha$ | = phase speed of an atmospheric wave (m $s^{-1}$) |
| $C_D$, $C_E$, $C_H$ | = bulk transfer coefficients for momentum, water vapor, energy (—) |
| $C_q$ | = mole concentration of gas $q$ (moles $cm^{-3}$) |
| $C'_{q,s,i}$ | = effective surface vapor concentration of gas $q$ over particles (moles $cm^{-3}$) |
| $C_r$ | = factor in expression for chemical reaction rate coefficient (K) |
| $C_{ratio}$ | = contrast ratio (—) |
| $C'_s$ | = saturation gas mole concentration (moles $cm^{-3}$) |

## d, D

| | |
|---|---|
| $d_c$ | = displacement height in a canopy (m) |
| $d_i$, $d_q$ | = diameter of particles in size bin $i$ and of a gas molecule (cm) |
| $D_{aq,q}$, $D_g$ | = diffusion coefficients of solute in water, water in soil ($cm^2$ $s^{-1}$) |
| $\bar{D}_A$, $\bar{D}_M$, $\bar{D}_N$, $\bar{D}_V$ | = geometric-mean area, mass, number, volume diameter (cm) |
| $D'_{q,i}$, $D'_{v,i}$ | = corrected diffusion coefficient of gas $q$, of water vapor over particles ($cm^2$ $s^{-1}$) |
| $D^{eff}_{q,i}$ | = effective diffusion coefficient of gas $q$ over particles ($cm^2$ $s^{-1}$) |
| $D_J$ | = Julian day of the year (days) |
| $D_{p,i}$, $D_q$, $D_v$ | = diffusion coefficient of particle, gas $q$, water vapor in air ($cm^2$ $s^{-1}$) |
| $D_h$ | = thermal diffusion coefficients ($cm^2$ $s^{-1}$) |
| $D_Y$ | = number of days in a year |

## e, E

| | |
|---|---|
| $\mathrm{e}_i$ | = basis function in finite-element method (—) |
| $e_\lambda$ | = actual radiance emitted by a body (W $m^{-2}$ $\mu m^{-1}$ $sr^{-1}$) |

| | |
|---|---|
| $E$ | = energy density (J m$^{-3}$) |
| E | = entrainment rate of outside air into a cloud (s$^{-1}$) |
| $E_{a,q}$, $E_{d,q}$ | = activation energy of adsorption, of desorption for species $q$ (J mole$^{-1}$) |
| $E_f$ | = turbulent moisture flux (evaporation rate) at ground surface (kg m$^{-2}$ s$^{-1}$) |
| $E_{i,j}$ | = coalescence efficiency (—) |
| $E_r$ | = activation energy of a chemical reaction (J mole$^{-1}$) |
| Ek | = Ekman number (—) |

## f, F

| | |
|---|---|
| $f$ | = Coriolis parameter (s$^{-1}$) |
| $f(\delta_*)$ | = anisotropic correction term (—) |
| $f_{l,i}$ | = fraction of a grid cell consisting of land-use type $i$ (—) |
| $f_r$ | = relative humidity of air (percent or fraction) |
| $f_{0,q}$ | = reactivity factor for gases (—) |
| $F_b$ | = infrared irradiant flux, summed over all wavelengths (W m$^{-2}$) |
| $\mathbf{F}_c$ | = Coriolis force vector (N) |
| $F_D$, $F_G$ | = drag force, gravitation force on falling particle (N) |
| $\mathbf{F}_g^*$, $\mathbf{F}_g$ | = true, effective gravitational force vector (N) |
| $F_{h,i}$, $F_{q,i}$ | = thermal and vapor ventilation factors (—) |
| $F_p$ | = irradiance of the sun's photosphere (W m$^{-2}$) |
| $\mathbf{F}_p$ | = pressure-gradient force vector (N) |
| $F_n$, $F_{n,g}$ | = solar + infrared flux summed over all wavelengths, net flux at ground (W m$^{-2}$) |
| $F_s$, $\bar{F}_s$ | = current total irradiance, mean solar constant (W m$^{-2}$) |
| $F_{s,\lambda}$, $\bar{F}_{s,\lambda}$ | = current, mean solar irradiance in a wavelength interval (W m$^{-2}$) |
| $\mathbf{F}_r$, $\mathbf{F}_R$ | = global, local apparent centrifugal force vector (N) |
| $\mathbf{F}_t$ | = turbulent flux divergence multiplied by air mass (N) |
| $\mathbf{F}_v$ | = viscous force vector (N) |
| $F_\lambda$ | = irradiance at wavelength $\lambda$ (W m$^{-2}$ μm$^{-1}$) |
| Fr | = Froude number (—) |

## g, G

| | |
|---|---|
| $g$, $g^*$, $g_0$ | = effective, true, globally averaged effective gravitational acceleration (m s$^{-2}$) |
| $g_a$, $g_a'$ | = asymmetry factor, asymmetry factor adjusted with delta function (—) |
| $g_M$ | = mean anomaly of the sun (deg) |
| G | = universal gravitational constant (m$^3$ kg$^{-1}$ s$^{-2}$) |
| G* | = Gibbs free energy (J) |
| $\Delta_f G_i^\circ$ | = standard molal Gibbs free energy of formation (J mole$^{-1}$) |
| $\Delta G$, $\Delta G_{a,q}$, $\Delta G_{d,q}$ | = change in molal Gibbs free energy of nucleation, adsorption, desorption (J mole$^{-1}$) |
| $G_i$ | = Cunningham slip-flow correction (—) |

**h, H**

| | |
|---|---|
| $h$ | = time step size (s) |
| h | = Planck's constant (J s) |
| $h_c$ | = canopy height (m) |
| $h_e$ | = equivalent depth of the atmosphere (m) |
| $H$ | = scale height (m) |
| $H_a$ | = local hour angle of the sun (deg) |
| $H_q$, $H_q^*$ | = Henry's constant, effective Henry's constant for species $q$ (moles $L^{-1}$ $atm^{-1}$) |
| $H_{r,i}$ | = radiative heating/cooling rate of particles of size $i$ (W) |
| $H_f$ | = vertical turbulent sensible heat flux in surface layer (W $m^{-2}$) |
| $H_c$ | = vertical flux of energy due to conduction (W $m^{-2}$) |
| $H^*$ | = enthalpy (J) |
| $\Delta_f H_i^\circ$ | = standard molal enthalpy of formation (J $mole^{-1}$) |
| $\Delta H_{a,q}$ | = change in molal enthalpy of adsorption (J $mole^{-1}$) |

**i, I**

| | |
|---|---|
| $\mathbf{i}$, $\mathbf{i}_\lambda$, $\mathbf{i}_R$ | = unit vector in Cartesian, spherical, cylindrical coordinates |
| $i_v$ | = van't Hoff factor (–) |
| I | = ionic strength (mole $kg^{-1}$) |
| $I_\lambda$, $I_B$ | = irradiance at wavelength $\lambda$, of background light (W $m^{-2}$ $\mu m^{-1}$ $sr^{-1}$) |

**j, J**

| | |
|---|---|
| $\mathbf{j}$, $\mathbf{j}_\varphi$, $\mathbf{j}_\theta$ | = unit vector in Cartesian, spherical, cylindrical coordinates |
| $J_{het}$, $J_{hom}$ | = heterogeneous, homogeneous nucleation rate (embryos $cm^{-2}$ $s^{-1}$, particles $cm^{-3}$ $s^{-1}$) |
| $J_q$ | = photolysis rate coefficient of species $q$ ($s^{-1}$) |
| $J_t$ | = Jacobian matrix of partial derivatives |

**k, K**

| | |
|---|---|
| $k$ | = von Kármán constant (—) |
| $\tilde{k}$ | = wavenumber in $x$-direction ($m^{-1}$) |
| $\mathbf{k}$, $\mathbf{k}_r$, $\mathbf{k}$ | = unit vectors in Cartesian, spherical, cylindrical coordinates |
| $k_{a,g,q,\lambda}$ | = mass absorption coefficient of gas $q$ at wavelength $\lambda$ ($cm^2$ $g^{-1}$) |
| $k_{a,q}$, $k_{d,q}$ | = rate of a gas adsorption to, desorption from a particle surface (molecules $mb^{-1}$ $s^{-1}$) |
| $k_F$, $k_S$, $k_T$ | = first-, second-, third-order rate coefficient ($s^{-1}$, $cm^3$ $molecule^{-1}$ $s^{-1}$, $cm^6$ $molecule^{-2}$ $s^{-1}$) |
| $k_B$ | = Boltzmann's constant (J $K^{-1}$) |
| $k_{q,i}$ | = mass-transfer coefficient for growth ($s^{-1}$) |
| $k_{s,q}$ | = rate coefficient for surface reaction of gas $q$ ($s^{-1}$) |
| K | = Stokes number (—) |
| $K$ | = order of matrix of partial derivatives |

$K_{eq}$ = equilibrium coefficient (various units)
$K_g$, $K_{g,sat}$ = hydraulic conductivity of soil, of soil at saturation ($m\ s^{-1}$)
$K_h$, $K_m$ = eddy diffusion coefficient for energy, momentum ($m^2\ s^{-1}$)
$\mathbf{K}_h$, $\mathbf{K}_m$ = eddy diffusion tensor for energy, momentum ($m^2\ s^{-1}$)
$\mathrm{Kn}_{a,i}$, $\mathrm{Kn}_{h,i}$, $\mathrm{Kn}_{q,i}$ = Knudsen number for air, energy, gas $q$ with respect to particles of size $i$ (—)

## l, L

$\tilde{l}$ = wavenumber in $y$-direction ($m^{-1}$)
$L$ = Monin–Obukhov length (m)
$L_c$ = sum of chemical loss terms (molecules $cm^{-3}\ s^{-1}$)
$L_e$, $L_m$, $L_s$ = latent heats of evaporation, melting, sublimation of water ($J\ kg^{-1}$)
$L_{e,q}$ = latent heat of evaporation of species $q$ ($J\ kg^{-1}$)
$L_M$ = mean longitude of the sun (deg)
$L_p$ = luminosity of the sun's photosphere (W)
LAI = leaf area index (—)

## m, M

$\tilde{m}$ = wavenumber in $z$-direction ($m^{-1}$)
$m_a$, $m_d$, $m_q$, $m_v$ = molecular weight of total air, dry air, gas $q$, water vapor ($g\ mole^{-1}$)
$\mathfrak{m}$, $\mathfrak{m}_a$ = molality, molality of species alone in solution ($mole\ kg^{-1}$)
$\mathsf{m}$ = mass of a single drop ($g\ particle^{-1}$)
$\mathsf{m}_i$ = mass concentration of particles in size bin $i$ (grams per cubic centimeter of air)
$m_p$, $m_s$ = molecular weight of a particle, of solute in a drop ($g\ mole^{-1}$)
$m_\lambda$ = complex index of refraction (—)
$M_a$, $M_v$, $M_d$, $M_c$ = mass of an air parcel, of water vapor, of dry air, of an entrainment region in a cloud (kg)
$\bar{M}$, $\bar{M}_q$, $\bar{M}_{p,i}$ = mass of an air molecule, of a molecule of gas $q$, of a particle of size $i$ (g)
$M_e$ = mass of earth (kg)
$\mathrm{M}_L$ = total mass concentration of particles in a lognormal mode ($\mu g\ m^{-3}$)
$M_s$, $M_w$ = mass of solute, liquid water in a drop (g)

## n, N

$n$, $n_d$, $n_q$, $n_v$ = number of moles of a gas, dry air, gas $q$, water vapor
$n_{a,\lambda}$, $n_\lambda$ = real part of index of refraction of air, particle (—)
$n_c$ = number of molecules in a nucleated cluster
$n_i$, $n_T$ = number concentration of particles of size $i$, in a size distribution ($particles\ cm^{-3}$)

| | |
|---|---|
| $n_m$ | = maximum number of adsorption sites on a surface (sites $cm^{-2}$) |
| $n_s$, $n_w$ | = number of moles of solute, of water in a drop |
| $n_{s,q}$ | = surface-area concentration of adsorbed gas $q$ (molecules $cm^{-2}$) |
| $\hat{N}$ | = array of gas number concentrations (molecules $cm^{-3}$) |
| $N, \bar{N}, N'$ | = precise, mean, and perturbation number concentration of a gas (molecules $cm^{-3}$) |
| $N_a$, $N_d$, $N_q$, $N_v$ | = number concentration of total air, dry air, gas $q$, water vapor (molecules $cm^{-3}$) |
| $N_A$, $N_C$ | = number of anions, cations in solution |
| $N_{ag}$ | = number of radiatively absorbing gases |
| $N_B$ | = number of particle size bins in a model size distribution |
| $\mathrm{N}_{Bo}$ | = Bond number (—) |
| $N_{bv}$ | = Brunt-Väisälä frequency ($s^{-1}$) |
| $N_c$ | = number of grid cells in a model |
| $N_D$ | = number concentration of particles in a size distribution (particles $cm^{-3}$) |
| $N_{e,h}$ | = number of diabatic energy sources or sinks |
| $N_{e,t}$ | = number of external sources or sinks of a trace species |
| $N_{e,x,t}$, $N_{f,x,t}$ | = exact, finite difference solution to partial differential equation (molecules $cm^{-3}$) |
| $N_g$ | = number of array values in a model |
| $N_G$ | = number of gases in a model |
| $N_{JD}$ | = number of days from the beginning of Julian year 2000 |
| $N_{loss,i}$, $N_{prod,i}$ | = number of chemical loss, production terms for species $i$ |
| $\mathrm{N}_L$ | = number concentration of particles in a lognormal mode (molecules $cm^{-3}$) |
| $N_L$ | = number of layer centers in a model |
| $N_M$ | = number of meteorological variables in a model |
| $N_{obs}$ | = number of observations of a variable at a given time |
| $N_P$ | = physical-property number (—) |
| $N_R$ | = number of radiative variables in a model |
| $N_T$ | = number of different particle size distributions in a model |
| $N_{tim}$ | = number of observation times of a variable |
| $N_V$ | = number of components in a particle |
| $N_X$ | = number of externally mixed particle types in a model |

## p, P

| | |
|---|---|
| $p$, $p_a$ | = pressure, pressure of total air (mb) |
| $\hat{p}_a$, $p_a''$ | = pressure in large-scale environment, perturbation to large-scale pressure (mb) |
| $p_d$, $p_q$, $p_v$ | = partial pressure exerted by dry air, by gas $q$, by water vapor (mb) |
| $p_{d,s}$ | = dry-air pressure at the ground surface (mb) |
| $p_{d,0}$ | = reference pressure for potential temperature equation (mb) |

$p_{q,s}$, $p_{v,I}$, $p_{v,s}$ = saturation vapor pressure of gas $q$, of water over ice, of water over liquid when surface is flat and dilute (mb)

$p'_{q,s,i}$, $p'_{v,I,i}$, $p'_{v,s,i}$ = effective saturation vapor pressure of gas $q$, of water over ice, of water over liquid (mb)

$p_{a,surf}$, $p_{a,top}$ = air pressure at the surface, top of a model atmosphere (mb)

$P$, $c_{p,d}P$ = potential temperature factor (—), Exner function (J kg$^{-1}$ K$^{-1}$)

**P** = predictor matrix for Gear's method

$P_g$ = flux of liquid water reaching soil surface in soil model (kg m$^{-2}$ s$^{-1}$)

$P_c$ = sum of chemical production terms (molecules cm$^3$ s$^{-1}$)

$P_s$ = scattering phase function (—)

$P_v$ = potential vorticity (s$^{-1}$ m$^{-1}$)

Pr = Prandtl number (—)

$\mathrm{Pr}_t$ = turbulent Prandtl number (—)

## q, Q

$q_q$ = moist-air mass mixing ratio of gas $q$ (kilograms per kilogram of moist air)

$q_{v,s}$ = saturation specific humidity at soil surface (kilograms per kilogram of moist air)

$q_v$, $\bar{q}_v$, $q'_v$ = total, mean, perturbation specific humidity (kilograms per kilogram of moist air)

$q_*$ = turbulent water-vapor scale (kg kg$^{-1}$)

$Q^*$, $Q$ = energy (J), energy per unit mass (J kg$^{-1}$) transferred to and from an air parcel

$Q_{a,i,\lambda}$, $Q_{f,i,\lambda}$, $Q_{s,i,\lambda}$ = single-particle absorption, forward-scattering, scattering efficiencies (—)

$Q_{c/e}$, $Q_{dp/s}$, $Q_{f/m}$ = change in energy due condensation, deposition, freezing (J)

$Q_{ir}$, $Q_{solar}$ = change of energy due to infrared heating/cooling, solar heating (J)

## r, R

$r^*$ = critical radius for growth of cloud drops (cm)

$r_c$ = critical radius of a nucleating cluster (cm)

$r_{c,g}$ = critical radius in modified gamma size distribution (μm)

$r_i$, $r_s$ = radius of particles in bin $i$, of a sphere for solid-angle calculations (cm)

$r_t$ = radius of a thermal (m)

$R$ = radial distance from the center of a drop (cm)

$R^*$ = universal gas constant

$R'$, $R_m$, $R_v$ = gas constants for dry air, moist air, water vapor

$R_a$, $R_b$, $R_s$ = resistance to turbulence, molecular diffusion, surface interactions (s m$^{-1}$)

$\mathbf{R}_c$, $R_c$ = vector, magnitude of radius of curvature (m)
$R_{c,n}$ = rate of reaction $n$ (molecules cm$^{-3}$ s$^{-1}$)
$R_{canp}$, $R_{conv}$, $R_{min}$ = canopy, buoyant-convection, minimum bulk canopy stomatal resistance (s m$^{-1}$)
$R_{cut}$, $R_{stom}$, $R_{meso}$ = cuticle, stomatal, mesophyll resistances (s m$^{-1}$)
$\mathbf{R}_e$, $R_e$ = vector, magnitude of radius of earth (m)
$R_{es}$, $\bar{R}_{es}$ = actual, mean earth–sun distance (m)
$R_b$ = radius of host particle for heterogeneous nucleation (cm)
$R_i$ = residual in finite-element method (variable units)
$R_n$ = rate of change of concentration due to process $n$ (variable units)
$R_p$ = radius of the sun, from center to photosphere (m)
$R_{soil}$, $R_{exp}$ = resistance to soil and leaf litter, to exposed surfaces (s m$^{-1}$)
Re = Reynolds number (—)
$\mathrm{Ri}_b$, $\mathrm{Ri}_c$, $\mathrm{Ri}_g$, $\mathrm{Ri}_T$ = bulk, critical, gradient, termination Richardson numbers (—)
Ro = Rossby number (—)

## s, S

$s$ = sigma-altitude level (fraction)
$\dot{s}$ = vertical velocity in the sigma-altitude coordinate (s$^{-1}$)
$S^*$ = critical saturation ratio (—)
$\mathrm{S}^*$, $\mathrm{S}$ = entropy (J K$^{-1}$), entropy per unit mass (J kg$^{-1}$ K$^{-1}$)
$\Delta S_{a,q}$ = entropy of adsorption on a surface (J mole$^{-1}$ K$^{-1}$)
$S_b$ = distance along path of electromagnetic wave (cm)
$S_f$ = net surface solar irradiance (W m$^{-2}$)
$S_p$ = stiffness predictor (s$^{-2}$)
$S_q$, $S'_{q,i}$ = ambient, equilibrium saturation ratio of gas $q$ (—)
$\mathrm{Sc}_q$, $\mathrm{Sc}_{p,i}$ = Schmidt number for gas $q$, particles of size $i$ (—)

## t, T

$t$ = time (s)
$t_s$ = time past local noon (s)
$T$, $T_c$ = air temperature in kelvin, Celsius
$T_D$ = dew point (K)
$T_e$ = effective temperature of the earth's surface without a greenhouse effect (K)
$T_g$, $T_{g,c}$ = ground surface temperature in kelvin, Celsius
$T_m$ = temperature of vapor molecules leaving the surface of a drop (K)
$T_0$ = reference temperature (298.15 K)
$T_p$ = effective temperature of the sun's photosphere (K)
$T_r$ = temperature at a liquid-drop surface (K)
$T_s$ = soil temperature below the ground surface (K)
$T_v$, $\hat{T}_v$, $T_{v,L}$ = virtual temperature of air, outside a cloud, at the lifting condensation level (K)

**u, U**

| | |
|---|---|
| $u$, $\bar{u}$, $u'$ | = precise, mean, perturbation west–east scalar velocity (m $s^{-1}$) |
| $u_*$ | = friction velocity (m $s^{-1}$) |
| $u_a$, $u_g$ | = west–east ageostrophic, geostrophic component of scalar velocity (m $s^{-1}$) |
| $\mathbf{u}_R$, $u_R$ | = radial velocity vector, scalar velocity in cylindrical coordinates (m $s^{-1}$) |
| U*, U | = internal energy (J), internal energy per unit mass (J $kg^{-1}$) of a gas |

**v, V**

| | |
|---|---|
| $v$, $\bar{v}$, $v'$ | = precise, mean, perturbation south–north scalar velocity (m $s^{-1}$) |
| $\mathbf{v}$, $\bar{\mathbf{v}}$, $\mathbf{v}'$ | = precise, mean, perturbation velocity vector (m $s^{-1}$) |
| $\lvert\mathbf{v}\rvert$, $\lvert\mathbf{v}_h\rvert$ | = wind speed, horizontal wind speed (m $s^{-1}$) |
| $\bar{v}_a$, $\bar{v}_{p,i}$, $\bar{v}_q$ | = thermal velocity of air, of particles of size $i$, of gas $q$ (cm $s^{-1}$) |
| $\mathbf{v}_A$ | = absolute velocity of a body on earth (m $s^{-1}$) |
| $v_a$, $v_g$ | = south–north ageostrophic, geostrophic component of scalar velocity (m $s^{-1}$) |
| $\mathbf{v}_g$ | = geostrophic velocity vector (m $s^{-1}$) |
| $\mathbf{v}_h$ | = horizontal velocity vector (m $s^{-1}$) |
| $v_i$, $v_{q,i}$ | = volume concentration of total particle, of component $q$ in a particle ($cm^3$ $cm^{-3}$) |
| $\mathbf{v}_\theta$, $v_\theta$ | = tangential velocity vector, scalar velocity in cylindrical coordinates (m $s^{-1}$) |
| $V$ | = volume of a gas ($m^3$) |
| $V_{d,gas}$, $V_{d,part,i}$ | = gas, particle dry-deposition velocity (cm $s^{-1}$) |
| $V_{f,i}$ | = fall (sedimentation) velocity of particle of size $i$ (cm $s^{-1}$) |
| $V_{G,a}$, $V_{G,c}$ | = average, cumulative interpolated parameter value in a spherical-coordinate cell |
| $V_L$ | = total volume concentration of particles in a lognormal mode ($cm^2$ $cm^{-3}$) |
| $V_{rat}$ | = volume ratio of two adjacent size bins (—) |

**w, W**

| | |
|---|---|
| $w$, $\bar{w}$, $w'$ | = precise, mean, perturbation vertical scalar velocities in altitude coordinate (m $s^{-1}$) |
| $w_g$, $w_{g,sat}$ | = volumetric, saturated water content of soil ($m^3$ $m^{-3}$) |
| $w_L$ | = liquid-water content of the air (g $m^{-3}$) |
| W*, W | = work (J), work per unit mass (J $kg^{-1}$) |
| $\mathrm{w}_p$ | = vertical velocity in the pressure coordinate (mb $s^{-1}$) |

### x, X

| | |
|---|---|
| $x$ | = west-east distance (m) |
| $x_h$ | = ratio of substrate radius to critical radius for nucleation (—) |
| $x_\lambda$ | = meteorological range at a given wavelength (km) |

### y, Y

| | |
|---|---|
| $y$ | = south–north distance (m) |
| $Y$ | = year A.D. |
| $Y_{q,p,\lambda,T}$ | = quantum yield (molecules photon$^{-1}$) |

### z, Z

| | |
|---|---|
| $z$ | = vertical distance (m) |
| $z_c$ | = altitude of center of cloud thermal above thermal base (m) |
| $z_{\text{LFC}}$ | = level of free convection (m) |
| $z_{\text{LNB}}$ | = level of neutral buoyancy in a cloud (m) |
| $z_r$ | = reference height of 10 m |
| $z_{ref}$ | = reference height (m) |
| $z_{surf}$, $z_{top}$ | = altitude above sea level of the ground surface, of the model top (m) |
| $\Delta z_t$ | = depth of atmosphere from surface to tropopause (m) |
| $z_{0,h}$, $z_{0,q}$, $z_{0,v}$ | = surface roughness length for energy, gas $q$, water vapor (m) |
| $z_{0,m}$, $z_{0,c}$ | = surface (aerodynamic) roughness length for momentum, averaged for a cell (m) |
| $Z$ | = absolute value of ionic charge (e.g., +1, +2) |
| $Z$ | = geopotential height (m) |
| $Z_n$ | = Zeldovich nonequilibrium correction factor (—) |
| $Z_t$ | = altitude difference between top of a column of air and the surface (m) |

### α, A (Alpha)

| | |
|---|---|
| $\alpha_a$, $\alpha_d$ | = specific volume of moist air, dry air (m$^3$ kg$^{-1}$) |
| $\alpha_c$ | = Charnock constant (—) |
| $\alpha_g$ | = wetness function for evaporation calculation (—) |
| $\alpha_g$ | = parameter in modified gamma size distribution |
| $\alpha_{q,i}$, $\alpha_h$ | = accommodation coefficient for gas $q$, for energy (—) |
| $\alpha_{i,\lambda}$ | = size parameter (—) |
| $\alpha_{s,j}$ | = coefficient used in Gear's method (—) |

### β, B (Beta)

| | |
|---|---|
| $\beta$ | = change of Coriolis parameter with latitude (s$^{-1}$ m$^{-1}$) |
| $\beta_g$ | = wetness function for evaporation calculation (—) |
| $\beta_h$, $\beta_m$ | = coefficients in dimensionless potential temperature, wind-shear equations (—) |
| $\beta_{i,j}$ | = coagulation kernel for particles $i$ and $j$ (cm$^3$ particle$^{-1}$ s$^{-1}$) |
| $\beta_s$ | = coefficient used in Gear's method (—) |

| | |
|---|---|
| $\beta_x$, $\beta_y$ | = rate at which molecules of gas $x$, $y$ strike a surface (molecules cm$^{-2}$ s$^{-1}$) |

## $\gamma, \Gamma$ (Gamma)

| | |
|---|---|
| $\gamma$ | $= c_{p,d}/c_{v,d}$ |
| $\gamma_q$ | = reaction probability of gas $q$ on a particle surface (—) |
| $\gamma_g$ | = parameter in modified gamma size distribution |
| $\gamma_h$, $\gamma_m$ | = coefficients in dimensionless potential temperature, wind shear equations (—) |
| $\gamma_+$, $\gamma_-$ | = activity coefficient of cation, of anion (—) |
| $\gamma_\pm$, $\gamma_{12b}$, $\gamma_{12m}$ | = mean, mean binary, mean mixed activity coefficient (—) |
| $\Gamma_d$, $\Gamma_s$, $\Gamma_v$, $\Gamma_w$ | = dry adiabatic, standard atmospheric, virtual-temperature, pseudoadiabatic lapse rate (K km$^{-1}$) |

## $\delta, \Delta$ (Delta)

| | |
|---|---|
| $\delta$ | = solar declination angle |
| $\delta_i$ | = mean distance from center of sphere for coagulation (cm) |

## $\epsilon$, E (Epsilon)

| | |
|---|---|
| $\epsilon$ | $= R'/R_v$ |
| $\epsilon_k$ | = dissipation rate of turbulent energy per gram of medium (cm$^2$ s$^{-3}$) |
| $\epsilon_{ob}$ | = obliquity of the ecliptic (deg) |
| $\epsilon_\lambda$, $\epsilon_{e,0}$ | = emissivity of a body at a given wavelength, average emissivity of earth (—) |

## $\zeta$, Z (Zeta)

| | |
|---|---|
| $\zeta_a$, $\zeta_{r,z}$ | = absolute vorticity, vertical component of relative vorticity (s$^{-1}$) |

## $\eta$, H (Eta)

| | |
|---|---|
| $\eta_a$ | = dynamic viscosity of air (kg m$^{-1}$ s$^{-1}$) |

## $\theta, \Theta$ (Theta)

| | |
|---|---|
| $\theta$ | = angle of curvature around a circle (deg) |
| $\theta$, $\theta_s$ | = zenith angle, solar zenith angle (deg) |
| $\theta_*$ | = turbulent potential temperature scale (K) |
| $\theta_c$ | = contact angle (deg) |
| $\theta_{f,q}$ | = surface coverage (fraction) |
| $\theta_p$, $\bar{\theta}_p$, $\theta'_p$ | = precise, mean, and perturbation potential temperature (K) |
| $\theta_{p,v}$ | = virtual potential temperature (K) |
| $\theta_{p,e}$ | = equivalent potential temperature (K) |
| $\theta_v$ | = potential virtual temperature (K) |
| $\Theta$ | = scattering angle (deg) |

**κ, K (Kappa)**

| | |
|---|---|
| $\kappa$ | $= R'/c_{p,d}$ |
| $\kappa_d$, $\kappa'_{d,i}$ | = thermal conductivity, corrected thermal conductivity of dry air (W $m^{-1}$ $K^{-1}$) |
| $\kappa_s$ | = thermal conductivity of soil–water–air mixture (W $m^{-1}$ $K^{-1}$) |
| $\kappa_\lambda$ | = imaginary part of the index of refraction |

**λ, Λ (Lambda)**

| | |
|---|---|
| $\lambda$, $\lambda_\alpha$ | = wavelength of an electromagnetic, of an atmospheric wave (μm) |
| $\lambda_a$, $\lambda_q$, $\lambda_{p,i}$ | = mean free path of air, of gas $q$, of a particle (cm) |
| $\lambda_h$ | = thermal mean free path (cm) |
| $\lambda_e$ | = longitude (deg) |
| $\lambda_{ec}$ | = ecliptic longitude of earth (deg) |
| $\lambda_p$ | = peak wavelength of emissions of radiation at a given temperature (μm) |
| $\lambda_R$ | = Rossby radius of deformation (m) |
| $\lambda_r$ | = empirical parameter in Marshall–Palmer rainfall distribution ($\mu m^{-1}$) |
| $\Lambda_c$ | = implicit loss coefficient for a chemical species ($s^{-1}$) |

**μ, M (Mu)**

| | |
|---|---|
| $\mu$, $\mu_s$ | = cosine of zenith angle, cosine of solar zenith angle (—) |
| $\mu_c$ | = Crank–Nicolson parameter |
| $\mu_i$, $\mu_i^\circ$ | = chemical potential of species $i$, of species $i$ at a reference temperature |

**ν, N (Nu)**

| | |
|---|---|
| $\nu$, $\nu_\alpha$ | = frequency of electromagnetic, of atmospheric wave ($s^{-1}$) |
| $\nu_c$ | = acoustic cutoff frequency ($s^{-1}$) |
| $\nu_0$ | = frequency of atomic vibrations on a surface ($s^{-1}$) |
| $\tilde{\nu}$ | = wavenumber of electromagnetic wave ($cm^{-1}$) |
| $\nu_a$ | = kinematic viscosity of air ($m^2$ $s^{-1}$) |
| $\nu_i$ | = stoichiometric coefficient in an equilibrium reaction |

**π, Π (Pi)**

| | |
|---|---|
| $\pi_a$ | = surface minus top pressure in a model (mb) |

**ρ, P (Rho)**

| | |
|---|---|
| $\rho_a$, $\rho_d$, $\rho_v$ | = density of moist air, dry air, water vapor (kg $m^{-3}$) |
| $\rho_g$, $\rho_s$, $\rho_w$ | = density of soil–water–air mixture, pure soil, liquid water (kg $m^{-3}$) |
| $\rho_p$, $\rho_{p,i}$ | = density of a particle, of a particle of size $i$ (g $cm^{-3}$) |
| $\rho_{p,q}$ | = density of condensed component $q$ in a particle (g $cm^{-3}$) |
| $\rho_{v,s}$ | = density of water vapor at saturation (g $cm^{-3}$) |

### σ, Σ (Sigma)

| | |
|---|---|
| $\sigma$ | = sigma-pressure level (fraction) |
| $\dot{\sigma}$ | = vertical velocity in the sigma-pressure coordinate ($s^{-1}$) |
| $\sigma_{a,\lambda}, \sigma_{s,\lambda}$ | = extinction coefficients due to total absorption, total scattering ($cm^{-1}$) |
| $\sigma_{a,g,\lambda}, \sigma_{s,g,\lambda}$ | = extinction coefficients due to gas absorption, gas scattering ($cm^{-1}$) |
| $\sigma_{a,a,\lambda}, \sigma_{s,a,\lambda}$ | = extinction coefficients due to aerosol absorption, aerosol scattering ($cm^{-1}$) |
| $\sigma_{a,c,\lambda}, \sigma_{s,c,\lambda}$ | = extinction coefficients due to cloud-drop absorption, cloud-drop scattering ($cm^{-1}$) |
| $\sigma_b$ | = extinction coefficient of background light ($cm^{-1}$) |
| $\sigma_B$ | = Stefan–Boltzmann constant ($W\ m^{-2}\ K^{-4}$) |
| $\sigma_{ext,\lambda}, \sigma_\lambda$ | = total extinction coefficients at wavelength λ ($cm^{-1}$) |
| $\sigma_g$ | = geometric standard deviation (—) |
| $\sigma_p$ | = surface tension of a particle ($J\ cm^{-2}$) |
| $\sigma_u, \sigma_u^2$ | = standard deviation, biased variance of a distribution |
| $\sigma_0$ | = molecular surface area of one surface adsorption site ($cm^2$ ) |

### τ, T (Tau)

| | |
|---|---|
| $\tau_{xz}$ | = shearing-stress ($dyn\ m^{-2}$) |
| $\tau'$ | = optical depth adjusted with delta function (—) |
| $\tau_{ad,q}$ | = characteristic time for aqueous diffusion of species $q$ (s) |
| $\tau_A, \tau_{(1/2)A}$ | = $e$-folding lifetime, half-life of reacting species A (s) |
| $\tau_{bv}$ | = period of Brunt–Väisälä oscillations (s) |
| $\tau_d$ | = number of seconds in a day (s) |
| $\tau_\lambda$ | = total optical depth at wavelength λ (—) |

### υ, Υ (Upsilon)

| | |
|---|---|
| $\upsilon, \upsilon_i, \upsilon_{q,i}$ | = volume of one particle, of one particle in bin $i$, of component $q$ in bin $i$ ($cm^3$) |
| $d\upsilon_i$ | = volume width of size bin $i$ ($cm^3$) |
| $\upsilon_p$ | = particle volume at peak diameter in self-preserving size distribution ($cm^3$) |

### φ, ϕ, Φ (Phi)

| | |
|---|---|
| $\varphi$ | = latitude (deg) |
| $\phi, \phi_s$ | = azimuth angle, solar azimuth angle (deg) |
| $\phi_{c_p}, \phi^{\circ}_{c_p}$ | = apparent molal heat capacity, apparent molal heat capacity at infinite dilution ($J\ mole^{-1}\ K^{-1}$) |
| $\phi_h, \phi_m$ | = dimensionless temperature gradient, dimensionless wind shear (—) |
| $\phi_L$ | = relative apparent molal enthalpy ($J\ mole^{-1}$) |
| $\mathbf{\Phi}, \Phi$ | = vector, magnitude of geopotential ($m^2\ s^{-2}$) |
| $\Phi_a, \Phi_g$ | = ageostrophic, geostrophic component of geopotential ($m^2\ s^{-2}$) |

### χ, Χ (Chi)

| | |
|---|---|
| $\chi$ | = volume mixing ratio of a gas (molecules molecule$^{-1}$) |
| $\chi_i$ | = electrical capacitance of a crystal of size $i$ (cm) |

### ψ, Ψ (Psi)

| | |
|---|---|
| $\psi_h, \psi_m$ | = influence function for energy, momentum (—) |
| $\psi_p, \psi_{p,sat}$ | = moisture potential, saturated moisture potential of soil (cm) |

### ω, Ω (Omega)

| | |
|---|---|
| $\omega$ | = mass mixing ratio (kg kg$^{-1}$) |
| $\omega_{h,i}, \omega_{q,i}$ | = adjustment to thermal conductivity, diffusion coefficient (—) |
| $\omega_I, \omega_L, \omega_T, \omega_v$ | = mass mixing ratio of ice, liquid water, total water, water vapor (kg kg$^{-1}$) |
| $\omega_L, \hat{\omega}_L$ | = mass mixing ratio of liquid water in a cloud, outside a cloud (kg kg$^{-1}$) |
| $\omega_{v,s}$ | = mass mixing ratio of water vapor at saturation (kg kg$^{-1}$) |
| $\omega_{s,\lambda}, \omega'_s$ | = single-scattering albedo, single-scattering albedo adjusted with delta function (—) |
| $\omega'_v$ | = mass mixing ratio of water vapor outside a cloud (kg kg$^{-1}$) |
| $\boldsymbol{\omega}_z, \omega_z$ | = angular-velocity vector, scalar in cylindrical coordinates (m s$^{-1}$) |
| $\boldsymbol{\Omega}, \Omega$ | = vector, magnitude of the earth's angular velocity (s$^{-1}$) |
| $\Omega_a$ | = solid angle (sr) |

# APPENDIX B

# Tables

## B.1. STANDARD ATMOSPHERIC VARIABLES VERSUS ALTITUDE

**Table B.1.** Variation of Geopotential Height ($Z$), Air Pressure, Air Temperature, and Air Density with Altitude in a Standard Atmosphere

| Alt. (km) | Geopot. Height (km) | Press. (mb) | Temp. (K) | Density (kg m$^{-3}$) | Alt. (km) | Geopot. Height (km) | Press. (mb) | Temp. (K) | Density (kg m$^{-3}$) |
|---|---|---|---|---|---|---|---|---|---|
| 0 | 0 | 1013.25 | 288.15 | 1.225 | 22 | 21.924 | 40.5 | 218.57 | 0.0645 |
| 0.1 | 0.1 | 1001.20 | 287.50 | 1.213 | 23 | 22.917 | 34.7 | 219.57 | 0.0550 |
| 0.2 | 0.2 | 989.45 | 286.85 | 1.202 | 24 | 23.910 | 29.7 | 220.56 | 0.0469 |
| 0.3 | 0.3 | 977.72 | 286.20 | 1.190 | 25 | 24.902 | 25.5 | 221.55 | 0.0401 |
| 0.4 | 0.4 | 966.11 | 285.55 | 1.179 | 26 | 25.894 | 21.9 | 222.54 | 0.0343 |
| 0.5 | 0.5 | 954.61 | 284.90 | 1.167 | 27 | 26.886 | 18.8 | 223.54 | 0.0293 |
| 0.6 | 0.6 | 943.22 | 284.25 | 1.156 | 28 | 27.877 | 16.2 | 224.53 | 0.0251 |
| 0.7 | 0.7 | 931.94 | 283.60 | 1.145 | 29 | 28.868 | 13.9 | 225.52 | 0.0215 |
| 0.8 | 0.8 | 920.77 | 282.95 | 1.134 | 30 | 29.859 | 12.0 | 226.51 | 0.0184 |
| 0.9 | 0.9 | 909.71 | 282.30 | 1.123 | 31 | 30.850 | 10.3 | 227.50 | 0.0158 |
| 1 | 1.0 | 898.80 | 281.65 | 1.112 | 32 | 31.840 | 8.89 | 228.49 | 0.0136 |
| 1.5 | 1.5 | 845.59 | 278.40 | 1.058 | 33 | 32.830 | 7.67 | 230.97 | 0.0116 |
| 2 | 1.999 | 795.0 | 275.15 | 1.007 | 34 | 33.819 | 6.63 | 233.74 | 0.00989 |
| 2.5 | 2.499 | 746.9 | 271.91 | 0.957 | 35 | 34.808 | 5.75 | 236.51 | 0.00846 |
| 3 | 2.999 | 701.2 | 268.66 | 0.909 | 36 | 35.797 | 4.99 | 239.28 | 0.00726 |
| 3.5 | 3.498 | 657.8 | 265.41 | 0.863 | 37 | 36.786 | 4.33 | 242.05 | 0.00624 |
| 4 | 3.997 | 616.6 | 262.17 | 0.819 | 38 | 37.774 | 3.77 | 244.82 | 0.00537 |
| 4.5 | 4.497 | 577.5 | 258.92 | 0.777 | 39 | 38.762 | 3.29 | 247.58 | 0.00463 |
| 5 | 4.996 | 540.5 | 255.68 | 0.736 | 40 | 39.750 | 2.87 | 250.35 | 0.00400 |
| 5.5 | 5.495 | 505.4 | 252.43 | 0.697 | 41 | 40.737 | 2.51 | 253.11 | 0.00346 |
| 6 | 5.994 | 472.2 | 249.19 | 0.660 | 42 | 41.724 | 2.20 | 255.88 | 0.00299 |
| 6.5 | 6.493 | 440.7 | 245.94 | 0.624 | 43 | 42.711 | 1.93 | 258.64 | 0.00260 |
| 7 | 6.992 | 411.1 | 242.70 | 0.590 | 44 | 43.698 | 1.69 | 261.40 | 0.00226 |
| 7.5 | 7.491 | 383.0 | 239.46 | 0.557 | 45 | 44.684 | 1.49 | 264.16 | 0.00197 |
| 8 | 7.990 | 356.5 | 236.22 | 0.526 | 46 | 45.669 | 1.31 | 266.93 | 0.00171 |
| 8.5 | 8.489 | 331.5 | 232.97 | 0.496 | 47 | 46.655 | 1.16 | 269.68 | 0.0015 |
| 9 | 8.987 | 308.0 | 229.73 | 0.467 | 48 | 47.640 | 1.02 | 270.65 | 0.00132 |
| 9.5 | 9.486 | 285.8 | 226.49 | 0.440 | 49 | 48.625 | 0.903 | 270.65 | 0.00116 |
| 10 | 9.984 | 265.0 | 223.25 | 0.414 | 50 | 49.610 | 0.798 | 270.65 | 0.00103 |
| 11 | 10.981 | 227.0 | 216.78 | 0.365 | 55 | 54.528 | 0.425 | 260.77 | $5.7 \times 10^{-4}$ |
| 12 | 11.977 | 194.0 | 216.65 | 0.312 | 60 | 59.439 | 0.220 | 247.02 | $3.1 \times 10^{-4}$ |
| 13 | 12.973 | 165.8 | 216.65 | 0.267 | 65 | 64.342 | 0.109 | 233.29 | $1.6 \times 10^{-4}$ |
| 14 | 13.969 | 141.7 | 216.65 | 0.228 | 70 | 69.238 | 0.0522 | 219.59 | $8.3 \times 10^{-5}$ |

*(cont.)*

**Table B.1.** (cont.)

| Alt. (km) | Geopot. Height (km) | Press. (mb) | Temp. (K) | Density (kg m$^{-3}$) | Alt. (km) | Geopot. Height (km) | Press. (mb) | Temp. (K) | Density (kg m$^{-3}$) |
|---|---|---|---|---|---|---|---|---|---|
| 15 | 14.965 | 121.1 | 216.65 | 0.195 | 75 | 74.125 | 0.0239 | 208.40 | $4.0 \times 10^{-5}$ |
| 16 | 15.960 | 103.5 | 216.65 | 0.166 | 80 | 79.006 | 0.0105 | 198.64 | $1.8 \times 10^{-5}$ |
| 17 | 16.955 | 88.5 | 216.65 | 0.142 | 85 | 83.878 | 0.0045 | 188.89 | $8.2 \times 10^{-6}$ |
| 18 | 17.949 | 75.7 | 216.65 | 0.122 | 90 | 88.744 | 0.0018 | 186.87 | $3.4 \times 10^{-6}$ |
| 19 | 18.943 | 64.7 | 216.65 | 0.104 | 95 | 93.601 | 0.00076 | 188.42 | $7.5 \times 10^{-7}$ |
| 20 | 19.937 | 55.3 | 216.65 | 0.0889 | 100 | 98.451 | 0.00032 | 195.08 | $5.6 \times 10^{-7}$ |
| 21 | 20.931 | 47.3 | 217.58 | 0.0757 | | | | | |

*Source:* NOAA (1976).

## B.2. SOLAR IRRADIANCE AT THE TOP OF THE ATMOSPHERE

**Table B.2.** Extraterrestrial Solar Irradiance at the Top of the Earth's Atmosphere versus Wavelength

| $\lambda$ (μm) | $\bar{F}_{s,\lambda}$ (W m$^{-2}$ μm$^{-1}$) | $\lambda$ (μm) | $\bar{F}_{s,\lambda}$ (W m$^{-2}$ μm$^{-1}$) | $\lambda$ (μm) | $\bar{F}_{s,\lambda}$ (W m$^{-2}$ μm$^{-1}$) | $\lambda$ (μm) | $\bar{F}_{s,\lambda}$ (W m$^{-2}$ μm$^{-1}$) | $\lambda$ (μm) | $\bar{F}_{s,\lambda}$ (W m$^{-2}$ μm$^{-1}$) |
|---|---|---|---|---|---|---|---|---|---|
| 0.105 | 0.055 | 0.355 | 1125 | 0.605 | 1773 | 0.855 | 909 | 3.1 | 26 |
| 0.110 | 0.050 | 0.360 | 1077 | 0.610 | 1722 | 0.860 | 953 | 3.2 | 22.6 |
| 0.115 | 0.039 | 0.365 | 1274 | 0.615 | 1671 | 0.865 | 896 | 3.3 | 19.2 |
| 0.120 | 1.168 | 0.370 | 1359 | 0.620 | 1721 | 0.870 | 933 | 3.4 | 16.6 |
| 0.125 | 0.371 | 0.375 | 1219 | 0.625 | 1665 | 0.875 | 928 | 3.5 | 14.6 |
| 0.130 | 0.060 | 0.380 | 1340 | 0.630 | 1658 | 0.880 | 907 | 3.6 | 13.5 |
| 0.135 | 0.080 | 0.385 | 1113 | 0.635 | 1639 | 0.885 | 904 | 3.7 | 12.3 |
| 0.140 | 0.061 | 0.390 | 1345 | 0.640 | 1632 | 0.890 | 894 | 3.8 | 11.1 |
| 0.145 | 0.063 | 0.395 | 1096 | 0.645 | 1601 | 0.895 | 892 | 3.9 | 10.3 |
| 0.150 | 0.096 | 0.400 | 1796 | 0.650 | 1557 | 0.9 | 891 | 4 | 9.5 |
| 0.155 | 0.194 | 0.405 | 1643 | 0.655 | 1502 | 0.91 | 880 | 4.1 | 8.7 |
| 0.160 | 0.206 | 0.410 | 1768 | 0.660 | 1562 | 0.92 | 869 | 4.2 | 7.8 |
| 0.165 | 0.372 | 0.415 | 1810 | 0.665 | 1570 | 0.93 | 858 | 4.3 | 7.1 |
| 0.170 | 0.607 | 0.420 | 1760 | 0.670 | 1539 | 0.94 | 847 | 4.4 | 6.5 |
| 0.175 | 0.885 | 0.425 | 1719 | 0.675 | 1556 | 0.95 | 837 | 4.5 | 5.92 |
| 0.180 | 1.90 | 0.430 | 1615 | 0.680 | 1526 | 0.96 | 820 | 4.6 | 5.35 |
| 0.185 | 2.53 | 0.435 | 1798 | 0.685 | 1481 | 0.97 | 803 | 4.7 | 4.86 |
| 0.190 | 3.88 | 0.440 | 1829 | 0.690 | 1460 | 0.98 | 785 | 4.8 | 4.47 |
| 0.195 | 5.35 | 0.445 | 1951 | 0.695 | 1491 | 0.99 | 767 | 4.9 | 4.11 |
| 0.200 | 7.45 | 0.450 | 2048 | 0.700 | 1453 | 1.0 | 748 | 5 | 3.79 |
| 0.205 | 10.7 | 0.455 | 2043 | 0.705 | 1420 | 1.05 | 668 | 6 | 1.82 |
| 0.210 | 23.4 | 0.460 | 2054 | 0.710 | 1407 | 1.1 | 593 | 7 | .99 |
| 0.215 | 36.3 | 0.465 | 2012 | 0.715 | 1376 | 1.15 | 535 | 8 | .585 |
| 0.220 | 44.7 | 0.470 | 2007 | 0.720 | 1351 | 1.2 | 485 | 9 | .367 |
| 0.225 | 55.0 | 0.475 | 2042 | 0.725 | 1358 | 1.25 | 438 | 10 | .241 |
| 0.230 | 50.5 | 0.480 | 2061 | 0.730 | 1331 | 1.3 | 397 | 11 | .165 |
| 0.235 | 49.5 | 0.485 | 1867 | 0.735 | 1322 | 1.35 | 358 | 12 | .117 |
| 0.240 | 47.0 | 0.490 | 1943 | 0.740 | 1282 | 1.4 | 337 | 13 | .0851 |
| 0.245 | 62.2 | 0.495 | 1993 | 0.745 | 1276 | 1.45 | 312 | 14 | .0634 |

*(cont.)*

Table B.2. (cont.)

| $\lambda$ ($\mu$m) | $\bar{F}_{s,\lambda}$ (W m$^{-2}$ $\mu$m$^{-1}$) | $\lambda$ ($\mu$m) | $\bar{F}_{s,\lambda}$ (W m$^{-2}$ $\mu$m$^{-1}$) | $\lambda$ ($\mu$m) | $\bar{F}_{s,\lambda}$ (W m$^{-2}$ $\mu$m$^{-1}$) | $\lambda$ ($\mu$m) | $\bar{F}_{s,\lambda}$ (W m$^{-2}$ $\mu$m$^{-1}$) | $\lambda$ ($\mu$m) | $\bar{F}_{s,\lambda}$ (W m$^{-2}$ $\mu$m$^{-1}$) |
|---|---|---|---|---|---|---|---|---|---|
| 0.250 | 55.2 | 0.500 | 1892 | 0.750 | 1272 | 1.5 | 288 | 15 | .0481 |
| 0.255 | 69.5 | 0.505 | 1941 | 0.755 | 1262 | 1.55 | 267 | 16 | .0371 |
| 0.260 | 111 | 0.510 | 1937 | 0.760 | 1241 | 1.6 | 245 | 17 | .0291 |
| 0.265 | 212 | 0.515 | 1805 | 0.765 | 1220 | 1.65 | 223 | 18 | .0231 |
| 0.270 | 255 | 0.520 | 1811 | 0.770 | 1195 | 1.7 | 202 | 19 | .0186 |
| 0.275 | 197 | 0.525 | 1850 | 0.775 | 1179 | 1.75 | 180 | 20 | .0152 |
| 0.280 | 186 | 0.530 | 1907 | 0.780 | 1189 | 1.8 | 159 | 25 | .00617 |
| 0.285 | 317 | 0.535 | 1894 | 0.785 | 1183 | 1.85 | 142 | 30 | .00297 |
| 0.290 | 546 | 0.540 | 1840 | 0.790 | 1151 | 1.9 | 126 | 35 | .0016 |
| 0.295 | 573 | 0.545 | 1866 | 0.795 | 1142 | 1.95 | 114 | 40 | .000942 |
| 0.300 | 493 | 0.550 | 1845 | 0.800 | 1126 | 2 | 103 | 50 | .000391 |
| 0.305 | 669 | 0.555 | 1854 | 0.805 | 1112 | 2.1 | 90 | 60 | .00019 |
| 0.310 | 711 | 0.560 | 1801 | 0.810 | 1080 | 2.2 | 79 | 80 | .0000416 |
| 0.315 | 765 | 0.565 | 1828 | 0.815 | 1073 | 2.3 | 69 | 100 | .0000257 |
| 0.320 | 777 | 0.570 | 1824 | 0.820 | 1049 | 2.4 | 62 | 120 | .0000126 |
| 0.325 | 935 | 0.575 | 1851 | 0.825 | 1050 | 2.5 | 55 | 150 | .00000523 |
| 0.330 | 1041 | 0.580 | 1833 | 0.830 | 1027 | 2.6 | 48 | 200 | .00000169 |
| 0.335 | 950 | 0.585 | 1838 | 0.835 | 1012 | 2.7 | 43 | 250 | .0000007 |
| 0.340 | 1035 | 0.590 | 1760 | 0.840 | 1006 | 2.8 | 39 | 300 | .00000023 |
| 0.345 | 980 | 0.595 | 1791 | 0.845 | 983 | 2.9 | 35 | 400 | .00000011 |
| 0.350 | 1019 | 0.600 | 1752 | 0.850 | 951 | 3 | 31 | 1000 | 0 |

$\lambda$ is the midpoint of a wavelength interval, and the irradiance at each $\lambda$ is integrated from the lower to the upper edge of the interval. For example, the irradiance in the interval from $\lambda = 0.1475$ to $0.1575$ $\mu$m, centered at $\lambda = 0.15$ $\mu$m is $\bar{F}_{s,\lambda}\,\Delta\lambda = 0.096$ W m$^{-2}$ $\mu$m$^{-1}$ $\times$ 0.005 $\mu$m = 0.00048 W m$^{-2}$. The sum of irradiance over all bins is the solar constant. The data were derived from Woods et al. (1996) for $\lambda < 0.275$ $\mu$m, Nicolet (1989) for $0.275 < \lambda < 0.9$ $\mu$m, and Thekaekara (1974) for $\lambda > 0.9$ $\mu$m.

## B.3. GAS-PHASE SPECIES

Table B.3. Inorganic and Organic Gases and Their Possible Chemical Structures

**Chemical Name**
**Molecular Formula**
**Chemical Structure**

**Inorganic**

**Hydrogen and Oxygen Species**

| Atomic hydrogen | Atomic oxygen (triplet) | Molecular oxygen | Ozone |
|---|---|---|---|
| H | O | $O_2$ | $O_3$ |
| H• |  | O=O | 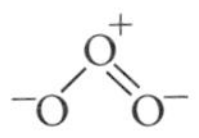 |

*(cont.)*

**Table B.3.** (cont.)

| Chemical Name<br>Molecular Formula<br>Chemical Structure | | | |
|---|---|---|---|
| **Inorganic** | | | |
| Hydroxyl radical<br>OH<br>$\dot{O}\text{–}H$ | Hydroperoxy radical<br>$HO_2$<br>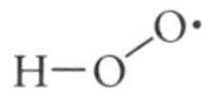 | Hydrogen peroxide<br>$H_2O_2$<br>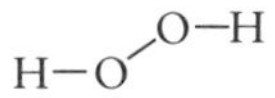 | Water vapor<br>$H_2O$<br>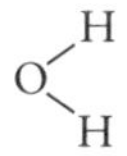 |
| **Nitrogen Species** | | | |
| Molecular nitrogen<br>$N_2$<br>$N\equiv N$ | Nitric oxide<br>NO<br>$\dot{N}=O$ | Nitrogen dioxide<br>$NO_2$<br>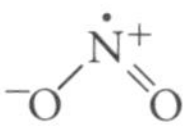 | Nitrate radical<br>$NO_3$<br>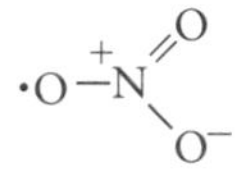 |
| Nitrous acid<br>HONO<br> | Nitric acid<br>$HNO_3$<br>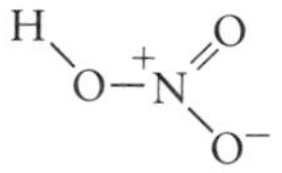 | Peroxynitric acid<br>$HO_2NO_2$<br>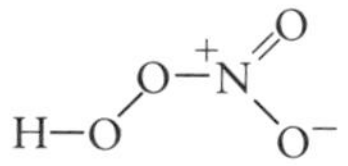 | Nitrous oxide<br>$N_2O$<br>$O=\overset{+}{N}=\overset{-}{N}$ |
| Dinitrogen pentoxide<br>$N_2O_5$<br>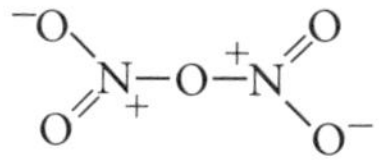 | | | |
| **Sulfur Species** | | | |
| Atomic sulfur<br>S<br>$:\ddot{S}\cdot$ | Sulfur monoxide<br>(sulfonyl or thionyl radical)<br>SO<br>$S=O$ | Sulfur dioxide<br>$SO_2$<br>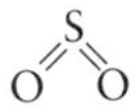 | Sulfur trioxide<br>$SO_3$<br>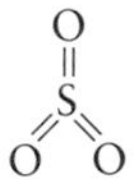 |
| Bisulfite<br>$HSO_3$<br>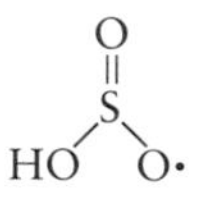 | Sulfuric acid<br>$H_2SO_4$<br>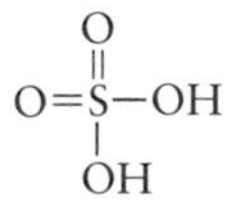 | Carbonyl sulfide<br>OCS<br>$O=C=S$ | Carbon monosulfide<br>CS<br>$^{-}C\equiv S^{+}$ |

*(cont.)*

### Table B.3. (cont.)

| **Chemical Name** |
| :---: |
| **Molecular Formula** |
| **Chemical Structure** |

## Inorganic

| Carbon disulfide | Hydrogen sulfide radical | Hydrogen sulfide | Methanethiol (Methyl sulfide) |
| :---: | :---: | :---: | :---: |
| $CS_2$ | HS | $H_2S$ | MeSH |
| | | | $CH_3SH$ |
| $S{=}C{=}S$ | (structure) | (structure) | (structure) |

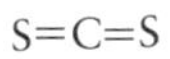

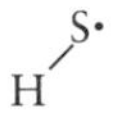

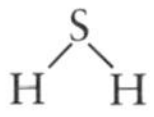

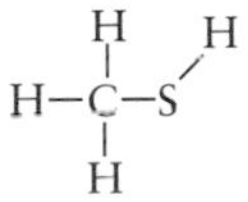

| Methanethiolate radical | Methanethiolate oxy radical | Methanethiolate peroxy radical $CH_3SO_2$ | Methanesulfenic acid |
| :---: | :---: | :---: | :---: |
| $CH_3S$ | $CH_3SO$ | | $CH_3SOH$ |
| (structure) | (structure) | (structure) | (structure) |

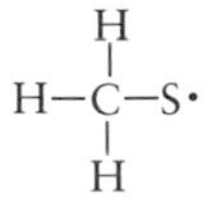

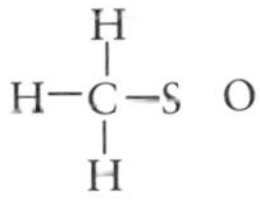

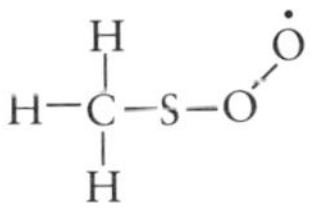

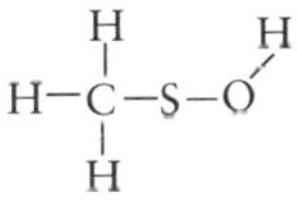

| Dimethyl sulfide | Dimethyl sulfide radical | Dimethyl sulfide–OH adduct | Dimethyl sulfone |
| :---: | :---: | :---: | :---: |
| DMS | $CH_3SCH_2$ | $CH_3S(OH)CH_3$ | $DMSO_2$ |
| $CH_3SCH_3$ | | | $CH_3S(O)_2CH_3$ |
| (structure) | (structure) | (structure) | (structure) |

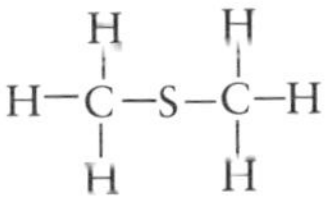

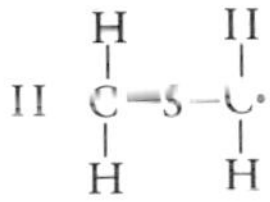

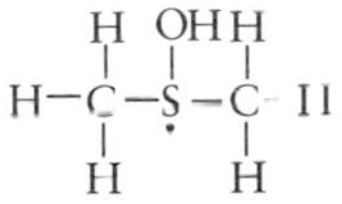

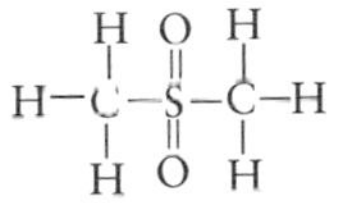

| Dimethyl disulfide | Methanesulfonic acid | Hydroxymethane-sulfonic acid | |
| :---: | :---: | :---: | :---: |
| DMDS | MSA | HMSA | |
| $CH_3SSCH_3$ | $CH_3S(O)_2OH$ | $HOCH_2S(O)_2OH$ | |
| (structure) | (structure) | (structure) | |

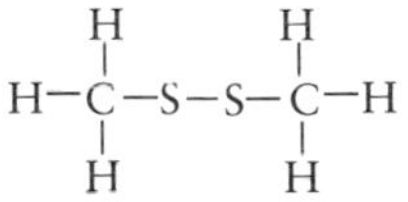

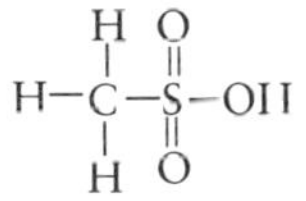

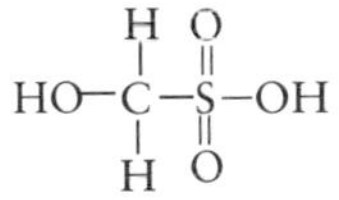

## Chlorine Species

| Atomic chlorine | Molecular chlorine | Hydrochloric acid | Chlorine monoxide |
| :---: | :---: | :---: | :---: |
| Cl | $Cl_2$ | HCl | ClO |
| $:\ddot{Cl}\cdot$ | Cl–Cl | H–Cl | $Cl{-}\dot{O}$ |

*(cont.)*

**Table B.3.** (cont.)

**Chemical Name**
**Molecular Formula**
**Chemical Structure**

**Inorganic**

Hypochlorous acid
HOCl
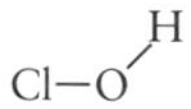

Chlorine peroxy radical
OClO
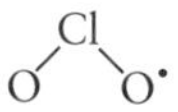

Chlorine peroxy radical
ClOO
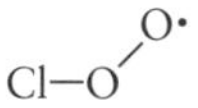

Dichlorine dioxide
$Cl_2O_2$
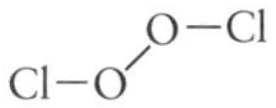

Chlorine nitrate
$ClONO_2$
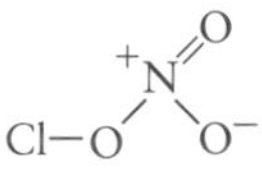

Chlorine nitrite
$ClNO_2$
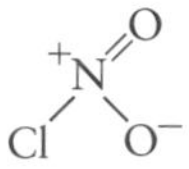

Methyl chloride
$CH_3Cl$
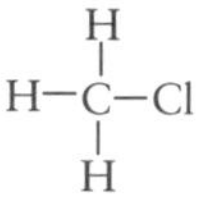

Methyl chloroform
$CH_3CCl_3$
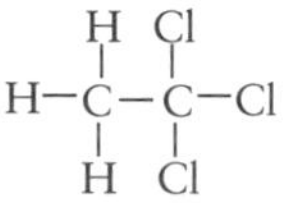

Trichlorofluoromethane
(CFC-11)
$CFCl_3$
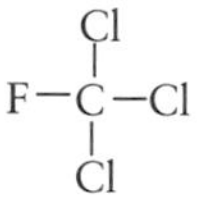

Dichlorodifluoromethane
(CFC-12)
$CF_2Cl_2$
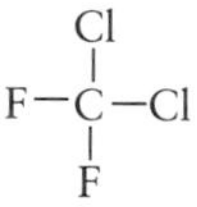

1-Fluorodichloro,
2-difluorochloroethane
(CFC-113)
$CFCl_2CF_2Cl$
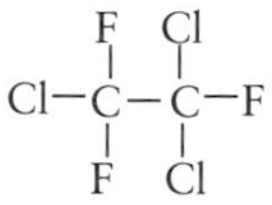

Chlorodifluoromethane
(HCFC-22)
$CF_2ClH$
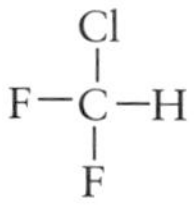

Carbon tetrachloride
$CCl_4$

**Bromine Species**

Atomic bromine
Br

Molecular bromine
$Br_2$
Br – Br

Hydrobromic acid
HBr
H—Br

Bromine chloride
BrCl
Br—Cl

Hypobromous acid
HOBr
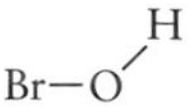

Bromine monoxide
BrO
$Br - \dot{O}$

Methyl bromide
$CH_3Br$
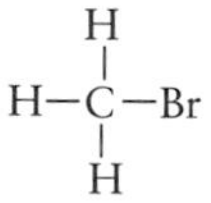

Bromine nitrate
$BrONO_2$
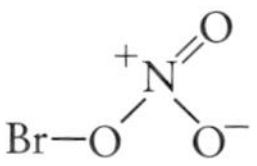

*(cont.)*

**Table B.3.** (cont.)

**Chemical Name**
**Molecular Formula**
**Chemical Structure**

## Inorganic

### Inorganic Carbon Species

Carbon monoxide
CO

$^{-}C{\equiv}O^{+}$

Carbon dioxide
$CO_2$

O=C=O

## Organic

### Alkanes

Methane
$CH_4$

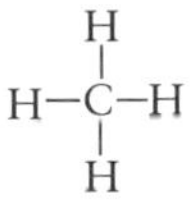

Ethane
$C_2H_6$

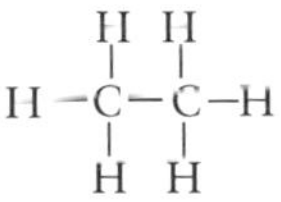

Propane
$C_3H_8$

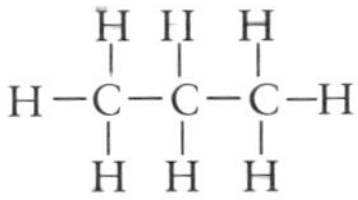

Butane
$C_4H_{10}$

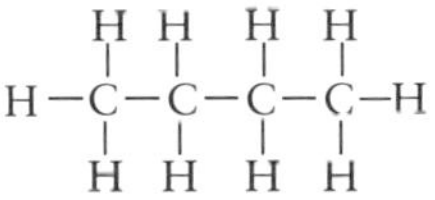

2,2-Dimethylpropane
(neopentane)
$C(CH_3)_4$

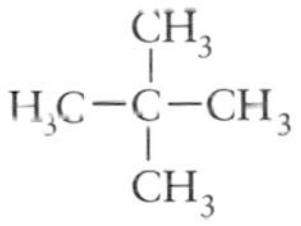

2,2,4-Trimethylpentane
(isooctane)
$(CH_3)_2CHCH_2C(CH_3)_3$

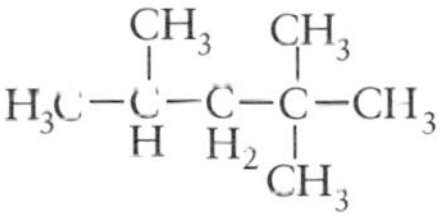

### Cycloalkanes

Cyclopropane
$(CH_2)_3$

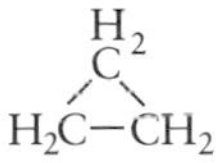

Cyclobutane
$(CH_2)_4$

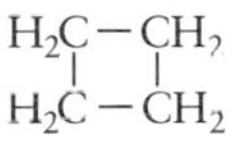

Cyclopentane
$(CH_2)_5$

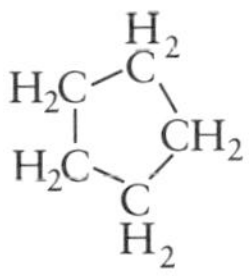

### Alkenes

Ethene
(ethylene)
$C_2H_4$

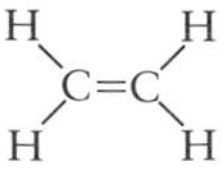

Propene
(propylene)
$C_3H_6$

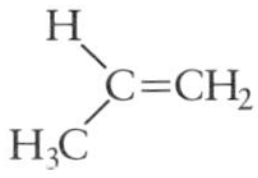

*trans*-2-Butene
$C_4H_8$

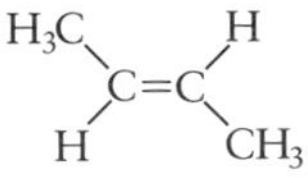

*cis*-2-Butene
$C_4H_8$

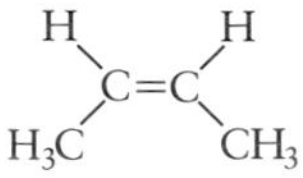

*(cont.)*

**Table B.3.** (cont.)

**Chemical Name**
**Molecular Formula**
**Chemical Structure**

## Organic

### Cycloalkenes

Cyclopentene
$C_5H_8$

Cyclohexene
$C_6H_{10}$

1-Methylcyclohexene
$C_7H_{12}$

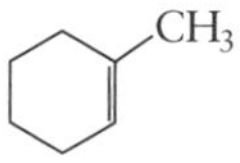


### Alkynes

Ethyne (acetylene)
$C_2H_2$

H−C≡C−H

Propyne
$C_3H_4$

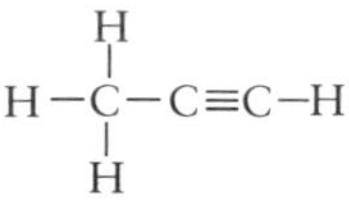


### Aromatics

Benzene
$C_6H_6$

Toluene
(methylbenzene)
**TOL**

$C_6H_5CH_3$

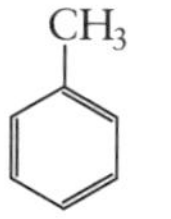


*o*-Xylene
(1,2-dimethylbenzene)
**XYL**

1,2-$(CH_3)_2C_6H_4$

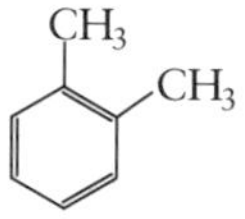


*m*-Xylene
(1,3-dimethylbenzene)
**XYL**

1,3-$(CH_3)_2C_6H_4$

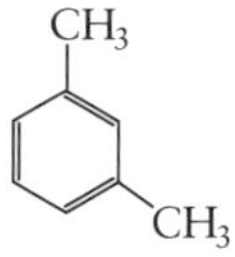


Ethylbenzene
$C_6H_5C_2H_5$

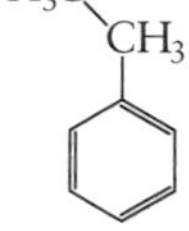


1,2,3-Trimethylbenzene
1,2,3-$(CH_3)_3C_6H_3$

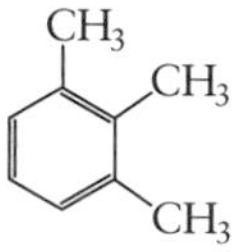


### Terpenes

Isoprene
(2-methyl-1,3-butadiene)
**ISOP**
$C_5H_8$

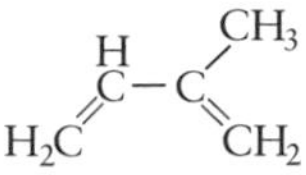


*d*-Limonene
$C_{10}H_{16}$

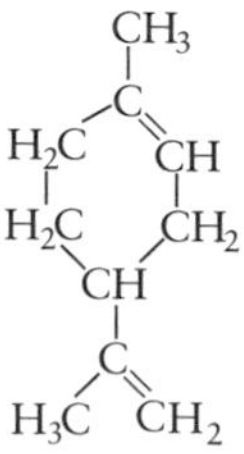


α-Pinene
$C_{10}H_{16}$

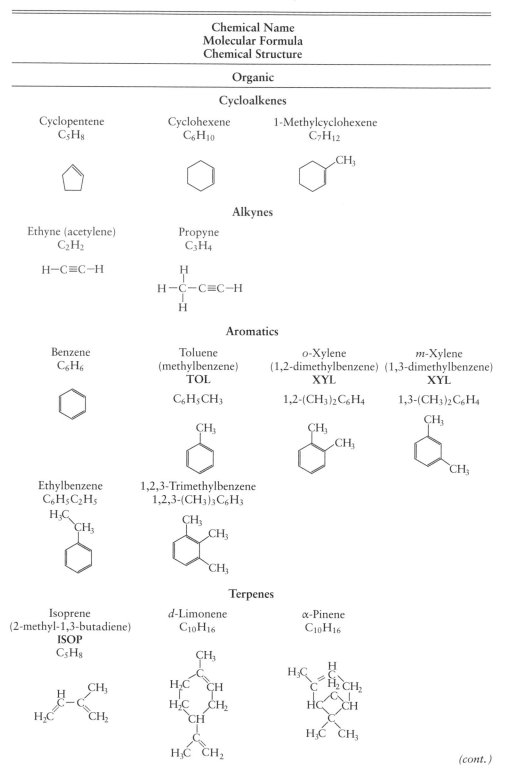

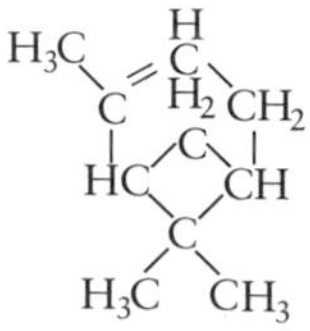


*(cont.)*

**Table B.3.** (cont.)

| Chemical Name<br>Molecular Formula<br>Chemical Structure | | | |
|---|---|---|---|
| **Organic** | | | |
| **Alcohols** | | | |
| Methanol<br>(methyl alcohol)<br>$CH_3OH$<br>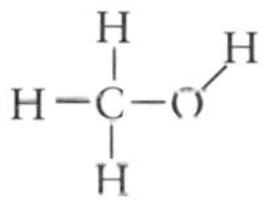 | Ethanol<br>(ethyl alcohol)<br>$C_2H_5OH$<br>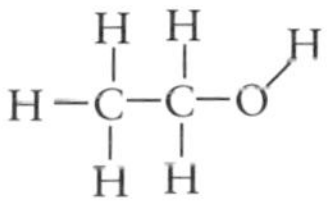 | *o*-Cresol<br>**CRES**<br>2-$CH_3C_6H_4OH$<br>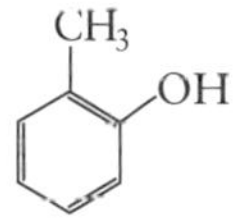 | Phenol<br>(hydroxybenzene)<br>$C_6H_5OH$<br>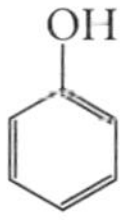 |
| **Aldehydes** | | | |
| Formaldehyde<br>(methanal)<br>HCHO<br>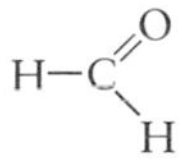 | Acetaldehyde<br>(ethanal)<br>$CH_3CHO$<br>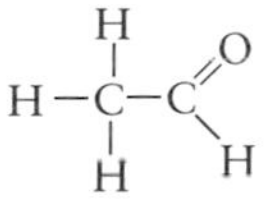 | Propionaldehyde<br>(propanal)<br>$CH_3CH_2CHO$<br>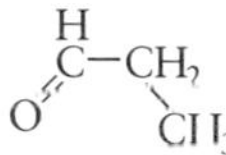 | Benzaldehyde<br>**BZA**<br>$C_6H_5CHO$<br>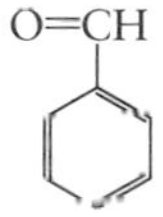 |
| Glycol aldehyde<br>$HOCH_2CHO$<br>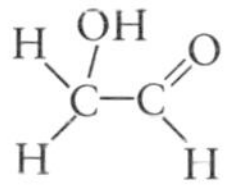 | Acrolein<br>(propenal)<br>$CH_2CH{=}CHO$<br>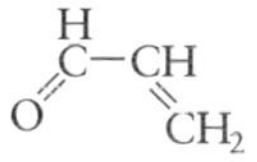 | Methacrolein<br>(2-methyl-2propenal)<br>**MACR**<br>$CH_2{=}C(CH_3)CHO$<br>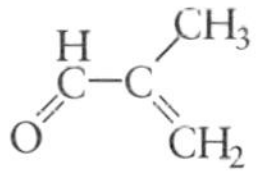 | |
| **Ketones** | | | |
| Acetone<br>(2-propanone)<br>$CH_3COCH_3$<br>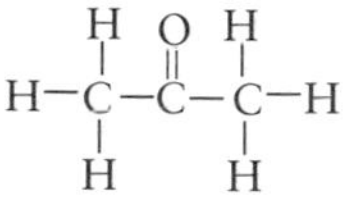 | Methylethylketone<br>(2-butanone)<br>$CH_3CH_2COCH_3$<br>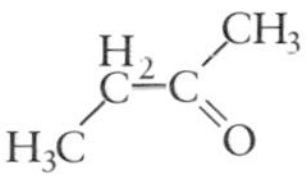 | Methylvinylketone<br>(3-buten-2-one)<br>**MVK**<br>$CH_2{=}CHCOCH_3$<br>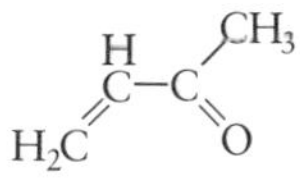 | |
| **Carboxylic Acids** | | | |
| Formic acid<br>(methanoic acid)<br>HCOOH<br>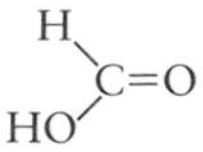 | Acetic acid<br>(ethanoic acid)<br>$CH_3COOH$<br>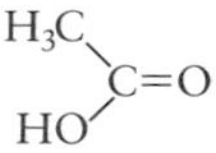 | Propionic acid<br>(propanoic acid)<br>$CH_3CH_2COOH$<br>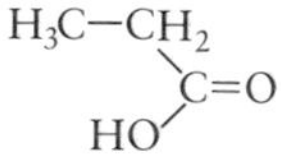 | |

*(cont.)*

**Table B.3.** (cont.)

**Chemical Name**
**Molecular Formula**
**Chemical Structure**

**Organic**

**Alkyl Radicals**

Methyl radical
$CH_3$

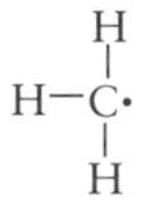

Ethyl radical
$C_2H_5$

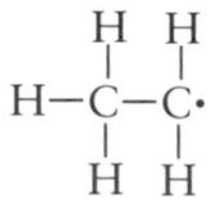

*n*-Propyl radical
$C_3H_7$

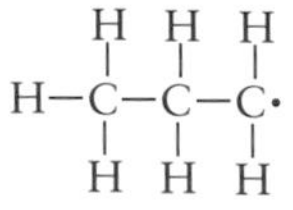

Isopropyl radical
$C_3H_7$

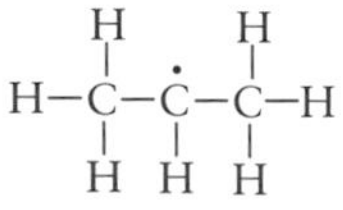

**Hydroxyalkyl Radicals**

Hydroxyethyl radical
(ethanyl radical)
$HOCH_2CH_2$

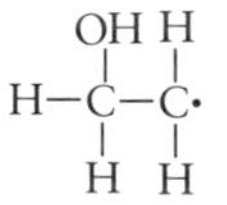

**Acyl (Alkanoyl) Radicals**

Formyl radical
(methanoyl radical)
HCO

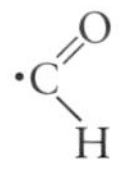

Acetyl radical
(ethanoyl radical)
$CH_3CO$

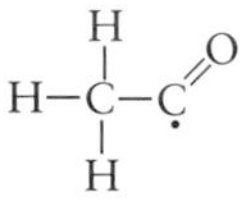

**Aryl Radicals**

Phenyl radical
$C_6H_5$

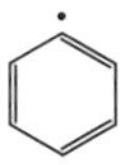

Methylphenyl radical
$C_6H_4(CH_3)$

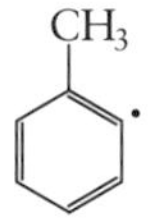

Benzyl radical
$C_6H_5CH_2$

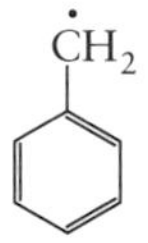

**Alkoxy Radicals**

Methoxy radical
$CH_3O$

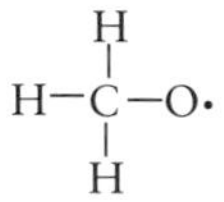

Ethoxy radical
$C_2H_5O$

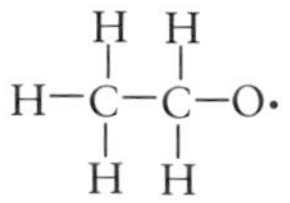

*n*-Propoxy radical
$C_3H_7O$

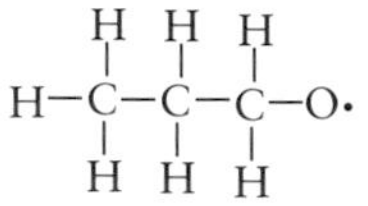

**Hydroxyalkoxy Radicals**

Hydroxyethyloxy radical
(ethanoloxy radical)
$HOCH_2CH_2O$

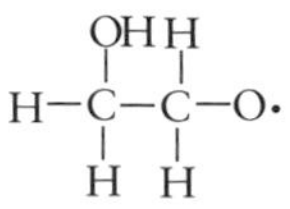

**Acyloxy (Alkanoyloxy) Radicals**

Acetyloxy radical
$CH_3C(O)O$

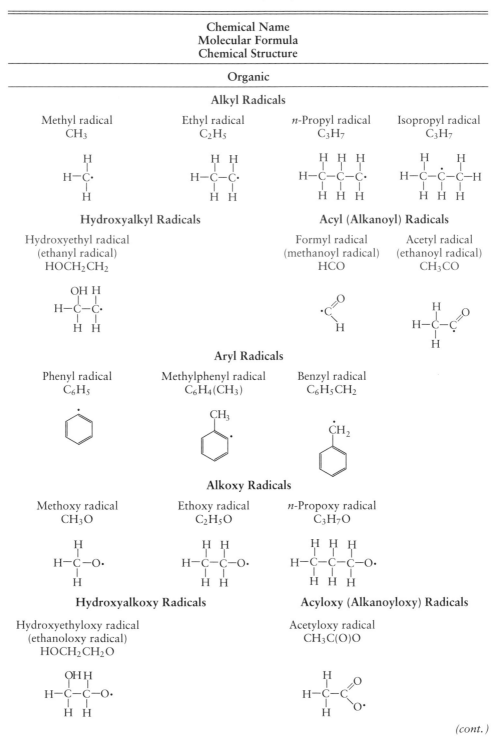

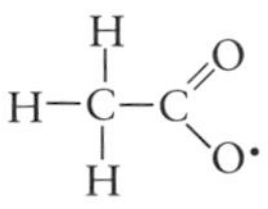

*(cont.)*

**Table B.3.** (cont.)

| Chemical Name<br>Molecular Formula<br>Chemical Structure | | | |
|---|---|---|---|
| **Organic** | | | |
| **Aryloxy Radicals** | | | |
| Phenoxy radical<br>**PHO**<br>$C_6H_5O$ | Methylphenoxy radical<br>**CRO**<br>$2\text{-}OC_6H_4CH_3$ | Benzoxy radical<br>$C_6H_5CH_2O$ | |
| 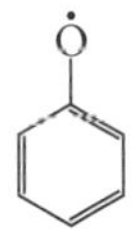 | 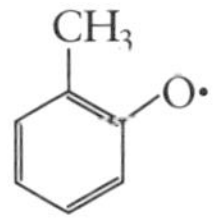 | 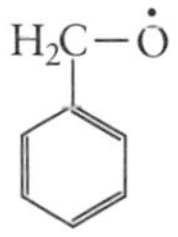 | |
| **Alkylperoxy Radicals** | | | |
| Methylperoxy radical<br>$CH_3O_2$ | Ethylperoxy radical<br>$C_2H_5O_2$ | *n*-Propylperoxy radical<br>$C_3H_7O_2$ | Alkylperoxy radical from OH addition to isoprene<br>**ISOH** |
| 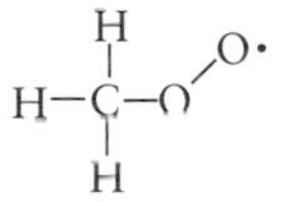 | 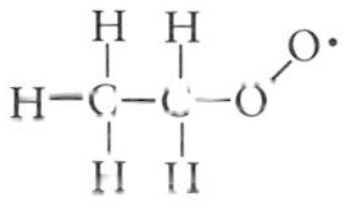 | 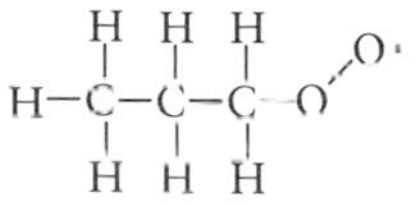 | 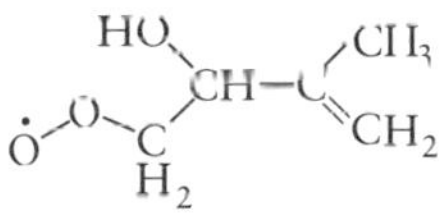 |
| **Hydroxyalkylperoxy Radicals** | | | |
| Hydroxymethylperoxy radical (methanolperoxy radical)<br>$HOCH_2O_2$ | Hydroxyethylperoxy radical (ethanolperoxy radical)<br>$HOC_2H_4O_2$ | Hydroxypropylperoxy radical (propanolperoxy radical)<br>$HOC_3H_6O_2$ | |
| 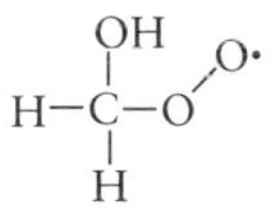 | 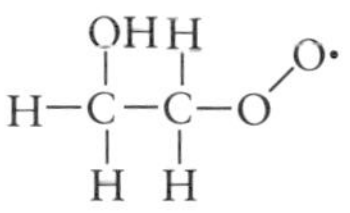 | 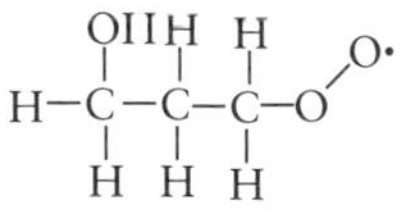 | |
| **Acylperoxy (Alkanoylperoxy) Radicals** | | | |
| Peroxyacetyl radical<br>$CH_3C(O)OO$ | Acetylmethylperoxy radical<br>$CH_3COCH_2OO$ | | |
| 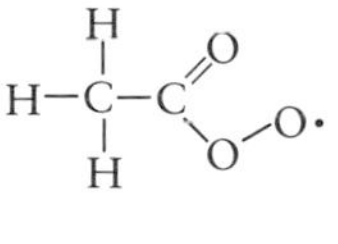 | 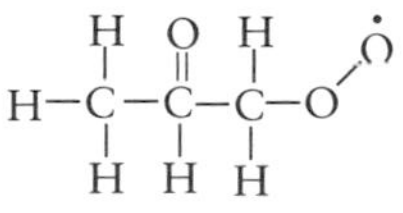 | | |

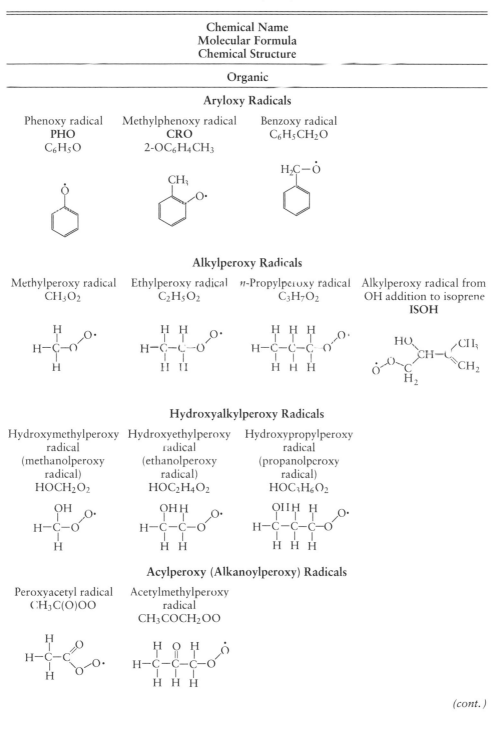

*(cont.)*

**Table B.3.** (cont.)

**Chemical Name**
**Molecular Formula**
**Chemical Structure**

**Organic**

**Arylperoxy Radicals**

| | | | |
|---|---|---|---|
| Phenylperoxy radical<br>**$PHO_2$**<br>$C_6H_5OO$<br>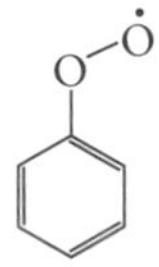 | Methylphenylperoxy radical<br>**$CRO_2$**<br>$2\text{-}O_2C_6H_4CH_3$<br>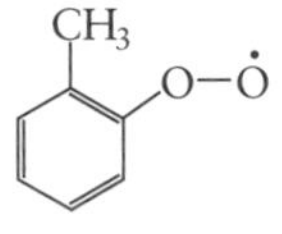 | Benzylperoxy radical<br>**$BO_2$**<br>$C_6H_5CH_2OO$<br>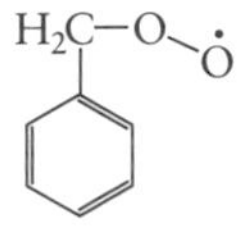 | Methylbenzylperoxy radical<br>**$XLO_2$**<br>$2\text{-}CH_2OOC_6H_5CH_3$<br>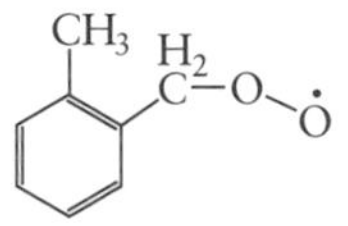 |

**Other Peroxy Radicals**

| | | | |
|---|---|---|---|
| Peroxybenzoyl radical<br>**$BZO_2$**<br>$C_6H_5OCOO$<br>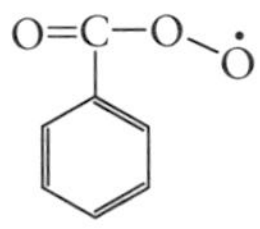 | MVK+$O_3$ product<br>$OOCCH_3CHCH_2$<br>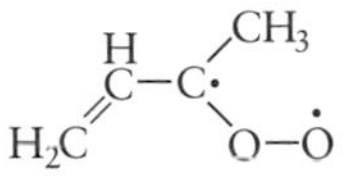 | 2-Methylperoxy propenoyl radical<br>$CH_2CCH_3C(O)OO$<br>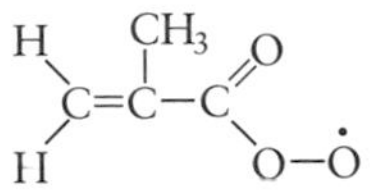 | MACR+$O_3$ product<br>$CH_2CCH_3CHOO$<br>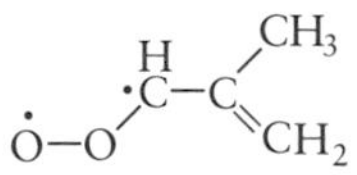 |
| Peroxy radical of methylglyoxal<br>**MGPX**<br>$CH_3COC(O)OO$<br>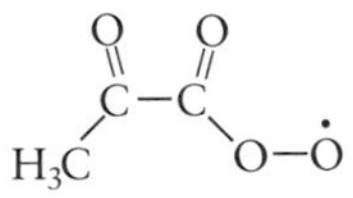 | MACR+OH product<br>$CH_2CCH_3C(O)OO$<br>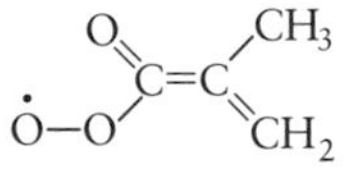 | Nitrated organic peroxy radical<br>**$PNO_2$**<br>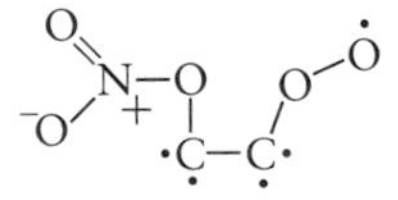 | |

**OH-Adducts**

| | | | |
|---|---|---|---|
| Methylvinylketone-OH adduct<br>**MV1**<br>$OOCH_2CH(OH)$<br>$C(CH_3)O$<br>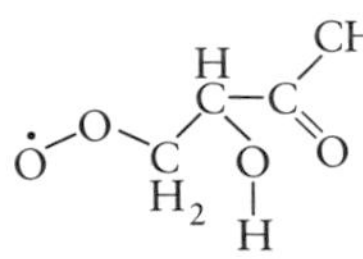 | Methylvinylketone-OH adduct<br>**MV2**<br>$HOCH_2CH(OO)$<br>$C(CH_3)O$<br>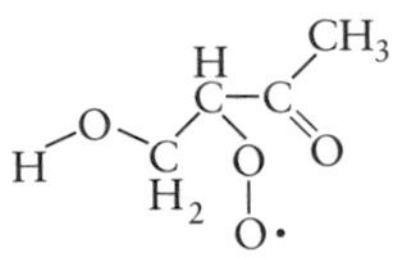 | Methacrolein-OH adduct<br>**MAC1**<br>$OCHC(O_2)(CH_3)$<br>$CH_2OH$<br>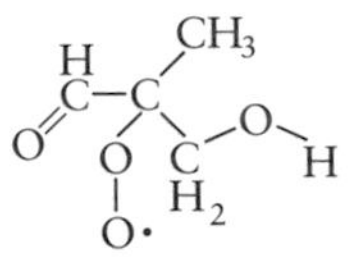 | Methacrolein-OH adduct<br>**MAC2**<br>$OCHC(OH)(CH_3)$<br>$CH_2OO$<br>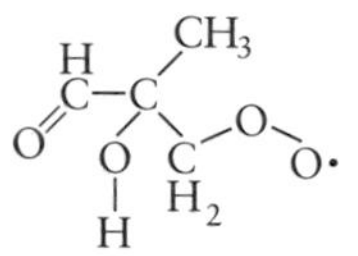 |

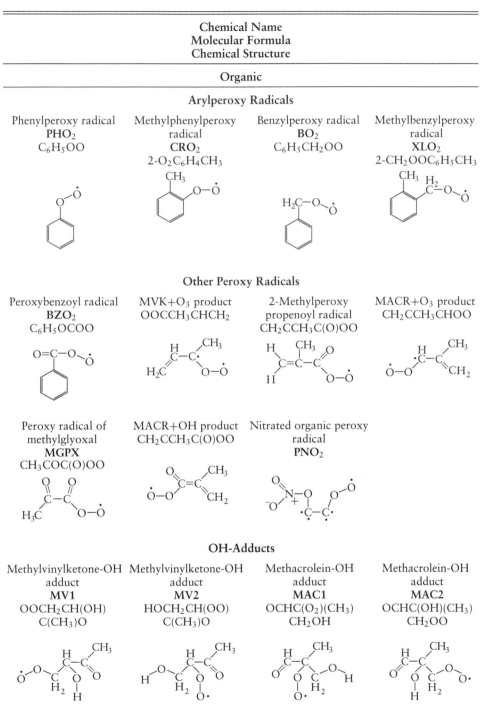

*(cont)*

**Table B.3.** (cont.)

**Chemical Name**
**Molecular Formula**
**Chemical Structure**

## Organic

Toluene–OH adduct
**$TO_2$**
2-$OHC_6H_5(O_2)CH_3$

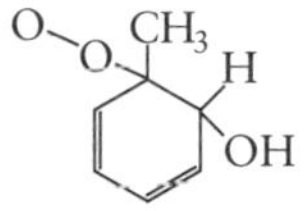

*o*-Xylene–OH adduct
**XINT**
4-OH-3-$O_2$-2-$CH_3C_6H_4CH_3$

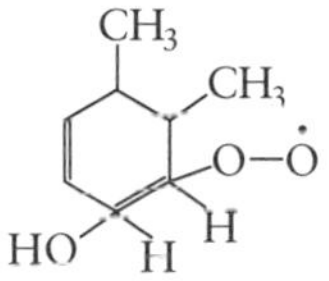

### Molozonides

Ethene molozonide
**OZD**
$CH_2O_3CH_2$

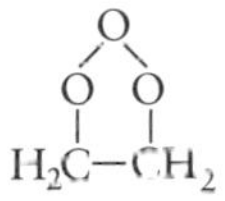

Propene molozonide
**OZD**
$CH_3CHO_3CH_2$

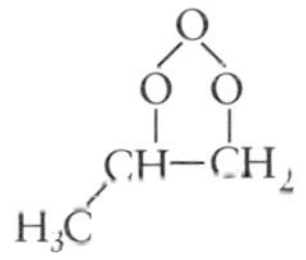

### Biradicals

Criegee biradical
$H_2COO$

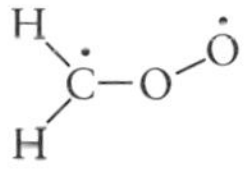

Methyl criegee biradical
$CH_3HCOO$

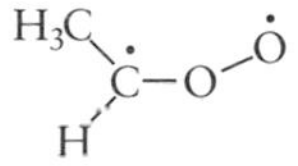

### Dicarbonyls

Glyoxal (ethanedial)
$(CHO)_2$

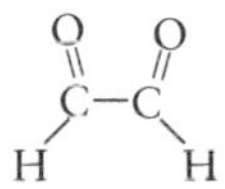

Methylglyoxal
**MGLY**
$CH_3COCHO$

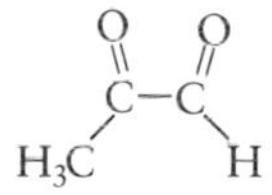

### Organic Peroxides

Methyl hydroperoxide
$CH_3OOH$

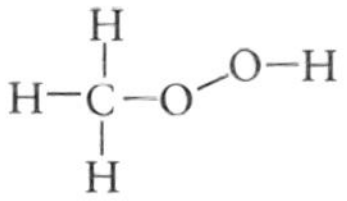

### Nitrates

Methyl nitrate
$CH_3ONO_2$

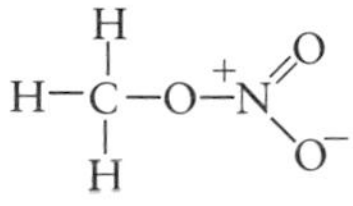

Ethyl nitrate
$C_2H_5ONO_2$

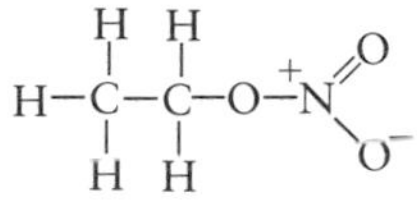

Propyl nitrate
$C_3H_7ONO_2$

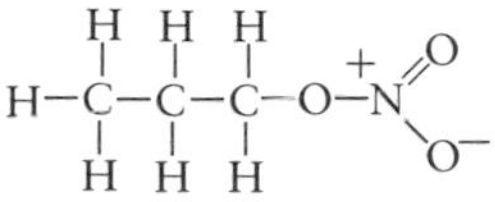

Benzyl nitrate
$C_6H_5CH_2ONO_2$

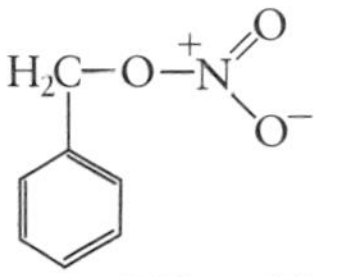

Peroxyacetyl nitrate
**PAN**
$CH_3C(O)OONO_2$

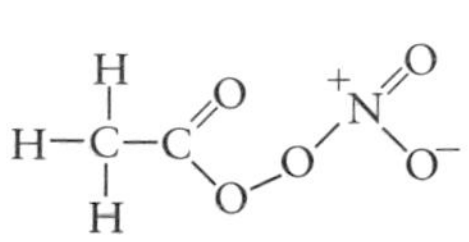

2-Methylperoxy propenoylnitrate
**MPAN**
$CH_2CCH_3C(O)OONO_2$

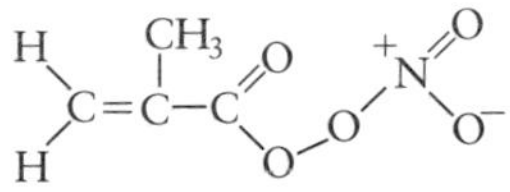

Peroxybenzoyl nitrate
**PBZN**
$C_6H_5C(O)OONO_2$

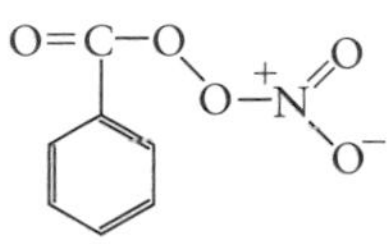

Isoprene-$NO_3$ adduct
**ISNT**
$C_5H_8ONO_2$

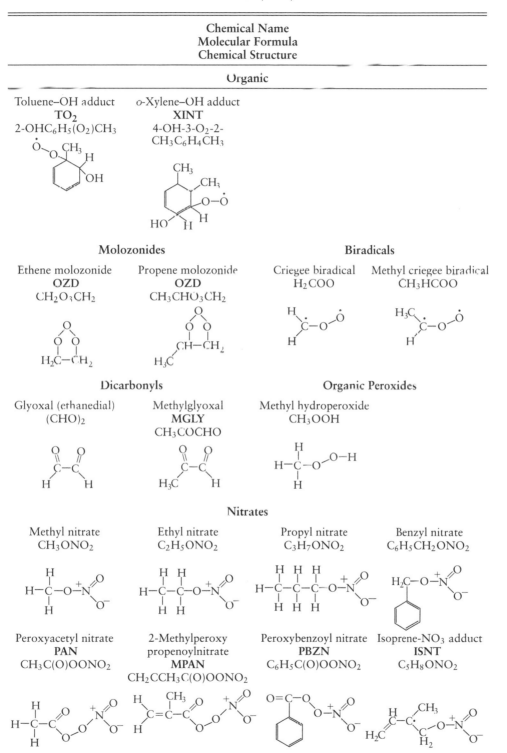

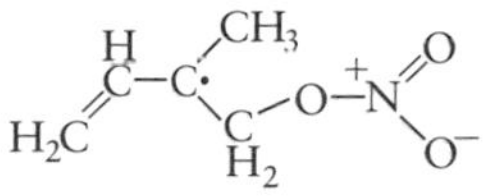

*(cont.)*

**Table B.3.** *(cont).*

| **Chemical Name**<br>**Molecular Formula**<br>**Chemical Structure** |
|---|
| **Organic** |

**Nitric Acids**

Methylperoxy nitric acid
$CH_3O_2NO_2$

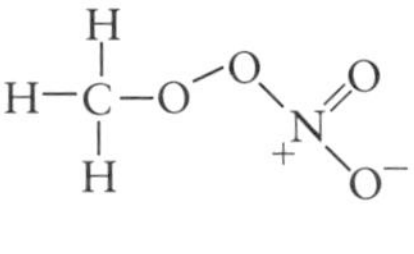

Ethylperoxy nitric acid
$C_2H_5O_2NO_2$

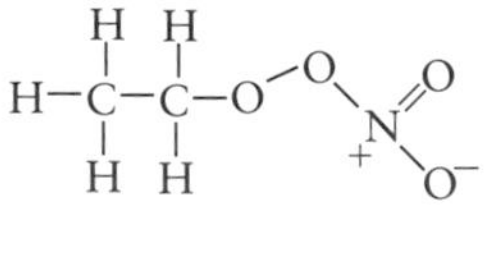

**Nitrites**

Methyl nitrite
$CH_3ONO$

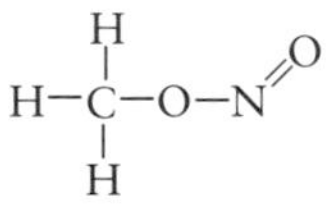

Ethyl nitrite
$C_2H_5ONO$

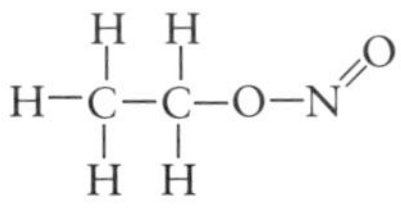

Propyl nitrite
$C_3H_7ONO$

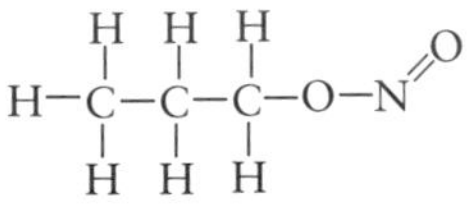

Benzyl nitrite
$C_6H_5CH_2ONO$

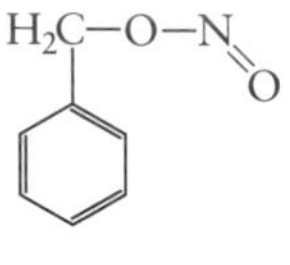

**Nitro Group**

Nitrobenzene
$C_6H_5NO_2$

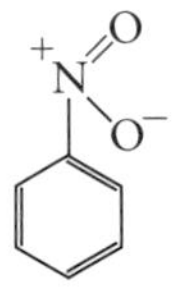

*m*-Nitrotoluene
3-$NO_2C_6H_4CH_3$

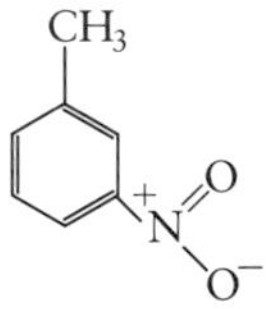

*o*-Nitrophenol
**NPHN**
2-$NO_2C_6H_4OH$

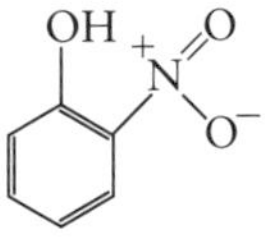

*m*-Nitrocresol
**NCRE**
3-$NO_2$-2-$OHC_6H_3CH_3$

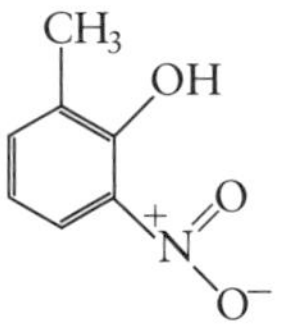

**Carbon-Bond Groups**

Paraffin carbon bond
**PAR**

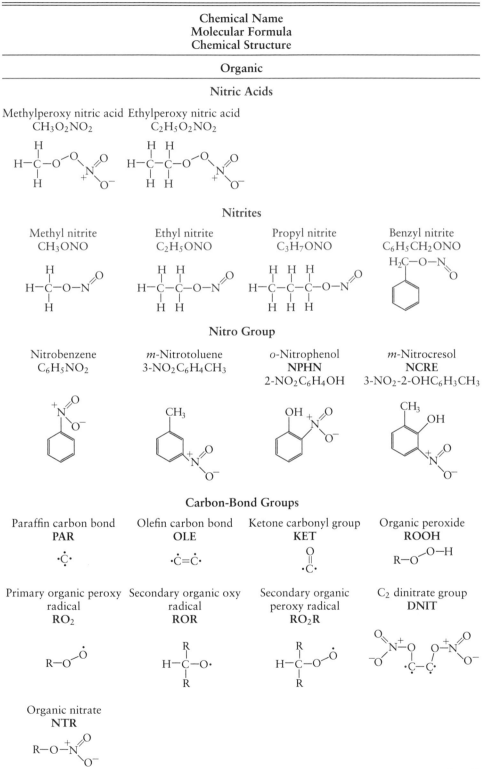

Olefin carbon bond
**OLE**

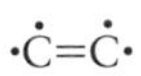

Ketone carbonyl group
**KET**

Organic peroxide
**ROOH**

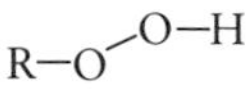

Primary organic peroxy radical
**$RO_2$**

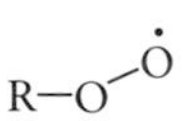

Secondary organic oxy radical
**ROR**

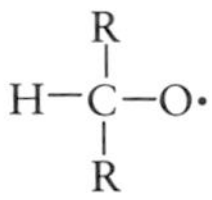

Secondary organic peroxy radical
**$RO_2R$**

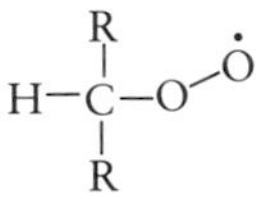

$C_2$ dinitrate group
**DNIT**

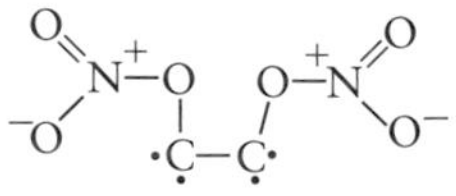

Organic nitrate
**NTR**

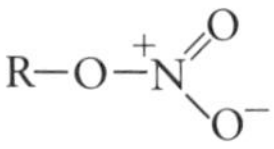

*(cont).*

Table B.3. (cont.)

| Chemical Name<br>Molecular Formula<br>Chemical Structure | | | |
|---|---|---|---|
| **Organic** | | | |
| **Miscellaneous Species and Groups** | | | |
| Dimethyl 2nd organic peroxide radical<br>**$AO_2$** | Peroxide radical of OPEN<br>**OPPX** | High-molecular-weight aromatic oxidation ring fragments<br>**OPEN** | Dinitrate of isoprene<br>**DISN** |
| Group of hydroxy carbonyl alkenes from isoprene–OH reaction<br>**IALD1** | Organic nitrates from OH addition to unsubstituted double bond<br>**ISNI1** | Organic nitrates from OH addition to substituted double bond and OH reactions of primary products<br>**ISNI2** | Alkyl peroxy radical from OH addition across the double bond of ISNI1 and ISNI2<br>**ISNIR** |
| Aromatic ring fragment acid<br>**ACID** | Organic peroxide from isoprene<br>**IPRX** | Paraffin to peroxy radical operator<br>**DOP** | Paraffin loss operator<br>**XOP** |

Molecular formulae and abbreviated names (boldface) used in Table B.4 are defined here. Organic sulfur-, chlorine-, and bromine-containing species are included in the inorganic categories for convenience.

## B.4. GAS-PHASE REACTIONS

Table B.4. Gas-Phase Chemical Kinetic Reactions, Reaction Rates, and Photoprocesses

| No. | Regions | Kinetic Reaction | $F_c$[a] | Rate Coefficient ($s^{-1}$, $cm^3\ s^{-1}$, or $cm^6\ s^{-1}$) | Refs.[b] |
|---|---|---|---|---|---|
| | | **Inorganic Chemistry** | | | |
| 1 | All | $O + O_2 + M \longrightarrow O_3 + M$ | | $6.00 \times 10^{-34}\ (300/T)^{2.3}$ | B |
| 2 | S | $O + O_3 \longrightarrow 2\ O_2$ | | $8.00 \times 10^{-12}\ e^{-2060/T}$ | A |
| 3 | S | $O(^1D) + O_3 \longrightarrow 2O_2$ | | $1.20 \times 10^{-10}$ | B |
| 4 | S | $O(^1D) + O_3 \longrightarrow O_2 + 2O$ | | $1.20 \times 10^{-10}$ | B |
| 5 | All | $O(^1D) + O_2 \longrightarrow O + O_2$ | | $3.20 \times 10^{-11}\ e^{67/T}$ | A |
| 6 | All | $O(^1D) + N_2 \longrightarrow O + N_2$ | | $1.80 \times 10^{-11}\ e^{107/T}$ | A |
| 7 | S | $O(^1D) + N_2 + M \longrightarrow N_2O + M$ | | $3.50 \times 10^{-37}(300/T)^{0.6}$ | B |
| 8 | S | $O(^1D) + N_2O \longrightarrow N_2 + O_2$ | | $4.40 \times 10^{-11}$ | A |
| 9 | S | $O(^1D) + N_2O \longrightarrow NO + NO$ | | $7.20 \times 10^{-11}$ | A |
| 10 | All | $O(^1D) + H_2 \longrightarrow OH + H$ | | $1.10 \times 10^{-10}$ | A |
| 11 | All | $O(^1D) + H_2O \longrightarrow OH + OH$ | | $2.20 \times 10^{-10}$ | A |
| 12 | All | $H + O_2 \xrightarrow{M} HO_2$ | (P) 0.85 | $5.40 \times 10^{-32}(300/T)^{1.8}$<br>$7.50 \times 10^{-11}$ | A |
| 13 | All | $H + O_3 \longrightarrow O_2 + OH$ | | $1.40 \times 10^{-10} e^{-470/T}$ | B |
| 14 | All | $H + HO_2 \longrightarrow H_2 + O_2$ | | $5.60 \times 10^{-12}$ | A |
| 15 | All | $H + HO_2 \longrightarrow OH + OH$ | | $7.20 \times 10^{-11}$ | A |
| 16 | All | $H + HO_2 \longrightarrow H_2O + O$ | | $2.40 \times 10^{-12}$ | A |
| 17 | All | $OH + O \longrightarrow H + O_2$ | | $2.30 \times 10^{-11} e^{110/T}$ | A |

*(cont.)*

Table B.4. (cont.)

| No. | Regions | Kinetic Reaction | $F_c$ [a] | Rate Coefficient ($s^{-1}$, $cm^3\ s^{-1}$, or $cm^6\ s^{-1}$) | Refs.[b] |
|---|---|---|---|---|---|
| | | **Inorganic Chemistry** | | | |
| 18 | All | $OH + O_3 \longrightarrow HO_2 + O_2$ | | $1.90 \times 10^{-12} e^{-1000/T}$ | A |
| 19 | All | $OH + H_2 \longrightarrow H_2O + H$ | | $7.70 \times 10^{-12} e^{-2100/T}$ | A |
| 20 | All | $OH + OH \longrightarrow H_2O + O$ | | $4.20 \times 10^{-12} e^{-240/T}$ | B |
| 21 | All | $OH + OH \xrightarrow{M} H_2O_2$ | (P) 0.5 | $6.90 \times 10^{-31}(300/T)^{0.8}$<br>$2.6 \times 10^{-11}$ | A |
| 22 | All | $OH + HO_2 \longrightarrow H_2O + O_2$ | | $4.80 \times 10^{11} e^{250/T}$ | A |
| 23 | All | $OH + H_2O_2 \longrightarrow HO_2 + H_2O$ | | $2.90 \times 10^{-12} e^{-160/T}$ | A |
| 24 | All | $OH + NO \xrightarrow{M} HONO$ | (P)0.90 | $7.40 \times 10^{-31}(300/T)^{2.4}$<br>$4.50 \times 10^{-11}$ | A |
| 25 | All | $OH + NO_2 \xrightarrow{M} HNO_3$ | (P)0.43 | $2.60 \times 10^{-30}(300/T)^{2.9}$<br>$6.70 \times 10^{-11}(300/T)^{0.6}$ | A |
| 26 | All | $OH + NO_3 \longrightarrow HO_2 + NO_2$ | | $2.00 \times 10^{-11}$ | A |
| 27 | All | $OH + HONO \longrightarrow H_2O + NO_2$ | | $1.80 \times 10^{-11} e^{-390/T}$ | A |
| 28 | All | $OH + HNO_3 \longrightarrow H_2O + NO_3$ | | *c* | A |
| 29 | All | $OH + HO_2NO_2 \longrightarrow H_2O + NO_2 + O_2$ | | $1.50 \times 10^{-12} e^{360/T}$ | A |
| 30 | All | $OH + CO \longrightarrow HO_2 + CO_2$ | | *d* | A |
| 31 | All | $HO_2 + O \longrightarrow OH + O_2$ | | $2.70 \times 10^{-11} e^{224/T}$ | A |
| 32 | All | $HO_2 + O_3 \longrightarrow OH + 2O_2$ | | $1.40 \times 10^{-14} e^{-600/T}$ | A |
| 33 | All | $HO_2 + HO_2 \longrightarrow H_2O_2 + O_2$ | | *e* | F |
| 34 | All | $HO_2 + NO \longrightarrow OH + NO_2$ | | $3.70 \times 10^{-12} e^{240/T}$ | A |
| 35 | All | $HO_2 + NO_2 \xrightarrow{M} HO_2NO_2$ | (P) 0.60 | $1.80 \times 10^{-31} 300/T^{3.2}$<br>$4.70 \times 10^{-12}(300/T)^{1.4}$ | A |
| 36 | All | $HO_2 + NO_3 \longrightarrow HNO_3 + O_2$ | | $4.00 \times 10^{-12}$ | A |
| 37 | All | $H_2O_2 + O \longrightarrow OH + HO_2$ | | $1.40 \times 10^{-12} e^{2000/T}$ | A |
| 38 | All | $NO + O \xrightarrow{M} NO_2$ | (P) 0.85 | $1.00 \times 10^{-31} 300/T^{1.6}$<br>$3.00 \times 10^{-11}(300/T)^{-0.3}$ | A |
| 39 | All | $NO + O_3 \longrightarrow NO_2 + O_2$ | | $1.80 \times 10^{-12} e^{-1370/T}$ | A |
| 40 | All | $NO_2 + O \longrightarrow NO + O_2$ | | $6.50 \times 10^{-12} e^{120/T}$ | A |
| 41 | All | $NO_2 + O \xrightarrow{M} NO_3$ | (P) 0.80 | $9.00 \times 10^{-31}(300/T)^{2.0}$<br>$2.20 \times 10^{-11}$ | A |
| 42 | All | $NO_2 + O_3 \longrightarrow NO_3 + O_2$ | | $1.20 \times 10^{-13} e^{-2450/T}$ | A |
| 43 | All | $NO_3 + O \longrightarrow NO_2 + O_2$ | | $1.70 \times 10^{-11}$ | A |
| 44 | All | $NO_3 + NO \longrightarrow 2\ NO_2$ | | $1.80 \times 10^{-11} e^{110/T}$ | A |
| 45 | All | $NO_3 + NO_2 \xrightarrow{M} N_2O_5$ | (P) 0.33 | $2.70 \times 10^{-30}(300/T)^{3.4}$<br>$2.00 \times 10^{-12}(300/T)^{-0.2}$ | A |
| 46 | All | $N_2O_5 \xrightarrow{M} NO_3 + NO_2$ | (P) 0.33 | $1.00 \times 10^{-3}(300/T)^{3.5}$<br>$\times e^{-11000/T}$<br>$9.70 \times 10^{14}(300/T)^{-0.1}$<br>$\times e^{-11080/T}$ | A |
| 47 | All | $N_2O_5 + H_2O \longrightarrow 2\ HNO_3$ | | $2.00 \times 10^{-21}$ | A |
| 48 | All | $HO_2NO_2 \xrightarrow{M} HO_2 + NO_2$ | (P) 0.60 | $5.00 \times 10^{-6} e^{-10000/T}$<br>$2.60 \times 10^{15} e^{-10900/T}$ | A |
| | | **Organic Chemistry** | | | |
| | | **Alkane, Alkene, and Aldehyde Chemistry** | | | |
| 49 | S | $CH_4 + O(^1D) \longrightarrow CH_3O_2 + OH$ | | $1.40 \times 10^{-10}$ | A |
| 50 | S | $CH_4 + O(^1D) \longrightarrow HCHO + H_2$ | | $2.30 \times 10^{-12}$ | A |

*(cont.)*

**Table B.4.** (cont.)

| No. | Regions | Kinetic Reaction | $F_c$ [a] | Rate Coefficient ($s^{-1}$, $cm^3\ s^{-1}$, or $cm^6\ s^{-1}$) | Refs.[b] |
|---|---|---|---|---|---|
| | | **Organic Chemistry** | | | |
| | | **Alkane, Alkene, and Aldehyde Chemistry** | | | |
| 51 | All | $CH_4 + OH \longrightarrow CH_3O_2 + H_2O$ | | $3.90 \times 10^{-12} e^{-1765/T}$ | A |
| 52 | All | $CH_3O + O_2 \longrightarrow HCHO + HO_2$ | | $7.20 \times 10^{-14} e^{-1080/T}$ | A |
| 53 | U/T | $CH_3O + NO \longrightarrow HCHO + HO_2 + NO$ | | $4.00 \times 10^{-12} (300/T)^{0.7}$ | A |
| 54 | U/T | $CH_3O + NO \xrightarrow{M} CH_3ONO$ | (P) 0.60 | $1.60 \times 10^{-29} (300/T)^{3.5}$<br>$3.60 \times 10^{-11} (300/T)^{0.6}$ | A |
| 55 | U/T | $CH_3O + NO_2 \xrightarrow{M} CH_3ONO_2$ | (P) 0.44 | $2.80 \times 10^{-29} (300/T)^{4.5}$<br>$2.00 \times 10^{-11}$ | A |
| 56 | U/T | $CH_3ONO_2 + OH \longrightarrow HCHO$<br>$+ NO_2 + H_2O$ | | $1.00 \times 10^{-14} e^{1060/T}$ | A |
| 57 | All | $CH_3O_2 + HO_2 \longrightarrow CH_3OOH + O_2$ | | $3.80 \times 10^{-13} e^{780/T}$ | A |
| 58 | All | $CH_3O_2 + NO \longrightarrow CH_3O + NO_2$ | | $4.20 \times 10^{-12} e^{180/T}$ | A |
| 59 | U/T | $CH_3O_2 + NO_2 \xrightarrow{M} CH_3O_2NO_2$ | (P) 0.36 | $2.50 \times 10^{-30} (300/T)^{5.5}$<br>$7.50 \times 10^{-12}$ | A |
| 60 | U/T | $CH_3O_2NO_2 \xrightarrow{M} CH_3O_2 + NO_2$ | (P) 0.36 | $9.00 \times 10^{-5} e^{-9690/T}$<br>$1.10 \times 10^{16} e^{-10560/T}$ | A |
| 61 | U/T | $CH_3O_2 + CH_3O_2 \longrightarrow 2\ CH_3O + O_2$ | | $5.90 \times 10^{-14} e^{-509/T}$ | A |
| 62 | U/T | $CH_3O_2 + CH_3O_2 \longrightarrow HCHO + CH_3OH$ | | $7.04 \times 10^{-14} e^{365/T}$ | A |
| 63 | U/T | $CH_3O_2 + CH_3C(O)OO \longrightarrow CH_3O_2$<br>$+ CH_3O + CO_2$ | | $5.10 \times 10^{-12} e^{272/T}$ | A |
| 64 | All | $CH_3OOH + OH \longrightarrow CH_3O_2 + H_2O$ | | $1.90 \times 10^{-12} e^{190/T}$ | A |
| 65 | U/T | $C_2H_6 + OH \longrightarrow C_2H_5O_2 + H_2O$ | | $7.90 \times 10^{-12} e^{-1030/T}$ | A |
| 66 | U/T | $C_2H_5O_2 + NO \longrightarrow C_2H_5O + NO_2$ | | $8.70 \times 10^{-12}$ | A |
| 67 | U/T | $C_2H_5O_2 + NO_2 \xrightarrow{M} C_2H_5O_2NO$ | (P) 0.31 | $1.30 \times 10^{-29} (300/T)^{6.2}$<br>$8.80 \times 10^{-12}$ | A |
| 68 | U/T | $C_2H_5O_2NO_2 \xrightarrow{M} C_2H_5O_2 + NO_2$ | (P) 0.31 | $4.80 \times 10^{-4} e^{-9285/T}$<br>$8.80 \times 10^{15} e^{-10440/T}$ | A |
| 69 | U/T | $C_2H_5O_2 + HO_2 \longrightarrow ROOH + O_2$ | | $2.70 \times 10^{-13} e^{1000/T}$ | A |
| 70 | U/T | $C_2H_5O + O_2 \longrightarrow CH_3CHO + HO_2$ | | $6.00 \times 10^{-14} e^{-550/T}$ | A |
| 71 | U/T | $C_2H_5O + NO \longrightarrow C_2H_5ONO$ | | $4.40 \times 10^{-11}$ | A |
| 72 | U/T | $C_2H_5O + NO \longrightarrow CH_3CHO + HO_2 + NO$ | | $1.30 \times 10^{-11}$ | A |
| 73 | U/T | $C_2H_5O + NO_2 \longrightarrow C_2H_5ONO_2$ | | $2.80 \times 10^{-11}$ | A |
| 74 | U/T | $C_3H_8 + OH \longrightarrow C_3H_7O_2 + H_2O$ | | $8.00 \times 10^{-12} e^{-590/T}$ | A |
| 75 | U/T | $C_3H_7O_2 + NO \longrightarrow C_3H_7O + NO_2$ | | $4.80 \times 10^{-12}$ | A |
| 76 | U/T | $C_3H_7O + O_2 \longrightarrow CH_3COCH_3 + HO_2$ | | $1.50 \times 10^{-14} e^{-200/T}$ | A |
| 77 | U/T | $C_3H_7O + NO \longrightarrow C_3H_7ONO$ | | $3.40 \times 10^{-11}$ | A |
| 78 | U/T | $C_3H_7O + NO \longrightarrow CH_3COCH_3 + HO_2 + NO$ | | $6.50 \times 10^{-12}$ | A |
| 79 | U/T | $C_3H_7O + NO_2 \longrightarrow C_3H_7ONO_2$ | | $3.50 \times 10^{-11}$ | A |
| 80 | U/T | $C_2H_4 + OH \xrightarrow{M} HOC_2H_4O_2$ | (P) 0.70 | $7.00 \times 10^{-29} (300/T)^{3.1}$<br>$9.00 \times 10^{-12}$ | A |
| 81 | U/T | $HOC_2H_4O_2 + NO \longrightarrow NO_2$<br>$+ 2\ HCHO + H$ | | $6.93 \times 10^{-12}$ | C |
| 82 | U/T | $HOC_2H_4O_2 + NO \longrightarrow NO_2$<br>$+ CH_3CHO + OH$ | | $2.07 \times 10^{-12}$ | C |
| 83 | U/T | $C_2H_4 + O_3 \longrightarrow HCHO + H_2COO$ | | $3.40 \times 10^{-15} e^{-2580/T}$ | A |
| 84 | U/T | $C_2H_4 + O_3 \longrightarrow HCHO + HCOOH^*$ | | $5.70 \times 10^{-15} e^{-2580/T}$ | A |
| 85 | U/T | $H_2COO + NO \longrightarrow NO_2 + HCHO$ | | $7.00 \times 10^{-12}$ | C |
| 86 | U/T | $H_2COO + H_2O \longrightarrow HCOOH + H_2O$ | | $4.00 \times 10^{-16}$ | C |

*(cont.)*

**Table B.4.** (cont.)

| No. | Regions | Kinetic Reaction | $F_c$[a] | Rate Coefficient ($s^{-1}$, $cm^3\ s^{-1}$, or $cm^6\ s^{-1}$) | Refs.[b] |
|---|---|---|---|---|---|
| | | **Organic Chemistry** | | | |
| 87 | U/T | $H_2COO + HCHO \longrightarrow OZD$ | | $2.00 \times 10^{-12}$ | C |
| 88 | U/T | $H_2COO + CH_3CHO \longrightarrow OZD$ | | $2.00 \times 10^{-12}$ | C |
| 89 | U/T | $HCOOH^* \longrightarrow CO_2 + H_2$ | | 0.21 | A |
| 90 | U/T | $HCOOH^* \longrightarrow CO + H_2O$ | | 0.60 | A |
| 91 | U/T | $HCOOH^* \longrightarrow OH + HO_2 + CO$ | | 0.19 | A |
| 92 | U/T | $C_3H_6 + OH \xrightarrow{M} HOC_3H_6O_2$ | (P) 0.50 | $8.00 \times 10^{-27}(300/T)^{3.5}$<br>$3.00 \times 10^{-11}$ | A |
| 93 | U/T | $HOC_3H_6O_2 + NO \longrightarrow NO_2 + CH_3CHO + HCHO + HO_2$ | | $6.00 \times 10^{-12}$ | C |
| 94 | U/T | $C_3H_6 + O_3 \longrightarrow HCHO + CH_3HCOO$ | | $4.10 \times 10^{-16}e^{-1880/T}$ | A |
| 95 | U/T | $C_3H_6 + O_3 \longrightarrow HCHO + CH_3HCOO^*$ | | $2.34 \times 10^{-15}e^{-1880/T}$ | A |
| 96 | U/T | $C_3H_6 + O_3 \longrightarrow CH_3CHO + H_2COO$ | | $1.03 \times 10^{-15}e^{-1880/T}$ | A |
| 97 | U/T | $C_3H_6 + O_3 \longrightarrow CH_3CHO + H_2COO^*$ | | $1.72 \times 10^{-15}e^{-1880/T}$ | A |
| 98 | U/T | $CH_3HCOO + NO \longrightarrow NO_2 + CH_3CHO$ | | $7.00 \times 10^{-12}$ | C |
| 99 | U/T | $CH_3HCOO + H_2O \longrightarrow CH_3COOH + H_2O$ | | $4.00 \times 10^{-16}$ | C |
| 100 | U/T | $CH_3HCOO + HCHO \longrightarrow OZD$ | | $2.00 \times 10^{-12}$ | C |
| 101 | U/T | $CH_3HCOO + CH_3CHO \longrightarrow OZD$ | | $2.00 \times 10^{-12}$ | C |
| 102 | U/T | $CH_3COOH^* \longrightarrow CH_4 + CO_2$ | | 0.16 | A |
| 103 | U/T | $CH_3COOH^* \longrightarrow CH_3O_2 + CO + OH$ | | 0.64 | A |
| 104 | U/T | $CH_3COOH^* \longrightarrow CH_3O + CO + HO_2$ | | 0.20 | A |
| 105 | All | $HCHO + OH \longrightarrow HO_2 + CO + H_2O$ | | $8.60 \times 10^{-12}e^{20/T}$ | A |
| 106 | All | $HCHO + O \longrightarrow OH + HO_2 + CO$ | | $3.40 \times 10^{-11}e^{-1550/T}$ | B |
| 107 | U/T | $HCHO + NO_3 \longrightarrow HNO_3 + HO_2 + CO$ | | $5.80 \times 10^{-16}$ | A |
| 108 | U/T | $HCHO + HO_2 \longrightarrow HOCH_2O_2$ | | $9.70 \times 10^{-15}e^{625/T}$ | A |
| 109 | U/T | $HOCH_2O_2 \longrightarrow HO_2 + HCHO$ | | $2.40 \times 10^{12}e^{-7000/T}$ | A |
| 110 | U/T | $HOCH_2O_2 + HO_2 \longrightarrow ROOH$ | | $5.60 \times 10^{-15}e^{2300/T}$ | A |
| 111 | U/T | $HOCH_2O_2 + NO \longrightarrow NO_2 + HO_2 + HCOOH$ | | $7.00 \times 10^{-12}$ | C |
| 112 | U/T | $CH_3CHO + O \longrightarrow CH_3C(O)OO + OH$ | | $1.80 \times 10^{-11}e^{-1100/T}$ | B |
| 113 | U/T | $CH_3CHO + OH \longrightarrow CH_3C(O)OO + H_2O$ | | $5.60 \times 10^{-12}e^{310/T}$ | A |
| 114 | U/T | $CH_3CHO + NO_3 \longrightarrow CH_3C(O)OO + HNO_3$ | | $1.40 \times 10^{-12}e^{-1860/T}$ | A |
| 115 | U/T | $CH_3C(O)OO + HO_2 \longrightarrow ROOH + O_2$ | | $1.14 \times 10^{-13}e^{1040/T}$ | A |
| 116 | U/T | $CH_3C(O)OO + HO_2 \longrightarrow CH_3O_2 + OH + CO_2$ | | $3.16 \times 10^{-13}e^{1040/T}$ | A |
| 117 | U/T | $CH_3C(O)OO + NO \longrightarrow NO_2 + CH_3O_2 + CO_2$ | | $2.00 \times 10^{-11}$ | A |
| 118 | U/T | $CH_3C(O)OO + NO_2 \xrightarrow{M} CH_3C(O)OONO_2$ | (P) 0.30 | $2.70 \times 10^{-28}(300/T)^{7.1}$<br>$1.20 \times 10^{-11}(300/T)^{0.9}$ | A |
| 119 | U/T | $CH_3C(O)OO + CH_3C(O)OO \longrightarrow 2\ CH_3O_2 + O_2$ | | $2.80 \times 10^{-12}e^{530/T}$ | A |
| 120 | U/T | $CH_3C(O)OONO_2 \xrightarrow{M} CH_3C(O)OO + NO_2$ | (P) 0.30 | $4.90 \times 10^{-3}e^{-12100/T}$<br>$5.40 \times 10^{16}e^{-13830/T}$ | A |
| 121 | U/T | $CH_3COCH_3 + OH \longrightarrow CH_3COCH_2OO + H_2O$ | | $2.80 \times 10^{-12}e^{-760/T}$ | A |
| 122 | U/T | $CH_3COCH_2OO + NO \longrightarrow CH_3C(O)OO + HCHO + NO_2$ | | $8.10 \times 10^{-12}$ | C |
| 123 | U/T | $CH_3OH + OH \longrightarrow HCHO + HO_2 + H_2O$ | | $2.63 \times 10^{-12}e^{-360/T}$ | A |
| 124 | U/T | $CH_3OH + OH \longrightarrow CH_3O + H_2O$ | | $4.66 \times 10^{-13}e^{-360/T}$ | A |
| 125 | U/T | $C_2H_5OH + OH \longrightarrow CH_3CHO + HO_2 + H_2O$ | | $3.69 \times 10^{-12}e^{-70/T}$ | A |
| 126 | U/T | $PAR + OH \longrightarrow RO_2 + H_2O$ | | $9.20 \times 10^{-14}$ | C |
| 127 | U/T | $PAR + OH \longrightarrow RO_2R + H_2O$ | | $7.20 \times 10^{-13}$ | C |
| 128 | U/T | $RO_2 + NO \longrightarrow NO_2 + HO_2 + CH_3CHO + XOP$ | | $7.70 \times 10^{-12}$ | C |
| 129 | U/T | $RO_2 + NO \longrightarrow NTR$ | | $4.40 \times 10^{-11}e^{-1400/T}$ | C |

*(cont.)*

**Table B.4.** (cont.)

| No. | Regions | Kinetic Reaction | $F_c$ [a] | Rate Coefficient ($s^{-1}$, $cm^3\ s^{-1}$, or $cm^6\ s^{-1}$) | Refs.[b] |
|---|---|---|---|---|---|
| | | **Organic Chemistry** | | | |
| 130 | U/T | $RO_2R + NO \longrightarrow NO_2 + ROR$ | | $7.00 \times 10^{-12}$ | C |
| 131 | U/T | $RO_2R + NO \longrightarrow NTR$ | | $1.20 \times 10^{-10}e^{-1400/T}$ | C |
| 132 | U/T | $ROR + NO_2 \longrightarrow NTR$ | | $1.50 \times 10^{-11}$ | C |
| 133 | U/T | $ROR \longrightarrow KET + HO_2$ | | $1.60 \times 10^{3}$ | C |
| 134 | U/T | $ROR \longrightarrow KET + DOP$ | | $2.10 \times 10^{14}e^{-8000/T}$ | C |
| 135 | U/T | $ROR \longrightarrow CH_3CHO + DOP + XOP$ | | $4.00 \times 10^{14}e^{-8000/T}$ | C |
| 136 | U/T | $ROR \longrightarrow CH_3COCH_3 + DOP + 2\ XOP$ | | $4.40 \times 10^{14}e^{-8000/T}$ | C |
| 137 | U/T | $XOP + PAR \longrightarrow$ | | $6.80 \times 10^{-12}$ | C |
| 138 | U/T | $DOP + PAR \longrightarrow RO_2$ | | $5.10 \times 10^{-12}$ | C |
| 139 | U/T | $DOP + PAR \longrightarrow AO_2 + 2\ XOP$ | | $1.50 \times 10^{-12}$ | C |
| 140 | U/T | $DOP + PAR \longrightarrow RO_2R$ | | $1.70 \times 10^{-13}$ | C |
| 141 | U/T | $DOP + KET \longrightarrow CH_3C(O)OO + XOP$ | | $6.80 \times 10^{-12}$ | C |
| 142 | U/T | $AO_2 + NO \longrightarrow NO_2 + CH_3COCH_3 + HO_2$ | | $8.10 \times 10^{-12}$ | C |
| 143 | U/T | $OLE + O \longrightarrow 2\ PAR$ | | $4.10 \times 10^{-12}e^{-324/T}$ | C |
| 144 | U/T | $OLE + O \longrightarrow CH_3CHO$ | | $4.10 \times 10^{-12}e^{-324/T}$ | C |
| 145 | U/T | $OLE + O \longrightarrow HO_2 + CO + RO_2$ | | $1.20 \times 10^{-12}e^{-324/T}$ | C |
| 146 | U/T | $OLE + O \longrightarrow RO_2 + XOP + CO + HCHO + OH$ | | $2.40 \times 10^{-12}e^{-324/T}$ | C |
| 147 | U/T | $OLE + OH \longrightarrow CH_3O_2 + CH_3CHO + XOP$ | | $5.20 \times 10^{-12}e^{504/T}$ | C |
| 148 | U/T | $OLE + O_3 \longrightarrow CH_3CHO + H_2COO + XOP$ | | $2.80 \times 10^{-15}e^{-2105/T}$ | C |
| 149 | U/T | $OLE + O_3 \longrightarrow HCHO + CH_3HCOO + XOP$ | | $2.80 \times 10^{-15}e^{-2105/T}$ | C |
| 150 | U/T | $OLE + O_3 \longrightarrow CH_3CHO + HCOOH^* + XOP$ | | $4.30 \times 10^{-15}e^{-2105/T}$ | C |
| 151 | U/T | $OLE + O_3 \longrightarrow HCHO + CH_3COOH^* + XOP$ | | $4.30 \times 10^{-15}e^{-2105/T}$ | C |
| 152 | U/T | $OLE + NO_3 \longrightarrow PNO_2$ | | $7.70 \times 10^{-15}$ | C |
| 153 | U/T | $PNO_2 + NO \longrightarrow DNIT$ | | $6.80 \times 10^{-13}$ | C |
| 154 | U/T | $PNO_2 + NO \longrightarrow HCHO + CH_3CHO + XOP + 2NO_2$ | | $6.80 \times 10^{-12}$ | C |
| | | **Aromatic Chemistry** | | | |
| 155 | U | $TOL + OH \longrightarrow BO_2 + H_2O$ | | $1.70 \times 10^{-13}e^{322/T}$ | C |
| 156 | U | $TOL + OH \longrightarrow CRES + HO_2$ | | $7.60 \times 10^{-13}e^{322/T}$ | C |
| 157 | U | $TOL + OH \longrightarrow TO_2$ | | $1.20 \times 10^{-12}e^{322/T}$ | C |
| 158 | U | $BO_2 + NO \longrightarrow NO_2 + BZA + HO_2$ | | $8.10 \times 10^{-12}$ | C |
| 159 | U | $BZA + OH \longrightarrow BZO_2 + H_2O$ | | $1.30 \times 10^{-11}$ | C |
| 160 | U | $BZO_2 + NO \longrightarrow NO_2 + PHO_2 + CO_2$ | | $2.50 \times 10^{-12}$ | C |
| 161 | U | $BZO_2 + NO_2 \longrightarrow PBZN$ | | $8.40 \times 10^{-12}$ | E |
| 162 | U | $PBZN \longrightarrow BZO_2 + NO_2$ | | $1.60 \times 10^{15}e^{-13033/T}$ | E |
| 163 | U | $PHO_2 + NO \longrightarrow NO_2 + PHO$ | | $8.10 \times 10^{-12}$ | C |
| 164 | U | $PHO + NO_2 \longrightarrow NPHN$ | | $1.30 \times 10^{-11}e^{300/T}$ | E |
| 165 | U | $CRES + OH \longrightarrow CRO + H_2O$ | | $1.60 \times 10^{-11}$ | C |
| 166 | U | $CRES + OH \longrightarrow CRO_2 + H_2O$ | | $2.50 \times 10^{-11}$ | C |
| 167 | U | $CRES + NO_3 \longrightarrow CRO + HNO_3$ | | $2.20 \times 10^{-11}$ | C |
| 168 | U | $CRO + NO_2 \longrightarrow NCRE$ | | $1.40 \times 10^{-11}$ | C |
| 169 | U | $CRO_2 + NO \longrightarrow NO_2 + OPEN + HO_2$ | | $4.00 \times 10^{-12}$ | C |
| 170 | U | $CRO_2 + NO \longrightarrow NO_2 + ACID + HO_2$ | | $4.00 \times 10^{-12}$ | C |
| 171 | U | $TO_2 + NO \longrightarrow NO_2 + OPEN + HO_2$ | | $7.30 \times 10^{-12}$ | C |
| 172 | U | $TO_2 + NO \longrightarrow NTR$ | | $8.10 \times 10^{-13}$ | C |
| 173 | U | $TO_2 \longrightarrow HO_2 + CRES$ | | $4.20$ | C |
| 174 | U | $XYL + OH \longrightarrow CRES + PAR + HO_2$ | | $3.32 \times 10^{-12}e^{116/T}$ | C |
| 175 | U | $XYL + OH \longrightarrow XLO_2 + H_2O$ | | $1.70 \times 10^{-12}e^{116/T}$ | C |
| 176 | U | $XYL + OH \longrightarrow TO_2$ | | $5.00 \times 10^{-12}e^{116/T}$ | C |
| 177 | U | $XYL + OH \longrightarrow XINT$ | | $6.60 \times 10^{-12}e^{116/T}$ | C |
| 178 | U | $XLO_2 + NO \longrightarrow NO_2 + HO_2 + BZA + PAR$ | | $8.10 \times 10^{-12}$ | C |

*(cont.)*

**Table B.4.** (cont.)

| No. | Regions | Kinetic Reaction | Rate Coefficient $F_c{}^a$ ($s^{-1}$, $cm^3\ s^{-1}$, or $cm^6\ s^{-1}$) | Refs.[b] |
|---|---|---|---|---|
| | | **Organic Chemistry** | | |
| 179 | U | $XINT + NO \longrightarrow NO_2 + HO_2 + 2\ CH_3COCHO + PAR$ | $8.10 \times 10^{-12}$ | C |
| 180 | U | $CH_3COCHO + OH \longrightarrow MGPX + H_2O$ | $1.50 \times 10^{-11}$ | A |
| 181 | U | $MGPX + NO \longrightarrow NO_2 + CH_3C(O)OO + CO_2$ | $8.10 \times 10^{-12}$ | C |
| 182 | U | $OPEN + OH \longrightarrow OPPX + CH_3C(O)OO + HO_2 + CO$ | $3.00 \times 10^{-11}$ | C |
| 183 | U | $OPEN + O_3 \longrightarrow CH_3CHO + MGPX + HCHO + CO$ | $1.60 \times 10^{-18} e^{-500/T}$ | C |
| 184 | U | $OPEN + O_3 \longrightarrow HCHO + CO + OH + 2\ HO_2$ | $4.30 \times 10^{-18} e^{-500/T}$ | C |
| 185 | U | $OPEN + O_3 \longrightarrow CH_3COCHO$ | $1.10 \times 10^{-17} e^{-500/T}$ | C |
| 186 | U | $OPEN + O_3 \longrightarrow CH_3C(O)OO + HCHO + HO_2 + CO$ | $3.20 \times 10^{-17} e^{-500/T}$ | C |
| 187 | U | $OPEN + O_3 \longrightarrow$ | $5.40 \times 10^{-18} e^{-500/T}$ | C |
| 188 | U | $OPPX + NO \longrightarrow NO_2 + HCHO + HO_2 + CO$ | $8.10 \times 10^{-12}$ | C |
| | | **Terpene Chemistry** | | |
| 189 | U/T | $ISOP + OH \longrightarrow ISOH$ | $2.55 \times 10^{-11} e^{410/T}$ | D |
| 190 | U/T | $ISOP + O_3 \longrightarrow 0.67 MACR + 0.26\ MVK + 0.3\ OH + 0.07\ PAR + 0.07\ OLE + 0.07\ H_2COO + 0.8\ HCHO + 0.06\ HO_2 + 0.15\ CO_2 + 0.05\ CO$ | $1.23 \times 10^{-14} e^{-2013/T}$ | D |
| 191 | U/T | $ISOP + O \longrightarrow 0.22\ MACR + 0.63\ MVK + 0.08\ ISOH$ | $3.30 \times 10^{-11}$ | D |
| 192 | U/T | $ISOP + NO_3 \longrightarrow ISNT$ | $7.80 \times 10^{-13}$ | D |
| 193 | U/T | $ISOH + NO \longrightarrow 0.364\ MACR + 0.477\ MVK + 0.840\ HCHO + 0.08\ ISNI1 + 0.08\ ISNI2 + 0.886\ HO_2 + 0.840\ NO_2$ | $1.22 \times 10^{-11} e^{-180/T}$ | D |
| 194 | U/T | $ISNT + NO \longrightarrow 1.1\ NO_2 + 0.8\ HO_2 + 0.80\ ISNI1 + 0.1\ MACR + 0.15\ HCHO + 0.05\ MVK + 0.05\ DISN$ | $1.39 \times 10^{-11} e^{-180/T}$ | D |
| 195 | U/T | $ISNI1 + OH \longrightarrow ISNIR$ | $3.35 \times 10^{-11}$ | D |
| 196 | U/T | $ISNI2 + OH \longrightarrow ISNIR$ | $1.88 \times 10^{-11}$ | D |
| 197 | U/T | $ISNIR + NO \longrightarrow 0.05\ DISN + 0.05\ HO_2 + 1.9\ NO_2 + 0.95 CH_3CHO + 0.95\ CH_3COCH_3$ | $1.39 \times 10^{-11} e^{-180/T}$ | D |
| 198 | U/T | $ISNI1 + O_3 \longrightarrow 0.2\ O + 0.08\ OH + 0.5\ HCHO + 0.5\ IALD1 + 0.5\ ISNI2 + 0.5\ NO_2$ | $5.00 \times 10^{-18}$ | D |
| 199 | U/T | $ISOH + ISOH \longrightarrow 0.6\ MACR + 0.6\ MVK + 1.2\ HCHO + 1.2\ HO_2$ | $2.00 \times 10^{-13}$ | D |
| 200 | U/T | $ISOH + HO_2 \longrightarrow IPRX$ | $6.15 \times 10^{-11} e^{-900/T}$ | D |
| 201 | U/T | $IPRX + OH \longrightarrow ISOH$ | $2.00 \times 10^{-11}$ | D |
| 202 | U/T | $IPRX + O_3 \longrightarrow 0.7\ HCHO$ | $8.00 \times 10^{-18}$ | D |
| 203 | U/T | $MACR + O_3 \longrightarrow 0.8\ CH_3COCHO + 0.7\ HCHO + 0.2\ O + 0.09\ H_2COO + 0.2\ CO + 0.275\ HO_2 + 0.215\ OH + 0.16\ CO_2 + 0.15\ CH_2CCH_3CHOO$ | $5.32 \times 10^{-15} e^{-2520/T}$ | D |
| 204 | U/T | $MVK + O_3 \longrightarrow 0.82\ CH_3COCHO + 0.8\ HCHO + 0.2\ O + 0.11\ H_2COO + 0.05\ CO + 0.06\ HO_2 + 0.08\ OH + 0.04\ CH_3CHO + 0.07\ OOCCH_3CHCH_2$ | $4.32 \times 10^{-15} e^{-2016/T}$ | D |
| 205 | U/T | $MACR + OH \longrightarrow 0.42\ MAC1 + 0.08\ MAC2 + 0.5\ CH_2CCH_3C(O)OO$ | $1.86 \times 10^{-11} e^{175/T}$ | D |
| 206 | U/T | $MVK + OH \longrightarrow 0.28\ MV1 + 0.72\ MV2$ | $4.11 \times 10^{-12} e^{453/T}$ | D |
| 207 | U/T | $MAC1 + NO \longrightarrow 0.95\ HO_2 + 0.95\ CO + 0.95\ CH_3COCH_3 + 0.95\ NO_2 + 0.05\ ISNI2$ | $1.39 \times 10^{-11} e^{-180/T}$ | D |

*(cont.)*

**Table B.4.** (cont.)

| No. | Regions | Kinetic Reaction | $F_c$[a] | Rate Coefficient ($s^{-1}$, $cm^3\ s^{-1}$, or $cm^6\ s^{-1}$) | Refs.[b] |
|---|---|---|---|---|---|
| | | **Organic Chemistry** | | | |
| 208 | U/T | $MAC2 + NO \longrightarrow 0.95\ HO_2 + 0.95\ HCHO + 0.95\ CH_3COCHO + 0.95\ NO_2 + 0.05\ ISNI2$ | | $1.39 \times 10^{-11} e^{-180/T}$ | D |
| 209 | U/T | $MV1 + NO \longrightarrow 0.95\ CH_3COCHO + 0.95\ HCHO + 0.05\ ISNI2 + 0.95\ NO_2 + 0.95\ HO_2$ | | $1.39 \times 10^{-11} e^{-180/T}$ | D |
| 210 | U/T | $MV2 + NO \longrightarrow 0.95\ CH_3CHO + 0.95\ CH_3C(O)OO + 0.05\ ISNI2 + 0.95\ NO_2$ | | $1.39 \times 10^{-11} e^{-180/T}$ | D |
| 211 | U/T | $MV1 + HO_2 \longrightarrow ROOH$ | | $6.15 \times 10^{-11} e^{-900/T}$ | D |
| 212 | U/T | $MV2 + HO_2 \longrightarrow ROOH$ | | $6.15 \times 10^{-11} e^{-900/T}$ | D |
| 213 | U/T | $MAC1 + HO_2 \longrightarrow ROOH$ | | $6.15 \times 10^{-11} e^{-900/T}$ | D |
| 214 | U/T | $MAC2 + HO_2 \longrightarrow ROOH$ | | $6.15 \times 10^{-11} e^{-900/T}$ | D |
| 215 | U/T | $CH_2CCH_3C(O)OO + NO_2 \longrightarrow MPAN$ | | $8.40 \times 10^{-12}$ | D |
| 216 | U/T | $MPAN \longrightarrow CH_2CCH_3C(O)OO + NO_2$ | | $1.58 \times 10^{16} e^{-13507/T}$ | D |
| 217 | U/T | $CH_2CCH_3C(O)OO + NO \longrightarrow C_2H_4 + CH_3O_2 + NO_2 + CO_2$ | | $1.40 \times 10^{-11}$ | D |
| | | **Sulfur Chemistry** | | | |
| 218 | All | $SO_2 + OH \xrightarrow{M} HSO_3$ | (P) 0.45 | $4.00 \times 10^{-31} (300/T)^{3.3}$ <br> $2.00 \times 10^{-12}$ | A |
| 219 | All | $HSO_3 + O_2 \longrightarrow SO_3 + HO_2$ | | $1.30 \times 10^{-12} e^{-330/T}$ | A |
| 220 | All | $SO_3 + H_2O \longrightarrow H_2SO_4$ | | $6.00 \times 10^{-15}$ | A |
| 221 | T | $CH_3SCH_3 + OH \longrightarrow CH_3SCH_2 + H_2O$ | | $1.13 \times 10^{-11} e^{-254/T}$ | A |
| 222 | T | $CH_3SCH_3 + OH \longrightarrow CH_3S(OH)CH_3$ | | $f$ | A |
| 223 | T | $CH_3SCH_2 + O_2 \longrightarrow CH_3SCH_2O_2$ | | $7.30 \times 10^{-13}$ | N |
| 224 | T | $CH_3SCH_2O_2 + NO \longrightarrow CH_3SCH_2O + NO_2$ | | $8.00 \times 10^{-12}$ | N |
| 225 | T | $CH_3SCH_2O \longrightarrow CH_3S + HCHO$ | | $1.00 \times 10^{1}$ | N |
| 226 | T | $CH_3S + O_2 \longrightarrow CH_3SOO^*$ | | $2.5 \times 10^{-18}$ | A |
| 227 | T | $CH_3SOO^* + NO \longrightarrow CH_3SO + NO_2$ | | $1.4 \times 10^{-11}$ | N |
| 228 | T | $CH_3SOO^* \longrightarrow CH_3S + O_2$ | | $6.0 \times 10^{2}$ | N |
| 229 | T | $CH_3SO \longrightarrow CH_3O_2 + SO$ | | $5.0 \times 10^{-5}$ | N |
| 230 | T | $CH_3SO + O_3 \longrightarrow CH_3SO_2 + O_2$ | | $2.0 \times 10^{-12}$ | N |
| 231 | T | $CH_3SO_2 \longrightarrow CH_3O_2 + SO_2$ | | $1.1 \times 10^{1}$ | N |
| 232 | T | $CH_3S(OH)CH_3 \longrightarrow CH_3SOH + CH_3O_2$ | | $5.0 \times 10^{5}$ | N |
| 233 | T | $CH_3S(OH)CH_3 + OH \longrightarrow CH_3S(O)_2CH_3 + 2HO_2$ | | $5.8 \times 10^{-11}$ | N |
| 234 | T | $CH_3SOH + OH \longrightarrow CH_3SO + H_2O$ | | $1.1 \times 10^{-10}$ | N |
| 235 | T | $CH_3SSCH_3 + OH \longrightarrow CH_3SOH + CH_3S$ | | $7.00 \times 10^{-11} e^{350/T}$ | A |
| 236 | T | $CH_3SH + OH \longrightarrow CH_3S + H_2O$ | | $9.90 \times 10^{-12} e^{356/T}$ | A |
| 237 | T | $H_2S + OH \longrightarrow HS + H_2O$ | | $6.30 \times 10^{-12} e^{-80/T}$ | A |
| 238 | T | $HS + O_2 \longrightarrow SO + OH$ | | $4.00 \times 10^{-19}$ | A |
| 239 | T/S | $SO + O_2 \longrightarrow SO_2 + O$ | | $1.60 \times 10^{-13} e^{-2280/T}$ | A |
| 240 | T/S | $OCS + OH \longrightarrow HS + CO_2$ | | $1.10 \times 10^{-13} e^{-1200/T}$ | A |
| 241 | T/S | $CS_2 + OH \longrightarrow HS + OCS$ | | $2.00 \times 10^{-15}$ | A |
| 242 | T/S | $CS + O_2 \longrightarrow OCS + O$ | | $2.90 \times 10^{-19}$ | A |
| 243 | T/S | $S + O_2 \longrightarrow SO + O$ | | $2.10 \times 10^{-12}$ | A |
| 244 | S | $Cl + O_2 + N_2 \longrightarrow ClOO + N_2$ | | $1.40 \times 10^{-33} (300/T)^{3.9}$ | A |
| 245 | S | $Cl + O_3 \longrightarrow ClO + O_2$ | | $2.90 \times 10^{-11} e^{-260/T}$ | A |

*(cont.)*

Table B.4. (cont.)

| No. | Regions | Kinetic Reaction | $F_c$[a] | Rate Coefficient ($s^{-1}$, $cm^3\ s^{-1}$, or $cm^6\ s^{-1}$) | Refs.[b] |
|---|---|---|---|---|---|
| | | **Chlorine Chemistry** | | | |
| 246 | S | $Cl + H_2 \longrightarrow HCl + H$ | | $3.70 \times 10^{-11} e^{-2300/T}$ | A |
| 247 | S | $Cl + HO_2 \longrightarrow HCl + O_2$ | | $1.80 \times 10^{-11} e^{170/T}$ | A |
| 248 | S | $Cl + H_2O_2 \longrightarrow HCl + HO_2$ | | $1.10 \times 10^{-11} e^{-980/T}$ | A |
| 249 | S | $Cl + NO_2 \xrightarrow{M} ClNO_2$ | (P) 0.60 | $1.80 \times 10^{-31} (300/T)^{2.0}$ <br> $1.00 \times 10^{-10} (300/T)^{1.0}$ | B |
| 250 | S | $Cl + HNO_3 \longrightarrow HCl + NO_3$ | | $2.00 \times 10^{-16}$ | A |
| 251 | S | $Cl + NO_3 \longrightarrow ClO + NO_2$ | | $2.40 \times 10^{-11}$ | A |
| 252 | S | $Cl + CH_4 \longrightarrow HCl + CH_3O_2$ | | $9.60 \times 10^{-12} e^{-1350/T}$ | A |
| 253 | S | $Cl + HCHO \longrightarrow HCl + CO + HO_2$ | | $8.20 \times 10^{-11} e^{-34/T}$ | A |
| 254 | S | $Cl + HOCl \longrightarrow Cl_2 + OH$ | | $2.50 \times 10^{-12} e^{-130/T}$ | B |
| 255 | S | $Cl + OClO \longrightarrow ClO + ClO$ | | $3.40 \times 10^{-11} e^{160/T}$ | A |
| 256 | S | $Cl + ClOO \longrightarrow Cl_2 + O_2$ | | $2.30 \times 10^{-10}$ | B |
| 257 | S | $Cl + ClOO \longrightarrow ClO + ClO$ | | $1.20 \times 10^{-11}$ | B |
| 258 | S | $ClO + O \longrightarrow Cl + O_2$ | | $3.80 \times 10^{-11} e^{70/T}$ | A |
| 259 | S | $ClO + O_3 \longrightarrow ClOO + O_2$ | | $1.50 \times 10^{-17}$ | A |
| 260 | S | $ClO + O_3 \longrightarrow OClO + O_2$ | | $1.00 \times 10^{-18}$ | A |
| 261 | S | $ClO + OH \longrightarrow Cl + HO_2$ | | $9.90 \times 10^{-12} e^{120/T}$ | B |
| 262 | S | $ClO + OH \longrightarrow HCl + O_2$ | | $1.10 \times 10^{-12} e^{120/T}$ | B |
| 263 | S | $ClO + HO_2 \longrightarrow HOCl + O_2$ | | $4.66 \times 10^{-13} e^{710/T}$ | B |
| 265 | S | $ClO + HO_2 \longrightarrow HCl + O_3$ | | $1.44 \times 10^{-14} e^{710/T}$ | B |
| 266 | S | $ClO + NO \longrightarrow Cl + NO_2$ | | $6.20 \times 10^{-12} e^{294/T}$ | A |
| 267 | S | $ClO + NO_2 \xrightarrow{M} ClONO_2$ | (P) 0.50 | $1.60 \times 10^{-31} (300/T)^{3.4}$ <br> $2.00 \times 10^{-11} (300/T)^{1.9}$ | A |
| 268 | S | $ClO + ClO \longrightarrow Cl + ClOO$ | | $3.00 \times 10^{-11} e^{-2450/T}$ | A |
| 269 | S | $ClO + ClO \xrightarrow{M} Cl_2O_2$ | (P) 0.60 | $1.70 \times 10^{-32} (300/T)^{4.0}$ <br> $5.40 \times 10^{-12}$ | A |
| 270 | S | $HCl + OH \longrightarrow Cl + H_2O$ | | $2.40 \times 10^{-12} e^{-330/T}$ | A |
| 271 | S | $ClONO_2 + O \longrightarrow Cl + NO_2$ | | $3.00 \times 10^{-12} e^{-800/T}$ | A |
| 272 | S | $OClO + O \longrightarrow ClO + O_2$ | | $2.40 \times 10^{-12} e^{-960/T}$ | A |
| 273 | S | $OClO + OH \longrightarrow HOCl + O_2$ | | $4.50 \times 10^{-13} e^{800/T}$ | A |
| 274 | S | $OClO + NO \longrightarrow ClO + NO_2$ | | $2.50 \times 10^{-12} e^{-600/T}$ | B |
| 275 | S | $ClOO \longrightarrow Cl + O_2$ | | $2.80 \times 10^{-10} e^{-1820/T}$ | A |
| 276 | S | $HOCl + O \longrightarrow ClO + OH$ | | $1.00 \times 10^{-11} e^{-1300/T}$ | A |
| 277 | S | $HOCl + OH \longrightarrow ClO + H_2O$ | | $3.00 \times 10^{-12} e^{-500/T}$ | A |
| 278 | S | $Cl_2O_2 \xrightarrow{M} ClO + ClO$ | (P) 0.60 | $1.00 \times 10^{-6} e^{-8000/T}$ <br> $4.80 \times 10^{15} e^{-8820/T}$ | A |
| | | **Bromine Chemistry** | | | |
| 279 | S | $Br + O_3 \longrightarrow BrO + O_2$ | | $1.70 \times 10^{-11} e^{-800/T}$ | A |
| 280 | S | $Br + HO_2 \longrightarrow HBr + O_2$ | | $1.40 \times 10^{-11} e^{-590/T}$ | A |
| 281 | S | $Br + H_2O_2 \longrightarrow HBr + HO_2$ | | $1.00 \times 10^{-11} e^{-3000/T}$ | B |
| 282 | S | $Br + HCHO \longrightarrow HBr + CO + HO_2$ | | $1.70 \times 10^{-11} e^{-800/T}$ | A |
| 283 | S | $BrO + O \longrightarrow Br + O_2$ | | $1.90 \times 10^{-11} e^{230/T}$ | A |
| 284 | S | $BrO + OH \longrightarrow Br + HO_2$ | | $7.5 \times 10^{-11}$ | B |
| 285 | S | $BrO + HO_2 \longrightarrow HOBr + O_2$ | | $3.40 \times 10^{-12} e^{540/T}$ | B |
| 286 | S | $BrO + NO \longrightarrow Br + NO_2$ | | $8.70 \times 10^{-12} e^{260/T}$ | A |
| 287 | S | $BrO + NO_2 \xrightarrow{M} BrONO_2$ | (P) 0.4 | $4.70 \times 10^{-31} (300/T)^{3.1}$ <br> $1.70 \times 10^{-11} (300/T)^{0.6}$ | A |
| 288 | S | $BrO + ClO \longrightarrow Br + OClO$ | | $1.60 \times 10^{-12} e^{430/T}$ | A |

*(cont.)*

**Table B.4.** (cont.)

| No. | Regions | Kinetic Reaction | $F_c{}^a$ | Rate Coefficient ($s^{-1}$, $cm^3\ s^{-1}$, or $cm^6\ s^{-1}$) | Refs.[b] |
|---|---|---|---|---|---|
| | | **Bromine Chemistry** | | | |
| 289 | S | $BrO + ClO \longrightarrow Br + ClOO$ | | $2.90 \times 10^{-12} e^{220/T}$ | A |
| 290 | S | $BrO + ClO \longrightarrow BrCl + O_2$ | | $5.80 \times 10^{-13} e^{170/T}$ | A |
| 291 | S | $BrO + BrO \longrightarrow 2Br + O_2$ | | $2.40 \times 10^{-12} e^{-40/T}$ | B |
| 292 | S | $BrO + BrO \longrightarrow Br_2 + O_2$ | | $2.80 \times 10^{-14} e^{860/T}$ | B |
| 293 | S | $BrO + O_3 \longrightarrow Br + 2O_2$ | | $1.00 \times 10^{-12} e^{-3200/T}$ | B |
| 294 | S | $HBr + OH \longrightarrow Br + H_2O$ | | $1.10 \times 10^{-11} (298/T)^{0.8}$ | A |
| 295 | S | $HOBr + O \longrightarrow BrO + OH$ | | $1.40 \times 10^{-10} e^{-430/T}$ | H |
| 296 | S | $BrCl + O \longrightarrow BrO + Cl$ | | $2.20 \times 10^{-10}$ | I |

**Photoprocesses**

| | Region | Photolysis Reaction | λ (μm) | J (25 km) ($s^{-1}$) | J (0 km) ($s^{-1}$) | Ref. |
|---|---|---|---|---|---|---|
| 297 | S | $O_2 + h\nu \longrightarrow O + O$ | <0.245 | $1.2 \times 10^{-11}$ | $4.36 \times 10^{-26}$ | B |
| 298 | All | $O_3 + h\nu \longrightarrow O(^1D) + O_2$ | <0.31 | $1.13 \times 10^{-4}$ | $5.08 \times 10^{-5}$ | B |
| 299 | All | $O_3 + h\nu \longrightarrow O + O_2$ | >0.31 | $4.95 \times 10^{-4}$ | $4.17 \times 10^{-4}$ | B |
| 300 | S | $H_2O + h\nu \longrightarrow H + OH$ | <0.21 | $6.61 \times 10^{-11}$ | 0 | A |
| 301 | All | $H_2O_2 + h\nu \longrightarrow 2\ OH$ | <0.355 | $1.18 \times 10^{-5}$ | $7.72 \times 10^{-6}$ | A |
| 302 | All | $NO_2 + h\nu \longrightarrow NO + O$ | <0.42 | $1.24 \times 10^{-2}$ | $8.82 \times 10^{-3}$ | B |
| 303 | All | $NO_3 + h\nu \longrightarrow NO_2 + O$ | 0.41–0.67 | $3.04 \times 10^{-1}$ | $2.84 \times 10^{-1}$ | A |
| 304 | All | $NO_3 + h\nu \longrightarrow NO + O_2$ | 0.59–0.63 | $2.65 \times 10^{-2}$ | $2.49 \times 10^{-3}$ | A |
| 305 | S | $N_2O + h\nu \longrightarrow N_2 + O(^1D)$ | <0.24 | $2.49 \times 10^{-8}$ | $1.81 \times 10^{-23}$ | B |
| 306 | All | $N_2O_5 + h\nu \longrightarrow NO_2 + NO_3$ | <0.385 | $4.73 \times 10^{-5}$ | $5.04 \times 10^{-5}$ | B |
| 307 | All | $HONO + h\nu \longrightarrow OH + NO$ | <0.40 | $2.75 \times 10^{-3}$ | $1.94 \times 10^{-3}$ | A |
| 308 | All | $HNO_3 + h\nu \longrightarrow OH + NO_2$ | <0.335 | $5.61 \times 10^{-6}$ | $8.23 \times 10^{-7}$ | A |
| 309 | All | $HO_2NO_2 + h\nu \longrightarrow HO_2 + NO_2$ | <0.33 | $1.14 \times 10^{-5}$ | $3.68 \times 10^{-6}$ | A |
| 310 | All | $HO_2NO_2 + h\nu \longrightarrow OH + NO_3$ | <0.33 | $5.67 \times 10^{-6}$ | $1.84 \times 10^{-6}$ | A |
| 311 | All | $HCHO + h\nu \longrightarrow 2\ HO_2 + CO$ | <0.334 | $6.17 \times 10^{-5}$ | $3.29 \times 10^{-5}$ | A |
| 312 | All | $HCHO + h\nu \longrightarrow CO + H_2$ | <0.37 | $7.64 \times 10^{-5}$ | $4.40 \times 10^{-5}$ | A |
| 313 | All | $CH_3OOH + h\nu \longrightarrow CH_3O + OH$ | <0.36 | $9.92 \times 10^{-6}$ | $5.73 \times 10^{-6}$ | A |
| 314 | U/T | $CH_3CHO + h\nu \longrightarrow CH_3O_2 + HO_2 + CO$ | <0.325 | $1.30 \times 10^{-5}$ | $6.35 \times 10^{-6}$ | A |
| 315 | U/T | $CH_3ONO_2 + h\nu \longrightarrow CH_3O + NO_2$ | <0.33 | $2.54 \times 10^{-6}$ | $1.25 \times 10^{-6}$ | A |
| 316 | U/T | $CH_3O_2NO_2 + h\nu \longrightarrow CH_3O_2 + NO_2$ | <0.325 | $1.76 \times 10^{-5}$ | $6.43 \times 10^{-6}$ | A |
| 317 | U/T | $C_2H_5ONO_2 + h\nu \longrightarrow C_2H_5O + NO_2$ | <0.330 | $1.49 \times 10^{-5}$ | $1.79 \times 10^{-6}$ | A |
| 318 | U/T | $C_3H_7ONO_2 + h\nu \longrightarrow C_3H_7O + NO_2$ | <0.330 | $1.94 \times 10^{-5}$ | $3.44 \times 10^{-6}$ | A |
| 319 | U/T | $CH_3CO_3NO_2 + h\nu \longrightarrow CH_3CO_3 + NO_2$ | <0.300 | $3.54 \times 10^{-6}$ | $7.23 \times 10^{-8}$ | A |
| 320 | U/T | $CH_3COCH_3 + h\nu \longrightarrow CH_3O_2 + CH_3C(O)OO$ | <0.335 | $1.92 \times 10^{-6}$ | $8.39 \times 10^{-7}$ | A |
| 321 | U/T | $KET + h\nu \longrightarrow CH_3C(O)OO + RO_2 + 2XOP$ | <0.33 | $1.92 \times 10^{-6}$ | $8.39 \times 10^{-7}$ | J |
| 322 | U/T | $MVK + h\nu \longrightarrow CH_3C(O)OO + C_2H_4 + HO_2$ | <0.325 | $8.11 \times 10^{-7}$ | $4.12 \times 10^{-7}$ | K |
| 323 | U/T | $MACR + h\nu \longrightarrow C_2H_4 + HO_2 + CO + CH_3O_2$ | <0.375 | $2.59 \times 10^{-6}$ | $1.64 \times 10^{-6}$ | A |
| 324 | U/T | $CH_3COCHO + h\nu \longrightarrow CH_3C(O)OO + CO + HO_2$ | <0.465 | $3.82 \times 10^{-4}$ | $2.94 \times 10^{-4}$ | A |
| 325 | U | $BZA + h\nu \longrightarrow PHO_2 + CO + HO_2$ | <0.385 | $6.45 \times 10^{-3}$ | $4.31 \times 10^{-5}$ | E |
| 326 | U | $OPEN + h\nu \longrightarrow CH_3C(O)OO + CO + HO_2$ | <0.33 | $5.13 \times 10^{-4}$ | $2.94 \times 10^{-4}$ | G |
| 327 | T | $CH_3SSCH_3 + h\nu \longrightarrow 2\ CH_3S$ | <0.40 | $1.30 \times 10^{-4}$ | $6.28 \times 10^{-5}$ | A |

*(cont.)*

**Table B.4.** (cont.)

| | | Photoprocesses | | | | |
|---|---|---|---|---|---|---|
| No. | Region | Photolysis Reaction | λ (μm) | J (25 km) ($s^{-1}$) | J (0 km) ($s^{-1}$) | Ref. |
| 328 | S | $OCS + h\nu \longrightarrow CO + S$ | <0.26 | $2.83 \times 10^{-8}$ | $1.72 \times 10^{-21}$ | A |
| 329 | T/S | $CS_2 + h\nu \longrightarrow CS + S$ | <0.34 | $2.46 \times 10^{-5}$ | $1.46 \times 10^{-5}$ | A |
| 330 | S | $HCl + h\nu \longrightarrow H + Cl$ | <0.22 | $2.13 \times 10^{-8}$ | $9.25 \times 10^{-24}$ | A |
| 331 | S | $ClO + h\nu \longrightarrow Cl + O$ | <0.305 | $1.30 \times 10^{-4}$ | $4.06 \times 10^{-5}$ | L |
| 332 | S | $ClOO + h\nu \longrightarrow ClO + O$ | <0.280 | $6.22 \times 10^{-8}$ | $2.81 \times 10^{-16}$ | B |
| 333 | S | $OClO + h\nu \longrightarrow ClO + O$ | <0.45 | $1.36 \times 10^{-1}$ | $9.86 \times 10^{-2}$ | A |
| 334 | S | $HOCl + h\nu \longrightarrow OH + Cl$ | <0.375 | $4.10 \times 10^{-4}$ | $2.59 \times 10^{-4}$ | A |
| 335 | S | $ClONO_2 + h\nu \longrightarrow Cl + NO_3$ | <0.40 | $6.22 \times 10^{-5}$ | $4.75 \times 10^{-5}$ | A |
| 336 | S | $Cl_2 + h\nu \longrightarrow Cl + Cl$ | <0.45 | $3.51 \times 10^{-3}$ | $2.37 \times 10^{-3}$ | A |
| 337 | S | $Cl_2O_2 + h\nu \longrightarrow Cl + ClOO$ | <0.36 | $2.02 \times 10^{-3}$ | $1.21 \times 10^{-3}$ | A |
| 338 | S | $ClNO_2 + h\nu \longrightarrow Cl + NO_2$ | <0.372 | $6.50 \times 10^{-4}$ | $3.94 \times 10^{-4}$ | A |
| 339 | S | $CH_3Cl + h\nu \longrightarrow CH_3O_2 + Cl$ | <0.22 | $1.31 \times 10^{-8}$ | $2.26 \times 10^{-24}$ | A |
| 340 | S | $CH_3CCl_3 + h\nu \longrightarrow 3\ Cl$ | <0.24 | $7.20 \times 10^{-7}$ | $5.12 \times 10^{-22}$ | A |
| 341 | S | $CCl_4 + h\nu \longrightarrow Cl + CCl_3$ | <0.25 | $8.51 \times 10^{-7}$ | $2.12 \times 10^{-21}$ | A |
| 342 | S | $CFCl_3 + h\nu \longrightarrow Cl + CFCl_2$ | <0.25 | $5.10 \times 10^{-7}$ | $4.34 \times 10^{-22}$ | A |
| 343 | S | $CF_2Cl_2 + h\nu \longrightarrow Cl + CF_2Cl$ | <0.226 | $4.60 \times 10^{-8}$ | $1.65 \times 10^{-23}$ | A |
| 344 | S | $CFCl_2CF_2Cl + h\nu \longrightarrow Cl + CFCl_2CF_2$ | <0.230 | $8.78 \times 10^{-8}$ | $2.63 \times 10^{-23}$ | B |
| 345 | S | $CF_2ClH + h\nu \longrightarrow Cl$ | <0.205 | $5.68 \times 10^{-10}$ | $3.04 \times 10^{-27}$ | B |
| 346 | S | $BrO + h\nu \longrightarrow Br + O$ | <0.375 | $5.88 \times 10^{-3}$ | $3.82 \times 10^{-3}$ | A |
| 347 | S | $HOBr + h\nu \longrightarrow Br + OH$ | <0.48 | $9.05 \times 10^{-4}$ | $6.12 \times 10^{-4}$ | A |
| 348 | S | $BrONO_2 + h\nu \longrightarrow Br + NO_3$ | <0.39 | $1.61 \times 10^{-3}$ | $1.06 \times 10^{-3}$ | A |
| 349 | S | $Br_2 + h\nu \longrightarrow 2Br$ | <0.60 | $3.56 \times 10^{-2}$ | $3.02 \times 10^{-2}$ | B |
| 350 | S | $CH_3Br + h\nu \longrightarrow CH_3O_2 + Br$ | <0.26 | $1.13 \times 10^{-6}$ | $3.97 \times 10^{-21}$ | A |
| 351 | S | $BrCl + h\nu \longrightarrow Br + Cl$ | <0.57 | $5.88 \times 10^{-3}$ | $4.61 \times 10^{-3}$ | M |

Species in the reactions are listed in Table B.3. Species above reaction arrows are second or third bodies included in pressure- dependent reactions (footnote a). $M = N_2 + O_2$ is total air. The "region" column indicates whether the reaction is important in urban areas (U), the free-troposphere (T), the stratosphere (S), urban areas and the free-troposphere (U/T), or all three regions (All). The "Refs." column refers to sources of data for reaction rate coefficients, absorption cross sections, and quantum yields. Photolysis coefficients at 0 and 25 km were calculated for a nonpolluted sky, zenith angle of 0°, and UV-surface abedo of 0.03. The wavelengths given are valid for the listed photoprocess.

[a](P) indicates a pressure-dependent reaction, for which the reaction rate is

$$k_r = \frac{k_{\infty,T}k_{0,T}[M]}{k_{\infty,T} + k_{0,T}[M]} F_c^{\left[1 + \left(\log_{10}\frac{k_{0,T}[M]}{k_{\infty,T}}\right)^2\right]^{-1}}$$

where $k_{0,T}$ is the temperature-dependent three-body, low-pressure limit rate coefficient (the first rate listed), $k_{\infty,T}$ is the two-body, high-pressure limit rate coefficient (the second rate listed), $[M] = [N_2] + [O_2]$ is the concentration (molecules $cm^{-3}$) of the third body, and $F_c$ is the broadening factor.

[b]A, Atkinson et al. (1997); B, DeMore et al. (1997); C, Gery et al. (1989); D, Paulson and Seinfeld (1992); E, Carter (1991); F, Stockwell (1995); G, Gery et al. (1988); H, Nesbitt et al. (1995); I, Clyne et al. (1976); J, assumed the same as for acetone; K, assumed the same as for methyl ethyl ketone; L, Watson (1977); M, Seery and Britton (1964); N, Yin et al. (1990).

[c]$k_r = k_1 + k_3[M]/(1 + k_3[M]/k_2)$, where $k_1 = 7.20 \times 10^{-15}e^{785/T}$, $k_2 = 4.10 \times 10^{-16}e^{1440/T}$, $k_3 = 1.90 \times 10^{-33}e^{725/T}$, and $[M] = [N_2] + [O_2]$ (molecules $cm^{-3}$).

[d]$k_r = 1.30 \times 10^{-13}(1 + 0.6p_{a.bar})(300/T)^{1.0}$, where $p_{a.bar}$ is the ambient air pressure in bars.

[e]$k_r = (2.30 \times 10^{-13}e^{600/T} + 1.70 \times 10^{-33}[M]e^{1000/T})(1. + 1.40 \times 10^{-21}[H_2O]e^{2200/T})$, where $[M] = [N_2] + [O_2]$ and $[H_2O]$ are in units of molecules $cm^{-3}$.

[f]$k_r = 1.7 \times 10^{-42}[M]e^{7810/T}/(1.0 + 5.5 \times 10^{-31}[M]\ e^{7460/T})$, where $[M] = [N_2] + [O_2]$ (molecules $cm^{-3}$).

## B.5. EQUILIBRIUM AND AQUEOUS-CHEMISTRY SPECIES

**Table B.5.** List of Species Involved in Equilibrium or Aqueous-Chemistry Reactions

| Chemical Name | Chemical Formula | Chemical Name | Chemical Formula |
|---|---|---|---|
| | **Gas-Phase Precursors** | | |
| Nitrogen dioxide | $NO_2(g)$ | Hydrochloric acid | $HCl(g)$ |
| Nitrate radical | $NO_3(g)$ | Carbon dioxide | $CO_2(g)$ |
| Nitrous acid | $HONO(g)$ | Formaldehyde | $HCHO(g)$ |
| Nitric acid | $HNO_3(g)$ | Formic acid | $HCOOH(g)$ |
| Peroxynitric acid | $HO_2NO_2(g)$ | Methanol | $CH_3OH(g)$ |
| Hydroxyl radical | $OH(g)$ | Methylperoxy radical | $CH_3O_2(g)$ |
| Hydrogen peroxide | $H_2O_2(g)$ | Methyl hydroperoxide | $CH_3OOH(g)$ |
| Water vapor | $H_2O(g)$ | Acetic acid | $CH_3COOH(g)$ |
| Ozone | $O_3(g)$ | Peroxyacetyl nitrate | $CH_3C(O)OONO_2(g)$ |
| Sulfur dioxide | $SO_2(g)$ | Peroxyacetic acid | $CH_3C(O)OOH(g)$ |
| Sulfuric acid | $H_2SO_4(g)$ | Methylglyoxal | $CH_3COCHO(g)$ |
| Ammonia | $NH_3(g)$ | Nitrocresol | $C_6H_3(CH_3)(OH)NO_2(g)$ |
| | **Dissociated Ionic Molecules** | | |
| Hydrogen ion | $H^+$ | Peroxysulfate radical ion | $SO_5^-$ |
| Ammonium ion | $NH_4^+$ | Peroxymonosulfate ion | $HSO_5^-$ |
| Sodium ion | $Na^+$ | Hydromethanesulfonate | $HOCH_2SO_3^-$ |
| Potassium ion | $K^+$ | Oxymethanesulfonate ion | $^-OCH_2SO_3^-$ |
| Magnesium ion | $Mg^{2+}$ | Nitrate ion | $NO_3^-$ |
| Calcium ion | $Ca^{2+}$ | Nitrite ion | $NO_2^-$ |
| Hydroxy ion | $OH^-$ | Chloride ion | $Cl^-$ |
| Hydroperoxy ion | $HO_2^-$ | Dichloride ion | $Cl_2^-$ |
| Peroxy ion | $O_2^-$ | Chlorine hydroxide radical ion | $ClOH^-$ |
| Bisulfate ion | $HSO_4^-$ | Bicarbonate ion | $HCO_3^-$ |
| Sulfate ion | $SO_4^{2-}$ | Carbonate ion | $CO_3^{2-}$ |
| Bisulfite ion | $HSO_3^-$ | Carbonate radical ion | $CO_3^-$ |
| Sulfite ion | $SO_3^{2-}$ | Formate | $HCOO^-$ |
| Sulfate radical ion | $SO_4^-$ | Acetate | $CH_3COO^-$ |
| | **Undissociated Molecules** | | |
| Nitrogen dioxide | $NO_2(aq)$ | Carbon monoxide | $CO(aq)$ |
| Nitrate radical | $NO_3(aq)$ | Carbon dioxide | $CO_2(aq)$ |
| Nitrous acid | $HONO(aq)$ | Formaldehyde | $HCHO(aq)$ |
| Nitric acid | $HNO_3(aq)$ | Methylene glycol | $H_2C(OH)_2(aq)$ |
| Peroxynitric acid | $HO_2NO_2(aq)$ | Formic acid | $HCOOH(aq)$ |
| Hydroxyl radical | $OH(aq)$ | Methanol | $CH_3OH(aq)$ |
| Hydroperoxy radical | $HO_2(aq)$ | Methylperoxy radical | $CH_3O_2(aq)$ |
| Hydrogen peroxide | $H_2O_2(aq)$ | Methyl hydroperoxide | $CH_3OOH(aq)$ |
| Water | $H_2O(aq)$ | Acetic acid | $CH_3COOH(aq)$ |
| Molecular oxygen | $O_2(aq)$ | Peroxyacetyl nitrate | $CH_3C(O)OONO_2(aq)$ |
| Ozone | $O_3(aq)$ | Peroxyacetic acid | $CH_3C(O)OOH(aq)$ |
| Sulfur dioxide | $SO_2(aq)$ | Methylgloxyal | $CH_3COCHO(aq)$ |
| Sulfuric acid | $H_2SO_4(aq)$ | Hydrated methylglyoxal | $CH_3COCH_3O_2(aq)$ |
| Ammonia | $NH_3(aq)$ | Nitrocresol | $C_6H_3(CH_3)(OH)NO_2(aq)$ |
| Hydrochloric acid | $HCl(aq)$ | Soluble manganese(II) ion and complexes | Mn(II) |

*(cont.)*

**Table B.5.** (cont.)

| Chemical Name | Chemical Formula | Chemical Name | Chemical Formula |
|---|---|---|---|
| Chlorine atom | Cl(aq) | Soluble iron(III) ion and complexes | Fe(III) |
| | | **Solids** | |
| Ammonium nitrate | $NH_4NO_3(s)$ | Potassium carbonate | $K_2CO_3(s)$ |
| Ammonium chloride | $NH_4Cl(s)$ | Magnesium nitrate | $Mg(NO_3)_2(s)$ |
| Ammonium bisulfate | $NH_4HSO_4(s)$ | Magnesium chloride | $MgCl_2(s)$ |
| Ammonium sulfate | $(NH_4)_2SO_4(s)$ | Magnesium sulfate | $MgSO_4(s)$ |
| Triammonium bisulfate | $(NH_4)_3H(SO_4)_2(s)$ | Magnesium carbonate | $MgCO_3(s)$ |
| Ammonium bicarbonate | $NH_4HCO_3(s)$ | Calcium nitrate | $Ca(NO_3)_2(s)$ |
| Sodium nitrate | $NaNO_3(s)$ | Calcium chloride | $CaCl_2(s)$ |
| Sodium chloride | NaCl(s) | Calcium sulfate | $CaSO_4{\cdot}2H_2O(s)$ |
| Sodium bisulfate | $NaHSO_4(s)$ | Calcium carbonate | $CaCO_3(s)$ |
| Sodium sulfate | $Na_2SO_4(s)$ | Manganese heptoxide | $Mn_2O_7(s)$ |
| Sodium bicarbonate | $NaHCO_3(s)$ | Silicon dioxide | $SiO_2(s)$ |
| Sodium carbonate | $Na_2Co_3(s)$ | Aluminium oxide | $Al_2O_3(s)$ |
| Potassium nitrate | $KNO_3(s)$ | Iron(III) oxide | $Fe_2O_3(s)$ |
| Potassium chloride | KCl(s) | Lead suboxide | $Pb_2O(s)$ |
| Potassium bisulfate | $KHSO_4(s)$ | Elemental carbon | EC (s) |
| Potassium sulfate | $K_2SO_4(s)$ | Organic carbon | OC (s) |
| Potassium bicarbonate | $KHCO_3(s)$ | | |

## B.6. THERMODYNAMIC DATA

**Table B.6.** Value of $\Delta_f G_i^\circ$, $\Delta_f H_i^\circ$, and $c_{p,i}^\circ$ for Several Substances

| Substance | $\Delta_f G_i^\circ$ (kJ mole$^{-1}$) | $\Delta_f H_i^\circ$ (kJ mole$^{-1}$) | $c_{p,i}^\circ$ (J mole$^{-1}$ K$^{-1}$) |
|---|---|---|---|
| $NH_3(g)$ | −16.45[A] | −46.11[A] | 35.06[A] |
| $HNO_3(g)$ | −74.72[A] | −135.06[A] | 53.35 [A] |
| HCl(g) | −95.299[A] | −92.307[A] | 29.126[A] |
| $SO_2(g)$ | −300.194[A] | −296.83[A] | 39.87[A] |
| $H_2SO_4$ (g) | −690.289[B] | −814.21[B] | — |
| $CO_2$ (g) | −394.359[A] | −393.509[A] | 37.11[A] |
| HCOOH(g) | −351.0[A] | −378.57[A] | — |
| $CH_3COOH(g)$ | −374.0[A] | −432.25[A] | 66.5[A] |
| $H_2O(aq)$ | −237.129[A] | −285.83[A] | 75.291[A] |
| $NH_3(aq)$ | −26.5[A] | −80.29[A] | 35.06[E] |
| $HNO_3(aq)$ | −111.25[A] | −207.36[A] | −86.6[A] |
| $H_2SO_3(aq)$ | −537.81[A] | −608.81[A] | — |
| $H_2SO_4(aq)$ | −690.003[A] | −813.989[A] | 138.91[A] |
| $CO_2(aq)$ | −385.98[A] | −413.8[A] | 277.64[H] |
| HCOOH(aq) | −372.3[A] | −425.43[A] | — |
| $CH_3COOH(aq)$ | −396.46[A] | −485.76[A] | — |
| $H^+$ | 0[A] | 0[A] | 0[A] |
| $NH_4^+$ | −79.31[A] | −132.51[A] | 79.9[A] |
| $Na^+$ | −261.905[A] | −240.12[A] | 46.4[A] |
| $K^+$ | −283.27[A] | −252.38[A] | 21.8[A] |
| $Mg^{2+}$ | −454.8[A] | −466.85[A] | — |
| $Ca^{2+}$ | −553.58[A] | −542.83[A] | — |

*(cont.)*

**Table B.6.** (cont.)

| Substance | $\Delta_f G_i^\circ$ (kJ mole$^{-1}$) | $\Delta_f H_i^\circ$ (kJ mole$^{-1}$) | $c_{p,i}^\circ$ (J mole$^{-1}$ K$^{-1}$) |
|---|---|---|---|
| $OH^-$ | $-157.244^A$ | $-229.994^A$ | $-148.5^A$ |
| $NO_3^-$ | $-111.25^F$ | $-207.36^F$ | $-86.6^F$ |
| $Cl^-$ | $-131.228^A$ | $-167.159^A$ | $-136.4^A$ |
| $HSO_4^-$ | $-755.91^A$ | $-887.34^A$ | $-84^A$ |
| $SO_4^{2-}$ | $-744.53^A$ | $-909.27^A$ | $-293.0^A$ |
| $HSO_3^-$ | $-527.73^A$ | $-626.22^A$ | — |
| $SO_3^{2-}$ | $-486.5^A$ | $-635.5^A$ | — |
| $HCO_3^-$ | $-586.77^A$ | $-691.99^A$ | $88.43^H$ |
| $CO_3^{2-}$ | $-527.81^A$ | $-677.14^A$ | $-234.52^H$ |
| HCOO | $-351.0^A$ | $-425.55^A$ | $-87.9^A$ |
| $CH_3COO-$ | $-369.31^A$ | $-486.01^A$ | $-6.3^A$ |
| $NH_4NO_3(s)$ | $-183.87^A$ | $-365.56^A$ | $139.3^A$ |
| $NH_4Cl(s)$ | $-203.167^B$ | $-314.86^B$ | $84.1^A$ |
| $NH_4HSO4(s)$ | $-823.0^C$ | $-1026.96^A$ | $127.5^D$ |
| $(NH_4)_2SO4(s)$ | $-901.67^A$ | $-1180.85^A$ | $187.49^A$ |
| $(NH_4)_3H(SO4)_2(s)$ | $-1730.0^C$ | $-2207.0^C$ | $315.0^D$ |
| $NH_4HCO_3(s)$ | $-665.9^A$ | $-849.4^A$ | — |
| $NaNO_3(s)$ | $367.0^A$ | $-467.85^A$ | $92.88^A$ |
| $NaCl(s)$ | $-384.24^B$ | $-411.26^B$ | $50.5^A$ |
| $NaHSO_4(s)$ | $-1003.81^B$ | $-1132.19^B$ | $85.0^D$ |
| $Na_2SO_4(s)$ | $-1270.16^A$ | $-1387.08^A$ | $128.2^A$ |
| $NaHCO_3(s)$ | $-851.0^A$ | $950.81^A$ | $87.61^A$ |
| $Na_2CO_3(s)$ | $-1044.44^A$ | $-1130.68^A$ | $112.3^A$ |
| $KNO_3(s)$ | $-394.86^A$ | $-494.63^A$ | $96.4^A$ |
| $KCl(s)$ | $-409.14^A$ | $-436.747^A$ | $51.3^A$ |
| $KHSO_4(s)$ | $-1031.3^A$ | $-1160.6^A$ | $87.16^G$ |
| $K_2SO_4(s)$ | $-1321.37^A$ | $-1437.79^A$ | $131.46^A$ |
| $KHCO_3(s)$ | $-863.5^A$ | $-963.2^A$ | $89.27^H$ |
| $K_2CO_3(s)$ | $-1063.5^A$ | $-1151.02^A$ | $114.43^A$ |
| $Mg(NO_3)_2(s)$ | $-589.4^A$ | $-790.65^A$ | $141.92^A$ |
| $MgCl_2(s)$ | $-591.79^A$ | $-641.32^A$ | $71.38^A$ |
| $MgSO_4(s)$ | $-1170.6^A$ | $-1284.9^A$ | $96.48^A$ |
| $MgCO_3(s)$ | $-1012.1^A$ | $-1095.8^A$ | $75.52^A$ |
| $Ca(NO_3)_2(s)$ | $-743.07^A$ | $-938.39^A$ | $149.37^A$ |
| $CaCl_2(s)$ | $-748.1^A$ | $-795.8^A$ | $72.59^A$ |
| $CaSO_4\text{-}2H_2O(s)$ | $-1797.28^A$ | $-2022.63^A$ | $186.02^A$ |
| $CaCO_3(s)$ | $-1128.79^A$ | $-1206.92^A$ | $81.88^A$ |

A. Wagman et al. (1982); B, Zaytsev and Aseyev (1992); C, Bassett and Seinfeld (1983); D, Wexler and Seinfeld, 1991); E, from A but for $NH_2^+$; F, from A but for $HNO_3$ since $NO_3^-$ data may be incorrect; G, Kim and Seinfield (1995); H, Meng et al. (1995).

## B.7. EQUILIBRIUM REACTIONS

**Table B.7.** Equilibrium Reaction Coefficients, and Coefficient Units

| No. | Reaction | | A | B | C | Units | Ref.[a] |
|---|---|---|---|---|---|---|---|
| 1. | $SO_2(g)$ | $\rightleftharpoons SO_2(aq)$ | 1.22 | 10.55 | | mol/kg-atm | A |
| 2. | $H_2O_2(g)$ | $\rightleftharpoons H_2O_2(aq)$ | $7.45 \times 10^4$ | 22.21 | | mol/kg-atm | B |
| 3. | $O_3(g)$ | $\rightleftharpoons O_3(aq)$ | $1.13 \times 10^{-2}$ | 7.72 | | mol/kg-atm | C |

*(cont.)*

**Table B.7.** (cont.)

| No. | Reaction | | A | B | C | Units | Ref.[a] |
|---|---|---|---|---|---|---|---|
| 4. | $NO_2(g)$ | $\rightleftharpoons NO_2(aq)$ | $1.00 \times 10^{-2}$ | 8.38 | | mol/kg-atm | D |
| 5. | $NO_3(g)$ | $\rightleftharpoons NO_3(aq)$ | $2.10 \times 10^{5}$ | 29.19 | | mol/kg-atm | E |
| 6. | $OH(g)$ | $\rightleftharpoons OH(aq)$ | $2.50 \times 10^{1}$ | 17.72 | | mol/kg-atm | E |
| 7. | $HO_2(g)$ | $\rightleftharpoons HO_2(aq)$ | $2.00 \times 10^{3}$ | 22.28 | | mol/kg-atm | E |
| 8. | $HONO(g)$ | $\rightleftharpoons HONO(aq)$ | $4.90 \times 10^{1}$ | 16.04 | | mol/kg-atm | F |
| 9. | $HO_2NO_2(g)$ | $\rightleftharpoons HO_2NO_2(aq)$ | $2.00 \times 10^{4}$ | | | mol/kg-atm | G |
| 10. | $HNO_3(g)$ | $\rightleftharpoons HNO_3(aq)$ | $2.10 \times 10^{5}$ | | | mol/kg-atm | D |
| 11. | $NH_3(g)$ | $\rightleftharpoons NH_3(aq)$ | $5.76 \times 10^{1}$ | 13.79 | −5.39 | mol/kg-atm | A |
| 12. | $HCHO(g)$ | $\rightleftharpoons HCHO(aq)$ | 3.46 | 8.9 | | mol/kg-atm | H |
| 13. | $HCOOH(g)$ | $\rightleftharpoons HCOOH(aq)$ | $5.39 \times 10^{3}$ | 18.9 | | mol/kg-atm | A |
| 14. | $CO_2(g)$ | $\rightleftharpoons CO_2(aq)$ | $3.41 \times 10^{-2}$ | 8.19 | −28.93 | mol/kg-atm | A |
| 15. | $CH_3OH(g)$ | $\rightleftharpoons CH_3OH(aq)$ | $2.20 \times 10^{2}$ | 16.44 | | mol/kg-atm | I |
| 16. | $CH_3O_2(g)$ | $\rightleftharpoons CH_3O_2(aq)$ | 6.00 | 18.79 | | mol/kg-atm | E |
| 17. | $CH_3OOH(g)$ | $\rightleftharpoons CH_3OOH(aq)$ | $2.27 \times 10^{2}$ | 18.82 | | mol/kg-atm | B |
| 18. | $CH_3COOH(g)$ | $\rightleftharpoons CH_3COOH(aq)$ | $8.60 \times 10^{3}$ | 21.58 | | mol/kg-atm | A |
| 19. | $CH_3C(O)OOH(g)$ | $\rightleftharpoons CH_3C(O)OOH(aq)$ | $4.73 \times 10^{2}$ | 20.70 | | mol/kg-atm | B |
| 20. | $CH_3C(O)OONO_2(g)$ | $\rightleftharpoons CH_3C(O)OONO_2(aq)$ | 2.90 | 19.83 | | mol/kg-atm | J |
| 21. | $CH_3COCHO(g)$ | $\rightleftharpoons CH_3COCHO(aq)$ | $3.70 \times 10^{3}$ | 25.33 | | mol/kg-atm | K |
| 22. | $HCHO(aq) + H_2O(aq)$ | $\rightleftharpoons H_2C(OH)_2(aq)$ | $1.82 \times 10^{3}$ | 13.49 | | $atm^{-1}$ | L |
| 23. | $SO_2\ (aq) + H_2O(aq)$ | $\rightleftharpoons H^+ + HSO_3^-$ | $1.71 \times 10^{-2}$ | 7.04 | | mol/kg | A |
| 24. | $CO_2(aq) + H_2O(aq)$ | $\rightleftharpoons H^+ + HCO_3^-$ | $4.30 \times 10^{-7}$ | −3.08 | 31.81 | mol/kg | A |
| 25. | $Cl(aq) + H_2O(aq)$ | $\rightleftharpoons H^+ + ClOH^-$ | $6.20 \times 10^{-8}$ | | | mol/kg | M |
| 26. | $NH_3(aq) + H_2O(aq)$ | $\rightleftharpoons NH_4^+ + OH^-$ | $1.81 \times 10^{-5}$ | −1.50 | 26.92 | mol/kg | A |
| 27. | $HNO_3(aq)$ | $\rightleftharpoons H^+ + NO_3^-$ | $1.20 \times 10^{1}$ | 29.17 | 16.83 | mol/kg | N |
| 28. | $HCl(aq)$ | $\rightleftharpoons H^+ + Cl^-$ | $1.72 \times 10^{6}$ | 23.15 | | mol/kg | O |
| 29. | $H_2O(aq)$ | $\rightleftharpoons H^+ + OH^-$ | $1.01 \times 10^{-14}$ | −22.52 | 26.92 | mol/kg | A |
| 30. | $H_2SO_4(aq)$ | $\rightleftharpoons H^+ + HSO_4^-$ | $1.00 \times 10^{3}$ | | | mol/kg | R |
| 31. | $H_2O_2(aq)$ | $\rightleftharpoons H^+ + HO_2^-$ | $2.20 \times 10^{-12}$ | −12.52 | | mol/kg | S |
| 32. | $HO_2(aq)$ | $\rightleftharpoons H^+ + O_2^-$ | $3.50 \times 10^{-5}$ | | | mol/kg | R |
| 33. | $HONO(aq)$ | $\rightleftharpoons H^+ + NO_2^-$ | $5.10 \times 10^{-4}$ | −4.23 | | mol/kg | F |
| 34. | $HCOOH(aq)$ | $\rightleftharpoons H^+ + HCOO^-$ | $1.86 \times 10^{-4}$ | −0.05 | | mol/kg | A |
| 35. | $CH_3COOH(aq)$ | $\rightleftharpoons H^+ + CH_3COO^-$ | $1.75 \times 10^{-5}$ | 0.10 | | mol/kg | A |
| 36. | $ClOH^-$ | $\rightleftharpoons Cl^- + OH(aq)$ | 1.43 | | | mol/kg | M |
| 37. | $Cl_2^-$ | $\rightleftharpoons Cl(aq) + Cl^-$ | $5.26 \times 10^{-6}$ | | | mol/kg | M |
| 38. | $HSO_3^-$ | $\rightleftharpoons H^+ + SO_3^{2-}$ | $5.99 \times 10^{-8}$ | 3.74 | | mol/kg | A |
| 39. | $HSO_4^-$ | $\rightleftharpoons 2\,H^+ + SO_4^{2-}$ | $1.02 \times 10^{-2}$ | 8.85 | 25.14 | mol/kg | A |
| 40. | $HCO_3^-$ | $\rightleftharpoons H^+ + CO_3^{2-}$ | $4.68 \times 10^{-11}$ | −5.99 | 38.84 | mol/kg | A |
| 41. | $HNO_3(g)$ | $\rightleftharpoons H^+ + NO_3^-$ | $2.51 \times 10^{6}$ | 29.17 | 16.83 | $mol^2/kg^2$-atm | A |
| 42. | $HCl\ (g)$ | $\rightleftharpoons H^+ + Cl^-$ | $1.97 \times 10^{6}$ | 30.19 | 19.91 | $mol^2/kg^2$-atm | A |
| 43. | $NH_3(g) + H^+$ | $\rightleftharpoons NH_4^+$ | $1.03 \times 10^{11}$ | 34.81 | −5.39 | $atm^{-1}$ | A |
| 44. | $NH_3(g) + HNO_3(g)$ | $\rightleftharpoons NH_4^+ + NO_3^-$ | $2.58 \times 10^{17}$ | 64.02 | 11.44 | $mol^2/kg^2$-$atm^2$ | A |
| 45. | $NH_3(g) + HCl(g)$ | $\rightleftharpoons NH_4^+ + Cl^-$ | $2.03 \times 10^{17}$ | 65.05 | 14.51 | $mol^2/kg^2$-$atm^2$ | A |
| 46. | $NH_4NO_3(s)$ | $\rightleftharpoons NH_4^+ + NO_3^-$ | $1.49 \times 10^{1}$ | −10.40 | 17.56 | $mol^2/kg^2$ | A |
| 47 | $NH_4Cl(s)$ | $\rightleftharpoons NH_4^+ + Cl^-$ | $1.96 \times 10^{1}$ | −6.13 | 16.92 | $mol^2/kg^2$ | A |
| 48. | $NH_4HSO_4(s)$ | $\rightleftharpoons NH_4^+ + HSO_4^-$ | $1.38 \times 10^{12}$ | −2.87 | 15.83 | $mol^2/kg^2$ | A |
| 49 | $(NH_4)_2SO_4(s)$ | $\rightleftharpoons 2\,NH_4^+ + SO_4^{2-}$ | 1.82 | −2.65 | 38.57 | $mol^3/kg^3$ | A |
| 50. | $(NH_4)_3H(SO_4)_2(s)$ | $\rightleftharpoons 3\,NH_4^+ + HSO_4^- + SO_4^{2-}$ | $2.93 \times 10^{1}$ | −5.19 | 54.40 | $mol^5/kg^5$ | A |
| 51. | $NH_4HCO_3(8)$ | $\rightleftharpoons NH_4^+ + HCO_3^-$ | 1.08 | −10.04 | | $mol^2/kg^2$ | A |
| 52. | $NaNO_3(s)$ | $\rightleftharpoons Na^+ + NO_3^-$ | $1.20 \times 10^{1}$ | −8.22 | 16.01 | $mol^2/kg^2$ | A |
| 53. | $NaCl(s)$ | $\rightleftharpoons Na^+ + Cl^-$ | $3.61 \times 10^{1}$ | −1.61 | 16.90 | $mol^2/kg^2$ | A |

*(cont.)*

**Table B.7.** (cont.)

| No. | Reaction | | A | B | C | Units | Ref.[a] |
|---|---|---|---|---|---|---|---|
| 54. | $NaHSO_4(s)$ | $\rightleftharpoons Na^+ + HSO_4^-$ | $2.84 \times 10^2$ | 1.91 | 14.75 | $mol^2/kg^2$ | A |
| 55. | $Na_2SO_4(s)$ | $\rightleftharpoons 2\ Na^+ + SO_4^{2-}$ | $4.80 \times 10^{-1}$ | 0.98 | 39.50 | $mol^3/kg^3$ | A |
| 56. | $NaHCO_3(s)$ | $\rightleftharpoons Na^+ + HCO_3^-$ | $3.91 \times 10^{-1}$ | −7.54 | −5.68 | $mol^2/kg^2$ | A |
| 57. | $Na_2CO_3(s)$ | $\rightleftharpoons 2Na^+ + CO_3^{2-}$ | $1.81 \times 10^1$ | 10.77 | 30.55 | $mol^3/kg^3$ | A |
| 58. | $KNO_3(s)$ | $\rightleftharpoons K^+ + NO_3^-$ | $8.72 \times 10^{-1}$ | −14.07 | 19.39 | $mol^2/kg^2$ | A |
| 59. | $KCl(s)$ | $\rightleftharpoons K^+ + Cl^-$ | 8.68 | −6.94 | 19.95 | $mol^2/kg^2$ | A |
| 60. | $KHSO_4(s)$ | $\rightleftharpoons K^+ + HSO_4^-$ | $2.40 \times 10^1$ | −8.42 | 17.96 | $mol^2/kg^2$ | A |
| 61. | $K_2SO_4(s)$ | $\rightleftharpoons 2\ K^+ + SO_4^{2-}$ | $1.57 \times 10^{-2}$ | −9.59 | 45.81 | $mol^3/kg^3$ | A |
| 62. | $KHCO_3(s)$ | $\rightleftharpoons K^+ + HCO_3^-$ | $1.40 \times 10^1$ | −7.60 | −2.52 | $mol^2/kg^2$ | A |
| 63. | $K_2CO_3(s)$ | $\rightleftharpoons 2\ K^+ + CO_3^{2-}$ | $2.54 \times 10^5$ | 12.46 | 36.73 | $mol^3/kg^3$ | A |
| 64. | $Mg(NO_3)_2(s)$ | $\rightleftharpoons Mg^{2+} + 2NO_3^-$ | $2.51 \times 10^{15}$ | | | $mol^3/kg^3$ | A |
| 65. | $MgCl_2$ | $\rightleftharpoons Mg^{2+} + 2Cl^-$ | $9.55 \times 10^{21}$ | | | $mol^3/kg^3$ | A |
| 66. | $MgSO_4(s)$ | $\rightleftharpoons Mg^{2+} + SO_4^{2-}$ | $1.08 \times 10^5$ | | | $mol^2/kg^2$ | A |
| 67. | $MgCO_3(s)$ | $\rightleftharpoons Mg^{2+} + CO_3^{2-}$ | $6.82 \times 10^{-6}$ | | | $mol^2/kg^2$ | A |
| 68. | $Ca(NO_3)_2(s)$ | $\rightleftharpoons Ca^{2+} + 2NO_3$ | $6.07 \times 10^5$ | | | $mol^3/kg^3$ | A |
| 69. | $CaCl_2$ | $\rightleftharpoons Ca^{2+} + 2Cl^-$ | $7.97 \times 10^{11}$ | | | $mol^3/kg^3$ | A |
| 70. | $CaSO_4\text{-}2H_2O(s)$ | $\rightleftharpoons Ca^{2+} + SO_4^{2-} + 2H_2O(aq)$ | $4.32 \times 10^{-5}$ | | | $mol^2/kg^2$ | A |
| 71. | $CaCO_3(s)$ | $\rightleftharpoons Ca^{2+} + CO_3^{2-}$ | $4.97 \times 10^{-9}$ | | | $mol^2/kg^2$ | A |

[a] A, Derived from Table B.6 using (18.37); B, Lind and Kok 1986; C, Kozac-Channing and Heltz 1983; D, Schwartz 1984; E, Jacob 1986; F, Schwartz and White 1981; G, Park and Lee 1987; H, Ledbury and Blair 1925; I, Snider and Dawson 1985; J, Lee 1984; K, Betterton and Hoffmann 1988; L, Le Henaff 1968; M, Jayson et al. 1973; N, derived from a combination of other rate coefficients in the table; O, Marsh and McElroy 1985; R, Perrin 1982; S, Smith and Martell 1976.

The equilibrium coefficient reads,

$$K_{eq}(T) = \mathrm{A} \exp\left\{\mathrm{B}\left(\frac{T_0}{T} - 1\right) + \mathrm{C}\left(1 - \frac{T_0}{T} + \ln\frac{T_0}{T}\right)\right\}$$

where

$$\mathrm{A} = K_{eq}(T_0),\ \mathrm{B} = -\frac{1}{R^* T_0}\sum_i k_i \nu_i \Delta_f H_i^\circ,\ \mathrm{C} = -\frac{1}{R^*}\sum_i k_i \nu_i c_{p,i}^\circ,\ \text{and } T_0 = 298.15\,\mathrm{K},$$

The terms in A, B, and C are defined in Chapter 18.

## B.8. AQUEOUS-PHASE REACTIONS

**Table B.8.** Aqueous-Phase Chemical Kinetic Reactions, Rate Coefficients, and Photoprocesses

| No. | Reaction | A | B | Ref.[a] |
|---|---|---|---|---|
| 1 | $S(IV) + H_2O_2(aq) + H^+ \longrightarrow S(VI) + 2\ H^+ + H_2O(aq)$ | | | |
| | $HSO_3^- + H_2O_2(aq) + H^+$ | $7.45 \times 10^7$ (1) | −15.96 | A |
| 2 | $S(IV) + HO_{2,T} \longrightarrow S(VI) + OH(aq) + H^+$ | | | |
| | $HSO_3^- + HO_2(aq)$ | $4.35 \times 10^5$ | | A |
| | $SO_3^{2-} + HO_2(aq)$ | $5.65 \times 10^5$ | | A |
| | $HSO_3^- + O_2^-$ | $4.35 \times 10^4$ | | A |
| | $SO_3^{2-} + O_2^-$ | $5.65 \times 10^4$ | | A |
| 3 | $S(IV) + OH(aq) \longrightarrow SO_5^- + H_2O(aq)$ | | | |
| | $HSO_3^- + OH(aq)$ | $4.20 \times 10^9$ | −5.03 | E |
| | $SO_3^{2-} + OH(aq)$ | $4.60 \times 10^9$ | −5.03 | E |

*(cont.)*

**Table B.8.** Aqueous-Phase Chemical Kinetic Reactions, Rate Coefficients, and Photoprocesses (cont.)

| No. | Reaction | A | B | Ref.[a] |
|---|---|---|---|---|
| 4 | $S(IV) + O_3(aq) \longrightarrow S(VI) + O_2(aq) + H^+$ | | | |
| | $SO_2(aq) + O_3(aq)$ | $2.40 \times 10^4$ | | A |
| | $HSO_3^- + O_3(aq)$ | $3.70 \times 10^5$ | −18.56 | A |
| | $SO_3^{2-} + O_3(aq)$ | $1.50 \times 10^9$ | −17.72 | A |
| 5 | $S(IV) + HCHO_T \longrightarrow HOCH_2SO_3^-$ | | | |
| | $HSO_3^- + HCHO(aq)$ | $7.90 \times 10^2$ | −16.44 | G |
| | $SO_3^{2-} + HCHO(aq)$ | $2.48 \times 10^7$ | −6.04 | G |
| 6 | $S(IV) + Cl_2^- \longrightarrow SO_5^- + 2\ Cl^- + H^+$ | | | |
| | $HSO_3^- + Cl_2^-$ | $3.40 \times 10^8$ | −5.03 | D |
| | $SO_3^{2-} + Cl_2^-$ | $1.60 \times 10^8$ | −5.03 | D |
| 7 | $S(IV) + SO_5^- \longrightarrow HSO_5^- + SO_5^-$ | | | |
| | $HSO_3^- + SO_5^-$ | $3.00 \times 10^5$ | −10.40 | D |
| | $SO_3^{2-} + SO_5^-$ | $1.30 \times 10^7$ | −6.71 | D |
| 8 | $S(IV) + SO_4^- \longrightarrow S(VI) + SO_5^- + H^+$ | | | |
| | $HSO_3^- + SO_4^-$ | $1.30 \times 10^9$ | −5.03 | H |
| | $SO_3^{2-} + SO_4^-$ | $5.30 \times 10^8$ | −5.03 | H |
| 9 | $S(IV) + HSO_5^- \longrightarrow 2\ S(VI) + 2H^+$ | | | |
| | $HSO_3^- + HSO_5^-$ | $7.10 \times 10^6$ | −10.47 | D |
| 10 | $S(IV) + CH_3OOH(aq) + H^+ \longrightarrow S(VI) + 2\ H^+ + CH_3OH(aq)$ | | | |
| | $HSO_3^- + CH_3OOH(aq) + H^+$ | $1.90 \times 10^7$ | −12.75 | A |
| 11 | $S(IV) + CH_3C(O)OOH(aq) \longrightarrow S(VI) + H^+ + CH_3COOH_T$ | | | |
| | $HSO_3^- + CH_3C(O)OOH(aq)$ | $3.60 \times 10^7$ (2) | −13.42 | E |
| 12 | $S(IV) + Fe(III) \longrightarrow S(VI) + H_2O_2(aq) + Fe(III)$ | | | |
| | $SO_3^{2-} + Fe(III)$ | $9.50 \times 10^7$ | −20.27 | B,C |
| 13 | $S(IV) + Mn(II) \longrightarrow S(VI) + H_2O_2(aq) + Mn(II) + H^+$ | | | |
| | $HSO_3^- + Mn(II)$ | $1.00 \times 10^3$ | −30.06 | Cc |
| 14 | $H_2O_2(aq) + OH(aq) \longrightarrow H_2O(aq) + HO_{2,T}$ | $2.70 \times 10^7$ | −5.70 | J |
| 15 | $H_2O_2(aq) + Cl_2^- \longrightarrow 2\ Cl^- + HO_{2,T} + H^+$ | $1.40 \times 10^5$ | −11.31 | Q |
| 16 | $H_2O_2(aq) + Cl(aq) \longrightarrow Cl^- + HO_{2,T} + H^+$ | $4.50 \times 10^7$ | | R |
| 17 | $H_2O_2(aq) + SO_4^- \longrightarrow HO_{2,T} + S(VI) + H^+$ | $1.20 \times 10^3$ | −6.71 | P |
| 18 | $H_2O_2(aq) + CO_3^- \longrightarrow HO_{2,T} + CO_{2,T}$ | $8.00 \times 10^5$ | −9.46 | O |
| 19 | $OH(aq) + HO_{2,T} \longrightarrow H_2O(aq) + O_2(aq)$ | | | |
| | $OH(aq) + HO_2(aq)$ | $7.00 \times 10^9$ | −5.03 | I |
| | $OH(aq) + O_2^-$ | $1.00 \times 10^{10}$ | −5.03 | I |
| 20 | $OH(aq) + HSO_5^- \longrightarrow SO_5^- + H_2O(aq)$ | $1.70 \times 10^7$ | −6.38 | H |
| 21 | $OH(aq) + CO_{2,T} \longrightarrow H_2O(aq) + CO_3^-$ | | | |
| | $OH(aq) + HCO_3^-$ | $1.50 \times 10^7$ | −6.41 | M |
| 22 | $OH(aq) + HCHO_T \longrightarrow HO_{2,T} + HCOOH_T + H_2O(aq)$ | | | |
| | $OH(aq) + H_2C(OH)_2(aq)$ | $2.00 \times 10^9$ | −5.03 | U |
| 23 | $OH(aq) + HOCH_2SO_3^- \longrightarrow SO_5^- + HCHO_T + H_2O(aq)$ | $1.40 \times 10^9$ | −5.03 | H |
| 24 | $OH(aq) + HCOOH_T \longrightarrow HO_{2,T} + CO_{2,T} + H_2O(aq)$ | | | |
| | $OH(aq) + HCOOH(aq)$ | $1.60 \times 10^8$ | −5.03 | V |
| | $OH(aq) + HCOO^-$ | $2.50 \times 10^9$ | −5.03 | W |
| 25 | $OH(aq) + CH_3COOH_T \longrightarrow HO_{2,T} + CO(aq) + HCOOH_T + H_2O(aq)$ | | | |
| | $OH(aq) + CH_3COOH(aq)$ | $2.00 \times 10^7$ | −6.25 | Y |
| | $OH(aq) + CH_3COO^-$ | $8.00 \times 10^7$ | −5.07 | Y |
| 26 | $OH(aq) + CH_3OH(aq) \longrightarrow HO_{2,T} + HCOOH_T + H_2O(aq)$ | $4.50 \times 10^8$ | −5.03 | W |

*(cont.)*

**Table B.8.** (cont.)

| No. | Reaction | A | B | Ref.[a] |
|---|---|---|---|---|
| 27 | $OH(aq) + CH_3OOH(aq) \longrightarrow CH_3O_2(aq) + H_2O(aq)$ | $2.70 \times 10^7$ | −5.70 | H |
| 28 | $OH(aq) + CH_3OOH(aq) \longrightarrow HO_{2,T} + HCOOH_T + H_2O(aq)$ | $1.90 \times 10^7$ | −6.04 | H |
| 29 | $HO_{2,T} + O_3(aq) \longrightarrow OH(aq) + 2\ O_2(aq)$ | | | |
| | $O_2^- + O_3(aq)$ | $1.50 \times 10^9$ | −5.03 | I |
| 30 | $HO_{2,T} + HO_{2,T} \longrightarrow H_2O_2(aq) + O_2(aq)$ | | | |
| | $HO_2(aq) + HO_2(aq)$ | $8.60 \times 10^5$ | −7.94 | K |
| | $HO_2(aq) + O_2^-$ | $1.00 \times 10^8$ | −5.03 | K |
| 31 | $HO_{2,T} + CH_3O_2(aq) \longrightarrow CH_3OOH(aq) + O_2(aq)$ | | | |
| | $HO_2(aq) + CH_3O_2(aq)$ | $4.30 \times 10^5$ | −10.07 | H |
| | $O_2^- + CH_3O_2(aq)$ | $5.00 \times 10^7$ | −5.37 | H |
| 32 | $HO_{2,T} + SO_4^- \longrightarrow S(VI) + H^+ + O_2(aq)$ | | | |
| | $HO_2(aq) + SO_4^-$ | $5.00 \times 10^9$ | −5.03 | H |
| | $O_2^- + SO_4^-$ | $5.00 \times 10^9$ | −5.03 | H |
| 33 | $HO_{2,T} + SO_5^- \longrightarrow HSO_5^- + O_2(aq)$ | | | |
| | $O_2^- + SO_5^-$ | $1.00 \times 10^8$ | −5.03 | H |
| 34 | $HO_{2,T} + Cl_2^- \longrightarrow 2\ Cl^- + O^2(aq)$ | | | |
| | $O_2^- + Cl_2^-$ | $1.00 \times 10^9$ | −5.03 | P |
| 35 | $HO_{2,T} + CO_{2,T} \longrightarrow HO_2^- + CO_3^-$ | | | |
| | $O_2^- + HCO_3$ | $1.50 \times 10^6$ | | N |
| 36 | $HO_{2,T} + CO_3^- \longrightarrow CO_{2,T} + O_2(aq) + OH^-$ | | | |
| | $O_2^- + CO_3^-$ | $4.00 \times 10^8$ | −5.03 | O |
| 37 | $SO_4^- + HCOOH_T \longrightarrow S(VI) + CO_{2,T} + HO_{2,T}$ | | | |
| | $SO_4^- + HCOO^-$ | $1.10 \times 10^8$ | −5.03 | AA |
| 38 | $SO_4^- + Cl^- \longrightarrow S(VI) + Cl(aq)$ | $2.60 \times 10^8$ | −5.03 | AA |
| 39 | $SO_5^- + HCOOH_T \longrightarrow HSO_5^- + CO_{2,T} + HO_{2,T}$ | | | |
| | $SO_5^- + HCOO^-$ | $1.40 \times 10^4$ | −13.42 | H |
| 40 | $SO_5^- + SO_5^- \longrightarrow 2\ SO_4^- + O_2(aq)$ | $6.00 \times 10^8$ | −5.03 | D |
| 41 | $HCOOH_T + CO_3^- \longrightarrow 2\ CO_{2,T} + HO_{2,T} + OH^-$ | | | |
| | $HCOO^- + CO_3^-$ | $1.10 \times 10^5$ | −11.41 | X |
| 42 | $HCOOH_T + Cl_2^- \longrightarrow CO_{2,T} + HO_{2,T} + 2\ Cl^-$ | | | |
| | $HCOO^- + Cl_2^-$ | $1.90 \times 10^6$ | −8.72 | Q |
| 43 | $H_2O_2(aq) + h\nu \longrightarrow 2\ OH(aq)$ | Radiation-dependent | | BB |
| 44 | $NO_3^- + h\nu \longrightarrow NO_2(aq) + OH(aq) + OH^-$ | Radiation-dependent | | BB |

S(IV), S(VI), $HO_{2,T}$, $CO_{2,T}$, $HCHO_T$, $HCOOH_T$, and $CH_3COOH_T$ are families, defined in (19.29)–(19.35), respectively. Rate coefficients are given for individual reactions (where reactants only are identified) that together constitute the family reaction above it. Rate coefficients are defined below. Use $O_2(aq)$, $H_2O(aq)$, or $OH^-$ to balance equations molecularly, if necessary.

[a] A, Hoffmann and Calvert (1985); B, Conklin and Hoffmann (1988); C, Martin and Hill (1987a); Cc, Martin and Hill (1987b); D, Huie and Neta (1987); E, Lind et al. (1987); G, Boyce and Hoffmann (1984); H, Jacob (1986); I, Sehested et al. (1968); J, Christensen et al. (1982); K, Bielski (1978); M, Weeks and Rabani (1966); N, Schmidt (1972); O, Behar et al. (1970); P, Ross and Neta (1979); Q, Hagesawa and Neta (1978); R, Graedel and Goldberg (1983); U, Bothe and Schulte-Frohlinde (1980); V, Scholes and Willson (1967); W, Anbar and Neta (1967); X, Chen et al. (1973); Y, Farhataziz and Ross (1977); AA, Wine et al. (1989); BB, Graedel and Weschler (1981).

Rate coefficients for individual reactions in each family reaction have the form,

$$K_{aq}(T) = A \exp\left[B\left(\frac{T_0}{T_1} - 1\right)\right] \qquad A = K_{aq}(T_0) \qquad B = -\frac{1}{R^* T_0}\sum_i k_i \nu_i \Delta_f H_i^\circ$$

where A has units of $s^{-1}$, $M^{-1}\ s^{-1}$, or $M^{-2}\ s^{-1}$, the terms in B are defined in Chapter 18, $T_0 = 298.15$ K, and $T$ is in kelvin. Specialized rate coefficients have the form

$$K_{aq}(T) = \frac{A \exp\{B[(T_0/T_1)-1]\}}{1+13[H^+]} \tag{1}$$

$$K_{aq}(T) = A \exp\left[B\left(\frac{T_0}{T}-1\right)\right] [H^+] + 7.0\times10^2 \tag{2}$$

where $K_{a1} = 0.0123$ mol $L^{-1}$, $K_{a2} = 6.61\times10^{-8}$ mol $L^{-1}$, and $[H^+]$ is in moles per liter. Rate coefficients for family reactions are described in (19.43)–(19.47).

## B.9. SOLUTE ACTIVITY COEFFICIENT DATA

**Table B.9.** Parameters for Calculating Electrolyte Mean Binary Activity Coefficients

| Electrolyte | $j$ | $B_j$ | $G_j$ | $H_j$ |
|---|---|---|---|---|
| HCl | | Hamer and Wu (1972) 16 m | Parker (1965) 55.5 m | Parker (1965) 15.9 m |
| | 0 | $-1.998104\times10^{-2}$ | 0 | 0 |
| | 1 | $-7.959068\times10^{-1}$ | $5.532198\times10^{-1}$ | $2.108728\times10^{0}$ |
| | 2 | $6.580198\times10^{-1}$ | $-2.571126\times10^{-1}$ | $8.542292\times10^{-1}$ |
| | 3 | $-7.409941\times10^{-2}$ | $2.790048\times10^{-1}$ | $-6.237459\times10^{-1}$ |
| | 4 | $1.345075\times10^{-2}$ | $-4.691631\times10^{-2}$ | $1.935911\times10^{-1}$ |
| | 5 | $-2.248651\times10^{-3}$ | $2.382485\times10^{-3}$ | $-2.037543\times10^{-2}$ |
| $HNO_3$ | | Hamer and Wu (1972) 28 m | Parker (1965) 55.5 m | Parker (1965) 55.5 m |
| | 0 | $-2.388378\times10^{-2}$ | 0 | 0 |
| | 1 | $-7.777787\times10^{-1}$ | $5.785894\times10^{-1}$ | $-4.785171\times10^{-1}$ |
| | 2 | $5.950086\times10^{-1}$ | $-9.860271\times10^{-1}$ | $6.521896\times10^{0}$ |
| | 3 | $-1.284278\times10^{-1}$ | $6.043012\times10^{-1}$ | $-2.605544\times10^{0}$ |
| | 4 | $1.291734\times10^{-2}$ | $-1.123169\times10^{-1}$ | $3.739984\times10^{-1}$ |
| | 5 | $-6.257155\times10^{-4}$ | $6.688134\times10^{-3}$ | $-1.832646\times10^{-2}$ |
| NaCl | | Hamer and Wu (1972) 6.1 m | Parker (1965) 6.1 m | Perron et al. (1981) 6.0 m |
| | 0 | $-6.089937\times10^{-3}$ | 0 | 0 |
| | 1 | $-1.015184\times10^{0}$ | $5.808744\times10^{-1}$ | $2.261834\times10^{0}$ |
| | 2 | $9.345503\times10^{-1}$ | $-1.163239\times10^{0}$ | $3.622494\times10^{0}$ |
| | 3 | $-4.615793\times10^{-1}$ | $5.136893\times10^{-1}$ | $-1.608598\times10^{0}$ |
| | 4 | $1.431557\times10^{-1}$ | $-1.029523\times10^{-1}$ | $2.092972\times10^{-1}$ |
| | 5 | $-1.700298\times10^{-2}$ | $1.401488\times10^{-2}$ | 0 |
| $NaNO_3$ | | Wu and Hamer (1980) 10.8 m | Parker (1965) 9.2 m | Parker (1965) 2.2 m |
| | 0 | $-6.638145\times10^{-3}$ | 0 | 0 |
| | 1 | $-1.024329\times10^{0}$ | $5.678220\times10^{-1}$ | $7.232987\times10^{-1}$ |
| | 2 | $6.877457\times10^{-1}$ | $-2.136826\times10^{0}$ | $1.918907\times10^{1}$ |
| | 3 | $-3.336161\times10^{-1}$ | $1.145031\times10^{0}$ | $-2.382164\times10^{1}$ |
| | 4 | $8.387414\times10^{-2}$ | $-2.585350\times10^{-1}$ | $1.367081\times10^{1}$ |
| | 5 | $-8.154844\times10^{-3}$ | $2.390815\times10^{-2}$ | $-3.064556\times10^{0}$ |
| $NaHSO_4$ | | Harvie et al. (1984) 6.0 m | Assumed same as for NaCl 6.1 m | Same as for NaCl 6.0 m |
| | 0 | $-8.890979\times10^{-3}$ | 0 | 0 |
| | 1 | $-9.559487\times10^{-1}$ | $5.808744\times10^{-1}$ | $2.261834\times10^{0}$ |
| | 2 | $8.758970\times10^{-1}$ | $-1.163239\times10^{0}$ | $3.622494\times10^{0}$ |

**Table B.9.** (cont.)

| Electrolyte | $j$ | $B_j$ | $G_j$ | $H_j$ |
|---|---|---|---|---|
| | 3 | $-4.607380 \times 10^{-1}$ | $5.136893 \times 10^{-1}$ | $-1.608598 \times 10^{0}$ |
| | 4 | $1.309144 \times 10^{-1}$ | $1.029523 \times 10^{-1}$ | $2.092972 \times 10^{-1}$ |
| | 5 | $-1.398546 \times 10^{-2}$ | $1.401488 \times 10^{-2}$ | 0 |
| $Na_2SO_4$ | | Goldberg (1981) | Wagman et al. (1982) | Holmes and Mesmer (1986) |
| | | 4.4 m | 3.1 m | 2.0 m |
| | 0 | $-2.323071 \times 10^{-2}$ | 0 | 0 |
| | 1 | $-3.321509 \times 10^{0}$ | $1.698182 \times 10^{0}$ | $9.410224 \times 10^{0}$ |
| | 2 | $3.388793 \times 10^{0}$ | $-5.160108 \times 10^{0}$ | $2.213823 \times 10^{1}$ |
| | 3 | $-2.402946 \times 10^{0}$ | $2.132810 \times 10^{0}$ | $-3.481895 \times 10^{1}$ |
| | 4 | $8.926764 \times 10^{-1}$ | $8.840108 \times 10^{-1}$ | $2.348397 \times 10^{1}$ |
| | 5 | $1.225933 \times 10^{-1}$ | $-5.143058 \times 10^{-1}$ | $-6.471345 \times 10^{0}$ |
| $NH_4Cl$ | | Hamer and Wu (1972) | Wagman et al. (1982) | Parker (1965) |
| | | 7.4 m | 7.0 m | 7.4 m |
| | 0 | $-5.022484 \times 10^{-3}$ | 0 | 0 |
| | 1 | $-1.037873 \times 10^{0}$ | $4.890513 \times 10^{-1}$ | $1.959107 \times 10^{0}$ |
| | 2 | $8.517483 \times 10^{-1}$ | $-7.013315 \times 10^{-1}$ | $9.894682 \times 10^{-1}$ |
| | 3 | $-4.225323 \times 10^{-1}$ | $4.682151 \times 10^{-1}$ | $-1.024499 \times 10^{-1}$ |
| | 4 | $1.214996 \times 10^{-1}$ | $-1.702461 \times 10^{-1}$ | $-2.354376 \times 10^{-1}$ |
| | 5 | $-1.471525 \times 10^{-2}$ | $2.502643 \times 10^{-2}$ | $6.600384 \times 10^{-2}$ |
| $NH_4NO_3$ | | Hamer and Wu (1972) | Vanderzee et al.(1980) | Roux et al. (1978) |
| | | 25.9 m | 25.0 m | 22.4 m |
| | 0 | $-1.044572 \times 10^{-2}$ | 0 | 0 |
| | 1 | $-1.004940 \times 10^{0}$ | $4.362921 \times 10^{-1}$ | $2.611682 \times 10^{0}$ |
| | 2 | $4.674064 \times 10^{-1}$ | $-1.455444 \times 10^{0}$ | $3.158677 \times 10^{0}$ |
| | 3 | $-1.750495 \times 10^{-1}$ | $6.282104 \times 10^{-1}$ | $-2.005748 \times 10^{0}$ |
| | 4 | $3.253844 \times 10^{-2}$ | $-1.123507 \times 10^{-1}$ | $4.113737 \times 10^{-1}$ |
| | 5 | $-2.276789 \times 10^{-3}$ | $7.438990 \times 10^{-3}$ | $-2.820677 \times 10^{-2}$ |
| $NH_4HSO_4$ | | Bassett and Seinfeld (1983) | Same as for $NH_4Cl$ | Same as for $NH_4Cl$ |
| | | 6.0 m | 7.0 m | 7.4 m |
| | 0 | $-2.708121 \times 10^{-3}$ | 0 | 0 |
| | 1 | $-1.095646 \times 10^{0}$ | $4.890513 \times 10^{-1}$ | $1.959107 \times 10^{0}$ |
| | 2 | $1.042878 \times 10^{0}$ | $-7.013315 \times 10^{-1}$ | $9.894682 \times 10^{-1}$ |
| | 3 | $-6.289405 \times 10^{-1}$ | $4.682151 \times 10^{-1}$ | $-1.024499 \times 10^{-1}$ |
| | 4 | $2.079631 \times 10^{-1}$ | $-1.702461 \times 10^{-1}$ | $-2.354376 \times 10^{-1}$ |
| | 5 | $-2.776957 \times 10^{-2}$ | $2.502643 \times 10^{-2}$ | $6.600384 \times 10^{-2}$ |
| $(NH_4)_2SO_4$ | | Filippov et al. (1985) | Wagman et al. (1982) | Sukhatme and Saikhedkar (1969) |
| | | 5.8 m | 5.5 m | 5.5 m |
| | 0 | $-2.163694 \times 10^{-2}$ | $2.297972 \times 10^{-1}$ | 0 |
| | 1 | $-3.377941 \times 10^{0}$ | $4.255129 \times 10^{-1}$ | $1.609902 \times 10^{-3}$ |
| | 2 | $3.118007 \times 10^{0}$ | $-2.220594 \times 10^{0}$ | $4.437758 \times 10^{0}$ |
| | 3 | $-1.920544 \times 10^{0}$ | $2.607601 \times 10^{0}$ | $6.101756 \times 10^{-3}$ |
| | 4 | $6.372975 \times 10^{-1}$ | $-1.243384 \times 10^{0}$ | $4.021805 \times 10^{-1}$ |
| | 5 | $-8.277292 \times 10^{-2}$ | $2.102563 \times 10^{-1}$ | $4.375833 \times 10^{-4}$ |

Parameters fit into (18.56), which fits into (18.55). B, which also fits into (18.51), is used to calculate binary activity coefficients at 298.15 K, $G$ is a heat-of-dilution parameter, and $H$ is a heat-capacity parameter. Molalities are the maximum molalities of the data used to derive the fits. From Jacobson et al. (1996b).

## B.10. WATER ACTIVITY DATA

**Table B.10.** Parameters for Calculating Molalities of Binary Electrolytes as a Function of Relative Humidity at 298.15 K

| | [A]$HCl$<br>0% r.h.; 18.5 m | [A]$HNO_3$<br>0% r.h.; 22.6 m | [B]$H^+/HSO_4^-$<br>0% r.h.; 30.4 m | [B]$2H^+/SO_4^{2-}$<br>0% r.h.; 30.4 m |
|---|---|---|---|---|
| $Y_0$ | $1.874637647 \times 10^{1}$ | $2.306844303 \times 10^{1}$ | $3.0391387536 \times 10^{1}$ | $3.0391387536 \times 10^{1}$ |
| $Y_1$ | $-2.052465972 \times 10^{1}$ | $-3.563608869 \times 10^{1}$ | $-1.8995058929 \times 10^{2}$ | $-1.8995058929 \times 10^{2}$ |
| $Y_2$ | $-9.485082073 \times 10^{1}$ | $-6.210577919 \times 10^{1}$ | $9.7428231047 \times 10^{2}$ | $9.7428231047 \times 10^{2}$ |
| $Y_3$ | $5.362930715 \times 10^{2}$ | $5.510176187 \times 10^{2}$ | $-3.1680155761 \times 10^{3}$ | $-3.1680155761 \times 10^{3}$ |
| $Y_4$ | $-1.223331346 \times 10^{3}$ | $-1.460055286 \times 10^{3}$ | $6.1400925314 \times 10^{3}$ | $6.1400925314 \times 10^{3}$ |
| $Y_5$ | $1.427089861 \times 10^{3}$ | $1.894467542 \times 10^{3}$ | $-6.9116348199 \times 10^{3}$ | $-6.9116348199 \times 10^{3}$ |
| $Y_6$ | $-8.344219112 \times 10^{2}$ | $-1.220611402 \times 10^{3}$ | $4.1631475226 \times 10^{3}$ | $4.1631475226 \times 10^{3}$ |
| $Y_7$ | $1.90992437 \times 10^{2}$ | $3.098597737 \times 10^{2}$ | $-1.0383424491 \times 10^{3}$ | $-1.0383424491 \times 10^{3}$ |
| | [C]$NaCl$<br>47% r.h.; 13.5 m | [C]$NaNO_3$<br>30% r.h.; 56.8 m | [D]$NaHSO_4$<br>1.9% r.h.; 158 m | [D]$Na_2SO_4$<br>58% r.h.; 13.1 m |
| $Y_0$ | $5.875248 \times 10^{1}$ | $3.10221762 \times 10^{2}$ | $1.8457001681 \times 10^{2}$ | $5.5983158 \times 10^{2}$ |
| $Y_1$ | $-1.8781997 \times 10^{2}$ | $-1.82975944 \times 10^{3}$ | $-1.6147765817 \times 10^{3}$ | $-2.56942664 \times 10^{3}$ |
| $Y_2$ | $2.7211377 \times 10^{2}$ | $5.13445395 \times 10^{3}$ | $8.444076586 \times 10^{3}$ | $4.47450201 \times 10^{3}$ |
| $Y_3$ | $-1.8458287 \times 10^{2}$ | $-8.01200018 \times 10^{3}$ | $-2.6813441936 \times 10^{4}$ | $-3.45021842 \times 10^{3}$ |
| $Y_4$ | $4.153689 \times 10^{1}$ | $7.07630664 \times 10^{3}$ | $5.0821277356 \times 10^{4}$ | $9.8527913 \times 10^{2}$ |
| $Y_5$ | | $-3.33365806 \times 10^{3}$ | $-5.5964847603 \times 10^{4}$ | |
| $Y_6$ | | $6.5442029 \times 10^{2}$ | $3.2945298603 \times 10^{4}$ | |
| $Y_7$ | | | $-8.002609678 \times 10^{3}$ | |
| | [A]$NH_4Cl$<br>47% r.h.; 23.2 m | [A]$NH_4NO_3$<br>62% r.h.; 28 m | $NH_4HSO_4$<br>6.5% r.h.; 165 m | [C]$(NH_4)_2SO_4$<br>37% r.h.; 29.0 m |
| $Y_0$ | $-7.110541604 \times 10^{3}$ | $3.983916445 \times 10^{3}$ | $2.9997156464 \times 10^{2}$ | $1.1065495 \times 10^{2}$ |
| $Y_1$ | $7.217772665 \times 10^{4}$ | $1.153123266 \times 10^{4}$ | $-2.8936374637 \times 10^{3}$ | $-3.6759197 \times 10^{2}$ |
| $Y_2$ | $-3.071054075 \times 10^{5}$ | $-2.13956707 \times 10^{5}$ | $1.4959985537 \times 10^{4}$ | $5.0462934 \times 10^{2}$ |
| $Y_3$ | $7.144764216 \times 10^{5}$ | $7.926990533 \times 10^{5}$ | $-4.5185935292 \times 10^{4}$ | $-3.1543839 \times 10^{2}$ |
| $Y_4$ | $-9.840230371 \times 10^{5}$ | $-1.407853405 \times 10^{6}$ | $8.110895603 \times 10^{4}$ | $6.770824 \times 10^{1}$ |
| $Y_5$ | $8.03407288 \times 10^{5}$ | $1.351250086 \times 10^{6}$ | $-8.4994863218 \times 10^{4}$ | |
| $Y_6$ | $-3.603924022 \times 10^{5}$ | $-6.770046794 \times 10^{5}$ | $4.7928255412 \times 10^{4}$ | |
| $Y_7$ | $6.856992393 \times 10^{4}$ | $1.393507324 \times 10^{5}$ | $-1.1223105556 \times 10^{4}$ | |

A, derived from Jacobson (1996b) and references therein; B, derived from Robinson and Stokes (1955); C, obtained from Tang (1997); D, derived from Tang and Munkelwitz (1994). *Y* coefficients fit into (18.66). The table also lists the lowest relative humidity (r.h.) and corresponding highest molality (**m**) for which each fit is valid.

## B.11. SURFACE RESISTANCE DATA

**Table B.11.** Parameters Used in Surface Resistance Equations of Chapter 20

| Parameter ($s\ m^{-1}$) | 1 | 2 | 3 | 4 | 5 | 6 | 7 | 8 | 9 | 10 | 11 |
|---|---|---|---|---|---|---|---|---|---|---|---|
| **Seasonal Category 1: Midsummer with lush vegetation** | | | | | | | | | | | |
| $R_i$ | $\infty$ | 60 | 120 | 70 | 130 | 100 | $\infty$ | $\infty$ | 80 | 100 | 150 |
| $R_{cut,0,q}$ | $\infty$ | 2000 | 2000 | 2000 | 2000 | 2000 | $\infty$ | $\infty$ | 2500 | 2000 | 4000 |
| $R_{canp}$ | 100 | 200 | 100 | 2000 | 2000 | 2000 | 0 | 0 | 300 | 150 | 200 |
| $R_{soil,SO_2}$ | 400 | 150 | 350 | 500 | 500 | 100 | 0 | 1000 | 0 | 220 | 400 |
| $R_{soil,O_3}$ | 300 | 150 | 200 | 200 | 200 | 300 | 2000 | 400 | 1000 | 180 | 200 |
| $R_{exp,SO_2}$ | $\infty$ | 2000 | 2000 | 2000 | 2000 | 2000 | $\infty$ | $\infty$ | 2500 | 2000 | 4000 |
| $R_{exp,O_3}$ | $\infty$ | 1000 | 1000 | 1000 | 1000 | 1000 | $\infty$ | $\infty$ | 1000 | 1000 | 1000 |
| **Seasonal Category 2: Autumn with unharvested cropland** | | | | | | | | | | | |
| $R_i$ | $\infty$ | $\infty$ | $\infty$ | $\infty$ | 250 | 500 | $\infty$ | $\infty$ | $\infty$ | $\infty$ | $\infty$ |
| $R_{cut,0,q}$ | $\infty$ | 9000 | 9000 | 9000 | 4000 | 8000 | $\infty$ | $\infty$ | 9000 | 9000 | 9000 |
| $R_{canp}$ | 100 | 150 | 100 | 1500 | 2000 | 1700 | 0 | 0 | 200 | 120 | 140 |
| $R_{soil,SO_2}$ | 400 | 200 | 350 | 500 | 500 | 100 | 0 | 1000 | 0 | 300 | 400 |
| $R_{soil,O_3}$ | 300 | 150 | 200 | 200 | 200 | 300 | 2000 | 400 | 800 | 180 | 200 |
| $R_{exp,SO_2}$ | $\infty$ | 9000 | 9000 | 9000 | 2000 | 4000 | $\infty$ | $\infty$ | 9000 | 9000 | 9000 |
| $R_{exp,O_3}$ | $\infty$ | 400 | 400 | 400 | 1000 | 600 | $\infty$ | $\infty$ | 400 | 400 | 400 |
| **Seasonal Category 3: Late autumn after frost, no snow** | | | | | | | | | | | |
| $R_i$ | $\infty$ | $\infty$ | $\infty$ | $\infty$ | 250 | 500 | $\infty$ | $\infty$ | $\infty$ | $\infty$ | $\infty$ |
| $R_{cut,0,q}$ | $\infty$ | $\infty$ | 9000 | 9000 | 4000 | 8000 | $\infty$ | $\infty$ | 9000 | 9000 | 9000 |
| $R_{canp}$ | 100 | 10 | 100 | 1000 | 2000 | 1500 | 0 | 0 | 100 | 50 | 120 |
| $R_{soil,SO_2}$ | 400 | 150 | 350 | 500 | 500 | 200 | 0 | 1000 | 0 | 200 | 400 |
| $R_{soil,O_3}$ | 300 | 150 | 200 | 200 | 200 | 300 | 2000 | 400 | 1000 | 180 | 200 |
| $R_{exp,SO_2}$ | $\infty$ | $\infty$ | 9000 | 9000 | 3000 | 6000 | $\infty$ | $\infty$ | 9000 | 9000 | 9000 |
| $R_{exp,O_3}$ | $\infty$ | 1000 | 400 | 400 | 1000 | 600 | $\infty$ | $\infty$ | 800 | 600 | 600 |
| **Seasonal Category 4: Winter, snow on ground and subfreezing** | | | | | | | | | | | |
| $R_i$ | $\infty$ | $\infty$ | $\infty$ | $\infty$ | 400 | 800 | $\infty$ | $\infty$ | $\infty$ | $\infty$ | $\infty$ |
| $R_{cut,0,q}$ | $\infty$ | $\infty$ | $\infty$ | $\infty$ | 6000 | 9000 | $\infty$ | $\infty$ | 9000 | 9000 | 9000 |
| $R_{canp}$ | 100 | 10 | 10 | 1000 | 2000 | 1500 | 0 | 0 | 50 | 10 | 50 |
| $R_{soil,SO_2}$ | 100 | 100 | 100 | 100 | 100 | 100 | 0 | 1000 | 100 | 100 | 50 |
| $R_{soil,O_3}$ | 600 | 3500 | 3500 | 3500 | 3500 | 3500 | 2000 | 400 | 3500 | 3500 | 3500 |
| $R_{exp,SO_2}$ | $\infty$ | $\infty$ | $\infty$ | 9000 | 200 | 400 | $\infty$ | $\infty$ | 9000 | $\infty$ | 9000 |
| $R_{exp,O_3}$ | $\infty$ | 1000 | 1000 | 400 | 1500 | 600 | $\infty$ | $\infty$ | 800 | 1000 | 800 |
| **Seasonal Category 5: transitional spring with partially green short annuals** | | | | | | | | | | | |
| $R_i$ | $\infty$ | 120 | 240 | 140 | 250 | 190 | $\infty$ | $\infty$ | 160 | 200 | 300 |
| $R_{cut,0,q}$ | $\infty$ | 4000 | 4000 | 4000 | 2000 | 3000 | $\infty$ | $\infty$ | 4000 | 4000 | 8000 |
| $R_{canp}$ | 100 | 50 | 80 | 1200 | 2000 | 1500 | 0 | 0 | 200 | 60 | 120 |
| $R_{soil,SO_2}$ | 500 | 150 | 350 | 500 | 500 | 200 | 0 | 1000 | 0 | 250 | 400 |
| $R_{soil,O_3}$ | 300 | 150 | 200 | 200 | 200 | 300 | 2000 | 400 | 1000 | 180 | 200 |
| $R_{exp,SO_2}$ | $\infty$ | 4000 | 4000 | 4000 | 2000 | 3000 | $\infty$ | $\infty$ | 4000 | 4000 | 8000 |
| $R_{exp,O_3}$ | $\infty$ | 1000 | 500 | 500 | 1500 | 700 | $\infty$ | $\infty$ | 600 | 800 | 800 |

Column headings are the following land-use types: 1, urban land; 2, agricultural land; 3, range land; 4, deciduous forest; 5, coniferous forest; 6, mixed forest including wetland; 7, water, both salt and fresh; 8, barren land, mostly desert; 9, nonforested wetland; 10, mixed agricultural and range land; and 11, rocky open areas with low-growing shrubs. Source: (Wesely 1989).

## B.12. MORE SURFACE RESISTANCE DATA

**Table B.12.** Parameters Used in Surface Resistance Calculations

| Chemical Formula | Chemical Name | $D_v/D_q$ | $H_q^*$(M atm$^{-1}$) | $f_{0,q}$ |
|---|---|---|---|---|
| $SO_2$ | Sulfur dioxide | 1.9 | $1 \times 10^5$ | 0 |
| $O_3$ | Ozone | 1.6 | $1 \times 10^{-2}$ | 1 |
| $NO_2$ | Nitrogen dioxide | 1.6 | $1 \times 10^{-2}$ | 0.1 |
| NO | Nitric oxide | 1.3 | $3 \times 10^{-3}$ | 0 |
| $HNO_3$ | Nitric acid | 1.9 | $1 \times 10^{14}$ | 0 |
| $H_2O_2$ | Hydrogen peroxide | 1.4 | $1 \times 10^5$ | 1 |
| $CH_3CHO$ | Acetaldehyde | 1.6 | $1.5 \times 10^1$ | 0 |
| HCHO | Formaldehyde | 1.3 | $6 \times 10^3$ | 0 |
| $CH_3OOH$ | Methyl hydroperoxide | 1.6 | $2.4 \times 10^2$ | 0.1 |
| $CH_3C(O)OOH$ | Peroxyacetic acid | 2.0 | $5.4 \times 10^2$ | 0.1 |
| HCOOH | Formic acid | 1.6 | $4 \times 10^6$ | 0 |
| $NH_3$ | Ammonia | 0.97 | $2 \times 10^4$ | 0 |
| $CH_3C(O)OONO_2$ | Peroxyacetyl nitrate | 2.6 | $3.6 \times 10^0$ | 0.1 |
| HONO | Nitrous acid | 1.6 | $1 \times 10^5$ | 0.1 |

Source: Wesely (1989).

# References


Adamson A. W. (1990) *Physical Chemistry of Surfaces*, 5th ed., John Wiley and Sons, Inc., New York.

Allen P. and Wagner K. (1992) 1987 California Air Resources Board emissions inventory, magnetic tapes ARA806, ARA807.

Al Nakshabandi G. and Konhke H. (1965) Thermal conductivity and diffusivity of soils as related to moisture tension and other physical properties. *Agric. Meteor.* **2**, 271–9.

Anbar M. and Neta P. (1967) A compilation of specific bimolecular rate constants for the reactions of hydrated electrons, hydrogen atoms, and hydroxyl radicals with inorganic and organic compounds in aqueous solution. *Int. J. Appl. Radiat. Isot.* **18**, 493–523.

Andreas E. L. (1992) Sea spray and the turbulent air–sea heat fluxes. *J. Geophys. Res.* **97**, 11,429–41.

Angell C. A., Guni M. O., and Sichina W. J. (1982) Heat capacity of water at extremes of supercooling and superheating. *J. Phys. Chem.* **86**, 998–1002.

Anthes R. A. (1977) A cumulus parameterization scheme utilizing a one-dimensional cloud model. *Mon. Wea. Rev.* **105**, 270–86.

Arakawa A. (1984) Boundary conditions in limited area models. Course notes, Dept. of Atmospheric Sciences, University of California, Los Angeles.

Arakawa A. (1988) Finite-difference methods in climate modeling. In *Physically-Based Modeling and Simulation of Climate and Climatic Change – Part 1*, M. E. Schlesinger, ed., Kluwer Acad. Publishers, 79–168.

Arakawa A. (1997) Adjustment mechanisms in atmospheric models. *J. Meteor. Soc. Japan* **75**, 155–79.

Arakawa A. and Konor C. S. (1995) Vertical differencing of the primitive equations based on the Charney–Phillips grid in hybrid $\sigma$-$p$ vertical coordinates. *Mon. Wea. Rev.* **124**, 511–28.

Arakawa A. and Lamb V. R. (1977) Computational design of the basic dynamical processes of the UCLA general circulation model. *Methods Comput. Phys.* **17**, 174–265.

Arakawa A. and Schubert W. H. (1974) Interaction of a cumulus cloud ensemble with large scale environment, Part I. *J. Atmos. Sci.* **31**, 674–701.

Arakawa A. and Suarez M. J. (1983) Vertical differencing of the primitive equations in sigma coordinates. *Mon. Wea. Rev.* **111**, 34–45.

Atkinson R., Lloyd A. C., and Winges L. (1982) An updated chemical mechanism for hydrocarbon/$NO_x$/$SO_2$ photooxidations suitable for inclusion in atmospheric simulation models. *Atmos. Environ.* **16**, 1341–55.

Atkinson R., Baulch D. L., Cox R. A., Hampson Jr. R. F., Kerr J. A., Rossi M. J., and Troe J. (1997) Evaluated kinetic, photochemical, and heterogeneous data for atmospheric chemistry. Supplement V. *J. Phys. Chem. Ref. Data* **26**, 521–1011.

Austin J. (1991) On the explicit versus family solution of the fully diurnal photochemical equations of the stratosphere. *J. Geophys. Res.* **96**, 12,941–74.

Avissar R. and Mahrer Y. (1988) Mapping frost-sensitive areas with a three-dimensional local-scale numerical model. Part I: Physical and numerical aspects. *J. Appl. Meteor.* **27**, 400–13.

Bader G. and Deuflhard P. (1983) A semi-implicit mid-point rule for stiff systems of ordinary differential equations. *Numer. Math.* **41**, 373–98.

Baldocchi D. D., Hicks B. B., and Camara P. (1987) A canopy stomatal resistance model for gaseous deposition to vegetated surfaces. *Atmos. Environ.* **21**, 91–101.

Bassett M. E. and Seinfeld J. H. (1983) Atmospheric equilibrium model of sulfate and nitrate aerosol. *Atmos. Environ.* **17**, 2237–52.

Bassett M. E. and Seinfeld J. H. (1984) Atmospheric equilibrium model of sulfate and nitrate aerosol-II. Particle size analysis. *Atmos. Environ.* **18**, 1163–70.

Bates T. S., Kiene R. P., Wolfe G. V., Matrai P. A., Chavez F. P., Buck K. R., Blomquist B. W., and Cuhel R. L. (1994) The cycling of sulfur in surface seawater of the Northeast Pacific. *J. Geophys. Res.* **99**, 7835–43.

Beard K. V. (1976) Terminal velocity and shape of cloud and precipitation drops aloft. *J. Atmos. Sci.* **33**, 851–64.

Behar D., Czapski G., and Duchovny I. (1970) Carbonate radical in flash photolysis and pulse radiolysis of aqueous carbonate solutions. *J. Phys. Chem.* **74**, 2206–10.

Berresheim H., Wine P. H., and Davis D. D. (1995) Sulfur in the atmosphere. In *Composition, Chemistry, and Climate of the Atmosphere*, H. B. Singh, ed., Van Nostrand Reinhold, New York, 251–307.

Betterton E. A. and Hoffmann M. R. (1988) Henry's law constants of some environmentally important aldehydes. *Environ. Sci. Technol.* **22**, 1415–8.

Betts A. K. (1986) A new convective adjustment scheme. Part I: Observational and theoretical basis. *Quart. J. Roy. Meteor. Soc.* **112**, 677–91.

Betts A. K. and Miller M. J. (1986) A new convective adjustment scheme. Part II: Single column tests using GATE wave, BOMEX, ATEX, and arctic air-mass data sets. *Quart. J. Roy. Meteor. Soc.* **112**, 693–709.

Bhumralkar C. M. (1975) Numerical experiments on the computation of ground surface temperature in an atmospheric general circulation model. *J. Appl. Meteor.* **14**, 67–100.

Bielski B. H. J. (1978) Reevaluation of the spectral and kinetic properties of $HO_2$ and $O_2^-$ free radicals. *Photochem. Photobiol.* **28**, 645–9.

Blackadar A. K. (1976) Modeling the nocturnal boundary layer. *Proc. Third Symp. on Atmospheric Turbulence, Diffusion and Air Quality*, Boston, American Meteorological Society, 46–9.

Blackadar A. K. (1978) Modeling pollutant transfer during daytime convection. *Proc. Fourth Symp. on Atmospheric Turbulence, Diffusion, and Air Quality*, Reno, American Meteorological Society, 443–7.

Blumthaler M. and Ambach W. (1988) Solar UVB-albedo of various surfaces. *Photochem. Photobiol.* **48**, 85–8.

Bojkov R. D. and Fioletov V. E. (1995) Estimating the global ozone characteristics during the last 30 years. *J. Geophys. Res.* **100**, 16,537–51.

Bolton D. (1980) The computation of equivalent potential temperature. *Mon. Wea. Rev.* **108**, 1046–53.

Bonsang B., Martin D., Lambert G., Kanakidou M., Le Roulley J. C., and Sennequier G. (1991) Vertical distribution of nonmethane hydrocarbons in the remote marine boundary layer. *J. Geophys. Res.* **96**, 7313–24.

Bothe E. and Schulte-Frohlinde D. (1980) Reaction of dihydroxymethyl radical with molecular oxygen in aqueous solution. *Z. Naturforsch. B, Anorg. Chem. Org. Chem.* **35**, 1035–9.

Bott A. (1989) A positive definite advection scheme obtained by nonlinear renormalization of the advective fluxes. *Mon. Wea. Rev.* **117**, 1006–15.

Bott A. and Carmichael G. R. (1993) Multiphase chemistry in a microphysical radiation fog model – a numerical study. *Atmos. Environ.* **27A**, 503–22.

Boubel R. W., Fox D. L., Turner D. B., and Stern A. C. S. (1994) *Fundamentals of Air Pollution.* Academic Press, Inc., San Diego.

Boyce S. D. and Hoffmann M. R. (1984) Kinetics and mechanism of the formation of hydroxymethanesulfonic acid at low pH. *J. Phys. Chem.* **88**, 4740–6.

Brock J. R., Zehavi D., and Kuhn P. (1986) Condensation aerosol formations and growth in a laminar coaxial jet: Experimental. *J. Aerosol Sci.* **17**, 11–22.

Bromley L. A. (1973) Thermodynamic properties of strong electrolytes in aqueous solutions. *AIChE J.* **19**, 313–20.

Brutsaert W. (1991) *Evaporation in the Atmosphere.* Kluwer Academic Publishers, Dordrecht, 299 pp.

Businger J. A., Wyngaard J. C., Izumi Y., and Bradley E. F. (1971) Flux-profile relationships in the atmospheric surface layer. *J. Atmos. Sci.* **28**, 181–9.

Calder K. L. (1949) Eddy diffusion and evaporation in flow over aerodynamically smooth and rough surfaces: A treatment based on laboratory laws of turbulent flow with special reference to conditions in the lower atmosphere. *Q. J. Mech. Appl. Math.* **2**, 153–76.

California Air Resources Board (CARB) (1988) Method used to develop a size-segregated particulate matter inventory. Technical Support Division, Emission Inventory Branch, California Air Resources Board, Sacramento, CA.

Campbell F. W. and Maffel L. (1974) Contrast and spatial frequency. *Sci. Am.* **231**, 106–14.

Carmichael G. R., Peters L. K., and Kitada T. (1986) A second generation model for regional-scale transport/chemistry/deposition. *Atmos. Environ.* **20**, 173–88.

Carter W. P. L. (1991) Development of ozone reactivity scales for volatile organic compounds, EPA 600/3-91-050. U.S. Environmental Protection Agency, Research Triangle Park, NC.

Cass G. R. (1979) On the relationship between sulfate air quality and visibility with examples in Los Angeles. *Atmos. Environ.* **13**, 1069–84.

Celia M. A. and Gray W. G. (1992) *Numerical Methods for Differential Equations.* Prentice-Hall, Englewood Cliffs, NJ.

Chameides W. L. (1984) The photochemistry of a remote marine stratiform cloud. *J. Geophys. Res.* **89**, 4739–55.

Chameides W. L. and Stelson A. W. (1992) Aqueous-phase chemical processes in deliquescent sea-salt aerosols: A mechanism that couples the atmospheric cycles of S and sea salt. *J. Geophys. Res.* **97**, 20,565–80.

Chang E., Nolan K., Said M., Chico T., Chan S., and Pang E. (1991) 1987 emissions inventory for the South Coast Air Basin: Average annual day. South Coast Air Quality Management District (SCAQMD), Los Angeles.

Chapman S. (1930) A theory of upper-atmospheric ozone. *Mem. Royal Met. Soc.* **3**, 104–25.

Chapman S. and Cowling T. G. (1970) *The Mathematical Theory of Nonuniform Gases.* Cambridge University Press, Cambridge, England.

Charlock T. P., Kondratyev K., and Prokofyev, M. P. (1993) Review of recent research on the climatic effect of aerosols. In *Aerosol Effects on Climate*, S. Gerard Jennings, ed., The University of Arizona Press, Tucson and London, Chapter 6, pp. 232–74.

Charlson R. J., Lovelock J. E., Andreae M. O., and Warren S. G. (1987) Oceanic phytoplankton, atmospheric sulphur, cloud albedo, and climate. *Nature* **326**, 655–61.

Charlson R. J., Langner J., Rodhe H., Leovy C. B., and Warren S. G. (1991) Perturbation of the northern hemisphere radiative balance by backscattering from anthropogenic sulfate aerosols. *Tellus* **43AB**, 152–63.

Charlson R. J., Schwartz S. E., Hales J. M., Cess R. D., Coakley Jr. J. A., Hansen J. E., and Hofmann D. J. (1992) Climate forcing by anthropogenic aerosols. *Science* **255**, 423–30.

Charney J. G. and Phillips N. A. (1953) Numerical integration of the quasigeostrophic equations for barotropic and simple baroclinic flows. *J. Meteor.* **10**, 71–99.

Charnock H. (1955) Wind stress on a water surface. *Quart. J. Roy. Met. Soc.* **81**, 639–40.

Chatfield R. B., Gardner E. P., and Calvert J. G. (1987) Sources and sinks of acetone in the troposphere: Behavior of reactive hydrocarbons and a stable product. *J. Geophys. Res.* **92**, 4208–16.

Chen C. (1991) A nested grid nonhydrostatic, elastic model using a terrain-following coordinate transformation: The radiative-nesting boundary conditions. *Mon. Wea. Rev.* **119**, 2852–69.

Chen S., Cope V. W., and Hoffman M. Z. (1973) Behavior of $CO_3^-$ radicals generated in the flash photolysis of carbonatoamines complexes of cobalt(III) in aqueous solution. *J. Phys. Chem.* 77, 1111–6.

Cheng M.-D. and Arakawa A. (1997) Inclusion of rainwater budget and convective downdrafts in the Arakawa–Schubert cumulus parameterization. *J. Atmos. Sci.* **54**, 1359–78.

Chock D. P. (1991) A comparison of numerical methods for solving the advection equation – III. *Atmos. Environ.* **25A**, 853–71.

Chock D. P. and Winkler S. L. (1994) A comparison of advection algorithms coupled with chemistry. *Atmos. Environ.* **28**, 2659–75.

Chock D. P., Sun P., and Winkler S. L. (1996) Trajectory-grid: An accurate sign-preserving advection–diffusion approach for air quality modeling. *Atmos. Environ.* **30**, 857–68.

Christensen H., Sehested K., and Corfitzen H. (1982) Reactions of hydroxyl radicals with hydrogen peroxide at ambient and elevated temperatures. *J. Phys. Chem.* **86**, 1588–90.

Chylek P. (1977) A note on extinction and scattering efficiencies. *J. Appl. Meteor.* **16**, 321–2.

Chylek P. and Coakley Jr. J. A. (1974) Aerosols and climate. *Science* **183**, 75–7.

Chylek P., Videen G., Ngo D., Pinnick R. G., and Klett J. D. (1995) Effect of black carbon on the optical properties and climate forcing of sulfate aerosols. *J. Geophys. Res.* **100**, 16,325–32.

Clapp R. B. and Hornberger G. M. (1978) Empirical equations for some soil hydraulic properties. *Water Resour. Res.* **14**, 601–4.

Cleaver B., Rhodes E., and Ubbelohde A. R. (1963) Studies of phase transformations in nitrates and nitrites I. Changes in ultra-violet absorption spectra on melting. *Proc. Roy. Soc. London* **276**, 437–53.

Clegg S. L. and Brimblecombe P. (1995) Application of a multicomponent thermodynamic model to activities and thermal properties of 0–40 mol $kg^{-1}$ aqueous sulphuric acid from <200 K to 328 K. *J. Chem. Eng. Data* **40**, 43–64.

Clegg S. L., Brimblecombe P., Liang Z., and Chan C. K. (1997) Thermodynamic properties of aqueous aerosols to high supersaturation: II – A model of the system $Na^+$–$Cl^-$–$NO_3^-$–$SO_4^{2-}$–$H_2O$ at 298.15 K. *Aerosol. Sci. Technol.* **27**, 345–66.

Clyne M. A. A., Monkhouse P. B., and Townsend L. W. (1976) Reactions of $O(^3P)$ atoms with halogens: The rate constants for the elementary reactions $O(^3P)$ + BrCl, $O(^3P)$ + $Br_2$ and $O(^3P)$ + $Cl_2$. *Int. J. Chem. Kinet.* **8**, 425–49.

Cohen M. D., Flagan R. C., and Seinfeld J. H. (1987a) Studies of concentrated electrolyte solutions using the electrodynamic balance. 1. Water activities for single-electrolyte solutions. *J. Phys. Chem.* **91**, 4563–74.

Cohen M. D., Flagan R. C., and Seinfeld J. H (1987b) Studies of concentrated electrolyte solutions using the electrodynamic balance. 2. Water activities for mixed-electrolyte solutions. *J. Phys. Chem.* **91**, 4575–82.

Comes F. J., Forberich O., and Walter J. (1997) OH field measurements: A critical input into model calculations on atmospheric chemistry. *J. Atmos. Sci.* **54**, 1886–94.

Conklin M. H. and Hoffmann M. R. (1988) Metal ion–S(IV) chemistry III. Thermodynamics and kinetics of transient iron(III)–sulfur(IV) complexes. *Environ. Sci. Technol.* **22**, 891–8.

Cotton W. R. and Anthes R. A. (1989) *Storm and Cloud Dynamics*. Academic Press, Inc., San Diego.

Courant R., Friedrichs K., and Lewy H. (1928) Über die partiellen Differenzengleichungen der mathematischen Physik. *Math. Ann.* **100**, 32–74.

Crank J. (1975) *The Mathematics of Diffusion*, 2nd ed. Clarendon Press, Oxford.

Crank J. and Nicolson P. (1947) A practical method for numerical evaluation of solutions of partial differential equations of the heat-conduction type. *Proc. Camb. Philos. Soc.* **43**, 50–67.

Crutzen P. J. (1971) Ozone production rates in an oxygen–hydrogen–nitrogen oxide atmosphere. *J. Geophys. Res.* **76**, 7311–27.

Cuenca R. H., Ek M., and Mahrt L. (1996). Impact of soil water property parameterization on atmospheric boundary layer simulation. *J. Geophys. Res.* **101**, 7269–77.

Cunningham E. (1910) On the velocity of steady fall of spherical particles through fluid medium. *Proc. Roy. Soc. London* **A83**, 357–65.

Curtiss C. F. and Hirschfelder J. O. (1952) Integration of stiff equations. *Proc. Nat. Acad. Sci. U.S.A.* **38**, 235–43.

Dabdub D. and Seinfeld J. H. (1994) Numerical advective schemes used in air quality models – sequential and parallel implementation. *Atmos. Environ.* **28**, 3369–85.

Davis E. J. (1983) Transport phenomena with single aerosol particles. *Aerosol Sci. Technol.* **2**, 121–44.

Dean, J. A. (1992) *Lange's Handbook of Chemistry*, McGraw-Hill, Inc., New York.

Deardorff J. W. (1977) A parameterization of ground surface moisture content for use in atmospheric prediction models. *J. Appl. Meteor.* **16**, 1182–5.

de Arellano J. V., Duynkerke P., and van Weele M. (1994) Tethered-balloon measurements of actinic flux in a cloud-capped marine boundary layer. *J. Geophys. Res.* **99**, 3699–705.

Deirmendjian D. (1969) *Electromagnetic Scattering on Spherical Polydispersions*. Elsevier, New York.

DeMore W. B., Sanders S. P., Golden D. M., Hampson R. F., Kurylo M. J., Howard C. J., Ravishankara A. R., Kolb C. E., and Molina M. J. (1997) Chemical kinetics and photochemical data for use in stratospheric modeling. Evaluation number 12, JPL Publ. 97-4. Jet Propulsion Laboratory, Pasadena, CA.

Ding P. and Randall D. A. (1998) A cumulus parameterization with multiple cloud-base levels. *J. Geophys. Res.* **103**, 11,341–53.

Donea J. (1984) A Taylor–Galerkin method for convective transport problems. *Int. J. Numer. Methods Engng.* **20**, 101–19.

Dorsch R. G. and Hacker P. (1951) Experimental values of surface tension of supercooled water, Tech. Note 2510. National Advisory Committee for Aeronautics (NACA).

Duce R. A. (1969) On the source of gaseous chlorine in the marine atmosphere. *J. Geophys. Res.* **70**, 1775–9.

Duce R. A. and Hoffman E. J. (1976) Chemical fractionation at the air/sea interface. *Annu. Rev. Earth Planet. Sci.* **4**, 187–228.

Dudhia J. (1993) A nonhydrostatic version of the Penn State NEAR mesoscale model: Validation less and simulations of an Atlantic cyclone and cold front. *Mon. Wea. Rev.* **21**, 1493–1513.

Dyer A. J. (1974) A review of flux-profile relationships. *Boundary Layer Meteor.* **7**, 363–72.

Dyer A. J. and Bradley E. F. (1982) An alternative analysis of flux-gradient relationships at the 1976 ITCE. *Boundary Layer Meteor.* **22**, 3–19.

Eddington S. A. (1916) On the radiative equilibrium of the stars. *Mon. Not. Roy. Astronom. Soc.* **77**, 16–35.

Edlen B. (1966) The refractive index of air. *Meteorology.* **2**, 71–80.

Eliasen E., Machenhauer B., and Rasmussen E. (1970) On a numerical method for integration of the hydrodynamical equations with a spectral representation of the horizontal fields, Report No. 2. Institut for Teoretisk Meteorologi, University of Copenhagen, 35 pp.

Elliott S., Turco R. P., and Jacobson M. Z. (1993) Tests on combined projection/forward differencing integration for stiff photochemical family systems at long time step. *Computers Chem.* **17**, 91–102.

Emanuel K. A. (1991) A scheme for representing cumulus convection in large-scale models. *J. Atmos. Sci.* **38**, 1541–57.

Encyclopedia Britannica committee, eds. (1980) *Encyclopedia Britannica* , 15th ed., Encyclopedia Britannica, Inc., Chicago.

Eriksson E. (1960) The yearly circulation of chloride and sulfur in nature; meteorological, geochemical and pedological implications. Part II. *Tellus* **12**, 63–109.

Erisman J. W., van Pul W. A. J., and Wyers P. (1994) Parameterization of surface resistance for the quantification of atmospheric deposition of acidifying pollutants and ozone. *Atmos. Environ.* **28**, 2595–607.

Ethridge D. M., Pearman G. I., and Fraser P. J. (1992) Changes in tropospheric methane between 1841 and 1978 from a high accumulation rate Antarctic ice core. *Tellus* **44B**, 282–94.

Farhataziz and Ross A. B. (1977) Selected specific rates of transients from water in aqueous solutions, III. Hydroxyl radical and perhydroxyl radical and their radical ions, Rep. NSRDBS-NBS 59. U.S. Dept. of Commerce, Washington, DC.

Farman J. C., Gardiner B. G., and Shanklin J. D. (1985). Large losses of total ozone in Antarctica reveal seasonal $ClO_x/NO_x$ interaction. *Nature* **315**, 207–10.

Fassi-Fihri A., Suhre K., and Rosset R. (1997) Internal and external mixing in atmospheric aerosols by coagulation: Impact on the optical and hygroscopic properties of the sulphate–soot system. *Atmos. Environ.* **31**, 1392–402.

Filippov V. K., Charykova M. V., and Trofimov Y. M. (1985) Thermodynamics of the system $NH_4H_2PO_4$–$(NH_4)_2SO_4$–$H_2O$ at 25°C. *J. Appl. Chem. U.S.S.R.* **58**, 1807–11.

Finlayson-Pitts B. J. and Pitts J. N. Jr. (1986) *Atmospheric Chemistry: Fundamentals and Experimental Techniques*. John Wiley & Sons, Inc., New York.

Fleming E. L., Chandra S., Schoeberl M. R., and Barnett J. J. (1988) Monthly mean global climatology of temperature, wind, geopotential height, and pressure for 1–120 km, Tech. Memo. 100697. NASA, 85 pp.

Fletcher N. H. (1958) Size effect in heterogeneous nucleation. *J. Chem. Phys.* **29**, 572–6.

Flubacher P., Leadbetter A. J., and Morrison J. A. (1960) Heat capacity of ice at low temperatures. *J. Chem. Phys.* **33**, 1751–5.

Foster V. G. (1992) Determination of the refractive index dispersion of liquid nitrobenzene in the visible and ultraviolet. *J. Phys. D* **25**, 525–9.

Fowler L. D., Randall D. A., and Rutledge S. (1996) Liquid and ice cloud microphysics in the CSU general circulation model. Part I: Model description and simulated microphysical processes. *J. Climate* 9, 489–529.

Frank W. M. and Cohen C. (1987) Simulation of tropical convective systems. Part I: A cumulus parameterization. *J. Atmos. Sci.* **44**, 3787–99.

Friedlander S. K. (1977) *Smoke, Dust, and Haze. Fundamentals of Aerosol Behavior.* John Wiley & Sons, Inc., New York.

Friedlander S. K. (1983) Dynamics of aerosol formation by chemical reaction. *Ann. N.Y. Acad. Sci.* **404**, 354–64.

Fritsch J. M. and Chappel C. F. (1980) Numerical prediction of convectively driven mesoscale pressure systems. Part I: Convective parameterization. *J. Atmos. Sci.* **37**, 1722–33.

Fuchs N. A. (1964) *The Mechanics of Aerosols* (translated by R. E. Daisley and M. Fuchs). Pergamon Press, New York.

Fuchs N. A. and Sutugin A. G. (1971). Highly dispersed aerosols. In *Topics in Current Aerosol Research*. G. M. Hidy and J. R. Brock, eds. Pergamon Press, New York, Vol. 2, pp. 1–60.

Ganzeveld L. and Lelieveld J. (1995) Dry deposition parameterization in a chemistry general circulation model and its influence on the distribution of reactive trace gases. *J. Geophys. Res.* **100**, 20,999–1,012.

Garcia R. R., Stordal F., Solomon S., and Kiehl J. T. (1992) A new numerical model of the middle atmosphere 1. Dynamics and transport of tropospheric source gases. *J. Geophys. Res.* **97**, 12,967–91.

Garratt J. R. (1992) *The Atmospheric Boundary Layer.* Cambridge University Press, Cambridge.

Garratt J. R. and Hicks B. B. (1973) Momentum, heat and water vapour transfer to and from natural and artificial surfaces. *Q. J. R. Meteorol. Soc.* **99**, 680–7.

Gazdag J. (1973) Numerical convective schemes based on accurate computation of space derivatives. *J. Comp. Phys.* **13**, 100–13.

Gear C. W. (1971) *Numerical Initial Value Problems in Ordinary Differential Equations.* Prentice-Hall, Englewood Cliffs, NJ.

Gelbard F. (1990) Modeling multicomponent aerosol particle growth by vapor condensation. *Aerosol Sci. Technol.* **12**, 399–412.

Gelbard F. and Seinfeld J. H. (1980) Simulation of multicomponent aerosol dynamics. *J. Colloid Interface Sci.* **78**, 485–501.

Gery M. W., Fox D. L. and Jeffries H. E. (1985) A continuous stirred tank reactor investigation of the gas-phase reaction of hydroxyl radicals and toluene. *Intl. J. Chem. Kinetics* **17**, 931–55.

Gery M. W., Whitten G. Z., and Killus J. P. (1988) Development and testing of the CBM-IV for urban and regional modeling, Rep. EPA-600/3-88-012. U.S. Environmental Protection Agency, Research Triangle Park, NC.

Gery M. W., Whitten G. Z., Killus J. P., and Dodge M. C. (1989) A photochemical kinetics mechanism for urban and regional scale computer modeling. *J. Geophys. Res.* **94**, 12,925–56.

Giauque W. F. and Stout J. W. (1936) The entropy of water and the third law of thermodynamics. The heat capacity of ice from 15 to 273 K. *J. Am. Chem. Soc.* **58**, 1144–50.

Gillette D. A., Patterson Jr. E. M., Prospero J. M., and Jackson M. L. (1993) Soil aerosols. In *Aerosol Effects on Climate.* S. G. Jennings, ed. University of Arizona Press, Tucson, 73–109.

Gittens G. J (1969) Variation of surface tension of water with temperature. *J. Coll. Interface Sci.* **30**, 406–12.

Goldberg R. N. (1981) Evaluated activity and osmotic coefficients for aqueous solutions: Thirty-six uni-bivalent electrolytes. *J. Phys. Chem. Ref. Data* **10**, 671–764.

Golding B. W. (1992) An efficient nonhydrostatic forecast model. *Meteor. Atmos. Phys.* **50**, 89–103.

Gong W. and Cho H.-R. (1993) A numerical scheme for the integration of the gas-phase chemical rate equations in three-dimensional atmospheric models. *Atmos. Environ.* **27A**, 2147–60.

Goodin W. R., McRae G. J., and Seinfeld J. H. (1979) A comparison of interpolations methods for sparse data: Application to wind and concentration fields. *J. Appl. Meteor.* **18**, 761–71.

Graedel T. E. and Goldberg K. I. (1983) Kinetic studies of raindrop chemistry, 1. Inorganic and organic processes. *J. Geophys. Res.* **88**, 10,865–82.

Graedel T. E. and Weschler C. J. (1981) Chemistry within aqueous atmospheric aerosols and raindrops. *Rev. Geophys.* **19**, 505–39.

Grell G. A. (1993) Prognostic evaluation of assumptions used by cumulus parameterizations. *Mon. Wea. Rev.* **121**, 764–87.

Groblicki P. J., Wolff G. T., and Countess R. J. (1981) Visibility-reducing species in the Denver "brown cloud" – I. Relationships between extinction and chemical composition. *Atmos. Environ.* **15**, 2473–84.

Hack J. J. (1992) Climate system simulation: Basic numerical and computational concepts. In *Climate System Modeling*, K. E. Trenberth, ed. Cambridge University Press, Cambridge, 283–318.

Hack J. J. (1994) Parameterization of moist convection in the National Center for Atmospheric Research community climate model (CCM2). *J. Geophys. Res.* **99**, 5551–68.

Hagesawa K. and Neta P. (1978) Rate constants and mechanisms of reaction for $Cl_2^-$ radicals. *J. Phys. Chem.* **82**, 854–7.

Hairer E. and Wanner G. (1991) *Solving Ordinary Differential Equations II. Stiff and Differential–Algebraic Problems*. Springer-Verlag, Berlin.

Hale G. M. and Querry M. R. (1973) Optical constants of water in the 200-nm to 200-μm wavelength region. *Appl. Opt.* **12**, 555–63.

Hamer W. J. and Wu Y.-C. (1972) Osmotic coefficients and mean activity coefficients of uni-univalent electrolytes in water at 25°C. *J. Phys. Chem. Ref. Data* **1**, 1047–99.

Hamill P., Turco R. P., Kiang C. S., Toon O. B., and Whitten R. C. (1982) An analysis of various nucleation mechanisms for sulfate particles in the stratosphere. *J. Aerosol Sci.* **13**, 561–85.

Hansen J. E. (1969) Radiative transfer by doubling very thin layers. *Astrophys. J.* **155**, 565–73.

Hansen J., Lacis A., Lee P., and Wang W.-C. (1980) Climatic effects of atmospheric aerosols presented at Conference on Aerosols: Urban and rural characteristics, source and transport studies. *Ann. N.Y. Acad. Sci.* **338**, 575–87.

Harned H. S. and Owen B. B. (1958) *The Physical Chemistry of Electrolyte Solutions*. Reinhold, New York, Chapter 8.

Harrington J. Y., Meyers M. P., Walko R. L., and Cotton W. R. (1995) Parameterization of ice crystal conversion process due to vapor deposition for mesoscale models using double-moment basis functions. Part I: Basic formulation and parcel model results. *J. Atmos. Sci.* **52**, 4344–66.

Hartman D. L. (1994) *Global Physical Climatology*. Academic Press, Inc., San Diego.

Harvey R. B., Stedman D. H., and Chameides W. (1977) Determination of the absolute rate of photolysis of $NO_2$. *J. Air Pol. Control Assn*, **27**, 663–6.

Harvie C. E., Moller N., and Weare J. H. (1984) The prediction of mineral solubilities in natural waters: The Na–K–Mg–Ca–H–Cl–$SO_4$–OH–$HCO_3$–$CO_3$–$CO_2$–$H_2O$ system to high ionic strengths at 25°C. *Geochim. Cosmochim. Acta* **48**, 723–51.

Henyey L. C. and Greenstein J. L. (1941) Diffuse radiation in the galaxy. *Astrophys. J.* **93**, 70–83.

Hering S. V. and Friedlander S. K. (1982) Origins of aerosol sulfur size distributions in the Los Angeles Basin. *Atmos. Environ.* **16**, 2647–56.

Hering S. V., Friedlander S. K., Collins J. C., and Richards L. W. (1979) Design and evaluation of a new low pressure impactor 2. *Environ. Sci. Technol.* **13**, 184–8.

Hertel O., Berkowicz R., and Christensen J. (1993) Test of two numerical schemes for use in atmospheric transport–chemistry models. *Atmos. Environ.* **27A**, 2591–611.

Hesstvedt E., Hov O., and Isaksen I. S. A. (1978) Quasi-steady-state approximations in air pollution modeling: Comparison of two numerical schemes for oxidant prediction. *Int. J. Chem. Kin.* **10**, 971–94.

Himmelblau P. M. (1964) Diffusion of dissolved gases in liquids. *Chem. Rev.* **64**, 527–50.

Hindmarsh A. C. (1983) ODEPACK, a systematized collection of ODE solvers. In *Scientific Computing*, R. S. Stepleman et al., eds., North-Holland, Amsterdam, 55–74.

Hinze J. O. (1975) *Turbulence: An Introduction to its Mechanism and Theory*, 2nd ed. McGraw-Hill, New York, 790 pp.

Hitchcock D. R., Spiller L. L., and Wilson W. E. (1980) Sulfuric acid aerosols and HCl release in coastal atmospheres: Evidence of rapid formation of sulfuric acid particulates. *Atmos. Environ.* **14**, 165–82.

Hoffmann M. R. and Calvert J. G. (1985) Chemical transformation modules for Eulerian acid deposition models, Vol. 2. The aqueous-phase chemistry, EPA/600/3-85/017, U.S. Environmental Protection Agency, Research Triangle Park, NC.

Hofmann D. J., Oltmans S. J., Harris J. M., Johnson B. J., and Lathrop J. A. (1997) Ten years of ozonesonde measurements at the South Pole: Implications for recovery of springtime Antarctic ozone. *J. Geophys. Res.* **102**, 8931–43.

Hogstrom U. (1987) Non-dimensional wind and temperature profiles in the atmospheric surface layer: A reevaluation. *Boundary-Layer Meteor.* **42**, 55–78.

Holmes H. F. and Mesmer R. E. (1986) Thermodynamics of aqueous solutions of the alkali metal sulfates. *J. Solution Chem.* **15**, 495–518.

Holton J. R. (1992) *An Introduction to Dynamical Meteorology*. Academic Press, Inc., San Diego.

Holton J. R., Haynes P. H., McIntyre M. E., Douglass A. R., Rood R. B., and Pfister L. (1995) Stratosphere–troposphere exchange. *Rev. Geophys.* **33**, 403–39.

Horvath H. (1995) Size segregated light absorption coefficient of the atmospheric aerosol. *Atmos. Environ.* **29A**, 875–83.

Houze R. A., Jr. (1993) *Cloud Dynamics*. Academic Press, Inc., San Diego.

Hughes T. J. R. and Brooks A. N. (1979) A multidimensional upwind scheme with no crosswind diffusion. In *Finite Element Methods for Convection Dominated Flows*, ( T. J. R. Hughes) ed., AMD Vol. 34. ASME, New York, 19–35.

Huie R. E. and Neta P. (1987) Rate constants for some oxidations of S(IV) by radicals in aqueous solutions. *Atmos. Environ.* **21**, 1743–7.

Hynes A. J., Wine P. H., and Semmes D. H. (1986) Kinetic mechanism of OH reactions with organic sulfides. *J. Phys. Chem.* **90**, 4148–56.

Ingalls M. N., Smith L. R., and Kirksey R. E. (1989) Measurement of on-road vehicle emission factors in the California South Coast Air Basin. Volume I: Regulated emissions, Report to the Coordinating Research Council under Project SCAQS-1. Southwest Research Institute, San Antonio, TX.

Intergovernmental Panel on Climate Change (IPCC) (1990). *Climate Change: The IPCC Scientific Assessment*, J. T. Houghton, G. J. Jenkins, and J. J. Ephraums, eds. Cambridge University Press, Cambridge.

Irvine W. M. (1968) Multiple scattering by large particles. II. Optically thick layers. *Astrophys. J.* **152**, 823–34.

Irvine W. M. (1975) Multiple scattering in planetary atmospheres. *Icarus* **25**, 175–204.

Jacob D. J. (1986) Chemistry of OH in remote clouds and its role in the production of formic acid and peroxymonosulfonate. *J. Geophys. Res.* **91**, 9807–26.

Jacob D. J., Gottlieb E. W., and Prather M. J. (1989a) Chemistry of a polluted cloudy boundary layer. *J. Geophys. Res.* **94**, 12,975–13,002.

Jacob D. J., Sillman S., Logan J. A., and Wofsy S. C. (1989b) Least independent variables method for simulation of tropospheric ozone. *J. Geophys. Res.* **94**, 8497–509.

Jacobson M. Z. (1994) Developing, coupling, and applying a gas, aerosol, transport, and radiation model to study urban and regional air pollution. Ph.D. Thesis, Dept. of Atmospheric Sciences, University of California, Los Angeles.

Jacobson M. Z. (1995) Computation of global photochemistry with SMVGEAR II. *Atmos. Environ.* **29A**, 2541–6.

Jacobson M. Z. (1997a) Development and application of a new air pollution modeling system. Part II: Aerosol module structure and design. *Atmos. Environ.* **31A**, 131–44.

Jacobson M. Z. (1997b) Development and application of a new air pollution modeling system. Part III: Aerosol-phase simulations. *Atmos. Environ.* **31A**, 587–608.

Jacobson M. Z. (1997c) Numerical techniques to solve condensational and dissolutional growth equations when growth is coupled to reversible aqueous reactions. *Aerosol Sci. Technol.* **27**, 491–8.

Jacobson M. Z. (1998a) Vector and scalar improvement of SMVGEAR II through absolute error tolerance control. *Atmos. Environ.* **32**, 791–6.

Jacobson M. Z. (1998b) Studying the effects of aerosols on vertical photolysis rate coefficient and temperature profiles over an urban airshed. *J. Geophys. Res.* **103**, 10,593–604.

Jacobson M. Z. (1998c) Effects of soil moisture on temperatures, winds, and pollutant cocentrations in Los Angeles. *J. Appl. Meteorol.*, in press.

Jacobson M. Z. and Turco R. P. (1994) SMVGEAR: A sparse-matrix, vectorized Gear code for atmospheric models. *Atmos. Environ.* **28A**, 273–84.

Jacobson M. Z. and Turco R. P. (1995) Simulating condensational growth, evaporation, and coagulation of aerosols using a combined moving and stationary size grid. *Aerosol Sci. Technol.* **22**, 73–92.

Jacobson M. Z., Turco R. P., Jensen E. J., and Toon O. B. (1994) Modeling coagulation among particles of different composition and size. *Atmos. Environ.* **28A**, 1327–38.

Jacobson M. Z., Lu R., Turco R. P., and Toon O. B. (1996a) Development and application of a new air pollution modeling system. Part I: Gas-phase simulations. *Atmos. Environ.* **30B**, 1939–63.

Jacobson M. Z., Tabazadeh A., and Turco R. P. (1996b) Simulating equilibrium within aerosols and nonequilibrium between gases and aerosols. *J. Geophys. Res.* **101**, 9079–91.

Jaenicke R. (1988) *Numerical Data and Functional Relationships in Science and Technology*, New Series Vol. 4, Meteorology Sub. Vol. b, Physical and Chemical Properties of Air, G. Fischer, ed. Springer-Verlag.

Jarvis P. G., James G. B., and Landsberg J. J. (1976) Coniferous forest. In *Vegetation and the Atmosphere*, Vol. 2, J. L. Monteight, ed. Academic Press, New York, 171–240.

Jayne J. T., Davidovits P., Worsnop D. R., Zahniser M. S., and Kolb C. E. (1990) Uptake of $SO_2$ by aqueous surfaces as a function of pH: The effect of chemical reaction at the interface. *J. Phys. Chem.* **94**, 6041–8.

Jayson G. G., Parsons B. J., and Swallow A. J. (1973) Some simple, highly reactive, inorganic chlorine derivatives in aqueous solution. *Trans. Faraday Soc.* **69**, 1597–607.

Jeans J. (1954) *The Dynamical Theory of Gases*. Dover, New York.

John W., Wall S. M., Ondo J. L., and Winklmayr W. (1989) Acidic aerosol size distributions during SCAQS. Final Report for the California Air Resources Board under Contract No. A6-112-32.

Joseph J. H., Wiscombe W. J., and Weinman J. A. (1976) The delta–Eddington approximation for radiative flux transfer. *J. Atmos. Sci.* **33**, 2452–9.

Junge C. E. (1961) Vertical profiles of condensation nuclei in the stratosphere. *J. Meteorol.* **18**, 501–9.

Kaimal J. C. and Finnigan J. J. (1994) *Atmospheric Boundary Layer Flows: Their Structure and Measurement*. Oxford University Press, New York.

Kao C.-Y. J. and Ogura Y. (1987) Response of cumulus clouds to large-scale forcing using the Arakawa–Schubert cumulus parameterization. *J. Atmos. Sci.* **44**, 2437–48.

Kaps P. and Rentrop P. (1979) Generalized Runge–Kutta methods of order four with stepsize control for stiff ordinary differential equations. *Numer. Math.* **33**, 55–88.

Kasahara A. (1974) Various vertical coordinate systems used for numerical weather prediction. *Mon. Wea. Rev.* **102**, 509–22.

Kasten F. (1968) Falling speed of aerosol particles. *J. Appl. Meteor.* **7**, 944–7.

Kawata Y. and Irvine W. M. (1970) The Eddington approximation for planetary atmospheres. *Astrophys. J.* **160**, 787–90.

Kerker M. (1969) *The Scattering of Light and Other Electromagnetic Radiation*. Academic Press, New York.

Kiehl J. T. and Briegleb B. P. (1993) The relative role of sulfate aerosols and greenhouse gases in climate forcing. *Science* **260**, 311–14.

Kim Y. P. and Seinfeld J. H. (1995) Atmospheric gas–aerosol equilibrium: III. Thermodynamics of crustal elements $Ca^{2+}$, $K^{+}$, and $Mg^{2+}$. *Aerosol Sci. and Technol.* **22**, 93–110.

Kim Y. P., Seinfeld J. H., and Saxena P. (1993a) Atmospheric gas–aerosol equilibrium I. Thermodynamic model. *Aerosol Sci. Technol.* **19**, 157–81.

Kim Y. P., Seinfeld J. H., and Saxena P. (1993b) Atmospheric gas–aerosol equilibrium II. Analysis of common approximations and activity coefficient calculation methods. *Aerosol Sci. Technol.* **19**, 182–98.

King M. D. (1993) Radiative properties of clouds. In *Aerosol–Cloud–Climate Interactions*, P. V. Hobbs, ed., Academic Press, San Diego, Chapter 5.

Klemp J. B. and Wilhelmson R. B. (1978) The simulation of three-dimensional convective storm dynamics. *J. Atmos. Sci.* **3**, 1070–96.

Knudsen M. and Weber S. (1911) Luftwiderstand gegen die langsame Bewegung kleiner Kugeln. *Ann. Phys.* **36**, 981–94.

Köhler H. (1936) The nucleus in the growth of hygroscopic droplets. *Trans. Faraday Soc.* **32**, 1152–61.

Kondo J., Saigusa N., and Sato T. (1990) A parameterization of evaporation from bare soil surfaces. *J. Appl. Meteor.* **29**, 385–9.

Koschmieder H. (1924) Theorie der horizontalen Sichtweite. *Beitr. Phys. Freien Atm.* **12**, 33–53, 171–81.

Kozac-Channing L. F. and Heltz G. R. (1983) Solubility of ozone in aqueous solutions of 0–0.6 M ionic strength at 5–30°C. *Environ. Sci. Technol.* **17**, 145–9.

Kreitzberg C. W. and Perkey D. (1976) Release of potential instability. Part I: A sequential plume model within a hydrostatic primitive equation model. *J. Atmos. Sci.* **33**, 456–75.

Krekov G. M. (1993) Models of atmospheric aerosols. In *Aerosol Effects on Climate*, S. G. Jennings, ed., University of Arizona Press, Tucson, 9–72.

Krishnamurti T. N. and Moxim W. J. (1971) On parameterization of convective and non-convective latent heat release. *J. Appl. Meteor.* **10**, 3–13.

Krishnamurti T. N., Bedi H. S., and Hardiker V. M. (1998) *An Introduction to Global Spectral Modeling*. Oxford University Press, New York.

Krishnamurti T. N., Pan H.-L., Pasch R. J., and Molinari J. (1980) Cumulus parameterization and rainfall rates I. *Mon. Wea. Rev.* **108**, 465–72.

Kritz M. A. and Rancher J. (1980) Circulation of Na, Cl, and Br in the tropical marine atmosphere. *J. Geophys. Res.* **85**, 1633–9.

Kuo H. L. (1965) On formation and intensification of tropical cyclones through latent heat release by cumulus convection. *J. Atmos. Sci.* **22**, 40–63.

Kuo H. L. (1974) Further studies of the parameterization of the influence of cumulus convection on largescale flow. *J. Atmos. Sci.* **31**, 1232–40.

Kurihara Y. (1973) A scheme of moist convective adjustment. *Mon. Wea. Rev.* **101**, 547–53.

Lakshmi V., Wook E. F., and Chodhury B. J. (1997) Evaluation of special sensor microwave/satellite data for regional soil moisture estimation over the Red River basin. *J. Appl. Meteorol.* **36**, 1309–28.

Lamb H. (1910) On atmospheric oscillations. *Proc. Roy. Soc. London* **84**, 551–72.

Langford A. O., Proffitt M. H., VanZandt T. E., and Lamarque J.-F. (1996) Modulation of tropospheric ozone by a propagating gravity wave. *J. Geophys. Res.* **101**, 26,605–13.

Larson S., Cass G., Hussey K., and Luce F. (1984) Visibility model verification by image processing techniques. Final report to the California Air Resources Board under Agreement A2-077-32.

Lary D. J. (1997) Catalytic destruction of stratospheric ozone. *J. Geophys. Res.* **102**, 21,515–26.

Lax P. D. and Wendroff B. (1964) Difference schemes with high order of accuracy for solving hyperbolic equations. *Commun. Pure. Appl. Math.* **17**, 381–98.

Ledbury W. and Blair E. W. (1925) The partial formaldehyde vapour pressure of aqueous solutions of formaldehyde, II. *J. Chem. Soc.* **127**, 2832–9.

Lee H. D. P., translator (1951) *Meteorologica* by Aristotle, T. E. Page, ed. Harvard University Press, Cambridge.

Lee K. W. (1985) Conservation of particle size distribution parameters during Brownian coagulation. *J. Colloid and Interface Sci.* **108**, 199–206.

Lee T. J. and Pielke R. A. (1992) Estimating the soil surface specific humidity. *J. Appl. Meteor.* **31**, 480–4.

Le Henaff P. (1968) Méthodes détude et propriétés des hydrates, hemiacétals et hemiacétals derivés des aldehydes et des cétones. *Bull. Soc. Chim. France*, **11**, 4687–700.

Lenschow D. H., Li X. S., Zhu C. J., and Stankov B. B. (1988) The stably stratified boundary layer over the great planes. *Boundary Layer Meteor.* **42**, 95–121.

Lettau H. H. (1969) Note on aerodynamic roughness-parameter estimation on the basis of roughness element description. *J. Appl. Meteor.* **8**, 828–32.

Liang J. and Jacob D. J. (1997) Effect of aqueous-phase cloud chemistry on tropospheric ozone. *J. Geophys. Res.* **102**, 5993–6002.

Liang J. and Jacobson M. Z. (1998) A study of sulfur dioxide oxidation pathways for a range of liquid water contents, pHs, and temperatures. *J. Geophys. Res.*, in review.

Lide D. R., ed.-in-chief (1993) *CRC Handbook of Chemistry and Physics*. CRC Press, Inc., Boca Raton, FL.

Lind J. A. and Kok G. L. (1986) Henry's law determinations for aqueous solutions of hydrogen peroxide, methylhydroperoxide, and peroxyacetic acid. *J. Geophys. Res.* **91**, 7889–95.

Lind J. A., Kok G. L., and Lazrus A. L. (1987) Aqueous phase oxidation of sulfur(IV) by hydrogen peroxide, methylhydroperoxide, and peroxyacetic acid. *J. Geophys. Res.* **92**, 4171–7.

Lindzen, R. S. (1981) Turbulence and stress due to gravity wave and tidal breakdown. *J. Geophys. Res.* **86**, 9707–14.

Liou K. N. (1974) Analytic two-stream and four-stream solutions for radiative transfer. *J. Atmos. Sci.* **31**, 1473–5.

Liou K.-N. (1976) On the absorption, reflection and transmission of solar radiation in cloudy atmospheres. *J. Atmos. Sci.* **33**, 798–805.

Liou K. N. (1992) *Radiation and Cloud Processes in the Atmosphere*. Oxford University Press, New York.

List R. J., ed. (1984) *Smithsonian Meteorological Tables*, 6th ed. Smithsonian Institution Press, Washington, DC.

Lord S. J. and Arakawa A. (1980) Interaction of a cumulus cloud ensemble with the large-scale environment. Part II. *J. Atmos. Sci.* **37**, 2677–92.

Lorenz E. N. (1960) Energy and numerical weather prediction. *Tellus* **12**, 364–73.

Louis J.-F. (1979) A parametric model of vertical eddy fluxes in the atmosphere. *Boundary Layer Meteor.* **17**, 187–202.

Lu R. (1994) Development of an integrated air pollution modeling system and simulations of ozone distributions over the Los Angeles Basin. Ph.D. Thesis, University of California, Los Angeles.

Ludlum F. H. (1980) *Clouds and Storms*. The Pennsylvania State University Press, University Park.

Lurmann F. W., Carter W. P. L., and Coyner L. A. (1987) *A Surrogate Species Chemical Reaction Mechanism for Urban Scale Air Quality Simulation Models. Volume I: Adaption of the Mechanism*, EPA-600/3-87/014a, U.S. Environmental Protection Agency, Research Triangle Park, NC.

Lurmann F. W., Main H. H., Knapp K. T., Stockburger L., Rasmussen R. A., and Fung K. (1992) Analysis of the ambient VOC data collected in the Southern California Air Quality Study. Final Report to the California Air Resources Board under Contract A832-130.

Madronich S. (1987) Photodissociation in the atmosphere 1. Actinic flux and the effects of ground reflections and clouds. *J. Geophys. Res.* **92**, 9740–52.

Mahfouf J.-F. and Noilhan J. (1991) Comparative study of various formulations of evaporation from bare soil using in situ data. *J. Appl. Meteor.* **30**, 1354–65.

Mahfouf J.-F. and Noilhan J. (1996) Inclusion of gravitational drainage in a land surface scheme based on the force-restore method. *J. Appl. Meteor.* **35**, 987–92.

Mahrt L., Heald R. C., Lenschow D. H., Stankov B. B., and Troen I. (1979) An observational study of the structure of the nocturnal boundary layer. *Boundary Layer Meteor.* **17**, 247–64.

Makar P. A. and Karpik S. R. (1996) Basis-spline interpolation on the sphere: Applications to semi-lagrangian advection. *Mon. Wea. Rev.* **124**, 182–99.

Manabe S. J., Smagorinsky J., and Strickler R. F. (1965) Simulated climatology of a general circulation model with a hydrological cycle. *Mon. Weather Rev.* **93**, 769–98.

Marsh A. R. W. and McElroy W. J. (1985) The dissociation constant and Henry's law constant of HCl in aqueous solution. *Atmos. Environ.* **19**, 1075–80.

Marshall J. S. and Palmer W. (1948) The distribution of raindrops with size. *J. Meteor.* **3**, 165–8.

Martens C. S., Wesolowski J. J., Hariss R. C., and Kaifer R. (1973) Chlorine loss from Puerto Rican and San Francisco Bay Area marine aerosols. *J. Geophys. Res.* **78**, 8778–92.

Martin L. R. and Hill M. W. (1987a) The iron-catalyzed oxidation of sulfur: Reconciliation of the literature rates. *Atmos. Environ.* **21**, 1487–90.

Martin L. R. and Hill M. W. (1987b) The effect of ionic strength on the manganese catalyzed oxidation of sulfur(IV). *Atmos. Environ.* **21**, 2267–70.

Mason B. J. (1971) *The Physics of Clouds*. Clarendon Press, Oxford.

Matsuno T. (1966) Numerical integrations of the primitive equations by simulated backward difference scheme. *J. Meteor. Soc. Japan* **44**, 76–84.

Maxwell J. C. (1890) *The Scientific Papers of James Clerk Maxwell*, Vol. II, W. D. Niven, ed. Cambridge University Press, 636–40.

McCumber M. C. and Pielke R. A. (1981) Simulation of the effects of surface fluxes of heat and moisture in a mesoscale numerical model. Part I: Soil layer. *J. Geophys. Res.* **86**, 9929–38.

McElroy M. B., Salawitch R. J., Wofsy S. C., and Logan J. A. (1986) Reduction of Antarctic ozone due to synergistic interactions of chlorine and bromine. *Nature* **321**, 759–62.

McRae G. J., Goodin W. R., and Seinfeld J. H. (1982) Development of a second-generation mathematical model for urban air pollution – I. Model formulation. *Atmos. Environ.* **16**, 679–96.

Meador W. E. and Weaver W. R. (1980) Two-stream approximations to radiative transfer in planetary atmospheres: A unified description of existing methods and a new improvement. *J. Atmos. Sci.* **37**, 630–43.

Mellor G. L. and Yamada T. (1982) Development of a turbulence closure model for geophysical fluid problems. *Rev. Geophys. and Space Phys.* **20**, 851–75.

Meng Z., Seinfeld J. H., Saxena P., and Kim Y. P. (1995) Atmospheric gas aerosol equilibrium: IV. Thermodynamics of carbonates. *Aerosol Sci. Technol.* **23**, 131–154.

Mesinger F. and Arakawa A. (1976) Numerical methods used in atmospheric models. GARP Publication Series. No. 17, 1. World Meteorological Organization, 64 pp.

Middleton P. and Brock J. R. (1976) Simulation of aerosol kinetics. *J. Colloid Interface Sci.* **54**, 249–64.

Middleton W. E. K. (1952) *Vision Through the Atmosphere*. University of Toronto Press, Toronto, Canada.

Mihailovic D. T., Rajkovic B., Lalic B., and Dekic L. (1995) Schemes for parameterizing evaporation from a non-plant covered surface and their impact on partitioning the surface energy in land–air exchange parameterization. *J. Appl. Meteor.* **34**, 2462–75.

Millikan R. A. (1923) The general law of fall of a small spherical body through a gas, and its bearing upon the nature of molecular reflection from surfaces. *Phys. Rev.* **22**, 1–23.

Mitchell A. R. (1969) *Computational Methods in Partial Differential Equations*. John Wiley, New York.

Miyakoda K., Smagorinsky J., Strickler R. F., and Hembree G. D. (1969) Experimental extended predictions with a nine-level hemispheric model. *Mon. Wea. Rev.* **97**, 1–76.

Molina L. T. and Molina M. J. (1986) Production of $Cl_2O_2$ by the self reaction of the ClO radical. *J. Phys. Chem.* **91**, 433–6.

Molina M. J. and Roland F. S. (1974) Stratospheric sink for chlorofluoromethanes: Chlorine atom catalysed destruction of ozone. *Nature* **249**, 810–2.

Molinari J. (1982) A method for calculating the effects of deep cumulus convection in numerical models. *Mon. Wea. Rev.* **11**, 1527–34.

Monahan E. C., Spiel D. E., and Davidson K. L. (1986) A model of marine aerosol generation via whitecaps and wave disruption, in *Oceanic Whitecaps and Their Role in Air–Sea Exchange Processes*, E. C. Monahan and G. MacNiocaill, eds., D. Reidel, Norwell, MA, 167–174.

Monin A. S. and Obukhov A. M. (1954) Basic laws of turbulent mixing in the ground layer of the atmosphere. *Trans. Geophys. Inst. Akad. Nauk USSR* **151**, 1963–87.

Monin A. S. and Yaglom (1971) *Statistical Fluid Mechanics*. MIT Press, Cambridge, MA.

Moorthi S. and Suarez M. J. (1992) Relaxed Arakawa–Schubert: A parameterization of moist convection for general circulation models. *Mon. Wea. Rev.* **120**, 978–86.

Mordy W. (1959) Computations of the growth by condensation of a population of cloud droplets. *Tellus* **11**, 16–44.

Mozurkewich M., McMurry P. H., Gupta A., and Calvert J. G. (1987) Mass accommodation coefficients for $HO_2$ radicals on aqueous particles. *J. Geophys. Res.* **92**, 4163–70.

Muller H. (1928) Zur allgemeinen Theorie der raschem Koagulation. Die koagulation von Stabchen- und Blattchen-kolloiden; die Theorie beliebig polydisperser Systeme und der Stromungskoagulation. *Kolloidbeihefte* **27**, 223–50.

Munger W. J., Collett J. Jr., Daube B. C., and Hoffmann M. R. (1989) Carboxylic acids and carbonyl compounds in southern California clouds and fogs. *Tellus* **41b**, 230–42.

Mylonas D. T., Allen D. T., Ehrman S. H., and Pratsinis S. E. (1991) The sources and size distributions of organonitrates in Los Angeles aerosols. *Atmos. Environ.* **25A**, 2855–61.

National Oceanic and Atmospheric Administration (NOAA) (1976) *U.S. Standard Atmosphere*. Washington, DC.

Nautical Almanac Office (NAO) and Her Majesty's Nautical Almanac Office (1993) *Astronomical Almanac*. U.S. Government Printing Office, Washington DC.

Nebeker F. (1995) *Calculating the Weather*. Academic Press, Inc., San Diego.

Nesbitt F. L., Monks P. S., Wayne W. A., Stief L. J., and Touni R. (1995) The reaction of $O(^3P)$ + HOBr: Temperature dependence of the rate constant and importance of the reaction as an HOBr loss process. *Geophys. Res. Lett.* **22**, 827–30.

Nicolet M. (1989) Solar spectral irradiances with their diversity between 120 and 900 nm. *Planet. Space Sci.* **37**, 1249–89.

Noilhan J. and Planton S. (1989) A simple parameterization of land surface processes for meteorological models. *Mon. Wea. Rev.* **117**, 536–49.

Noll K. E., Fang K. Y. P., and Khalili E. (1990) Characterization of atmospheric coarse particles in the Los Angeles Basin. *Aerosol Sci. Technol.* **12**, 28–38.

Ogura Y. and Phillips N. A. (1962) Scale analysis of deep and shallow convection in the atmosphere. *J. Atmos. Sci.* **19**, 173–179.

Olscamp P. J., translator (1965) *Discourse on Method, Optics, Geometry, and Meteorology* by René Descartes. Bobbs-Merrill Company, Inc., Indianapolis.

Oke T. R. (1978) *Boundary Layer Climates*. Methuen, London.

Ooyama V. K. (1971) A theory on parameterization of cumulus convection. *J. Meteor. Soc. Japan* **49**, 744–56.

Orszag S. A. (1970) Transform method for calculation of vector coupled sums: Application to the spectral form of the vorticity equation. *J. Atmos. Sci.* **27**, 890–5.

Orszag S. A. (1971) Numerical simulation of incompressible flows within simple boundaries. I. Galerkin (spectral) representations. *Stud. Appl Math.* **50**, 293–326.

Osborne N. S., Stimson H. F., and Ginnings D. C. (1939) Measurements of heat capacity and heat of vaporization of water in the range of 0 degrees to 100 degrees celsius. *J. Res. Nat. Bur. Stand.* **23**, 197–260.

Pandis S. N. and Seinfeld J. H. (1989) Sensitivity analysis of a chemical mechanism for aqueous-phase atmospheric chemistry. *J. Geophys. Res.* **94**, 1105–26.

Pandis S. N., Harley R. A., Cass G. R., and Seinfeld J. H. (1992) Secondary organic aerosol formation and transport. *Atmos. Environ.* **26A**, 2269–82.

Park J.-Y. and Lee Y.-N. (1987) Aqueous solubility and hydrolysis kinetics of peroxynitric acid. Paper presented at 193rd Meeting, Am. Chem. Soc., Denver, CO, April 5–10.

Parker V. B. (1965) *Thermal Properties of Aqueous Uni-univalent Electrolytes*. National Standard Reference Data Series – NBS 2. U.S. Government Printing Office, Washington, DC.

Pasquill F. (1962) *Atmospheric Diffusion*. Van Nostrand, London.

Paulson S. E. and Seinfeld J. H. (1992) Development and evaluation of a photooxidation mechanism for isoprene. *J. Geophys. Res.* **97**, 20,703–15.

Penner J. E., Dickinson R. E., and O'Neill C. A. (1992) Effects of aerosol from biomass burning on the global radiation budget. *Science* **256**, 1432–3.

Pepper D. W., Kern C. D., and Long P. E. Jr. (1979) Modeling the dispersion of atmospheric pollution using cubic splines and chapeau functions. *Atmos. Environ.* **13**, 223–37.

Perrin D. D. (1982) *Ionization Constants of Inorganic Acids and Bases in Aqueous Solution*, 2nd ed. Pergamon, New York.

Perron G., Roux A., and Desnoyers J. E. (1981) Heat capacities and volumes of NaCl, $MgCl_2$, $CaCl_2$, and $NiCl_2$ up to 6 molal in water. *Can. J. Chem.* **59**, 3049–54.

Petersen R. L. (1997) A wind tunnel evaluation of methods for estimating surface roughness length at industrial facilities. *Atmos. Environ.* **31**, 45–57.

Philip J. R. (1957) Evaporation, and moisture and heat fields in the soil. *J. Meteor.* **14**, 354–66.

Phillips N. A. (1957) A coordinate system having some special advantages for numerical forecasting. *J. Meteor.* **14**, 184–5.

Pielke R. A. (1984) *Mesoscale Meteorological Modeling*. Academic Press, Inc., San Diego.

Pilinis C. and Seinfeld J. H. (1987) Continued development of a general equilibrium model for inorganic multicomponent atmospheric aerosols. *Atmos. Environ.* **21**, 2453–66.

Pilinis C. and Seinfeld J. H. (1988) Development and evaluation of an Eulerian photochemical gas–aerosol model. *Atmos. Environ.* **22**, 1985–2001.

Pitzer K. S. (1991) Ion interaction approach: Theory and data correlation. In *Activity Coefficients in Electrolyte Solutions*, 2nd ed., K. S. Pitzer, ed., CRC Press, Boca Raton, 75–153.

Pitzer K. S. and Mayorga G. (1973) Thermodynamics of electrolytes II. Activity and osmotic coefficients for strong electrolytes with one or both ions univalent. *J. Phys. Chem.* **77**, 2300–8.

Potter J. F. (1970) The delta-function approximation in radiative transfer theory. *J. Atmos. Sci.* **27**, 943–9.

Press W. H., Flannery B. P., Teukolsky, S. A., and Vetterling W. T. (1992). *Numerical Recipes: The Art of Scientific Computing*. Cambridge University Press, Cambridge.

Price G. V. and MacPherson A. K. (1973) A numerical weather forecasting method using cubic splines on a variable mesh. *J. Appl. Meteor.* **12**, 1102–13.

Pruppacher H. R. and Klett J. D. (1997) *Microphysics of Clouds and Precipitation*, 2nd rev. and enl. ed. Kluwer Academic Publishers, Dordrecht.

Purnell D. K. (1976) Solution of the advection equation by upstream interpolation with a cubic spline. *Mon. Weath. Rev.* **104**, 42–8.

Rao N. P. and McMurry P. H. (1989) Nucleation and growth of aerosol in chemically reacting systems. *Aerosol Sci. Technol.* **11**, 120–33.

Rasool S. I and Schneider S. H. (1971) Atmospheric carbon dioxide and aerosols. Effects of large increases on global climate. *Science* **173**, 138–41.

Reynolds D. W., Vonder Har T. H., and Cox S. K. (1975) The effect of solar radiation in the tropical troposphere. *J. Appl. Meteor.* **14**, 433–44.

Reynolds S. D., Roth P. M., and Seinfeld J. H. (1973) Mathematical modeling of photochemical air pollution – I: Formulation of the model. *Atmos. Environ.* **7**, 1033–61.

Richardson L. F. (1922) *Weather Prediction by Numerical Process*. Cambridge University Press, reprinted 1965, 236 pp.

Robert A. (1982) A semi-Lagrangian and semi-implicit numerical integration scheme for the primitive meteorological equations. *Japan. Meteor. Soc.* **60**, 319–25.

Robinson R. A. and Stokes R. H. (1955) *Electrolyte Solutions*. Academic Press, New York.

Rogers R. R. and Yau M. K. (1989) *A Short Course in Cloud Physics*. Pergamon Press, Oxford.

Rosenbaum J. S. (1976) Conservation properties of numerical integration methods for systems of ordinary differential equations. *J. Comp. Phys.* **20**, 259–67.

Ross A. B. and Neta P. (1979) Rate constants for reactions of inorganic radicals in aqueous solutions. NSRDS- NBS 65. National Bureau of Standards, U.S. Department of Commerce, Washington, DC.

Rossby C. and collaborators (1939) Relation between variations in the intensity of the zonal circulation of the atmosphere and the displacements of the semi-permanent centers of action. *J. Marine Res.* **2**, 38–55.

Roux A., Musbally G. M., Perron G., Desnoyers E., Singh P. P., Woolley E. M., and Hepler L. G. (1978) Apparent molal heat capacities and volumes of aqueous electrolytes at 25°C: $NaClO_3$, $NaClO_4$, $NaNO_3$, $NaBrO_3$, $NaIO_3$, $KClO_3$, $KBrO_3$, $KIO_3$, $NH_4NO_3$, $NH_4Cl$, and $NH_4ClO_4$. *Can. J. Chem.* **56**, 24–8.

Russell A. G., Winner D. A., Harley R. A., McCue K. F., and Cass G. R. (1993) Mathematical modeling and control of the dry deposition flux of nitrogen-containing air pollutants. *Environ. Sci. Technol.* **27**, 2772–82.

Russell L. M., Pandis S. N., and Seinfeld J. H. (1994) Aerosol production and growth in the marine boundary layer. *J. Geophys. Res.* **99**, 20,989–21,003.

Saffman P. G. and Turner J. S. (1956) On the collision of drops in turbulent clouds. *J. Fluid Mech.* **1**, 16–30.

Salby M. L. (1996) *Fundamentals of Atmospheric Physics*. Academic Press, Inc., San Diego.

Sander R., Lelieveld J., and Crutzen P. J. (1995) Modelling of nighttime nitrogen and sulfur chemistry in size resolved droplets of an orographic cloud. *J. Atmos. Chem.* **20**, 89–116.

Sandu A., Verwer J. G., van Loon M., Carmichael G. R., Potra F. A., Dabdub D., and Seinfeld J. H. (1997) Benchmarking stiff ODE solvers for atmospheric chemistry problems I: Implicit versus explicit. *Atmos. Environ.* **31**, 3151–66.

San Jose R., Casanova J. L., Viloria R. E., and Casanova J. (1985) Evaluation of the turbulent parameters of the unstable surface boundary layer outside Businger's range. *Atmos. Environ.* **19**, 1555–61.

Saxena P., Hudischewskyj A. B., Seigneur C., and Seinfeld J. H. (1986) A comparative study of equilibrium approaches to the chemical characterization of secondary aerosols. *Atmos. Environ.* **20**, 1471–83.

Saxena P., Mueller P. K., and Hildemann L. M. (1993) Sources and chemistry of chloride in the troposphere: A review. In *Managing Hazardous Air Pollutants: State of the Art*, W. Chow and K. K Connor, eds., Lewis Publishers, Boca Raton, FL, 173–190.

Schmidt K. H. (1972) Electrical conductivity techniques for studying the kinetics of radiation-induced chemical reactions in aqueous solutions. *Int. J. Radiat. Phys. Chem.* **4**, 439–68.

Schneider W., Moortgat G. K., Tyndall G. S., and Burrows J. P. (1987) Absorption cross-sections of $NO_2$ in the UV and visible region (200–700 nm) at 298 *K. J. Photochem. Photobiol, A: Chem.* **40**, 195–217.

Scholes G. and Willson R. L. (1967) $\gamma$-radiolysis of aqueous thymine solutions. Determination of relative reaction rates of OH radicals. *Trans. Faraday Soc.* **63**, 2982–93.

Schwartz S. E. (1984) Gas- and aqueous-phase chemistry of $HO_2$ in liquid water clouds. *J. Geophys. Res.* **89**, 11,589–98.

Schwartz S. E. (1986) Mass-transport considerations pertinent to aqueous phase reactions of gases in liquid-water clouds. In *Chemistry of Multiphase Atmospheric Systems*, NATO ASI Series, Vol. G6, W. Jaeschke, ed., Springer-Verlag, Berlin, 415–471.

Schwartz S. E. and White W. H. (1981) Solubility equilibria of the nitrogen oxides and oxyacids in aqueous solution. *Adv. Environ. Sci. Eng.* **4**, 1–45.

Seaman N. L., Ludwig F. L., Donall E. G., Warner T. T., and Bhumralkar C. M. (1989) Numerical studies of urban planetary boundary-layer structure under realistic synoptic conditions. *J. Appl. Meteor.* **28**, 760–81.

Seery D. J. and Britton D. (1964) The continuous absorption spectra of chlorine, bromine, bromine chloride, iodine chloride, and iodine bromide. *J. Phys. Chem.* **68**, 2263–6.

Sehested K., Rasmussen O. L., and Fricke H. (1968) Rate constants of OH with $HO_2$, $O_2^-$, and $H_2O_2^+$ from hydrogen peroxide formation in pulse-irradiated oxygenated water. *J. Phys. Chem.* **72**, 626–31.

Sehmel G. A. (1980) Particle and gas dry deposition: A review. *Atmos. Environ.* **14**, 983–1011.

Sellers W. D. (1965) *Physical Climatology*. University of Chicago Press, Chicago, 272 pp.

Sheridan P. J., Brock C. A., and Wilson J. C. (1994) Aerosol particles in the upper troposphere and lower stratosphere: Elemental composition and morphology of individual particles in northern midlatitudes. *Geophys. Res. Lett.* **21**, 2587–90.

Sherman A. H. and Hindmarsh A. C. (1980) GEARS: A package for the solution of sparse, stiff ordinary differential equations. Report UCRL-84102, Lawrence Livermore Laboratory.

Shimazaki T. and Laird A. R. (1970) A model calculation of the diurnal variation in minor neutral constituents in the mesosphere and lower thermosphere including transport effects. *J. Geophys. Res.* **75**, 3221–35.

Shir C. C. and Bornstein R. D. (1976) Eddy exchange coefficients in numerical models of the planetary boundary layer. *Boundary Layer Meteorol.* **11**, 171–85.

Shuttleworth W. J. (1989) Micrometeorology of temperate and tropical forest. *Phil. Trans. Roy. Soc. London* **B324**, 299–334.

Siegel R. and Howell J. R. (1992) *Thermal Radiation Heat Transfer*. Taylor and Francis, Washington, DC.

Singh H. B. (1995) Halogens in the atmospheric environment. In *Composition, Chemistry, and Climate of the Atmosphere*, H. B. Singh, ed., Van Nostrand Reinhold, New York.

Singh H. B., Kanakidou M., Crutzen P. J., and Jacob D. J. (1995) High concentrations and photochemical fate of oxygenated hydrocarbons in the global troposphere. *Nature* **378**, 50–4.

Singh H. B., Herlth D., Kolyer R., Salas L., Bradshaw J. D., Sandholm T., Davis D. D., Crawford J., Kondo Y., Koike M., Talbot R., Gregory G. L., Sachse G. W., Browell E., Blake D. R., Rowland F. S., Newell R., Merrill J., Heikes B., Liu S. C., Crutzen P. J., and Kanakidou M. (1996) Reactive nitrogen and ozone over the western Pacific: Distributions, partitioning, and sources. *J. Geophys. Res.* **101**, 1793–1808.

Singh H. B., Viezee W., and Salas L. J. (1988) Measurements of selected $C_2$–$C_5$ hydrocarbons in the troposphere: Latitudinal, vertical, and temporal variations. *J. Geophys. Res.* **93**, 15,861–78.

Slinn W. G. N., Hasse L., Hicks B. B., Hogan A. W., Lal D., Liss P. S., Munnich K. O., Sehmel G. A., and Vittori O. (1978) Some aspects of the transfer of atmospheric trace constituents past the air–sea interface. *Atmos. Environ.* **12**, 2055–87.

Smith R. M. and Martell A. E. (1976) *Critical Stability Constants, Vol. 4: Inorganic Complexes*. Plenum, New York.

Smolarkiewicz P. K. (1983) A simple positive definite advection scheme with small implicit diffusion. *Mon. Weather Rev.* **111**, 479–86.

Smoluchowski M. V. (1918) Versuch einer mathematischen Theorie der Koagulationskinetik kolloider Lösungen. Z. *Phys. Chem.* **92**, 129–68.

Snider J. R. and Dawson G. A. (1985) Tropospheric light alcohols, carbonyls, and acetonitrile: Concentrations in the southwestern United States and Henry's law data. *J. Geophys. Res.* **90**, 3797–805.

Sokolik I., Andronova A., and Johnson C. (1993) Complex refractive index of atmospheric dust aerosols. *Atmos. Environ.* **27A**, 2495–502.

Sommer L. (1989) *Analytical Absorption Spectrophotometry in the Visible and Ultraviolet*. Elsevier, Amsterdam.

Spencer J. W. (1971) Fourier series representation of the position of the sun. *Search* **2**, 172–2.

Staniforth A. and Cote J. (1991) Semi-Lagrangian integration schemes for atmospheric models – a review. *Mon. Wea. Rev.* **119**, 2206–23.

Stelson A. W., Bassett M. E., and Seinfeld J. H. (1984) Thermodynamic equilibrium properties of aqueous solutions of nitrate, sulfate and ammonium. In *Chemistry of Particles, Fogs and Rain*, J. L. Durham, ed., Ann Arbor Publication, Ann Arbor, MI, 1–52.

Stephens G. L. and Tsay S. C. (1990) On the cloud absorption anomaly. *Quart. J. Roy. Meteor. Soc.* **116**, 671–704.

Stockwell W. R. (1986) A homogeneous gas-phase mechanism for use in a regional acid deposition model. *Atmos. Environ.* **20**, 1615–32.

Stockwell W. R. (1995) On the $HO_2 + HO_2$ reaction: Its misapplication in atmospheric chemistry models. *J. Geophys. Res.* **100**, 11,695–8.

Stoer J. and Bulirsch R. (1980) *Introduction to Numerical Analysis*. Springer-Verlag, New York.

Stokes R. H. and Robinson R. A. (1966) Interactions in aqueous nonelectrolyte solutions. I. Solute–solvent equilibria. *J. Phys. Chem.* **70**, 2126–30.

Stommel H. (1947) Entrainment of air into a cumulus cloud. Part I. *J. Appl. Meteor.* **4**, 91–4.

Strom J., Okada K., and Heintzenber J. (1992) On the state of mixing of particles due to Brownian coagulation. *J. Aerosol Sci.* **23**, 467–80.

Stull R. B. (1988) *An Introduction to Boundary Layer Meteorology*. Kluwer Academic Publishers, Dordrecht.

Suck S. H. and Brock J. R. (1979) Evolution of atmospheric aerosol particle size distributions via Brownian coagulation: Numerical simulation. *J. Aerosol Sci.* **10**, 581–90.

Sukhatme S. P. and Saikhedkar N. (1969) Heat capacities of glycerol–water mixtures and aqueous solutions of ammonium sulfate ammonium nitrate and strontium nitrate. *Ind. J. Technol.* **7**, 1–4.

Tabazadeh A. and Turco R. P. (1993) A model for heterogeneous chemical processes on the surfaces of ice and nitric acid trihydrate particles. *J. Geophys. Res.* **98**, 12,727–40.

Tabazadeh A., Turco R. P., Drdla K., and Jacobson M. Z. (1994) A study of Type I polar stratospheric cloud formation. *Geophys. Res. Lett.* **21**, 1619–22.

Tang I. N. (1980) Deliquescence properties and particle size change of hygroscopic aerosols in generation of aerosols. In *Generation of Aerosols and Facilities for Exposure Experiments*, K. Willeke, ed., Ann Arbor Science, Ann Arbor, MI, 153–167.

Tang I. N. (1996) Chemical and size effects of hygroscopic aerosols on light scattering coefficients. *J. Geophys. Res.* **101**, 19,245–50.

Tang I. N. (1997) Thermodynamic and optical properties of mixed-salt aerosols of atmospheric importance. *J. Geophys. Res.* **102**, 1883–93.

Tang I. N. and Munkelwitz H. R. (1993) Composition and temperature dependence of the deliquescence properties of hygroscopic aerosols. *Atmos. Environ.* **27A**, 467–73.

Tang I. N. and Munkelwitz H. R. (1994) Water activities, densities, and refractive indices of aqueous sulfates and sodium nitrate droplets of atmospheric importance. *J. Geophys. Res.* **99**, 18,801–8.

Tang I. N., Wong W. T., and Munkelwitz H. R. (1981) The relative importance of atmospheric sulfates and nitrates in visibility reduction. *Atmos. Environ.* **15**, 2463–71.

Targuay M., Robert A., and Leprise R. (1990). A semiimplicit semilangrangian fully compressible regional forecast model. *Mon. Wea. Rev.* **118**, 1970–80.

Tapp M. C. and White P. W. (1976) A nonhydrostatic mesoscale model. *Quart J. Roy. Meteor. Soc.* **102**, 277–98.

Tesche T. W. (1988) Accuracy of ozone air quality models. *J. Environ. Engng.* **114**, 739–52.

Thekaekara M. P. (1974) Extraterrestrial solar spectrum, 3000–6100 Å at 1–Å intervals. *Appl. Opt.* **13**, 518–22.

Tjernstrom M. (1993) Turbulence length scales in stably stratified free shear flow analyzed from slant aircraft profiles. *J. Appl. Meteor.* **32**, 948–63.

Tiedtke M. (1989) A comprehensive mass flux scheme for cumulus parameterization in large-scale models. *Mon. Wea. Rev.* **117**, 1779–1800.

Toon O. B. and Pollack J. (1976) A global average model of atmospheric aerosols for radiative transfer calculations. *J. Appl. Meteor.* **15**, 225–46.

Toon O. B. and Ackerman T. P. (1981) Algorithms for the calculation of scattering by stratified spheres. *Appl. Opt.* **20**, 3657–60.

Toon O. B., Hamill P., Turco R. P., and Pinto J. (1986) Condensation of $HNO_3$ and HCl in the winter polar stratospheres. *Geophys. Res. Lett. Nov. Supp.* **13**, 1284–7.

Toon O. B., Turco R. P., Westphal D., Malone R., and Liu M. S. (1988) A multidimensional model for aerosols: Description of computational analogs. *J. Atmos. Sci.* **45**, 2123–43.

Toon O. B., McKay C. P., and Ackerman T. P. (1989a) Rapid calculation of radiative heating rates and photodissociation rates in inhomogeneous multiple scattering atmospheres. *J. Geophys. Res.* **94**, 16,287–301.

Toon O. B., Turco R. P., Jordan J., Goodman J., and Ferry G. (1989b) Physical processes in polar stratospheric ice clouds. *J. Geophys Res.* **94**, 11,359–80.

Tremback C. J., Powell J., Cotton W. R., and Pielke R. A. (1987) The forward-in-time upstream advection scheme: Extension to higher orders. *Mon. Wea. Rev.* **115**, 540–55.

Tsang T. H. and Brock J. R. (1982) Aerosol coagulation in the plume from a cross-wind line source. *Atmos. Environ.* **16**, 2229–35.

Tsang T. H. and Brock J. R. (1986) Simulation of condensation aerosol growth by condensation and evaporation. *Aerosol Sci. Technol.* **5**, 385–8.

Tsang, T. H. and Huang L. K. (1990) On a Petrov-Galerkin finite element method for evaporation of polydisperse aerosols. *Aerosol Sci. Technol.* **12**, 578–97.

Tsang T. H. and Korgaonkar N. (1987) Effect of evaporation on the extinction coefficient of an aerosol cloud. *Aerosol Sci. Technol.* **7**, 317–28.

Tung K. K. and Yang H. (1988) Dynamical component of seasonal and year-to-year changes in Antarctic and global ozone, *J. Geophys. Res.* **93**, 12,537–59.

Turco R. P. and Whitten R. C. (1974) A comparison of several computational techniques for solving some common aeronomic problems. *J. Geophys. Res.* **79**, 3179–85.

Turco R. P., Hamill P., Toon O. B., Whitten R. C., and Kiang C. S. (1979) The NASA–Ames Research Center stratospheric aerosol model: I. Physical processes and computational analogs. NASA Tech. Publ. (TP) 1362, iii-94.

Turco R. P., Toon O. B., and Hamill P. (1989) Heterogeneous physiochemistry of the polar ozone hole. *J. Geophys. Res.* **94**, 16,493–510.

Twomey S. A. (1976) Computations of the absorption of solar radiation by clouds. *J. Atmos. Sci.* **33**, 1087–91.

Twomey S. A. (1977) The effect of cloud scattering on the absorption of solar radiation by atmospheric dust. *J. Atmos. Sci.* **29**, 1156–9.

Twomey S. A. (1991) Aerosols, clouds, and radiation. *Atmos. Environ.* **25A**, 2435–42.

Tyndall G. S. and Ravishankara A. R. (1991) Atmospheric oxidation of reduced sulfur species. *Int. J. Chem. Kinet.* **23**, 483–527.

U.S. Department of the Army (1958) Universal Transverse Mercator grid. Tables for transformation of coordinates from grid to geographic; Clarke 1866 Spheroid. U.S. Government Printing Office, Washington, DC.

U.S. Environmental Protection Agency (USEPA) (1978) Air quality criteria for ozone and other photochemical oxidants. Report No. EPA-600/8-78-004.

van de Hulst H. C. (1957) *Light Scattering by Small Particles.* John Wiley and Sons, Inc., New York.

van den Hurk B. J. J. M., Bastiaansen W. G. M., Pelgrum H., and van Meijgaard E. V. (1997) A new methodology for assimilation of initial soil moisture fields in weather prediction models using meteosat and NOAA data. *J. Appl. Met.* **36**, 1271–83.

Vanderzee C. E., Waugh D. H., and Haas N. C. (1980) Enthalpies of dilution and relative apparent molar enthalpies of aqueous ammonium nitrate. The case of a weakly hydrolysed (dissociated) salt. *J Chem. Thermodynam.* **12**, 21–5.

van Doren J. M, Watson L. R., Davidovits P., Worsnop D. R., Zahniser S., and Kolb C. E. (1990) Temperature dependence of the uptake coefficients of $HNO_3$, HCl, and $N_2O_5$ by water droplets. *J. Phys. Chem.* **94**, 3256–69.

van Geuchten M. T. (1980) A closed-form equation for predicting the hydraulic conductivity of unsaturated soils. *Ann. Geophys.* **3**, 615–28.

van Weele M. and Duynkerke P. G. (1993) Effects of clouds on the photodissociation of $NO_2$: Observation and modelling. *J. Atmos. Chem.* **16**, 231–55.

van Zandt T. E. and Fritts D. C. (1989) A theory of enhanced saturation of the gravity wave spectrum due to increases in atmospheric stability. *Pure Appl. Geophys. Pageoph.* **130**, 399–420.

Varoglu E. and Finn W. D. L. (1980) Finite elements incorporating characteristics for one-dimensional diffusion–convection equation. *J. Comp. Phys.* **34**, 371–89.

Venkataraman C. and Friedlander S. K. (1994) Size distributions of polycyclic aromatic hydrocarbons and elemental carbon. 2. Ambient measurements and effects of atmospheric processes. *Environ. Sci. Technol.* **28**, 563–72.

Venkataraman C., Lyons J. M., and Friedlander S. K. (1994) Size distributions of polycyclic aromatic hydrocarbons and elemental carbon. 1. Sampling, measurement methods, and source characterization. *Environ. Sci. Technol.* **28**, 555–62.

Verwer J. G. (1994) Gauss–Seidel iteration for stiff ODEs from chemical kinetics. *SIAM J. Sci. Comput.* **15**, 1243–50.

Villars D. S. (1959) A method of successive approximations for computing combustion equilibria on a high speed digital computer. *J. Phys. Chem.* **63**, 521–5.

Waggoner A. P., Weiss R. E., Ahlquist N. C., Covert D. S., Will S., and Charlson R. J. (1981) Optical characteristics of atmospheric aerosols. *Atmos. Environ.* **15**, 1891–1909.

Wagman D. D., Evans W. H., Parker V. B., Schumm R. H., Halow I., Bailey S. M., Churney K. L., and Nuttall R. L. (1982) The NBS tables of chemical thermodynamic properties: Selected values for inorganic and $C_1$ and $C_2$ organic substances in SI units. *J. Phys. Chem. Ref. Data* **11**, Suppl. 2.

Walcek C. J., Brost R. A., and Chang J. S. (1986) $SO_2$, sulfate and $HNO_3$ deposition velocities computed using regional landuse and meteorological data. *Atmos. Environ.* **20**, 949–64.

Walcek C. J., Yuan H.-H., and Stockwell W. R. (1997) The influence of aqueous-phase chemical reactions on ozone formation in polluted and nonpolluted clouds. *Atmos. Environ.* **31**, 1221–37.

Walmsley J. L. and Wesely M. L. (1996) Modification of coded parameterizations of surface resistances to gaseous dry deposition. *Atmos. Environ.* **30A**, 1181–8.

Washington W. M. and Parkinson C. L. (1986) *An Introduction to Three-Dimensional Climate Modeling*. University Science Books, Mill Valley, CA.

Watson R. T. (1977) Rate constants for reactions of $ClO_x$ of atmospheric interest. *J. Phys. Chem. Ref. Data* **6**, 871–917.

Weare B. C., Temkin R. L., and Snell F. M. (1974) Aerosols and climate: Some further considerations. *Science* **186**, 827–8.

Weeks J. L. and Rabani J. (1966) The pulse radiolysis of deaerated aqueous carbonate solutions. *J. Phys. Chem.* **70**, 2100–6.

Weisman M. L., Skamarock W. C., and Klemp J. B. (1997) The resolution dependence of explicitly modeled convective systems. *Mon. Wea. Rev.* **125**, 527–48.

Welch R. M., Cox S. K., and Davis J. M. (1980) *Solar Radiation and Clouds*, Meteorological Monograph 17. American Meteorological Society.

Wendisch M. S., Mertes S., Ruggaber A., and Nakajima T. (1995) Vertical profiles of aerosol and radiation and the influence of a temperature inversion: Measurements and radiative transfer calculations. *J. Appl. Meteor.* **35**, 1703–15.

Wengle H. and Seinfeld J. H. (1978) Pseudospectral solution of atmospheric diffusion problems. *J. Comp. Phys.* **26**, 87–106.

Wesely M. L. (1989) Parameterization of surface resistances to gaseous dry deposition in regional-scale numerical models. *Atmos. Environ.* **23**, 1293–1304.

Wesely M. L. and Hicks B. B. (1977) Some factors that affect the deposition rates of sulfur dioxide and similar gases on vegetation. *J. Air Pollut. Control Ass.* **27**, 1110–6.

Wetzel P. J. and Chang J. (1987) Concerning the relationship between evapotranspiration and soil moisture. *J. Climate Appl. Meteor.* **26**, 18–27.

Wexler A. S. and Seinfeld J. H. (1990) The distribution of ammonium salts among a size and composition dispersed aerosol. *Atmos. Environ.* **24A**, 1231–46.

Wexler A. S. and Seinfeld J. H. (1991) Second-generation inorganic aerosol model. *Atmos. Environ.* **25A**, 2731–48.

Whitby E. R. (1985) The model aerosol dynamics model. Part I. Report to the U.S. Environmental Protection Agency. Dept. of Mechanical Engineering, University of Minnesota, Minneapolis.

Whitby K. T. (1978) The physical characteristics of sulfur aerosols. *Atmos. Environ.* **12**, 135–59.

Whitten G. Z., Hogo H., and Killus J. P. (1980) The carbon bond mechanism: A condensed kinetic mechanism for photochemical smog. *Environ. Sci. Technol.* **14**, 690–700.

Wigley T. M. L. and Raper S. C. B. (1992) Implications for climate and sea level of revised IPCC emissions scenarios. *Nature* **357**, 293–300.

Wilke C. R. and Chang P. (1955) Correlation of diffusion coefficients in dilute solutions. *Am. Inst. Chem. Eng. J.* **1**, 264–70.

Wine P. H., Tang Y., Thorn R. P., Wells J. R., and Davis D. D. (1989) Kinetics of aqueous-phase reactions of the $SO_4^-$ radical with potential importance in cloud chemistry. *J. Geophys. Res.* **94**, 1085–94.

Wiscombe W. (1977) The delta-M method: Rapid yet accurate radiative flux calculations for strongly asymmetric phase functions. *J. Atmos. Sci.* **34**, 1408–22.

Wiscombe W. J., Welch R. M., and Hall W. D. (1984) The effects of very large drops on cloud absorption. Part I: Parcel models. *J. Atmos. Sci.* **41**, 1336–55.

Wolf M. E. and Hidy G. M. (1997) Aerosols and climate: Anthropogenic emissions and trends for 50 years. *J. Geophys. Res.* **102**, 11,113–21.

Woodcock A. H. (1953) Salt nuclei in marine air as a function of altitude and wind force. *J. Meteorol.* **10**, 362–71.

Woods T. N., Prinz D. K., Rottman G. J., London J., Crane P. C., Cebula R. P., Hilsenrath E., Bruecker G. E., Andrews M. D., White O. R., VanHoosier M. E., Floyd L. E., Herring L. C., Knapp B. G., Pankratz C. K., and Reiser P. A. (1996) Validation of the UARS solar ultraviolet irradiances: Comparison with the ATLAS 1 and 2 measurements. *J. Geophys. Res.* **101**, 9541–69.

World Meteorological Organization (WMO) (1975) *Manual on the Observation of Clouds and Other Meteors*. World Meteorological Organization, Geneva.

World Meteorological Organization (WMO) (1995) Scientific assessment of ozone depletion: 1994. Rep. 25, Global Ozone Res. and Monit. Proj. World Meteorological Organization, Geneva.

Worsnop D. R., Fox L. E., Zahniser M. S., and Wofsy S. C. (1993). Vapor pressures of solid hydrates of nitric acid: Implications for polar stratospheric clouds. *Science* **259**, 71–4.

Wu J. (1993) Production of spume drops by the wind tearing of wave crests: The search for quantification. *J. Geophys. Res.* **98**, 18,221–7.

Wu Y.-C. and Hamer W. J. (1980) Revised values of the osmotic coefficients and mean activity coefficients of sodium nitrate in water at 25°C. *J. Phys. Chem. Ref. Data* **9**, 513–8.

Yanenko N. A. (1971) *The Method of Fractional Steps*. Springer-Verlag, 160 pp.

Yin F., Grosjean D., and Seinfeld J. H. (1990) Photooxidation of dimethyl sulfide and dimethyl disulfide. I: Mechanism development. *J. Atmos. Chem.* **11**, 309–64.

Young A. T. (1980) Revised depolarization corrections for atmospheric extinction. *Appl. Opt.* **19**, 3427–8.

Young T. R. and Boris J. P. (1977) A numerical technique for solving stiff ordinary differential equations associated with the chemical kinetics of reactive-flow problems. *J. Phys. Chem.* **81**, 2424–7.

Zawadski I., Torlaschi E., and Sauvageau R. (1981) The relationship between mesoscale thermodynamic variables and convective precipitation. *J. Atmos. Sci.* **38**, 1535–40.

Zaytsev I. D. and Aseyer G. G., eds. (1992) Properties of aqueous solutions of electrolytes (translated by M. A. Lazarev and V. R. Sorochenka). CRC Press, Boca Raton.

Zdunkowski W. G. and McQuage N. D. (1972) Short-term effects of aerosol on the layer near the ground in a cloudless atmosphere. *Tellus* **24**, 237–54.

Zdunkowski W. G., Welch R. M., and Paegle J. (1976) One-dimensional numerical simulation of the effects of air pollution on the planetary boundary layer. *J. Atmos. Sci.* **33**, 2399–414.

Zeldovich Y. B. (1942) Theory of new phase formation: cavitation. *J. Exp. Theor. Phys. (USSR)* **12**,525–38.

Zhang D. and Anthes R. A. (1982) A high-resolution model of the planetary boundary layer – sensitivity tests and comparisons with SESAME-79 data. *J. Appl. Meteor.* **21**, 1594–609.

# Index